T0094171

Exascale Scientific Applications

Chapman & Hall/CRC
Computational Science Series

Series Editor: Horst Simon

Deputy Director

Lawrence Berkeley National Laboratory

Berkeley, California, U.S.A.

PUBLISHED TITLES

Combinatorial Scientific Computing

Edited by Uwe Naumann and Olaf Schenk

Contemporary High Performance Computing: From Petascale Toward Exascale

Edited by Jeffrey S. Vetter

Contemporary High Performance Computing: From Petascale Toward Exascale, Volume Two

Edited by Jeffrey S. Vetter

Data-Intensive Science

Edited by Terence Critchlow and Kerstin Kleese van Dam

Elements of Parallel Computing

Eric Aubanel

The End of Error: Unum Computing

John L. Gustafson

Exascale Scientific Applications: Scalability and Performance Portability

Edited by Tjerk P. Straatsma, Katerina B. Antypas, and Timothy J. Williams

From Action Systems to Distributed Systems: The Refinement Approach

Edited by Luigia Petre and Emil Sekerinski

Fundamentals of Multicore Software Development

Edited by Victor Pankratius, Ali-Reza Adl-Tabatabai, and Walter Tichy

Fundamentals of PARALLEL Multicore Architecture

Yan Solihin

The Green Computing Book: Tackling Energy Efficiency at Large Scale

Edited by Wu-chun Feng

Grid Computing: Techniques and Applications

Barry Wilkinson

High Performance Computing: Programming and Applications

John Levesque with Gene Wagenbreth

High Performance Parallel I/O

Prabhat and Quincey Koziol

High Performance Visualization: Enabling Extreme-Scale Scientific Insight
Edited by E. Wes Bethel, Hank Childs, and Charles Hansen

Industrial Applications of High-Performance Computing: Best Global Practices
Edited by Anwar Osseyran and Merle Giles

Introduction to Computational Modeling Using C and Open-Source Tools
José M. Garrido

Introduction to Concurrency in Programming Languages
Matthew J. Sottile, Timothy G. Mattson, and Craig E. Rasmussen

Introduction to Elementary Computational Modeling: Essential Concepts, Principles, and Problem Solving
José M. Garrido

Introduction to High Performance Computing for Scientists and Engineers
Georg Hager and Gerhard Wellein

Introduction to Modeling and Simulation with Matlab and Python
Steven I. Gordon and Brian Guilfoos

Introduction to Reversible Computing
Kalyan S. Perumalla

Introduction to Scheduling
Yves Robert and Frédéric Vivien

Introduction to the Simulation of Dynamics Using Simulink®
Michael A. Gray

Peer-To-Peer Computing: Applications, Architecture, Protocols, and Challenges
Yu-Kwong Ricky Kwok

Performance Tuning of Scientific Applications
Edited by David Bailey, Robert Lucas, and Samuel Williams

Petascale Computing: Algorithms and Applications
Edited by David A. Bader

Process Algebra for Parallel and Distributed Processing
Edited by Michael Alexander and William Gardner

Programming for Hybrid Multi/Many-Core MPP Systems
John Levesque and Aaron Vose

Scientific Data Management: Challenges, Technology, and Deployment
Edited by Arie Shoshani and Doron Rotem

Software Engineering for Science
Edited by Jeffrey C. Carver, Neil P. Chue Hong, and George K. Thiruvathukal

Exascale Scientific Applications
Scalability and Performance Portability

Edited by
Tjerk P. Straatsma, Katerina B. Antypas,
and Timothy J. Williams

CRC Press
Taylor & Francis Group
Boca Raton London New York

CRC Press is an imprint of the
Taylor & Francis Group, an **informa** business

A CHAPMAN & HALL BOOK

CRC Press
Taylor & Francis Group
6000 Broken Sound Parkway NW, Suite 300
Boca Raton, FL 33487-2742

© 2018 by Taylor & Francis Group, LLC
CRC Press is an imprint of Taylor & Francis Group, an Informa business

No claim to original U.S. Government works

Printed on acid-free paper

International Standard Book Number-13: 978-1-138-19754-1 (Hardback)

This book contains information obtained from authentic and highly regarded sources. Reasonable efforts have been made to publish reliable data and information, but the author and publisher cannot assume responsibility for the validity of all materials or the consequences of their use. The authors and publishers have attempted to trace the copyright holders of all material reproduced in this publication and apologize to copyright holders if permission to publish in this form has not been obtained. If any copyright material has not been acknowledged, please write and let us know so we may rectify in any future reprint.

Except as permitted under U.S. Copyright Law, no part of this book may be reprinted, reproduced, transmitted, or utilized in any form by any electronic, mechanical, or other means, now known or hereafter invented, including photocopying, microfilming, and recording, or in any information storage or retrieval system, without written permission from the publishers.

For permission to photocopy or use material electronically from this work, please access www.copyright.com (http://www.copyright.com/) or contact the Copyright Clearance Center, Inc. (CCC), 222 Rosewood Drive, Danvers, MA 01923, 978-750-8400. CCC is a not-for-profit organization that provides licenses and registration for a variety of users. For organizations that have been granted a photocopy license by the CCC, a separate system of payment has been arranged.

Trademark Notice: Product or corporate names may be trademarks or registered trademarks, and are used only for identification and explanation without intent to infringe.

Visit the Taylor & Francis Web site at
http://www.taylorandfrancis.com

and the CRC Press Web site at
http://www.crcpress.com

Printed and bound in the United States of America by
Edwards Brothers Malloy on sustainably sourced paper

Contents

Foreword xi

Preface xiii

About the Editors xvii

Contributors xix

Chapter 1 Portable Methodologies for Energy Optimization on Large-Scale
Power-Constrained Systems 1

Kevin J. Barker and Darren J. Kerbyson

Chapter 2 Performance Analysis and Debugging Tools at Scale 17

*Scott Parker, John Mellor-Crummey, Dong H. Ahn, Heike Jagode,
Holger Brunst, Sameer Shende, Allen D. Malony, David Lecomber,
John V. DelSignore, Jr., Ronny Tschüter, Ralph Castain, Kevin Harms,
Philip Carns, Ray Loy, and Kalyan Kumaran*

Chapter 3 Exascale Challenges in Numerical Linear and Multilinear Algebras 51

Dmitry I. Lyakh and Wayne Joubert

Chapter 4 Exposing Hierarchical Parallelism in the FLASH Code for Supernova
Simulation on Summit and Other Architectures 95

Thomas Papatheodore and O. E. Bronson Messer

Chapter 5 NAMD: Scalable Molecular Dynamics Based on the Charm++ Parallel
Runtime System 119

Bilge Acun, Ronak Buch, Laxmikant Kale, and James C. Phillips

Chapter 6 Developments in Computer Architecture and the Birth and Growth
of Computational Chemistry 145

Wim Nieuwpoort and Ria Broer

Chapter 7 On Preparing the Super Instruction Architecture and Aces4 for
Future Computer Systems 151

Jason Byrd, Rodney Bartlett, and Beverly A. Sanders

Chapter 8 Transitioning NWChem to the Next Generation of Manycore Machines 165

*Eric J. Bylaska, Edoardo Aprà, Karol Kowalski, Mathias Jacquelin,
Wibe A. de Jong, Abhinav Vishnu, Bruce Palmer, Jeff Daily, Tjerk P. Straatsma,
Jeff R. Hammond, and Michael Klemm*

Chapter 9   Exascale Programming Approaches for Accelerated Climate Modeling
for Energy 187

   Matthew R. Norman, Azamat Mametjanov, and Mark Taylor

Chapter 10   Preparing the Community Earth System Model for Exascale Computing 207

   *John M. Dennis, Christopher Kerr, Allison H. Baker, Brian Dobbins, Kevin Paul,
   Richard Mills, Sheri Mickelson, Youngsung Kim, and Raghu Kumar*

Chapter 11   Large Eddy Simulation of Reacting Flow Physics and Combustion 231

   Joseph C. Oefelein and Ramanan Sankaran

Chapter 12   S3D-Legion: An Exascale Software for Direct Numerical Simulation
of Turbulent Combustion with Complex Multicomponent Chemistry 257

   *Sean Treichler, Michael Bauer, Ankit Bhagatwala, Giulio Borghesi,
   Ramanan Sankaran, Hemanth Kolla, Patrick S. McCormick, Elliott Slaughter,
   Wonchan Lee, Alex Aiken, and Jacqueline Chen*

Chapter 13   Data and Workflow Management for Exascale Global Adjoint Tomography 279

   *Matthieu Lefebvre, Yangkang Chen, Wenjie Lei, David Luet, Youyi Ruan,
   Ebru Bozdağ, Judith Hill, Dimitri Komatitsch, Lion Krischer, Daniel Peter,
   Norbert Podhorszki, James Smith, and Jeroen Tromp*

Chapter 14   Scalable Structured Adaptive Mesh Refinement with Complex Geometry 307

   Brian Van Straalen, David Trebotich, Andrey Ovsyannikov, and Daniel T. Graves

Chapter 15   Extreme Scale Unstructured Adaptive CFD for Aerodynamic Flow Control 319

   *Kenneth E. Jansen, Michel Rasquin, Jed Brown, Cameron Smith,
   Mark S. Shephard, and Chris Carothers*

Chapter 16   Lattice Quantum Chromodynamics and Chroma 345

   Bálint Joó, Robert G. Edwards, and Frank T. Winter

Chapter 17   PIC Codes on the Road to Exascale Architectures 375

   Henri Vincenti, Mathieu Lobet, Remi Lehe, Jean-Luc Vay, and Jack Deslippe

Chapter 18   Extreme-Scale *De Novo* Genome Assembly 409

   *Evangelos Georganas, Steven Hofmeyr, Leonid Oliker, Rob Egan,
   Daniel Rokhsar, Aydin Buluc, and Katherine Yelick*

Chapter 19   Exascale Scientific Applications: Programming Approaches for
Scalability, Performance, and Portability: KKRnano 431

   Paul F. Baumeister, Marcel Bornemann, Dirk Pleiter, and Rudolf Zeller

Chapter 20   Real-Space Multiple-Scattering Theory and Its Applications at Exascale 449

Markus Eisenbach and Yang Wang

Chapter 21   Development of QMCPACK for Exascale Scientific Computing 461

*Anouar Benali, David M. Ceperley, Ed D'Azevedo, Mark Dewing,
Paul R. C. Kent, Jeongnim Kim, Jaron T. Krogel, Ying Wai Li, Ye Luo,
Tyler McDaniel, Miguel A. Morales, Amrita Mathuria, Luke Shulenburger,
and Norm M. Tubman*

Chapter 22   Preparing an Excited-State Materials Application for Exascale 481

*Jack Deslippe, Felipe H. da Jornada, Derek Vigil-Fowler, Taylor Barnes,
Thorsten Kurth, and Steven G. Louie*

Chapter 23   Global Gyrokinetic Particle-in-Cell Simulation 507

William Tang and Zhihong Lin

Chapter 24   The Fusion Code XGC: Enabling Kinetic Study of Multiscale Edge
Turbulent Transport in ITER 529

*Eduardo D'Azevedo, Stephen Abbott, Tuomas Koskela, Patrick Worley,
Seung-Hoe Ku, Stephane Ethier, Eisung Yoon, Mark Shephard, Robert Hager,
Jianying Lang, Jong Choi, Norbert Podhorszki, Scott Klasky,
Manish Parashar, and Choong-Seock Chang*

Index 553

Foreword

Over the past quarter century, computational modeling and simulation have become an integral part of the fabric of science and engineering research and innovation. Simulation science has advanced our understanding of the creation of the universe, mechanisms of chemical and biochemical processes, impacts of natural disasters (tornados, hurricanes, and earthquakes) and changes in the earth's climate, optimization of combustion and fusion energy processes, and many, many more. Digital technologies are now spreading to the observational sciences, which are being revolutionized by the advent of powerful new sensors that can detect and record a wide range of physical, chemical, and biological phenomena—from the massive digital cameras in a new generation of telescopes to sensor arrays for characterizing ecological and geological processes and new sequencing instruments for genomics research and precision medicine. Data science, as one of the newest applications of digital technologies, is rapidly developing.

In large part, the advances in simulation and data science are driven in a synergistic loop by the continuing advances in computing technologies. From the 1970s onward, increases in computing power were driven by Moore's law and Dennard scaling, with a doubling in computing power occurring every 18–24 months. Thus, the end of the 1980s saw the deployment of computer systems capable of performing a billion arithmetic operations per second. Ten years later, computing technology had advanced to the point that it was possible to perform a trillion arithmetic operations per second. In the 2010s, computer systems capable of a quadrillion operations per second were being fielded. Exascale computers, 1000 times more powerful than petascale computers, will arrive in the next few years. Because of the enormous potential of exascale computers for addressing society's most pressing problems as well as advancing science and engineering, China, Europe, Japan, and the United States are pressing forward with the development of exascale computer systems.

But, the transformation in computing technologies required to attain the exascale poses significant challenges. One must advance the entire computational ecosystem to take advantage of advanced exascale technologies such as applications, algorithms, and software development tools. With the demise of Dennard scaling in the mid-2000s, increases in the performance of computational modeling and simulation codes can only be achieved through the use of a larger and larger number of processors (or compute cores). Although this "scalability" problem has been with us for 25 years, for much of that time, its impact was only lightly felt because of the dramatic increases in the performance of single compute cores—a factor of two orders of magnitude from 1989 to 2004 alone. With single core performance now stalled, computational scientists and engineers must confront the scalability problem head on.

The need for ever more scalability has dramatically increased the difficulty of developing science and engineering applications for leading-edge computers. At the heart of the problem is the discovery of new algorithms that scale to large numbers of compute cores for a broad range of science and engineering applications. This problem can only be solved through innovative research that combines the talents of computational scientists and engineers, computer scientists, and applied mathematicians. But, even given scalable algorithms, the development of science and engineering applications for computers with tens of millions of compute cores, tens of petabytes of memory, and hundreds of petabytes of disk storage is challenging. The software must be written, debugged, optimized, and, to the extent possible, made resilient to computer faults (e.g., the loss of a compute core or a memory block) and be portable among different exascale computer architectures—none of this is easy or straightforward. Progress will require the creation of new software development tools and libraries and/or substantial reformulation of existing tools, all of which must be integrated into a robust, easy-to-use application development environment.

The authors of the chapters in this book are the pioneers who will explore the exascale frontier. The path forward will not be easy for the reasons noted above. These authors, along with their colleagues who will produce these powerful computer systems, will, with dedication and determination, overcome the scalability problem, discover the new algorithms needed to achieve exascale performance for the broad range of applications that they represent, and create the new tools needed to support the development of scalable and portable science and engineering applications. Although the focus is on exascale computers, the benefits will permeate all of science and engineering because the technologies developed for the exascale computers of tomorrow will also power the petascale servers and terascale workstations of tomorrow. These affordable computing capabilities will empower scientists and engineers everywhere.

This Foreword has touched primarily on the issue of computational modeling and simulation, which is the focus of the current book, but data-driven discovery also requires advanced computing systems to collect, transport, store, manage, integrate, and analyze increasingly large amounts of invaluable data. The knowledge gained from data-driven discovery is already transforming our understanding of many natural phenomena and the future is full of promise. We expect many new data science applications to arise as this field advances.

Thom H. Dunning, Jr.
Northwest Institute for Advanced Computing
Pacific Northwest National Laboratory and University of Washington
Seattle, Washington

Preface

Scientific computing has become one of the fundamental pillars of science, combining theory and experiment. Computing is providing capabilities allowing theoretical concepts to be cast in computational modeling and simulation methods for the interpretation, prediction, and design of experiments or for providing unique and detailed understanding of physical systems that are impossible or prohibitively difficult, expensive, or dangerous to study experimentally. Computing also plays an increasingly important role in the analysis of large-scale observational and experimental data with the objective of validating or improving the theoretical models of the underlying physical phenomena, as well as informing and guiding new experiments. The scientific enterprise is depending on computing to address many of the fundamental intellectual challenges for understanding the natural world including the evolution of life, the properties and reactivity of materials that make up our environment, and the formation and expansion of the universe. Computing has an increasingly transformational role in practically every aspect of society as well, including economic competitiveness, advanced manufacturing, health care, environmental sustainability, natural disaster recovery, social media and entertainment, national security, and energy security.

The enormous advances in the integration of computing into virtually everything we do is in part the result of the rapid technological developments of the last decades. The largest computers available have become faster by almost three orders of magnitude roughly every decade. Current leadership computing facilities provide systems capable of providing tens of petaflops and exaflops-capable systems are expected in the 2021–2023 timeframe. The computer architectures that have made these increases in processing power possible have gone through a number of significant conceptual changes, from fast scalar processors in the 1970s, vector processors in the 1980s, parallel systems in the 1990s and 2000s, to the current transition from massively parallel homogeneous computer systems to the highly complex systems with extensive hierarchies in processors and accelerators, volatile and nonvolatile memory, and communication networks.

With each new generation of technologies, the system software designers, programming environments and tool providers, and application software developers are faced with the challenge of adapting or rewriting their codes in such a way as to take full advantage of the capabilities offered by the new computer systems, as well as to be portable between different concurrently available architectures. This book presents twenty-four chapters by software development teams from a variety of scientific disciplines focusing on the programming practices to achieve scalability on high-end computer systems while at the same time maintaining architectural and performance portability for different computer technologies. The premise of this publication is that scientific application developers learn best by example, and this volume intends to document and disseminate the strategies being developed and used, the experiences obtained and best practices followed in these early scientific application porting efforts, especially those with the goal of achieving high scalability with minimal loss in performance portability.

For the current volume, contributions from developers of highly scalable applications in a wide variety of scientific domains were invited. Many of these developers are participating in application readiness programs focused on readying applications for the next generation exascale architectures, including the Center for Accelerated Application Readiness at the Oak Ridge Leadership Computing Facility, the NERSC Exascale Science Application Program at the National Energy Research Scientific Computing Center, and the Early Science Program at the Argonne Leadership Computing Facility. This publication is organized in a section on general aspects of portable application development, followed by sections that highlight modeling and simulation application developments in specific scientific domains: astrophysics and cosmology, biophysics, chemical physics, climate science, combustion science, earth science, engineering, high-energy physics, informatics, materials science, and plasma physics.

In the development of large applications for high-performance computing systems, in addition to scalability and parallel performance, much attention is given to architectural and performance portability. In Chapter 1, Barker and Kerbyson discuss energy efficiency as one of the key metrics for performance on exascale systems, and introduce the concept of power-portability as the ability of applications to effectively use power-saving features across a variety of architectures. In Chapter 2, Parker et al. explore performance and debugging tools available on current high-performance resources for the development of applications for exascale systems and the underlying techniques, and the architectural features that they exploit. In Chapter 3, Lyakh and Joubert describe the challenges in developing performance portable linear and multilinear algebra libraries for large heterogeneous exascale architectures.

The remainder of this volume is dedicated to the development of specific scientific applications. In Chapter 4, Papatheodore and Messer outline the exploitation of hierarchical parallelism in the adaptive-mesh, multiphysics simulation framework FLASH that is extensively used in the field of astrophysics for the simulation of supernovae explosions. In Chapter 5, Acun et al. describe the programming approach based on the Charm++ parallel framework for the highly scalable biophysics application NAMD, used by thousands of users for molecular dynamics simulations of biomolecular systems. In Chapter 6, Nieuwpoort and Broer give a historical perspective of the advancement of electronic computing technologies and the development of computational chemistry and describe how the concurrent development of chemical physics applications have always adapted to changes in computer technologies, allowing for a projection of programming issues going forward toward exascale. In Chapter 7, Byrd et al. describe their efforts to prepare ACES-4 a theoretical chemistry application based on the use of the domain-specific language SIAL for future computer systems. In Chapter 8, Bylaska et al. highlight the development of the parallel computational chemistry application NWChem and their strategy to transition this application to the next generation massively threaded and many-core exascale systems. Two related applications in climate science are represented in this volume. In Chapter 9, Norman et al. describe the programming approaches taken for the effective use of the Accelerated Climate Modeling for Energy (ACME) and their focus on performance portability. In Chapter 10, Dennis et al. focus on the preparation of the Community Earth System Model (CESM) for exascale computational resources. Combustion science applications are represented by two applications. In Chapter 11, Oefelein and Sankaran describe the development of the highly scalable code RAPTOR for the large eddy simulations of reactive flow physics and combustion in complex geometries. In Chapter 12, Treichler et al. illustrate the use of Legion in the development of the exascale software for direct numerical simulation of turbulent combustion with complex multicomponent chemistry. In the field of earth science, two applications are highlighted. In Chapter 13, Lefebvre et al. focus on the data and workflow management aspects of their exascale application SpecFEM for global adjoint tomography and the simulation and analysis of seismic events in the earth's crust. In Chapter 14, Van Straalen et al. highlight different aspects of optimizing scalability and portability of adaptive mesh refinement with complex geometries as implemented in ChomboCrunch and used for the simulation of unsteady flows found in pore scale reactive transport processes associated with subsurface problems including carbon sequestration. In Chapter 15, Jansen et al. describe the development of extreme-scale unstructured adaptive computational fluid dynamics for aerodynamic flow control in the engineering application PHASTA. In Chapter 16, Joó et al. describe their highly scalable implementation of lattice chromodynamics in the code CHROMA, as an example of a high energy physics application. In Chapter 17, Vincenti et al. outline the development of particle-in-cell methodologies as found in the particle accelerator modeling code WARP for exascale architectures. In Chapter 18, Georganas et al. explore the informatics challenges of *de novo* genome assembly at an extreme scale, as implemented by the Meraculous algorithm. The field of material science is represented with four applications. In Chapter 19, Baumeister et al. highlight aspects of scalability, performance, and portability for the KKRnano application. In Chapter 20, Eisenbach and Wang describe the development of LSMS, a linear scaling code for real-space multiple-scattering theory and application to first principles study of magnetic and crystal structure phase transitions in solid-state materials. In Chapter 21, Benali et al. describe aspects of the implementation of quantum Monte Carlo techniques in the QMCPACK, using statistical sampling techniques to solve the many-body Schrödinger equation for solid-state systems. In Chapter 22, Deslippe et al. describe BerkeleyGW, a software package for evaluation of electron excited-state properties and optical responses of materials based on the many-body perturbation theory as formulated by GW and Bethe–Salpeter equation methodology. In the last two chapters, two plasma physics applications are highlighted. In Chapter 23, Tang and Lin describe their implementation for exascale architectures of GTC, a global gyrokinetic particle-in-cell code designed to address turbulent transport of plasma particles and associated confinement, from present generation devices to much larger ITER-size plasmas. In Chapter 24, D'Azevedo et al. illustrate the approach of using nested OpenMP parallelism, adaptive parallel I/O, data reduction, and load balancing based on dynamic repartitioning in XGC, an application enabling kinetic studies of multiscale edge turbulent transport in ITER.

A few other people have had a significant role in completing this book. First, we thank Randi Cohen, senior acquisitions editor for Computer Science at CRC Press/Taylor & Francis Group, who has patiently guided us through the entire process from the initial discussions about the concept for this book to the production of the final product. Her experience and expert advice throughout the nearly two-year long process has been crucial in the successful completion of this publication. It has been a pleasure working with her. Peyton Ticknor, administrator at Oak Ridge National Laboratory, provided the day-to-day administrative support for this project. Her attention to detail and organizational skills made it easy to keep this project on track and review and deliver all manuscripts in a timely fashion. The discussions with and advice from our colleagues who have helped develop the concept for this book should be acknowledged, in particular Bronson Messer and Mike Papka.

I owe a great thank you to my colleagues Katie Antypas and Tim Williams who served as coeditors for this book. It has been a great experience to develop the plans for this book together and execute the process of communicating with authors and reviewing the collection of submitted manuscripts. Finally, our spouses, Ineke Straatsma, Mark Palatucci, and Jane Archer deserve our appreciation for their support that allowed us the additional time and energy that went into the preparation of this book.

This publication was developed using resources of the Oak Ridge Leadership Computing Facility, which is a DOE Office of Science User Facility supported under Contract DE-AC05-00OR22725; the Argonne Leadership Computing Facility, supported by the Office of Science, U.S. Department of Energy, under Contract DE-AC02-06CH11357; and National Energy Research Scientific Computing Center, a DOE Office of Science User Facility supported by the Office of Science of the U.S. Department of Energy under Contract No. DE-AC02-05CH11231.

This book provides an overview of the practical approaches that scientific application developers from a variety of scientific domains are using to prepare their application for the next generation of pre-exascale and exascale architectures. The description of their efforts to achieve scalability, architectural and performance portability, maintainability and usability for their applications—covering a variety of algorithms, implementations, and programming approaches—is intended to serve as examples for development efforts in other scientific domains.

Dr. Tjerk P. Straatsma
Oak Ridge National Laboratory

About the Editors

Tjerk P. Straatsma is the group leader of the Scientific Computing Group in the National Center for Computational Sciences, a division that houses the Oak Ridge Leadership Computing Facility, at Oak Ridge National Laboratory, Oak Ridge, Tennessee, and adjunct faculty member in the Chemistry Department of the University of Alabama in Tuscaloosa, Alabama. He earned his PhD in mathematics and natural sciences from the University of Groningen, the Netherlands. After a postdoctoral associate appointment, followed by a faculty position in the Department of Chemistry at the University of Houston, Texas, he moved to Pacific Northwest National Laboratory (PNNL) where, as core developer of the NWChem computational chemistry software, he established a program in computational biology and was group leader of the computational biology and bioinformatics group. Straatsma served as Director for the Extreme Scale Computing Initiative at PNNL, focusing on developing science capabilities for emerging petascale computing architectures. He was promoted to Laboratory Fellow, the highest scientific rank at the laboratory.

In 2013, he joined Oak Ridge National Laboratory, where, in addition to being group leader of the Scientific Computing Group, he is also the director of the Center for Accelerated Application Readiness and the Applications Working Group in the Institute for Accelerated Data Analytics and Computing focusing on preparing scientific applications for the next generation pre-exascale and exascale computer architectures.

Straatsma has been a pioneer in the development, efficient implementation, and application of advanced modeling and simulation methods as key scientific tools in the study of chemical and biomolecular systems, complementing analytical theories and experimental studies. His research focuses on the development of computational techniques that provide unique and detailed atomic level information that is difficult or impossible to obtain by other methods, and that contributes to the understanding of the properties and function of these systems. In particular, his expertise is in the evaluation of thermodynamic properties from large-scale molecular simulations, having been involved since the mid-1980s, in the early development of thermodynamic perturbation and thermodynamic integration methodologies. His research interests also include the design of efficient implementations of these methods on modern, complex computer architectures, from the vector processing supercomputers of the 1980s to the massively parallel and accelerated computer systems of today.

Since 1995, he has been a core developer of the massively parallel molecular science software suite NWChem and is responsible for its molecular dynamics simulation capability.

Straatsma has coauthored nearly 100 publications in peer-reviewed journals and conferences, was recipient of the 1999 R&D 100 Award for the NWChem molecular science software suite, and was recently elected Fellow of the American Association for the Advancement of Science.

Katerina B. Antypas is the data department head at the National Energy Research Scientific Computing (NERSC) Center, Berkeley, California, which includes the Data and Analytics Services Group, Data Science Engagement Group, Storage Systems Group, and Infrastructure Services Group. The department's mission is to pioneer new capabilities to accelerate large-scale data-intensive science discoveries as the Department of Energy Office of Science workload grows to include more data analysis from experimental and observational facilities such as light sources, telescopes, satellites, genomic sequencers, and particle colliders. Antypas is also the project manager for the NERSC-8 system procurement, a project to deploy NERSC's next generation HPC supercomputer in 2016, named Cori, a system comprised of the Cray interconnect and Intel Knights Landing manycore processor. The processor features on-package, high bandwidth memory and more than 64 cores per node with four hardware threads each. These technologies offer applications with great performance potential, but require users to make changes to applications in order to take advantage of multilevel memory and a large number of hardware threads. To address this concern, Antypas and the NERSC-8 team launched the NERSC Exascale Science Applications Program (NESAP), an initiative to prepare approximately 20 application teams for the Knights Landing architecture through close partnerships with vendors, science application experts, and performance analysts.

Antypas is an expert in parallel I/O application performance, and for the past six years has given a parallel-I/O tutorial at the SC conference. She also has expertise in parallel application performance, HPC architectures, and HPC user support and Office of Science user requirements. Antypas is also a PI on a new ASCR Research Project, "Science Search: Automated MetaData Using Machine Learning."

Before coming to NERSC, Antypas worked at the ASC Flash Center at the University of Chicago supporting the FLASH code, a highly scalable, parallel, adaptive mesh refinement astrophysics application. She wrote the parallel I/O modules in HDF5 and Parallel-NetCDF for the code. She has an MS in computer science from the University of Chicago, Illinois, and a bachelor's in physics from Wellesley College, Massachusetts.

Timothy J. Williams is deputy director of science at the Argonne Leadership Computing Facility at Argonne National Laboratory, Lemont, Illinois. He works in the Catalyst team—computational scientists who work with the large-scale projects using ALCF supercomputers. Williams manages the Early Science Program (ESP). The goal of the ESP is to prepare a set of scientific applications for early, preproduction use of next-generation computers such as ALCF's most recent Cray-Intel system based on second generation Xeon Phi processors, Theta; and their forthcoming pre-exascale system, Aurora, based on third generation Xeon Phi. Williams received his BS in Physics and Mathematics from Carnegie Mellon University, Pittsburgh, Pennsylvania, in 1982; he earned his PhD in physics in 1988 from the College of William and Mary, Williamsburg, Virginia, focusing on numerical study of a statistical turbulence theory using Cray vector supercomputers. Since 1989, he has specialized in the application of large-scale parallel computation to various scientific domains, including particle-in-cell plasma simulation for magnetic fusion, contaminant transport in groundwater flows, global ocean modeling, and multimaterial hydrodynamics. He spent 11 years in research at Lawrence Livermore National Laboratory (LLNL) and Los Alamos National Laboratory. In the early 1990s, Williams was part of the pioneering Massively Parallel Computing Initiative at LLNL, working on plasma PIC simulations and dynamic alternating direction implicit (ADI) solver implementations on the BBN TC2000 computer. In the late 1990s, he worked at Los Alamos' Advanced Computing Laboratory with a team of scientists developing the POOMA (Parallel Object Oriented Methods and Applications) framework—a C++ class library encapsulating efficient parallel execution beneath high-level data-parallel interfaces designed for scientific computing. Williams then spent nine years as a quantitative software developer for the financial industry at Morgan Stanley in New York, focusing on fixed-income securities and derivatives, and at Citadel in Chicago, focusing most recently on detailed valuation of subprime mortgage-backed securities. Williams returned to computational science at Argonne in 2009.

Contributors

Stephen Abbott
Oak Ridge National Laboratory
Oak Ridge, Tennessee

Bilge Acun
Department of Computer Science
University of Illinois at Urbana-Champaign
Champaign, Illinois

Dong H. Ahn
Lawrence Livermore National Laboratory
Livermore, California

Alex Aiken
Stanford University
Stanford, California

Edoardo Aprà
William R. Wiley Environmental Molecular
  Sciences Laboratory
Pacific Northwest National Laboratory
Richland, Washington

Allison H. Baker
National Center for Atmospheric Research
Computational Information Systems Laboratory
Boulder, Colorado

Kevin J. Barker
High Performance Computing
Pacific Northwest National Laboratory
Richland, Washington

Taylor Barnes
National Energy Research Scientific Computing
  Center
Lawrence Berkeley National Laboratory
Berkeley, California

Rodney Bartlett
Department of Chemistry
University of Florida
Gainesville, Florida

Michael Bauer
NVIDIA Research
Santa Clara, California

Paul F. Baumeister
Jülich Supercomputing Centre
Forschungszentrum Jülich
Jülich, Germany

Anouar Benali
Argonne National Laboratory
Lemont, Illinois

Ankit Bhagatwala
Lawrence Berkeley National Laboratory
Berkeley, California

Giulio Borghesi
Sandia National Laboratories
Livermore, California

Marcel Bornemann
Peter-Grünberg Institut
Forschungszentrum Jülich
Jülich, Germany

Ebru Bozdağ
Department of Geophysics
Colorado School of Mines
Golden, Colorado

Ria Broer
Department of Theoretical Chemistry
Zernike Institute for Advanced Materials
University of Groningen
Groningen, the Netherlands

Jed Brown
Argonne National Laboratory
Lemont, Illinois

Holger Brunst
Technische Universität Dresden
Dresden, Germany

Ronak Buch
Department of Computer Science
University of Illinois at Urbana-Champaign
Champaign, Illinois

Aydin Buluc
Computational Research Division
Lawrence Berkeley National Laboratory
Berkeley, California

Eric J. Bylaska
William R. Wiley Environmental Molecular
    Sciences Laboratory
Pacific Northwest National Laboratory
Richland, Washington

Jason Byrd
ENSCO, Inc.
Melbourne, Florida

Philip Carns
Argonne National Laboratory
Lemont, Illinois

Chris Carothers
Department of Computer Science
Rensselaer Polytechnic Institute
Troy, New York

Ralph Castain
Intel Corporation
Bend, Oregon

David M. Ceperley
Department of Physics and National Center for
    Supercomputing Applications
University of Illinois at Urbana-Champaign
Urbana, Illinois

Choong-Seock Chang
Princeton Plasma Physics Laboratory
Princeton, New Jersey

Jacqueline Chen
Sandia National Laboratories
Livermore, California

Yangkang Chen
Oak Ridge National Laboratory
Oak Ridge, Tennessee

Jong Choi
Oak Ridge National Laboratory
Oak Ridge, Tennessee

Felipe H. da Jornada
Department of Physics
University of California at Berkeley
and
Materials Sciences Division
Lawrence Berkeley National Laboratory
Berkeley, California

Jeff Daily
Advanced Computing, Mathematics and Data
    Division
Pacific Northwest National Laboratory
Richland, Washington

Eduardo D'Azevedo
Oak Ridge National Laboratory
Oak Ridge, Tennessee

Wibe A. de Jong
Computational Research Division
Lawrence Berkeley National Laboratory
Berkeley, California

John V. DelSignore, Jr.
Rogue Wave Software Inc.
Louisville, Colorado

John M. Dennis
Computational & Information Systems
    Laboratory
National Center for Atmospheric Research
Boulder, Colorado

Jack Deslippe
National Energy Research Scientific
    Computing
Lawrence Berkeley National Laboratory
Berkeley, California

Mark Dewing
Argonne National Laboratory
Lemont, Illinois

Brian Dobbins
National Center for Atmospheric Research
Boulder, Colorado

Robert G. Edwards
Jefferson Laboratory
Norfolk, Virginia

Rob Egan
Joint Genome Institute
Lawrence Berkeley National Laboratory
Berkeley, California

Markus Eisenbach
Oak Ridge National Laboratory
Oak Ridge, Tennessee

Stephane Ethier
Princeton Plasma Physics Laboratory
Princeton, New Jersey

Evangelos Georganas
Computational Research Division
Lawrence Berkeley National Laboratory
Berkeley, California

Daniel T. Graves
Lawrence Berkeley National Laboratory
Berkeley, California

Robert Hager
Princeton Plasma Physics Laboratory
Princeton, New Jersey

Jeff R. Hammond
Data Center Group, Intel Corporation
Hillsboro, Oregon

Kevin Harms
Argonne National Laboratory
Lemont, Illinois

Judith Hill
Oak Ridge National Laboratory
Oak Ridge, Tennessee

Steven Hofmeyr
Computational Research Division
Lawrence Berkeley National Laboratory
Berkeley, California

Mathias Jacquelin
Computational Research Division
Lawrence Berkeley National Laboratory
Berkeley, California

Heike Jagode
University of Tennessee
Knoxville, Tennessee

Kenneth E. Jansen
University of Colorado
Boulder, Colorado

Bálint Joó
Jefferson Laboratory
Norfolk, Virginia

Wayne Joubert
National Center for Computational
Sciences
Oak Ridge National Laboratory
Oak Ridge, Tennessee

Laxmikant Kale
Department of Computer Science
University of Illinois at Urbana-Champaign
Champaign, Illinois

Paul R. C. Kent
Oak Ridge National Laboratory
Oak Ridge, Tennessee

Darren J. Kerbyson
High Performance Computing
Pacific Northwest National Laboratory
Richland, Washington

Christopher Kerr
Consultant

Jeongnim Kim
Intel Corporation
Portland, Oregon

Youngsung Kim
National Center for Atmospheric Research
Boulder, Colorado

Scott Klasky
Oak Ridge National Laboratory
Oak Ridge, Tennessee

Michael Klemm
Software and Services Group
Intel Deutschland GmbH
Feldkirchen, Germany

Hemanth Kolla
Sandia National Laboratories
Albuquerque, New Mexico

Dimitri Komatitsch
Aix-Marseille University
Centrale Marseille
Marseille Cedex, France

Tuomas Koskela
Lawrence Berkeley National Laboratory
Berkeley, California

Karol Kowalski
William R. Wiley Environmental Molecular
    Sciences Laboratory
Pacific Northwest National Laboratory
Richland, Washington

Lion Krischer
ETH Zürich
Institute of Geophysics
Zürich, Switzerland

Jaron T. Krogel
Oak Ridge National Laboratory
Oak Ridge, Tennessee

Seung-Hoe Ku
Princeton Plasma Physics Laboratory
Princeton, New Jersey

Raghu Kumar
National Center for Atmospheric Research
Boulder, Colorado

Kalyan Kumaran
Argonne National Laboratory
Lemont, Illinois

Thorsten Kurth
National Energy Research Scientific Computing
    Center
Lawrence Berkeley National Laboratory
Berkeley, California

Jianying Lang
Intel Corporation
Santa Clara, California

David Lecomber
ARM Ltd.
Oxford, United Kingdom

Wonchan Lee
Stanford University
Stanford, California

Matthieu Lefebvre
Department of Geosciences
Princeton University
Princeton, New Jersey

Remi Lehe
Lawrence Berkeley National Laboratory
Berkeley, California

Wenjie Lei
Department of Geosciences
Princeton University
Princeton, New Jersey

Ying Wai Li
Oak Ridge National Laboratory
Oak Ridge, Tennessee

Zhihong Lin
University of California
California, Irvine

Mathieu Lobet
Lawrence Berkeley National Laboratory
Berkeley, California

Steven G. Louie
Department of Physics
University of California at Berkeley
and
Materials Sciences Division
Lawrence Berkeley National Laboratory
Berkeley, California

Ray Loy
Argonne National Laboratory
Lemont, Illinois

David Luet
Department of Geosciences
Princeton University
Princeton, New Jersey

Ye Luo
Argonne National Laboratory
Lemont, Illinois

Dmitry I. Lyakh
National Center for Computational Sciences
Oak Ridge National Laboratory
Oak Ridge, Tennessee

Allen D. Malony
Department of Computer and Information
  Science
University of Oregon
Eugene, Oregon

Azamat Mametjanov
Argonne National Laboratory
Lemont, Illinois

Amrita Mathuriya
Intel Corporation
Portland, Oregon

Patrick S. McCormick
Los Alamos National Laboratory
Los Alamos, New Mexico

Tyler McDaniel
Los Alamos National Laboratory
Los Alamos, New Mexico

John Mellor-Crummey
Department of Computer Science
Rice University
Houston, Texas

O. E. Bronson Messer
Oak Ridge National Laboratory
Oak Ridge, Tennessee

Sheri Mickelson
Argonne National Laboratory
Lemont, Illinois

Richard Mills
Argonne National Laboratory
Lemont, Illinois

Miguel A. Morales
Lawrence Livermore National Laboratory
Livermore, California

Wim Nieuwpoort
Department of Theoretical Chemistry
Zernike Institute for Advanced Materials
University of Groningen
Groningen, the Netherlands

Matthew R. Norman
Oak Ridge National Laboratory
Oak Ridge, Tennessee

Joseph C. Oefelein
Combustion Research Facility
Sandia National Laboratories
Livermore, California

Leonid Oliker
Computational Research Division
Lawrence Berkeley National Laboratory
Berkeley, California

Andrey Ovsyannikov
Lawrence Berkeley National Laboratory
Berkeley, California

Bruce Palmer
Advanced Computing, Mathematics and Data
  Division
Pacific Northwest National Laboratory
Richland, Washington

Thomas Papatheodore
Oak Ridge National Laboratory
Oak Ridge, Tennessee

Manish Parashar
Rutgers University
Piscataway, New Jersey

Scott Parker
Argonne National Laboratory
Lemont, Illinois

Kevin Paul
National Center for Atmospheric Research
Boulder, Colorado

Daniel Peter
Extreme Computing Research Center
King Abdullah University of Science and
  Technology (KAUST)
Thuwal, Saudi Arabia

James C. Phillips
Beckman Institue and National Center for
    Supercomputing Applications
University of Illinois Urbana-Champaign
Champaign, Illinois

Dirk Pleiter
Jülich Supercomputing Centre
Forschungszentrum Jülich
Jülich, Germany

Norbert Podhorszki
Oak Ridge National Laboratory
Oak Ridge, Tennessee

Michel Rasquin
Cenaero, Universite Libre de Bruxelles
Charleroi, Belgium

Daniel Rokhsar
Joint Genome Institute
Lawrence Berkeley National Laboratory
Berkeley, California

Youyi Ruan
Department of Geosciences
Princeton University
Princeton, New Jersey

Beverly A. Sanders
Department of Computer & Information
    Science & Engineering
University of Florida
Gainesville, Florida

Ramanan Sankaran
Center for Computational Sciences
Oak Ridge National Laboratory
Oak Ridge, Tennessee

Sameer Shende
University of Oregon
Eugene, Oregon

Mark S. Shephard
Rensselaer Polytechnic Institute
Troy, New York

Luke Shulenburger
Sandia National Laboratory
Albuquerque, New Mexico

Elliott Slaughter
Stanford University
Stanford, California

Cameron Smith
Rensselaer Polytechnic Institute
Troy, New York

James Smith
Department of Geosciences
Princeton University
Princeton, New Jersey

Tjerk P. Straatsma
National Center for Computational Sciences
Oak Ridge National Laboratory
Oak Ridge, Tennessee

William Tang
Princeton University
Princeton Plasma Physics Laboratory
Princeton, New Jersey

Mark Taylor
Sandia National Laboratory
Albuquerque, New Mexico

David Trebotich
Lawrence Berkeley National Laboratory
Berkeley, California

Sean Treichler
Stanford University
Stanford, California

Jeroen Tromp
Department of Geosciences
and
Program in Applied & Computational
    Mathematics
Princeton University
Princeton, New Jersey

Ronny Tschüter
Technische Universität Dresden
Dresden, Germany

Norm M. Tubman
University of California-Berkeley
Berkeley, California

Brian Van Straalen
Lawrence Berkeley National Laboratory
Berkeley, California

Jean-Luc Vay
Lawrence Berkeley National Laboratory
Berkeley, California

Derek Vigil-Fowler
National Renewable Energy Laboratory
Boulder, Colorado

Henri Vincenti
Lawrence Berkeley National Laboratory
Berkeley, California
and
Lasers Interactions and Dynamics Laboratory
CEA Saclay
Saclay, France

Abhinav Vishnu
Advanced Computing, Mathematics and Data
    Division
Pacific Northwest National Laboratory
Richland, Washington

Yang Wang
Pittsburgh Supercomputing Center
Carnegie Mellon University
Pittsburgh, Pennsylvania

Frank T. Winter
Jefferson Laboratory
Norfolk, Virginia

Patrick Worley
Oak Ridge National Laboratory
Oak Ridge, Tennessee

Katherine Yelick
Computational Research Division
Lawrence Berkeley National Laboratory
Berkeley, California

Eisung Yoon
Rensselaer Polytechnic Institute
Troy, New York

Rudolf Zeller
Institute for Advanced Simulation
Forschungszentrum Jülich
Jülich, Germany

1 Portable Methodologies for Energy Optimization on Large-Scale Power-Constrained Systems

Kevin J. Barker and Darren J. Kerbyson

CONTENTS

1.1 Introduction 1
1.2 Background: How Architectures Drive the ASET Approach 3
1.3 The ASET Approach 4
    1.3.1 Optimizing Per-Core Energy 5
    1.3.2 Optimizing Power Allocation across a Parallel System 6
1.4 ASET Implementation 8
    1.4.1 Example: Wave-Front Algorithms 8
    1.4.2 Example: Load-Imbalanced Workloads 10
1.5 Case Study: ASETs versus Dynamic Load Balancing 11
    1.5.1 Power Measurements and Analysis 12
1.6 Conclusions 15
References 15

1.1 INTRODUCTION

The high-performance computing (HPC) landscape is evolving rapidly on the way toward exascale computing. Whereas ultimate performance was previously the sole metric for computing platform success, future systems will be required to achieve unprecedented levels of performance within tightly constrained power budgets. This emphasis on energy efficiency within the context of high performance will necessitate new approaches to optimizing power usage at all levels, from the underlying technology through the system architecture and software stackto and including application software.

Various hardware manufacturers have developed controls, such as fine-grained power scaling and power gating of components, that enable compute resources to make more efficient use of the available power. In HPC, it is often that these power-saving controls go unused, or at best observe and react to application activities locally on each node [1–3]. While architectural approaches to energy optimization have been and continue to be explored [4,5], further energy savings can potentially be made available by incorporating application information into intelligent runtime decision-making. In this way, opportunities for energy savings may be predictively analyzed and acted upon, allowing runtime software to move beyond merely being reactive and instead to proactively exploit fine-grained power saving.

Placing the burden of power monitoring and allocation on the application software is undesirable especially when considering that in many cases, developing the functionality and scaling applications to exascale are challenging by themselves. Incorporating power optimizations into applications would require developers to be additionally concerned with possibly arcane architectural features of each potential execution platform. This poses a severe limitation to *power portability*, and hence the ability of application software to utilize the available power-saving features across a range of platforms.

A key requirement is a suitable set of abstractions that facilitate the gathering of power usage data and the use of the power controls provided by hardware both locally, on each node, as well as globally across a system. Further, intelligent runtime software that can incorporate application-specific behavioral information is necessary so that the available power resources can be utilized most effectively. This runtime software can be made more intelligent by incorporating predictive and quantitative models of application behavior, allowing opportunities for power savings to be identified and optimized so that power delivery closely matches application requirements.

In this chapter, we describe the *Application-Specific Energy Template (ASET)* approach to energy optimization developed at Pacific Northwest National Laboratory (PNNL). The key concept of an ASET is the separation of concerns, that is, the separation of the power-saving controls from the application software that allows application information to be used for power optimization. An energy template can be considered as the encapsulation of application-specific information into an actionable model that identifies opportunities for proactive energy optimization [6]. These models generally describe per-core behavior and can identify optimization opportunities that cannot be found without application knowledge. The ASET approach has been effectively demonstrated in facilitating power and energy optimizations across a range of workloads and systems. They are *application driven*; the techniques that we target are driven by information provided by the application and have been developed in the context of scientific workloads that execute on today's large-scale systems.

ASETs make use of application-specific information that can be encapsulated in behavioral models. The predictions made by these models take as inputs observable metrics in performance and power utilization. In turn, the predictions are used by runtime software to control the power consumption of various system components. As systems continue to evolve with an eye toward energy optimization, we expect the fidelity of both observable and controllable power characteristics to improve both spatially (e.g., the granularity of functional units) and temporally (i.e., at higher frequencies). A key driver for ASET development is to isolate the application software from this continually evolving architectural landscape. ASETs, therefore, provide a needed level of indirection (and portability) between the application software and underlying hardware.

ASETs work in two primary ways, based on the source of the application information they utilize

1. *Locally:* Per-core information can be used to identify periods of *slack* in which processor cores are not performing useful work and can therefore be moved into a lower power state (p-state). This may happen during long-latency events, such as global synchronization operations or waiting for incoming messages. Periods of inactivity resulting from algorithmic data-dependence can also reveal opportunities for energy optimization and are harder to automatically detect. Application information, in the form of a behavioral model, is necessary to not only identify when such algorithmic opportunities will exist, but also to determine their duration and calculate the cost/benefit of processor core p-state transitioning.

2. *Globally:* Information on application load across a system can be used to optimize power distribution across the parallel system. In this way, ASETs can optimize execution time by increasing the power allocation to compute resources that lie on the parallel performance critical path. This can be thought of as an alternative to dynamic load balancing, in which data (and by association, computation) are moved away from overloaded resources to underloaded resources in an attempt to more evenly distribute a dynamically changing workload.

However, unlike dynamic load balancing, dynamically allocating power (which we term dynamic power steering [DPS]) minimizes or eliminates the cost of data movement and maintains any locality built into the initial load distribution.

In a *power-constrained* system, an ASET with global information can also assist when a system is power constrained, that is, when the full system cannot draw a given amount of power at a specific time. ASETs incorporating a power model in addition to a performance model can alter the p-states of processor cores such that the global power constraint is satisfied while improving the performance of those resources on the performance critical path.

In the following sections, we describe the ASET approach in more detail. Beginning with some background on current and potential future architectures, we build the case for why a unified and abstract approach to application-driven power and energy management is necessary. We then describe the ASET approach and give some brief insight into how ASETs can be implemented in runtime software. Finally, we provide some results that demonstrate the effectiveness of the ASET approach on realistic workloads and parallel systems.

1.2 BACKGROUND: HOW ARCHITECTURES DRIVE THE ASET APPROACH

There are many mechanisms that are used in the current hardware to reduce power consumption. Many of these have their origins in embedded computing systems where both power and energy are limited and where devices are often overprovisioned with hardware resources that cannot all be used at the same time [7]. Additionally, they are targeted at many if not all subsystems within a computing system including the processor cores, the cache subsystem, memory controllers, external memories, motherboard chipsets, fans, network interface cards (NICs), network switches, as well as power supplies and power distribution across a data center. When coupled with the different types of power control mechanisms, their domain of influence, and their frequency of operation and the idiosyncratic differences from system to system, it is clear that attempting to save power (or energy) is not easy, especially if the expectation is on the application programmer to tackle such a task.

Power savings generally arise from three distinct mechanisms:

1. Power scaling: In which the p-state of the target component is changed. The p-state is typically lowered to reduce power consumption but this also reduces the performance of the component—this is especially true for processor cores, and it may or may not actually save energy as this depends upon the activity of the component and whether this activity is in the critical processing path for the application.

2. Power gating: In which a component, or part of it, is effectively turned off. This can significantly save power draw as long as the component is not required, but power gating a device can take significant time to reestablish an operational state.

3. Power shifting: In which the total power consumption of a set of components typically remains constant but power is shifted between components so that the more active components receive more power. This is aimed at reducing the critical path within normal execution.

The most widely used power-scaling mechanism is dynamic voltage–frequency scaling (DVFS) and is typically available on most processors from major vendors including Intel, AMD, IBM, and ARM. Different p-states are established at different voltage–frequency operating points. Often changing just the operating frequency can be done quickly, whereas additionally changing the voltage can take significantly longer. In complementary metal–oxide–semiconductor (CMOS), the dynamic power consumption is related to frequency (f) and voltage (v) as fv^2, and thus a change in voltage has a larger proportional impact than changes in frequency. Though reductions in voltage are becoming harder to achieve as devices operate at near-threshold voltages [8], or in some cases subthreshold [9], when reliability of components become increasingly challenging.

In addition, the domain of influence of DVFS can vary significantly, from a single core on a processor socket to the entire processor socket including the on-chip cache and memory controllers. Other examples of power scaling include multifrequency memories whose frequencies impact on the memory data rates.

Power gating is advantageous when the components are not going to be used for some time. From a coarse level, this could include nodes that are not utilized within a computing system (or spare-nodes that are added to, or taken from a running job). It can be applied to an accelerated system, power-gating accelerators (e.g., graphics processing units [GPUs]) when not used by a particular application. Increased power gating is expected as more heterogeneous resources are added to a processor, resources which may either only be used some of the time or resources that may only be used for specific types of applications.

There have been several cases in which power shifting have been explored, that is, transferring power where it is needed most to speed up the components that are most actively in use. These tend to focus locally within a processing node. An example of this is the shifting of power between the processor and external memories (dynamic random-access memory [DRAM]), or their p-state being changed in combination, to minimize energy consumption [10]. Power shifting can also be applied at a higher level, across nodes within a system, changing the p-state of resources in multiple nodes while satisfying an overall system power constraint. This is an approach we use in our case study later.

It is a complex and challenging endeavor to optimize power consumption as can be gained from this brief discussion of possible power-saving mechanisms. In addition, on a given system, some of these may be in existence and smaller subset may actually be available for use at a user level. Many are used just within the hardware itself, for instance, to ensure that the thermal design point (TDP) is never exceeded and that the component(s) stay(s) within a safe operating range.

Our approach of separating the power-saving mechanisms, from the application, using an ASET is described below. This, we feel provides an approach for the optimization of power and is not tied to a particular system, power-saving mechanism, nor implementation. It does, however, rely on application information being exposed as input to an ASET, and in response the ASET will utilize the power-saving mechanisms available on a particular system.

1.3 THE ASET APPROACH

The key to power portability across disparate systems is the use of ASETs, that provide the linkage between application software and runtime software capable of manipulating system p-states. ASETs encompass application-specific behavioral models that can be used to guide runtime systems to make informed decisions on when to change the p-states of system resources. ASETs are populated by information provided by the applications, providing a per-process view of the application's current state and can be viewed as a state vector containing usage information covering a set of local resources.

The ASET approach is illustrated in Figure 1.1. ASETs are defined in terms of a state machine encapsulating a view of application state as well as the transitions between states. Application state is defined behaviorally, including the level of activity on various resources available within the system. States may capture application behavior at any level of granularity, such as at the loop or function level or between interprocessor communication events. System resources may refer to any component for which power use can be observed and where it can be controlled (e.g., processor core, memory

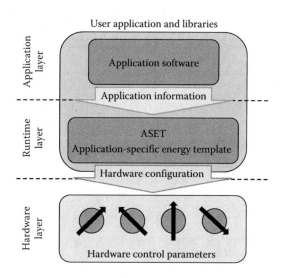

FIGURE 1.1 The ASET captures application-specific behavior and is parameterized in terms of dynamic application information. The ASET's predictions are then used to configure hardware parameters defining power utilization.

controller, or network interface). In the case of systems available today, the ability to observe and control power is limited. However, it is envisioned that emerging and future systems will have measurement and control capabilities at a much finer level of granularity in both space and time, allowing power-optimizing runtime software a greater degree of freedom to exploit opportunities for power savings.

ASETs are defined strictly in terms of application behavior and are parameterized in terms of the power required for each state. This can be determined empirically, by observing the behavior of previous executions of a particular code region, or can be modeled using predictive power models. ASET model states also capture performance information that indicates the amount of time the application is expected to remain in each state. This information is critical in evaluating the cost versus benefit analysis that will enable runtime software to determine which power or energy optimizations to utilize.

ASETs will not only convey information from the application to the runtime software, but also can serve as the mechanism for feeding back information to the application regarding potential optimizations (e.g., whether or not dynamic load balancing involving data migration is appropriate). In this way, we envision that ASETs may themselves be dynamic, adapting over time to reflect application adaptation.

1.3.1 OPTIMIZING PER-CORE ENERGY

Communication-based ASETs (described in greater detail in [6]) make use of application-specific information to identify *a priori* when processor cores will wait for incoming data and thus may be placed in a low *p*-state to save energy. Such states arise in many commonly occurring parallel constructs used by applications in a variety of domains and include cases such as work starvation where work is handed out by a central queue which gradually drains over time, and load imbalance in which some processor cores will finish their processing in advance of others.

These long-latency events result in what is often called *slack time*. Often, this slack time can be automatically identified, and the *p*-state of the processor core reduced, using a mechanism such as DVFS and/or idling the processor cores for a brief period of time. Such cases are shown in Figure 1.2a. Alternatively, the speed of processing prior to a synchronization event can be altered

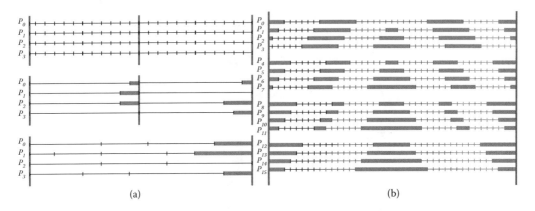

FIGURE 1.2 (See color insert.) Example application communication templates showing periods of idleness (green) suitable for energy savings. Synchronous patterns are shown in (a), while asynchronous patterns requiring application knowledge for energy optimization are shown in (b).

using DVFS, ensuring that all processor cores arrive at the synchronization point simultaneously. However, for a large class of applications no such global synchronization points exist and so these techniques are insufficient. Such a case is the wave-front algorithm shown in Figure 1.2b. In this case, all processor cores undertake the same amount of work and have the same amount of idle time, but at different points during their execution. Idle periods in this case are a result of data dependencies and not load imbalance. Identifying such areas of slack is made even more difficult when this behavior varies over time.

ASETs address this difficulty by capturing a description of application activities in order to guide the runtime software in making energy-optimizing decisions. ASETs make use of the following principles:

- An ASET represents a sequence of active and idle *states* for each processor core and may vary from core to core.
- The rules associated with the *transition* from one state to another uses predetermined application information that represents expected parallel activity.
- The transition between states makes use of *triggers* that monitor application activity. For parallel activities built using the message passing interface (MPI), these triggers may be calls to the parallel activities provided by the MPI runtime.
- Minimal changes to an application source are needed other than to enable/disable a particular template.
- The MPI runtime is augmented to be able to idle a processor core for a certain period of time guided by the information supplied by the template.

Changing the p-state of the processor cores during the identified idle phases can save energy. Whereas the role of the ASET is to identify when this will occur, it is the job of the underlying runtime system to act on this information and change the p-state. The mechanisms to accomplish this are potentially platform-specific; by isolating these platform-specific mechanisms to the runtime software, power portability is maintained for the application.

1.3.2 OPTIMIZING POWER ALLOCATION ACROSS A PARALLEL SYSTEM

While a per-core ASET as described above has proven successful in reducing the power and energy consumption of parallel workloads, decisions are made at a strictly local (per-core) level. However, it is the case that further optimizations in power and energy efficiency may be realized if information describing the state of the entire parallel computation is taken into consideration. To achieve this, ASETs incorporate the concept of *DPS*; using dynamic application-specific information, power can be routed to those computational resources that lie along the performance critical path and potentially away from those that do not. DPS can therefore achieve the following two important goals.

1. Performance improvement via selective power boosting. In this manner, DPS can be used to improve the performance of applications that are load imbalanced and whose load distribution varies over time. Dynamic load balancing is often employed to achieve near-equal quantities of work on each processor allowing processors to progress through application steps at the same speed. However, this often requires significant movement of data across the system from one memory domain to another. In successful load balancing techniques, the cost (in terms of time) of both evaluating a load balancing decision-making algorithm as well as data movement is less than the idle time lost to load imbalance and thus represents an overall reduction in runtime. The complexity for determining optimum, or even reasonable, balanced distributions, as well as determining which data to migrate, can be increasingly complex for irregular data found in a many scientific applications.
2. Performance optimization within power budget constraints. Restrictive power budgets for the largest scale systems imply that it may be the case that not all architectural components may be utilized to their full capabilities simultaneously. As a result, the parallel system may be used in a partially *throttled-down* configuration, or, alternatively, an asymmetric power distribution may be employed across the machine. Using DPS, an ASET has the capability to route power to overloaded resources while simultaneously routing power away from underutilized resources in order to maintain a fixed global power budget.

DPS uses application-specific information as enabled by the ASETs. It is most suited to calculations that are naturally load imbalanced and whose degree of load imbalance varies over time. For example, consider the case of a particle-in-cell (PIC) application in which charged particles are initially uniformly distributed within a space, as shown in Figure 1.3a. The application of an external electric field could significantly perturb the arrangement of the particles leading to a natural load imbalance (Figure 1.3b), which, using conventional load balancing could result in Figure 1.3c. With DPS, there would be no data movement, but rather the resources which hold a higher number of particles would receive more power than those with fewer (Figure 1.3d).

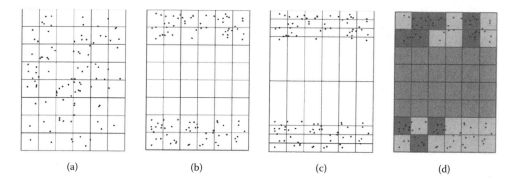

(a)   (b)   (c)   (d)

FIGURE 1.3 Example distribution of particles in a particle-in-cell code showing domain boundaries before and after load balancing and with DPS (gray shading indicates power level supplied to each domain). (a) Initial distribution. (b) During processing. (c) With load balancing. (d) With DPS.

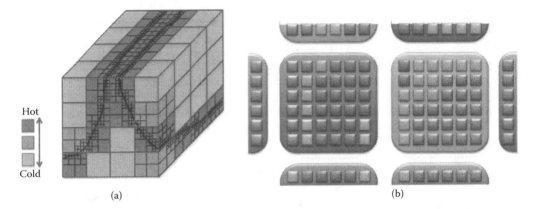

FIGURE 1.4 Example of adaptive mesh refinement application showing notional hot and cold data regions (a) as well as an example of hot and cold cores/sockets in a system (b).

DPS is not limited to PIC codes; it also benefits codes that use complex irregular data sets, including adaptive mesh refinement (AMR) (Figure 1.4) in which cells are at different levels of refinement as the result of physical simulation (e.g., combustion wave fronts that move over time). Many such applications undertake periodic load balancing, but the cost of such actions increase with scale. Data movement operations for load balancing can be reduced or eliminated with DPS.

1.4 ASET IMPLEMENTATION

The use of ASETs has been demonstrated for a number of applications exhibiting multiple types of load balance patterns. Below we describe how ASETs have been used for two examples: the first being for wave-front applications and the second being for load-imbalanced workloads.

1.4.1 Example: Wave-Front Algorithms

As an example, consider a wave-front algorithm whose processing flow is shown in Figure 1.5. Wave-front algorithms are characterized by a dependency in the order of processing grid points within a data grid. Each grid point can only be processed when previous grid points in the direction of processing flow have been processed. Typically, a computation starts at a corner grid point, flows through the grid, and exits at the opposite corner. This flow is considered as a wave front as it passes through the entire data grid. Complexities arise when the direction of wave-front travel varies from phase to phase. The available parallelism within a wave-front application, that is, the number of grid points that can be processed simultaneously, is equal to the dimensionality of the grid minus one; for example, a three-dimensional grid is typically decomposed in only two dimensions.

The most important features that need to be characterized for use in an ASET are the expected amount of delay each processor core will experience prior to processing a subdomain of the data grid and when to transition from one processing phase to another. The ASET for a wave-front algorithm consists of 10 phases. There are four *active* phases (shown as phases 1, 3, 5, and 7) in Figure 1.6 that correspond to the four wave-front directions (wave fronts originating from each corner of the two-dimensional grid), as shown in Figure 1.5, and the last active phases consists of a global reduction collective operation (implemented, for instance, with an MPI_Allreduce operation provided by the MPI runtime). There is a *wait* phase that occurs prior to each active phase. The state transition diagram is shown in Figure 1.6 and the parameters associated with each state as well as the rules for state transitions are listed in Table 1.1.

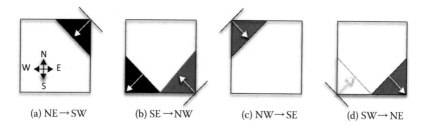

(a) NE→SW   (b) SE→NW   (c) NW→SE   (d) SW→NE

FIGURE 1.5 Ordering of wave-front travel for a two-dimensional domain discretization showing the first steps in each direction: from North-East to South-West (a), from South-East to North-West (b), from North-West to South-East (c), and from South-West to North-East (d). Phases (b) and (d) can proceed concurrently once the wave-front has progressed beyond the initial point of the next wave.

Active ● Idle

FIGURE 1.6 States and transitions in the wave-front ASET.

TABLE 1.1

Summary of the Parameters of the Wave-Front ASET

State	Type	Trigger	Value
0	Wait	Elapsed time	$(P_X - C_X) + (P_Y - C_Y)$
1	Active	MPI_Recv	$N_{\text{Blocks}} \times (!N + !E)$
2	Wait	Elapsed time	$2 \times (C_Y - 1)$
3	Active	MPI_Recv	$N_{\text{Blocks}} \times (!S + !E)$
4	Wait	Elapsed time	$2 \times ((C_X - 1) + (P_Y - C_Y))$
5	Active	MPI_Recv	$N_{\text{Blocks}} \times (!N + !W)$
6	Wait	Elapsed time	$2 \times (C_Y - 1)$
7	Active	MPI_Recv	$N_{\text{Blocks}} \times (!S + !W)$
8	Wait	Elapsed time	$(P_X - C_X) + (P_Y - C_Y)$
9	Active	MPI_AllReduce	1

The transition from a wait phase to a subsequent active phase takes place automatically after an expected wait time. This is the amount of time before the first activity in the active phase will occur. It is given by the state's value, which is defined in terms of the processor core's location in the processor grid and is the specified as the number of blocks to wait for. For example, before entering phase one, the active phase for the NE→SW direction from Figure 1.4a, the distance of the processor core from the NE corner determines the amount of wait time for phase 0. In comparison, the wait time in phase 2, that is, the delay between a processor core completes phase 1 (the NW→SW direction) to starting phase 3 (the SW→NW), depends only on the distance from the South boundary. Note that the processor grid is indexed from 1 to C_X cores in the X-dimension, and 1 to C_Y in the Y-dimension. This corresponds to that used in Sweep3D with the first processor core ($C_X = 1$, $C_Y = 1$) being located at the SW corner, and processor core ($C_X = P_X$, $C_Y = P_Y$) located at the NE corner.

The transition from an active phase to a subsequent wait phase is achieved after a number of the trigger events have been seen. For the first four active phases, the trigger is the reception of a message (implemented as a call to `MPI_Recv`), and the last is an `MPI_Allreduce`. The number of receives that will occur in any of the active phases is nominally two per block for a processor core as indicated by the earlier pseudocode but is different for cores on the edge of the processor grid. *N, S, E, W* are used in Table 1.1 to denote whether a core is on the respective processor edge (set if on the edge). For example, the number of receives in phase one for the *NE* corner will be zero (it initiates the wave-front, and hence has no receives), whereas cores on either the *N* or *E* edge will see only one receive per block and all other cores will see two per block.

1.4.2   Example: Load-Imbalanced Workloads

The ASET-enabled DPS approach optimizes power consumption in two primary ways:

1. Minimization of the power associated with load balancing by eliminating data movement between computational resources for the purposes of redistributing computational load
2. Determining how power can be assigned to those resources that have more work to perform

DPS is most suited where the static calculation of an ideal power distribution is impossible, such as those applications that exhibit dynamic and input-dependent load imbalance that varies over time. Further, applications whose performance is impacted by changes to the node or core *p*-state (i.e., *p*-state) are most amenable to this approach; routing more power to overloaded resources should cause a significant improvement in performance.

DPS results in a *power-optimized* system in which power is directed to the work being performed, enabling applications to optimize performance within global power constraints. A heuristic is used to identify those processor cores that lie along the performance critical path and improve their performance though an additional power allocation. In order to prevent the total power draw across the parallel system from exceeding a prescribed threshold, processor cores that are not on the performance critical path are identified for a reduction in power (in cases without a power constraint, the *p*-states of processor cores not on the performance critical path may not need to be reduced). As long as any power reduction does not then place these processor cores on the performance critical path, the new power distribution is valid.

The heuristic shown in Heuristic 1.1 is used to select the *p*-state for all processor cores in a single application iteration or phase. The maximum amount of work is calculated over all processors (Step 3) and its associated time cost is derived using a performance model (Step 4). The *p*-state for all other processor cores is calculated as being the slowest that does not impact the overall execution

---

Heuristic 1.1   Power assignment within ASET-enabled dynamic power steering

---

Start

1: $PWR_{max} = maximum\ globally\ available\ power$
2: p-state$_{max} = fastest$ p-state
3: $N_{work\_max} = max(N_{work\_i})\ \forall\ i \in \{P_i\}$
4: $t_{work\_max} = N_{work\_max} \cdot t_{work}(\text{p-state}_{max})$
5: $\forall\ i \in \{P_i\ |P_i <> P_{work\_max}\} find\ the\ slowest$ p-state such that $t_{work\_i} < t_{work\_max}$
6: $PWR_i = t_{work\_i}(\text{p-state}_i)$
7: $PWR_{global} = \sum_{i=0}^{p} PWR(\text{p-state}_i)$
8: *If* $PWR_{global} > PWR_{max}$ *then reduce* p-state$_{max}$ *and repeat from Step 3*
9: *Assign* p-state *calculated to each processor core*

End

---

time (Step 5). If the global power budget is exceeded, then the p-state of the highest loaded processor is reduced and the assignment heuristic is repeated from Step 3.

As in the wave-front example, this heuristic is embedded in the ASET for a load-imbalanced workload. The exact mechanism for altering processor-core p-state is determined by the underlying hardware. However, the ASET provides a layer of isolation between the application and the hardware. Changes mandated by alternative methods for either observing or controlling power distribution will be confined to the ASET and will not require changes to the application source code.

1.5   CASE STUDY: ASETS VERSUS DYNAMIC LOAD BALANCING

The efficacy of both the local and global optimization methodologies has been described in previous publications [6,11,12]. However, it is of interest to compare the performance and energy efficiency gains made by the ASET approach to the more traditional dynamic load balancing approach. For generality a synthetic benchmark was designed and implemented for use in this analysis.

The following three parameters describe the execution of the benchmark:

Compute intensity describes the ratio of computation to memory access. In this case, we configure
the benchmark to be either high in compute intensity (the computation contained within
the benchmark executes with memory found exclusively in the lowest level of on-chip
cache) or high in memory intensity (all operands are fetched from main memory).
Load imbalance is used to describe the disparity between what are initially the most and least loaded
processor cores.
Initial data distribution parameter describes how the work is initially allocated to the processor
cores.

There are two configurations explored here: a *blocked* distribution pattern in which all processor cores within a single power domain (a single socket on the system under test) are assigned the same initial amount of work. This pattern aims to minimize the idle time caused by load imbalance within a single power domain, as all processor cores within each domain must be set to the same p-state. A *two-dimensional* distribution pattern initially assigns work to each processor core in proportion to the distance of that processor core from the northwest corner of a two-dimensional processor grid. This pattern can be thought of as corresponding to the initial step in a wave-front algorithm in which the wave is beginning from the northwest corner of the domain.

The synthetic benchmark begins by assigning tasks to each processor core. Each task is configured to reflect the degree of compute or memory intensity specified by the input parameters. Initially, the same number of tasks is assigned to each processor core; load imbalance is affected by altering the amount of computation performed by each task. In a reflection of the idealized parameters of the benchmark, there is no communication between tasks during execution. A lack of data dependencies between tasks means that tasks may be executed in any order. The benchmark concludes when all tasks are completed.

Dynamic load balancing requires migrating tasks between memory domains. Although this is not a parameter we explore here, the amount of data that must be transferred with each task is a tunable parameter. Migrating large amounts of data acts as a detriment to dynamic load balancing performance. However, for this analysis we consider the tasks to be relatively small at only 16 KB.

A number of load balancing algorithms were evaluated, ranging from a single global work queue from which tasks are stolen by each idle processor core, to load balancing *neighborhoods* mirroring the topology of the load distribution. The most ideal configuration (i.e., the configuration resulting in the best performance) was determined for each processor core to select some fixed number of random neighbors. Note that for applications that rely on data locality to achieve performance, this may not be the most effective algorithm. Tasks are migrated using a *pull* methodology:

idle processor cores attempt to steal available tasks from their neighbors. Once tasks have been migrated a single time, they will execute on their new processor cores. This is to prevent thrashing that may occur as tasks migrate multiple times. Tasks are migrated in small batches to reduce the overheads.

To parameterize the performance and power models that are used within the ASET to predict the result of ASET-enabled DPS power allocation, it is necessary to know the power draw of the node components under load for each available p-state. The system used for testing consisted of 32 two-socket nodes each with 10-core Intel Xeon E5-2680 CPUs. Each socket is a single power domain, meaning all cores within a single socket must be set to the same p-state. For this system, p-states are specified by CPU clock frequency in approximately 100 MHz increments. As systems are developed with increased spatial and temporal power control, it will be the case that much finer grained power domains will be available. The system was instrumented using Penguin Computing's Power Insight 2.0 [13], which is able to measure up to 13 power measurements as well as eight thermal measurements within each node at an aggregate rate of 1 kHz. Power measurements for the processor sockets and memory banks were used in the following case study.

1.5.1 POWER MEASUREMENTS AND ANALYSIS

Figure 1.7 shows the CPU and memory idle and active power draws for both compute-intensive (all floating point operations utilize operands that are found in Level-1 on-chip cache) and memory-intensive (all floating point operations utilize operands that are fetched from main memory) workloads for a single node. As shown, the CPU power draw increases with CPU p-state but is not impacted by the difference in compute/memory intensity (Figure 1.7a). This contrasts with a larger difference in power draw by the memory for different compute/memory intensities (Figure 1.7b).

DVFS was used to adjust only the p-state of processor cores on our test system. Because DVFS only impacts CPU frequency, memory power adjustment was not an available option. However, this may be a possibility in other systems. Additionally, because the difference in CPU power consumption between compute-intensive and memory-intensive workloads is minimal, we use an approximate value to represent power consumption in Heuristic 1.1.

Figure 1.8 contains measured per-core execution time for the synthetic benchmark for both compute-intensive and memory-intensive task specifications, as well as for both blocked and two-dimensional initial data distribution on an 8-node system with 160 total processor cores. The default p-state for each CPU core is defined to be 2.0 GHz; this is the clock frequency for both the default (i.e., no load balancing) runs as well as the dynamic load balancing runs. The power consumed by the eight nodes in this configuration is taken to be the maximum power budget that cannot be exceeded during DPS power balancing. This p-state was selected as the default because it gives some opportunity for performance improvement at the cost of increased per-core power consumption. The x-axis is core ID, while the y-axis is execution time. Each plot contains execution time using both ASET-enabled DPS, as well as traditional dynamic load balancing. Execution time for the default case is also shown.

In all cases, both ASET-enabled DPS as well as dynamic load balancing reduced the overall execution time (denoted by the longest-running processor core). In the case of a blocked initial load distribution (Figure 1.8a and c), both ASET-enabled DPS and dynamic load balancing achieve comparable performance, with ASET-enabled DPS leading to the best (lowest) execution time. However, for the two-dimensional initial load distribution, dynamic load balancing outperforms ASET-enabled DPS. This is due to the poor matching of initial load levels and hardware-supported power domains. Because of the coarse-grained power distribution on the system under test, there remains significant idle time on each CPU socket due to the varying load across the socket. Finer grained power control of each core would counter this issue.

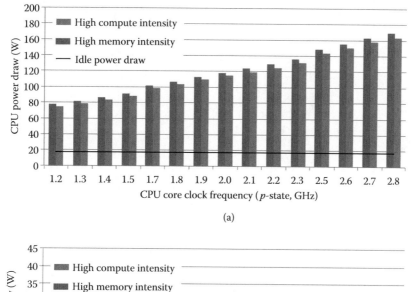

(a)

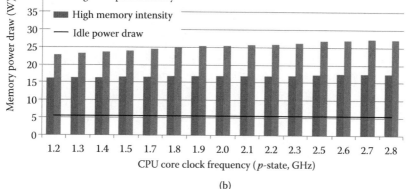

(b)

FIGURE 1.7 Single-node power draw for both CPU and memory for compute-intensive and memory-intensive workloads across all available p-states: (a) CPU power draw and (b) memory power draw.

Figure 1.9 presents total execution time, average power consumption, and total energy consumption for both blocked and two-dimensional initial load distributions, as well as for compute-intensive and memory-intensive benchmark configurations. The data is normalized to the runtime, power consumption, and energy usage when using identical input parameters to the synthetic benchmark without using dynamic load balancing or ASET-enabled DPS. The data shows that in all cases, both dynamic load balancing and ASET-enabled DPS improve both runtime and energy efficiency. However, ASET-enabled DPS outperforms dynamic load balancing in those cases in which the load distribution closely aligns with the hardware-supported power domains (i.e., the blocked initial load distribution). This indicates that ASET-enabled DPS can provide benefit beyond what is typically found for dynamic load balancing in those cases in which fine-grained power control is possible (the blocked data distribution maps workload levels to available power domains), or when the workload patterns of the application map nicely to power domains defined by the hardware. In addition, because this workload is highly idealized, it is believed that more realistic workloads that depend on data locality or involve a high cost in computing and implementing dynamic data redistribution patterns will benefit from ASET-enabled DPS.

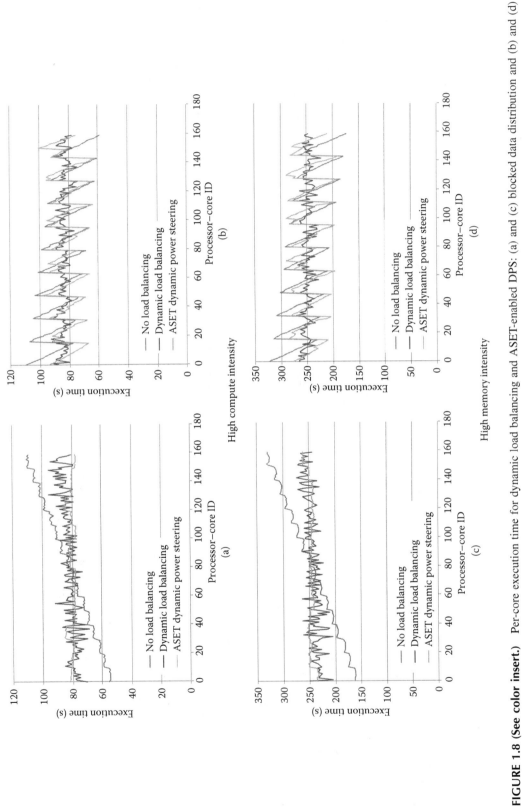

FIGURE 1.8 (See color insert.)   Per-core execution time for dynamic load balancing and ASET-enabled DPS: (a) and (c) blocked data distribution and (b) and (d) two-dimensional data distribution.

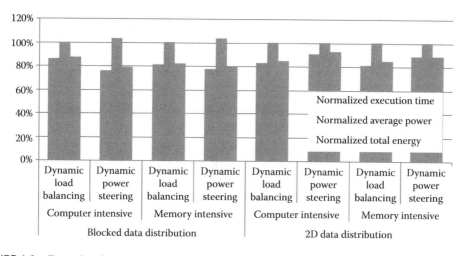

FIGURE 1.9 Execution time, average power consumption, and total energy normalized to an execution not employing dynamic load balancing or ASET-enabled DPS.

1.6 CONCLUSIONS

In response to requirements for providing heretofore-unseen levels of performance within increasingly strict power bounds, HPC architectures are providing features allowing for both the observation and tailoring of power distribution across system resources. However, this presents a problem for application developers and software engineers: in order to structure software to make use of these features, developers find themselves locking into specific architectures that provide specific functionality. Not only does this mean that software must be designed and implemented with a particular feature set in mind, but that porting software to new systems becomes increasingly difficult.

The ASET approach enables developers to target and make use of intelligent runtime software and decision-making algorithms with the goal of optimizing power usage to improve parallel application performance. ASETs provide an abstraction of the underlying platform-specific power management capabilities, giving applications a consistent interface and removing the burden of understanding complex power optimization methods encountered when targeting state-of-the-art architectures and systems. Per-core optimization strategies making use of application-specific behavioral models to exploit algorithmic periods of slack work in conjunction with global system models that exploit load imbalance to route power to where it can be most effectively utilized. ASETs therefore are able to incorporate a number of sophisticated optimization strategies that may be beyond the capabilities of domain scientists and application developers; with ASETs, developers can quickly develop application software that fully utilizes each architecture's power optimization tools.

The ASET approach has proven to be able to exploit opportunities for energy optimization that are not visible without the use of application-specific information. By incorporating application-driven models into the runtime decision-making process, runtime software can be *proactive* at exploiting upcoming behavioral patterns. ASETs have demonstrated their effectiveness using both real and synthetic workloads in which a number of runtime parameters may be adjusted to capture the processing and memory access characteristics of a wide range of large-scale workloads.

REFERENCES

1. V. W. Freeh and D. K. Lowenthal. Using multiple energy gears in MPI programs on a power-scalable cluster. In *ACM Symposium on Principles and Practice of Parallel Programming (PPoPP)*, pp. 164–173, Chicago, IL: ACM, 2005.

2. C. Liu, A. Sivasubramaniam, M. Kandemir, and M. J. Irwin. Exploiting barriers to optimize power consumption of CMPs. In *Parallel and Distributed Processing Symposium (IPDPS)*, Washington, DC: IEEE, 2005.

3. N. Kappiah, V. W. Freeh, and D. K. Lowenthal. Just in time dynamic voltage scaling: Exploiting internode slack to save energy in MPI programs. In *IEEE/ACM Conference for High-Performance Computing, Networking, Storage, and Analysis*, Washington, DC: IEEE/ACM, 2005.

4. B. Rountree, D. K. Lowenthal, B. R. deSupinski, M. Schulz, V. W. Freeh, and T. Bletsch. Adagio: Making DVS practical for complex HPC applications. In *23rd International Conference on Supercomputing (ICS)*, pp. 460–469, New York, NY: ACM, 2009.

5. A. Vishnu, S. Song, A. Marquez, K. J. Barker, D. J. Kerbyson, K. Cameron, and P. Balaji. Designing energy efficient communication runtime systems for data centric programming models. In *IEEE International Conference on Green Computing and Communications*, pp. 229–236. Washington, DC: IEEE, 2010.

6. D. J. Kerbyson, A. Vishnu, and K. J. Barker. Energy templates: Exploiting application information to save energy. In *IEEE International Conference on Cluster Computing*, Washington, DC: IEEE, 2011.

7. S. Mittal. A survey of techniques for improving energy efficiency in embedded computing systems. *International Journal of Computer Aided Engineering and Technology* 6, no. 4 (2014): 440–459.

8. S. Mittal. A survey of architectural techniques for near-threshold computing. *ACM Journal on Emerging Technologies in Computing Systems (JETC)* 12, no. 4 (2016): 46.

9. L. Tan, S. Song, P. Wu, Z. Chen, R. Ge, and D. J. Kerbyson. Investigating the interplay between energy efficiency and resilience in high performance computing. In *Parallel and Distributed Processing Symposium (IPDPS), 2015 IEEE International*, pp. 786–796. Washington, DC: IEEE, 2015.

10. V. Sundriyal and M. Sosonkina. Joint frequency scaling of processor and DRAM. *Journal of Supercomputing* 72, no. 4 (2016): 1549–1569.

11. K. J. Barker, D. J. Kerbyson, and E. Anger. On the feasibility of dynamic power steering. In *2nd International Workshop on Energy Efficient Supercomputing (E2SC)*, pp. 60–69. Piscataway, NJ: IEEE Press, 2014.

12. K. J. Barker and D. J. Kerbyson. Modeling the performance and energy impact of dynamic power steering. In *Workshop on Large-Scale Parallel Processing (LSPP)*, pp. 1380–1389. Washington, DC: 2016.

13. J. H. Laros, P. Pokorny, and D. DeBonis. PowerInsight—A commodity power measurement capability. In *International Green Computing Conference (IGCC)*, pp. 1–6, IEEE, 2013.

2 Performance Analysis and Debugging Tools at Scale

Scott Parker, John Mellor-Crummey, Dong H. Ahn,
Heike Jagode, Holger Brunst, Sameer Shende,
Allen D. Malony, David Lecomber, John V. DelSignore Jr.,
Ronny Tschüter, Ralph Castain, Kevin Harms,
Philip Carns, Ray Loy, Kalyan Kumaran

CONTENTS

2.1	Introduction	18
2.2	Tool and Debugger Building Blocks	19
	2.2.1 Hardware Performance Counters	19
	2.2.2 Sampling	20
	2.2.2.1 Event-Based Sampling	20
	2.2.2.2 Instruction-Based Sampling	20
	2.2.2.3 Data-Centric Sampling	21
	2.2.3 Call Stack Unwinding	21
	2.2.4 Instrumentation	22
	2.2.4.1 Source-Code Instrumentation	23
	2.2.4.2 Compiler-Based Instrumentation	23
	2.2.4.3 Binary Instrumentation	23
	2.2.5 Library Interposition	23
	2.2.6 Tracing	24
	2.2.7 GPU Performance Tools and Interfaces	25
	2.2.8 MPI Profiling, Tools, and Process Acquisition Interfaces	26
	2.2.9 OMPT—A Performance Tool Interface for OpenMP	27
	2.2.10 Process Management Interface—Exascale	28
	2.2.10.1 Architecture and Infrastructure	28
	2.2.10.2 Requirements	28
2.3	Performance Tools	28
	2.3.1 Performance Application Programming Interface	29
	2.3.2 HPCToolkit	30
	2.3.3 TAU	31
	2.3.4 Score-P	32
	2.3.5 Vampir	33
	2.3.6 Darshan	35
2.4	Debugging Tools	36
	2.4.1 Allinea DDT	37
	2.4.2 The TotalView Debugger	39
	2.4.2.1 Asynchronous Thread Control	39
	2.4.2.2 Reverse Debugging	39
	2.4.2.3 Heterogeneous Debugging	39

        2.4.2.4   Architecture and Infrastructure...40
        2.4.2.5   Multicast and Reduction...40
        2.4.2.6   Debugger Requirements 40
    2.4.3   Valgrind and Memory Debugging Tools 41
    2.4.4   Stack Trace Analysis Tool 42
    2.4.5   MPI and Thread Debugging 43
2.5   Conclusions...44
References 45

2.1   INTRODUCTION

While high-performance computing (HPC) systems are often assessed in terms of their raw computing power, it is their ability to efficiently and correctly perform large-scale scientific simulations that makes them invaluable tools for modern scientific research. Performance tools and debuggers are critical components that enable computational scientists to fully exploit the computing power of these systems. Performance tools help developers of applications, frameworks, and libraries use HPC platforms efficiently by collecting information to analyze and model the performance of their code on a given architecture and they help to identify costly program regions and quantify their impact on code performance and scalability.

Debugging tools help developers to identify and correct a variety of correctness problems in applications including logic errors, data corruption, and parallel race conditions. Debugging tools enable observation, exploration, and control of program state, such that a developer can verify that execution correlates to what is intended. Both performance tools and debuggers rely on a well-developed ecosystem of supporting components from both the system hardware, such as hardware performance counters, and system software, such as the CUDA Debugger API and MPIR API.

A variety of challenges exist in enabling performance tools and debuggers on today's petascale systems. Tools and debuggers typically have low-level dependencies on system components, including instruction sets and operating systems calls. Therefore, the unique architectures of HPC systems can impose significant cost on tool and debugger development. For example, Blue Gene/Q vector instructions have presented porting challenges to tools such as Valgrind, and limitations of the CUDA Debugger API have made fine grain control of graphical processing unit (GPU) threads difficult for debuggers. The introduction of new architectural features, such as deeper memory hierarchies, has added to the analysis requirements for performance tools. The degree of parallelism on today's systems, reaching to over a million-way concurrency, demands that debuggers and tools be able to interact with jobs in a scalable way in order to control, collect, output, and visualize the large volume of data generated. Tools and debuggers require visibility into hardware and software that is not always provided, the hardware does not always provide adequate hardware counters for performance tools, and higher level programming models such as OpenMP may not provide access to information on their internal state.

This chapter explores present-day challenges and those likely to arise as new hardware and software technologies are introduced on the path to exascale. A variety of performance and debugging tools that have been successfully employed on today's largest HPC systems will be discussed. The focus is on third-party tools, those not provided directly by system vendors. While vendor tools can provide important functionality, in many cases the third-party tools represent the leading edge of tool development and often face the most hurdles on a new platform as they are often developed with limited or no vendor support. Third-party tools have also had the most cross-platform availability and therefore the largest base and history in the user community. In addition to covering specific tools, this chapter covers some of the underlying hardware (such as hardware counters), software (such as the PMPI interface), and techniques (such as sampling) that enable tools and debuggers.

2.2 TOOL AND DEBUGGER BUILDING BLOCKS

Critical underlying and enabling technological components and techniques upon which performance and debugging tools rely are presented in this section. This includes hardware performance counters (Section 2.2.1), sampling (Section 2.2.2), call stack unwinding (Section 2.2.3), instrumentation (Section 2.2.4), library interposition (Section 2.2.5), tracing (Section 2.2.6), GPU tool and debugger interfaces (Section 2.2.7), message-passing interface (MPI) profiling and tools interfaces (Section 2.2.8), the OpenMP tools interface (Section 2.2.9), and scalable startup and process attachment (Section 2.2.10).

2.2.1 Hardware Performance Counters

Modern processors contain a small set of registers known as hardware counters onto which a larger set of performance-related events can be mapped. Hardware counters can collect accurate and detailed information about a wide range of hardware performance metrics. The kind of counters that are available is highly hardware dependent; even across the CPUs of a single vendor, each CPU generation has its own implementation. Also, the number of events, as well as which events that can be counted simultaneously by the underlying hardware, varies with each processor architecture. The PAPI API (Section 2.3.1) is a common way through which tools and applications access these counters.

About a decade ago, the primary method of boosting the computational capacity of a chip shifted focus to increasing the number of processing units (cores). These cores share performance-critical chip resources called off-core, uncore, or sometimes NorthBridge, which we refer to under the unified moniker of *inter-core* resources. Included is everything but the core itself (e.g., memory hierarchy, memory controllers, on-chip networks, on-chip GPUs, Peripheral Component Interconnect [PCI] bus, and power). Interactions between cores and contention for shared inter-core resources become increasingly important origins of potential performance bottlenecks. The current situation with inter-core counters is similar to the one with core counters, in that many different platform-specific interfaces exist, with some poorly documented or not documented at all, which makes determining what is being measured very difficult.

The following example highlights the complexity of *meaningful* system-wide performance measurements: today's chip architectures, as they are found in most modern systems ranging from workstations up to TOP500 supercomputers (including all modern Intel Xeon, Xeon-Phi, and Advanced Micro Device [AMD] processors), intend to facilitate heterogeneous computing. The heterogeneous system architecture allows any on-chip GPU (or other accelerators) to operate in the same memory space as the CPU cores, which results in the sharing of various levels of the cache (inter-core resource), while a single cache-coherent address space is provided and needs to be maintained. A number of existing studies show that (1) cache coherency creates substantial additional on-chip traffic (Schuchardt et al. 2013; Keramidas and Kaxiras 2010); (2) hierarchical caching does not improve the probability of finding a cache line locally (Xu et al. 2011); and (3) conventional hardware coherence creates too much long-distance communication (Ros et al. 2012). Given the increased on-chip parallelism, the complexity of correlating on-core events with shared inter-core events, together with the currently limited vendor and kernel support, the fact that shared hardware resources are currently the limiting factor of performance is no surprise.

Over time, other system components beyond the processing chip have gained performance interfaces (e.g., GPUs, I/O, network interfaces). For instance, many network switches and network interface cards (NICs) contain counters that can monitor various events related to performance and reliability. Possible events include checksum errors, dropped packets, and packets sent and received by a node, and the number of packets originating from a node as opposed to being passed through a node. Furthermore, other system health measurements, such as chip- or board-level temperature sensors, are available and useful to monitor, preferably in a portable manner. Although the set of hardware counters for various system components is necessarily dependent on the underlying hardware, the ability to monitor these system-side events is the only path forward for fully understanding and improving application efficiency and scalability on future exascale hardware.

2.2.2 SAMPLING

Sampling is a technique utilized by many tools, including HPCToolkit (Section 2.3.2), TAU (Section 2.3.3), and Score-P (Section 2.3.4). Monitoring all events in program executions on extreme-scale parallel systems is impractical. To reduce the cost of monitoring, performance tools typically use *sampling* to monitor only a subset of events. Within a parallel program execution, one can use sampling at a coarse grain by opting to collect information about only a subset of processes or threads. At a finer grain, one may employ *synchronous* or *asynchronous* sampling to monitor a subset of events that occur during execution of threads selected for monitoring.

To use synchronous sampling, one augments a program with instrumentation (also know as a probe; see Section 2.2.4) that is encountered as part of its normal control flow but then uses this instrumentation to collect information only a subset of the times that it is encountered. For example, Photon MPI (Vetter 2002) uses this approach to monitor a subset of the instrumented messaging operations encountered during execution of an MPI program.

In contrast, asynchronous sampling uses timer expiration or hardware counter (Section 2.2.1) unit triggers to enable a tool to associate a metric with a thread's execution context. A strength of asynchronous sampling is that a tool can use it to measure and attribute fine-grain performance metrics with low overhead. In the rest of this section, we survey mechanisms for asynchronous sampling.

2.2.2.1 Event-Based Sampling

For decades, Unix profiling tools, e.g., *gprof* (Graham et al. 1982), have employed interval timers to periodically interrupt an executing program to determine where a program spends execution time. Today's microprocessors include hardware counters that can be used to count hundreds of different kinds of events in addition to time. Each counter can be configured to trigger an interrupt when a target event count threshold is reached. Typically, a tool can concurrently count a handful of events for each hardware thread. Using such counters, one can count events that represent work (e.g., instructions executed), resource consumption (e.g., cycles), or inefficiency (e.g., stalls).

On microprocessors that support out-of-order execution, it is difficult to precisely attribute an event to the instruction that caused it. For out-of-order cores, metrics attributed using event-based sampling (EBS) are too imprecise to support fine-grain execution analysis at the instruction level. Even so, metrics gathered using EBS can provide meaningful aggregate information at the loop or routine level.

To avoid imprecise attribution using EBS, Intel processors provide special hardware support for *precise event-based sampling* (PEBS) (Sprunt 2002), which uses a microassist to precisely attribute events to instructions. Until Intel's Goldmont microarchitecture, not all kinds of events supported precise EBS (Intel Corporation 2016). To reduce the overhead of collecting and processing event-based samples, Intel's PEBS hardware can log machine state associated with a sample event in a buffer so that the processor need not be interrupted until the buffer fills.

2.2.2.2 Instruction-Based Sampling

In the late 1990s, DEC developed instruction-based sampling (IBS) to support detailed performance monitoring and precise attribution of metrics on out-of-order cores (Dean et al. 1997). At regular intervals, the hardware selects an instruction for monitoring dispatching it for execution. As a monitored instruction moves through the execution pipeline, the core records a detailed record of significant events (e.g., cache misses), latencies associated with pipeline stages, the effective address of a branch target or memory operand, and whether the instruction completed successfully or was aborted. When a sampled instruction completes execution, an interrupt is triggered and a tool can inspect the detailed account of the instruction's execution. The execution record for a sampled instruction supports detailed analysis of pipeline and memory hierarchy utilization.

IBM's Power5 and its successors refine the idea of IBS with marked instructions. Rather than collecting samples for arbitrary instructions, marked instructions enable a tool to randomly sample instructions that satisfy a predicate condition that can focus monitoring, e.g., monitor loads that filled a primary cache block from memory that was not in any cache or instructions whose delay at a particular stage in the pipeline exceeded a specified threshold (IBM Corporation 2015). Marked instructions can increment event counts and then be used to trigger a sample when a particular threshold value is exceeded.

In 2015, NVIDIA introduced hardware support for periodically sampling instructions being executed by each streaming multiprocessor (SM) in its GPUs (Matwankar). At the expiration of each sampling period, a sampling agent inspects an active warp from an SM and records its program counter (PC) and execution state. The execution state can indicate whether instruction is ready to issue or blocked waiting for another instruction to complete execution, a cache miss or one or more pending memory accesses to complete, a synchronization to complete, or functional units to become available. This sampling capability enables a tool to monitor the execution of code on GPU architectures, see what kinds of delays occur during execution, and associate them with the code incurring the delays.

2.2.2.3 Data-Centric Sampling

With the exception of NVIDIA's GPUs, all of the IBS implementations support data-centric sampling (DCS). For a sampled instruction, the effective address of a data access (if any) is available in addition to the address of the instruction itself. This enables a tool to associate access latency or access to particular levels of the memory hierarchy with both instructions and data. On Intel processors, by using precise EBS in conjunction with a load latency performance monitoring facility, one can measure the latency of a load from dispatch to its final write back from the memory subsystem (Intel Corporation 2016). Using these capabilities, a tool can associate the effective address of a data access with its load latency and identify the data source for the load.

Using performance monitoring unit capabilities for IBS or EBS, one can obtain detailed insight into the execution behavior of programs and how they leverage various capabilities of a microprocessors and graphics processing units, e.g., functional units and the memory hierarchy. A challenge for tools is translating measurements using a collection of hardware counters into high-level insight about what aspects of a program's behavior need attention and how to address the problems observed. With hundreds of performance counters available, tool developers need a top-down methodology for analyzing performance data, such as the one for Intel processors (Yasim 2014) or IBM's CPI stack model (Srinivas et al. 2011). Armed with such a methodology, a sampling-based performance tool can collect the set of metrics necessary to diagnose any inefficiencies present, and pinpoint code and/or data associated with problems discovered.

A requirement for successful diagnosis of performance problems on forthcoming exascale systems is a capable set of hardware performance counters for processor cores and accelerators (if any) along with a top-down methodology to use them. The next section addresses call stack unwinding, an approach often used to associate a performance metric gathered using sampling with an applications program's calling context.

2.2.3 CALL STACK UNWINDING

Parallel applications typically invoke functions such as message-passing communication primitives from more than one context. Understanding the performance of such applications requires associating each functions' costs with the contexts in which it is invoked. Call-graph profilers such as *gprof* (Graham et al. 1982) apportion the cost of each function among its callers by measuring the time spent in each function using asynchronous sampling, counting the number of times the function is called by each caller using instrumentation at function entry, and then dividing the function's costs among its callers according to the proportion of calls from each. This approach can produce misleading results when the cost of functions, e.g., MPI_Wait, depends on the context in which they are invoked (Ponder and Fateman 1988).

To more accurately attribute the costs of context-dependent function invocations, many modern profilers associate the cost of a function invocation with multiple levels of *calling context*—the stack of procedures active in a thread at a particular point in time. While some performance tools gather calling context information by instrumenting each function's entry and exit (Muller et al. 2007; Shende and Malony 2006), the runtime cost of such instrumentation can be excessive and require mitigation approaches such as *function filtering* (TU Dresden Center for Information Services and High Performance Computing) or *throttling* (Department of Computer and Information Science). The cost of tracking calling contexts using instrumentation is proportional to the number of function invocations in an execution. Rather than instrumenting functions to obtain calling context, other tools use *call stack unwinding*. Call stack unwinding is performed by determining the base pointer, stack pointer, and PC in the caller of the active function and then repeating this process for every other function on the call stack. Call stack unwinding for a function can be performed by using a library such as *libunwind* (Mosberger-Tang), which uses recipes encoded by the compiler as DWARF Frame Description Entry (FDE) and Common Information Entry (CIE) records (DWARF Debugging Information Format Committee), or by using binary analysis to determine unwinding recipes directly from a function's machine code (Tallent et al. 2009a). The unwinding recipe that applies for a code location indicates how to compute the value of the stack pointer for the caller and where to find values for caller's base pointer and PC, which may be in registers or on the stack at offsets relative to the current frame's base pointer or stack pointer. The cost of unwinding the call stack grows with the call chain depth.

Call stack unwinding is commonly used in conjunction with asynchronous sampling to attribute the cost associated with a sample event to the calling context where the event was triggered. The cost of using call stack unwinding in conjunction with asynchronous sampling can be reduced when temporally adjacent sample events occur in contexts that share common procedure frames. Since all procedure frames on the call stack for the previous sample are already known, there is no reason to unwind through these frames again. One can identify a common unwind prefix with the previous sample by using a *sample bit* to mark each procedure frame seen by the previous unwind (Whaley 2000), or using a *trampoline* to mark the innermost procedure frame in the call stack that has not returned since the last sample (Arnold and Sweeney 1999). Using either of these approaches, a call stack unwinder need only unwind the new call-path suffix and append it to the prefix in common with the previous sample. This approach reduces the average cost of unwinding to the average number of new procedure frames seen at each sample event.

Call stack unwinding is used by HPCToolkit (Section 2.3.2) (Adhianto et al. 2010b; Rice University) as well as by debugging tools such as the Stack Trace Analysis tool (STAT) (Section 2.4.4) (Arnold et al. 2007), and other performance tools such as Open|Speedshop (Open|Speedshop) and TAU (Section 2.3.3) (Shende and Malony 2006).

2.2.4 INSTRUMENTATION

Instrumentation is the insertion of code (or *probes*) to perform measurement in a program. It is a vital step in performance analysis, especially for parallel programs. Program instrumentation can take place at several levels of program transformation where events of interest to measure are defined. This can span several levels from source code (manual, preprocessor), compiler, library interposition, linker, binary/dynamic, interpreter, to virtual machine. A complete performance view of the parallel execution sometimes requires contribution of event-specific information from multiple instrumentation levels. Tools can use instrumentation to make measurements for different types of events, such as *atomic*, *context*, or *interval* events. An *atomic event* denotes a single action that is triggered with data associated with it. For instance, an atomic event could indicate that some memory was allocated of a particular size. A *context event* is an atomic event with additional information about the executing context at the point where it is triggered. For example, a tool might maintain an event stack to capture the nesting relationship between interval events. The context event could include the event stack with the event measurement. In contrast to atomic events, an *interval event* is comprised of a pair of *begin* and *end* events to measure code regions such as loops and routines.

2.2.4.1 Source-Code Instrumentation

Tools typically provide an API for manual instrumentation of source code. Source-code instrumentation is a robust and portable instrumentation option that easily extends to support all compilers for a given language. However, manually inserting timer calls in the source code can be tedious and error-prone. Source-code instrumentation is simplified by using a source-to-source preprocessor based on a static analysis system. The degree of instrumentation can be specified to exclude or include routines and files for instrumentation or specify loops in routines for instrumentation. Source instrumentation offers the flexibility to control the instrumentation granularity, for example a nesting level may be specified to specify instrumentation of outer or inner loops.

2.2.4.2 Compiler-Based Instrumentation

Modern compilers used in HPC systems provide special flags to enable instrumentation of routine entry and exit by triggering a pair of calls with a function address and its callsite location. A performance measurement tool that implements these calls then maps the address to the source-code location using symbol and line map information recorded in an executable or shared library by a compiler. The Binary File Descriptor (BFD) interface provided by the GNU binutils library (Binutils) provides an interface that can be used to perform this mapping. The overhead of compiler-based instrumentation is slightly higher than source-code instrumentation, and the tool needs to take special care of instrumentation of dynamic shared objects (DSOs) while mapping source-code locations and demangling C++ symbol names. Additionally, the use of compiler-based instrumentation alone limits the granularity of instrumentation to the function level which precludes performance measurement at level of lines or loops.

2.2.4.3 Binary Instrumentation

Instrumentation of codes after the compiler has produced that binary is also possible. This can enable, for example, probes for interval events at routine and loop boundaries. Performance tools can instrument binaries using binary rewrite tools such as the Dyninst API and MAQAO.

2.2.5 LIBRARY INTERPOSITION

Library wrapping is a form of source instrumentation whereby the original library routines are replaced by instrumented versions, which in turn call the original routines. The problem is how to avoid modifying the library calling interface. Some libraries provide support for interposition, where an alternative name-shifted interface to the native library is provided and weak bindings are used for application code linking. The advantage of this approach is that library-level instrumentation can be implemented by defining a wrapper interposition library layer that inserts instrumentation calls before and after calls to the native routines. Many tools use MPI's support for interposition, PMPI (Section 2.2.8), for performance instrumentation of the MPI libraries and similarly some tools make use of OpenSHMEM's PSHMEM interface.

Another approach to routine-level instrumentation is runtime interception of calls. This can be accomplished through preloading a DSO. For instance, tools can leverage the Linux LD_PRELOAD environment variable to preload a shared object that intercepts MPI calls using the PMPI profiling interface. Likewise, it is also possible to preload replacements for libraries like the POSIX I/O, CUDA, and OpenACC libraries that invoke instrumented routines. In the case of CUDA, the CUPTI interface is used. Runtime interposition of libraries is a technique that can be applied in general with any shared library that is loaded by the program. TAU (Section 2.3.3) provides a tool (called *tau_gen_wrapper*) that can read the library's interface by parsing the header files in Program Database Toolkit (PDT) and automatically generate a library that redefines the routines specified in the header file. The first routine invoked from this wrapped library loads the given library using the dlopen() system call. Internally, each routine in the wrapper library searches for the corresponding call from the given library and calls the routine with the appropriate arguments. The wrappers can implement tool instrumentation calls before and after the calls to the library being wrapped.

In order to allow for preloading DSOs applications must be dynamically linked. However, on many HPC platforms static linking is the default and preferred option. On Linux systems, the linker supports function wrapping when statically linking by using the wrap, name flag for each routine to be wrapped. Since there may be a several of these flags (one for each routine), they can be stored in a file and passed using the special @argfile syntax supported by most linkers. When @argfile is passed on the command line, the linker will read command-line options from a file named *argfile*. This allows tools to use interposition libraries that can be linked in or preloaded in the context of the executing application to instrument the code transparently without requiring any modifications to the source code or the build system.

2.2.6 TRACING

Tracing is a commonly applied method for analyzing details of parallel program execution and is utilized by a variety of tools including Score-P (Section 2.3.4), Vampir (Section 2.3.5), TAU (Section 2.3.3), and HCPToolkit (Section 2.3.2). Unlike profiling, the tracing approach records a program's state or state change as timed samples or events, i.e., as a combination of timestamp, and state specific data. This creates a data stream, which allows very detailed observations of parallel programs. With this technology, synchronization and communication patterns of parallel program runs can be traced and analyzed in terms of performance and correctness. Normally, the analysis is done in a postmortem step, i.e., after completion of the program. The recorded data's high level of detail also enables software tools for automatic performance analysis and estimation. Program traces can also be used to calculate performance profiles. Computing profiles from trace data allows arbitrary time intervals and process groups to be specified. This is in contrast to *fixed* profiles or phase profiles accumulated during runtime.

The concept of tracing can offer a high degree of detail. However, the consequence of an increasing degree of detail can be a literal flood of information. This information needs to be stored, visualized, analyzed, and archived. It scales with the number of processes and the observation time. The volume of data easily exceeds the size of today's computer main memories. Clearly, the data volume depends on the frequency at which data is measured and recorded. Depending on the source of information, two different cases are relevant in tracing. Case one collects program state samples asynchronously at a predefined rate, which results in a predictable data volume that can be estimated by the simple formula: size = *number of processes * sample rate * sample size * time*. The resulting data needs to be analyzed statistically. Case two records individual synchronous state changes (events) of a parallel program, e.g., entering and leaving of synchronization or communication phases, whenever they occur. This scenario does not allow estimating the size of the data prior to executing the parallel program as the corresponding data rate depends on implementation details of the observed parallel program itself. Figure 2.1 illustrates the difference between sample- and event-based tracing graphically. While event-based tracing records every incarnation of a predefined event, sample-based tracing records frequent samples, which can be used to derive statistical properties of the given event only. Tracing often deploys the event-driven collection methodology, presumably because its resulting data stream enables a very descriptive representation of a program's behavior over time. Likewise, the measurement overhead during program tracing obeys the same causality.

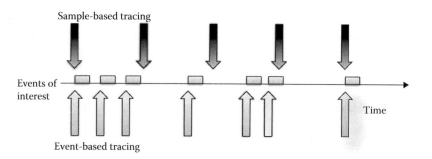

FIGURE 2.1 Difference between sample-based and event-based tracing.

Recent developments in tracing focus on intelligent performance data preselection, filtering, and processing. Likewise, parallelism and data distribution are introduced into performance data processing.

2.2.7 GPU PERFORMANCE TOOLS AND INTERFACES

General-purpose graphical processing units (GPGPUs) have emerged as viable exascale-computing elements. In fact, two of the three Department of Energy's (DOE) CORAL pre-exascale machines (CORAL) are underpinned by NVIDIA Volta GPUs, and a majority of their FLOPs come from these devices. Keeping abreast of this architectural trend, applications are being increasingly migrated to GPUs, but find that realizing the full potentials of GPUs can often require significant efforts. On the path to exascale computing lies a high demand for an effective GPU code-development tools ecosystem.

The performance-analysis tools of hardware vendors, in particular NVIDIA's, are generally well respected. Tools like nvprof (Profiler User's Guide) and NVIDIA Visual Profiler offer key capabilities needed to reason about the application's performance on a single GPU-enabled node. nvprof is a command-line-based profiler, which can summarize the overall performance of a GPU code, but also trace activities such as each kernel execution and memory copy/set instance and CUDA runtime API calls. NVVP is a graphical tool that shows such information in a unified time line and allows a further drill down into raw data and GPU performance metrics.

NVIDIA's proprietary tools, however, do not provide adequate support for analyzing an MPI program running on multiple compute nodes. Their strategy is to fill this gap by enabling open-source tools including HPCToolkit (Section 2.3.2), TAU (Section 2.3.3), and Vampir (Section 2.3.5). In fact, both proprietary and open-source tools build on the same performance-analysis infrastructure called CUPTI. Continuous evolution of this infrastructure, therefore, is essential to providing an effective performance-tools environment.

With respect to debugging applications running on GPUs, parallel debuggers such as TotalView (Section 2.4.2) and DDT (Section 2.4.1) already have this capability. They build either directly on NVIDIA's low-level debug interface called CUDA Debugger API or CUDA-GDB, NVIDIA's single-process debugger, which uses the same API internally. While these debuggers offer standard debugging operations such as stopping the program at certain points in execution and allowing users to inspect GPU variables across different nodes, some of the fundamental deficiencies of this API currently limit its effectiveness for HPC application.

One deficiency of the CUDA Debugger API is a hardware-centric interface that allows the debugger to control only threads that are active and currently reside on the GPU. With such semantics, the debuggers cannot control threads at the logical grid/block level. Further, asynchronous control of threads so that they all can reach the same execution point is not supported. Such idioms are becoming increasingly important for high-level programming models such as OpenMP. Other deficiencies include a lack of debugging support for Multi-Process Service (MPS) and suboptimal performance on applications that launch millions of kernels. Compilers and runtimes also have deficiencies: they generally lack debugging support for high-level programming models such as OpenMP (Protze et al. 2015).

With respect to certain classes of GPU memory correctness errors such as misaligned or out-of-range accesses to either shared or local memory, CUDA-MEMCHECK (CUDA-MEMCHECK) provides handy ways to detect and locate them in the GPU code. However, its data-race detection is significantly limited as it can only detect races occurring in shared memory.

2.2.8 MPI Profiling, Tools, and Process Acquisition Interfaces

Most programs for large-scale parallel systems employ message passing communication expressed using the Message Passing Interface (MPI). To support monitoring of message passing applications, the MPI standard defines a profiling interface, commonly referred to as the PMPI interface, to be provided as part of any MPI library implementation. This interface is commonly utilized by tools, including HPCToolkit (Section 2.3.2), TAU (Section 2.3.3), and Score-P (Section 2.3.4) to collected MPI-related performance information. The interface allows tools to add instrumentation (Section 2.2.4) to the MPI library by way of transparent interception of MPI calls and is provided through a requirement that all MPI routines be available with both an MPI_ and a PMPI_ prefix. This enables wrapper libraries to redefine the MPI_ entry point with added instrumentation and then call the associated PMPI_ function from within the modified routine. This provides for a simple and portable mechanism for tools to interface with MPI without requiring access to the underlying MPI implementation source. In environments that support weak symbols, such as Linux, the MPI_ routines are typically defined as weak symbols, which allows the linker to preferentially link to any MPI_ routines defined in a tool wrapper library as standard strong symbol. If no such strong symbols exist the linker will default to using the weak symbols to resolve function references. For environments without weak symbols, two MPI libraries implementing the routines with the two different prefixes may be required. In either case, the profiling interface allows for link time substitution of MPI_ routines with instrumented tool wrapper library routine. References are resolved to an interposing library by simply including the library at link time. Major limitations of the MPI profiling interface is that it restricts the interception of MPI calls to a single tool wrapper library and it does not provide a mechanism to access internal performance or state information of an MPI implementation.

This last limitation has been alleviated as of the MPI 3.0 standard through the introduction of the MPI tool interface. This interface provides a mechanism for MPI implementations to expose variables, each of which represents a particular property, settings, or performance measurement from within the implementation. The interface consists of two parts, with the first part providing information about. control variables through which the MPI implementation tunes its configuration along with the ability to set them. The second part provides access to performance variables that can provide insight into internal performance information of the implementation. The specification does not define any control or performance variables and instead allows the implementation to specify the specific variables that it will provide. The specification does, however, provide the necessary routines, prefixed with MPI_T_, to find all of the variables provided by a particular implementation, to query their properties, to retrieve their descriptions, and to access and alter their values. The variables provided through the tool interface are scoped to provide information for the general MPI environment of a process, or for individual MPI objects, such as communicators or data types.

To allow debuggers and other tools to attach to an MPI job, the MPIR process acquisition interface was created (DelSignore). While this interface is not part of the official MPI standard, it has become a de facto standard which is utilized by a wide variety of MPI implementations. MPIR is not an API but is instead a rendezvous protocol between the MPI implementation and the tool which allows tools to obtain a list of MPI process descriptors, receive event signals, and locate the message queue display library. An MPI process descriptor contains information on the node address, executable name, and process ID for each MPI process in the job. This enables a tool to locate the MPI processes that belong to a job, which is needed when launching MPI jobs under tool control or attaching to already running MPI jobs. The interface also allows for event signaling from the MPI runtime to the tool to indicate events such as process spawn and abort and provides a mechanism to specify the path to the MPIR Message Queue Display library, which allows tools to access the state of the message queues in the MPI processes.

2.2.9 OMPT—A Performance Tool Interface for OpenMP

Unlike MPI (Section 2.2.8), the OpenMP specification has historically not provided a standardized interface to support the collection of performance information. A challenge when analyzing the performance of parallel applications that employ OpenMP to exploit node-level parallelism is that there is a substantial semantic gap between a program's OpenMP source code and its implementation. OpenMP compilers separate regions associated with certain OpenMP directives from the functions in which they appear by *outlining* each such region into its own function. Such radical program transformations make it difficult for performance tools to attribute measurements of an OpenMP program's execution back to its source code.

To address this problem and support development of high-quality, portable, *first-party* performance tools (a first-party performance tool is one that runs within the address space of the program that is being monitored), the Preview 1 release of OpenMP 5.0 specification includes a performance tools interface known as OMPT (OpenMP Language Working Group 2016). The design for OMPT includes support for building performance tools that monitor execution on one or more OpenMP devices using sampling and/or tracing. Once the OpenMP 5.0 standard is finalized, any tool that uses the OMPT interface to interact with an OpenMP runtime system should be able to monitor and analyze the performance of any OpenMP program developed using any compliant OpenMP implementation.

The design for OMPT includes a specification for the following interfaces:

- An initialization interface through which a tool can register a tool initializer and finalizer
- A routine that enables a tool to query the state of an OpenMP thread to determine what the thread is doing (e.g., serial work, parallel work, waiting for synchronization, idle, etc.)
- Routines that enable a tool to examine the sequence of nested parallel regions and tasks (if any) surrounding the current context
- A routine that helps a tool map implementation-level call stacks back to their source-level representations by identifying which stack frames belong to the OpenMP runtime system
- A callback interface that enables a tool to receive notification of OpenMP *events* that occur as OpenMP regions execute
- Callback signatures for tool-provided routines that may be invoked by the runtime's callback interface
- An interface that enables a tool to trace operations (e.g., data allocation and deallocation, data copies, kernel execution) on OpenMP target devices
- A runtime library routine that an application can invoke to control a tool

At this writing, several OpenMP runtime systems, including the LLVM OpenMP runtime and IBM's LOMP OpenMP runtime, implement prototypes of the OMPT interface. Several tools, including HPCToolkit (Section 2.3.2) (Adhianto et al. 2010b; Rice University) and TAU (Section 2.3.3) (Shende and Malony 2006; University of Oregon), have used prototypes of the OMPT interface to help gather and interpret performance measurements of OpenMP node program executions as part of parallel applications.

2.2.10 Process Management Interface—Exascale

Program startup on an exascale system will present challenges due to the sheer number of processes and threads involved. The need to launch tool and debugger daemons—which themselves will be massively parallel—only adds to the difficulty. The process management interface (PMI) has been used for quite some time as a means of exchanging wireup information needed for interprocess communication. PMI-Exascale (PMIx) provides an extended version of the PMI definitions specifically designed to both support next generation clusters up to and including exascale sizes, and to provide an abstraction interface by which tools and applications can portably interact with the system management software stack (SMS). The overall objective of the PMIx community is to (1) define a set of agnostic APIs (not affiliated with any specific programming model code base) to support interactions between application processes and the SMS, (2) establish a standards-like process for maintaining the definitions, and (3) provide a reference implementation of the PMIx standard that demonstrates the desired level of scalability.

At the heart of the PMIx project is the concept of *instant on* job startup—i.e., that processes in a job are provided with all the information required for interprocess communication during initialization, without requiring any peer-to-peer data exchange. This allows applications to be started and ready to communicate as fast as the system can spawn them, independent of the number of nodes, and/or processes being launched. PMIx offers this same capability to tools and debuggers to launch and wireup any required daemons, including co-spawning of debugger daemons at application launch, subject to the capabilities of the underlying SMS.

In addition, PMIx provides a channel by which tools can query the SMS and applications for information and capabilities. This includes the ability to obtain information on process location, pid, and state of executing applications; available scheduler queues and their status; network topology, process location relative to the network, and network traffic status; and file system statistics.

2.2.10.1 Architecture and Infrastructure

PMIx uses a dll-based plug-in architecture that supports third-party runtime additions, as well as proprietary binary modules. Communications are constrained to occur between client processes and the PMIx server on their node—tools, however, can connect across nodes to the server of their choice. Information collected by each server is shared with its clients via shared memory, thus minimizing the memory footprint at extreme scales. Cross-node communications (e.g., to obtain diagnostic information from a remote resource) are provided by the host SMS or by third-party libraries accessed by the PMIx server.

2.2.10.2 Requirements

PMIx itself requires only that a thread-enabled version of libevent be available. User-available capabilities are dependent on supported features of the SMS on the target system.

2.3 PERFORMANCE TOOLS

This section presents a number of the performance tools that are currently being used to analyze the performance of applications running at scale on the largest pre-exascale systems. These tools are Performance Application Programming Interface (PAPI) (Section 2.3.1), HPCToolkit (Section 2.3.2), TAU (Section 2.3.3), Score-P (Section 2.3.4), Vampir (Section 2.3.5), and Darshan (Section 2.3.6). All of these tools have development plans that extend to exascale machines and it is necessary that future exascale platforms enable the underlying elements in order to provide a base on which scalable performance tools can be built.

2.3.1 PERFORMANCE APPLICATION PROGRAMMING INTERFACE

The use of hardware counters (Section 2.2.1) to measure and improve the performance of scientific applications has become an integral part of the software ecosystem needed to successfully exploit the current and emerging generations of HPC systems. The PAPI provides tool designers and application scientists with a consistent interface and methodology to the performance counters that are found across the entire computing system. The HPC community has relied on PAPI to track low-level hardware operations for more than 15 years. While it can be used independently as a performance monitoring library and tool for application analysis, PAPI finds its greatest utility as a middleware component for a number of third-party profiling, tracing, sampling, and autotuning toolkits—such as TAU (Section 2.3.3) (Shende and Malony 2006), Scalasca (Geimer et al. 2010), Vampir (Section 2.3.5) (Knuepfer and Brunst 2011), Score-P (Section 2.3.4) (Knüpfer et al. 2012), HPC-Toolkit (Section 2.3.2) (Adhianto et al. 2010b), CrayPat (The CrayPat Performance Analysis Tool n.d.), PerfExpert (Burtscher et al. 2010), Active Harmony (Tapus et al. 2002)—making it the de facto standard for hardware counter analysis. More specifically, PAPI enables users to collect performance counter information from various hardware and software components—including most major CPUs (Terpstra et al. 2009), GPUs and accelerators (Malony et al. 2011), networks (McCraw et al. 2013; Using the PAPI Cray NPU Component n.d.), I/O systems (Terpstra et al. 2009), and power interfaces (McCraw et al. 2014; Dongarra et al. 2012), as well as virtual cloud environments (Weaver et al. 2012; Weaver et al. 2013). The collected information can then be utilized by computational scientists to analyze, model, and productively maximize the performance of their applications on a wide variety of architectures.

The basic underlying components upon which PAPI relies are the physical hardware counters (Section 2.2.1) and a recent Linux kernel version (at least 2.6.31 or newer) to support various CPUs using the "perf_event" subsystem. This last requirement is relaxed, however, for distributions like Red Hat Enterprise Linux (RHEL) that backport newer CPU support to older kernels. That said, not everything PAPI supports relies on Linux "perf_event." In fact, most of the PAPI components (other than the CPU core and uncore components) in the src/components directory provide access to various performance counters from user space through third-party drivers. For instance, the PAPI InfiniBand component uses the sysfs interface to access InfiniBand performance counters from user space; the PAPI CUDA component uses the CUPTI drivers (CUDA Tools SDK) to access performance counters for GPUs (Section 2.2.7); and the PAPI NVML component uses the "nVidia Management Library" to enable monitoring of temperature on GPUs.

PAPI is an ongoing project with the mission to continuously broaden its monitoring power to meet the performance analysis needs of new and advanced technologies that prove fundamental for both state-of-the-art and forthcoming computing capabilities. As software and hardware evolve, it has become increasingly difficult for application scientists to simultaneously address two software engineering challenges: software complexity and hardware complexity.

To address hardware complexity, the PAPI team continues to work closely with hardware vendors (1) to develop new components to monitor performance-critical resources of new compute units, deep and heterogeneous memory hierarchies, and novel interconnect technologies; and (2) to ensure seamless integration of PAPI with new architectures at or near their release.

To address software complexity, the PAPI team works on enabling cross-layer and integrated modeling and analysis of the entire hardware–software ecosystem by extending PAPI with the capability to expose performance metrics for key software components that are currently treated as black boxes, such as numerical libraries, application frameworks, and task runtime systems.

Enabling the monitoring of both types of performance events—hardware and software related—uniformly, through one consistent PAPI interface, is fundamental to PAPI's continuous development and will be mission critical to achieving the performance and efficiency goals of next-generation applications as the scale and complexity of platforms continue to grow toward exascale.

2.3.2   HPCTₒₒₗₖᵢₜ

HPCToolkit (Adhianto et al. 2010b) is a suite of multiplatform tools for measurement, analysis, attribution, and presentation of performance data for parallel programs. HPCToolkit helps developers of applications, frameworks, and libraries use parallel platforms efficiently by identifying costly program regions, assessing inefficiencies, and quantifying their impact on code performance and scalability within and across nodes. HPCToolkit supports analysis of program executions that employ both multithreaded and distributed-memory parallelism.

To gather metrics during a program execution, HPCToolkit collects data *asynchronously* when an event trigger fires (e.g., a hardware performance monitoring unit triggers an interrupt) or *synchronously* when a wrapped function (e.g., an MPI operation) is called. At the point of collection, HPCToolkit unwinds the application's call stack to identify the calling context where collection was initiated and augments one or more metric counts associated with that context. Typically, asynchronous sampling is used to collect application performance metrics and synchronous data collection initiated by instrumentation is used to collect semantic information, e.g., bytes communicated. Both sampling and stack unwinding are described in further detail Sections 2.2.2 and 2.2.3.

To attribute performance metrics to source code, HPCToolkit employs binary analysis to precisely attribute performance metrics to program features (Tallent et al. 2009a). To do so, HPCToolkit recovers program structure from application binaries using Dyninst (Williams et al. 2015)—a toolkit for analysis, instrumentation, and execution control of binary programs. HPCToolkit uses Dyninst to parse machine code into an internal machine-independent representation of instructions, identify functions, build a control flow graph (CFG) for each function, and perform interval analysis in each CFG to recover information about loops and their nesting. HPCToolkit combines this with information about inlining available from an executable's symbol table to accurately map each machine instruction back to source code. This enables HPCToolkit to attribute costs to source lines, loops, inlined C++ templates and functions, procedures, and call chains.

To help pinpoint bottlenecks in parallel program executions, HPCToolkit supports differential profiling (Coarfa et al. 2007; Tallent et al. 2009a). With this technique, one compares multiple executions at different scales to quantify scalability losses within and across nodes on parallel systems and pinpoint their causes. HPCToolkit also supports a collection of novel problem-focused techniques to quantify and attribute performance losses due to inefficient parallelization (Tallent and Mellor-Crummey 2009a,b), resource contention (Tallent et al. 2010b), load imbalance (Tallent et al. 2010a), and exposed memory latency (Liu and Mellor-Crummey 2011, 2013, 2014). To help users analyze performance data collected about one or more executions, HPCToolkit provides effective user interfaces that present code-centric and time-centric views of application performance (Adhianto et al. 2010a; Tallent et al. 2011).

Plans to prepare HPCToolkit for exascale call for extending it in several different ways.

- First, HPCToolkit currently records one or more files of measurement data for each thread in an execution; for exascale, HPCToolkit must employ collective I/O to aggregate measurement data into a smaller and more manageable number of files.
- Second, performance-analysis tools for HPC platforms typically ignore kernel activity on compute nodes. Experience has shown that one can't understand the root causes of performance bottlenecks without understanding activity within the kernel and device drivers (Chabbi et al. 2013). To address this issue, HPCToolkit will be extended to use Linux perf_events to sample within the kernel and device drivers.
- Third, to make it possible to map performance measurement data from an implementation-level view to a user-level view, HPCToolkit will rely on lightweight support at various levels in the software stack, including communication libraries and threaded programming models; the emerging OMPT performance tools interface for OpenMP 5.0 is an exemplar of such support (OpenMP Language Working Group 2016).

- Fourth, to support better measurement of accelerated computations, HPCToolkit is being extended to collect data using PC sampling on GPUs.
- Fifth, HPCToolkit's user interfaces will be extended with a data-centric view that attributes execution metrics to static variables and dynamic allocations identified by their calling contexts, as well as a resource-centric view to support bottleneck analysis, especially in the memory hierarchy.
- Other planned extensions include improving measurement and analysis to cope with extremely large thread counts, and explaining performance bottlenecks at the microarchitecture level by using top-down models, e.g., Srinivas et al. (2011) and Yasim (2014).

Finally, since program executions on exascale platforms will generate massive amounts of data, automated techniques for analyzing performance data, pinpointing and quantifying bottlenecks will be necessary as well.

2.3.3   TAU

TAU is an open-source framework and tools suite for performance instrumentation, measurement, and analysis of scalable parallel applications and systems. The TAU project began at the University of Oregon in 1990s with the goal of creating a framework that could produce robust, portable, and scalable performance tools for use in all parallel programs and systems over several technology generations. Today, the TAU Performance System® (Shende and Malony, 2006) is a ubiquitous suite of tools for supporting all major HPC development languages (e.g., C, C++, Fortran, Python, Chapel, UPC), operating systems, communication libraries (e.g., MPI, Global Arrays, DMAPP, OpenSH-MEM), runtime systems (e.g., CUDA, OpenMP, OpenACC, OpenCL, pthread), and application libraries (e.g., PETSC, HDF5). TAU provides robust support for observing parallel performance on a broad range of platforms, for managing multiexperiment performance data, and for characterizing performance properties and mining performance features. Research and development of the TAU Performance System has been supported by the National Science Foundation (NSF), DOE, and Department of Defense (DOD). TAU is widely used in HPC research centers throughout the United States and around the world.

TAU's architecture has three components: instrumentation, measurement, and analysis. TAU provides both direct-code instrumentation (Section 2.3.3), where probes are inserted in the program to measure performance metrics (e.g., time, hardware counts), and indirect EBS (Section 2.2.2), where interrupts (e.g., caused by periodic timer or hardware performance counters) are used to enable measurement. TAU utilizes several mechanisms for source to binary-code instrumentation, and it provides access to multiple sources of performance data. In particular, it utilizes the PAPI (Section 2.3.1) library to get access to processor specific hardware performance counter (Section 2.2.1) data. The measurement component of TAU provides robust parallel profiling and tracing (Section 2.3.3). Parallel profiling aggregates performance metrics for all instrumented events in a parallel program at the resolution of individual execution threads. TAU supports multiple types of parallel profiling: flat, callpath/callgraph, and phase. In contrast, TAU's parallel tracing supports logging time-stamped events with event-specific parameters to thread-specific trace buffers. TAU allows online and hierarchical trace merging and conversion. As well as performance data, TAU can collect and store in TAUdb metadata about the hardware/system environment, its runtime configuration, how the application was built and compiled, its execution parameters and input files, and any application-specific information.

The performance-analysis technology in TAU primarily focuses on parallel profiles. The TAUdb performance data management framework has been used extensively to provide a managed repository of profile results from parallel application experiments and a robust access mechanism for TAU's analysis tools. The ParaProf tool provides scalable analysis and rich interaction with a TAU parallel

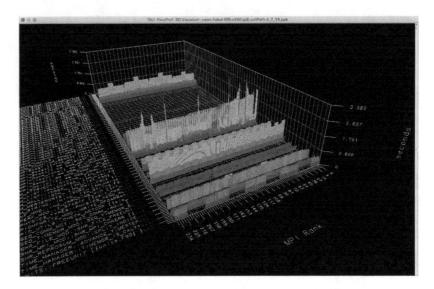

FIGURE 2.2 (See color insert.)   ParaProf.

profiles, including 2D and 3D visualizations (Figure 2.2). The PerfExplorer tools were developed for multiexperiment analysis and performance data mining. It enables analysis workflows to be created and supports higher level reasoning and meta-analysis through the use of inference rules.

The TAU Performance System continues to be enhanced, most recently with the integration of its measurement infrastructure with parallel software layers. For instance, TAU can access low-level performance statistics from the MPI layer using the MPI-T tool interface (Section 2.2.8) and can set performance critical control variables to tune MPI's performance at runtime. TAU also provides unique support for runtime monitoring by interfacing with external tools at runtime using the BEACON event and control notification backplane. Improvements are also being made to simplify TAU's use. The TAU Commander environment from ParaTools, Inc. builds upon TAU by providing a simplified interface to all aspects of TAU, from installation, to instrumentation, to launching the executable and analysis tasks, using a single command.

2.3.4   SCORE-P

Score-P is a measurement infrastructure that can utilize both instrumentation (Section 2.2.4) and sampling (Section 2.2.2). It is a highly scalable and easy-to-use tool suite for recording fine-grained performance events with special focus on parallel applications. It supports a wide range of HPC platforms and programming models. Score-P provides core measurement services for a range of widely recognized performance tools, such as Vampir (Section 2.3.5), TAU (Section 2.3.3), and Scalasca. Typical questions Score-P helps to answer are

- Which call paths in my program consume most of the time?
- How much time is spent in communication or synchronization?
- What is the performance of kernels that execute on accelerators such as GPUs?
- Where does my code allocate or free memory and are there any leaks?

Figure 2.3 depicts an overview of the Score-P instrumentation and measurement infrastructure, the produced data formats, and the analyzing tools from the VI-HPS ecosystem. Supported programming models and other event sources are modularized at the lowest level.

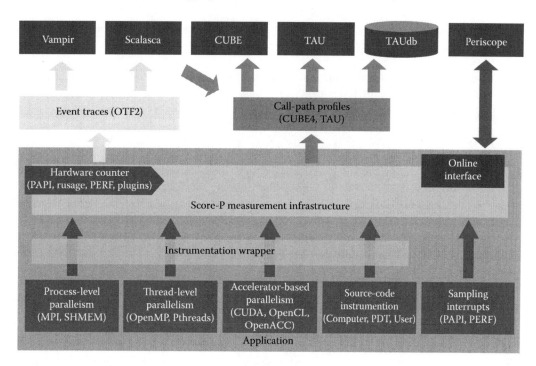

FIGURE 2.3 The Score-P instrumentation and measurement framework at a glance.

Score-P collects event data during the execution of an instrumented application. The Score-P measurement system offers various instrumentation methods at compile/link time, including automatic compiler instrumentation, manual source-code instrumentation, the instrumentation of OpenMP pragmas or the interception of MPI, SHMEM, and Pthreads library calls. Monitoring of accelerator-based paradigms like CUDA, OpenACC, and OpenCL is also supported by Score-P. As an alternative to automatic compiler instrumentation, events can be generated using a sampling approach.

At runtime, the instrumented program can be configured to record an event trace (Section 2.2.6) in the OTF2 format or produce a CUBE4 call-path profile. In profiling mode, the performance events are summarized at runtime separately for each call path. In tracing mode, each individual event with its corresponding timestamp is recorded in temporal order. Because of the use of standardized open-source data formats, multiple analysis tools can work on the same data from a single measurement run. Call-path profiles can be examined in TAU or the Cube profile browser. Event traces can be examined in Vampir or used for automatic bottleneck analysis with Scalasca. Alternatively, online analysis with periscope is possible. Filtering techniques allow precise control over the amount of data to be collected. Optionally, metrics from an extensive set of performance data sources like PAPI (Section 2.3.1), rusage, and perf can be recorded. Furthermore, there exists a metric plug-in interface. It enables users to implement routines to gather metric data from machine specific sources. Measurement mode and any external metric sources are specified at runtime without recompilation. Score-P is available as open-source under a 3-clause BSD License.

2.3.5 VAMPIR

The Vampir performance visualizer works with Score-P (Section 2.3.4) trace files (Section 2.2.6) and allows studying a program's runtime behavior at a fine level of detail. This includes the display of detailed performance event recordings over time in timelines and aggregated profiles. Interactive navigation and zooming are the key features of the tool, which help to quickly identify inefficient or faulty parts of a program. Vampir helps to answer questions such as the following:

- How well does my program make progress over time?
- When and where does my program suffer from load imbalances and why?
- Why is the time spent in communication or synchronization higher than expected?
- Are I/O operations delaying my program?
- Does my hybrid program interplay well with the given accelerator?

Before using Vampir, an application program needs to be measured with Score-P. Running the program with Score-P produces a bundle of trace files in the Open Source OTF2 community format. When opening the bundle with Vampir, a timeline thumbnail of the data is presented. This thumbnail allows preselecting a subset or the total data volume for a detailed inspection. The program behavior over time is presented to the user in an interactive chart called master timeline. Further charts with different analysis focus can be added as depicted in Figure 2.4. These charts can be divided into timeline charts and statistical charts.

Timeline charts show program events *en detail* over time. The chain of events of monitored processing elements is displayed on a horizontal time axis. Detailed information about the invocation of subroutines, communication, synchronization, and hardware counter measurements is provided. Figure 2.4 shows different timeline types on the left-hand side from top to bottom. The *Master Timeline* (upper left) represents multiple processes as individual rows, whereas a *Process Timeline* (not depicted) focuses on a single process and its stack levels. Counter data is addressed in *Counter Timelines* (bottom left) and the *Performance Radar*, which illustrates multiple processes at the same time. Statistical charts (right) reveal accumulated measures, which are computed from the corresponding event data. An overview of the phases of the entire program run is given in the *Zoom Toolbar* (top right), which can also be used to zoom and shift to the program phases of interest.

VampirServer is the parallel, distributed version of Vampir. It implements parallelized event analysis algorithms, which enable fast and interactive rendering of very complex performance monitoring data. Large data volumes can be analyzed due to the close integration with state-of-the-art parallel production environments. VampirServer follows a client/server approach, which allows bulky performance data to be kept close to its origin. The client-side visualization of the data can be carried

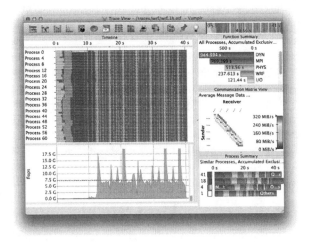

FIGURE 2.4 An typical analysis session in Vampir with a selection of timeline and statistical charts.

out from any Vampir instance on a laptop or desktop PC connected to the Internet. The server can efficiently handle program traces with hundreds of thousands of processes.

2.3.6 DARSHAN

I/O operations on HPC systems typically involve a high degree of parallelism both within the application and within the filesystem itself. This leads to significant challenges in identifying I/O bottlenecks at scale. Darshan (Carns et al. 2009, 2011) is a scalable, lightweight I/O profiling tool that helps to address this challenge by characterizing and summarizing the I/O behavior of production applications. This characterization data includes operation counts, request offsets, request sizes, and basic timestamp information for the most commonly used I/O APIs. The data is written into a compact compressed log file once execution is complete. Darshan's lightweight design makes it possible to monitor all jobs running on a large-scale HPC system; the measurement overhead is significantly smaller than typical I/O variability.

Application users can use the utilities provided with Darshan to produce a report summarizing key details of an application's I/O behavior, such as operation counts by interface, file counts, request size histograms, and I/O time lines. Figure 2.5 shows an example of Darshan summary report output. This information is then used to guide optimization and performance debugging efforts. HPC system providers can also analyze system-wide Darshan logs in aggregate to understand the overall I/O workload of their system, gauge the effectiveness of new technologies, and identify applications in need of guidance for I/O.

Darshan version 3.x (Snyder et al. 2016) introduced a modular architecture which allows new types of instrumentation to be added to Darshan without modifying the log structure or internal software core. Figure 2.6 illustrates the modular structure of Darshan and associated log files. Existing Darshan modules, including those for POSIX I/O, MPI-IO, Parallel NetCDF, and HDF5 I/O interfaces, the Lustre file system, and the Blue Gene/Q runtime environment, each collect data using methods appropriate for that module. Modules may use the MPI profiling interface

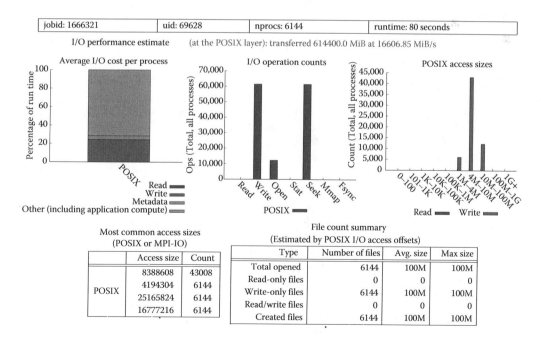

FIGURE 2.5 Example output from Darshan analysis.

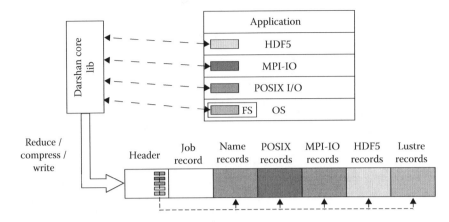

FIGURE 2.6 Darshan 3.x modular core and log format.

(Section 2.2.8) (MPI 2012), link-time function wrappers (GNU wrap), run-time dynamic function wrappers (LD_PRELOAD), environment variables, or system APIs for this purpose.

Computing in the Exascale era calls for increasing levels of parallelism throughout the software stack, heterogeneous architectures, and multilevel I/O subsystems. These trends all serve to make the overall I/O path more complex and more challenging to understand and tune. This is true not only for existing I/O libraries and programming models, but also for new alternatives that may be deployed at Exascale. The modular Darshan architecture is designed to help address this challenge by enabling Darshan to incorporate new data sources as platforms evolve. Exascale system designers can facilitate this effort by providing robust monitoring capability via well-defined APIs, at multiple levels of the I/O stack, with minimal perturbation of production execution. It is also increasingly important to capture both application-specific behavior and system-wide behavior in order to correlate activity and produce a holistic view of I/O performance. This type of monitoring capability can be integrated into Darshan and other portable performance-analysis tools to improve efficiency across a broad range of computing platforms.

2.4 DEBUGGING TOOLS

This section surveys a number of the state-of-the-art debugging tools that enable debugging of applications running at scale on the largest pre-exascale systems. Two full-featured debuggers commonly used are DDT (Section 2.4.1) and TotalView (Section 2.4.2). Tools like Valgrind (Section 2.4.3), STAT (Section 2.4.4), and MPI and thread debuggers (Section 2.4.5) provide more narrow yet still critical specialized functionality.

The essence of a debugging tool is enabling observation, exploration, and control of program state, such that a developer can, for example, verify that what is currently occurring correlates to what is intended. Such tools can also encompass automation—such as memory debugging or MPI deadlock detection.

Core requirements for a debugger include the abilities:

- To discover and attach to (or launch) processes rapidly
- To control individual processes and execution contexts
- To query individual execution contexts for their state (PC, stack, memory, registers)
- To convey program state in the developer's model: variables, memory, source code

At the heart of this are standards appropriate to the non-HPC world:

- The ptrace system call—providing the ability to access another process's registers and memory (peeking and poking) and single stepping.
- Support for interrupts—the ability of a user to pause a process, or to insert a breakpoint or a data watchpoint.
- Calling conventions (part of an *ABI*) that enable a stack to be unwound.
- A debug symbols format—DWARF is today's standard—a highly expressive format that enables the debugger to understand variables, structures, classes, and source locations. It also often includes information that enables stack unwinding.
- Other utility helpers such as libthread_db which provides debuggers the ability to query for each thread within a process.

What makes a parallel debugger—and one suited to extreme scale is

- The ability to achieve each of the items of a regular debugger—en-masse in node counts—and in per-node execution contexts.

Parallel debuggers also have needs related to the programming model—for example, MPI message queue debugging, which is currently implemented as a helper library similar to libthread_db, or the ability of the debugger to calculate state based on multiple individual processes—such as when examining partitioned global address space (PGAS) data.

2.4.1 ALLINEA DDT

One debugger used at petascale today is Allinea DDT. Its design exploits the available parallelism by running significant tool components at the actual compute resources. This offloading of computation required for debugging to the location of the actual executing processes provides fast individual access, but also the foundation of ability to scale freely as the number of nodes increases. To provide that actual scalability, a scalable tree overlay control network—with the ability to broadcast and reduce—is used to rapidly control processes and to gather and merge data from those compute nodes. This architecture enables multi-petascale jobs to have responsive collective operations often taking around 100–200 ms to complete on many cluster-based systems.

The *entire scalability* experience is a defining challenge of tool developers: from data acquisition and process (or context) control through to the eye of the developer. It is not sufficient to rapidly query or control execution context—if the information retrieved cannot be readily understood or acted on. DDT has features—for example, parallel stack views, smart highlighting, and sparklines—which address visual scalability (Figure 2.7).

The *parallel stack view* gathers stack traces from every execution context and merges the information—with reduction occurring within the control tree—to provide a tree view of the stack traces and the number of occurrences. This enables the developer to rapidly establish outliers and similarities, and to select a nominee process (or group of processes) at each stack frame to explore.

Smart highlighting uses the reduction network to identify min, max, the number of values equal to a particular value, and whether an expression has changed and applying color highlighting as a visual cue.

Sparklines take this further—and plot a 64-pixel square bitmap—of the graph of a value across every process. The 64-pixels width/height enables patterns to be seen clearly and can be calculated using the reduction network efficiently. The information conveyed is easily processed by our minds—with patterns, or the absence of patterns, being almost instantaneously recognized.

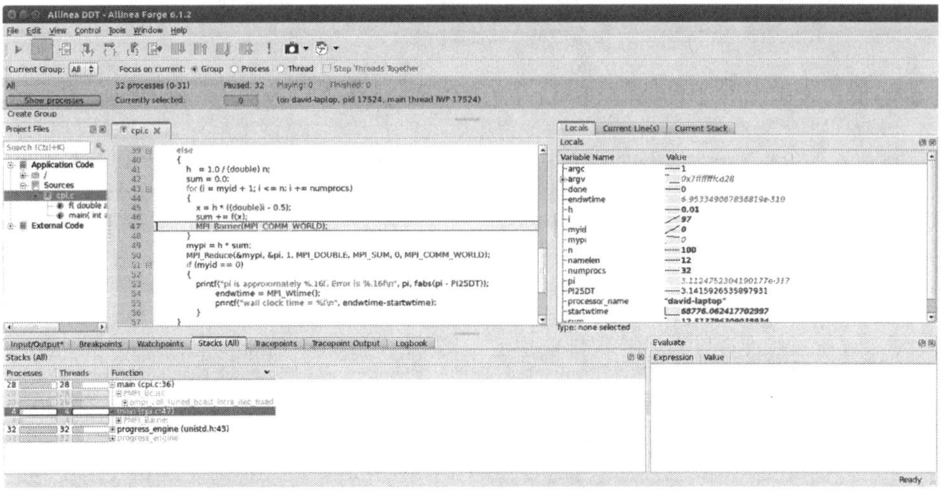

FIGURE 2.7   AllineaDDT.

As we scale up, we observe that debugging 1,000,000 cores is no less tractable than debugging 1,000 cores from a user-interface perspective—both quantities rely on good information reduction. Supporting this must be a major focus: Inter- and intra-node scalability is necessary to provide the capabilities.

Some of the prominent challenges to the future include the following:

- The capabilities of accelerators and specialized devices such as GPUs or field-programmable gate arrays (FPGAs) to support debugging at speed. Such devices can present limitations in the ability of a user to control execution or to receive data in a timely or consistent manner. Where it proves impossible to control or query, then computation risks becoming a *black box* without more sophisticated data validation—enabling the developer only to see the before and after effects of calculation.
- Similar to the use of specialized devices, reductions in resources made available to tools can cause issues. Debuggers, like most tools, rely on the ability of systems to provide scalable launch mechanisms (Section 2.2.10) and to provide resource and performance by which a debugger can perform its tasks. On systems with design-limited compute resources for debugging (e.g., one Blue Gene/Q IO node was solely responsible for debugging 128 compute nodes), the available performance is insufficient to support sparklines. If it is not possible to rapidly query stack traces from every context, even parallel stack traces would become impossible.
- Changes to models of parallelism and the degree of state exposed to tools. The MPI+X programming approaches clearly change the MPI hegemony of consistent (often bulk-synchronous style) coding. This hegemony has enabled simplification in the past—will it continue? Equally, will runtime systems used by task-based parallel models expose their internal graph of uncompleted calculations to debuggers? If not, how will a developer know what computation is complete and what is not?

It's evident that much of the tool capabilities that benefit developers today depend on favorable choices at the system and programming model level. While the word codesign has clearly seen overuse, in the context of tools this is one area that codesign stands to bring a significant benefit.

2.4.2 The TotalView Debugger

The TotalView for HPC debugger from Rogue Wave Software has a 30-year history of HPC debugging excellence. Built from day 1 as the parallel debugger for the BBN Butterfly, TotalView has supported most major HPC systems from Cray, HP, IBM, Intel, SGI, Sun, and other vendors, running Unix, Linux, or custom kernels. Over time, TotalView has continually adapted to the ever-changing HPC landscape by supporting new HPC architectures, languages, and technologies. Its innovative techniques have aided debugging across all major sectors: commercial, government, and academia.

Today, TotalView is a mature, robust, scalable, and highly feature-rich debugger supporting the C, C++, and Fortran programming languages, hybrid mixtures of parallel programming paradigms, such as MPI, OpenMP, pthreads, CUDA, OpenACC, UPC, CAF, Global Arrays, SHMEM, and more.

TotalView provides a graphical user interface (GUI) and command-line interface (CLI) for interactive debugging that allows users to debug large numbers of processes and threads consisting of multiple languages and parallel programming paradigms, all from a single point of control. Standard debugger features include starting and stopping threads, processes, and groups of threads or processes; source and instruction-level step and step-over; breakpoints and conditional breakpoints; display of source code, variables, arrays, memory blocks, and registers; and modifying program state. The CLI is a programmable Tcl interpreter that supports the creation of user-defined utilities to extend debugger functionality and provides a scripting language that allows for automated operation. TVScript and MemScript are frameworks for noninteractive batch debugging. Using an *event-action* model, a user defines a series of *events* that may occur within the program and the actions to take when an event occurs. Data is logged to a set of output files for review when the batch job has completed.

TotalView offers many other features, including a C, C++, and Fortran expression evaluator; data watchpoints; data visualizers; the ability to view the elements of Standard Template Library (STL) collection classes (map, set, array, list, string, etc.); display aggregation; memory debugging; core file debugging; asynchronous thread control; reverse debugging; heterogeneous debugging; and extensive help and documentation.

2.4.2.1 Asynchronous Thread Control

TotalView supports a thread-debugging feature called *asynchronous thread control*, particularly useful for debugging multithreaded applications. TotalView leverages system-level execution-control mechanisms (ptrace(), /proc, or other process-control APIs) that allow it to control the execution of individual threads. The asynchronous thread-control features of TotalView give the user fine-grained control to stop and start individual threads, create stop-thread breakpoints, hold and release individual threads, synchronize the execution of a team of threads at the same program location, single-step threads in lock-step, and more.

2.4.2.2 Reverse Debugging

The TotalView ReplayEngine reverse debugging feature on Linux x86_64 systems records the execution history of a program and allows repeatedly running or single stepping both backward and forward within the history. The ability to work backward from a failure, error, or crash to its root cause eliminates the need to restart the program. Execution within recorded history is completely deterministic, so once a *bug* is captured, it remains in the history. The Replay on Demand feature allows enabling reverse debugging during a debugging session. The save and restore functionality allows saving the execution history to a file and loading it later to continue debugging.

2.4.2.3 Heterogeneous Debugging

TotalView supports debugging heterogeneous applications involving a mix of processor architectures (e.g., PowerPC, x86_64, GPU, etc.) down to single-thread granularity. TotalView's address space model supports processes containing multiple heterogeneous address spaces, where various threads can run in their own address space (e.g., CUDA contexts), or share a process address space with other threads (e.g., pthreads). As programming languages are extended to accommodate accelerator devices—for example, OpenMP 4 TARGET constructs—this address space model allows TotalView to more accurately reconstruct the conceptual model in which a process is a *container* for CPU, GPU, and other devices.

2.4.2.4   Architecture and Infrastructure

TotalView's architecture is extremely resource efficient in terms of memory and file I/O utilization. The debugger client running on the front-end node reads and stores exactly one copy of the executable and shared library symbol tables. The symbols are *shared* across all processes under the control of the debugger client. The backend debugger servers do not read or store symbol table information, and use very little compute node memory, allowing the application to use nearly all available memory (see Figure 2.8). Further, the lightweight servers allow TotalView to run in resource-constrained environments, such as IBM's Blue Gene/Q IO nodes.

2.4.2.5   Multicast and Reduction

An MRNet (MRNet: A Multicast/Reduction Network) tree of client, communication, and server processes forms a highly scalable debugger infrastructure. The client multicasts operations to the servers, and the results are reduced (aggregated) as they return up the tree. The fan-out and depth of the tree is user-configurable and can be adapted to accommodate any scale.

Display aggregation in the GUI and CLI uses reduction techniques that group processes and threads by common properties or values using user-specified *group-by* terms. Processes or threads are deemed equivalent when, for example, they have the same data values (e.g., a variable value), program state (running, stopped, at breakpoint, etc.), or call frames (meaning they are inside the same function or at the same line number). Aggregation reduces display clutter, improves performance, and makes it easier to find *outlier* processes or threads that are not behaving like their peers.

2.4.2.6   Debugger Requirements

TotalView has a modest list of requirements, including X11 support for the GUI client; a full-featured process and thread tracing interface such as ptrace() or the /proc file system; accurate debugging information, such as ELF, DWARF, or Stabs; the MPIR interface for MPI codes (Section 2.2.8);

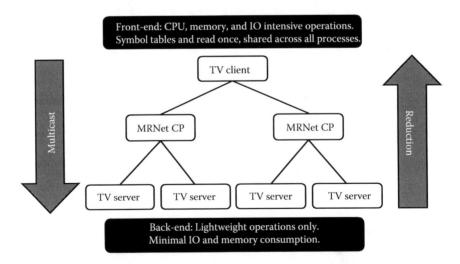

FIGURE 2.8   TotalView's architecture.

thread-level tracing interface (libpthread_db) for threads; TCP/IP sockets and remote shell support (e.g., ssh) for debugging multiple nodes; and the NVIDIA CUDA Debug API (Section 2.2.7) for CUDA applications.

2.4.3 VALGRIND AND MEMORY DEBUGGING TOOLS

Program errors related to the misuse of memory can be especially difficult to identify and resolve. These errors include the use of uninitialized data, access to unallocated memory regions, buffer overflows, improper memory deallocation, and memory leaks. These problems can be hard to reproduce and can manifest in ways that do not provide a clear connection to the underlying error making them challenging to debug. The resolution of these errors can be facilitated through the use of tools that provide memory debugging capabilities.

A number of tools are available to help identify and correct memory errors including libraries such as Electric Fence, the AddressSanitizer (Serbryany et al. 2012) feature of the GNU and LLVM compilers, full-featured debuggers such as TotalView (Section 2.4.2) and DDT (Section 2.4.1), and tools such as Valgrind (Nethercote and Seward 2007). These tools differ in performance, types of bugs detected, and likelihood of bug detection but generally employ some subset of a common set of techniques to monitor memory accesses, monitor memory allocation/deallocation operations, and identify unreachable allocated regions of memory.

Valgrind, a frequently used tool in Linux environments, provides a relatively complete set of memory debugging features including memory leak detection, detection of invalid heap memory operations including buffer overflow and use of unallocated memory, detecting the use of uninitialized memory, identification of overlapping function arguments, and identification of invalid or mismatched deallocation operations. This functionality is provided by the Memcheck component of Valgrind, which is implemented, along with other components, within a virtual environment provided by the Valgrind core through which program instructions are passed. The Valgrind core translates program instructions into an intermediate representation (IR) which is then made available to components for modification or instrumentation. The modified IR is then translated back into native machine code which is executed on the CPU. While tools such as AddressSanitizer provide memory debugging without the use of a virtual environment, the virtual environment provided by Valgrind provides a general framework for implementing Memcheck features and allows for binary analysis and instrumentation of executables that can be used for a variety of purposes in other Valgrind components.

To spot memory errors, Memcheck employs techniques common to full-featured memory debuggers: the use of shadow memory, instrumentation of memory access instructions, and tracking of memory allocations/deallocations through a replacement memory management library. The shadow memory maintained by Memcheck contains information about the validity and addressability of each byte in regular memory. Valid bytes are those that have been initialized by the program and addressable bytes are those that can be legally read or written. Instrumentation is inserted by Memcheck around most program instructions to check the validity and addressability of each byte accessed. Validity checks are performed when generating memory addresses, for control flow decisions, and for systems calls. For all other operations, validity information is computed and tracked but does not trigger an error report. Byte-level addressability checks are performed for each read and write operation. A program's heap memory is initially marked as unaddressable, when an allocation is performed the appropriate bytes are marked addressable and then marked unaddressable again when released. These operations are performed by a Memcheck library that replaces the standard system allocators and intercepts calls to malloc, calloc, and other memory allocation routines. To detect buffer overflows, the library also brackets allocated heap blocks with red zone regions that will trigger an error if any byte in the region is accessed. In addition to monitoring heap memory accesses, the accessibility of the stack region is tracked using the stack pointer location. While Memcheck has the capability to detect buffer overruns in heap memory, it cannot detect these errors in the stack or static memory. Valgrind, however, provides an experimental component, SGcheck, that provides this capability. Memcheck optionally identifies memory leaks at program termination by determining whether any of the memory blocks that remain allocated by the program are unreachable.

The collection of techniques consisting of shadow memory, instruction-level instrumentation, allocator library replacement, and an assessment of block reachability at program termination is employed to varying degrees by a variety of memory debugging tools. The GNU and LLVM Address-Sanitizer component employs all of these techniques and provides many similar capabilities to Valgrind. AddressSanitizer, however, utilizes a different approach to instruction instrumentation, performing instrumentation at compile time, and not employing a virtual environment as Valgrind does. As a result, the overhead of AddressSanitizer can be significantly lower than that of Valgrind, which can produce over an order of magnitude increase in code execution time and a 25% increase in memory usage. This overhead can be a major impediment to the use of memory debuggers for codes with long runtimes and large memory usage. Additionally, a particular concern for HPC platforms is that while most memory debuggers can work with large-scale parallel applications, output is generated on a per-process and per-thread basis which can quickly make large-scale debugging impractical. Cross-platform availability for memory debuggers can be limited as they typically have strong dependencies on low-level system features. Tight coupling to both processor instruction sets and operating system calls makes porting full-featured memory debugging tools a significant undertaking on a new platform.

2.4.4 STACK TRACE ANALYSIS TOOL

Exascale computing will reach a new peak in terms of the number of threads of execution used by an application, and along with it will grow the scales at which bugs manifest themselves. STAT (Arnold et al. 2007) is a highly scalable, lightweight debugger that has proven effective in isolating elusive errors that only emerge in production computing.

STAT helps to isolate bugs by gathering stack traces from each individual process of a parallel application and merges them into a global, yet compact representation. Each stack trace, as depicted in Figure 2.9a, captures the function calling sequence of an individual process, which informs the user of what part of the code the process is currently executing. The merged traces then form a call-graph prefix tree, which can be seen in Figure 2.9b. STAT labels the edges of these merged traces with the count and set of MPI ranks that exhibited that calling sequence. Further, nodes in the prefix tree that are visited by the same set of processes are given the same color, providing the user with a quick means of identifying the various process equivalence classes.

In fact, process equivalence classes are STAT's main idiom to help users focus only on a manageable number of processes. Many applications typically demonstrate only a small number of differently behaving classes with a few processes taking an anomalous call path, a similarly small number of processes waiting for point-to-point communication, and the remaining processes stuck in a collective operation. Thus, users can effectively debug a large-scale application, even with scales over extremely large numbers of MPI processes, by focusing on a small subset of the representative tasks.

STAT builds on a highly portable and scalable, open-source tools infrastructure, including Launch-MON (Ahn et al. 2008) for tool daemon launching, which then relies on the MPIR (Section 2.2.8) debug interface (DelSignore), MRNet (MRNet: A Multicast/Reduction Network) for scalable communication, and the StackwalkerAPI (Dyninst API) for obtaining stack traces.

Since its inception, STAT has continued to evolve for higher scalability (Lee et al. 2008) and with additional features (Ahn et al. 2009). Its current efforts include adding support for a wider range of programming models to be more effective on exascale architectural and programming models trends. For instance, the developers are currently studying the trade-off between scalability and capabilities when debugging MPI+OpenMP applications. For this purpose, they are closely working with the OpenMP language committee to standardize and leverage the OpenMP debug interface called OMPD (Protze et al. 2015).

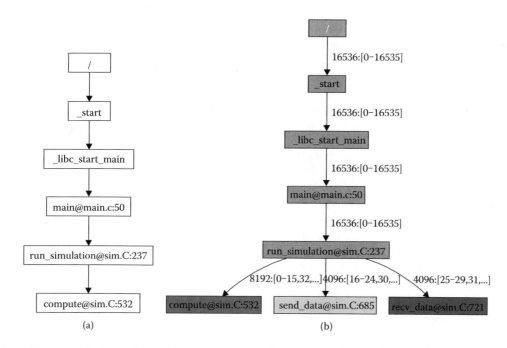

FIGURE 2.9   A single stack trace (a) and a STAT merged call-graph prefix tree (b).

Exascale computing will require increasing dimensions in complexity, including diverse architectures, performance portability middlewares, and high-level programming models (e.g., Legion 2016. Developing standardized interfaces to debug these aspects would be greatly beneficial over requiring each tool to implement its own hooks. Indeed, the ad-hoc approach has proven itself to be unproductive and fragile, and may become unsustainable if not intractable in the exascale landscape. STAT uses the current MPI+OpenMP efforts to establish a viable and sustainable model for supporting a wide range of exascale programming models.

2.4.5   MPI AND THREAD DEBUGGING

Parallelism, both at the thread and process level can present a unique set of challenges for insuring program correctness. MPI is widely used in implementing parallelism at the process level and the MPI profiling interface described earlier (Section 2.2.8) can be utilized by debugging and testing tools to identify MPI-related problems. In particular, record-and-replay tools and noise injectors can build on this interface to help debugging and testing nondeterministic MPI applications and to expose unintended message races. As an example, ReMPI (Sato et al. 2015) is a recently released tool that records and deterministically replays MPI program execution by reproducing message-receive orders. Because the large amount of data that traditional record-and-reply techniques record precludes its practical applicability to extreme-scale parallel applications, ReMPI (Sato et al. 2015) uses a new compression algorithm called Clock Delta Compression (CDC) (Sato et al. 2015) to solve this scalability challenge. CDC defines a reference order of message receives based on a totally ordered relation using Lamport clocks, and only records the differences between this reference logical-clock order and an observed order. At the time of writing, ReMPI has been used by Lawrence Livermore National Lab (LLNL) users and already proven effective in debugging highly elusive nondeterministic bugs. For example, it recently helped ParaDiS, a dislocation dynamics application developed at LLNL, to debug a previously intractable bug by reproducing nondeterministic fatal errors at the exactly same physics cycles.

ReMPI developers also developed a novel noise injection tool called NINJA (Sato et al. 2017), which can easily expose unintended MPI message races. Tools such as NINJA can complement record-and-replay tools because it can become even harder to observe a nondeterministic bug in record mode (i.e., a *heisenbug*). A study (NINJA) has shown that NINJA can manipulate the network timing such that a rarely observed bug can manifest itself easily and quickly.

At the thread level, the correct implementation of multithreaded code is challenging and can lead to latent bugs due to race conditions. Nevertheless, large applications have begun to transition to an MPI+X model, with X being some kind of threading model such as OpenMP. Many "X" models exist and most of them advertise easy and high-level exploitation of on-node resources. However, despite these claims, there is often only rudimentary experience with these models, which raises the chance for bugs associated with multithreading and the level of difficulty in coping with them. In particular, data races are notoriously difficult to catch, in part because they can be difficult to spot and reproduce through traditional testing.

A race detector can be a significant aid in exposing these bugs. In fact, considerable progress has been made over the years in data-race detection applied to low-level threading models. Unfortunately, though, very few practical tools are available for high-level "X" programming models important to HPC, such as OpenMP. For example, mature open-source tools like the Helgrind (Jannesari et al. 2009) component of Valgrind (Section 2.4.3) and ThreadSanitizer (Serebryany and Iskhodzhanov 2009) only target the underlying POSIX threads and do not support high-level models. This lack of knowledge of the high-level model can lead to many false positives and large overheads. Intel Inspector XE does provide OpenMP support, but still suffers from accuracy and overhead problems (Atzeni et al. 2016) as well as portability issues due to hardware vendor lock-in.

Several recent efforts aim at overcoming these challenges. Specifically, Archer (Atzeni et al. 2016) paves a path via which mature open-source dynamic race checkers can be adapted to HPC's high-level models for high accuracy, low overheads, and portability.

Archer combines well-layered modular static and dynamic analysis stages to detect races in OpenMP applications. Its static analysis passes classify any given OpenMP code region into one of two categories: guaranteed race-free and potentially racy. By applying dynamic analyses only to potentially racy regions, it ameliorates performance and memory overheads, which is especially important for large production applications. For its dynamic analysis, Archer employs ThreadSanitizer unmodified, but enhances it by annotating the OpenMP runtime library to convey the OpenMP specific *happens-before* relations to ThreadSanitizer.

The initial version of Archer used a direct annotation approach: modifying the OpenMP runtime at the source-code level. While its annotation patch has already been merged into the Clang/LLVM OpenMP runtime and has already proven effective, this kind of direct approach is limited to work only with specific OpenMP runtimes annotated at the source-code level.

To extend the benefit to a wider array of OpenMP runtimes, Archer has been working with the OpenMP community to define the standard tools interface OMPT (Section 2.2.9), which provides a portable abstraction for tools across runtimes, and started migrating its annotation scheme to it. The new approach enables Archer to be portable across OpenMP runtime implementations.

On our continuing path to exascale computing, we expect that programmers will demand tools like Archer not only for OpenMP but also for other high-level models including emerging tasking models such as Legion. Lessons learned from Archer indicate that critical codesign steps must be taken to meet this demand.

2.5 CONCLUSIONS

A rich and varied ecosystem of performance and debugging tools will be a critical component of a successful exascale platform. These systems are expected to present a variety of challenges for tool development, including increasing systems complexity, diverse and heterogeneous architectures, new performance portability middleware components, and new high-level programming models. In order for exascale systems to provide the tools environment needed to maximize the productivity of both the system and the computational scientists using them, it is critical that the developers of these systems engage in codesign with the tools community.

For tool developers, a number of challenges exist to adapt to future exascale platforms:

- Tools often have low-level dependencies on system components, including instruction sets and operating system calls. New system architectures can require significant redevelopment work to support these new architectures.
- Heterogeneous systems require tools and debuggers to work with multiple and often very different hardware components, increasing tool complexity, and development effort.
- Increasing system parallelism presents ongoing scaling challenges for tools and debuggers.
- New hardware technologies developed for exascale systems may require tools to develop correspondingly new data collection and analysis capabilities.
- The increased complexity and volume of performance and debugging data likely to be seen on exascale systems risks overwhelming tool users. Tools and debuggers may need to develop advanced techniques such as automated filtering and analysis to reduce the complexity seen by the user.

The tools discussed in this chapter are all in production at scale on today's petascale machines and are evolving on a path toward exascale. This experience has led to the identification of a number of possible issues that need to be addressed during the design of future exascale machines:

- Hardware designs need to provide the fundamental components necessary to enable performance and debugging tools, such as relevant on-chip hardware performance counters that allow attribution of on-core, off-core, and network events to software along with a top-down methodology to use them.
- Exascale systems need to provide necessary compute, software, and IO resources to enable tools to function at scale on the system. Tools and debuggers rely on a variety of system features, including the ability of systems to provide scalable launch mechanisms and to provide compute and memory resources necessary for them to perform their tasks.
- Specialized devices have in the past limited the ability of tools and debugger to control execution or to receive data in a timely and consistent manner.
- System software must be designed to support tools and debuggers. Low-level software infrastructure components should provide tools full access to hardware resources.
- New and existing programming models, runtimes, and libraries should provide standardized hooks for tools and debuggers that expose a sufficient degree of state. The MPI_T and OpenMP Tools interfaces are examples of the types of interface capabilities that are needed.
- System hardware and software components should be sufficiently documented to allow third-party tools to fully utilize and analyze the system hardware and software.

REFERENCES

L. Adhianto, S. Banerjee, M. Fagan, M. Krentel, G. Marin, J. Mellor-Crummey, and N. R. Tallent. HPC-TOOLKIT: Tools for performance analysis of optimized parallel programs. *Concurrency and Computation: Practice and Experience* 22(6) (2010b): 685–701.

L. Adhianto, J. Mellor-Crummey, and N. R. Tallent. Effectively presenting call path profiles of application performance. In *Workshop on Parallel Software Tools and Tool Infrastructures, in with the 2010 International Conference on Parallel Processing*, Philadelphia, PA: IEEE, 2010a.

D. H. Ahn, D. C. Arnold, B. R. de Supinski, G. L. Lee, B. P. Miller, and M. Schulz. Overcoming scalability challenges for tool daemon launching. In *Proceedings of International Conference on Parallel Processing (ICPP)*. Portland, OR, September 2008. Piscataway, NJ: IEEE.

D. H. Ahn, B. R. de Supinski, I. Laguna, B. Liblit, G. L. Lee, B. P. Miller, and M. Schulz. Scalable temporal order analysis for large scale debugging. In *Proceedings of the IEEE/ACM Supercomputing Conference (SC|09)*. Portland, OR, November 2009. New York, NY: ACM.

D. C. Arnold, D. H. Ahn, B. R. de Supinski, G. Lee, B. P. Miller, and M. Schultz. Stack trace analysis for large scale debugging. In *Proceeding of the 21st IEEE International Parallel and Distributed Processing Symposium (IPDPS)*. Long Beach, CA, March, April 2007. 1–10. Piscataway, NJ: IEEE.

M. Arnold and P. F. Sweeney. Approximating the calling context tree via sampling. Technical Report 21789, IBM, 1999.

S. Atzeni, G. Gopalakrishnan, Z. Rakamaric, D. H. Ahn, I. Laguna, M. Schulz, G. L. Lee, J. Protze, M. S. Müller. ARCHER: Effectively spotting data races in large OpenMP applications. In *Proceedings of the 30th IEEE International Parallel and Distributed Processing Symposium (IPDPS)*. Chicago, IL, May 2016. Piscataway, NJ: IEEE.

M. Bauer, S. Treichler, E. Slaughter, and A. Aiken. Legion: Expressing locality and independence with logical regions. In *Proceedings of the International Conference on High Performance Computing, Networking, Storage and Analysis (SC '12)*. Los Alamitos, CA: IEEE Computer Society Press, 2012. Article 66, 11 pages.

M. Burtscher, B.-D. Kim, J. Diamond, J. McCalpin, L. Koesterke, and J. Browne. PerfExpert: An easy-to-use performance diagnosis tool for HPC applications. In *Proceedings of the ACM/IEEE International Conference for High Performance Computing, Networking, Storage and Analysis, SC*. Washington, DC: IEEE Computer Society, 2010. 1–11.

P. Carns, K. Harms, W. Allcock, C. Bacon, S. Lang, R. Latham, and R. Ross. Understanding and improving computational science storage access through continuous characterization. *Scientific Data Storage*. New Orleans, LA: IEEE, 2009.

P. Carns, R. Latham, R. Ross, K. Iskra, S. Lang, and K. Riley. 24/7 characterization of petascale I/O workloads. In *Proceedings of 2009 Workshop on Interfaces and Architectures for Scientific Data Storage*. New Orleans, LA, USA: IEEE, 2009.

M. Chabbi, K. Murthy, M. Fagan, and J. Mellor-Crummey. Effective sampling-driven performance tools for GPU-accelerated supercomputers. In *International Conference for High Performance Computing, Networking, Storage and Analysis (SC)*, New York, NY: ACM, November 2013. 1–12.

C. Coarfa, J. Mellor-Crummey, N. Froyd, and Y. Dotsenko. Scalability analysis of SPMD codes using expectations. In *Proceeding of the 21st International Conference on Supercomputing*, New York, NY: ACM, 2007. 13–22.

CORAL. Retrieved from https://asc.llnl.gov/CORAL/. Last accessed December 5, 2016.

The CrayPat Performance Analysis Tool. (n.d.). Retrieved from http://docs.cray.com/books/S-2376-63/, 2015.

CUDA-GDB. Retrieved from https://docs.nvidia.com/cuda/cuda-gdb. Last accessed December 5, 2016.

CUDA-MEMCHECK. Retrieved from https://developer.nvidia.com/cuda-memcheck. Last accessed December 5, 2016.

CUPTI. Retrieved from https://docs.nvidia.com/cuda/cupti/. Last accessed December 5, 2016.

J. Dean, J. E. Hicks, C. A. Waldspurger, W. E. Weihl, and G. Chrysos. ProfileMe: hardware support for instruction-level profiling on out-of-order processors. In *MICRO 30: Proceedings of the 30th annual ACM/IEEE International Symposium on Microarchitecture*. Washington, DC: IEEE Computer Society. 1997, 292–302.

Debugger API. Retrieved from https://docs.nvidia.com/cuda/debugger-api/index.html. Last accessed December 5, 2016.

J. DelSignore, D. Ahn, R. Castiain, J. Squires, M. Schulz. (n.d.). The MPIR Process Acquisition Interface Version 1.0. Retrieved from http://mpi-forum.org/docs/mpir-specification-10-11-2010.pdf. Last accessed December 5.

Department of Computer and Information Science, University of Oregon Advanced Computing Laboratory. TAU User's Guide. 2017. Retrieved from https://www.cs.uoregon.edu/research/tau/tau-usersguide.pdf. Last accessed November 11, 2016.

J. Dongarra, H. Ltaief, P. Luszczek, and V. Weaver. Energy footprint of advanced dense numerical linear algebra using tile algorithms on multicore architectures. In *2nd International Conference on Cloud and Green Computing (CGC)*. 2012. 274–281.

DWARF Debugging Information Format Committee. DWARF Debugging Information Format Version 4. Retrieved from http://www.dwarfstd.org/doc/DWARF4.pdf, Last accessed: June 10, 2010.

Dyninst API. (n.d.) Retrieved from http://www.dyninst.org/dyninst. Last accessed December 5, 2016.

M. Geimer, F. Wolf, B. J. N. Wylie, E. Ábrahám, D. Becker, and B. Mohr. The Scalasca performance toolset architecture. *Concurrency and Computation: Practice and Experience* 22(6) (April 2010): 702–719.

S. Graham, P. Kessler, and M. McKusick. Gprof: A call graph execution profiler. In *SIGPLAN '82 Symposium on Compiler Construction*. New York, NY: ACM. June 1982, 120–126.

IBM Corporation. Power ISA Version 2.07B. April 9, 2015.

Intel Corporation. Intel 64 and IA-32 architectures software developers manual: Combined volumes: 1, 2A, 2B, 2C, 2D, 3A, 3B, 3C and 3D. Order number: 325462-060US, September 2016.

A. Jannesari, K. Bao, V. Pankratius, and W. F. Tichy. Helgrind+: An efficient dynamic race detector. In *Proceedings of the 23rd international Parallel & Distributed Processing Symposium (IPDPS'09)*. Rome, Italy: IEEE, 2009.

S. Keramidas and G. Kaxiras. Sarc coherence: Scaling directory cache coherence in performance and power. *Micro (IEEE)* 30(5) (September 2010): 54–65.

H. Knuepfer and A. Brunst. Vampir. In: Brunst H. and Knüpfer A. (eds) *Encyclopedia of Parallel Computing*. Boston, MA: Springer US. 2011:2125–2129.

A. Knüpfer et al. Score-P: A joint performance measurement run-time infrastructure for periscope, Scalasca, TAU, and Vampir. In: Brunst H., Müller M., Nagel W., Resch M. (eds) *Tools for High Performance Computing 2011*, Berlin, Heidelberg: Springer, 2012.

G. L. Lee, D. H. Ahn, D. C. Arnold, B. R. de Supinski, M. Legendre, B. P. Miller, M. Schulz, and B. Liblit. Lessons learned at 208K: Towards debugging millions of cores. In *the Proceedings of the IEEE/ACM Supercomputing Conference (SC|08)*. Austin, TX, November 2008. Piscataway, NJ: IEEE

Legion. Retrieved from http://legion.stanford.edu. December 5, 2016.

X. Liu and J. Mellor-Crummey. Pinpointing data locality problems using data-centric analysis. In *CGO '11: Proceeding of the 2011 International Symposium on Code Generation and Optimization*. Chamonix, France, Washington, DC: IEEE Copmuter Society, April 2011.

X. Liu and J. Mellor-Crummey. A data-centric profiler for parallel programs. In *Proceeding of the International Conference on High Performance Computing, Networking, Storage and Analysis, SC '13*. New York, NY: ACM, 2013. 28:1–28:12.

X. Liu and J. Mellor-Crummey. A tool to analyze the performance of multithreaded programs on NUMA architectures. In *Proceedings of the 19th ACM SIGPLAN Symposium on Principles and Practice of Parallel Programming, PPoPP '14*. New York, NY: ACM, 2014. 259–272.

A. D. Malony, S. Biersdorff, S. Shende, H. Jagode, S. Tomov, G. Juckeland, R. Dietrich, D. Poole, and C. Lamb. Parallel performance measurement of heterogeneous parallel systems with GPUs. In *Proceedings of the 2011 International Conference on Parallel Processing, ICPP '11*. Washington, DC: IEEE Computer Society, 2011. 176–185.

S. Matwankar. CUDA 7.5: Pinpoint Performance Problems with Instruction-Level Profiling. Retrieved from https://devblogs.nvidia.com/parallelforall/cuda-7-5-pinpoint-performance-problems-instruction-level-profiling. Last accessed December 5, 2016.

H. McCraw, J. Ralph, A. Danalis, and J. Dongarra. Power monitoring with PAPI for extreme scale architectures and dataflow-based programming models. In *Workshop on Monitoring and Analysis for High Performance Computing Systems Plus Applications (HPCMASPA 2014), IEEE Cluster*. Madrid, Spain: IEEE, 2014. 385–391.

H. McCraw, D. Terpstra, J. Dongarra, K. Davis, and R. Musselman. Beyond the CPU: Hardware performance counter monitoring on blue gene/Q. In *Proceedings of the International Supercomputing Conference 2013, ISC'13*. Heidelberg, Germany: Springer. 2013. 213–225.

D. Mosberger-Tang. (n.d.) libunwind 1.1. Retrieved from http://nongnu.org/libunwind. Last accessed October 6, 2012.

MPI: A Message-Passing Interface Standard Version 3.0. 2012. Message Passing Interface Forum.

MRNet: A Multicast/Reduction Network. Retrieved from http://www.paradyn.org/mrnet/. Last accessed December 5, 2016.

M. S. Muller, A. Knupfer, M. Jurenz, M. Lieber, H. Brunst, H. Mix, and W. E. Nagel. Developing scalable applications with Vampir, Vampirserver and VampirTrace. In *Parallel Computing: Architectures, Algorithms and Applications, ParCo*, Germany: Forschungszentrum Julich and RWTH Aachen University, September 4–7, 2007. 637–644.

Multi-Process Service. Retrieved from https://docs.nvidia.com/deploy/pdf/CUDA_Multi_ Process_Service _Overview.pdf, 2015 Last accessed December 5, 2016.

N. Nethercote and J. Seward. Valgrind: A framework for heavyweight dynamic binary instrumentation. In *Proceeding of the ACM SIGPLAN Conference on Programming Language Design and Implementation (PLDI '07)*, June 2007. 89–100.

NVIDIA Visual Profiler. Retrieved from https://developer.nvidia.com/nvidia-visual-profiler, 2015. Last accessed December 5, 2016.

OpenMP Language Working Group. OpenMP Technical Report 4: Version 5.0 Preview 1. Retrieved from http://www. openmp.org/wp-content/uploads/openmp-tr4.pdf, November 2016. Last accessed December 27, 2016.

Open|Speedshop. Retrieved from https://openspeedshop.org, 2016. Last accessed December 5, 2016.

ParaDiS. Retrieved from http://paradis.stanford.edu. December 5, 2016.

C. Ponder and R. J. Fateman. Inaccuracies in program profilers. *Software. Practice. Experience* 18(5) (May 1988): 459–467.

Profiler User's Guide. Retrieved from https://docs.nvidia.com/cuda/profiler-users-guide. December 5, 2016.

J. Protze, I. Laguna, D. H. Ahn, J. DelSignore, A. Burton, M. Schulz, M. S. Müller. Lessons learned from implementing OMPD: A debugging interface for OpenMP. In *11th International Workshop on OpenMP (IWOMP)*. Aachen, Germany: Springer International Publishing, October 2015.

Rice University. HPCToolkit Performance Tools Project. Retrieved from http://hpctoolkit.org. Last accessed December 5, 2016.

A. Ros, B. Cuesta, M. Gomez, A. Robles, and J. Duato. Cache miss characterization in hierarchical large-scale cache-coherent systems. In *IEEE 10th International Symposium on Parallel and Distributed Processing with Applications (ISPA)*, Madrid, Spain, 2012. Piscataway, NJ: IEEE 691–696.

K. Sato, D. H. Ahn, I. Laguna, G. L. Lee, M. Schulz. Clock delta compression for scalable order-replay of non-deterministic parallel applications. In *Proceedings of IEEE/ACM Supercomputing Conference (SC|15)*. Austin, TX. Piscatawwy, NJ: IEEE, November 2015.

K. Sato, D. H. Ahn, I. Laguna, G. L. Lee, M. Schulz. Chris chambreau noise injection techniques for reproducing subtle and unintended message races. In *Proceeding of the 22nd ACM SIGPLAN Symposium on Principles and Practice of Parallel Programming*. Austin, TX. New York, NY: ACM, February 2017.

M. Schuchhardt, A. Das, N. Hardavellas, G. Memik, and A. Choudhary. The impact of dynamic directories on multicore interconnects. *Computer* 46(10) (2013): 32–39.

K. Serbryany, D. Bruening, A. Potapenko, and D. Vyukov. AddressSanitizer: A fast address sanity checker. In *USENIX Annual Technical Conference*, Boston, MA. Berkeley, CA: USENIX Association, 2012.

K. Serebryany and T. Iskhodzhanov. ThreadSanitizer: Data race detection in practice. In *Proceedings of the Workshop on Binary Instrumentation and Applications WBIA '09*. New York, NY: ACM, 2009. 62–71.

S. Shende and A. D. Malony. The TAU parallel performance system. *International Journal of High Performance Computing Applications* 20(2) (Summer 2006): 287–331.

S. Snyder, P. Carns, K. Harms, R. Ross, G. K. Lockwood, N. J. Wright. 2016. Modular HPC I/O characterization with Darshan. In *Workshop on Extreme-Scale Programming Tools (ESPT 2016)*. Piscataway, NJ: IEEE Press, 2016.

B. Sprunt. Pentium 4 performance-monitoring features. *IEEE Micro* 22(4) (2002): 72–82.

M. Srinivas et al. IBM POWER7 performance modeling, verification, and evaluation. *IBM Journal of Research and Development* 55(3) (May 2011):4:1–4:19.

N. R. Tallent, L. Adhianto, and J. M. Mellor-Crummey. Scalable identification of load imbalance in parallel executions using call path profiles. In *Proceeding of the 2010 ACM/IEEE Conference on Supercomputing*, Washington, DC: IEEE 2010a.

N. R. Tallent and J. Mellor-Crummey. Effective performance measurement and analysis of multithreaded applications. In *Proceeding of the 14th ACM SIGPLAN Symposium on Principles and Practice of Parallel Programming*. New York, NY: ACM, 2009a. 229–240.

N. R. Tallent and J. M. Mellor-Crummey. Identifying performance bottlenecks in work-stealing computations. *Computer* 42(12) (2009a): 44–50.

N. R. Tallent, J. M. Mellor-Crummey, L. Adhianto, M. W. Fagan, and M. Krentel. Diagnosing performance bottlenecks in emerging petascale applications. In *Proceeding of the ACM/IEEE Conference on Supercomputing*. New York, NY: ACM, 2009b. 1–11.

N. R. Tallent, J. Mellor-Crummey, and M. W. Fagan. Binary analysis for measurement and attribution of program performance. In *Proceeding of the 30th ACM SIGPLAN Conference on Programming Language Design and Implementation*. New York, NY: ACM, 2009b. 441–452.

N. R. Tallent, J. M. Mellor-Crummey, M. Franco, R. Landrum, and L. Adhianto. Scalable fine-grained call path tracing. In *Proceeding of the 25th International Conference on Supercomputing*. Tucson, AZ, New York, NY: ACM June 2011.

N. R. Tallent, J. M. Mellor-Crummey, and A. Porterfield. Analyzing lock contention in multithreaded applications. In *Proceeding of the 15th ACM SIGPLAN Symposium on Principles and Practice of Parallel Programming*. New York, NY: ACM, 2010b, 269–280.

C. Tapus, I.-H. Chung, and J. K. Hollingsworth. Active harmony: Towards automated performance tuning. In *Proceedings of the 2002 ACM/IEEE Conference on Supercomputing, SC '02*. Los Alamitos, CA: IEEE Computer Society Press, 2002. 1–11.

D. Terpstra, H. Jagode, H. You, and J. Dongarra. Collecting performance data with PAPI-C. In *Tools for High Performance Computing*. Berlin, Heidelberg: Springer. 2009, 157–173.

TU Dresden Center for Information Services and High Performance Computing. Vampirtrace 5.14.4 User Manual. Retrieved from https://tu-dresden.de/zih/forschung/ressourcen/dateien/projekte/vampirtrace/dateien/VT-UserManual-5.14.4.pdf Last accessed: August 9, 2017

University of Oregon. TAU Portable Profiling. Retrieved from http://tau.uoregon.edu. Last accessed December 5, 2016.

Using the PAPI Cray NPU Component. (n.d.). Retrieved from http://docs.cray.com/books/S-0046-10//S-0046-10.pdf.

J. Vetter. Dynamic statistical profiling of communication activity in distributed applications. In *Proceedings of the 2002 ACM SIGMETRICS International Conference on Measurement and Modeling of Computer Systems*. New York, NY: ACM, 2002. 240–250.

V. Weaver et al. PAPI 5: Measuring power, energy, and the cloud. In *IEEE International Symposium on Performance Analysis of Systems and Software (ISPASS)*, Austin, TX. April 21-23, 2013. 124–125.

V. Weaver, M. Johnson, K. Kasichayanula, J. Ralph, P. Luszczek, D. Terpstra, and S. Moore. Measuring energy and power with PAPI. In *ICPPW '12 Proceedings of the 2012 41st International Conference on Parallel Processing Workshops*, Pittsburgh, PA, September 10-13, 2012. 262–268.

J. Whaley. A portable sampling-based profiler for Java virtual machines. In *Proceedings of the ACM 2000 conference on Java Grande*. New York, NY: ACM, 2000. 78–87.

W. R. Williams, X. Meng, B. Welton, and B. P. Miller. Tools for High Performance Computing 2015. Proceedings of the 9th International Workshop on Parallel Tools for High Performance Computing, Dresden, Germany, September 2015. Switzerland: Springer International, 2016. 2015.

Y. Xu, Y. Du, Y. Zhang, and J. Yang. A composite and scalable cache coherence protocol for large scale CMPs. In *Proceedings of the International Conference for High Performance Computing, Networking, Storage and Analysis*, Seattle, WA, November, 2011. New York, NY: ACM, 2011. 285–294.

A. Yasim. A top-down method for performance analysis and counters architecture. In *IEEE International Symposium on Performance Analysis of Systems and Software (ISPASS)*. Monterey, CA, 2014, IEEE March 2014, 35–44.

3 Exascale Challenges in Numerical Linear and Multilinear Algebras

Dmitry I. Lyakh and Wayne Joubert

CONTENTS

3.1 Introduction 52
3.2 Linear Algebra 52
    3.2.1 Applications 52
    3.2.2 Linear Algebra Operations: State of Practice 53
        3.2.2.1 Dense Linear Algebra Operations 53
        3.2.2.2 Sparse Linear Algebra Operations 55
    3.2.3 Parallel and Accelerated Algorithms 55
        3.2.3.1 Hardware Considerations 55
        3.2.3.2 Dense Linear Algebra Algorithms 58
        3.2.3.3 Sparse Linear Algebra Algorithms 60
    3.2.4 Extreme Scale Issues 63
        3.2.4.1 Higher Thread Count 63
        3.2.4.2 Changing Memory Hierarchies 64
        3.2.4.3 Communication Network Developments 64
        3.2.4.4 Growing Resilience Concerns 64
    3.2.5 Software 65
        3.2.5.1 Third-Party Libraries 65
        3.2.5.2 Vendor Libraries 66
    3.2.6 Conclusion 66
3.3 Tensor Algebra 67
    3.3.1 Tensors in Different Scientific Disciplines 67
    3.3.2 Basic Tensor Algebra Operations 71
    3.3.3 Tensor Decompositions and Higher Level Operations 73
    3.3.4 Parallel Algorithms for Basic Tensor Operations 77
    3.3.5 Extreme Scale Solutions 79
        3.3.5.1 Projected Exascale Computing Hardware Roadmap 79
        3.3.5.2 Hardware Abstraction Scheme and Virtual Processing 80
        3.3.5.3 HPC Scale Abstraction 84
        3.3.5.4 Hierarchical Task-Based Parallelism via Recursive Data Placement
                and Work Distribution 85
3.4 Conclusions 89
Acknowledgment 90
References 90

3.1 INTRODUCTION

Scientific and mathematical libraries play a universally important role in leadership class high-performance computing (HPC) (Dongarra et al., 2015). Many science applications require the use of specific high-performance libraries in order to function. These libraries embody many person-years of algorithmic development and code optimization effort, resulting in a more efficient use of HPC resources and better allocation of precious developer time.

As we progress to exascale and beyond, the need for libraries, which raise the abstraction level and thus address a larger scope and granularity of operations needed by the software developer, is becoming more acute. HPC nodes are becoming increasingly complex and harder to program for high performance. At the same time, developer resources remain roughly constant. Therefore, the opportunity to reuse highly tuned code in the form of portable scientific libraries across multiple projects becomes increasingly valuable.

A set of seven computational motifs (dense and sparse linear algebra, structured and unstructured grids, fast Fourier transforms, particle and Monte Carlo methods) has been posited as a framework for thinking about the scope of algorithm types that are prevalent in HPC applications (Colella, 2005). Over the course of time, these have never hardened into *seven libraries* or *seven operation categories* due to the fundamentally amorphous and variegated nature of the contents of each class as evidenced in the way computational scientists and developers have needed to implement them. This being said, certain of these motifs, in particular, dense and sparse linear algebra as well as the fast Fourier transform, have been more amenable to formalization and encapsulation in highly successful libraries. Yet, the challenge of performance portability across different, often heterogeneous, computer architectures remains largely open. Exascale scientific computing systems will likely be characterized by a diverse set of computer architectures as the Moore's law era is approaching its critical point. Consequently, the choice of a portable programming model (or programming approach) becomes critical for scalable implementation of scientific applications at extreme scale, revitalizing the importance of efficient and portable scientific libraries, in particular, linear algebra libraries.

Apart from the decades-long development history of linear algebra algorithms and libraries, the efforts toward library-based implementation of tensor (or multilinear) algebra algorithms were much less pronounced until very recently. The push toward the latter was mostly stimulated by the need to perform large-scale electronic structure simulations and run large-scale data analysis workflows in various fields, including the prominent field of deep learning. Scalable numerical tensor algebra algorithms are currently under active development, and their efficient portable implementation in domain-specific libraries is in very high demand.

This chapter covers both linear and multilinear algebra fields. Section 3.1 highlights linear algebra algorithms and the challenges on their path to exascale. It provides a generic overview of the current state of this rather extensive field. Section 3.2 describes numerical tensor algebra algorithms and theorizes on a specific solution for performance portability on large-scale heterogeneous HPC architectures that originates from the design of the ExaTENSOR library being developed by the first author at the Oak Ridge Leadership Computing Facility (OLCF).

3.2 LINEAR ALGEBRA

3.2.1 APPLICATIONS

Usage patterns at HPC facilities indicate that dense and sparse linear algebra operations are among the most commonly used in scientific applications. For example, in reference (Joubert and Su, 2012), a study of the application workloads for a 2-year period on the OLCF Jaguar system revealed that of the top 20 most heavily used applications, 11 used dense linear algebra and 12 required sparse linear algebra. Within the Center for Accelerated Application Readiness (CAAR) effort to prepare applications for the pre-exascale summit system, a full 9 of the 13 focus applications, or 70%, depend on mathematical libraries to supply dense and/or sparse linear algebra operations. Furthermore, usage of libraries, as reported in Anantharaj et al. (2013), on the OLCF Jaguar and Titan systems over a period of time showed the predominant use of Cray LibSci, which is typically employed by applications for its linear algebra support. Indeed, both the HPL and high-performance geometric multigrid (HPGMG) benchmarks, highly differing measures of performance of the top HPC systems and intended to be very rough proxies of real application performance, are themselves based on linear algebra algorithms (https://www.top500.org/project/linpack; https://hpgmg.org).

This trend is expected to continue, not only because science applications by the inherent nature of their models and algorithms require highly optimized linear algebra operations but also for architectural reasons. As shown repeatedly, for example, in Kogge (2008), a primary limiter for exascale computing will be power, and in this regime, the primary power cost is data movement; thus, operations with high computational intensity as measured by operations per memory reference (i.e., operations that allow many operations on a data element before requiring to read a new data element into the processing unit) will run most efficiently. This favors computations exhibiting this property such as dense linear algebra operations and, specifically, basic linear algebra subprograms (BLAS-3) kernels. Furthermore, the regularity of structure of dense linear algebra is perfectly suited to vector processing, a defining characteristic of the anticipated exascale node architecture.

Due to the widespread use of dense and sparse linear algebra in HPC applications—which in actual practice tends to be limited to only a fairly narrow scope of required function calls—efforts to implement and optimize these operations in the form of libraries, whether by vendors or third-party developers, are the efforts well spent in terms of potential impact on increased code speed, increases of cost and energy efficiency at facilities, improvements to developer productivity, and increased science output to reach mission goals.

3.2.2 Linear Algebra Operations: State of Practice

Clearly, the scope and variety of linear algebra operations is very broad. Numerical linear algebra, due to its integral importance to scientific simulations, has been the subject of intense scrutiny by many research communities for decades, with many of the results finding their way into production applications. However, the focus of the current study is much more limited; specifically, we focus on linear algebra operations expected to be most commonly required at exascale. From our experience at a leadership computing user facility with large systems and a diverse user base using some of the world's most scalable high-performance applications, we have observed over time a remarkably consistent picture of the kinds of linear algebra operations needed. Though there will always be codes with highly specialized operations out of the norm, we focus here on the most common operations, which are important in themselves but also paradigmatic of the performance issues faced by the broader scope of linear algebra methods.

In the following, we focus on general computational characteristics of algorithms most commonly used at leadership scale rather than detailed descriptions of the algorithms (for the latter, see standard references, e.g., Golub and Van Loan, 2013).

3.2.2.1 Dense Linear Algebra Operations

A given matrix A may possess the property that many or most of its entries are zero, that is, the matrix is sparse. When computationally advantageous, the data structures and algorithms used to operate on this matrix can be selected to take advantage of this sparsity. Otherwise, the matrix may be considered dense, with all matrix entries stored and operated upon.

It is useful to classify linear algebra operations as BLAS-1 (scalar, vector), BLAS-2 (vector, matrix), and BLAS-3 (matrix, matrix) operations, based on the kind of operation requested and the computational complexity (Dongarra et al., 1990), as a generalization of the original BLAS standard (Lawson et al., 1979). Of particular interest are the dense BLAS-3 operations, since these, in particular, are able to yield the high computational intensities leading to efficient computations on modern processors. The BLAS-1, BLAS-2, and BLAS-3 operations can, in turn, be composed with each other to form a vast array of higher level operations of high value to scientific application codes.

Though the range of operations implemented in linear algebra libraries and potentially useful to science applications is large (Anderson et al., 1999) in practice, the number of linear algebra operations in common use is, in fact, relatively small. We consider here a small set of commonly used operations representative of those typically used in practice on leadership class systems. In the following subsections, we give a survey of our findings based on use cases prevalent on OLCF leadership class HPC systems.

3.2.2.1.1   GEMM

The matrix–matrix multiplication for general dense matrices, implemented as single-precision general matrix-matrix multiplication (SGEMM), double-precision general matrix-matrix multiplication (DGEMM), single-precision complex general matrix-matrix multiplication (CGEMM), or double-precision complex matrix-matrix multiplication (ZGEMM), is required by many application codes. In its simplest form, this operation is $C = AB$, where A is dimension $M \times K$ and B is $K \times N$. Though it is desirable from a performance standpoint that M, N, and K be as large as possible to maximize computational intensity and amortize overheads, in some applications, these operations may be small, for example, $M = K = 32$. For science applications in practice, the matrix entries are most commonly double precision or double complex and in some cases single precision.

3.2.2.1.2   LU Factor/Solve

The companion operations of factoring a matrix as $A = LU$ lower and upper triangular factors and then solving the associated linear system $Ax = b$ via forward and backward solves is a requirement of some applications. The factorization operation itself may attain high computational intensity by employing GEMM operations; the forward and backward solves, however, are of lower computational intensity and generally less parallel due to recursions but are also generally less costly than the factorization, thus less performance-critical. In some cases, the factorization might be used once and then discarded; alternatively, the factorization might be used multiple times to solve repeated linear systems with different right-hand sides. Other types of factorizations such as Cholesky and QR factorizations are also sometimes used by applications.

3.2.2.1.3   Matrix Diagonalization

The eigendecomposition of a matrix as $A = PJP^{-1}$ with J diagonal or an upper bidiagonal Jordan normal form is also needed by some science codes. In addition to the diagonal entries of J (eigenvalues), the columns of P (eigenvectors) are, in some cases, required. Iterative eigensolvers can be employed if approximations to only a few eigenvalues and possibly eigenvectors are needed; if all are needed, a full dense eigensolver is typically used. Applications may require eigensolvers applicable to a symmetric matrix or alternatively to a nonsymmetric matrix, leading to different algorithmic variants. In some cases, generalized eigenvalue problem solves or singular value decompositions (SVDs) are needed.

3.2.2.1.4   Batched Operations

Often, it is necessary to perform simultaneously many repeated instances of one of the above operations for a set of matrices $\{A_i\}$, possibly all of the same size. Though computational intensity is no greater from operating on many matrices compared to one, efficiency gains are still possible from batching these operations together, by increasing the available parallelism and potentially amortizing the costs of transfer latencies. Though not in the original BLAS and linear algebra package (LAPACK) standards, batched operations are increasingly supported by linear algebra libraries.

3.2.2.2 Sparse Linear Algebra Operations

3.2.2.2.1 Sparse Linear Solvers

The solution of the linear system $Ax=b$ for sparse A is an expensive component of many codes, for example, those performing semi-implicit or implicit time-dependent simulations of partial differential equations. For some simulations, the linear solve problem is tractable only by sparse direct solvers based on factorization techniques; however, for leadership class HPC, it is usually preferable to attempt use of iterative methods, which typically have much better computational complexity and scalability characteristics; we thus focus primarily on iterative methods.

The calculation of iterates x_i, which converge with increasing i to the true solution x, is performed by different choices of iterative method based on whether the matrix A is symmetric in its entries across the main diagonal or conversely nonsymmetric. For the symmetric case, the conjugate gradient method is generally used, whereas for the nonsymmetric case, the restarted generalized minimal residual (GMRES) method is popular. These Krylov solver methods make use of matrix–vector products with A as well as BLAS-1 and other BLAS-2 operations; the low-computational intensities of these operations generally result in low-computational intensities for the solve except in special cases when A can be represented in a way that enables use of BLAS-3 computations.

It is general practice to solve systems of the form $Ax=b$ by first preconditioning with a matrix M that approximates A, so that the preconditioned system $M^{-1}Ax=M^{-1}b$ is much easier to solve with the aforementioned iterative methods and also the required multiplication of M^{-1} by a vector is comparatively inexpensive. For leadership scale applications, several preconditioners are common. First, letting M be a diagonal matrix composed of the main diagonal entries of A, the Jacobi preconditioning, is a weak preconditioner but has the virtue of being inexpensive, straightforwardly parallelized, and easy to implement. Second, incomplete factorization methods, such as incomplete LU factorization (ILU), based on approximately factoring A by discarding nonzero entries during the factorization process, are effective for a wide class of problems but are difficult to parallelize due to their recursions. Third, multigrid methods based on solving a hierarchical sequence of problems similar to the original problem offer optimal order-N solution time for many problems and are parallelizable but are limited in their scope of efficient applicability. Finally, physics-based preconditioners can be custom-developed on a case-by-case basis when the physical problem that gives rise to the linear system is suggestive of a simpler problem that is easier to solve.

3.2.2.2.2 Sparse Eigensolvers

Sparse eigenproblems can also be classified as symmetric or nonsymmetric based on the properties of the matrix A. The Lanczos method and the Arnoldi method, respectively, are generally used, analogously to the conjugate gradient and GMRES methods for linear systems. These methods are typically used to compute approximations to a subset of the eigenvalues and, optionally, eigenvectors. Both the GMRES linear solver and the Arnoldi eigensolver suffer from the property that both storage requirements and work per iteration increase linearly as the number of iterations increases. Restarting techniques can be used to mitigate this problem at the expense of some degradation of convergence rate. It is also possible to improve the convergence of these methods by means of preconditioning, for example, with the generalized Davidson's method (Morgan and Scott, 1986).

3.2.3 PARALLEL AND ACCELERATED ALGORITHMS

3.2.3.1 Hardware Considerations

It has always been a challenge to map algorithms to hardware to enable codes to solve problems as quickly and economically as possible. The history of computing has seen many changes in computer architecture, necessitating repeated refactoring and rewriting of application codes and libraries. In more recent years, advanced hardware has trended unmistakably toward increasing node complexity, exacerbating the burden of code optimization for developers, who must now implement increasingly complex code to extract top performance for their algorithms.

These changes have assuredly impacted developers of algorithms, software, and libraries for dense and sparse linear algebra. In the following subsections, a short, selective list of key issues that must be faced in the effort to develop highly efficient linear algebra software is discussed.

3.2.3.1.1 Processor Instruction Sets

The actual machine instructions executed on the processors of HPC systems are typically generated by compilers, though occasionally developers resort to assembly language, an approach currently experiencing a revival in the broader developer community (http://www.i-programmer.info/news/98-languages/9904-assembler-in-the-top-ten-languages-for-july.html). In lieu of this, ensuring that the compiler generates highly efficient machine instructions can require careful crafting of the source code and might mean code tweaks to optimize code to the specific compiler, though this practice, when taken to the extreme, can lead to code that is difficult to maintain across compiler versions and test cases. Ensuring that specific high-performing instructions are used, such as fused multiply-add (FMA) or special function instructions, can require attention and in some cases requires use of hardware-specific intrinsics.

3.2.3.1.2 Vectorization

Vector units in modern processors have grown in length to 8, 16, or 32 elements. Vector hardware is a fundamental aspect of pre-exascale and exascale compute nodes. The first challenge to porting linear algebra to vector processors is adapting the algorithm components to a form suitable to vectorization, for example, stride-1 aligned memory accesses with operations that are supported by the vector units. The next concern is actually applying vectorization in the code, whether by prescriptive use of directives (Intel Phi), explicit reference to vector lanes (NVIDIA CUDA) or higher level approaches. In some cases, one must verify that the compiler, in fact, properly vectorizes the code, for example, via user inspection of compilation listings, and if vectorization is not successful then adjust the code accordingly. Attention must also be paid to operations that hinder vectorization, for example, the need to perform a reduction across vector lanes in the middle of a larger computation.

3.2.3.1.3 Threading

Threading is recognized as an essential requirement for attaining high performance on pre-exascale and exascale systems. Algorithmically, decisions must be made regarding the problem dimensions to be mapped to the compute threads and how the mapping is made. The manner of binding of threads to processor cores impacts performance. Allowing for dynamic rather than static scheduling generally allows for better use of hardware resources by allowing the code to submit work immediately to threads as they become idle. Hyperthreading to execute concurrent multiple threads per processor core is a key to high performance; tuning on a case-by-case basis may be required to determine optimal thread counts.

3.2.3.1.4 Memory

Programming for optimal use of the memory hierarchy of modern processors has become an issue of paramount importance, insofar as the number of levels of the hierarchy has continued to increase, the latency and bandwidth characteristics of the levels are significant and vary heavily, and issues affecting performance at each level are complex. We treat different components of the memory hierarchy in turn.

3.2.3.1.4.1 Registers and Caches

Compute intensive operations perform best if all data stays resident in registers for as long as possible; however, long, complex computational kernels can require many registers, leading to register spillage and underutilization of vector units. Caches (L1, L2, L3, and on GPUs texture cache and constant cache) are either programmable or automatic with various eviction policies and associativity characteristics; direct mapped caches require special attention since certain data alignments can unexpectedly result in thrashing and attendant performance losses. The mapping from cache to registers can have performance-related impacts, for example, bank conflicts for graphics processing unit (GPU) shared memory. Caches favor stride-1 access; cache line size can affect performance, for example, causing performance losses for unstructured memory accesses in sparse linear algebra operations.

3.2.3.1.4.2   Main Memory   Nonuniform memory access (NUMA) issues affect performance when different cores on a node have different memory latency or bandwidth characteristics depending on which memory chip or bank on the node is being accessed. This interacts with threading, which, in some cases, assigns by a first-touch policy the memory pages to memory units, potentially resulting in nonlocal siting of pages and thus performance loss. More generally, the fact that multiple processor cores may need to access a single memory unit results in competition among the cores for the available bandwidth for memory intensive codes. Memory page size also affects performance depending on the problem. Prefetching, either done by the compiler or manually, can improve performance with respect to caches by initiating memory requests early and allowing overlap of memory operations and computations.

3.2.3.1.4.3   Accelerator Memory   For discrete accelerated processors used in conjunction with conventional CPUs, transfers between CPU and GPU memories must be performed. This can be done explicitly with staging to overlap with computations for efficiency. NVLINK hardware on some pre-exascale systems will allow for automatic data migration; in some cases, it will likely still be necessary to explicitly manage transfers for high performance.

3.2.3.1.4.4   High Bandwidth Memory   Pre-exascale systems will locate large quantities of high-speed memory directly on the processor chip. This will again raise questions as to whether it is best to program transfers explicitly or let them occur automatically. Programming models are also an issue, insofar as developers want a performance portable solution across vendors but standards for memory management have yet to be developed and standardized.

3.2.3.1.4.5   Nonvolatile RAM   Nonvolatile RAM (NVRAM), to be available in pre-exascale systems, offers the potential for solving much larger problems than is possible, using only main memory. Issues must be resolved regarding the use of explicit versus automatic programming approaches, programming models based on explicit bulk transfer versus addressability of the NVRAM as extended memory, and the need for standardized application programming interfaces (APIs) across vendors for performance portability.

3.2.3.1.4.6   Out-of-Core   Occasionally, an application may require use of the spinning-disk file system as extended memory. Proper staging and optimization of these operations are key for performance.

3.2.3.1.5   Internode Parallelism

For computations across many compute nodes, the most immediate determiners of performance are communication bandwidth, which affects transfer speeds for codes with large messages, and communication latency, which is operative for small-message codes, for example, in strong scaling regimes. Message injection characteristics of the interconnect hardware are also significant when a single node must send and/or receive many smaller messages possibly from different nodes at the same time. Node execution configuration characteristics, for example, whether one or multiple processes per node send messages, affect performance. For bandwidth-limited codes, use of asynchronous communications overlapped with computations can significantly improve performance. One-sided messaging via shared memory (SHMEM)-type programming models and direct transfer between GPU main memories using GPUDirect can enhance performance. Finally, interconnect topology is sometimes significant, and methods for mapping communication patterns strategically to the network can be useful for reducing slowdowns caused by network contention (Sankaran et al., 2015).

3.2.3.1.6   Scheduling

An overarching theme becoming increasingly important on the path to exascale is scheduling of asynchronous operations. To use as much of the needed hardware of an exascale node as possible at any given time, including CPUs, discrete accelerators, interconnects, memories, and input and output (I/O) subsystems, it is advantageous to launch asynchronous nonblocking work units to perform operations with these hardware components. This enables overlap of operations and also keeps idle hardware busy by scheduling work to begin as soon as resources become available. This can be accomplished using a dynamic task-based programming model. Task-based programming models will additionally be helpful if resiliency concerns become critical at exascale, since task-based programming will potentially allow failed work tasks to be relaunched.

3.2.3.1.7   Summary

It is clear that HPC compute nodes have become highly complex, with many interlocking performance determiners that play a decisive role in the realizable performance of codes. It is essential that, to serve project needs on exascale as well as later post-exascale systems, linear algebra operations, which play a central role in the success of many leadership class science applications, be well optimized for these architectures going forward.

3.2.3.2   Dense Linear Algebra Algorithms

We now consider how the exascale hardware characteristics described above impact the performance of fundamental dense linear algebra algorithms.

3.2.3.2.1   GEMM

The dense matrix–matrix product operation is a core service, which, due to its high compute intensity, enables high efficiency for many higher level linear algebra operations that depend on it. To make GEMM operations on modern compute nodes execute with highly optimized performance, every one of the previously mentioned performance considerations must be taken into account.

3.2.3.2.1.1   Instruction Sets   Modern processors offer FMA instructions, which execute in fewer clock cycles than individual multiply and add instructions used together. Code efficiently implementing GEMM must use FMA instructions for this fundamental required operation. This might be possible through a high-level language directly or might require calls to custom intrinsics or use of assembly code. Optimization techniques, such as loop unrolling, are typically required in order to ensure high density of FMA instructions.

3.2.3.2.1.2   Vectorization   The sequence of FMAs for a GEMM computation must map effectively to the processor's vector units. Factors such as alignment in memory, data layout, blocking factors, order of computations, and loop unroll depth are to be considered.

3.2.3.2.1.3   Threading   Work at the low level must be divided so that it can be distributed efficiently to multiple compute threads. If there are many GEMMs of nearly the same size, for example, from batched GEMM requests, another axis of parallelism is thus available for threading; this is much needed in modern processors that require huge numbers of compute threads in order to approach peak performance. Batched GEMMs with different matrix sizes may require further considerations to address load balancing issues.

3.2.3.2.1.4   Memory: Registers   Since FMAs operate on registers, it is important to perform many FMAs in the form of register-to-register operations between the intervening register load and store operations to and from memory to keep the FMA density high and minimize the impact of costly memory references.

3.2.3.2.1.5   Memory: Caches   The full GEMM operation must be subdivided into blocks that can be staged to fit within the caches for fast transfer into registers. This applies to either conventional caches or programmable scratchpad memories such as GPU shared memory. When appropriate, prefetching can be used so that the processor memory controller can work on satisfying memory references while at the same time FMA operations are taking place. Memory layout and access patterns require consideration to avoid performance variances.

3.2.3.2.1.6   Main memory: The same kinds of blocking and staging techniques are used to load the last-level cache from main memory. NUMA issues and thread binding to processor cores may need to be considered to optimize effective bandwidth.

3.2.3.2.1.7   Accelerator Memory   On current GPU hardware, the latency and bandwidth costs of transferring matrix data to and from the GPU can take a toll on the performance of pure GEMM operations performed on the GPU. Thus, it is desirable to overlap transfers and computations in a pipelined fashion when possible. The GEMM can be broken into blocks, and at any given time, a transfer to the GPU of a matrix block, a GEMM operation on the GPU, and a transfer of a result matrix block back from the GPU can all be in flight at the same time. NVLINK hardware may make this more automatic, but scheduling will be critical; transfer requests must not be posted *just in time* as the data are needed but rather must be posted beforehand by a prefetching technique so that the data are ready for computation at the right time.

3.2.3.2.1.8   High Bandwidth Memory, NVRAM, Out-of-Core   The same considerations are apropos for other layers of the memory hierarchy. Computations must be appropriately blocked, and computations and transfers must be staged, typically in an explicit fashion, in order to maintain high performance.

3.2.3.2.1.9   Internode Parallelism   In practice, most applications require individual GEMMs to be performed in an isolated fashion per-node. However, some applications require multinode GEMMs. In such cases, the same principles apply of blocking the computations and overlapping transfers with computations by use of asynchronous protocols. In practice, the GEMMs on different nodes typically have the same size; if not, special considerations are needed to address load balancing concerns.

3.2.3.2.1.10   Scheduling   All of the above considerations require close attention be paid to optimal scheduling of computations and data transfers. When possible, this is generally best done by using dynamic scheduling to get highest performance.

3.2.3.2.1.11   Practical Considerations   Vendors and third-party library developers deliver general-purpose libraries for performing GEMM calculations efficiently. However, specific projects may have requirements that are not met by these libraries. For example, we have encountered codes that always generate matrices with one very large dimension and another dimension that is small, for which library codes are typically not well optimized (cf. Joubert et al., 2015). In another example, the double precision DGEMM may be well optimized but the more infrequent double complex ZGEMM less so. When feasible, requests can be made to the library developers to optimize the cases needed for critical tasks. Also, GEMM software must be able to take advantage of all CPU cores and not just the attached accelerators such as GPUs. In other cases, users may require their own control to be able to allocate some compute threads, cores, GPUs, or parts of GPUs to a GEMM operation and others to other operations.

3.2.3.2.2   LU Factor/Solve

LU factorizations rely on well-optimized lower level kernels to achieve high performance, such as GEMM operations, triangular matrix forward and backward solves and low-rank updates. LU factorization implementations benefit from formulations rich in GEMM operations so as to leverage in-depth code optimizations deployed to GEMM library kernels. To achieve high performance on modern compute nodes, LU factorizations must use all node compute resources, for example, a heterogeneous mix of CPUs and GPUs or Intel Phi processors. Formulation of base kernel operations and efficient scheduling of these operations become of paramount importance, and it is convenient to formulate the dependencies between these operations as a directed acyclic graph (DAG) submitted to a runtime scheduler (see e.g., Haidar, 2016). The lower/upper triangular matrix factorization (LU) solve process making use of the factorization takes much less time than the factorization process itself for very large problems yet possess its own challenges. The recursions inherent to the triangular solves are not naturally vectorized or threaded with the efficiency of other operations such as GEMMs; techniques such as imposing a block structure on the matrix are required to increase available parallelism. When appropriate, mixed precision techniques can speed up the LU factorization and solve process. The need to perform batched LU factorizations and solves by some applications puts emphasis on code optimization for a different part of the problem space; the addition of more linear systems provides more parallelism suitable for a threaded environment, but the individual problems may be smaller resulting in potentially lower computational intensity and more overheads and thus a greater challenge to performance optimization.

3.2.3.2.3   Matrix Diagonalization

The full eigendecomposition of a symmetric matrix A is typically done in two steps: (1) reduction of the matrix to tridiagonal form by similarity transformations in the form of Householder reflections and (2) the eigendecomposition of the resulting tridiagonal matrix. The former step typically requires most of the computational work. To map the tridiagonal reduction to accelerated nodes, it is possible to reformulate the computation as a series of blocked Householder transformations based on low-rank updates using GEMMs in addition to matrix–vector products, the latter tending to limit performance due to their lower computational intensity (see, e.g., Yamazaki, 2013 for details).

The full decomposition of general nonsymmetric matrices into eigenvectors and eigenvalues is composed of three parts. First, the matrix is transformed to an upper Hessenberg matrix via block Hessenberg reduction, which is mainly GEMM dominated. Second, the upper Hessenberg matrix is transformed into an upper triangular matrix by QR iteration; each iteration is composed of mixed BLAS-1 and BLAS-3 operations, and the iteration count is variable depending on the problem. Third, eigenpairs are computed for the tridiagonal matrix using a blocked method employing GEMM operations (see, e.g., Gates, 2014 for details).

Acceleration of symmetric and nonsymmetric eigensolvers for modern compute nodes is an area of active research that is likely to require innovations in algorithms as well as implementations. Recurring themes of this work include reformulation of algorithms to convert as much work as possible to high computational intensity BLAS-3 GEMM operations, optimizing the remaining BLAS-1 and BLAS-2 bottlenecks as much as possible, blocking of computations to map work and data to the memory hierarchy and compute hardware, and careful scheduling to make best use of many-core and heterogeneous node hardware.

3.2.3.3   Sparse Linear Algebra Algorithms

Sparse linear algebra occupies a very different computational regime compared to dense linear algebra. When the matrices are sparse, data transfer costs within the memory hierarchy take a much more front-and-center role in determining performance, simply because there is much less reuse of any given data element for performing multiple computations with that element at the time when it is resident within the compute unit. Performance factors such as good use of instruction sets, vectorization, and threading are still important but are overshadowed by the heavy costs of transferring data across the memory hierarchy.

For the purposes of this discussion, it is most natural to consider performance issues of sparse solvers in three parts: the base iterative solver algorithms, the sparse matrix–vector product (SpMV) operations, and the preconditioning operations.

3.2.3.3.1 Iterative Solvers

The most popular iterative linear solvers and eigensolvers, for example, the Krylov methods conjugate gradients, GMRES, Lanczos, and Arnoldi, are mathematically related and take similar forms. These can be classified as short recurrence methods with fixed work per iteration (conjugate gradients, Lanczos) and long recurrence methods with work per iteration and storage requirements linear in the iteration count (GMRES, Arnoldi). For the short recurrence methods, the cost of the matrix–vector product and preconditioner generally dominate the cost of the iterative solve in all but the simplest problems; for long recurrence methods, the time and storage costs of the Krylov solver itself can be problematic. Long recurrence methods for nonsymmetric problems can be periodically restarted to reduce expense, though this weakens convergence of the methods, acutely so for very hard problems. Short recurrence methods such as BiCGSTAB and TFQMR are alternatives, though their convergence characteristics are often uncertain.

For long recurrence methods, a potential modification to improve performance on exascale architectures is to restructure the BLAS-1 vector orthogonalizations to instead take the form of BLAS-2 matrix–vector operations to attain some reduction in memory traffic; care must be taken here to avoid adverse impacts on numerical properties of the algorithms. To further improve performance of long recurrence methods, algorithm variants based on unrolling the iteration loop (Swanson and Chronopoulos, 1992; Joubert and Carey, 1992a; Demmel et al., 2009) provide fruitful opportunities. For these methods, several matrix–vector product and preconditioning steps are performed in sequence without intervening vector orthogonalization operations, these being deferred and done all at once to reduce latencies and other costs. These methods can improve performance in three ways. First, the number of global reductions is reduced; this can be helpful in some situations, for example, strong scaling grid-based codes, though the benefit from this may be limited depending on the problem because of vendors' attempts thus far to keep global reduction latency costs low. Second, some opportunity is created to convert BLAS-2 vector orthogonalizations to BLAS-3 operations, thereby increasing computational intensity. Finally, for both the long recurrence and the short recurrence cases, temporal locality can be increased across multiple matrix–vector product and preconditioning operations, this being beneficial for memory traffic depending on the structure of the operations (Joubert and Carey, 1992b).

3.2.3.3.2 Sparse Matrix–Vector Products

In practice, the compute time for SpMVs is sometimes small compared to preconditioning costs. However, SpMV costs are significant when strong, expensive preconditioners are not effective, necessitating the use of the inexpensive Jacobi preconditioning. SpMV costs are also significant for the case when Jacobi iteration is a computationally expensive component of other solvers, such as Jacobi smoothers for multigrid preconditioners. SpMV is generally of low-computational intensity since it is BLAS-2; furthermore, if the underlying problem is unstructured or has no exploited structure, it requires irregular memory accesses, which are costly. For some problem-specific cases, the matrix–vector product operator might be reformulated in terms of BLAS-3 operations; presently, this case is not usual, but such methods are currently being pursued for exascale systems. Also, some simulation problems result in a full sparse matrix composed of many small dense matrices, for example, due to multiple degrees of freedom per grid cell for a grid-based problem; for situations for which it makes sense, inefficiencies due to irregularity of memory access can be reduced by exploiting this structure, for example, using a block compressed sparse row storage format, and to some extent vector elements can be reused, lowering memory traffic costs. In other cases, zero padding might be added to the sparse matrix to create a dense block structure, though this artificially adds extra computations and as such is generally of limited value. In still other cases, the science problem or parts of the problem may have constant coefficients, so that the sparse matrix may have many repeated identical elements; exploiting this can result in less memory traffic. More generally, matrix element deduplication methods (Stevenson et al., 2012) can, in some cases, drastically reduce the number of loads of matrix elements into the processor and thus improve performance.

For structured problems derived, for example, from finite difference discretizations, the SpMV can, in some cases, be significantly accelerated by use of stencil-based techniques to increase spatial, and in some cases, temporal, locality to reduce memory traffic (Datta et al., 2009; Nguyen et al., 2010). Analogously, for unstructured problems, reorderings of the matrix and vectors, e.g, using space-filling curve orderings, can be applied so as to reduce the incidence of cache misses and thus improve performance in a similar way.

For solving very large problems out-of-core by using disk storage, matrix blocking methods can be used in conjunction with block preconditioners and unrolled iteration loop iterative methods where appropriate (Hoemmen, 2011). The same kinds of methods can be used in principle for other parts of the memory hierarchy, for example, NVRAM and on-processor high bandwidth memory (HBM).

3.2.3.3.3   Preconditioners

Frequently, finding a good preconditioner is fundamental to solving the required problem effectively. For some special cases, the specific nature of the problem to be solved suggests a custom preconditioner that resembles the original problem but is simpler to solve, thus resulting in an effective physics-based preconditioner. Though these are effective in some special cases, often the targeted problem calls for a general-purpose preconditioner. We consider two representative classes in common use: incomplete factorizations and multigrid methods.

3.2.3.3.3.1   Incomplete Factorizations   Incomplete factorization preconditioners such as incomplete Cholesky (IC) and ILU are based on performing an approximate factorization of the matrix A into factors L and U to be used for forward and backward solves. This is done by modifying the standard factorization method so as to discard entries that would result in fill-in of the zero entries of the matrix outside of the original sparsity pattern of A or alternatively by keeping only selected entries according to some other criterion (see e.g., Saad, 2003). This is an effective general-purpose preconditioner for many problems in practice. However, it is not algorithmically scalable, in the sense that for classes of problems of interest, for example, steady-state solutions for various types of physical simulations, the computational complexity of solving the problems to a given tolerance is worse than $O(N)$ in the matrix dimension. The method does have the strength that the triangular matrices L^{-1} and U^{-1}, which are implicit in the method though not explicitly computed, may be dense, thus the method is, in principle, able to represent global, long-distance coupling characteristics that may be present in the operator A^{-1}. However, by the same token, the algorithm, for orderings of practical interest, is difficult to parallelize due to long-distance, recursive couplings inherent in the forward and backward solves as well as the factorization itself. Reorderings of the unknowns can be applied to reduce the amount of recursion, but this impacts the convergence rate. Alternatively, parallelism can be increased by eliminating connections in the matrices, also impacting convergence. If this is not done, then the incomplete factorization problem can be solved in parallel by resorting to wavefront methods, in which the recursion is organized into a sequence of parallel steps such that within a step the problem can be fully parallelized (Li and Saad, 2013; Joubert and Oppe, 1994). All of the above-described methods have performance stress points similar to the SpMV operation, such as

memory locality, vectorization, memory access patterns, and effective use of the memory hierarchy. The wavefront methods have additional challenges: less available parallelism per wavefront than is available for the full SpMV; the need for synchronizations or communications between one wavefront and the next; and substantially less available parallelism for the early and late wavefronts. Performance can be improved by expanding L^{-1} and U^{-1} as partial Neumann expansions and pipelining the computation to increase memory locality (Chow and Patel, 2015); however, the work per iteration in this approach is increased, and convergence is, in general, weakened due to truncation of the expansion.

3.2.3.3.3.2 Multigrid Methods   Multigrid methods are based on generating a hierarchy of subproblems resembling the original problem that are used to accelerate the solution of the original problem. Multigrid solvers often have the highly desirable property of algorithmic scalability; for important classes of problems, the operation count for a solve up to desired accuracy is proportional to the matrix dimension, $O(N)$. A multigrid solve is composed of a setup phase in which the subproblems and transfer operators are formed and then an iterative solve process in which for each solver iteration the sequence of subproblems is traversed to recursively improve the approximate solution. The general class of methods can be subdivided into structured grid forms (geometric multigrid), for which assumptions can be made based on regular stencil structure and characteristics of the physics, and unstructured forms (algebraic multigrid), which incur additional costs as can be expected from irregular computations as described earlier (cf. Briggs et al., 2000). The multigrid setup time is generally dominated by sparse matrix triple product operations, whereas the multigrid solve proper is, depending on the choice of smoother, dominated by SpMV operations. The sparse matrix–matrix computations for the setup phase, though decidedly more complex than SpMV, have similar performance concerns to the SpMV computation, viz., memory locality, data reuse, optimization to the memory hierarchy, memory access patterns, and to a lesser degree vectorization and threading. Multigrid methods inherit all of the performance challenges endemic to the SpMV computation, with some additional challenges. The subproblems used to accelerate solution of the original problem are smaller and thus have less available parallelism. These subproblems are typically solved in sequence through the subproblem hierarchy; they can alternatively be solved simultaneously by breaking the sequential dependencies yielding more parallelism; however, this results in slower convergence in general. An effective technique is to remap the smaller subproblems to smaller numbers of compute nodes for more efficient solution with respect to internode communication (see Goedecker and Hoisie, 2001; Joubert and Cullum, 2006).

3.2.4   Extreme Scale Issues

On the march from petascale systems to pre-exascale systems such as the CORAL systems and then to exascale, it has not yet become entirely clear what the final configuration of a truly power efficient practical exascale system will look like. Nevertheless, the same factors understood as early as (Kogge, 2008) are still relevant: the relentless increase in the number of compute threads, the cost of moving data through the node memory hierarchy as well as off-node, and growing concerns about resiliency. We consider each of these, in turn, with their impacts on linear algebra algorithms.

3.2.4.1   Higher Thread Count

As leadership systems grow in capability, the number of nodes may grow modestly or in some cases may in fact decrease. The more dramatic change, however, is the increasing quantity of compute threads per node, per package, and per core. This impacts both dense and sparse linear algebra algorithms, especially dense linear algebra due to the high computational intensity of the calculations. For weak scaling regimes for which the required computations grow in concert with the growth in number of nodes or threads, the added work must be mapped effectively to the additional threads. The added threads may increase opportunities for hiding latencies for the computations to improve efficiency. Also, the increased quantity of thread hardware on the node is expected to make it even more important to use dynamic scheduling to keep as many threads busy as much as possible. For strong scaling regimes, the amount of work available to keep the threads busy will eventually run out. More parallelism may be extracted for some operations, for example, by parallelizing the SpMV not only across matrix rows but also across nonzeros of each row. When parallelism is then exhausted, scientists must find new problem formulations or new problems that are suited to the expanding number of threads, for example, more degrees of freedom per gridcell or running of ensemble calculations so that more instances of small weakly coupled or independent problems can be run to generate worthwhile science results.

3.2.4.2 Changing Memory Hierarchies

Only recently have users begun to understand the performance and effective use of HBM and NVRAM on leadership system compute nodes. Additional forthcoming disruptions may have further impact on code optimization approaches on the road to exascale systems, such as, for example, hierarchical register files (Dally, 2015), three-dimensional (3D) Xpoint memory and Nano-RAM, not to mention the fall of NVRAM pricing leading to its increasing prevalence. The presence, however, of a complex memory hierarchy itself is sure to endure. The changing characteristics of the hierarchy, for example, bandwidths, latencies, protocols, and user interface as well as differences between different systems and different APIs make it difficult for linear algebra software developers to keep codes up-to-date. For dense linear algebra, it will be necessary to continue to stage data through the memory hierarchy to attain high computational intensity by adapting known methods such as blocking, prefetching, and dynamic scheduling in new ways. Sparse linear algebra operations are much more memory intensive, and optimizations for the memory hierarchy will be critical. Methods described earlier such as stencil computation optimization, blocking, reordering for better locality, and unrolled iteration loop methods will become increasingly valuable to improve performance and will need to be adapted to the changes in the memory hierarchy. More algorithm innovations in addition to attempts to reformulate science problems to increase computational intensity in meaningful ways will be driving forces for making better use of exascale nodes.

3.2.4.3 Communication Network Developments

The recent trend has been for the ratio of off-node communication bandwidth to node processing power to decrease substantially. Developments in silicon photonics may remedy this. However, off-node communications will continue to be expensive. For dense linear algebra, if the problem size per node is high, computational intensity considerations will cause computation to dominate communication. However, for cases requiring a smaller quantity of work per node, communication will dominate, and alternative formulations will be needed. Sparse linear algebra will be even more vulnerable to growing communication costs in a manner similar to the challenge from memory hierarchy issues. Communication-avoidant iterative methods (Hoemmen, 2011) as well as subdomain-based preconditioners, which do more on-node work per unit of communication, will grow in appeal. Reformulation of computational approaches to science problems that minimize communications will continue to be valuable. Methods for mapping algorithms and codes to the underlying interconnect topology have already shown performance benefits for some codes (Sankaran et al., 2015) and are likely to be relevant for sparse linear algebra algorithms at exascale.

3.2.4.4 Growing Resilience Concerns

It is not yet entirely clear that what will be required for handling resilience issues at exascale, beyond the commonly used checkpoint/restart approach, most recently supported by burst buffer memories.

However, the drive to 10-nm feature sizes and lower suggests that it makes sense to be prepared for the eventuality of frequent hardware errors (Heroux, 2014). Dense and sparse linear algebra codes formulated in terms of a tasked-based programming model are able to readily resubmit tasks that fail due to detected hardware failures. Recasting of existing linear algebra algorithms into a form that is resilient to the incidence of silent errors will make it possible to execute these algorithms with confidence in environments with moderate probabilities of such errors (see e.g., Du et al., 2012; Heroux and Hoemmen, 2013; Herault and Robert, 2015).

3.2.5 SOFTWARE

We now discuss some of the available libraries for performing linear algebra operations on HPC systems. The list is not exhaustive; we focus on the main libraries that are available and most heavily used on leadership class systems. Based on the ongoing changes in node architectures in the lead-up to exascale, most of these codes in varying degrees are being modernized and adapted to take advantage of modern node hardware. As compute hardware grows in complexity going forward to exascale, the user community will look to processor manufacturers, third-party library developers and system integrators to provide unified solutions that deliver high-performance linear algebra services to application codes.

3.2.5.1 Third-Party Libraries

LAPACK (http://www.netlib.org/lapack), the well-known successor to the prior LINPACK library, provides dense linear algebra support for conventional CPUs, including a BLAS implementation as well as many higher level dense linear algebra routines. It is the reference code used as the basis for performance optimization and threading for multiple conventional single- and multicore CPUs produced by vendors.

OpenBLAS (http://www.openblas.net) is a successor to the highly optimized GotoBLAS2 library for conventional multicore processors. It is threaded with vectorization and supports x86/x86-64, ARM/ARM64, and POWER processors.

ScaLAPACK (http://www.netlib.org/scalapack) is a distributed parallel library depending on BLAS and LAPACK for its per-processor computations. It contains distributed parallel BLAS (PBLAS) and distributed dense solvers and eigensolvers. Efforts are underway to update these capabilities for exascale hardware.

Matrix algebra for GPU and multicore architectures (MAGMA; http://icl.cs.utk.edu/magma) contains BLAS and higher level dense linear algebra routines, including batched operations, for GPUs. It supports multicore and multi-GPU nodes and uses static scheduling. It also has some support for sparse linear algebra iterative solvers, in part through use of the NVIDIA cuSPARSE library.

Parallel linear algebra software for multicore architectures (PLASMA; https://bitbucket.org/icl/plasma) is a software package for solving dense linear algebra problems using homogeneous multicore processors and Xeon Phi coprocessors. It supports BLAS and higher level dense linear algebra. Using a task-based compute model with dynamic DAG scheduling, it can deliver significantly higher performance than LAPACK.

Distributed parallel linear algebra software for multicore architectures (DPLASMA) (http://icl.utk.edu/dplasma) is designed for distributed systems with multicore multisocket nodes and, if available, accelerators including GPUs or Intel Phi. It contains distributed BLAS-3 and distributed higher level dense linear solvers and eigensolvers. It is based on a distributed parallel runtime with dynamic DAG task scheduling. It can yield significantly higher performance than ScaLAPACK.

Trilinos (https://trilinos.org) is a multipurpose mathematical and scientific library with support for many computation types including sparse linear algebra. The Tpetra subpackage, successor to Epetra, provides parallel sparse solver services for accelerated (GPU or Intel Phi) nodes through the Kokkos package, with additional services provided by the Teuchos subpackage. Some of the linear algebra subpackages built on this infrastructure are Belos (Krylov solvers), Amesos2 (direct solvers), Ifpack2 (incomplete factorizations and other preconditioners), MueLu (algebraic multigrid), and others.

HYPRE (http://computation.llnl.gov/projects/hypre-scalable-linear-solvers-multigrid-methods) provides parallel Krylov solvers for structured, semistructured, and unstructured grids. It also provides a variety of parallel preconditioners including incomplete factorizations and algebraic multigrid. It supports threading but currently does not support accelerated nodes with GPUs; future support for exascale node hardware seems likely.

Portable, extensible toolkit for scientific computation (PETSc) (https://www.mcs.anl.gov/petsc) contains a variety of sparse solvers. Based on its infrastructure for distributed vectors and sparse matrices, the KSP subpackage of PETSc contains Krylov solvers and incomplete factorization and algebraic multigrid preconditioners, among others. It also has interfaces to other libraries such as the HYPRE preconditioner library and the Trilinos multilevel (ML) preconditioner. It currently does not support threads or GPU acceleration; efforts have begun to improve support for exascale nodes.

3.2.5.2   Vendor Libraries

NVIDIA GPU-accelerated libraries (https://developer.nvidia.com/gpu-accelerated-libraries) provide support for dense and sparse linear algebra on GPU-accelerated nodes. The collection includes cuBLAS (BLAS functionality for multi-GPU nodes), cuSOLVER (dense and sparse direct solvers for linear equations and eigenproblems), cuSPARSE (sparse matrix support and incomplete factorization preconditioners), and AmgX (scalable multinode algebraic multigrid solves). These libraries are expected to be well optimized for generations of GPU accelerators going forward. However, they will need to be integrated into full solutions that manage operations in the context of all available node hardware, for example, managing the staging of computations from NVRAM when appropriate and marshaling available compute resources of the CPU.

Intel Math Kernel Library (MKL; https://software.intel.com/en-us/intel-mkl) is Intel's vehicle for delivering high-performance linear algebra and other functionality to conventional multicore and Intel Phi processors. It provides BLAS functionality with vectorization and threading capabilities, linear solvers and eigensolvers for dense systems, optimized ScaLAPACK functionality, sparse direct solvers for clusters, and iterative sparse solvers.

Cray Scientific and Math Libraries (CSML; http://docs.cray.com/books/S-2529-116/S-2529-116.pdf) support a variety of mathematical capabilities including dense and sparse linear algebra. The LibSci library supports BLAS, LAPACK, and ScaLAPACK functionality, and more recently, the LibSci_ACC library supports dense linear algebra functionality using all node resources including CPU and GPU, optimized by autotuning. The iterative refinement toolkit aids with mixed precision dense linear algebra. Cray also supplies optimized Trilinos and PETSc (with embedded HYPRE) libraries, optimized to use high-performance autotuned SpMV product operations via Cray Adaptive Sparse Kernel (CASK) framework.

IBM's Engineering and Scientific Subroutine Library (ESSL) and Parallel ESSL (PESSL) libraries deliver linear algebra and other functionality, optimized for POWER processors and connected GPU accelerators. ESSL provides BLAS and LAPACK support for a single compute node. PESSL provides ScaLAPACK support. The codes are heavily optimized with threading and vectorization for POWER processors. Recent releases include library versions that support offloading of some operations to GPUs.

3.2.6   Conclusion

The ascent to exascale computing will perhaps be the most difficult hurdle ever faced by attempts to deliver truly high performing scientific applications. Deployment of highly performant linear algebra algorithms and software will play a central role in the successful use of these systems by many science application codes.

3.3 TENSOR ALGEBRA

3.3.1 Tensors in Different Scientific Disciplines

Tensors are mathematical objects ubiquitously present in many scientific disciplines. In the most general case, a *tensor* can be defined as a multidimensional array of objects (tensor elements), that is, a map $T^{j_1 \cdots j_n}_{i_1 \cdots i_m} \to \mathbf{S}$, where $\mathbf{S}$ is a set of objects (possible values of tensor elements), and each tensor index has a span over a certain integer range, either finite or infinite. The characteristic integer pair (m, n) is called the *tensor valence* while the total number of tensor dimensions, $(n + m)$, is called either the *tensor order* (in mathematical literature) or the *tensor rank* (in physical literature). We will adopt the physical nomenclature in this text because the term *tensor order* has a different meaning in some physical disciplines (on the other hand, the term *tensor rank* has a different meaning in math). By requiring $\mathbf{S}$ to be a ring, we obtain a more specific definition of a tensor, enhanced with an additional algebraic structure (we can now multiply and add tensor elements). Ultimately, when $\mathbf{S}$ is a field, or even more specifically, the field of real ($\mathbb{R}$) or complex ($\mathbb{C}$) numbers, we end up with a more custom, yet generic enough, definition of a tensor, which will be assumed throughout. The grouping of tensor indices into two rows is important when the tensor obeys additional linear transformation rules, which we will clarify later.

In math and science, there are three main situations in which the notion of a tensor emerges naturally. The first one, although not historically, is the formation of a direct (tensor) product of two vector spaces, $\mathbf{V}^{(1)}$ and $\mathbf{V}^{(2)}$. If the n-dimensional space $\mathbf{V}^{(1)}$ has a basis $\left\{ \vec{e}^{(1)}_k, k = [1, n] \right\}$ and the m-dimensional space $\mathbf{V}^{(2)}$ has a basis $\left\{ \vec{e}^{(2)}_k, k = [1, m] \right\}$, then the direct product basis $\left\{ \vec{e}^{(1)}_k \vec{e}^{(2)}_l, k = [1, n], l = [1, m] \right\}$ will span the tensor product space $\mathbf{W}$, which is a vector space itself (an arrow on top of a small letter designates a vector). A vector from the space $\mathbf{V}^{(1)}$ is specified by its n components $\vec{a} \equiv \left\{ a^k, k = [1, n] \right\}$, a vector from the space $\mathbf{V}^{(2)}$ is specified by its m components $\vec{b} \equiv \left\{ b^k, k = [1, m] \right\}$, while a vector from the tensor product space $\mathbf{W}$ is specified by its $(n \cdot m)$ components $\vec{w} \equiv \left\{ w^{kl}, k = [1, n], l = [1, m] \right\}$ arranged in a two-index tuple (*multi-index*). By taking tensor products of multiple vector spaces, one can construct higher order tensor product spaces. In case $\mathbf{W}$ is constructed from a repeated application of the tensor product operation to the same *inner-product* vector space $\mathbf{V}$, the *metric tensor*

$$g_{ij} \equiv \sum_k \left(e^k_i \right)^* e^k_j \tag{3.1}$$

and its inverse (in a matrix sense)

$$g^{ij} \equiv (g_{ij})^{-1} \tag{3.2}$$

will play an important role in tensor algebra operations (in Equation 3.2, the inverse operation is applied to the matrix g_{ij} as a whole). In Equation 3.1, e^k_j is the k-th component of the basis vector $\vec{e}_j$ expanded in the *standard* orthonormal basis, and the star in the superscript means complex conjugation (hereafter, we assume a vector space over the complex field). The basis vectors of the parental vector space $\mathbf{V}$ may or may not be orthogonal in general. Although, in some cases, the vector space $\mathbf{V}$ may be spanned by a *frame* instead of a basis, that is, an overcomplete set of vectors with linear dependencies, we will not consider this case in the following discussion. Importantly, the metric tensor and its inverse can be used for *lowering/raising* tensor indices, the two operations defined as follows:

- *Index lowering:* $T^{ij}_{k} = g_{kl} T^{ilj} \equiv \sum_l g_{kl} T^{ilj}$
- *Index raising:* $T^k_{ij} = g^{kl} T_{ilj} \equiv \sum_l g^{kl} T_{ilj}$

In the definitions above, we adopt the implicit summation convention where we run a summation over those index labels that appear twice on the right-hand side. Upper indices of the tensor are called *contravariant* indices; lower indices of the tensor are called *covariant* indices. Index *variance* determines whether the index is associated with the covariant, $(\vec{e}_i)$, or contravariant (dual) basis, $(\vec{e}^i \equiv g^{ij}\vec{e}_j)$, which, in turn, determines whether the index transforms with the basis transformation matrix or its inverse during a nondegenerate basis transformation in the vector space associated with the index. The duality requirement mandates $\left(e_i^k\right)^* e^{jk} \equiv g_i^j = \delta_i^j$, where δ_i^j is the Kronecker delta. When raising/lowering a tensor index, its horizontal position also matters, in general, since it associates the tensor index with a specific vector space from the underlying tensor product. However, in practice, this can often be ignored (as in our equations so far).

The second situation, which the notion of a tensor actually takes its origin from, is rooted in differential geometry where tensors represent linear and multilinear relations between geometric objects on differential manifolds. The classical *tensor calculus* was widely adopted in physics where it provides a covariant description of physical laws that does not depend on the choice of the coordinate system. In particular, the requirement of the invariance of a scalar with respect to the basis introduces a distinction between the covariant and contravariant tensor indices with their specific linear transformation rules stated above. Also, in many physical applications, tensors appear in a form of *tensor fields*, that is, tensor-valued functions of multiple variables, including scalar fields, vector fields, and higher rank (higher order) tensor fields, as found in theory of electromagnetism, nonlinear optics, continuum mechanics, special and general relativity, etc.

Finally, the third situation where tensors emerge is connected to the recent raise of big data analytics where the underlying (big) data sets are inherently multidimensional in many cases, being represented by tensors. Here, one may encounter the data produced by scientific instruments in materials science, medicine, and astronomy, to name a few scientific disciplines. Other examples include social data, financial data, and any other data arranged as a multidimensional array of numbers. The entire field of deep learning is based on tensor operations used in both the training and inference phases. Additionally, the notion of a *hypergraph* as a generalization of a *graph* is an important mathematical construct for studying complex relations between objects, and its representation is essentially a tensor as well. We should note that, in data analytics, the underlying vector spaces do not necessarily possess an inner product, thus having no metric tensor as well (in this case, the co- and contravariance of tensor indices becomes irrelevant such that all tensor indices can be placed either up or down persistently, simply specifying a multidimensional array of data).

Notwithstanding the importance of the classical tensor calculus in physics, in this text, we will mostly focus on relatively recent large-scale applications of tensors in quantum many-body theory (QMBT) and general data analytics. In both domains, the tensor size and the compute power demand grow rapidly with the problem size, making these applications ideally suited for extreme scale high-performance computing. Additionally, the quantum many-body methods, used in electronic and nuclear structure theories, are often compute intensive, that is, the flop/byte ratio is high, favoring flop-oriented modern HPC architectures such as Cray XK7 Titan and IBM/Nvidia Summit supercomputers deployed (or to be deployed) by the U.S. Department of Energy (DOE) at the Oak Ridge National Laboratory (ORNL). In fact, the DOE Office of Science allocated special funding for the preparation of electronic and nuclear structure community codes for the leadership HPC systems within the CAAR program at ORNL (https://www.olcf.ornl.gov/caar/). In particular, the electronic structure codes NWChem, LS-DALTON, and DIRAC and the nuclear structure code NUCCOR are part of this program. Although the three electronic structure codes address slightly different aspects of quantum-chemical simulations and thus require somewhat different capabilities, they all heavily rely on numerical tensor algebra. The same is true for the nuclear structure code.

Let us briefly describe the numerical essence of QMBT since it has been our main use case so far, motivating the development of scalable tensor algebra algorithms. Using the language of *second quantization* ubiquitously employed in QMBT, the quantum many-body Hamiltonian operator can be written as

$$H \equiv \hat{a}_p^+ H_q^p \hat{a}^q + \hat{a}_p^+ \hat{a}_r^+ H_{qs}^{pr} \hat{a}^s \hat{a}^q, \tag{3.3}$$

where $\hat{a}_p^+$ and $\hat{a}^q$ are the (quasi)-particle creation and annihilation operators, respectively, H_q^p and H_{qs}^{pr} are the 1- and 2-body Hamiltonian tensors (rank-2 and rank-4, respectively), and an implicit summation over repeated indices is assumed. The tensor indices run over some basis in which each basis vector represents a quantum state of a single quantum particle. In general, the Hamiltonian may contain higher than 2-body terms (in nuclear structure theory, it is necessary to include at least the 3-body term represented by rank-6 tensors in order to obtain accurate results). The goal is to obtain the eigenvalues and eigenvectors of the above Hamiltonian operator, which is normally done with some approximate numeric scheme (a QMBT method). In principle, the exact solution for N interacting quantum particles, called a *wavefunction*, can be written in the *hole-particle* representation in the following form:

$$|\Psi\rangle = \left(1 + \hat{a}_{a_1}^+ C_{i_1}^{a_1} \hat{a}^{i_1} + \frac{1}{2!2!} \hat{a}_{a_1}^+ \hat{a}_{a_2}^+ C_{i_1 i_2}^{a_1 a_2} \hat{a}^{i_2} \hat{a}^{i_1} + \frac{1}{3!3!} \hat{a}_{a_1}^+ \hat{a}_{a_2}^+ \hat{a}_{a_3}^+ C_{i_1 i_2 i_3}^{a_1 a_2 a_2} \hat{a}^{i_3} \hat{a}^{i_2} \hat{a}^{i_1} + \dots \right) |0\rangle, \tag{3.4}$$

where $C_{i_1 \dots}^{a_1 \dots}$ are the so-called *configuration interaction* (CI) amplitude tensors, and $|0\rangle$ is the reference state of N quantum particles in which each particle occupies its own quantum state independently of others. This N-particle wavefunction is expanded in the Hilbert space constructed as a tensor product of single-particle Hilbert spaces. The exact expansion for $|\Psi\rangle$ that describes a state of N interacting quantum particles will contain tensors with ranks up to $2N$ in the hole-particle second quantized formalism. Consequently, an exponentially large number of tensor elements needs to be determined (and stored) when N is growing. This is exactly where quantum theory hits the wall of an exponentially large Hilbert space on which the N-particle wavefunction is supported. Fortunately, the number of significant tensor elements required for achieving a reasonable accuracy of the solution is much smaller in many practical cases. Consequently, *the main goal of computational QMBT is to approximate the exact wavefunction by only low-rank tensors while exploiting tensor sparsity as much as possible*. A straightforward approximation, called the CI method, simply ignores all tensors in the wavefunction expansion that have ranks higher than some predefined value, which is normally 2, 3, or 4. Unfortunately, this approach has serious theoretical drawbacks, leading to unphysical results in some cases. A much more clever approximation is based on the *coupled-cluster* (CC) theory, a *de facto* standard of accuracy in quantum chemistry. The CC wavefunction has the following form

$$|\Psi\rangle = \exp(\hat{T}) = \left(1 + \hat{T} + \frac{1}{2!} \hat{T}^2 + \frac{1}{3!} \hat{T}^3 + \frac{1}{4!} \hat{T}^4 + \dots \right) |0\rangle \tag{3.5}$$

where the so-called *cluster operator* $\hat{T}$ is parameterized by low-rank tensors, for example,

$$\hat{T} \equiv \hat{a}_{a_1}^+ T_{i_1}^{a_1} \hat{a}^{i_1} + \frac{1}{2!2!} \hat{a}_{a_1}^+ \hat{a}_{a_2}^+ T_{i_1 i_2}^{a_1 a_2} \hat{a}^{i_2} \hat{a}^{i_1} \tag{3.6}$$

Equation 3.6 corresponds to the CC approach with singles and doubles (CCSD), the most widely used approximation in which only rank-2 and rank-4 tensors need to be computed. Due to the exponentiation of the cluster operator $\hat{T}$ higher rank tensors are implicitly present in the wavefunction expansion, being approximated by low-rank T tensors, for example,

$$C_{i_1 i_2 i_3}^{a_1 a_2 a_3} = \left\{ T_{i_1}^{a_1} T_{i_2 i_3}^{a_2 a_3} \right\}_A + \left\{ T_{i_1}^{a_1} T_{i_2}^{a_2} T_{i_3}^{a_3} \right\}_A \tag{3.7}$$

where $\{\dots\}_A$ generates all possible products, obtained by unique index permutations, and (anti-) symmetrizes the obtained expansion in accordance with proper quantum statistics (fermionic/

bosonic). Consequently, the CC wavefunction contains all the terms present in the exact wavefunction but approximates higher rank tensors as products of low-rank tensors.

From the computational point of view, conventional CC approximations are rather compute intensive. For example, the basic CCSD approach has the computational cost $O(N^6)$ flops and memory requirements $O(N^4)$ bytes, where N is the number of simulated quantum particles. There has been quite a bit of effort recently directed toward scaling and adaptation of CC codes on leadership HPC systems. All three electronic structure codes participating in the DOE CAAR program at ORNL, namely NWChem, LS-DALTON, and DIRAC, have the goal to improve the performance and scalability of their CC modules. Additionally, the LS-DALTON code implements a linear-scaling variant of CC theory, in which the sparsity of the cluster amplitude tensors, $T_{i_1}^{a_1}$ and $T_{i_1 i_2}^{a_1 a_2}$, is explicitly exploited, reducing the memory requirements and computational cost to $O(N)$ in the asymptote of large N.

In practice, the CCSD approach works extremely well when the wavefunction is dominated by the reference state $|0\rangle$ (see Equation 3.5) such that the rest of the exact wavefunction can be efficiently approximated using the cluster amplitude tensors, $T_{i_1}^{a_1}$ and $T_{i_1 i_2}^{a_1 a_2}$, and their products, like the one in Equation 3.7. For more complicated cases, there exist generalizations of the ordinary CC theory (Lyakh et al., 2012), which are not as efficient unfortunately. However, in general, the CC way of approximating higher rank tensors by simple cluster products of low-rank tensors is not accurate when the exact wavefunction is dominated by many components from the many-particle Hilbert space. In this case, a more efficient approach is offered by the *tensor network* (TN) theory. In essence, the key idea behind TN theory is to approximate higher rank tensors as *contracted* products of lower rank tensors, a so-called tensor network. To give an example, let us rebase the exact wavefunction given by Equation 3.4 upon the true vacuum reference state $|\rangle$ (no particles) instead of the N-particle reference state $|0\rangle$:

$$|\Psi\rangle = \frac{1}{N!} \hat{a}_{p_1}^+ \dots \hat{a}_{p_N}^+ C^{p_1 \cdots p_N} |\rangle \qquad (3.8)$$

where the product of N creation operators, $\hat{a}_{p_1}^+ \dots \hat{a}_{p_N}^+$, creates N quantum particles in states $p_1 \dots p_N$ (for fermions, no two particles are allowed to occupy the same quantum state). In this (equivalent) representation, the exact wavefunction is encoded by a rank-N tensor $C^{p_1 \cdots p_N}$. The exact solution, that is, the tensor $C^{p_1 \cdots p_N}$, can be obtained by a variational optimization of the Hamiltonian expectation value. In TN theory, this tensor is expanded as a network of low-rank tensors that can be done in many ways. One of the simplest TNs is called the *tensor train* (TT) (Oseledets and Tyrtyshnikov, 2009; Dolgov et al., 2014) or *matrix product state* (MPS) (Affleck et al., 1987; Chan et al., 2016):

$$C^{p_1 \cdots p_N} = T_{\alpha_1 \alpha_2}^{p_1} T_{\alpha_2 \alpha_3}^{p_2} \dots T_{\alpha_N \alpha_1}^{p_N} \qquad (3.9)$$

where the original rank-N tensor is approximated by a contracted product of N rank-3 tensors. Based on this approximation, the Hamiltonian expectation value can be optimized using the density matrix renormalization group (DMRG) procedure (White, 1993). In this way, the original exponential computational cost and memory requirements with respect to N are replaced by polynomial bounds (while sacrificing the exactness). The MPS-based tensor compression is one of the simplest ones, yet very powerful in many cases. There exist more sophisticated TNs, for example, tree tensor networks (TTN) (Shi et al., 2006; Grasedyck, 2010; Nakatani and Chan, 2013; Murg et al., 2015), projected entangled pair states (PEPS) (Verstraete and Cirac, 2004; Orus, 2014), multiscale entanglement renormalization ansatz (MERA) (Vidal, 2008), etc. The advantage of the TN theory over the CC approach is the absence of a bias towards any specific reference state of N quantum particles, whereas the CC approach is computationally more superior in reconstructing *small* parts of the wavefunction around the dominating reference state.

Currently, the methodological part of both the CC approach and the TN theory is relatively well developed, although there are still some open theoretical problems that need to be addressed. In contrast, the computational body of both theories is suffering from the lack of scalable implementations capable of running efficiently on leadership HPC systems, especially those based on accelerated node architectures (with few case-specific exceptions (Valiev et al., 2010; Lotrich et al., 2008). Existing codes normally provide their own implementation of specific CC or TN approximations, which have limited scalability and accelerator support in most cases. The code reuse between different scientific software projects tends to be low. The software development community does not seem to share a common set of abstractions with well-defined key primitives for numeric tensor algebra, making development and deployment of tensor algebra libraries somewhat challenging. Yet, with the steady increase of architectural complexity at both the node and HPC system levels, a sustainable way toward larger-scale and more accurate simulations of quantum systems will require hardware abstraction and a separation of the higher level algorithm expression from the hardware-specific implementation details, thus demanding a feature-rich massively parallel software framework for numeric tensor algebra. To date, there have been few works addressing specific computational challenges in numeric tensor algebra, for example, scalability of dense tensors contractions (Baumgartner et al., 2005; Lai et al., 2013; Solomonik et al., 2013, 2014; Rajbhandari et al., 2014a), the support of (block-)sparse tensor algebra (Calvin et al., 2015; Riplinger et al., 2016), the support of accelerators (https://github.com/DmitryLyakh/TAL_SH.git), and task-based runtimes (Danalis et al., 2015). However, none of these efforts is complete enough by itself in order to address the challenge of exascale computing. In attempt to prepare for such a challenge (and a great opportunity), the ExaTENSOR project was initiated at the ORNL in 2014. The goal of this project is to design and implement scalable algorithms for a number of key tensor algebra primitives that can run efficiently on vastly different hardware and HPC size. Since the work is still ongoing, below we will mostly describe the design of the ExaTENSOR framework, its anticipated functionality and key innovations that are expected to enable efficient utilization of exascale resources in the near future.

3.3.2 BASIC TENSOR ALGEBRA OPERATIONS

The basic tensor algebra consists of a few primitive operations in particular:

1. *Multiplication by a scalar (scaling)*: $D^{a_1 \ldots a_n}_{i_1 \ldots i_m} = \alpha \cdot L^{a_1 \ldots a_n}_{i_1 \ldots i_m}$

2. *Tensor addition*: $D^{a_1 \ldots a_n}_{i_1 \ldots i_m} = L^{a_1 \ldots a_n}_{i_1 \ldots i_m} + R^{a_1 \ldots a_n}_{i_1 \ldots i_m}$

3. *Tensor product*: $D^{a_1 \ldots a_n b_1 \ldots b_k}_{i_1 \ldots i_m j_1 \ldots j_l} = L^{a_1 \ldots a_n}_{i_1 \ldots i_m} R^{b_1 \ldots b_k}_{j_1 \ldots j_l}$

4. *Tensor contraction*: $D^{a_1 \ldots a_n b_1 \ldots b_k}_{i_1 \ldots i_m j_1 \ldots j_l} = L^{a_1 \ldots a_n c_1 \ldots c_r}_{i_1 \ldots i_m d_1 \ldots d_s} R^{b_1 \ldots b_k d_1 \ldots d_s}_{j_1 \ldots j_l c_1 \ldots c_r}$

In all above tensor operations, the index order in each tensor may be arbitrarily permuted in general. Additionally, two tensor norms are custom:

1. *1-norm*: $\eta_1 = |D|^{a_1 \ldots a_n}_{i_1 \ldots i_m} E^{a_1 \ldots a_n}_{i_1 \ldots i_m}$

2. *2-norm*: $\eta_2 = \left(\bar{D}^{a_1 \ldots a_n}_{i_1 \ldots i_m} D^{a_1 \ldots a_n}_{i_1 \ldots i_m} \right)^{1/2}$

Here, $|D|^{a_1 \ldots a_n}_{i_1 \ldots i_m}$ is a tensor obtained from $D^{a_1 \ldots a_n}_{i_1 \ldots i_m}$ by taking absolute values of the tensor elements, $\bar{D}^{a_1 \ldots a_n}_{i_1 \ldots i_m}$ is a complex-conjugate tensor obtained from $D^{a_1 \ldots a_n}_{i_1 \ldots i_m}$ by taking complex conjugates of the tensor elements, and $E^{a_1 \ldots a_n}_{i_1 \ldots i_m}$ is a tensor with all elements equal to 1. These few tensor operations form the numerical body of QMBT at the primitive level. They also serve as basic primitives in other scientific disciplines in which tensor algebra is the main numerical framework. To a large extent, generic higher level tensor algebra algorithms can be composed from the above primitive operations, thus demanding an efficient scalable implementation of the latter (here, we mostly refer to the *tensor contraction* operation as the other operations are rather trivial to parallelize).

Among the above operations, only the tensor contraction may have high compute (arithmetic) intensity, that is, the flop/byte ratio (number of floating point operations divided by the data size). Here, the data size is the total number of elements in all three tensors participating in a tensor contraction. We will call the number of elements in a tensor $D_{i_1 \ldots i_m}^{a_1 \ldots a_n}$ its volume, $\text{vol}\left(D_{i_1 \ldots i_m}^{a_1 \ldots a_n}\right)$. Furthermore, we will also define the *volume of a multi-index* $\{p_1 \ldots p_n\}$ as the number of unique combinations of the index values. In the simplest case, the multi-index volume is equal to the product of index extents (each index is associated with a specific dimension of the tensor and the extent of that dimension is the index extent). In these terms, for example, $\text{vol}\left(D_{i_1 \ldots i_m}^{a_1 \ldots a_n}\right) = \text{vol}\left(\{i_1 \ldots i_m\}\right) \cdot \text{vol}\left(\{a_1 \ldots a_n\}\right)$. The indices participating in a tensor contraction

$$D_{i_1 \ldots i_m j_1 \ldots j_l}^{a_1 \ldots a_n b_1 \ldots b_k} = L_{i_1 \ldots i_m d_1 \ldots d_s}^{a_1 \ldots a_n c_1 \ldots c_r} R_{j_1 \ldots j_l c_1 \ldots c_r}^{b_1 \ldots b_k d_1 \ldots d_s} \tag{3.10}$$

are classified as either *contracted* (those summed over) or *uncontracted* (those belonging to the resulting tensor $D_{i_1 \ldots i_m j_1 \ldots j_l}^{a_1 \ldots a_n b_1 \ldots b_k}$). We define the total number of floating point operations in a tensor contraction as twice the volume of all uncontracted indices times the volume of all contracted indices (doubling comes from considering multiplication and addition as two separate floating point operations). The total number of floating point operations, F, can also be expressed in terms of the tensor volumes as

$$F = 2\sqrt{\text{vol}(D) \cdot \text{vol}(L) \cdot \text{vol}(R)} \tag{3.11}$$

where we omitted tensor indices for the sake of clarity. The compute (arithmetic) intensity of the tensor contraction is

$$C = \frac{2\sqrt{\text{vol}(D) \cdot \text{vol}(L) \cdot \text{vol}(R)}}{\text{vol}(D) + \text{vol}(L) + \text{vol}(R)} \tag{3.12}$$

There are two main routes how a tensor contraction can be evaluated: direct and indirect. In the direct approach, a tensor contraction is evaluated as a loop nest over the contracted indices, with subsequent optimizations. In the indirect approach, a tensor contraction is transformed into a matrix–matrix multiplication via transposing the input/output tensors (bold capital letters designate matrices):

- $\mathbf{L}_{\{i_1 \ldots i_m a_1 \ldots a_n\}, \{c_1 \ldots c_r d_1 \ldots d_s\}} = L_{i_1 \ldots i_m d_1 \ldots d_s}^{a_1 \ldots a_n c_1 \ldots c_r}$
- $\mathbf{R}_{\{c_1 \ldots c_r d_1 \ldots d_s\}, \{j_1 \ldots j_l b_1 \ldots b_k\}} = R_{j_1 \ldots j_l c_1 \ldots c_r}^{b_1 \ldots b_k d_1 \ldots d_s}$
- $\mathbf{D}_{\{i_1 \ldots i_m a_1 \ldots a_n\}, \{j_1 \ldots j_l b_1 \ldots b_k\}} = \mathbf{L}_{\{i_1 \ldots i_m a_1 \ldots a_n\}, \{c_1 \ldots c_r d_1 \ldots d_s\}} \cdot \mathbf{R}_{\{c_1 \ldots c_r d_1 \ldots d_s\}, \{j_1 \ldots j_l b_1 \ldots b_k\}}$
- $D_{i_1 \ldots i_m j_1 \ldots j_l}^{a_1 \ldots a_n b_1 \ldots b_k} = \mathbf{D}_{\{i_1 \ldots i_m a_1 \ldots a_n\}, \{j_1 \ldots j_l b_1 \ldots b_k\}}$

As one can see, tensor indices are permuted (this step is called a *tensor transpose*), multiple tensor indices enclosed within curly brackets are flattened into a single matrix index, and a matrix–matrix multiplication is performed. We will call the uncontracted indices of the left/right tensor argument (L/R) *left/right uncontracted*, respectively. Thus, in each tensor contraction, there are three groups of indices: contracted, left uncontracted, and right uncontracted. This grouping helps to establish the correspondence with the matrix–matrix multiplication. Such a close relation of the (dense) tensor contraction operation to the (dense) matrix–matrix multiplication also leads to similar algorithmic bounds (Solomonik et al., 2013, 2014; Rajbhandari et al., 2014a). Importantly, similarly to the matrix–matrix multiplication, a tensor contraction has a lot of inherent parallelism and data reuse. Indeed (1) the summation over contracted indices in Equation 3.10 can be split into multiple sub-summations in an arbitrary way and those sub-summations can be evaluated in parallel, (2) different parts of the destination tensor (result) can be evaluated in parallel, and (3) in many cases, each tensor element is accessed many times during a tensor contraction.

So far, we have considered a generic case, however, in practice, tensors often have an additional structure and/or sparsity that can provide sometimes enormous savings in the required storage and the number of floating point operations to be performed in tensor operations. At the same time, an efficient computer implementation becomes more challenging in this case. In QMBT, tensors often have *permutational symmetry*, meaning that the tensor indices are grouped into one or more groups and within each group all index permutations refer to the same value, possibly multiplied by the permutation parity (sign). For example, $D^{a_1 a_2}_{i_1 i_2} = -D^{a_1 a_2}_{i_2 i_1} = -D^{a_2 a_1}_{i_1 i_2} = D^{a_2 a_1}_{i_2 i_1}$. As a consequence, the required storage space for such a tensor can reduce drastically (especially for higher rank tensors) but mapping many tensor elements to the same memory location, while maintaining a proper sign and high compute efficiency, becomes a challenge.

Another origin of tensor sparsity in physics may come from symmetry associated with each tensor index value (or, equivalently, with the corresponding basis vector of the corresponding vector space). In this case, each index value (integer) may be associated with a certain irreducible representation of some symmetry group. Then, normally due to some restrictions imposed by the physics of the problem, certain combinations of tensor indices will refer to the tensor elements that are strictly zero because the product of the corresponding irreducible representations does not transform according to the required symmetry representation. By reordering basis vectors in each vector space in a tensor product in accordance with the symmetry labels, the entire blocks of the tensor can be identified to be strictly zero (to be ignored in tensor contractions). To give an illustration, let us consider some tensor $D^{a_1 a_2}_{i_1 i_2}$ and suppose there are disjoint subranges of i_1, i_2, a_1, a_2 labeled by symmetries A and B. Then, for example, it may happen that only the following blocks of the tensor $D^{a_1 a_2}_{i_1 i_2}$ can be nonzero: D^{AA}_{AA} and D^{BB}_{BB}.

In electronic structure theory, there is yet another kind of tensor sparsity originating from the form of the interaction potential acting on quantum particles (physical interaction potentials normally decay with the interaction distance). For example, the CC tensors from Section 3.2.1, $T^{a_1}_{i_1}$ and $T^{a_1 a_2}_{i_1 i_2}$, describe correlations between hole-particle excitations (Lyakh and Bartlett, 2014), where vector spaces associated with the lower indices of these tensors describe quantum holes, while those associated with the upper indices describe quantum particles (holes and particles are *virtual* quantum states widely utilized in the second-quantization formalism, each represented by a vector from a Hilbert space $\mathcal{H}(\mathbb{R}^3)$). If these individual hole/particle states (complex functions of three spatial variables) have bounded supports localized in some regions of the 3D real space $\mathbb{R}^3$ and these spatial regions are sufficiently distant from each other such that the interaction is weak, then the corresponding tensor elements, which show the strength of the correlation, are likely to have negligible values which can be neglected in tensor operations. By properly reordering spatially localized basis vectors in the underlying vector spaces, entire blocks of the tensor can be identified to be negligible due to such a physical reason.

In our numerical tensor algebra framework, we consider all these cases at the level of block sparsity, that is, we assume that our tensors have a *block-sparse* structure in general, being composed of (nonnegligible) tensor blocks (a *tensor block* is a tensor slice obtained by restricting index ranges to specific subranges). Furthermore, we also consider *scale-adaptive* tensor algebra (Lyakh, 2014), a formalism originally suggested for large-scale *ab initio* electronic structure calculations, in which tensor dimensions can shrink due to a projection of the corresponding vector spaces into smaller (principal) subspaces. In this formalism, a tensor is polymorphic in the level of resolution of its dimensions; the resolution may vary in different tensor operations, depending on the other tensor operands.

3.3.3   TENSOR DECOMPOSITIONS AND HIGHER LEVEL OPERATIONS

Having defined the basic tensor operations (primitives), we can proceed to higher level tensor operations that can be composed from these basic primitives. As we briefly mentioned above, tensors are multidimensional objects tending to be very large in size in many interesting scientific applications, thus making efficient *tensor compression* the most important higher level numerical operation to consider. One can distinguish two scenarios here. First, a given (large) tensor needs to be compressed into a smaller tensor or a specific mathematical construct consisting of smaller tensors. We will call this case an *a posteriori* tensor compression. In the second scenario, a certain internal structure can be imposed on a tensor from the beginning, thus providing a real-time approximation of that tensor, parameterized by a smaller number of variables. We will call this scenario an *a priori* tensor compression. In general, higher level tensor algorithms may include both an *a priori* and *a posteriori* tensor compression steps. The examples of an *a priori* tensor compression from QMBT include the CC formalism, where the higher rank tensors are approximated by a sum of products of lower rank tensors, the MPS ansatz used in the DMRG algorithm, and other wave function ansaetze provided by the tensor network state (TNS) theory. At the same time, the MPS-based DMRG algorithm also involves the SVD step applied to a matrix in order to approximate it by its principal components (an *a posteriori* tensor compression). There are also examples of an SVD-based *a posteriori* tensor compression in CC theory (Kinoshita et al., 2003).

In order to compress a tensor, one may choose either to reduce its rank or to decrease the extent of some/all tensor dimensions, or to do both. In order to decrease the extent of a tensor dimension, a projection to a lower dimensional subspace of the original vector space associated with that dimension is needed. At the same time, the information loss (tensor norm reduction) should be minimized. For rank-2 tensors, a well-defined procedure for the dimensionality reduction is provided by the SVD, where the subspace spanned by the singular vectors with the largest singular values is optimal in the 2-norm. For higher rank tensors, unfortunately, no straightforward unique generalization exists. However, one can still construct reasonable lower dimensional subspaces for each tensor dimension by matricizing the tensor with respect to the chosen dimension. For example, focusing on dimension i_2 of tensor $T_{i_1 i_2}^{a_1 a_2}$, the matricized form will be

$$M_{i_2,J} \equiv M_{i_2,\{i_1 a_1 a_2\}} = T_{i_1 i_2}^{a_1 a_2} \tag{3.13}$$

where the remaining three dimensions of the tensor were flattened into a single dimension, thus forming a matrix $M_{i_2,J}$. By multiplying matrix $M_{i_2,J}$ by its Hermitian conjugate $\bar{M}_{J,i_2'}$, we obtain a Hermitian, positive-semidefinite matrix $\tilde{M}_{i_2,i_2'}$, which can be diagonalized. The eigenspace corresponding to the leading largest eigenvalues can be used as a reasonable approximation subspace for the dimension i_2.

The compression of individual tensor dimensions is a useful technique, but it does not provide a remedy to the curse of dimensionality, that is, it does not reduce the tensor rank (number of tensor dimensions). A rather general solution to the latter problem is provided by the TN theory, in which a higher dimensional tensor is represented as a contraction of a number of lower dimensional tensors (or, in a more general case, as a sum of such contracted products). The exact way how the lower dimensional tensors are contracted to form the original higher dimensional tensor does matter, that is, the structure of the TN defines its class and affects the quality of the approximation. It is custom in TN theory to represent tensor contractions graphically as graphs. A vertex of such a graph is a tensor and the edges (or arrows) associated with this vertex are tensor indices. The graph may contain *open* edges/arrows, that is, the edges attached to a vertex only from one side (with the other end open). The open edges/arrows represent uncontracted indices of a tensor whereas the *closed* edges/arrows, which connect two vertices, represent the contracted indices. Figure 3.1 illustrates such a graphical representation of tensors. Since the structure of a given TN can be represented as a graph, different classes of TNs can be distinguished, based on the kinds of the corresponding graphs. To give a more concrete example, we will briefly consider few classes of TNs below.

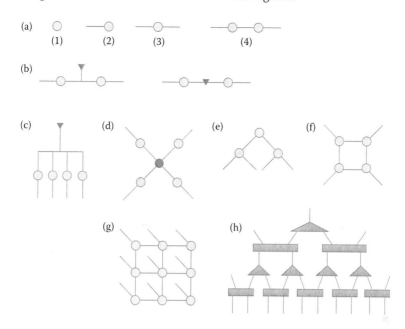

FIGURE 3.1 Graphical representation of tensors, tensor operations, and tensor ansaetze. (a) Tensors and tensor contractions: (1) scalar, (2) vector, (3) matrix, and (4) matrix–matrix multiplication. (b) Singular value decomposition (SVD). (c) Canonical polyadic (CP)-decomposition. (d) Tucker decomposition. (e) Tree tensor network (TTN) ansatz. (f) Tensor train (TT) ansatz. (g) Projected entanglement pair state (PEPS) ansatz. (h) Multiscale entanglement renormalization ansatz (MERA).

One of the most broadly known tensor decompositions is the *canonical polyadic* (CP) *decomposition* (Hitchcock, 1927; Carroll and Chang, 1970), also called CANDECOMP/PARAFAC or simply CP:

$$T_{i_1 i_2}^{a_1 a_2} \approx C_r X_r^{a_1} Y_r^{a_2} U_{i_1}^r V_{i_2}^r \tag{3.14}$$

where a summation over the index r is implied. The corresponding graph is shown in Figure 3.1. By varying the extent of this internal index r, one can control the quality of the approximation. In the following discussion, we will call the extent of an internal index its *valence*. Internal indices are closed indices that define the structure (connectivity) of the TN graph. In the context of the CP tensor decomposition, the extent of the index r may also be called the *resolution rank*, the notion closely related to the matrix rank (in a mathematical sense). It is important to note that the original rank-4 tensor is approximated by a contraction of four rank-2 tensors with one rank-1 tensor. This is a generalized tensor contraction, in which more than two tensors can share a contracted dimension (the corresponding graph is a generalized graph in which more than two vertices can be connected with an edge). Provided that the valence of index r is small enough, an efficient compression of the original tensor can be achieved. However, in general, the valence of r can easily grow large, rendering the CP ansatz inefficient.

The Tucker tensor decomposition (Tucker, 1966), also called the higher order SVD, represents another way of tensor compression:

$$T_{i_1 i_2}^{a_1 a_2} \approx X_{a_1'}^{a_1} Y_{a_2'}^{a_2} t_{i_1' i_2'}^{a_1' a_2'} U_{i_1}^{i_1'} V_{i_2}^{i_2'} \tag{3.15}$$

where $t^{a_1' a_2'}_{i_1' i_2'}$ is the *Tucker kernel*. Figure 3.1d shows the corresponding graph. The indices of the Tucker kernel are expected to have much smaller extent than the indices of the original tensor for the Tucker approximation to be efficient. Yet, the number of dimensions in the Tucker kernel stays the same, making it challenging to apply the Tucker decomposition to higher rank tensors. To overcome this problem, a hierarchical Tucker decomposition (Grasedyck, 2010) can be introduced in the form of TTNs, depicted in Figure 3.1e. The TT TN shown in Figure 3.1f, also called the MPS (Affleck et al., 1987; Chan et al., 2016) (see Equation 3.9), can be viewed as a special case of the TTN.

More sophisticated TNs have also been suggested, including the PEPS (Verstraete and Cirac, 2004) and MERA (Vidal, 2008) (see Figure 3.1g and h). In general, the choice of a particular structure of the TN is dictated by (a) the internal structure of the tensor being compressed, and (b) the memory requirements and computational cost of the compression scheme (in the context of the problem being solved). However, one can separate out several key concepts used for the construction of TNs. The first concept is the *isometric compression* defined by a tensor $S^{j_1 \ldots j_m}_{i_1 \ldots i_n}$ with n *input* lines (indices) and m *output* lines (indices), where $m < n$, with the examples shown in Figure 3.2. The requirement of isometry can be stated as

$$\bar{S}^{i_1 \ldots i_n}_{j_1 \ldots j_m} S^{j_1' \ldots j_m'}_{i_1 \ldots i_n} = \delta^{j_1'}_{j_1} \cdot \ldots \cdot \delta^{j_m'}_{j_m} \qquad (3.16)$$

where $\delta^{j_1'}_{j_1}$ is the Kronecker delta, and the bar on top of the first argument of the tensor contraction means complex conjugation (Figure 3.2 provides the corresponding graphical representation). In essence, the isometric tensor $S^{j_1 \ldots j_m}_{i_1 \ldots i_n}$ maps a larger tensor product space represented by the dimensions $i_1 \ldots i_n$ into a smaller tensor product space represented by the dimensions $j_1 \ldots j_m$ with an orthonormal basis (tensor $S^{j_1 \ldots j_m}_{i_1 \ldots i_n}$ essentially specifies a set of orthonormal vectors spanning the smaller subspace). By contracting a given tensor with the optimal (n, m) isometric tensor, one reduces the rank of the tensor by $(n-m)$. The optimal isometric tensor is the one that minimizes the tensor norm loss during such compression.

Another fundamental concept is the adaptivity of the valence (extent) of internal indices of a TN. By increasing/decreasing the valence of internal indices, one is increasing/decreasing the quality of the compression, with the computational cost following the same direction. In the most general case, a TN should be fully adaptive, both in structure (connections between graph vertices) and the valence of internal indices (this is still an active area of research). The numerical optimization of generic TNs is mostly based on two primitives, the tensor contraction and the SVD, thus requiring them to run efficiently at scale.

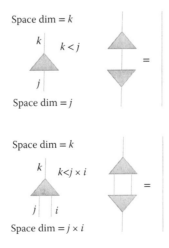

FIGURE 3.2   Isometric tensor compression.

3.3.4 PARALLEL ALGORITHMS FOR BASIC TENSOR OPERATIONS

The dense, block-sparse, and even more generally sparse basic tensor algebra operations all have a high degree of parallelism. For example, the most important basic tensor algebra primitive, the tensor contraction operation, consists of many elementary multiplications/accumulates, each multiplying a number from one input tensor by a number from another input tensor with a subsequent accumulation of the result in the corresponding element of the output tensor (generalized tensor contractions may have more than two input tensors). The elementary multiplications/accumulates are all independent of each other as long as the potential race conditions are avoided during the accumulation step. Consequently, there is a lot of flexibility how a given basic tensor operation can be decomposed into smaller pieces to be distributed among available computing units. For example, looking at Equation 3.10, the computational work can be decomposed into smaller chunks (*tasks*) by (a) splitting the destination (output) tensor into a number of smaller tensor slices in an arbitrary way, (b) splitting the sum over contracted indices into multiple subsums in an arbitrary way. Then the general algorithm optimization problem consists of two coupled subproblems:

1. *Granularity*: What is the optimal granularity of decomposition of data (tensors) and computations (tensor operations)?
2. *Mapping*: What is the optimal way of distribution of data (tensor slices) and computations (tasks) among available computing resources (memory and computing units) such that each computing unit is kept maximally occupied?

Here, we assume that the granularity of tasks is bounded and the bound does not depend on the problem size, which implies that the data size per task is bounded with the bound independent of the problem size. It is also important to note that the data distribution granularity is generally not directly related to the task granularity.

In the recent years, there have been a number of efforts put toward scalable numerical tensor algebra, mostly dealing with dense tensors (Baumgartner et al., 2005; Lotrich et al., 2008; Solomonik et al., 2013, 2014; Rajbhandari et al., 2014a,b) (see Calvin et al., 2015; Riplinger et al., 2016 regarding distributed sparse tensor algebra algorithms). In these works, the tensor contraction operation was the key tensor algebra primitive to be implemented efficiently. In practice, there have been two common choices for the distributed tensor storage layout: (1) the *blocked* distribution of tensors and (2) the *cyclic* distribution of tensors. The hybrid of the two is called the *block-cyclic* distribution. In the blocked distribution, also called the tiled distribution because of its analogy to the tiled matrix distribution, each tensor dimension may split into multiple segments and the tensor slices (blocks) specified by the combinations of different segments in different tensors dimensions are distributed among the nodes of an HPC system. The best distributed dense tensor contraction algorithms provide the interconnect topology aware regular distribution of tensor slices (Solomonik et al., 2013, 2014; Rajbhandari et al., 2014a,b). So far, the blocked distribution has been the prevalent scheme for distributed tensor storage (Baumgartner et al., 2005; Lotrich et al., 2008; Calvin et al., 2015). In contrast, the cyclops tensor framework (CTF) (Solomonik et al., 2013, 2014) utilized the cyclic distribution of tensors. In this scheme, the interconnect topology aware data mapping is based on the redistribution of individual tensor elements across all nodes. The cyclic distribution is better suited for storing tensors with permutational symmetry as compared to the blocked distribution, although some progress has been made recently with the use of the block-cyclic storage layout (Rajbhandari et al., 2014b). The cyclic distribution also provides better static load balancing in performing dense tensor contractions in a communication-optimal way. The price to pay is the potential necessity of the global (all-to-all) redistribution of tensor elements between different tensor contractions. Despite the prevalence of the two aforementioned tensor storage layouts, in general, the proper exascale tensor algebra framework should be flexible in terms of supported data storage layouts as there is no single ideal layout for all possible use cases.

The computational aspects of the distributed tensor contraction operation performed on dense tensors are well understood nowadays (Solomonik et al., 2013, 2014; Rajbhandari et al., 2014a,b), largely due to its close resemblance of the distributed matrix–matrix multiplication (in fact, any tensor contraction can be mapped to the matrix–matrix multiplication via tensor transposes, see Section 3.2.2). Both the tensor contraction and matrix–matrix multiplication have the same lower bounds for the necessary communication volume and both can benefit from an extra memory used for a partial data replication in the same way (Solomonik et al., 2013, 2014; Rajbhandari et al., 2014a). The corresponding task placement and data communication patterns are static and regular, making possible their efficient mapping on a given interconnect topology (Solomonik et al., 2013, 2014; Rajbhandari et al., 2014a). Provided that the tensor contraction has a sufficiently high arithmetic intensity (see Equation 3.12), the data transfers can be efficiently overlapped with ongoing computations, thus keeping each HPC node busy at all times (compute-bound case) (Solomonik et al., 2013, 2014; Rajbhandari et al., 2014a). Tensor contractions with a low arithmetic intensity are communication bandwidth bound. In some advanced electronic structure applications based on dense tensor algebra, the memory size can also become the limiting factor, thus requiring an out-of-core algorithm (Lotrich et al., 2008) and data compression whenever possible. This case has got much less attention in the computer science literature.

Things become more complicated when the tensors are *block sparse*, each tensor being a sparse collection of nonzero tensor blocks (slices). In this setting, a distributed algorithm can no longer assume some static regularity in the data storage, data transfers, and the computational work, unless it is focusing on a very specific regular case. In general, a block-sparse tensor can be stored as a distributed associative map in which the explicit exploitation of the interconnect topology for data/work placing becomes challenging and case dependent. Taking into account the irregularity of the block-sparse tensor algebra workloads, the task-based programming model is the most natural candidate for the parallel algorithm expression. To the best of our knowledge, to date, there are only two distributed tensor algebra frameworks being actively developed that specifically target the scalable, efficient implementation of block-sparse tensor computations: TiledArray (Calvin et al., 2015) and ExaTENSOR (the subject of this chapter). In particular, the TiledArray framework leverages the available task-based runtimes PaRSEC (Bosilca et al., 2013; Wu et al., 2015) and MADNESS (https://github.com/m-a-d-n-e-s-s/madness). The ExaTENSOR framework is less developed in terms of the implementation, but its complete design is presented in the following sections where the novel algorithmic aspects are explicitly highlighted. Besides, we should mention another framework for sparse tensor computations, called SparseMaps (Riplinger et al., 2016), which provides numeric primitives for a specific class of methods from electronic structure theory (Riplinger and Neese, 2013).

Apart from the distributed algorithms, there has also been a considerate effort put toward the implementation of basic tensor algebra primitives on a single (generally heterogeneous) node, including the support of multicore CPU, NVIDIA GPU, and Intel Xeon Phi. Not surprisingly, the tensor contraction operation has got the most attention here as well, as the most important basic tensor algebra primitive. Both dense and block-sparse cases were considered, where the latter is eventually reduced to the dense case (each individual nonzero block in a block-sparse tensor is dense). One can distinguish two dominating approaches in the implementation of tensor contractions on shared-memory systems: direct and indirect. In the direct approach, the tensor contraction is implemented as an explicit nest of loops over all involved indices, with subsequent optimizations involving loop reordering, loop blocking, etc. (Baumgartner et al., 2005). This approach is best implemented via an automated code generator, as each distinct tensor contraction may require a different code for optimal performance. In the indirect approach, the tensor contraction is transformed into the matrix–matrix multiplication via tensor transposes performed on the input/output tensor arguments as described in Section 3.2.2. The matrix–matrix multiplication is then executed by one of the highly optimized vendor-provided BLAS libraries. In the indirect approach, the key is to optimize the tensor transpose step, that is, an arbitrary permutation of tensor dimensions. A general algorithm for performing arbitrary tensor transposes efficiently on multicore CPU, NVIDIA GPU, and Intel Xeon Phi was presented in Lyakh (2015), but the implementation of that algorithm was suboptimal and could not deliver the maximally possible memory bandwidth, especially on accelerators. Later, a highly optimized implementation for NVIDIA GPU was suggested in the form of the cuTT library (https://github.com/ap-hynninen/cutt.git), which was shown to achieve close to the maximally possible GPU memory bandwidth (Hynninen and Lyakh, 2016). Recently, a more general approach was introduced in the form of the tensor transpose compiler (TTC) infrastructure (https://github.com/HPAC/TTC.git), a framework capable of automatically generating an optimized tensor transpose code for different hardware architectures, including x86 CPU, NVIDIA GPU, and Intel Xeon Phi, based on a higher level specification of the tensor transpose of interest (Springer et al., 2016a,b). Despite achieving a high memory bandwidth, the above algorithms are still out-of-place, thus still potentially introducing a 100% data size overhead for general tensor contractions. Very recently, this problem has been resolved with the use of the TTC framework, where instead of transposing the entire tensor in one step with the general tensor–tensor contraction (GETT) kernel (Springer and Bientinesi, 2016) is transposing smaller parts of the tensors, one at a time, immediately performing the matrix–matrix multiplication. This approach keeps the efficiency at a high level while only having a constant memory overhead when performing general (dense) tensor contractions. However, it is still based on a case-specific automated code generation. There were also other notable efforts directed toward a direct implementation of the tensor contraction operation (Nielson et al., 2015; Matthews, 2016). Despite a tremendous progress made in the optimization of the dense tensor contraction algorithms based on autotuning and code generation, an efficient universal tensor contraction kernel (one per hardware architecture) is still of demand, as in some use cases, the specific tensor contractions to be executed may not be known at compile time (alternatively, this problem could be solved by just-in-time code generation and compilation if the underlying programing language supports that).

To summarize, basic tensor algebra operations have been implemented in a number of software libraries, each of which is focusing on specific computational features for specific scientific domains, including distributed dense (Lai et al., 2013; Lotrich et al., 2010; Solomonik et al., 2013, 2014; Rajbhandari et al., 2014a,b) and block-sparse (Calvin et al., 2015) tensor operations for QMBT, block-sparse tensor computations on a single node (Epifanosky et al., 2013) for quantum chemistry, general-purpose dense tensor operations on heterogeneous nodes equipped with multicore CPU and NVIDIA GPU (https://github.com/DmitryLyakh/TAL_SH.git), and single/multinode dense tensor computations for deep learning (https://developer.nvidia.com/cudnn; https://www.tensorflow.org; https://deeplearning4j.org; https://github.com/amznlabs/amazon-dsstne.git). However, no single framework is general enough to address a wide variety of use cases while being able to run efficiently on vastly different hardware and at different scale, from a single workstation up to a leadership HPC system. Below, we present a prototypical design of such a framework under the name ExaTENSOR, where we focus on scalable task-based algorithms and performance-portable software solutions for numerical tensor algebra.

3.3.5 EXTREME SCALE SOLUTIONS

3.3.5.1 Projected Exascale Computing Hardware Roadmap

Although the uncertainty in the future exascale HPC hardware architectures still remains, there are some clear trends that are likely to persist. First of all, the computing power of processors (Flop/s) continues to grow rapidly, mostly due to increasing the number of computing units on a chip, extending the vector size, and supporting a concurrent execution of multiple threads per core at the hardware level. However, the existing lithographic chip fabrication techniques are currently approaching their limit, without having a mature technology to replace them yet. Consequently, one may expect simpler but more parallel processor architectures in the future, that is, a chip will contain more cores of a simpler design as compared to the current state-of-the-art out-of-order processors, like x86-64 Intel Xeon. This trend is confirmed by the success of the GPU and many-core HPC hardware architectures, like Titan at OLCF (https://www.olcf.ornl.gov/titan/) and Cori at NERSC (http://www.nersc.gov/users/computational-systems/cori/), which are optimized for computational throughput as opposed to the instruction execution latency optimization in modern CPU. Also, it is likely that we will be observing a growth in the heterogeneous processor architectures where different cores available on a chip will specialize in different kinds of workloads, as no single core architecture is optimal for all possible computational workloads. A bright example is the latest Chinese supercomputer Sunway TaihuLight (https://www.top500.org/system/178764). Another big effort in this direction is the heterogeneous system architecture (HSA) project from AMD (http://developer.amd.com/resources/heterogeneous-computing/what-is-heterogeneous-system-architecture-hsa/). Besides, taking into account, the importance of energy efficiency for large-scale computing, one could envision a growing interest in low-energy generic (https://www.arm.com/products/processors/instruction-set-architectures/index.php) as well as domain-specific processor architectures specializing in a given computational workload, for example, the architectures based on the field-programmable gate arrays (FPGA) (https://en.wikipedia.org/wiki/Field-programmable_gate_array).

Contrary to the steep increase in the processing power, memory bandwidth, network interconnect bandwidth, and especially its latency were not improving at the same fast pace. The most recent progress in these directions is related to the introduction of the on-chip 3D stacked memory (http://www.amd.com/en-us/innovations/software-technologies/hbm; http://www.hybridmemorycube.org), on-chip integration of the network interface (http://www.intel.com/content/www/us/en/high-performance-computing-fabrics/omni-path-architecture-fabric-overview.html), and the deployment of fast interconnects for communicating data between CPU and (NVIDIA GPU) accelerators (http://www.nvidia.com/object/nvlink.html). Clearly, to realize the full power of all computing units on flop-oriented node architectures, it is extremely important to feed them with data fast enough. This is especially a problem in some data analysis workflows in which the arithmetic intensity (and data reuse) is low. Additionally, since a data transfer consumes much more energy than an elementary computation, moving memory closer to the compute units becomes a priority for energy efficiency, especially at extreme scale. As a consequence, the memory system is becoming increasingly more hierarchical. Besides all these, the amount of memory per core has not been growing for some time (and even shrinking), although the total amount of memory per HPC node has been increasing, but not as fast as the computational power per node.

We expect that the parallel algorithm optimization and its efficient, performance-portable implementation will be the most time consuming and critical tasks in the exascale computing era, that is, one can no longer expect a free boost in performance when faster processors come on the market. Below, we present our ideas on the performance portability strategy for numerical tensor algebra, which can be generalized to other computational domains as well. Subsequently, we detail the design of the ExaTENSOR software framework which will serve as the computational (tensor algebra) engine in the CC module of the quantum-chemistry code DIRAC on the next generation supercomputer Summit to be deployed by the OLCF in 2018 (our main use case so far). At this point, only some (lower level) components of the ExaTENSOR framework have been implemented and the developmental work is still in active progress. Consequently, we will not provide much of benchmarking data here. Instead, we will present our conceptual approach to the performance portability problem at extreme scale (e.g., for tensor algebra workloads) and highlight the novel elements of our design.

3.3.5.2 Hardware Abstraction Scheme and Virtual Processing

In one of possible architectural scenarios, the complexity of the node architecture is likely to grow on the way to exascale, that is, an HPC node is likely to become denser and more heterogeneous. For example, the node of the future pre-exascale OLCF supercomputer Summit will be equipped with multiple IBM Power CPU and multiple NVIDIA GPU accelerators, in contrast to a single NVIDIA GPU on the current OLCF supercomputer Titan. Also, the memory hierarchy on a node is likely to become more complex on future systems. This increase in hardware complexity and heterogeneity poses serious challenges for programming model portability. For example, the future exascale HPC software written once would be expected to be portable across multiple node architectures, and ideally also perform well on all those architectures. To achieve this goal, the hardware specifics should be abstracted away by a proper portable programming model (a portable way of algorithm expression). The most common approach to date was based on the use of parallel programming *directives* incorporated in one of the general-purpose programming languages used for scientific HPC computing, such as Fortran, C, and C++. The two major directive-based standards for on-node parallel programming are OpenMP and OpenACC, both relying on the decomposition and assignment of the computational work to individual *threads* that can be executed in parallel on the computing units available on the node (the current OpenACC 2+ standards do not use the term thread explicitly but the OpenACC parallelization scheme is essentially based on nested multi-threading). In general, a thread is a code that can be executed by a computing unit concurrently with other threads executed on other computing units. The current OpenACC standard as well as the latest OpenMP 4+ standards allow placing threads on different devices (in principle), for example, multicore CPU, NVIDIA GPU, AMD GPU, Intel Xeon Phi, etc. These standards provide a unified general-purpose framework for on-node parallel programming (the execution of each thread can additionally be vectorized in both standards). Yet, the concept of a thread is rather low level (possibly hardware specific) and static, making it difficult to express irregular parallel algorithms in terms of threads. A better, higher level logical abstraction for concurrent execution is a *task*, specified by the input/output data and a specific computation to perform on the input data to produce the output data. The concept of a task is the most general parallel programming primitive such that any parallel computation can be logically decomposed into tasks (not uniquely in most cases). Within the task-based programming paradigm, a parallel code generates tasks, schedules them for concurrent execution on different computing units by binding them to the corresponding threads (via a task-scheduling runtime system), checks for the completion of each task, and respects possible task dependencies specified by the corresponding *DAG*. Tasks are explicitly supported by the OpenMP standard, but not yet by the OpenACC standard. The C++11 language and later versions explicitly provide their own primitives for task-based concurrent programming, as the C++ object-oriented philosophy does not align well with directive-based programming.

Although the task-based programming model is already general enough for direct expression of any irregular parallel algorithm in terms of tasks that are subsequently mapped to all available computing units across the entire HPC system, in our design of the ExaTENSOR framework, we introduce an additional level of abstraction at the node level for better hardware encapsulation, portability, and fault tolerance. Namely, each physical HPC node is virtualized as a domain-specific *software processor* that can accept domain-specific instructions (DSIs) (tensor algebra instructions in our case), decode them, fetch the input data, dispatch the instructions to appropriate computing devices available on the physical node, and upload the result back to its main storage location. In a sense, this approach is somewhat analogous to the Java virtual machine, but for a domain-specific case. In fact, the domain-specific code interpretation in the domain of electronic structure theory was pioneered in the ACES III software project (Lotrich et al., 2008), with its built-in super-instruction language and processing runtime (Sanders et al., 2010). Very recently, a modernized version of that effort was released under the name ACES IV (Sanders et al., 2017; https://github.com/UFParLab/aces4). The approach used in our work is somewhat similar in spirit, being based on similar general principles. Yet, our specific design and implementation are unique in a number of aspects and can be considered as a further elaboration of the concept of domain-specific hardware virtualization. We introduce a number of novel design elements, for example, hierarchical virtual processing, in order to achieve better performance portability in separating the hardware-agnostic algorithm specification and hardware-specific task scheduling. We also provide a generic formalization of the concept of domain-specific virtual processing.

The general architecture of a *domain-specific virtual processor* (DSVP) running on a physical node is illustrated in Figure 3.3. In general, the physical node may run more than one DSVP, thus providing a convenient means for node resource sharing. Regardless of the specific nature of the *domain-specific instruction set* (DSIS) that can be processed by a given DSVP, the DSVP execution pipeline is based on the following basic primitives:

- Fetch a block of DSIs from the higher level.
- Decode a DSI.
- Allocate/deallocate *local* memory for storing data.
- (Pre-)fetch the *remote* input data for each DSI.
- Transfer the data between different memory spaces accessible *locally*.
- Dispatch the *ready* DSIs to specific computing units.
- Check the *completion* (or *fault*) of a DSI.
- Upload the *remote* output data back to its persistent storage location.
- Respect the data *dependencies* between *local* DSIs.
- Retire a DSI.
- Upload the block of retired DSIs back to the higher level.

The above primitives are sufficient for setting up a functional DSVP. As one can see, in general, the DSVP receives instructions from a higher level of the domain-specific hardware virtualization scheme, thus enabling a *hierarchical virtualization* of an HPC system (hierarchical domain-specific virtual processing). This is a novel feature of our domain-specific hardware virtualization design that

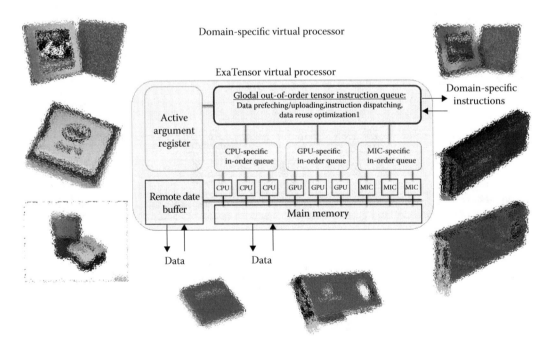

FIGURE 3.3 General architecture of a domain-specific virtual processor (DSVP).

will be extensively exploited in the ExaTENSOR framework for numerical tensor algebra workloads. We will describe it later in the text.

The DSIs executed by a DSVP are essentially the elementary tasks the parallel algorithm is decomposed into, plus some auxiliary instructions. Thus, a DSVP provides a runtime system for a *dynamic* execution of task-based algorithms. A cluster of DSVPs forms the basis for the distributed task-based runtime system. The orchestration of task scheduling within a cluster of DSVPs is done at a higher level of virtualization (system-wide HPC virtualization) described later. Importantly, the cluster of DSVPs consists of structurally similar virtual nodes (DSVPs) as the underlying (diverse) physical hardware is hidden by a uniform node virtualization. Thus, the distributed task-scheduling decisions are solely based on abstract characteristics of the virtual nodes, for example, the computational throughput (Flop/s), memory size (Bytes), network interface bandwidth and latency, etc. By abstracting away the specifics of the node hardware, one can better decouple the data/task distribution logic on a system-wide level from that on a node level, thus simplifying the algorithm parallelization. Essentially, the physical nodes become plug-and-play suppliers of computational resources for the virtual nodes running on them. Furthermore, a DSVP running on a physical node as an additional layer of abstraction is beneficial for hardware codesign efforts as the focus is shifted on efficient mapping of domain-specific primitives onto physical hardware resources. That is, instead of mapping an algorithm directly to available hardware, it is mapped onto a virtual (domain-specific) computing architecture, and then the (virtual) components of that virtual architecture are subsequently mapped to the actual hardware, thus decoupling the algorithm from the hardware details.

From the implementation point of view, a DSVP contains a multichannel out-of-order instruction queue (MC-OOO-IQ) implemented as a vector of linked lists. Each incoming DSI (task) is placed into one of the linked lists, based on its formal characteristics. In our case (numerical tensor algebra), these are (a) number of flops to perform (task compute granularity); (b) arithmetic intensity (flop-to-byte ratio); (c) data locality (size of the remote arguments to be fetched/uploaded remotely). These characteristics influence the priority of the instruction (task) and the choice of the specific device to be executed on (e.g., NVIDIA GPUs favor higher granularity and arithmetic intensity). Additionally, the choice of the specific device on which the instruction is to be executed will be affected by the local data residence (e.g., if one of the arguments of a tensor operation already resides in memory of some device, that device will have higher chances to be selected as the execution device since it requires less data movement). Furthermore, an instruction operating on remote data that has already been fetched for another (previously decoded) instruction will simply reuse that locally present data. Essentially, in such caching mechanism, the locally present remote data will not be discarded while it is still referenced by at least one of the unfinished instructions, as long as there is enough memory. Finally, the DSVP must respect data dependencies between DSIs in its MC-OOO-IQ. The easiest way to implement this is to register locally all instruction operands and supply a "READY" flag to those output operands which are inputs for subsequent instructions scheduled by the DSVP (the flag is set once the operand is ready, thus enabling the progress of the dependent instructions). In our design, each DSVP can only respect local data dependencies, that is, those within its own instruction queue. The global data dependencies, that is, data dependencies between instructions scheduled on different DSVPs, are controlled by higher levels of domain-specific HPC virtualization described later in the text. This drastically simplifies the task-scheduling logic on each DSVP and makes DSVPs more autonomous, which is good for fault tolerance.

The life cycle of the DSVP consists of appending new DSIs in the MC-OOO-IQ, repeatedly traversing the MC-OOO-IQ, registering instruction operands, initiating remote data fetches/uploads, dispatching instructions to specific compute devices by placing them into device-specific instruction queues, executing the instructions asynchronously on specific compute devices, checking for the completion/error, and finally retiring the completed (or failed) instructions, removing them from the MC-OOO-IQ and sending them back to the higher level of control. In a generic implementation, the classes of DSIs are derived from the abstract DSI class, thus providing a common interface for executing the DSVP workflow. The classes of DSI operands are also derived from the abstract DSI operand class for the same reason. In this way, the DSIs can be dispatched to any computing device present on node via the use of a device-unified API interface. Underneath, each specific device kind will have its own backend implemented as a library driver capable of executing elementary DSIs on that device kind, for example, multicore CPU, NVIDIA GPU, Intel Xeon Phi, AMD GPU, etc. Each new device kind will only require an implementation of the new library driver for the specific elementary operations to be executed on that device kind. Nothing else needs to be changed in the code, that is, all higher layers are hardware agnostic. The device-unified API interface for basic tensor algebra operations in the ExaTENSOR framework is provided by the TAL-SH library (https:// github.com/DmitryLyakh/TAL_SH.git) which currently supports multicore CPU and NVIDIA GPU (the Intel Xeon Phi support is planned in future).

Besides the multichannel out-of-order instruction queue, a DSVP needs a local memory allocator capable of dealing with a hierarchical memory organization. Namely, the memory allocator should be able to *quickly* allocate memory which is close to a specific computing device, if requested. The memory allocator should have a portable interface, that is, it should wrap device-specific memory allocators in a portable way. The same requirements also apply to the local data transfer engine used by a DSVP for transferring data between two arbitrary locations in local memory (within a node). A portable data transfer API interface should hide the device-specific data transfer API functions underneath. Clearly, the data transfer API must provide an asynchronous execution mode.

To fetch/upload blocks of DSIs as well as to fetch/upload the remote data the instructions operate on, a DSVP must have remote communication capabilities. The DSIs can be fetched/uploaded via two-sided communications, that is, the DSVP is persistently checking for new blocks of DSIs, coming from the higher level of control, and periodically uploading back the blocks of retired DSIs. In contrast, the fetch/upload of the remote data required/produced by the DSIs executed by a DSVP is done via a one-sided communication model as it introduces less synchronization and can benefit from advanced hardware features, like the remote direct memory access (RDMA) capability provided by upper class network interface cards (NICs). In the current implementation, ExaTENSOR relies on MPI-3 API for both types of remote communication. Alternatively, one-sided communications can be implemented on top of one of the available partitioned global address space (PGAS) libraries, for example, OpenSHMEM. Yet, the communication API interface used by a DSVP must be portable regardless of the chosen communication backend.

3.3.5.3   HPC Scale Abstraction

Having delegated the on-node task scheduling to a DSVP, we are still left with the problem of efficient data/work decomposition and data/work distribution among DSVP nodes. As we emphasized earlier, the basic tensor operations, for example, the tensor contraction operation, are very flexible in terms of how the data (tensors) and computations (tensor operations) can be decomposed into smaller pieces. Such a flexibility suggests looking for the most natural parallelization scheme which would preferably be independent of the HPC system scale (the total number of DSVP nodes). For this, we need to depart from a traditional flat representation of an HPC system and introduce a more structured view. That is, while a DSVP provides a hardware-agnostic abstraction of a physical compute node, we need to introduce another, higher layer of abstraction to hide the HPC system scale, thus becoming both hardware and machine scale agnostic. To accomplish this goal, we suggest a *recursive virtualization* of the HPC system illustrated in Figure 3.4. At the top level, the entire HPC system is considered as a single DSVP, level-0 DSVP, capable of executing DSIs (basic tensor algebra operations in our case). Then, by splitting the entire HPC system into several subsystems, we obtain a set of interconnected level-1 DSVPs constituting the level-0 DSVP. By recursively applying this decomposition procedure, we obtain level-2 DSVPs by decomposing each level-1 DSVP, level-3 DSVPs by decomposing each level-2 DSVP, etc., up to the *last-level* DSVPs, which are just our basic DSVPs introduced in the previous section. We will call these last-level DSVPs *elementary*

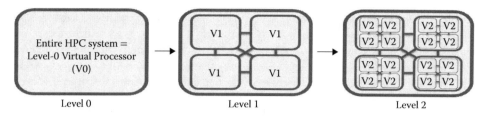

FIGURE 3.4 Hierarchical representation of a high-performance computing (HPC) system.

DSVPs in order to distinguish them from the *composite* DSVPs obtained by aggregation of elementary DSVPs (all higher level DSVPs except the last-level DSVPs are composite). One should note that each composite DSVP is a virtual entity instantiated on some physical node without a pre-defined mapping (it is not hard-wired to a specific physical node). Also, composite DSVPs run a different workflow than elementary DSVPs, which will be described in the next section (composite DSVPs are grouped together to form a hierarchical distributed task-scheduling manager). In our formal recursive decomposition scheme, we impose a tree structure on the initially flat view of the HPC system, which will be called the *HPC system decomposition tree*. The specific tree characteristics, that is, branching in each tree node and the composition of each tree node (a list of constituent physical nodes), will be affected by the topology of the HPC interconnect. Our guiding principle, here, is to decompose each level-k DSVP into a set of level-$(k+1)$ DSVPs by maximizing the aggregated communication bandwidth within each level-$(k+1)$ DSVP, which is likely to minimize the aggregated communication bandwidth in-between the level-$(k+1)$ DSVPs (*dense* packing of nodes). Having fully constructed such a domain-specific HPC system virtualization scheme, we will have two kinds of processes in our runtime, which will then be mapped to specific physical nodes:

1. Elementary DSVP, that is, the intranode task-scheduling runtime
2. Composite DSVP, that is, the internode task-scheduling runtime

The number of elementary DSVP processes, that is, the processes which perform actual work, will significantly outweigh the number of composite DSVP processes that are solely responsible for higher level control and global runtime optimization. The elementary DSVP processes will form groups controlled by their dedicated composite DSVP processes that are, in turn, merged into groups controlled by higher level composite DSVP processes, etc., up to the top level-0 DSVP process (root). Using the above scheme, on each level of hierarchy, each (composite) DSVP consists of a bounded number of lower level DSVPs, thus simplifying the data/work distribution decisions. That is, instead of globally distributing the data/work among $O(N)$ processes, we always distribute data/work locally among $O(1)$ processes, but do it recursively ($O(\log(N))$ recursions will be necessary). This recursive data/work distribution heuristic is in no way guaranteed to be optimal for an arbitrary tensor algebra workload, but it is scalable and, for many tensor algebra workloads, it is expected to deliver a reasonable performance, provided that we take an additional care of load balancing and communication pattern regularization, as described in the next section.

3.3.5.4 Hierarchical Task-Based Parallelism via Recursive Data Placement and Work Distribution

Having structured a given HPC system into a *hierarchical domain-specific virtual processor* (H-DSVP) with multiple levels of compute/storage granularity, we need to decompose the tensors (our domain-specific data) and basic tensor operations (our DSIs) into smaller pieces that can subsequently be mapped onto available DSVPs at each level of hierarchy. Essentially, we need to introduce a recursive data/task decomposition scheme in which the data/task granularity at each refinement level is tuned with respect to the storage/compute granularity of the DSVPs at the corresponding HPC scale refinement level which the data/tasks are distributed over. Such recursive data/ task mapping is illustrated in Figure 3.5. At the top level (level 0), which consists of a single level-0 DSVP, the original DSIs (basic tensor operations) are read from a user-defined program written in a domain-specific language (DSL). Each new tensor is registered as a sparse distributed collection of tensor slices, each tensor slice being a tensor itself. That is, the recursive definition of a tensor is: *A tensor is a generally sparse collection of tensors, up to the individual tensor elements*. Based on the granularity of the level-1 DSVPs that the level-0 DSVP splits into, the original tensor at level 0 is formally decomposed into multiple (smaller) tensors of appropriate granularity (only those smaller tensors are retained which are estimated to be nonzero). These smaller tensors are distributed among the level-1 DSVPs, each level-1 DSVP obtaining its own subset. Then, each level-1 DSVP performs exactly the same decomposition step, splitting its tensors into smaller tensors and distributing those among its children level-2 DSVPs. The entire process is recursively repeated until reaching the last level of the hierarchy (leaves of the HPC system decomposition tree). It may happen that at some level of such data decomposition, the tensors can become small enough such that their further decomposition would be unnecessary, at least at that level. In this case, these (already small) tensors can simply be replicated at the next level of the HPC system granularity refinement, that is, each level-$(k+1)$ child DSVP will obtain the same tensors as its parent level-k DSVP. Such a partial replication of data will only happen if the size of the tensors is considerably smaller than the storage granularity at that specific level of refinement. The partially replicated data can then decrease the communication volume during the execution of tasks.

Since we are interested in general block-sparse tensor algebra workloads, the structure of the tensor may not be known *a priori* in some cases, for example, when the tensor is formed as the result of a tensor operation operating on generic block-sparse tensor inputs. In this case, the tensor will have a *dynamic storage layout*, which will be finalized after the defining tensor operation has completed. Clearly, without any *a priori* information on the sparsity structure of the underlying tensors, the above-described data distribution scheme may easily result in unbalanced memory footprints. To prevent this, a data stealing mechanism will be activated to equilibrate the memory consumption

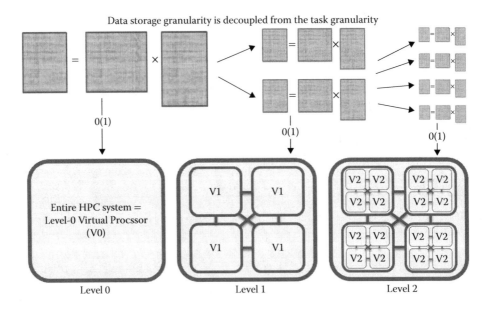

FIGURE 3.5 Recursive data/task mapping.

across different DSVPs on each level of the HPC system scale refinement. The data stealing process is also executed recursively, first from the bottom of the HPC system decomposition tree all the way up to the top level, and then back to the bottom. That is, the HPC system decomposition tree (layers of DSVPs organized in a tree) can record memory consumption for each tree node (a tree node represents a specific DSVP at a specific level of the HPC system scale refinement). Then, if unbalanced, the tree can rebalance its memory resource utilization by pushing the excess data up the hierarchy first and then propagating it down into underutilized tree branches, such that at the end the memory utilization at each leaf is more-or-less uniform.

Proceeding further, the recursive data decomposition naturally induces a recursive work decomposition. That is, the original basic tensor operations read at the top level are split into smaller scale operations (tasks) recursively by decomposing each tensor operand in accordance with its recursive storage decomposition, subsequently distributing thus obtained smaller scale tasks among the children DSVPs at each level of the HPC system scale refinement. Again, for general block-sparse tensor algebra workloads, this data-driven recursive work decomposition scheme can easily lead to a load imbalance in terms of the processing unit utilization across different last-level DSVPs. Similarly to balancing the memory resource utilization across last-level DSVPs, the compute resources can also be balanced via a recursive work stealing mechanism in which the overutilized leaves of the HPC system decomposition tree propagate their excess work up to the higher levels of the tree until reaching the parent nodes of underutilized tree branches to where the work will be redirected.

It is important to note that although the recursive work decomposition is driven by the recursive data decomposition, it is not tightly coupled to the latter. This means that the operands of tensor operations at a specific level of granularity refinement can, in general, be aggregates of multiple chunks of data stored separately (e.g., when the recursive decomposition stops before reaching the elementary storage granularity) or, conversely, they can be parts of the stored data chunks (e.g., when the recursive decomposition continues after reaching the elementary storage granularity). In other words, a tensor is physically stored as a generally sparse collection of specific (nonzero) tensor slices distributed over the elementary DSVPs, but the tensor operands in an elementary task are allowed to be both the aggregates of those elementary storage blocks as well as slices of them. This adds more flexibility for achieving a better overlap of remote data transfers and ongoing computations.

The workflow of a composite DSVP at any level of hierarchy consists of accepting the data specification (tensors) and work specification (tensor operations) from the higher level of hierarchy (from its master DSVP), then formally splitting the data and instructions into smaller pieces and distributing those among its children (lower level) DSVPs, up to the last level of hierarchy. As a result, each elementary (last-level) DSVP will be given its data to store and instructions to execute (the actual data allocation and instruction execution happens solely within the elementary DSVPs). The composite DSVPs only store metadata, for example, the tensor decomposition tables at each level of hierarchy, the work decomposition tables at each level of hierarchy, the resource utilization at each level of hierarchy, etc.

Since the hierarchical tensor decomposition tables are distributed among DSVPs, there is a need in information exchange in order to set the actual location of tensor operands in each individual task. This is accomplished by horizontal rotations within the HPC system decomposition tree. Namely, at any given tree level (granularity refinement level) each tree node first collects a list of tasks from the upper level. Then each tree node at that level sends its list of tasks to the tree node on the right and receives a list of tasks from the tree node on the left in a cyclic fashion, repeating this procedure until each tree node gets its original list of tasks back. During each such rotation, each tree node traverses the received list of tasks and sets the physical location of the task operands in case it owns the corresponding metadata. Yet, it may happen that at the end of the rotation phase the physical location of some (or even all) task operands will still be unknown, which means that it will be set on one of the lower levels of granularity refinement. Having reached the granularity level right before the last level, the location of all task operands will be set (no further granularity refinement is planned as the last level corresponds to the elementary DSVPs that will perform the actual work at that final level of granularity). Thus, each nonleave level in the HPC system decomposition tree first collects lists of tasks from the upper level, then rotates the obtained lists among the tree nodes on the same level and sets the data location (if known), and then distributes the tasks over the lower tree level (each tree node distributes its tasks among its children nodes). Additionally, the rotation phase will also perform resource utilization balancing, including workload balancing, memory footprint balancing, and communication load balancing. Essentially, each node of the HPC system decomposition tree (a DSVP) will contain the information on the current workload (flops consumed by the scheduled tasks), current memory footprint (bytes consumed by the data), and communication load (bytes to be transferred to/from DSVP). During the rotation phase, the tasks may migrate from their original lists to the lists belonging to other DSVPs in order to balance the resource consumption at each DSVP at a given level of granularity refinement. The probability of task migration will be dictated by the unbalance in the resource utilization at the specific granularity level as well as the locality of task operands. Note that by balancing the communication load, we can dynamically regularize the communication pattern, thus ensuring better scalability of irregular computational workloads (block-sparse tensor algebra is one of them).

As an ultimate result of the hierarchical task-scheduling process, each leave of the HPC system decomposition tree, that is, an elementary DSVP, will obtain its own list of tasks to execute and data to store. The location of each task operand will be known at that time, so the elementary DSVP can immediately proceed to the data prefetching step, thus starting the task execution workflow. However, we still need to manage data dependencies between the tasks. The data dependencies between the tasks scheduled on the same elementary DSVP will be taken care of by that specific DSVP as described earlier. In general, the data dependencies between the tasks scheduled on different elementary DSVPs will require remote synchronization between DSVPs, thus increasing the communication traffic as well as complicating the mitigation of potential node failures; consequently, we devised a simplified mechanism. Specifically, the lists of tasks assigned to each elementary DSVP are split into chunks called *task packets*, which are dispatched to elementary DSVPs as a whole. The tasks from any currently executed task packet on a specific elementary DSVP are not allowed to have data dependencies to the tasks contained in any other currently executed task packet on a different DSVP. That is, all tasks with cross-DSVP data dependencies are deferred until the corresponding data is ready. In this scheme, the global synchronization is done solely within the composite DSVP layer that consists of a relatively small group of processes. In such a setting, a failure of the node on which an elementary DSVP is running will simply lead to discarding the corresponding data and currently executed tasks and subsequently assigning them to a different (healthy) DSVP. However, a failure of the node on which a composite DSVP is running will have more severe consequences, in particular metadata loss, which can be mitigated by replicating metadata on multiple (at least two) composite DSVPs. In any case, since the number of composite DSVP processes is much smaller than the number of elementary DSVP processes, the probability of a failure of a composite DSVP process is proportionally smaller as well.

Finally, since some (dense) tensor algebra workloads are heavy on memory utilization, a possibility to run *out-of-core* algorithms will be required in general. For exascale HPC systems, we should generally assume a hierarchical memory model, including on-chip memory, on-package memory, regular DRAM, NVRAM, and finally disk memory. To run out-of-core algorithms, we need to be able to efficiently utilize NVRAM and disk memory. The presence of NVRAM on HPC nodes is highly beneficial for out-of-core algorithms as it has a higher bandwidth than disk I/O. The most straightforward approach suggests using NVRAM for swapping memory pages (operating system level) when a tensor algebra algorithm runs out of DRAM memory. In a more elaborate scheme, NVRAM can be used as a cache between the disk and DRAM, managed either at the system level or at the user level (or by a middleware layer, like MPI I/O). If implemented at the user level, our portable design philosophy, based on the domain-specific resource virtualization, suggests introducing a virtual service responsible for such cached disk I/O that will include all operations involving persistent memory (NVRAM, disk, etc.). We can call it the *domain-specific virtual I/O service* (DSVIOS). The basic primitives of DSVIOS are

1. An asynchronous prefetch of a block of data (tensor slice) from persistent memory into DRAM
2. An asynchronous upload of a block of data (tensor slice) from DRAM to persistent memory
3. Synchronization primitives (test, wait)
4. Importing an entire tensor from persistent memory
5. Exporting an entire tensor to persistent memory

The first three primitives operate on individual tensor slices, that is, the specific parts of a distributed tensor, in an asynchronous manner. They can be used by an elementary DSVP to issue requests to load/store our-of-core data from/to persistent memory. The DSVIOS processes will take care of these requests and provide a caching mechanism via NVRAM (if present). In contrast, the last two primitives from the above list involve all composite DSVPs as they operate on entire tensors (let us remind that a tensor is recursively defined as a generally sparse collection of tensors, up to individual tensor elements). When importing an entire tensor from persistent memory, the metadata tables will be created by composite DSVPs, the tensor will be distributed among DSVPs, followed by the actual memory allocation performed by elementary DSVPs. When exporting an entire tensor to persistent memory, the corresponding distributed metadata tables will be traversed over and all constituent tensor slices the tensor is stored in terms of will be offloaded to persistent memory for permanent storage, together with their metadata. Essentially, we are storing a distributed dictionary (associative map) here, in which each entry is defined by the tensor signature (key) and the tensor data (value). The *tensor signature* is a unique identifier of the tensor (tensor slice). The *tensor data* includes the tensor shape, data kind, tensor storage layout tables, other attributes, and the tensor body (tensor elements). Clearly, in general, importing (reading) a tensor may require data redistribution in case the execution configuration of the job differs from the one which exported (stored) the tensor.

Ideally, it would be convenient to offload the NVRAM disk caching feature to MPI I/O or other suitable middleware, thus simplifying the implementation of the DSVIOS. If such an option is not available, one has to rely on a custom memory allocator for NVRAM and implement the caching logic explicitly within the DSVIOS, subsequently providing the optimized DSVIOS service API to DSVPs. While taking more effort, a user-defined disk caching mechanism via NVRAM can be explicitly optimized for domain-specific workloads, for example, tensor algebra workloads in our case. That is, more complex hardware and more complex workloads peculiar to exascale computing provide an excellent opportunity for hardware–system–middleware–application codesign, where different features may migrate between those layers, based on the ease of implementation, cost and energy efficiency, and other criteria. We believe that, in this flexible scheme, the domain-specific virtualization of HPC resources provides a convenient abstraction for portability and codesign of the HPC systems targeting the corresponding domain-specific workloads.

3.4 CONCLUSIONS

Extreme scale computing presents an enormous opportunity for breakthrough scientific simulations as well as a grand challenge of writing scalable, efficient, portable, and maintainable software for performing these simulations. Although not without exceptions, it is hard to expect the majority of domain scientists to also become computer science experts capable of properly handling the complexity of diverse computer architectures and large-scale software design. Ideally, many domain scientists would prefer to focus on a high-level algorithm expression in terms of domain-specific primitives, letting the underlying runtime system execute these algorithms efficiently on any kind of computing resources they may have. Thus, an efficient implementation of domain-specific and generic math primitives in portable software libraries is crucial for the success of this high-productivity science model. The specific libraries with well-defined interfaces can then be hierarchically aggregated into larger toolkits or frameworks, providing a proper level of abstraction to scientific software developers.

In general, the chosen portable programming model, which is expected to deliver performance portability as well, can be different in different scientific HPC projects, to name few possibilities (we restrict ourselves to standard HPC languages here, although there is no guarantee that new alternatives will not take over in future):

- Fortran/C/C++ with MPI and directives (OpenMP, OpenACC)
- Fortran/C/C++ with a PGAS library, like GlobalArrays or OpenSHMEM, and directives (OpenMP, OpenACC)
- C++/UPC++ with a templated toolkit for accelerator programming, like Kokkos
- Explicit task-based programming models and runtimes
- Domain-specific, hardware-specific code generation from a high-level specification that can be used together with one of the above portability strategies
- Domain-specific virtual processing, advocated in this text for numeric tensor algebra workloads, which can also be used with one of the above portability strategies

However, regardless of the choice, the challenge of good quality software design will be growing due to the growth in algorithm and computer architecture complexity, thus again presenting an excellent opportunity for reusable, high-quality, high-performance software libraries, toolkits, and frameworks.

Another grand challenge, which has not been discussed in much detail in this text, is the fault tolerance of the exascale software stack. To this point, checkpointing has been the major fault mitigation strategy in HPC scientific computing. Unfortunately, the current middleware running on most scientific HPC platforms (e.g., MPI) does not really provide any reliable mechanism for tolerating hardware faults yet. Fault tolerance is an active research topic in computer science these days, and we optimistically hope that proper solutions will come to the market of scientific HPC computing in the near future.

ACKNOWLEDGMENT

This research used resources of the Oak Ridge Leadership Computing Facility, which is a DOE Office of Science User Facility supported under Contract DE-AC05-00OR22725.

REFERENCES

Affleck, I., T. Kennedy, E. H. Lieb, and H. Tasaki. (1987). Rigorous results on valence-bond ground states in antiferromagnets. *Physical Review Letters* 59, 799.
Anantharaj, V., F. Foertter, W. Joubert, and J. Wells. (2013). Approaching exascale: Application requirements for OLCF leadership computing. https://www.olcf.ornl.gov/wp-content/uploads/2013/01/OLCF_Requirements_TM_2013_Final1.pdf.
Anderson, E., Z. Bai, C. Bischof, S. Blackford, J. Demmel, J. Dongarra, J. Du Croz, A. Greenbaum, S. Hammarling, A. McKenney, and D. Sorensen. (1999). *LAPACK Users' Guide* (3rd ed.). Philadelphia, PA: Society for Industrial and Applied Mathematics. ISBN 0-89871-447-8.

Baumgartner, G., A. Auer, D. E. Bernholdt, A. Bibireata, V. Choppella, D. Cociorva, X. Gao, R. J. Harrison, S. Hirata, S. Krishnamoorthy, S. Krishnan, C-C. Lam, Q. Lu, M. Nooijen, R. M. Pitzer, J. Ramanujam, P. Sadayappan, and A. Sibiryakov. (2005). Synthesis of high-performance parallel programs for a class of ab initio quantum chemistry models. *Proceedings of the IEEE* 93, 276–292.

Bosilca, G., A. Bouteiller, A. Danalis, M. Faverge, M. Herault, and J. Dongarra. (2013). PaRSEC: Exploiting heterogeneity to enhance scalability. *Proceedings of the 2013 IEEE Computing in Science and Engineering* 15(6), 36–45.

Briggs W.L., V. E. Henson, and S. F. McCormick. (2000). *A Multigrid Tutorial* (2nd ed.). Philadelphia, PA: SIAM.

Calvin, J. A., C. A. Lewis, and E. F. Valeev. (2015). Scalable task-based algorithm for multiplication of block-rank-sparse matrices, IA3 '15 *Proceedings of the 5th Workshop on Irregular Applications*: Architectures and Algorithms, Austin TX, Nov 15, 2015, ISBN: 978-1-4503-4001-4.

Carroll, J. and J.-J. Chang. (1970). Analysis of individual differences in multidimensional scaling via an N-way generalization of 'Eckart-Young' decomposition. *Psychometrika* 35, 283–319.

Chan, G. K.-L., A. Keselman, N. Nakatani, Z.-D. Li, and S. R. White. (2016). Matrix product operators, matrix product states, and ab initio density matrix renormalization group algorithms. *Journal of Chemical Physics* 145, 014102.

Chow, E. and A. Patel. (2015). Fine-grained parallel incomplete LU factorizations. *SIAM Journal on Scientific Computing* 37(2), C169–C193.

Colella, P. (2005). Defining Software Requirements for Scientific Computing. http://www.lanl.gov/orgs/hpc/salishan/salishan2005/davidpatterson.pdf.

Dally, B. (20150. Challenges for Future Computing Systems. https://www.cs.colostate.edu/~cs575dl/Sp2015/Lectures/Dally2015.pdf.

Danalis, A., H. Jagode, G. Bosilca, and J. Dongarra. (2015). PaRSEC in practice: Optimizing a legacy chemistry application through distributed task-based execution. *Proceedings of the 2015 IEEE International Conference on Cluster Computing*, IEEE, Chicago IL, Sep 8-11, 2015, pp. 304–313.

Datta, K., S. Kamil, S. Williams, L. Oliker, J. Shalf, and K. Yelick. (2009). Optimization and performance modeling of stencil computations on modern microprocessors. *SIAM Review* 51(1), 129–159.

Demmel, J., M. Hoemmen, M. Mohiyuddin, and K. Yelick. (2009). Minimizing communication in sparse matrix solvers. *Proceedings of the 2009 ACM/IEEE Conference on Supercomputing*, New York, NY, Nov 2009.

Dolgov, S. V., B. N. Khoromskij, I. V. Oseledets, and D. V. Savostyanov. (2014). Computation of extreme eigenvalues in higher dimensions using block tensor train format. *Computer Physics Communications* 185, 1207–1216.

Dongarra, J. J., J. Du Croz, S. Hammarling, and I. S. Duff. (1990). A set of level 3 basic linear algebra subprograms. *ACM Transactions on Mathematical Software* 16(1), 1–17. doi:10.1145/77626.79170. ISSN 0098-3500.

Dongarra, J., J. Kurzak, P. Luszczek, T. Moore, and S. Tomov. (2015). Jack Dongarra et al. on Numerical Algorithms and Libraries at Exascale. https://www.hpcwire.com/2015/10/19/ numerical-algorithms-and-libraries-at-exascale.

Du, P., P. Lusczek, and J. Dongarra. (2012). High Performance Dense Linear System Solver with Resilience to Multiple Soft Errors. *Procedia Computer Science* 9, 216–225. Proceedings of the International Conference on Computational Science, ICCS 2012, http://www.sciencedirect.com/science/article/pii/S1877050912001445.

Epifanosky, E., M. Wormit, T. Kus, A. Landau, D. Zuev, K. Khistyaev, P. Manohar, I. Kaliman, A. Dreuw, and A. I. Krylov. (2013). New implementation of high-level correlated methods using a general block tensor library for high-performance electronic structure calculations. *Journal of Computational Chemistry* 34, 2293–2309 (2013).

Gates, M., A. Haidar, and J. Dongarra. (2014). Accelerating Eigenvector Computation in the Nonsymmetric Eigenvalue Problem, VECPAR 2014, Eugene, OR, June 2014.

Goedecker, S. and A. Hoisie. (2001). *Performance Optimization of Numerically Intensive Codes*. Philadelphia, PA: SIAM.

Golub, G. H. and C. F. Van Loan. (2013). *Matrix Computations* (4th ed.). Baltimore, MD: Johns Hopkins University Press.

Grasedyck, L. (2010). Hierarchical singular value decomposition of tensors. SIAM. *Journal on Matrix Analysis and Applications* 31, 2029–2054 (2010).

Haidar, A., S. Tomov, K. Arturov, M. Guney, S. Story, and J. Dongarra. (2016). LU, QR, and Cholesky factorizations: Programming model, performance analysis and optimization techniques for the Intel Knights Landing Xeon Phi. *IEEE High Performance Extreme Computing Conference (HPEC'16)*. Waltham, MA, IEEE, Sept 2016.

Herault, T. and Y. Robert, (Eds.). (2015). *Fault-Tolerance Techniques for High-Performance Computing*. Berlin, Germany: Springer.

Heroux, M. A. (2014). Toward Resilient Algorithms and Applications. https://arxiv.org/abs/1402.3809.

Heroux, M. A. and M. Hoemmen. (2013). Toward Resilient Algorithms and Applications. http://www.sandia.gov/~maherou/docs/HerouxTowardResilientAlgsAndApps.pdf.

Hitchcock, F. L. (1927). Multiple invariants and generalized rank of a p-way matrix or tensor. *Journal of Mathematics and Physics* 7, 39–79.

Hoemmen, M. (2011). A communication-avoiding, hybrid-parallel, rank-revealing orthogonalization method. *2011 IEEE International Parallel and Distributed Processing Symposium*, Anchorage, AK. http://ieeexplore.ieee.org/document/6012905/?arnumber=6012905.

Hynninen, A.-P. and D. I. Lyakh. (2016). cuTT: A high-performance tensor transpose library for CUDA compatible GPUs. Arxiv: 1705.01598 (2017), submitted to *ACM Transactions on Mathematical Software*.

Joubert, W., R. Archibald, M. Berrill, M. W. Brown, M. Eisenbach, R. Grout, J. Larkin, J. Levesque, B. Messer, M. Norman, B. Philip, R. Sankaran, A. Tharrington, and J. Turner. (2015). Accelerated application development: The ORNL Titan experience. *Computers and Electrical Engineering* 46, 123–138.

Joubert, W. and J. Cullum. (2006). Scalable algebraic multigrid on 3500 processors. *ETNA. Electronic Transactions on Numerical Analysis* 23, 105–128.

Joubert, W. and T. Oppe. (1994). Improved SSOR and incomplete Cholesky solution of linear equations on shared memory and distributed memory parallel computers. *Journal of Numerical Linear Algebra with Applications* 1(3), 287–311.

Joubert, W. and S. Su. (2012). An analysis of computational workloads for the ORNL Jaguar system, ICS'12: *Proceedings of the 26th ACM International Conference on Supercomputing*, Venice, Italy.

Joubert, W. D. and G. F. Carey. (1992a). Parallelizable restarted iterative methods for nonsymmetric linear systems. I: Theory. *International Journal of Computer Mathematics* 44(1–4), 243–267.

Joubert, W. D. and G. F. Carey. (1992b). Parallelizable restarted iterative methods for nonsymmetric linear systems. II: Parallel implementation. *International Journal of Computer Mathematics* 44(1–4), 269–290.

Kinoshita, T., O. Hino, and R. J. Bartlett. (2003). Singular value decomposition approach for the approximate coupled-cluster method. *Journal of Chemical Physics* 119, 7756.

Kogge P. (Ed.), (2008). *ExaScale Computing Study: Technology Challenges in Achieving Exascale Systems*, P. Kogge (ed.). Defense Advanced Research Projects Agency. http://users.ece.gatech.edu/~mrichard/ExascaleComputingStudyReports/ECS_reports.htm.

Lai, P.-W., K. Stock, S. Rajbhandari, S. Krishnamoorthy, and P. Sadayappan. (2013). A framework for load balancing of tensor contraction expressions via dynamic task partitioning. *Proceedings of the SC'13: International Conference on High-Performance Computing, Networking, Storage and Analysis*, Denver, CO.

Lawson, C., R. Hanson, D. Kincaid, and F. Krogh. (1979). Algorithm 539: Basic linear algebra subprograms for fortran usage. *ACM Transactions on Mathematical Software* 5(3), 308–323.

Li, R., and Y. Saad. (2013). GPU-accelerated preconditioned iterative linear solvers. *Journal of Supercomputing* 63(2), 443–466. http://link.springer.com/article/10.1007%2Fs11227-012-0825-3.

Lotrich, V., N. Flocke, M. Ponton, A. Yau, A. Perera, E. Deumens, and R. J. Bartlett. (2008). Parallel implementation of electronic structure energy, gradient and Hessian calculations. *Journal of Chemical Physics* 128, 194104.

Lotrich, V. F., J. M. Ponton, A. S. Perera, E. Deumens, R. J. Bartlett, and B. A. Sanders. (2010). Super instruction architecture for petascale electronic structure software: The story. *Molecular Physics* 108, 3323.

Lyakh, D. I. (2014). Scale-adaptive tensor algebra for local many-body methods of electronic structure theory. *International Journal of Quantum Chemistry* 114, 1607–1618.

Lyakh, D. I. (2015). An efficient tensor transpose algorithm for multicore CPU, Intel Xeon Phi, and NVIDIA Tesla GPU. *Computer Physics Communications* 189, 84–91.

Lyakh, D. I. and R. J. Bartlett. (2014). Algebraic connectivity analysis molecular electronic structure theory II: Total exponential formulation of second-quantised correlated methods. *Molecular Physics* 112, 213–260.

Lyakh, D. I., M. Musial, V. F. Lotrich, and R. J. Bartlett. (2012). Multireference nature of chemistry: The coupled-cluster view. *Chemical Reviews* 112, 182–243.

Matthews, D. A. (2016). High-performance tensor contraction without BLAS. *arXiv*:1607.00291.

Morgan, R. B. and D. S. Scott. (1986). Generalizations of Davidson's method for computing eigenvalues of sparse symmetric matrices. *SIAM Journal on Scientific and Statistical Computing* 7(3), 817–825.

Murg, V., F. Verstraete, R. Schneider, P. R. Nagy, and O. Legeza. (2015). Tree tensor network state with variable tensor order: An efficient multireference method for strongly correlated systems. *Journal of Chemical Theory and Computation* 11, 1027–1036.

Nakatani, N. and G. K.-L. Chan. (2013). Efficient tree tensor network states (TTNS) for quantum chemistry: Generalizations of the density matrix renormalization group algorithm. *Journal of Chemical Physics* 138, 134113.

Nguyen, A., N. Satish, J. Chhugani, C. Kim, and P. Dubey. (2010). 3.5-D blocking optimization for stencil computations on modern CPUs and GPUs. *Proceedings of the 2010 ACM/IEEE International Conference for High Performance Computing, Networking, Storage and Analysis*, New Orleans, LA. IEEE Computer Society.

Nielson, T., A. Rivera, P. Balaprakash, M. Hall, P. D. Hovland, E. Jessup, and B. Norris. (2015). Generating efficient tensor contractions for GPUs. *ICPP'15 Proceedings of the 2015 44th International Conference on Parallel Processing*, Beijing, China, pp. 969–978.

Orus, R. (2014). A practical introduction to tensor networks: Matrix product states and projected entangled pair states. *Annals of Physics* 349, 117–158.

Oseledets, I. V. and E. E. Tyrtyshnikov. (2009). Breaking the curse of dimensionality, or how to use SVD in many dimensions. *SIAM Journal on Scientific Computing* 31, 3744–3759.

Rajbhandari, S., A. Nikam, P.-W. Lai, K. Stock, S. Krishnamoorthy, and P. Sadayappan. (2014a). A communication-optimal framework for contracting distributed tensors. *Proceedings of the SC'14: International Conference for High-Performance Computing, Networking, Storage and Analysis*. ISBN:978-1-4799-5500-8. doi:10.1109/SC.2014.36.

Rajbhandari, S., A. Nikam, P.-W. Lai, K. Stock, S. Krishnamoorthy, and P. Sadayappan. (2014b). CAST: Contraction algorithm for symmetric tensors. *Proceedings of the 2014 43rd International Conference on Parallel Processing*. doi:10.1109/ICPP.2014.35.

Riplinger, C. and F. Neese. (2013). An efficient and near linear scaling pair natural orbital based local coupled cluster method. *Journal of Chemical Physics* 138, 034106.

Riplinger, C., P. Pinski, U. Becker, E. F. Valeev, and F. Neese. (2016). Sparse maps—A systematic infrastructure for reduced-scaling electronic structure methods. II. Linear scaling domain based pair natural orbital coupled cluster theory. *Journal of Chemical Physics* 144, 024109.

Rogers, D. M. (2016). Efficient primitives for standard tensor linear algebra. *Proceedings of XSEDE16*, July 17–21, 2016, Miami, FL. doi:10.1145/2949550.2949580.

Saad, Y. (2003). *Iterative Methods for Sparse Linear Systems* (2nd ed.). Philadelphia, PA: Society for Industrial and Applied Mathematics.

Sanders, B., R. Bartlett, E. Deumens, V. Lotrich, and M. Ponton. (2010). A block-oriented language and runtime system for tensor algebra with very large arrays. *Proceedings of the 2010 ACM/IEEE International Conference for High-Performance Computing, Networking, Storage, and Analysis*, IEEE Computer Society, Washington DC, ISBN 978-1-4244-7558-9.

Sanders, B. A., J. N. Byrd, N. Jindal, V. F. Lotrich, D. I. Lyakh, A. Perera, and R. J. Bartlett. (2017). ACES 4: A platform for computational chemistry calculations with extremely large block-sparse arrays. *Proceedings of the 2017 IEEE International Parallel and Distributed Processing Symposium (IPDPS)*, Orlando FL, ISSN: 1530-2075.

Sankaran, R., J. Angel, and M. Brown. (2015). Genetic algorithm based task reordering to improve the performance of batch scheduled massively parallel scientific applications. *Concurrency and Computation, Practice and Experience* 27(17), 4763–4783. http://dx.doi.org/10.1002/cpe.3457.

Shi, Y.-Y., L.-M. Duan, and G. Vidal. (2006). Classical simulation of quantum many-body systems with a tree tensor network. *Physical Review* A 74, 022320.

Solomonik, E., D. Matthews, J. R. Hammond, and J. Demmel. (2013). Cyclops tensor framework: Reducing communication and eliminating load imbalance in massively parallel contractions. Technical Report No. UCB/EECS-2013-11, 2013. http://www.eecs.berkeley.edu/Pubs/TechRpts/2013/EECS-2013-11.html.

Solomonik, E., D. Matthews, J. R. Hammond, J. F. Stanton, and J. Demmel. (2014). A massively parallel tensor contraction framework for coupled-cluster computations. *Journal of Parallel and Distributed Computing* 74, 3176–3190.

Springer, P. and P. Bientinesi. (2016). Design of a high-performance GEMM-like tensor-tensor multiplication. *arXiv*:1607:00145.

Springer, P., J. R. Hammond, and P. Bientinesi. (2016a). TTC: A high-performance compiler for tensor transpositions. *arXiv*:1603.02297.

Springer, P., A. Sankaran, and P. Bientinesi. (2016b). TTC: A tensor transposition compiler for multiple architectures. *Proceedings of the ARRAY'16*, June 14, 2016, Santa-Barbara, CA. ISBN:978-1-4503-4384-8/16/06, doi:10.1145/2935323.2935328.

Stevenson, J. P., A. Firoozshahian, A. Solomatnikov, M. Horowitz, and D. Cheriton. (2012). Sparse matrix-vector multiply on the HICAMP architecture. *ICS '12: Proceedings of the 26th ACM international conference on Supercomputing*, Venice, Italy, pp. 195–204.

Swanson, C. and A. T. Chronopoulos. (1992). Orthogonal s-step methods for nonsymmetric linear systems of equations. *Proceedings of ICS '92, The 6th ACM International Conference on Supercomputing*, Washington, DC, pp. 456–465.

Tucker, L. (1966). Some mathematical notes on three-mode factor analysis. *Psychometrika* 31, 279–311.

Valiev, M., E. J. Bylaska, N. Govind, K. Kowalski, T. P. Straatsma, H. J. J. Van Dam, D. Wang, J. Nieplocha, E. Apra, T. L. Windus, and W. A. de Jong. (2010). NWChem: A comprehensive and scalable open-source solution for large scale molecular simulations. *Computer Physics Communications* 181, 1477–1489.

Verstraete, F. and J. I. Cirac. (2004). Renormalization algorithms for quantum many-body systems in two and higher dimensions. *arXiv:cond-mat*/0407066.

Vidal, G. (2008). Class of quantum many-body states that can be efficiently simulated. *Physical Review Letters* 101, 110501.

White, S. R. (1993). Density-matrix algorithms for quantum renormalization groups. *Physical Review* B 48, 10345.

Wu, W., A. Bouteiller, G. Bosilca, M. Faverge, and J. Dongarra. (2015). Hierarchical DAG scheduling for hybrid distributed systems. *Proceedings of the 29th IEEE International Parallel and Distributed Processing Symposium*, May 2015, Hyderabad, India.

Yamazaki, I., T. Dong, R. Solcà, S. Tomov, J. Dongarra, and T. C. Schulthess. (2013). Tridiagonalization of a dense symmetric matrix on multiple GPUs and its application to symmetric eigenvalue problems. *Concurrency and Computation: Practice and Experience* 26(16), 2652–2666.

4 Exposing Hierarchical Parallelism in the FLASH Code for Supernova Simulation on Summit and Other Architectures

Thomas Papatheodore and O. E. Bronson Messer

CONTENTS

4.1 Background and Scientific Methodology 95
    4.1.1 Type Ia Supernovae 96
    4.1.2 Core-Collapse Supernovae 98
4.2 FLASH Algorithmic Details 99
    4.2.1 FLASH Physics Modules 100
    4.2.2 Multiphysics Implementation 102
4.3 Programming Approach 104
    4.3.1 Nuclear Burning Module 104
        4.3.1.1 OpenMP Threading on Titan 105
        4.3.1.2 GPU Optimization on Titan 105
    4.3.2 EoS Module 108
4.4 Benchmarking Results 109
    4.4.1 Nuclear Burning Module 109
        4.4.1.1 OpenMP Threading 111
        4.4.1.2 GPU Optimization on Titan 113
    4.4.2 EoS Module 114
4.5 Summary 115
Acknowledgments 116
References 116

4.1 BACKGROUND AND SCIENTIFIC METHODOLOGY

Since roughly 100 million years after the big bang, the primordial elements hydrogen (H), helium (He), and lithium (Li) have been synthesized into heavier elements by thermonuclear reactions inside of the stars. The change in stellar composition resulting from these reactions causes stars to evolve over the course of their lives. Although most stars burn through their nuclear fuel and end their lives quietly as inert, compact objects, whereas others end in explosive deaths. These stellar explosions are called supernovae and are among the most energetic events known to occur in our universe. Supernovae themselves further process the matter of their progenitor stars and distribute this material into the interstellar medium of their host galaxies. In the process, they generate $\sim 10^{51}$ ergs of kinetic energy by sending shock waves into their surroundings, thereby contributing to galactic dynamics as well.

The fate of individual stars depends on their initial masses. Stars that are born with less than eight solar masses (solar mass = mass of sun = 2×10^{33} g) of material will first burn their H fuel into He and subsequently burn the He into carbon (C) and oxygen (O). If sufficiently massive, these stars can continue to burn some of the newly formed C into O, neon (Ne), and magnesium (Mg), but, due to their low masses, this is their final burning stage before the nuclear reactions cease. Without energy generation from the nuclear reactions, there is no longer a way to support the stars against their own self-gravity and they begin to collapse. This continues until densities become large enough to force electrons into higher energy states as demanded by the Pauli exclusion principle. These *degenerate* electrons become the dominant contribution to the pressure of the collapsing star, and it is this electron degeneracy pressure that finally halts the collapse. What remains after such an occurrence is a white dwarf (WD); a hot, dense, inert ball of (mostly) C and O (supported by degenerate electrons) that slowly radiates its existing thermal energy (generated from previous burning epochs) out into its surroundings. These objects compress roughly a solar mass of material into a volume of the size of the Earth, with an average density and temperature of about 10^6 g cm^{-3} and 10^7 K. Although they are inert, we will see in the upcoming sections that these WDs can end their lives in thermonuclear explosions (type Ia supernovae [SNe Ia]) under the right conditions. On the other hand, stars that are born with more than eight solar masses of material can have central densities and temperatures high enough to fuse elements as heavy as iron (Fe). After exhausting their fuel, these stars also begin to collapse; however, the electron degeneracy pressure is not sufficient to halt the inward fall of their massive cores. In such cases, the collapse continues until the inner core is compressed slightly beyond nuclear densities (here, it is the neutron degeneracy pressure that halts the collapse), where it then rebounds by accelerating the inward falling material near the core outward into the remainder of the star. This *bounce*, along with subsequent neutrino heating, powers these *core-collapse* supernovae.

4.1.1 Type Ia Supernovae

SNe Ia are commonly attributed to thermonuclear explosions of C/O white dwarfs (C/O WDs) in binary stellar systems. These thermonuclear supernovae are unique among all other supernovae because they derive their explosion energies from nuclear energy release instead of the gravitational collapse of massive stellar cores. They are responsible for approximately 50% of the Fe production in the cosmos, and therefore play a significant role in galactic chemical evolution. Additionally, they can be used as distance indicators in observational cosmology by providing unparalleled insight into the geometry and evolution of our universe.

Observations of SNe Ia show a roughly homogeneous group of events, both spectroscopically and photometrically. Their light curves rise to peak brightness in ∼15–20 days by reaching an average magnitude of −19.3 with a variation of ∼0.3 between events (Hillebrandt et al., 2013). This rise is followed by a similarly steep decline that transitions into an exponential decrease. The spectra of SNe Ia show large amounts of high-velocity (10–20,000 km s^{-1}) intermediate-mass elements (IMEs; elements between the C/O fuel and the Fe peak) at early times (Filippenko, 1997), and if detected early enough, even higher velocity (∼25,000 km s^{-1}) carbon features can be seen (Parrent et al., 2011; Silverman and Filippenko, 2012; Zheng et al., 2013). As the observations continue, the composition evolves toward heavier elements, until about 2 weeks after peak when the spectra are dominated by lower velocity (5–10,000 km s^{-1}) Fe-group elements (IGEs). It is important to note that these IGEs are dominated by ^{56}Ni (∼0.3–0.9 solar masses for normal events), and it is the radioactive decay of the ^{56}Ni (^{56}Ni $\rightarrow ^{56}$Co $\rightarrow ^{56}$Fe) that is thought to power the optical light curves (see Churazov et al., 2014 for recent evidence). The homogeneous observational properties just outlined only hold for *normal* (Branch et al., 1993) SNe Ia, which make up about 70%–75% of all events. There are also *peculiar* subclasses of SNe Ia that account for the remainder of the population (Li et al., 2011).

Although there seems to be general agreement that SNe Ia arise from exploding C/O WDs, there is still no consensus on the pathway(s) that lead to explosion. Recall that C/O WDs are very stable, inert objects with no nuclear reactions. So what causes them to explode? As mentioned earlier, it appears that a companion star is required to drive them toward explosion, and it is the nature of the companion that divides possible explosion scenarios into different categories.

In the *single-degenerate scenario* (Wheeler and Hansen, 1971; Whelan and Icko, 1973), a C/O WD accretes matter (H or He) from a nondegenerate companion star (main sequence, giant, or He-burning star). This accreted matter is burned stably to C and O on the surface of the WD, thereby increasing its mass and central density. As the WD approaches the Chandrasekhar mass, densities and temperatures near its center reach values where C-burning ignites. This ignition is followed by a phase of convective C burning that lasts for centuries (Woosley et al., 2004) before eventually leading to a runaway that incinerates the entire object. The sequence of events that follows the onset of nuclear runaway is still a matter of debate, but the mechanism that best supports the observational data seems to be a turbulent deflagration (subsonic burning front) which transitions into a detonation (supersonic burning front) at some point during the explosion (Hillebrandt and Niemeyer, 2000). Parameterized models of this scenario have been shown to reliably match light curves and elemental abundances (Kasen et al., 2009), but further investigation is needed to help establish a theoretical understanding.

In the *double-degenerate scenario*, the companion star is an another WD, and it is the eventual merger of the two WDs (via energy loss through gravitational wave emission) that leads to a thermonuclear explosion in this case (Webbink, 1984; Iben and Tutukov, 1984). Recent simulations have shown that double WD systems with mass ratios near unity can lead to violent mergers (Pakmor et al., 2010, 2011, 2012) where the secondary (lower mass) WD is torn apart just before the objects make contact, and the in-falling material directly impacts the surface of the primary. This leads to thermodynamic conditions suitable for a detonation to occur during the merger that incinerates the entire object. It is important to note that the primary WD is essentially unaltered by the merger, so the detonation occurs in a sub-Chandrasekhar mass WD (densities $\sim 5 \times 10^7$ g cm^{-3}) surrounded by the disrupted matter of the secondary. Although binary population studies (Ruiter et al., 2012) have shown this scenario to agree with the observed SNe Ia rate (assuming all mergers lead to SNe Ia), detonations in (even the most likely) violent merger simulations must be manually ignited (due to unresolved length scales) once thermodynamic conditions are believed to be suitable.

In the *double-detonation scenario*, a C/O WD stably accretes He from either a He-burning star or a He-rich WD, which builds up a layer of He on its surface. Compressional heating near the bottom of this layer, or direct impact of the accretion stream onto the WD, might lead to a He detonation. This surface detonation is believed to be capable of inducing a subsequent detonation in the underlying C/O core, which then incinerates the entire object before reaching the Chandrasekhar mass by producing a SNe Ia. This secondary detonation can be ignited either directly, as the He detonation transitions into a C detonation at the He–C/O interface (*edge-lit* case), or due to compression of the core as a converging shockwave (produced as the He detonation travels around the WD surface) meets somewhere off-center (*core-compression* case). Unlike the single-degenerate scenario, which involves a Chandrasekhar mass WD, pure detonations in these sub-Chandrasekhar mass WDs (with lower densities) are capable of producing IMEs in the outer ejecta. However, a remaining obstacle for this explosion model is the overproduction of ^{44}Ti, which is not seen in normal SNe Ia (Kromer et al., 2010).

Simulating thermonuclear supernovae is complicated by a range of length scales that spans more than 10 orders of magnitude. The progenitor WD that is to be exploded has a radius of $\sim 10^8$ cm while the detonation fronts can be less than 10^{-1} cm. Even with our most advanced computing capabilities, it is impossible to resolve these microscopic burning fronts in full-star simulations. Therefore, the shock interface is typically modeled as a discontinuity between fuel and ashes that is spread out numerically over several computational zones. The resolution of these zones is ~ 1 km in the most expensive models. In order to evolve the unresolved nuclear burning that occurs in these simulations one can employ a nuclear reaction network (of various sizes) that is solved in situ during the calculations or use an approximate model of the burning. The latter approach is computationally less expensive but the fidelity of the results can be compromised due to inaccurate feedback of nuclear energy into the fluid and the altered abundances that this implies.

4.1.2 CORE-COLLAPSE SUPERNOVAE

Core-collapse supernovae (CCSNe), the explosive final moments of massive stars, are complex, dynamic, multiphysics problems coupling all four of the fundamental forces to produce a bright explosion from the birth of a neutron star (NS) or black hole (BH). The central engine of a CCSN generates rare transient signals in gravitational waves and neutrinos, and the explosion creates and ejects many chemical elements including the primary constituents of the Earth. The CCSN problem has been an evolving computational challenge for several decades and today, we are entering an era where the well-resolved, symmetry-free, three-dimensional (3D) simulations that are required to understand these complex stellar explosions and their byproducts are now possible.

After several million years of evolution and nuclear energy release, a massive star's core is composed of iron (and similar *iron-peak* elements) from which no further nuclear energy can be released by fission or fusion. Outside the Fe-core are shells representative of the previous burning stages—a silicon shell, oxygen shell, etc., out to a helium shell surrounded by an envelope of hydrogen. At the base of the Si-shell, nuclear burning continues to grow the Fe-core below. When the mass of the Fe-core reaches the limiting Chandrasekhar mass, it starts to collapse. This collapse takes only milliseconds to compress a region, initially thousands of kilometers in size to a few tens of kilometers. The result is to raise the density of the inner core to super-nuclear densities. The equation of state (EoS) of super-nuclear mater is very stiff and resisting further compression. A shock is formed at the edge of the inner core when the realization of this density causes a sharp seizure of the collapse. This shock will eventually lead to the violent expulsion of the rest of the star, through the transfer of a modest portion of the $\sim 10^{53}$ ergs ($= 10^{46}$ J $= 100$ Bethe) change in core gravitational binding energy into the outer parts of the start to produce a 1 Bethe explosion.

Aside from their role in producing compact objects (i.e., NSs and BHs) and their role as gravitational wave sources, CCSNe are an important link in our chain of origin from the big bang to the present. They are the dominant source of elements in the periodic table between oxygen and iron (Woosley and Weaver, 1995; Thielemann et al., 1996), and there is a growing evidence that they, or one of the related deaths of massive stars (O-Ne core collapse, collapsar, etc.), are indeed responsible for producing half the elements heavier than iron (Argast et al., 2004). Observations of nuclear abundances allow nucleosynthesis calculations to place powerful constraints on conditions deep in the interior of supernovae and their progenitors, places hidden from direct observation.

Spherically symmetric models that have found above the innermost iron and nickel dominated regions; the passage of the shock leaves a layer rich in alpha isotopes: 40Ca, 36Ar, 32S, and 28Si, the products of incomplete silicon burning. Above this is a layer of 16O, in the outer portions of which, significant fractions of 20Ne, 24Mg, and 12C are found. Finally comes the helium layer and hydrogen envelope, assuming they were not driven off as part of the stellar wind. Though the passage of the shock does not grossly alter the composition of the outer layers, appreciable amounts of many rarer isotopes can be produced by the combination of the shock passage and the neutrino flux.

Convection and hydrodynamic instabilities greatly complicate this picture by destroying strict compositional layering in the progenitor. The implication is that gross asymmetries must be present in the core and be part of the mechanism itself. However, few simulations to date have directly considered the impact of this multidimensional behavior on the nucleosynthesis. For example, Nagataki et al. (1998) and Maeda et al. (2002) employed parameterized *thermal bomb* models, which demonstrated that in aspherical models a larger fraction of the ejecta experiences alpha-rich freeze out (where matter expanding outward cools, allowing the light nuclei to recombine into iron, nickel, and neighboring nuclei). Unfortunately, these simulations tracked the composition only via postprocessing, that is, no evolution of the composition was included within the hydrodynamic simulation.

While the central engine plays its role in initiating the explosion perhaps in 1 s and nucleosynthesis is largely complete within a few seconds; thereafter, the spatial and velocity distribution of the newly made isotopes continues to develop over many minutes as the shock propagates through the rest of the star. However, performing fully coupled, neutrino radiation hydrodynamics (RHD) simulations beyond roughly 1 s is computationally infeasible in multiple spatial dimensions. An alternative is to stop the RHD simulations at O (1 s) and continue the adiabatic evolution of the ejecta using only a reactive flow solver (i.e., neglecting radiation).

Both of these (and other) problems in explosive stellar astrophysics and nucleosynthesis require multiphysics approaches, where solvers for hydrodynamics, nuclear kinetics, radiation transport, and other physics are coupled self-consistently and evolved in time. Furthermore, the disparate length scales involved in the multiphysics processes in stellar astrophysics requires the resolution of fine details (e.g., small-scale instabilities) in the context of Earth and larger sized bodies. Adaptive mesh refinement (AMR), where additional resolution is dynamically allocated where and when it is needed to resolve gradients is the preferred solution for this problem. We use the FLASH code to accomplish all of these aims.

4.2 FLASH ALGORITHMIC DETAILS

FLASH is an adaptive-mesh, multiphysics simulation code that is widely used in astrophysics and is the part of the Center for Accelerated Application Readiness (CAAR) program at Oak Ridge National Laboratory (ORNL). FLASH was developed initially under the aegis of the Department of Energy (DOE)/Advanced Simulation and Computing (ASC) Academic Strategic Alliance Program (ASAP), under the Advanced Scientific Computing Initiative (ASCI, now ASC) at the Center from Astrophysical Thermonuclear Flashes (FLASH Center) at the University of Chicago (Antypas et al., 2006; Dubey et al., 2009.) The first public release of FLASH was in 2000 (Fryxell et al., 2000). Since then, the code has been used for simulations in astrophysics, cosmology, terrestrial combustion, and high-energy density experiments involving intense lasers, among other applications. According to statistics compiled by the Flash Center (http://flash.uchicago.edu/site/publications/flash_pubs.shtml), the code has been used in over 1100 publications since 2000.

The list of physics capabilities required to span all these fields is large, and FLASH is best described as a framework for multiphysics simulation that allows a user to compose an appropriate set of physics modules, tailored to the problem at hand. FLASH implements a block-structured adaptive grid, adding resolution elements in areas of complex flow. This adaptive grid capability forms the heart of the framework and is currently provided by an AMR package (PARAMESH) and a set of mixed hyperbolic–parabolic–elliptic partial differential equation (PDE) solvers on that AMR mesh. These PDE solvers are augmented by a set of ordinary differential equation (ODE) solvers for local physics (e.g., for nuclear reaction networks). All of these physics modules are incorporated via operator splitting, where the overall evolution of the fully coupled system is accomplished as a sum of individual, simpler evolution operators. These suboperators are implemented in FLASH using a variety of numerical techniques, each one appropriate to the subtask: finite-volume methods are used to evolve the hydrodynamics in the code; mutligrid and multipole methods are used to solve the Poisson equation for gravity; N-body evolution methods are used to track Lagrangian tracer particles in problems of interest. Core-collapse and thermonuclear supernova phenomenology are two problems where many (if not essentially all) of these physics modules must be brought to bear to perform realistic simulations capable of confronting observations with quantitative predictions.

FLASH contains approximately 500,000 lines of code, though not every module is compiled for each distinct application. FLASH is written primarily in Fortran, with some C and a Python script for building applications and utilizes the Message Passing Interface (MPI) library for distributed-memory parallelism. Aside from MPI, FLASH has always relied on a conspicuously small number of external libraries. This lack of external dependencies is intentional, as the code has often been used on cutting-edge platforms early in their deployment, a circumstance where libraries are often lacking. Specifically, to perform a SNe Ia simulation with FLASH requires only MPI, HDF5, and some version of LAPACK/BLAS is present on a platform.

FLASH has been under strict version control since its initial creation, including nightly regression tests for more than a decade. We believe that no other astrophysical simulation code has been subject to such careful attention to software engineering practices (Dubey et al., 2013). The FLASH code is also essentially unique among astrophysics application codes in that it has been subjected to a formal verification and validation program (Calder et al., 2002; Weirs et al., 2005). Two of the features of FLASH that have proved to be very popular among users are the quick start guide and a collection of well-documented example application setups. Many users often start with an example setup related to their particular problem. The guide and this *head start* often lead to users being able to quickly customize these example setups for their own use, which considerably reduces their initial effort. In general, FLASH development has led to the adoption of a set of software engineering practices that have been successfully used to ensure a productive software development environment (Dubey et al., 2013, 2015; Dubey and Van Straalen, 2014). These practices include strong version control, a comprehensive testing regime that includes unit, integration and system level testing, and continuous integration.

FLASH also makes extensive use of OpenMP throughout many modules. One notable exception to this—and, perhaps, the one module that would benefit most from extensive threading and vectorization on Summit—is the module designed to evolve kinetic equations governing composition and energy release (the so-called Burn unit). Among other applications, this module is used to model interstellar chemistry (i.e., to determine what we see when we look out to the Universe) and to model nuclear burning in stars and explosive events like supernovae (i.e., to determine what we and the universe around us are made of). Because the solution of these equations requires a large number of spatially localized FLOPs, threading this module via OpenMP for use on summit has the possibility of increasing simulation performance considerably. This is specifically the module that we have targeted for most of our development work under the aegis of CAAR.

We have developed two prototype setups that serve as *performance laboratories* for our optimization work. The first problem setup is the central detonation of a C/O WD in hydrostatic equilibrium, which is typical of SN Ia simulations (WD problem hereafter). The second problem setup is the late-time evolution of a CCSN that is mapped from the results of a previous early time simulation (CCSN problem hereafter). Each of these problem setups involves some of the most important and computationally intense modules of FLASH, namely hydrodynamics, self-gravity, nonpolytropic equations of state, and nuclear kinetics. We explore some of the details of each of these modules next.

4.2.1 FLASH PHYSICS MODULES

FLASH solves the compressible equations of hydrodynamics (Euler equations),

$$\frac{\partial \rho}{\partial t} + \nabla \cdot (\rho v) = 0 \tag{4.1}$$

$$\frac{\partial \rho v}{\partial t} + \nabla \cdot (\rho v v) + \nabla P = \rho g \tag{4.2}$$

$$\frac{\partial \rho E}{\partial t} + \nabla \cdot \left[(\rho E + P) v\right] = \rho v \cdot g \tag{4.3}$$

along with an advection equation for each nuclear species,

$$\frac{\partial \rho X_l}{\partial t} + \nabla \cdot \left(\rho X_l v\right) = 0 \tag{4.4}$$

In these equations, ρ, v, P, g, E, and X_l are the density, velocity, pressure, gravitational acceleration, energy, and mass fraction of the lth nuclear species of the fluid, respectively. The energy E is the sum of the internal energy and kinetic energy per unit mass. The solution is calculated using a second-order directionally split piecewise-parabolic method (PPM; Colella and Woodward, 1984), which is a higher order Godunov method (Godunov, 1959). In this method, space and time are discretized using finite volume and explicit forward differencing, respectively. The fluxes through zone boundaries are updated by solving a Riemann problem at each boundary. PPM is a higher order method because it models zone variables with piecewise-parabolic functions as opposed to piecewise-linear functions used in lower order methods. PPM is a popular choice for astrophysical hydrodynamics because it allows accurate capturing of shocks (ubiquitous in astrophysical settings) at a reasonable computational cost.

The Euler equations are closed with the addition of an EOS, which relates, for instance, the pressure of the fluid to its density and temperature. A polytropic EOS can be used to describe simple gasses:

$$P = (\gamma - 1)\,\rho \epsilon \tag{4.5}$$

where γ is the adiabatic index, ρ is the fluid density, and ϵ is the specific internal energy. However, for the problems described in this work, the use of a simple EoS is insufficient. In these dense astrophysical environments, the pressure is dominated by the contribution of degenerate electrons (and even neutrons in the case of a CCSN), so more complicated expressions must be used. Because the EOS is called many times during stellar explosion problems, the time spent in the EOS routines can become large.

A common way to reduce this time is by creating a table of data using the full, detailed EOS and then using a simpler EOS that interpolates thermodynamic quantities from the table. In FLASH, the primary example of this is the tabular Helmholtz EoS (Timmes and Swesty, 2000). It accounts for an electron–positron plasma where the electrons and positrons can have relativistic velocities and an arbitrary degree of degeneracy. The Helmholtz EOS makes use of a high-order interpolation (i.e., quintic Hermite polynomials) to determine arbitrary values of the Helmholtz free energy—along with its first and second derivatives—in order to calculate all thermodynamic quantities of interest.

The nuclear burning that occurs in stellar explosions is tracked by evolving a nuclear reaction network. A reaction network consists of a set of ODEs of the form

$$\frac{dY_i}{dt} = R_i = f\left(Y_i\right) \tag{4.6}$$

where Y_i and R_i represent the abundance and total reaction rate of the isotope i. After a single time step, the abundance of isotope i is

$$Y_i^{n+1} = Y_i^n + dY_i \tag{4.7}$$

Using backward Euler time differencing (first order), these equations can be combined to derive an expression for dY:

$$\left(\frac{1}{\Delta t} - \frac{dF(Y^n)}{dY}\right) * dY = f(Y^n) \tag{4.8}$$

which is in the form $Ax = b$. Therefore, evolving the nuclear kinetics involves building and solving a $n \times n$ matrix equation in each computational zone on the grid. This yields a change in abundances of the nuclear species along with the corresponding energy release.

The gravitational acceleration,

$$g(x) = -\nabla \varphi(x) \tag{4.9}$$

is self-consistently calculated by solving the Poisson equation,

$$\nabla^2 \varphi(\boldsymbol{x}) = 4\pi G \rho(\boldsymbol{x})$$ (4.10)

Although FLASH can solve the Poisson equation using either a multipole or multigrid algorithm, we employ the former due to the near-spherical symmetry in our problems. The multipole solver determines the gravitational potential by calculating the center of mass across the computational domain,

$$\boldsymbol{x}_{cm} = \frac{\int d^3 \boldsymbol{x} \boldsymbol{x} \rho(\boldsymbol{x})}{\int d^3 \boldsymbol{x} \rho(\boldsymbol{x})}$$ (4.11)

computing the multipole moments,

$$\frac{1}{|\boldsymbol{x} - \boldsymbol{x}'|} = 4\pi \sum_{l=0}^{\infty} \sum_{m=-l}^{l} \frac{1}{2l+1} \frac{r_<^l}{r_>^{l+1}} Y_{lm}^* \left(\theta', \varphi'\right) Y_{lm}(\theta, \varphi)$$ (4.12)

used to create the density distribution, and using the moments to determine the potential field,

$$\varphi(\boldsymbol{x}) = -\frac{\alpha}{4\pi} \int d^3 \boldsymbol{x}' \frac{\rho(\boldsymbol{x}')}{|\boldsymbol{x} - \boldsymbol{x}'|}$$ (4.13)

Each of these steps involves global MPI communication and therefore can limit the scaling above ~10K MPI ranks.

4.2.2 MULTIPHYSICS IMPLEMENTATION

Among the most important reasons to study stellar explosions are the nuclear transmutations that they effect by forging the building blocks of ourselves and our planet. The thermonuclear burning that occurs in stellar explosion models is evolved by integrating the set of Equations 4.6 through 4.8 above that represent the abundances of a set of nuclear species. For each computational zone, this involves building an $n \times n$ (n = number of species in network) set of ODEs (i.e., Jacobian and right-hand-side matrices) and solving the resulting set of equations.

A common weakness of current multidimensional models is the use of small thermonuclear reaction networks, consisting of 10–20 nuclear species, allowing the cost of the network to be comparable to the cost of the hydrodynamics. For example, the current public version of FLASH includes nuclear reaction networks with 7, 13, and 19 isotopes. In contrast, parameterized explosion models have long used much more realistic thermonuclear reaction networks, containing hundreds or even thousands of species (see, e.g., Rauscher et al., 2002), where physical fidelity in the thermonuclear kinetics is realized by reducing the fidelity of the explosion model itself. While the use of the small networks has been partially ameliorated by postprocessing analysis of the nucleosynthesis (generally based on tracer particles) using realistic networks, such postprocessing studies cannot redress errors in the energy generation rate from the small network nor include the full effects of mixing.

Simplified networks fail to accurately describe both the composition and energy distribution of supernovae ejecta as directly observed. This deficiency has been recognized for some time, leading to the development of postprocessing schemes to obtain detailed abundances. In postprocessing, a thermodynamic profile generated by tracer particles and a reduced network (in-situ with the neutrino RHD) is used to evolve a larger nuclear network. A major limitation of this approach is the accuracy of the rate of nuclear energy released by the smaller in-situ network within the hydrodynamics. Since the nucleosynthesis depends on the thermodynamic conditions, and consequently the nuclear energy generation, a feedback exists that cannot be captured with postprocessing, significantly affecting the abundances of species such as 44Ti 57Fe, 58Ni, and 60Zn (Hix and Thielemann, 1999). Another principal limitation is realized when using one of the most popular types of reduced network, an alpha network. In an alpha network, only species that are built from units of two protons and two neutrons from 4He to 56Fe (i.e., so-called alpha nuclei, e.g., 12C, 16O, 28Si, etc.) are included. The primary problem with this network model is the inability to follow the effects of electron or neutrino capture in neutronization, wherein an electron and proton combine to form a neutron and release neutrinos. This causes the electron fraction (Ye = Z/A), and, concomitantly, the electron pressure to be miscalculated. Finally, postprocessing is not capable of capturing the observed mixing of the chemical elements due to the lack of coupling to the hydrodynamics.

A network size of approximately 150 is the next logical step in nucleosynthesis calculations (see Figure 4.1), as it encompasses a significant fraction of elemental abundances and energy-producing reactions important to both the CCSNe and SNe Ia problems, allowing proper neutronization and a much more accurate rate of nuclear energy generation. The inclusion of 150-species networks in the FLASH burners is the primary focus of the initial part of our code porting efforts. However, without *accelerated* nuclear networks, the evolution of the supernova to days and weeks, using in situ networks, will simply not be possible. Instead, we will have to fall back on the postprocessing of Lagrangian tracer particles with large networks. It has recently been shown (J. A. Harris, private communication) that such postprocessing in the CCSN problem can lead to significant differences in the amount of predicted 44Ti produced in the event. The production of this radioactive element is important, as it is a prime marker of supernova nucleosynthesis given the narrow gamma-ray lines produced by its decay. Small networks significantly overproduce this isotope (i.e., far in excess of what is actually observed), as the flow in the network must pass through 44Ti on the way to the iron peak. A larger 150-species network allows flows around 44Ti to populate the iron peak, markedly reducing the simulated abundance of this easily observed species.

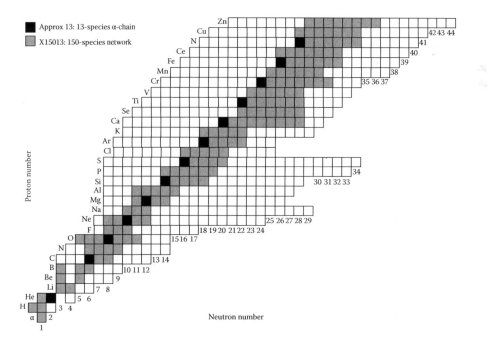

FIGURE 4.1   The participating species in an alpha network (black) and an expanded 150-species network (gray) in the N-Z plane.

4.3 PROGRAMMING APPROACH

As mentioned above, FLASH makes extensive use of OpenMP in most parts of the code, along with MPI throughout. Important for our purposes, the current data layout used in FLASH for the variables representing the fluid fields (i.e., those variables evolved with the hydro unit) inhibits vectorization, as the fields are next to each other in memory. A straightforward remedy would be to restructure the data for the hydro module so that the unit subroutines operate on multiple grid points at a time. This would effectively change the data layout from array of structures (AofS) to structure of arrays (SofA) by using a separate array for each fluid field. This tension between what might be referred to as a *logical* data ordering (i.e., one where all the data for a given block of grid zones is held contiguously in memory) and a *vectorizable* ordering (i.e., an ordering where individual field variables across spatial zones are stored contiguously) is a common one throughout FLASH. All FLASH physics modules are designed to operate on grid blocks as a fundamental unit, making the "logical" ordering natural in some sense. Regardless, for most modules, this can be overcomed and data can be placed in "vectorizable" order through local memory copies. Effecting a change to the *global* data layout is really only important for physics modules such as hydrodynamics, where the computations done in each grid are not independent (i.e., they depend on neighboring grid zones).

FLASH was designed from inception to rely on as few libraries as possible, in an effort to increase code portability. Nevertheless, for our work on the burn unit, an important exception to this rule is the use of optimized linear algebra subroutines, for example, accelerated versions of LAPACK and BLAS routines (e.g., MAGMA or cuBLAS). It should be noted here that FLASH has long exploited only a limited set of linear solvers, as the number of available burners has been small. Because of this, many users have made use of the GIFT or MA28 (sparse) solvers packaged directly with the code.

We seek to expand and generalize the list of available burners, relying on accelerated versions of DGESV to maximize performance. Given the relatively small size—but large number—of linear solves associated with the burn unit, we have chosen to use OpenMP to build the Jacobians on the central processing unit (CPU), and rely on batched versions of the linear solvers on the graphics processing unit (GPU) to achieve good performance. We have ported the nuclear burning module to run on GPUs and installed it into the FLASH code. We have also obtained preliminary results from accelerating the Helmholtz EoS module. In what follows, we will discuss the restructuring that was required in each of these modules as well as the methods used for offloading work to the GPUs.

4.3.1 Nuclear Burning Module

The nuclear burning module is responsible for evolving the composition and calculating the energy release within each grid zone in FLASH. In this module (and throughout many of the physics modules in FLASH), there is a set of nested DO loops used to iterate over these grid zones within all blocks local to an MPI rank (see Figure 4.2). Each iteration of this set of loops sends a single zone of thermodynamic (temperature, density) and composition (mass fractions of all species) data to the burner. Inside the burner, these data are used to build an $n \times n$ (n: number of species in the network*) system of ODEs representing the evolution of the n species. The solution of this system gives the updated composition and nuclear energy release for the zone. Without any threading, this process of building and solving each system of ODEs is performed sequentially for each zone. As we will see in the next section, when using large nuclear reaction networks ($n >$ tens of species), the time spent in the nuclear burning routines can dominate the runtime, so it is necessary to speed up the calculations.

---

* As mentioned above, $n = 150$ for all simulations in this work.

```
do blockID = 1, blockCount

  do k = zzone_min, zzone_max
    do j = yzone_min, yzone_max
      do i = xzone_min, xzone_max

        Call burner for a single zone within a block

      end do
    end do
  end do

end do
```

FIGURE 4.2   Nested loop structure used to iterate over all zones in each of the blocks local to an MPI rank.

4.3.1.1   OpenMP Threading on Titan

In order to reduce the time spent computing the nuclear kinetics (and therefore the overall simulation time when using large networks), we added OpenMP threading to the set of nested loops in Figure 4.2. We tested two threading models, where one we threaded the block loop (threadBlockList_CPU) and where in other, we threaded the zone loop corresponding to the dimensionality of the problem (e.g., j loop for our 2D setup; threadWithinBlock_CPU). For the former, each thread iterates over all zones within a single block, *burning* each zone in turn. The number of threads determines the number of blocks iterated over concurrently. For the latter, the zones within each block are distributed among the available threads, so each thread *burns* its portion of the zones sequentially (but in parallel with other threads). For example, if 16 OpenMP threads are used, each thread is responsible for burning 16 of the 256 zones* in the block (i.e., a row of zones).

4.3.1.2   GPU Optimization on Titan

In order to take advantage of the GPUs on Titan, we used a combination of OpenMP threading, CUDA for memory management, and batched cuBLAS library calls. Each OpenMP thread uses a batched library call to solve multiple systems of ODEs (i.e., burn multiple zones) in parallel on the GPU. We tested three different models by combining these layers of node-level parallelism. For each model, we initialize our simulations by allocating (for each thread) memory on the host and pinning it to a corresponding memory allocation on the GPU using cudaHostMalloc and cudaMalloc, respectively (pinned memory is required to allow asynchronous data transfers). This block of memory is allocated for the data structure that holds the num_zones (number of zones of thermodynamic and composition data sent to the burner each hydrodynamic time step—which differs for each of our methods) $n \times n$ systems of ODEs that will be solved (in batches) on the GPUs during the evolution.

For our first model, we used OpenMP threading over the block loop (threadBlockList_GPU). In this version, each thread fills arrays with thermodynamic and composition data for all zones in a block and sends these arrays (entire *block of zone data*) to the burner (Figure 4.3). Inside the burner, this data is used to build a system of ODEs (representing the evolution of the composition) for each zone, yielding an array of systems of ODEs. This array is the data structure, that is, the allocated memory for (on host and device) at the beginning of the simulation.† Each OpenMP thread then loops over its

---

* In 2D, each block contains 16×16 zones.

† Because we send an entire block of zones to the burner (256 zones in 2D), the size of this data structure is num_zones $\times$ $n \times n \times$ size of double $= 256 \times 150 \times 150 \times 8$ bytes ~46 MB. Recall, however, that the initial allocation of memory was performed by each thread so that each thread can work on a different block. Therefore, the total size of these data structures (per MPI rank) in this method is OMP_NUM_THREADS $\times$ 46 MB–737 MB when using 16 OpenMP threads.

```
!$omp parallel do
do blockID = 1, blockCount

  do k = zzone_min, zzone_max
    do j = yzone_min, yzone_max
      do i = xzone_min, xzone_max

          Fill arrays with thermodynamic
          and composition data for all zones
          within a single block

      end do
    end do
  end do

  Send entire "block of zones" (arrays) to burner

    In burner, array (zone) data are used to build
    system of ODEs for each zone. Then, the (256)
    systems are sent to the GPU in batches (of 16),
    solved using batched LU decomposition, and the
    results for each block are sent back to the host.

  do k = zzone_min, zzone_max
    do j = yzone_min, yzone_max
      do i = xzone_min, xzone_max

          Update main grid data structure with
          output from burner for a single block

      end do
    end do
  end do

end do
!$omp end parallel do
```

FIGURE 4.3   ThreadBlockList_GPU loop structure.

array with a stride of length batch_size, and each "batch of ODEs" is (on separate CUDA streams) sent to the GPU, solved using a batched LU decomposition (cuBLAS), and the results are sent back to the host.

The optimal number of OpenMP threads versus batch size was determined by measuring the performance when using different combinations of these parameters for this version of the burner (threadBlockList_GPU). The results of these tests showed that using 16 OpenMP threads and a batch size of 16 yielded the shortest time to completion. Using these values, we can give a more specific example of the batching within the burner for this model. The data structure is an array holding 256 150×150 matrices (systems of ODEs). Each thread sends batches of 16 matrices from its host data structure to the corresponding position within the data structure in GPU memory, solves each batch of systems on the GPU, and sends the results back to the host. Each of these sequences of send, solve, and receive (for a batch of matrices) are carried out asynchronously on separate CUDA streams by each OpenMP thread (i.e., each of the 16 OpenMP threads uses 16 CUDA streams to send, solve, and receive a batch of 16 systems of ODEs; 256 total CUDA streams). Using asynchronous data transfers on multiple CUDA streams allows us to overlap computation on the GPU with data transfers between the host and device. We also tested this version of the burner using dynamic OpenMP thread scheduling (threadBlockList_GPU_dyn) instead of the default static scheduling.

For our next method (threadWithinBlock_GPU), we again fill arrays with thermodynamic and composition data for all zones in a block, but instead of sending the entire arrays to the burner, we send batches of the arrays based on OpenMP thread ID (Figure 4.4). Here we adopt the optimal value of 16 for both batch size and number of OpenMP threads based on our previous results. Inside the burner, this data is used to build a system of ODEs for each zone, yielding an array

```
do blockID = 1, blockCount

  do k = zzone_min, zzone_max
    do j = yzone_min, yzone_max
      do i = xzone_min, xzone_max

        Fill arrays with thermodynamic
        and composition data for all zones
        within a single block

      end do
    end do
  end do

end do

!$omp parallel
Send batches of the arrays to the burner
based on thread ID

  In burner, array(zone) data are used to build
  System of ODEs for each zone. Then, the (16)
  Systems are sent to the GPU in a batch (of 16),
  Solved using batched LU decomposition, and the
  Results for each batch are sent back to the host.

!$omp end parallel

do blockID = 1, blockCount

  do k = zzone_min, zzone_max
    do j = yzone_min, yzone_max
      do i = xzone_min, xzone_max

        Update main grid data structure with
        output from burner for a single block

      end do
    end do
  end do

end do
```

FIGURE 4.4   ThreadWithinBlock_GPU loop structure.

of 16 $n \times n$ systems of ODEs.* After building its 16 $n \times n$ systems of ODEs, each thread asynchronously sends these systems from the host to the GPU, solves them on the GPU, and receives the results back on the host. Here, we used one CUDA stream per OpenMP thread (16 total CUDA streams).

Because the threadWithinBlock_GPU version only sends one batch of (16) zones to the GPU per thread, dynamic thread scheduling is not advantageous. Therefore, in our next version of the burner (threadWithinBlock_GPU_dyn), we instead fill arrays with thermodynamic and composition data from *all* zones within *all* blocks local to an MPI rank (Figure 4.5). We then iterate over the arrays with a stride of batch_size, sending each batch of (16) zones to the burner to be treated in the same manner as in threadWithinBlock_GPU. Here, however, we gain the advantage of effectively collapsing the block loop and being able to use dynamic thread scheduling.

4.3.2 EoS Module

In order to maintain thermodynamic consistency during each time step, the EoS must be called after any change is made to the thermodynamic or composition variables in a zone. This means that within each physics module, there are many calls to the EOS. Due to the number of calls to the EOS throughout simulations of astrophysical events, reducing the time spent in the EOS is important. This is especially true for simulations that do not require large reaction networks or perhaps do not require any nuclear burning. Collaborators at Stony Brook University have already developed a version of the Helmholtz EOS accelerated through the use of OpenACC directives, which we are installing into FLASH (Jacobs et al., 2016). The source code for this version of the EOS is part of a shared repository of microphysics modules (Starkiller) that can run in FLASH, as well as BoxLib-based codes such as CASTRO and MAESTRO.

In order to understand the best way to use this accelerated version of the EOS in FLASH, we created a driver program that mimics the time stepping and AMR block/zone structure used in FLASH. The EOS itself stores thermodynamic and composition data (temperature, density, pressure, composition, etc.) for a single zone within a Fortran structure. The driver program uses an array of these structures to store the conditions in many grid zones. For each time step, the driver fills each element (structure representing a single zone) of the array with mock grid data and then calls the EOS on each element (zone), which updates each structure (zone) with other thermodynamic quantities. These quantities are computed by using biquintic interpolation in a table of the Helmholtz free energy.

Because each zone can be interpolated independently of the others, the parallelization strategy is to calculate one zone per CUDA thread. The main parts of the program flow are as follows. At the beginning of the run, we allocate the main data array on the host and device (this array will remain in memory for the duration of the program, and will be used to pass zone data back and forth between host and device). We then read in the tabulated Helmholtz free energy data and make a copy that resides in device memory. Then, for each time step, we update the device with new mock grid data, call the EOS (launch the kernel) on each array element (zone), and update the host with the newly calculated quantities. Within the kernel (EOS call), the terms needed to construct the interpolation are calculated in separate functions decorated with acc routine. This maps to the GPU so that each interpolation of a zone occurs on a separate CUDA thread.

---

* In this case, because we only send a batch of 16 (of the total 256) zones to the burner, the size of the data structure is num_zones $\times n \times n \times$ size of double $= 16 \times 150 \times 150 \times 8$ bytes $\sim$3 MB. Recall, however, that the initial allocation of memory was performed by each thread so each thread can work on a different batch of the block. Therefore, the total size of these data structures (per MPI rank) in this model is OMP_NUM_THREADS $\times$ 3 MB $\sim$46 MB when using 16 OpenMP threads.

```
do blockID = 1, blockCount

  do k = zzone_min, zzone_max
    do j = yzone_min, yzone_max
      do i = xzone_min, xzone_max

          Fill arrays with thermodynamic
          and composition data for all zones
          of all blocks local to an MPI rank

      end do
    end do
  end do

end do

!$omp parallel do
do m = 1, total_num_zones, batch_size

  Send batches of the arrays to the burner

    In burner, array (zone) data are used to build
    system of ODEs for each zone. Then, the (16)
    systems are sent to the GPU in a batch (of 16),
    solved using batched LU decomposition, and the
    results for each batch are sent back to the host.

end do
!$omp end parallel do

do blockID = 1, blockCount

  do k = zzone_min, zzone_max
    do j = yzone_min, yzone_max
      do i = xzone_min, xzone_max

          Update main grid data structure with
          output from burner for all zones of all
          blocks local to an MPI rank

      end do
    end do
  end do

end do
```

FIGURE 4.5   ThreadWithinBlock_GPU_dyn loop structure.

4.4   BENCHMARKING RESULTS

4.4.1   NUCLEAR BURNING MODULE

In order to benchmark the performance of the accelerated burner in the context of the WD problem, we first ran a 2D axisymmetric simulation from 0 to 1.5 s of simulation time (i.e., a typical timescale for a SN Ia explosion). Because the number of adaptive-mesh blocks, as well as the amount and locality of burning within these blocks changes over time, we then ran several shorter simulations at different epochs of the explosion. These simulations were each started from a checkpoint file written out during the initial full simulation (at 0.2, 0.4, 0.6, 0.8, 1.0, and 1.2 s of simulation time) and ran for 20 hydrodynamic (global) time steps. Figure 4.6 shows the performance results from the simulation started at 0.6 s with no threading. Here, we can see that the nuclear burning module accounts for ~99% of the total run time, demonstrating how using large reaction networks can set the timescale

```
===================================================================================
    seconds in monitoring period :           9938.890
       number of evolution steps :                 20
-----------------------------------------------------------------------------------
accounting unit                      time sec  num calls   secs avg   time pct
-----------------------------------------------------------------------------------
initialization                         15.800          1     15.800      0.159
 IO_readCheckpoint                      2.744          1      2.744      0.028
 amr_refine_derefine                    0.054          1      0.054      0.001
 guardcell Barrier                      0.001          1      0.001      0.000
 guardcell internal                     0.246          1      0.246      0.002
 restrictAll                            0.277          1      0.277      0.003
 writePlotfile                          0.149          1      0.149      0.002
writePlotfile                           0.148          1      0.148      0.001
evolution                            9922.941          1   9922.941     99.840
 cosmology                              0.000         80      0.000      0.000
 hydro                                 75.929         40      1.898      0.764
  guardcell Barrier                     0.542         80      0.007      0.005
  guardcell internal                   18.010         80      0.225      0.181
  hy_ppm_sweep                         57.376         80      0.717      0.577
   eos gc                               0.008       6080      0.000      0.000
   hy_block                            21.150       6080      0.003      0.213
   hy_ppm_updateSoln                    1.877      12160      0.000      0.019
   Grid_conserveFluxes                 12.913         80      0.161      0.130
   eos                                 20.560       6080      0.003      0.207
 sourceTerms                         9825.216         40    245.630     98.856
  burn                               9825.216         40    245.630     98.856
   guardcell Barrier                    0.483         40      0.012      0.005
   guardcell internal                   9.665         40      0.242      0.097
   bn_burner                         9799.192     777282      0.013     98.594
 Particles_advance                      0.000         40      0.000      0.000
 gravity Barrier                        0.001         40      0.000      0.000
 gravity                                0.487         40      0.012      0.005
  Looped Moment All Reduce              0.171         40      0.004      0.002
 IO_output                             13.936         40      0.348      0.140
  diagnostics                           0.168         40      0.004      0.002
  checkpointing                        13.714         10      1.371      0.138
   writeCheckpoint                     13.713         10      1.371      0.138
    restrictAll                         2.647         10      0.265      0.027
 Grid_updateRefinement                  7.259         20      0.363      0.073
  tree                                  7.258         20      0.363      0.073
   amr_refine_derefine                  1.691         10      0.169      0.017
   markRefineDerefine                   3.231         10      0.323      0.033
    guardcell Barrier                   0.000         10      0.000      0.000
    guardcell internal                  3.207         10      0.321      0.032
     eos gc                             0.799         10      0.080      0.008
   updateData                           0.000         10      0.000      0.000
  updateParticleRefinement              0.000         10      0.000      0.000
 compute dt                             0.113         20      0.006      0.001
===================================================================================
```

FIGURE 4.6   Performance results from a simulation started at 0.6 s and ran for 20 hydrodynamic time steps, showing that the nuclear kinetics (marked as burn) account for ~99% of the total simulation time.

of the overall simulation. In the following subsections, we will discuss the results of the same set of simulations (started from checkpoint files at 0.2, 0.4, 0.6, 0.8, 1.0, and 1.2 s) performed with our OpenMP and GPU-accelerated versions of the burner.

4.4.1.1 OpenMP Threading

A comparison of our two threading models using OpenMP (CPU only) showed that the threadWithinBlock_CPU version of the burner was faster at all epochs of the simulations, so we chose this version to further optimize (but see Section 4.4.1.2 for comparison with all other models). The solid line in Figure 4.7 shows the results of the simulation that was started at 0.6 s and used the threadWithinBlock_CPU version of the burner. Here, the same simulation was run using 1, 2, 4, 8, and 16 OpenMP threads. Using this model, we obtained a speedup of ~8× when using 16 threads.

To further decrease the evolution time, we also improved the load balancing scheme used in FLASH (Calder et al., 2000). When a block is refined, the new child blocks are temporarily placed at the end of rank's block list. When a block is de-refined, sibling blocks are simply removed from the structure. Memory locations of those removed blocks are overwritten by packing the list of remaining on-rank blocks. After all refinements and de-refinements are completed, the redistribution of blocks is performed using a Morton space-filling curve (Warren and Salmon, 1993). (Other space-filling curves, which in principle have better spatial locality properties [Hilbert and Peano curves; Sagan, 1994], have been tried in the FLASH code with no measurable improvement in performance.) The Morton number is computed by interleaving the bits of the integer coordinates of each block. The Morton ordering in FLASH is such that no pattern of refinement or de-refinement can result in a block with a Morton number lower than its parent nor larger than the Morton number of its neighbor. Thus, a global sort to restore a Morton ordered list is unnecessary, and the redistribution can be done entirely with local sorts and nearest neighbor communications (Calder et al., 2000). Each block can be assigned a work value that can change as a function of time and/or position and can be used to weight the blocks. The total list, in Morton order, is cut into a number of pieces equal to the number of MPI ranks, with each piece having roughly the same amount of work. Blocks are

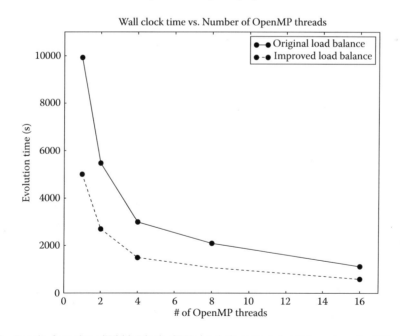

FIGURE 4.7 Results from threadWithinBlock_CPU simulations started at 0.6 s and run for 20 hydrodynamic time steps, showing the evolution time when using 1, 2, 4, 8, and 16 OpenMP threads—with and without improved load balancing.

then moved between MPI ranks such that each rank receives roughly equal amounts of work. The amount of communication is highly problem dependent. Simulations with quasi-static features can result in almost no data movement during the redistribution step. Conversely, problems where sharp gradients propagate over a large part of the computational domain (e.g., a quasi-spherical shock front propagating from a central explosion) can result in more than 50% of the blocks being moved in a redistribution event.

When simulating stellar explosions, there is often a natural load imbalance imposed by the presence of a burning front moving through the stellar material. Burning that occurs within this front takes many more subtime steps* than burning that occurs behind it or ahead of it (where no burning occurs). However, in the public version of FLASH all blocks are weighted evenly (i.e., the weight function is 1) so they are distributed evenly among the MPI ranks. Therefore, if one MPI rank is assigned several blocks where a significant amount of burning takes place and another MPI rank is assigned only blocks where no burning occurs, then a load imbalance will be present. To improve this load balancing, we weighted each block based on the maximum number of burning time steps taken by any zone within it. By doing so, MPI ranks assigned blocks with many burning time steps are given fewer total blocks to compute. Figure 4.8 illustrates this improvement by showing the time that each of the 8 MPI ranks spent in the burning routines versus the total number of burning time steps taken by each rank. We can see that without the improvement (circles in Figure 4.8) some MPI ranks took many more burning time steps than others (because they were assigned many blocks within and behind the burning front). We can also see that with the improved load balancing (triangles in Figure 4.8), we were able to reduce the maximum number of subtime steps taken by any one MPI rank, which resulted in a shorter runtime. With this improvement, we gained an additional ~2× speedup relative to the original load balancing (dashed line in Figure 4.7).

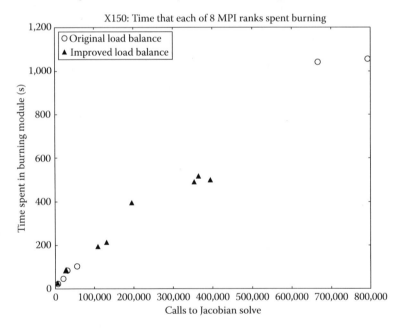

FIGURE 4.8 Time spent in nuclear burning routines versus the number of calls to the ODE solver (proxy for number of burning time steps). Simulation started from 0.6 s using the threadWithinBlock_CPU version of the burner—with and without improved load balancing. Each point represents an MPI rank (each with 16 threads)—i.e., 8 MPI ranks per run.

---

* The burning module takes many subtime steps (burning time steps) to evolve through a single hydrodynamic (global) time step.

4.4.1.2   GPU Optimization on Titan

Here, we will compare the results of our GPU-accelerated versions of the burner with the CPU-only (OpenMP) versions. Figure 4.9 shows a comparison of the time spent in the burner when evolving through 20 hydrodynamic time steps—at the six different simulation epochs—when using different versions of the burner. We will refer to this figure throughout this subsection. It is probably easiest to read from this figure by noting that solid lines are CPU only, dashed lines are GPU with static OpenMP thread scheduling, and dotted lines are GPU with dynamic OpenMP thread scheduling.

As mentioned above, the zone-threaded CPU-only version of the burner (threadWithin-Block_CPU; solid line with square markers) was faster than the block-threaded CPU-only version (threadBlockList_CPU; solid line with circle markers) at all epochs of the explosion. In fact, thread-WithinBlock_CPU was comparable with some of the GPU-enhanced versions. The block-threaded GPU version (threadBlockList_GPU; dashed line with triangle markers) was faster than its CPU counterpart (threadBlockList_CPU) but was the slowest of the GPU versions. However, by using dynamic OpenMP thread scheduling, the performance of the block-threaded GPU version (thread-BlockList_GPU_dyn) was much better. This performance increase is due to ameliorating a finer grained aspect of load imbalance inherent in these problems. Because the zones within some blocks take many more burning time steps than others, an OpenMP thread can stall an entire MPI rank if one of its blocks takes the longest to complete. However, when using dynamic threading, if a thread gets stalled by one block, the remaining threads can pick up the additional blocks while the stalled thread finishes. This version has mixed success at outperforming threadWithinBlock_CPU at different explosion epochs. The zone-threaded GPU version (threadWithinBlock_GPU; dashed line with star markers) had similar performance to threadWithinBlock_CPU, while its counterpart with dynamic thread scheduling was faster than threadWithinBlock_CPU at almost all epochs (same timing at 0.2 s). Recall that this variation of the burning module collapses the block/zone loop when using along with dynamic OpenMP thread scheduling. By collapsing the block loop, it is only possible for a thread to get held up within a batch of zones as opposed to an entire block of zones. The

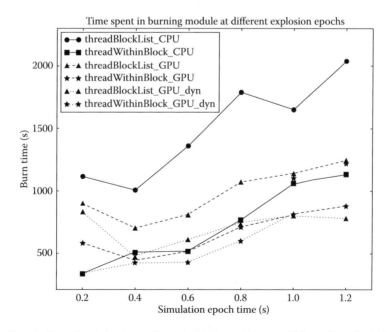

FIGURE 4.9   Results from simulations started at 0.2, 0.4, 0.6, 0.8, 1.0, and 1.2 s and run for 20 hydrodynamic time steps using different versions of the nuclear burning module.

dynamic thread scheduling ensures that if a thread does stall due to a large number of burning time steps within a zone, the other threads can pick up the remaining work.

4.4.2   EoS Module

In order to benchmark the EOS test problem, we ran simulations using a single CPU thread for increasing problem sizes.

The problem size was altered by using 32, 64, 128, or 256 blocks (range of blocks per MPI rank that might be encountered in a typical 2D supernova simulation using FLASH), each with 256 zones per block. We chose to use a single CPU thread for our benchmark because the public FLASH code (with no threading) uses a single core to compute all of the EOS calls (i.e., in all the zones) local to an MPI rank. However, we also wanted to know how our GPU-accelerated versions of the EOS compared to a CPU-threaded version that uses the full CPU capability of a node, so we ran the same set of simulations using 16 OpenMP threads to compute the interpolations in all zones. The results of these CPU-only simulations can be seen as the solid lines in Figure 4.10.

We then ran the set of simulations (32, 64, 128, 256 blocks) using the OpenACC version of the EOS (128 CUDA threads per thread block were used for all GPU versions of the EOS—Note: these *thread blocks* should not be confused with the *AMR blocks*). The results of these simulations can be seen as the dashed lines in Figure 4.10. Our original OpenACC implementation can be seen as the dashed line with circle markers. Figure 4.11 (top panel) shows that data transfers dominate the runtime for this problem. In order to reduce the data transfer times, we allocated pinned memory for the main data structure in the next implementation, resulting in a ~2× speedup relative to the first version with pageable memory (dashed line with square markers). In addition, only a subset of the structure variables (for each zone) are needed to interpolate values from the table, so we added an array that held a reduced set of variables to pass data to the device. All structure variables (within each element of the main data array) are filled for each zone within the EOS (kernel call) so we need to send the entire array back from the GPU to the host. By sending only a limited set of the thermodynamic and composition data to the GPU for each zone, we were able to speed up the

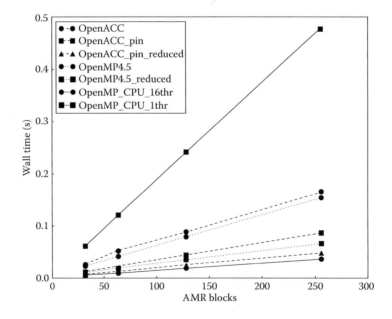

FIGURE 4.10   Comparison of timing results when using the CPU-only (1 and 16 threads), OpenACC, and OpenMP4.5 (with offloading) versions of the EOS code for different problem sizes.

```
FULL DATA SET --------------

Time(%)        Time    Calls        Avg        Min        Max  Name
 46.74%    40.870ms       96   425.73us   2.4960us   8.0196ms  [CUDA memcpy HtoD]
 40.59%    35.491ms        5   7.0983ms   7.0292ms   7.1722ms  [CUDA memcpy DtoH]
 12.68%    11.086ms        5   2.2171ms   2.2148ms   2.2202ms  eos_driver_26_gpu


REDUCED DATA SET ----------

Time(%)        Time    Calls        Avg        Min        Max  Name
 72.05%    35.481ms        5   7.0962ms   7.0295ms   7.1663ms  [CUDA memcpy DtoH]
 21.92%    10.795ms        5   2.1589ms   2.1541ms   2.1641ms  eos_driver_29_gpu
  6.03%     2.9690ms      97   30.607us   2.4960us   361.41us  [CUDA memcpy HtoD]
```

FIGURE 4.11 Timing results when sending full data set to and from the GPU (top) and when sending a reduced set of data to GPU and the full data set back from the GPU for the 256 block case. Host-to-device transfers are marked as HtoD, device-to-host transfers are marked as DtoH, and computing the kernel is labeled as eos_driver.

program by an additional ~2× (dashed line with triangle markers in Figure 4.10). Figure 4.11 shows the timing for host-to-device data transfers (HtoD), device-to-host data transfers (DtoH), and kernel execution. We can see that before reducing the amount of data sent to the GPU, the HtoD time was the largest percentage of time. By sending only a subset of data to the GPU, we reduced the HtoD time by 90%, leaving it as the smallest percentage of time.

In addition to our OpenACC implementation, we have also ported this EOS code to target GPUs using OpenMP4.5. We successfully compiled and ran the program with both the xlf and xlflang compilers on Power8+NVIDIA Pascal machines. In order to compare with our OpenACC version, we ran the same set of simulations (32, 64, 128, 256 blocks) using OpenMP4.5 on Titan, where we used the cray compiler. Figure 4.11 shows the results of the simulations run on Titan when using the full and reduced data sets (dotted lines). We were unable to make comparisons using pinned memory because an OpenMP4.5 implementation was not yet available.

Here, we should note that although we did not outperform the 16-thread (CPU-only) OpenMP version of the code, we have not utilized the full capability of the node (16 CPUs + GPU). In the final implementation of the GPU-accelerated EOS that will be used in FLASH, we will use all CPU cores on a node, each launching kernel calls from separate CUDA streams—similar to the method used in the nuclear burning module. However, even with a single core hosting the GPU, we have improved performance to within ~30% of the runtime of the CPU-OpenMP version that uses all of the *CPU* compute power of the node. Furthermore, the summit architecture will allow for additional reduction in runtime by decreasing the data transfer times (increased bandwidth and unified memory) and increased computational performance (multiple later generation GPUs per node).

4.5 SUMMARY

Accelerated versions of a large-network burning module and a degenerate matter EoS have put multi-petaflop scale, high-fidelity simulations of the nucleosynthesis from both core-collapse and thermonuclear supernovae nearly within reach. Work remains to be done on the other two major multiphysics components for these problems, namely the stellar hydrodynamics module and the multipole gravity solver. Though the gravity solver's performance is dominated by network performance (particularly the performance of collective operations), considerable performance increase can be realized at the node level in the hydrodynamics through the introduction of an accelerated Riemann solver for the PPM solution. However, close attention will have to be paid to memory layout for this module, as indicated earlier.

Though accelerating the hydrodynamics solver will have a modest effect on a large-network reactive flow simulation for supernovae (i.e., the network solve will likely continue to dominate the computational intensity), such acceleration would likely prove useful in other problem contexts. Precisely, this realization—that is, FLASH is used by many users for many other kinds of multiphysics problems—provides strong motivation for us to ensure that our performance improvements are as portable as possible. FLASH will be run on other near-future platforms besides summit (e.g., Intel Phi-based machines), as well as dozens to hundreds of small clusters in institutes and universities worldwide. As we carry our improvements forward, our initial experiences with OpenMP 4.5 have convinced us that using this particular programming model for future improvements is prudent. However, OpenMP still lacks a standardized way to perform memory placement and layout on GPUs in a portable manner. We hope that this situation will resolve in the near future, helping to ensure that our efforts in this regard will not be hamstrung by a lack of portable memory management.

ACKNOWLEDGMENTS

This work used resources of the Oak Ridge Leadership Computing Facility (OLCF) at the Oak Ridge National Laboratory, which is supported by the Office of Science of the U.S. Department of Energy under Contract No. DE-AC05-00OR22725. We thank IBM for their hospitality, expertise, and access to early hardware during a Hackathon hosted at the T.J. Watson Research Center. The authors acknowledge the contributions of our CAAR coinvestigators and other FLASH collaborators, including Sean Couch, Alan Calder, Petros Tzeferacos, and Anshu Dubey. We also thank our collaborators in the Starkiller shared microphysics project: Mike Zingale, Adam Jacobs, and Max Katz.

REFERENCES

Antypas, K. B., Calder, A. C., Dubey, A., Gallagher, J. B., Joshi, J., Lamb, D. Q., Linde, T., et al. FLASH: Applications and future. In A. Deane et al., *Parallel Computational Fluid Dynamics 2005: Theory and Applications* (2006). Amsterdam: Elsevier. 235.

Argast, D., Samland, M., Thielemann, F. -K., and Qian, Y. -Z. Neutron star mergers versus core-collapse supernovae as dominant r-process sites in the early galaxy. *Astron. Astrophys.*, 416 (2004): 997–1011.

Branch, D., Fisher, A., and Nugent, P. On the relative frequencies of spectroscopically normal and peculiar Type Ia supernovae. *Astrophys. J.*, 106 (1993): 2383.

Calder, A. C., Curtis, B. C., Dursi, L. J., Fryxell, B., MacNeice, P., Olson, K., Ricker, P., et al. High performance reactive fluid flow simulations using adaptive mesh refinement on thousands of processors. In *Proceedings of the 2000 ACM/IEEE conference on Supercomputing (SC '00). 56.* Washington, DC: IEEE Computer Society (2000).

Calder, A. C., Fryxell, B., Plewa, T., Rosner, R., Dursi, L. J., Weirs, V. G., Dupont, T., et al. On validating an astrophysical simulation code. *Astrophys. J., Suppl. Ser.*, 143 (2002): 201.

Churazov, E., Sunyaev, R., Isern, J., Knödlseder, J., Jean, P., Lebrun, F., Chugai, N., et al. Cobalt-56 γ-ray emission lines from the Type Ia supernova 2014j. *Nature*, 512 (2014): 406–408.

Colella, P. and Woodward, P. The piecewise parabolic method (PPM) for gas-dynamical simulations. *J. Comp. Phys.*, 54 (1984): 174.

Dubey, A., Antypas, K., Calder, A., Daley, C., Fryxell, B., Gallagher, J., Lamb, D., et al. Evolution of FLASH, a multiphysics scientific simulation code for high performance computing. *Int. J. High Perform. Comput. Appl.*, 28 (2013): 225–237.

Dubey, A., Antypas, K., Ganapathy, M. K., Reid, L. B., Riley, K., Sheeler, D., et al. Extensible component-based architecture for FLASH, a massively parallel, multiphysics simulation code. *Parallel Comp.*, 35(10) (2009): 512–522.

Dubey, A. and Van Straalen, B. Experiences from software engineering of large scale AMR multiphysics code frameworks. *J. Open Res. Softw.*, 2 (2014): e7.

Dubey, A., Weide, K., Lee, D., Bachan, J., Daley, C., Olofin, S., Taylor, N., et al. Ongoing verification of a multiphysics community code: FLASH. *Softw. Pract. Exp.* 45 (2015): 233–244.

Filippenko, A. V. Optical spectra of supernovae. *Annu. Rev. Astron. Astrophys.*, 35(1) (1997): 309–355.

Fryxell, B., Olson, K., Ricker, P., Timmes, F. X., Zingale, M., Lamb, D. Q., MacNeice, P., et al. FLASH: An adaptive mesh hydrodynamics code for modeling astrophysical thermonuclear flashes. *Astrophys. J. Suppl. Ser.*, 131(1) (2000): 273–334.

Godunov, S. K. A difference method for numerical calculation of discontinuous solutions of the equations of hydrodynamics. *(Russian) Mat. Sb. (N.S.)*, 47(89) (1959): 271–306.

Hillebrandt, W., Kromer, M., Röpke, F. K., and Ruiter, A. J. Towards an understanding of Type Ia supernovae from a synthesis of theory and observations. *Front. Phys.*, 8(2) (2013): 116–143.

Hillebrandt, W. and Niemeyer, J. C. Type Ia supernova explosion models. *Annu. Rev. Astron. Astrophys.*, 38(1) (2000): 191–230.

Hix, W.R. and Thielemann, F.K. Silicon burning II: Quasi-equilibrium and explosive burning. *Astrophys. J.*, 511 (1999): 862–875.

Iben, I. J. and Tutukov, A. V. Supernovae of type I as end products of the evolution of binaries with components of moderate initial mass (m not greater than about 9 solar masses). *Astrophys. J., Suppl. Ser.*, 54 (1984): 335.

Jacobs, A., Zingale, M., and Katz, M. private communication, (2016) in prep.

Kasen, D., Röpke, F. K., and Woosley, S. E. The diversity of Type Ia supernovae from broken symmetries. *Nature*, 460(7257) (2009): 869–872.

Kromer, M., Sim, S. A., Fink, M., Röpke, F. K., Seitenzahl, I. R., and Hillebrandt, W. Double-detonation sub-Chandrasekhar supernovae: Synthetic observables for minimum helium shell mass models. *Astrophys. J.*, 719(2) (2010): 1067–1082.

Li, W., Leaman, J., Chornock, R., Filippenko, A. V., Poznanski, D., Ganeshalingam, M., Wang, X., Modjaz, M., Jha, S., Foley, R. J., and Smith, N. Nearby supernova rates from the lick observatory supernova search-II. The observed luminosity functions and fractions of supernovae in a complete sample. *Mon. Not. R. Astron. Soc.*, 412(3) (2011): 1441–1472.

Maeda, K., Nakamura, T., Nomoto, K., Mazzali, P. A., Patat, F., and Hachisu, I. Explosive nucleosynthesis in aspherical hypernova explosions and late-time spectra of SN 1998bw. *Astrophys. J.*, 565 (2002): 405.

Nagataki, S., Shimizu, T. M., and Sato, K. Matter mixing from axisymmetric supernova explosion. *Astrophys. J.*, 495 (1998): 413.

Pakmor, R., Hachinger, S., Röpke, F. K., and Hillebrandt, W. Violent mergers of nearly equal-mass white dwarf as progenitors of subluminous Type Ia supernovae. *Astron. Astrophys.*, 528 (2011): A117.

Pakmor, R., Kromer, M., Röpke, F. K., Sim, S. A., Ruiter, A. J., and Hillebrandt, W. Sub-luminous Type Ia supernovae from the mergers of equal-mass white dwarfs with mass ~0.9 solar masses. *Nature*, 463(7277) (2010): 61–64.

Pakmor, R., Kromer, M., Taubenberger, S., Sim, S. A., Röpke, F. K., and Hillebrandt, W. Normal Type Ia supernovae from violent mergers of white dwarf binaries. *Astrophys. J.*, 747(1) (2012): L10.

Parrent, J. T., Thomas, R. C., Fesen, R. A., Marion, G. H., Challis, P., Garnavich, P. M., Milisavljevic, D., et al. A study of carbon features in Type Ia supernova spectra. *Astrophys. J.*, 732(1) (2011): 30.

Rauscher, T., Heger, A., Hoffman, R. D. and Woosley, S.E. Nucleosynthesis in massive stars with improved nuclear and stellar physics. *Astrophys. J.*, 576 (2002): 323–348.

Ruiter, A. J., Sim, S. A., Pakmor, R., Kromer, M., Seitenzahl, I. R., Belczynski, K., Fink, M., et al. On the brightness distribution of Type Ia supernovae from violent white dwarf mergers. *Mon. Not. R. Astron. Soc.*, 429(2) (2012): 1425–1436.

Sagan, H. *Space-Filling Curves*. Berlin: Springer-Verlag (1994).

Silverman, J. M. and Filippenko, A. V. Berkeley supernova Ia program–IV. Carbon detection in early-time optical spectra of Type Ia supernovae. *Mon. Not. R. Astron. Soc.*, 425(3) (2012): 1917–1933.

Thielemann, F. -K., Nomoto, K., and Hashimoto, M. Core-collapse supernovae and their ejecta. *Astrophys. J.*, 460 (1996): 408.

Timmes, F. X. and Swesty, F. D. The accuracy, consistency, and speed of an electron-positron equation of state based on table interpolation of the helmholtz free energy. *Astrophys. J.*, 126 (2000): 501–516.

Warren, M. S. and Salmon, J. K. *Proceeding of Supercomputing 93.* Washington, DC: IEEE Computer Society (1993).

Webbink, R. F. Double white dwarfs as progenitors of r coronae borealis stars and type I supernovae. *Astrophys. J. Lett.*, 277 (1984): 355.

Weirs, G., Dwarkadas, V., Plewa, T., Tomkins, C., and Marr-Lyon, M. Validating the flash code: Vortex-dominated flows. *Astrophys. Space Sci.*, 298 (2005): 341.

Wheeler, J. C. and Hansen, C. J. Thermonuclear detonations in collapsing white dwarf stars. *Astrophys. Space Sci.*, 11 (1971): 373.

Whelan, J. and Icko, J. I. Binaries and supernovae of type I. *Astrophys. J. Lett.*, 186 (1973): 1007.

Woosley, S. E. and Weaver, T. A. The evolution and explosion of massive stars. II. Explosive hydrodynamics and nucleosynthesis. *Astrophys. J., Suppl. Ser.*, 101 (1995): 181.

Woosley, S. E., Wunsch, S., and Kuhlen, M. Carbon ignition in Type Ia supernovae: An analytic model. *Astrophys. J.*, 607(2) (2004): 921–930.

Zheng, W., Silverman, J. M., Filippenko, A. V., Kasen, D., Nugent, P. E., Graham, M., Wang, X., et al. The very young Type Ia supernova 2013dy: Discovery, and strong carbon absorption in early-time spectra. *Astrophys. J.*, 778(1) (2013): L15.

5 NAMD: Scalable Molecular Dynamics Based on the Charm++ Parallel Runtime System

Bilge Acun, Ronak Buch, Laxmikant Kale,
and James C. Phillips

CONTENTS

5.1 Introduction..120
5.2 Scientific Methodology...120
5.3 Algorithmic Details 123
5.4 Programming Approach...126
    5.4.1 Performance and Scalability 127
        5.4.1.1 Dynamic Load Balancing 127
        5.4.1.2 Topology Aware Mapping 127
        5.4.1.3 SMP Optimizations...128
        5.4.1.4 Optimizing Communication 129
        5.4.1.5 GPU Manager and Heterogeneous Load Balancing....................129
        5.4.1.6 Parallel I/O 130
    5.4.2 Portability 131
    5.4.3 External Libraries 131
        5.4.3.1 FFTW...131
        5.4.3.2 Tcl..132
        5.4.3.3 Python..132
5.5 Software Practices 132
    5.5.1 NAMD 132
    5.5.2 Charm++..134
5.6 Benchmarking Results...135
    5.6.1 Extrapolation to Exascale 137
        5.6.1.1 Science Goals 137
        5.6.1.2 Runtime System Enhancements Needed 138
        5.6.1.3 Supporting Fine-Grain Computations 138
        5.6.1.4 Optimizations Related to Wide Nodes 138
5.7 Reliability and Energy-Related Concerns 139
    5.7.1 Fault Tolerance 139
    5.7.2 Energy, Power, and Variation..140
        5.7.2.1 Thermal-Aware Load Balancing..140
        5.7.2.2 Speed-Aware Load Balancing 140
        5.7.2.3 Power-Aware Job Scheduling with Malleable Applications 140
        5.7.2.4 Hardware Reconfiguration...141

5.8 Summary...141
Acknowledgments 141
References 142

5.1 INTRODUCTION

Nanoscale Molecular Dynamics (NAMD) is a scalable molecular dynamics (MD) application designed for high-performance atomic-level simulation of large biomolecular systems at a *femtosecond* time step resolution [1]. Used by tens of thousands of scientists—85,000 users since 2000—on everything from laptops and desktops to supercomputers and graphical processing units (GPUs) (even iPads!), NAMD has enabled significant breakthroughs in understanding the structure and the behavior of viruses and cellular organelles.

The fixed size nature of the molecular systems requires fine-grained parallelization techniques and *strong scaling* to achieve efficient simulation of long timescales. NAMD is built on top of the parallel framework Charm++ [2,3], which provides a robust infrastructure that enables simulations of molecular systems that has billions to scale up to hundreds of thousands of cores, while providing portability with performance across different supercomputing architectures. The NAMD and Charm++ collaboration was started in 1992 by principal investigators Klaus Schulten, Laxmikant V. Kale, and Robert Skeel. NAMD's performance needs have motivated Charm++ to develop new methods and abstractions resulting in successful codevelopment and interdisciplinary collaboration. In 2012, the IEEE Computer Society's Sidney Fernbach Award was jointly awarded to Kale and Schulten "for outstanding contributions to the development of widely used parallel software for large biomolecular systems simulation." One example of this is the petascale simulation of the HIV capsid with 64 million atoms on the Blue Waters Cray XE6 System at NCSA, enabling the precise determination of the chemical structure of the HIV capsid for the first time. This groundbreaking study was published and featured in *Nature* in 2013 and "petascale simulations are now being used to explore the interactions of the full capsid with drugs and with host cell factors critical to the infective cycle" [4]. Meanwhile, our efforts continue to prepare NAMD for exascale simulations not only in terms of performance and portability, but also in terms of reliability and energy efficiency as well.

This chapter discusses the scientific methodology behind NAMD (Section 5.2), algorithmic details (Section 5.3), NAMD's Charm++ based programming approach (Section 5.4), software practices used in maintaining NAMD and Charm++ (Section 5.5), scaling results (Section 5.6), and finally reliability and energy considerations for future architectures (Section 5.7).

5.2 SCIENTIFIC METHODOLOGY

NAMD is one of many programs enabling what are generally called MD simulations, distinguished from molecular mechanics (MM) calculations. MM provides energies and forces associated with a particular molecular configuration (three-dimensional [3D] coordinates for all atoms), which can combine with a minimization algorithm to obtain an optimized low-energy configuration. MD simulation, in contrast, calculates a trajectory of atomic coordinates evolved from initial atomic positions and velocities via classical dynamics as expressed by Newton's second law, $F = ma$.

It is important to distinguish the properties of dynamics simulations as applied to molecular systems from other *n-body problems* such as planetary dynamics under gravitational interactions. In planetary dynamics, the initial positions and velocities of the bodies being simulated are known with high accuracy, allowing the simulation to be run for millennia not only forwards, but also backwards to reproduce historical observations. The numerical integrators employed for such calculations must be of comparable high accuracy.

In molecular simulations, the initial positions are either of low accuracy, as from X-ray crystallography of proteins, or are randomly generated representative configurations of bulk materials such as water and lipid membranes, and hence only the statistical behavior of the simulated system is of interest rather than the specific trajectory. The choice of an appropriate integrator for MD is governed not by short-time accuracy but by long-time properties such as conservation of energy, and therefore the symplectic and reversible Verlet integrator is most often employed, with the trajectory generated at fixed time steps and stored at fixed intervals for later analysis.

Atomic velocities are generated randomly based on classical thermodynamics for the target temperature of the simulation, and the simulation must be run for an equilibration period in order to establish an equipartition between position and velocity degrees of freedom. During the equilibration period, and often for the duration of the simulation, the temperature of the system is steered toward a target by either random or deterministic adjustments to the atomic velocities.

Simulations of biomolecular systems in particular are often performed under periodic boundary conditions, in which the simulated particles interact not only directly with each other, but also with an infinite regular 3D lattice of images. This lattice does not typically correspond to an actual crystal structure, but is arbitrarily sized so as to contain the simulated structure plus a suitable quantity of solvent (water and ions). Such boundary conditions eliminate edge effects such as surface tension and evaporation and thus provide an illusion of infinite bulk solvent. Despite the infinite domain available, the volume per image is that of a single periodic cell. If the volume is too large bubbles will form during the simulation, whereas if the volume is small and incompressible liquid-phase simulation will experience high pressure. For this reason, the pressure of the system must be equilibrated as well by adjustments to the cell volume. The individual cell dimensions may be adjusted independently for anisotropic systems such as a bilayer membrane, or uniformly for bulk solvent. Again, both random and deterministic methods are available.

Temperature and pressure control enable simulations to be conducted in three common thermodynamic ensembles: NVE (constant particle count, volume, and energy), NVT (constant particle count, volume, and temperature), and NPT (constant particle count, pressure, and temperature). Advanced methods may enable simulations in a constant chemical potential ensemble where the number of particles varies, the most common being a constant pH ensemble in which protons (hydrogen ions) are added and removed from specific titratable locations in the molecule.

The forces (negated energy gradients) employed in an MD simulation may derive from quantum mechanical (QM) calculations, in which the electronic forces on the atomic nuclei are classically integrated to propagate nuclear coordinates. QM–MD calculations can accurately model electronic polarization and excitation effects, and the breaking of chemical bonds, but the expense of such calculations precludes their use for long-timescale calculations. Fortunately, life as we know it is limited to moderate temperatures and pressures where chemical reactions only occur at the catalytic sites of enzyme proteins and most processes consist of low-energy conformational transformations and molecular migrations. Therefore, biomolecular applications can generally substitute low-cost classical potential functions for quantum calculations, although hybrid QM–MM methods may be employed to model chemical reactions, modeling only the few atoms near the region via QM.

Classical MM potential functions employ additive terms for bonded and nonbonded interactions. Bonded terms represent the covalent bonding structure of the simulated molecules, with directly bonded atoms linked by simple springs of the form $U(r) = k(r - r_0)^2$ where r is the interatomic distance and the constants k and r_0 are parameterized for each pair of atoms based on QM calculations of either the actual molecule or appropriate analogues. This parameterization is greatly simplified by the fact that all proteins employ the same 20 amino acids, and likewise all DNA the same four bases, and hence a single set of parameters can be employed for all biomolecular simulations. New parameters need be developed only for small novel ligands such as drugs. A similar energy term $k(\theta - \theta_0)^2$ is applied to the angles between all pairs of bonds to a common atom, and also to planar sets of three bonds to a common atom. Finally, one or more torsion terms of the form $k(1 - \cos(n(\varphi - \varphi_0)))$ are applied to all dihedral angles between pairs of angles with an overlapping bond.

Nonbonded terms apply to all pairs of atoms in the simulation, including images from periodic boundary conditions, with the exception of pairs of atoms that are bonded to each other or to a common atom because the interactions of these closely bonded pairs are fully represented by the bonded terms in the potential function. Nonbonded terms represent electrostatic forces between partial charges on the atoms (due to uneven distribution of the molecular electron cloud relative to the nuclear charge), van der Waals attractive interactions, and short-range repulsive interactions. Electrostatic interactions follow Coulomb's law $U(r) = Cq_1q_2/r$, whereas the van der Waals r^{-6} attractive potential is traditionally coupled with r^{-12} repulsion in the Lennard–Jones form $U(r) = 4\epsilon((\sigma/r)^{12} - (\sigma/r)^6)$. As with the bonded terms above, partial charges q and Lennard–Jones parameters ϵ and σ are fitted based on QM calculations. For simplicity and efficiency, several atom types are defined for each element present in the simulation to represent the variety of chemical bonding environments, and these types are used to assign bonded and Lennard–Jones parameters. Lennard–Jones parameters for pairs of atoms with different types are approximated by geometric averaging of the individual ϵ energy terms and (most commonly) algebraic averaging of the atomic radii σ. The formalism for deriving parameters specified by the force-field developer must be adhered to and combining parameters developed via different methods is not advised.

The simplified form of the potential functions necessarily limits the accuracy and generality of the simulation. Parameters are typically developed to reproduce structures at standard temperature and pressure. Water models, in particular, are designed as solvents to model proteins, lipids, and DNA, and to maintain correct structure at 1 atm and 300 K. Water is a highly complex fluid that maintains a network of hydrogen bonds both with hydrophilic solutes and between molecules of water itself. It is unreasonable to expect a simple three-atom water model (or even more complex five-point models employing additional charge centers) to serve the needs of biomolecular simulations while also correctly reproducing all properties of water, for example, the various phases of ice.

It is common for the great majority of atoms in a biomolecular simulation to be solvent (i.e., water and a few ions), in particular when a single protein or protein complex is being studied. The explicit representation of water molecules is then sometimes replaced with an implicit solvent model, in which more complex and expensive multibody interaction terms are used to approximate the averaged effect of the possible solvent configurations on the solvent. The solvent thus experiences a smooth mean-force potential that enables greater conformational exploration, useful in the study of large domain motions and protein folding. Implicit solvent cannot, however, reproduce specific interactions with small numbers of water molecules, such as inside channels. The reduction in atom count and resulting performance increase is also reduced or eliminated for larger simulations.

The nonbonded electrostatic and Lennard–Jones interactions are nominally $O(N^2)$, which would make large-scale simulations prohibitively expensive. This is addressed by imposing a short-range cutoff of 8–12 A on the Lennard–Jones terms, reducing them to $O(N)$, whereas the long-range $1/r$ electrostatic interactions are most commonly calculated in an efficient manner by the particle-mesh Ewald (PME) method for periodic simulations with explicit solvent. PME smoothly separates the Coulomb interaction into a $1/r$-like but exponentially decaying short-range potential that is obtained by direct pairwise calculation, and a long-range potential that is smooth and finite at short ranges, allowing its efficient representation via a 1A grid. Atomic charges are spread with $O(N)$ complexity onto small regions of the grid ($4 \times 4 \times 4$ to $8 \times 8 \times 8$), convolved with the long-range potential via $O(N \log N)$ 3D fast Fourier transform (FFTs), and atomic forces interpolated from the grid again with $O(N)$ complexity. Other methods such as fast multipole and multilevel summation can achieve $O(N)$ complexity for the full electrostatics calculation, but in practice the FFTs are a small fraction of the PME calculation and the grid can be coarsened to 2A by employing higher-order interpolation. The actual performance limitation of the PME FFT is the global communication required for its parallelization.

Simulations must be long enough to provide sufficient sampling to observe the molecular process of interest to the scientist. A standard all-atom MD simulation can be integrated stably with a 1 fs time step via the Verlet *leap-frog* integrator (equivalent to Velocity verlet). Constraining the lengths of all bonds to hydrogen atoms, and the H–O–H angle of water molecules, allows a 2 fs time step (or longer with new *geodesic* integration methods that calculate constraint forces at several subtime steps). In addition, multiple time-stepping allows the slowly changing long-range electrostatics calculation to be done only every three steps. Beyond this point, longer time steps require coarse-grained models, in which groups of atoms are represented by larger particles with smoothed interactions that foreclose the study of many processes that rely on atomic detail.

A billion 1 fs time steps must be executed in sequence to achieve a microsecond of simulated time, reasonable by modern standards, but requiring 12 days even with excellent performance of a millisecond per step. Multiple independent simulations can and should be executed simultaneously with different initial velocities to increase sampling and statistical confidence, but other methods can be used to enhance sampling. The simplest enhanced sampling method is simulated annealing— raising the temperature of the simulation to increase barrier hopping followed by a slow cooling. More sophisticated methods such as accelerated MD directly lower barrier heights by scaling calculated forces when total system energies are above the thermally accessible level. When the exact mechanism under study is known but too slow to be observed during the simulation, time-dependent biases or steering forces can be used to induce the required transitions. Biased calculations can be used to efficiently extract the conformational free energy profile along a transition path. It is even possible to calculate the free energy difference between chemical states of the system by gradually creating, destroying, or transmuting atoms.

Multicopy algorithms enhance and direct sampling by loosely coupling otherwise independent simulations. The most common multicopy algorithms employ replica-exchange protocols, in which simulations along a range of temperatures or bias parameters periodically attempt to exchange configurations with their neighbors based on the Metropolis criterion, with overlapping bias windows analyzed to produce free energy profiles. Transition paths can be found via the string method, which tracks swarms of trajectories released along a candidate path in order to reach a lower energy path. Milestoning uses large numbers of independent trajectories to measure transition rates and first passage times between conformational hypersurface *milestones* in order to calculate both free energy profiles and transition rates. Finally, independent trajectories can be launched from known states to construct and extend Markov state models of biomolecular dynamics, including protein folding.

The types of scientific questions that can be asked and answered about biomolecular systems varies with the scale and complexity of the system. Alanine dipeptide has but two nontrivial degrees of freedom (rotatable dihedral angles) and can be sampled completely, thus making it a necessary but insufficient test for method development. Single protein systems, fully solvated, comprise around 100,000 atoms, and all but their slowest processes are studied with modern enhanced sampling methods. The ribosome, a complex aggregate of multiple proteins and nucleic acids responsible for protein synthesis in the cell, requires three million atoms to be simulated and remains a topic of cutting-edge research. Finally, the HIV capsid, at 64 million atoms, was assembled by MD simulations combining individual capsid protein crystal structures with cryoelectron microscopy density profiles, yielding an atomic-resolution capsid structure that is being simulated binding to both drugs and naturally occurring molecules in the cell to better understand the function of the capsid in the viral infection process. Simulations of the HIV capsid are run on NCSA Blue Waters, one of the most powerful supercomputers in the world.

5.3   ALGORITHMIC DETAILS

The basic design of NAMD's parallel algorithm dates to the early 1990s. The fact that it has survived this long, with only relatively modest changes to its basic parallel architecture, is due to the following factors:

1. The original design was done with a careful *isoefficiency analysis*, that is, analysis of asymptotic scalability.

2. Use of message driven execution to facilitate communication/computation overlap and enable flexible runtime scheduling.
3. Separation of concerns between what to do in parallel and where/when to execute computations, thus enabling powerful adaptive runtime system (RTS) support that could improve and change over the years in response to new architectural and application-related challenges without changing the basic structure of NAMD itself.

We explain these factors below.

Simply put, isoefficiency analysis deals with the question: for a given parallel algorithm (and, in the original formulation, characteristics of a parallel machine), at what rate must the problem size grow in order to maintain the same parallel efficiency when the number of processors is increased. Here, parallel efficiency is simply the speedup obtained divided by the number of processors. Suppose an algorithm is running on 10 processors with a speedup of 8. Running the same problem on 20 processors, the speedup should ideally be 16, but due to increased costs (typically, and for our purposes here, mainly, communication costs) it falls to a lower value. If you run a large size problem, for a properly designed algorithm, the communication-to-computation ratio *may* decrease and then it becomes possible to get a speedup of 16. Such algorithms are said to be scalable, and the rate at which the problem size must increase with respect to the number of processors is the isoefficiency of the algorithm. Thus, an algorithm with quadratic isoefficiency is one for which the problem size (defined as sequential execution time) must increase four times when the number of processors doubles to maintain the same parallel efficiency.

In the 1990s, many parallel MD programs were developed by converting an existing sequential program. Methods such as atom replication (where the array of all atoms was available on all processors) or simple static atom partitioning combined with force-decomposition [5] naturally arose in this scenario. However, our early analysis showed that these techniques were not scalable in the sense of isoefficiency analysis, that is, no amount of increase in problem size could bring the communication/computation ratio down, when increasing the number of processors. Along with a few peers [6], NAMD pioneered the idea of *spatial decomposition*, which was tedious but scalable. Atoms were decomposed into cubes, and they had to be migrated every K steps to the correct cubes based on their coordinates. Cut-off based electrostatic force calculation, the dominant part of the MD time step, becomes scalable with this technique.

The second technique was somewhat more radical at that time. Which processor should do the task of calculating forces between the atoms of two neighboring cubes? Prevailing processor-centric thinking would typically answer this by making the processor that housed one of the cubes do the computations. NAMD stipulated that the work will be done by another object, called the *compute* object denoted by a diamond in Figure 5.1, and that the placement of this object to a processor be left to the RTS. In particular, the placement could be (and often was) on a third processor which housed neither of the two cubes involved. This creates a degree of freedom for load balancing and allows utilization of a large number of processors (e.g., many more processors than the number of cubes). A similar idea, with more specific choices for the *third processor* has since been independently invented with names such as mid-point method, by Shaw et al., [7], Snir et al., [8], and Blue Matter team from IBM [9]. One can think of this technique as a hybrid between spatial decomposition and force decomposition.

Until the time of NAMD design, Charm++ (and Charm, its precursor) was an experimental programming model used mainly for divide-and-conquer and state-space search problems, which were representative of the problems addressed in the era of the so-called Fifth-Generation computing systems. Its main feature was the idea that the computation could be broken down into logical objects by the programmer but the assignment of those objects to processors was to be automated by a *RTS*. The related idea was that of message-driven execution: given that there are many objects on a processor, the object to be selected for execution should be the one for which a message (or method invocation) is already available. This allowed automatic, adaptive overlap of computation and communication (see the description in the next section).

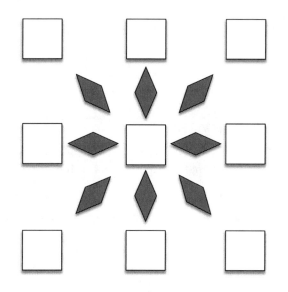

FIGURE 5.1   NAMD hybrid decomposition of patches (squares) and nonbonded computes (diamonds).

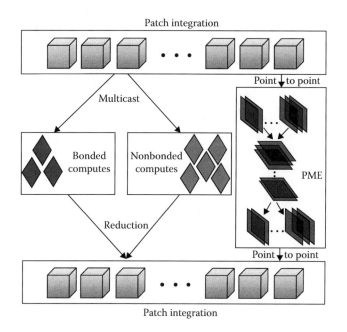

FIGURE 5.2   Compute flow of NAMD.

The above idea of hybrid decomposition might be novel in the context of a processor-centric programming model, but was natural in Charm++. The computation was logically decomposed into multiple collections of objects (see Figure 5.2). The PME summation, or the fast-multipole based algorithm for aperiodic systems, was used to calculate the long-range contributions of electrostatic forces. Because Charm++ was in the early stages of development, NAMD was initially implemented in Parallel Virtual Machine (PVM) as well as Charm++. However, the message-driven execution was a natural fit for the NAMD algorithm. For example, it could easily handle interleaving of PME messages, coordinate messages, and force messages, which is tricky to express modularly in PVM (or for that matter, in message passing interface [MPI]).

Thus, Charm++ and NAMD coevolved synergistically, in a process that we may call codesign today. Charm++ development was further helped by codevelopment of applications in astronomy and quantum chemistry as well as several engineering applications that arose at the rocket simulation center at Illinois.

5.4 PROGRAMMING APPROACH

NAMD is built upon the Charm++ parallel programming framework, which is a C++ based model with data-driven objects [2]. The Charm++ adaptive RTS has three main properties:

- **Overdecomposition:** This property consists in dividing the application data into objects, where the number of objects is typically much more than the number of processors. In Charm++ these are C++ objects called *chares*. Programmers make this division logically without having to worry about the underlying low-level components, that is, cores, and the RTS controls mapping of the objects to the processors. Overdecomposition carries many benefits; the logical division of data both increases programmer productivity and also application performance. It enables the overlap of computation and communication by increasing the likelihood that there are objects ready to execute in a processor while others are waiting for messages to be received over the network. Furthermore, division into small units may result in better cache performance as well. Charm++ also provides indexed collections of *chares,* which are called *chare arrays.* The programmer has the flexibility to create multiple types of *chares* and *chare arrays.*
- **Asynchronous message-driven execution:** This attribute means that execution of work units is driven by availability of messages, instead of a preprogrammed sequence. Messages are simply function calls on objects that can reside on local or remote processors. When the RTS detects the arrival of a message, it puts it on the message queue of the corresponding object and schedules it when possible. *Asynchronous* implies that the sender does not block the execution for any immediate response from the receiver side. The function calls, also called *entry* methods, look like standard C++ methods; the only addition is that the programmer specifies these functions in an interface file, called the *ci* file. Charm++ uses a simple translator to read the functions in the interface file and generate the necessary code in the background to be able to make these function calls execute on remote objects. The programmer does not need to worry about object locations, the RTS takes care of transferring the function call and the argument data to remote or local objects transparently.

Combined with the overdecomposition property, asynchronous execution can help reduce the communication related delays that can significantly hurt performance of applications. After the send operation, the sender can continue execution and does not have to wait for a response from the receiver. Also, the receiver can work on executing other messages as long as there are messages in its queue and does not need to be blocked waiting for a specific message. This scheme hides the latency of communication and can reduce the network pressure by spreading it over time.

- **Migratability:** This property is the ability to move objects from processor to processor. It allows the RTS to map and redistribute objects dynamically during execution to improve performance. For example, the RTS measures the load of the processors during application execution and moves objects around to keep the amount of work on every processor balanced. Charm++ provides a *packing* and *unpacking* (PUP) framework to enable migration of objects. The programmer specifies which data fields of the *chare* object need to be migrated and the RTS handles serialization and deserialization. Dynamic migratability enables many features in Charm++:

- Dynamic load balancing
- Recovery from hard and soft failures
- Temperature and power aware work distribution
- Malleable job scheduling with shrink and expand number of processors

These are discussed later in more detail in Sections 5.4.1 and 5.7.

NAMD benefits from these properties of Charm++. Other features of Charm++ that improve performance of NAMD are described further in this section.

5.4.1 PERFORMANCE AND SCALABILITY

Charm++ provides multiple methods to improve performance and scalability for both intranode and internode contexts. These methods are dynamic load balancing, TopoManager application programming interface (API) for topology-aware mapping, symmetric multi-processing (SMP)-specific mode, communication optimization methods such as prioritized messages and persistent communication, GPU Manager API for heterogeneous architectures, and Parallel I/O support.

5.4.1.1 Dynamic Load Balancing

The Charm++ runtime has a rich dynamic load-balancing framework that is based on *overdecomposition*. Each processing element (PE) typically has multiple chares that can be migrated from one PE to another to achieve better balance. The load balancing framework stands on the *principle of persistence*, which means that the load of objects in the past is often a good indicator of their load in the future. During execution, the runtime collects load information in terms of execution time of chares and the communication graph between them. This information is used to make load balancing decisions and decide where to migrate chares. Further, applications can add custom metrics to implement application-specific load balancing strategies to achieve better load balance.

NAMD has multiple different kinds of chares: patches, bonded computes, nonbonded computes, and three types of PME computes. All of these chares are instrumented in the runtime to collect their load information and the total load of each processor is calculated by summation of the object loads on that processor. Although all object loads are considered for load balancing, only nonbonded computes can migrate from one processor to another. This is partly because the nonbonded computes occupy a significant portion of the total execution time, and partly to minimize migrations and to simplify load balancing decision-making. The other object locations are statically assigned at startup.

The load balancing strategy described is centralized, requiring load and communication information of all objects to be collected on one processor. However, when running at large scale, this can create a serialization bottleneck and the data may become too large to fit in memory. To prevent this bottleneck, support for hierarchical load balancing was implemented. This way, processors are organized into a tree with customizable levels and number of children per tree node. First, load information is exchanged in a bottom up fashion and load balancing is done top down. At every level, each node and its children form a group where the previously explained refinement-based load balancing strategy is invoked within each group. This effectively restricts collection of information inside the group. Hierarchical load balancing has been shown to perform 100 times better than centralized load balancing in terms of the time it takes to load balance on petascale systems [10].

5.4.1.2 Topology Aware Mapping

Up to a few thousands of nodes, NAMD performs well on toroidal network topologies that are commonly found in supercomputing systems without worrying about the underlying network topology. However, scaling up to tens of thousands of nodes requires topology-aware placement of the work units to avoid network contention. Charm++ provides an API, TopoManager that provides physical topology information to the application to be able to map the PEs to the underlying topology. Topology aware placement schemes can deal with noncontiguous allocations (e.g., a random subset of processors within a 3D torus, or a denser torus with *holes,* i.e., unavailable nodes, in the middle).

The *replicas* in multicopy NAMD algorithms [11] do not exchange much information except some control parameters and aggregate information such as the total energy of the simulation or the potential energy of an applied bias; the majority of the communication happens within the replicas, or partitions. However, the overall performance of the simulation is limited by the slowest replica, therefore, it is important to have the same simulation rates for the replicas. There can be performance variations between partitions for various reasons; data-dependent algorithmic reasons, or network/system noise. The network noise can be solved by having equal, uniform, and compact partitioning of the physical nodes as much as possible. NAMD uses a recursive bisection strategy to do topology-aware partition mapping and topology-aware placement of compute objects within each partition [4].

5.4.1.3   SMP Optimizations

Multicore compute nodes in parallel architectures have driven the need for a multithreaded SMP runtime mode of Charm++. Charm++ offers two different build modes: SMP and non-SMP. In the non-SMP mode, each Charm++ PE runs as an OS process. Each process can send and receive messages from other processes. In SMP mode, by contrast, PEs run as threads, and all threads that belong to the same OS process form a *logical node,* or a *Charm++ node.* Each logical node has a communication thread with its own dedicated core that is responsible for sending and receiving internode communication messages. Other threads, whose count per logical node is customizable, simply act as workers. This mode, where the PEs on the same node share a single memory address space, provides multiple benefits on multicore nodes. These are improved memory footprint, more efficient intranode communication, fine-grained parallelism, and reduced launch time.

- **Reduced memory footprint:** In SMP mode, Charm++ PEs on the same node can now share common read-only and write-once data structures. This not only reduces the memory footprint, but it also improves the cache performance as well, compared to the non-SMP mode. Optimizations are done in NAMD to share certain information such as objects that contains static physical attributes of atoms. This has shown 3–10 times reduction in the memory usage in NAMD and better cache performance that is directly attributed to reduction in the memory footprint [3,12].
- **Improved intra- and internode communication**: Usage of the shared address space among the PEs in the same node improves intranode performance. When sending messages within a node, the runtime only needs to send a pointer to the message, instead of making a copy of it as in the non-SMP mode. In addition to intranode benefits, this optimization provides better communication performance in internode communication as well, such as in broadcasts and multicasts within the Charm++ runtime and in NAMD's multicast operations [12]. In such cases, when a remote PE sends a message to multiple PEs on the same node, only one internode message is sufficient to be sent, then the message can be forwarded to the intranode PEs locally. This removes the need to send multiple expensive internode messages, replacing them with cheaper intranode communication.
- **Reduction in launch time:** Because the SMP mode creates fewer OS processes than the non-SMP mode, it uses less system resources and significantly reduces the launch time of the applications at large scales. For example; in the non-SMP mode, it takes 6 minutes to launch NAMD with *mpirun*, when running on 224,076 cores on Jaguar supercomputer

with a Charm++ build that uses MPI for communication. Whereas in the SMP mode with 1 process and 12 threads per physical node, *mpirun* takes around 1 minute to launch NAMD [3]. This effect is all the more important on cluster-like machines, such as Texas Advanced Computing Center (TACC)'s Stampede.

- **Enabling fine-grained parallelism:** During the PME phase of NAMD, it has been observed that whereas some cores are busy with computation, the neighbor cores are idle. To exploit more fine grained parallelism, the *CkLoop* library was developed in Charm++. CkLoop offers OpenMP-like shared memory multithreading where Charm++ PEs are reused to execute the spawned tasks, and it has been shown to improve performance [13]. Using OpenMP for this purpose with Charm++ on the same cores is not easy because the two runtimes are not aware of each other. However, we have an ongoing effort to realize this because it is more programmer friendly to use existing well-known libraries like OpenMP instead of a custom one to achieve the same task.

5.4.1.4   Optimizing Communication

Communication constitutes a major part of the PME phase in NAMD, therefore, optimizing PME is critical to achieve scaling to a large number of nodes. Prioritized messages and persistent communication are two ways to optimize PME communication.

- **Prioritized messages:** Charm++ runtime provides the ability to assign priority to messages. Commonly, applications send different types of messages because they have different phases or functions. Some of those messages might be on the critical path and their delay might affect overall performance. For example, in NAMD arrival of PME messages drives the FFT computation. Another type of message in NAMD triggers nonbonded computation when two messages from the patches arrive at the compute object. Because PME messages are more critical, we assign a higher priority to them. The runtime queues honor priorities and use first-in-first-out (FIFO) order among messages of identical priority.

- **Persistent communication:** In the default mode of Charm++, when messages arrive at the receiver, the corresponding memory is allocated and the message is placed there. The memory destination is not known beforehand. On the other hand, most scientific applications are iterative and have *persistent* communication patterns that repeat over the execution of the application. Exploiting this knowledge gives the opportunity to save the time to allocate and free memory on the receiver side, and utilize remote direct memory access (RDMA) more efficiently. Implementing PME with persistent messages, NAMD's performance improved by 10% when running a 100-million-atom simulation on the Titan machine at Oak Ridge National Laboratory (ORNL).

5.4.1.5   GPU Manager and Heterogeneous Load Balancing

Charm++ supports GPUs on a basic level via an extension called GPU Manager. GPU Manager handles the delegation and execution of CUDA kernels in the context of the asynchronous message-driven runtime of Charm++. GPU Manager operates by registering GPU kernels to be managed by the RTS. The runtime asynchronously invokes kernels when data is available on the device, automating the overlap of data movement and execution. GPU Manager also automatically copies data to and from the device before and after kernel execution.

To use GPU Manager, the user supplies an explicit CUDA kernel and denotes buffers to be moved to and from the device. A callback must also be registered with the runtime, which is called after the kernel has finished and data has been copied back. This step is necessary because the call to GPU Manager returns once the runtime has copied the CUDA buffers; it does not block until the kernel has finished. GPU Manager coordinates data movement and kernel invocations through a FIFO queue. When a processor on the node is idle, the runtime invokes a function to issue new requests to the GPU.

Although GPU Manager allows GPUs to be utilized by the Charm++ runtime, it does not provide much control over execution. Namely, load balancing, a hallmark of Charm++, cannot be done because GPU load information is not used and there is no mechanism to move or retarget work. In order to solve this issue, Charm++ offers some experimental constructs on top of GPU Manager. The first of these is the Accel Framework [14], which can dynamically generate CUDA kernels from host code and dynamically decide where particular methods should be executed. The Accel Framework allows programmers to annotate methods with the "accel" keyword to indicate that a method can be targeted at multiple hardware platforms. Methods can contain multiple implementations for different hardware targets separated by keywords. The Accel Framework has a variety of strategies to determine where to execute particular entry methods. The strategy can be specified as a runtime argument. These strategies can specify a static division between the various available hardware targets (e.g., 30% of invocations go to CPU, 70% go to GPU) or use dynamic measurement techniques to automatically balance work. In order to maximize GPU utilization and avoid serialization, the Accel Framework tries to batch multiple device method calls into a single kernel launch. Object data used inside *accel* methods must be annotated to mark it for transfer in, out, or both. Additional annotations can specify the lifecycle of a variable, such as indicating that it should remain on the device for later use. After execution, a callback is called just as in GPU Manager. One deficiency of the Accel Framework is that load information is only used to balance work across the different hardware resources of a single node. It can run in a distributed job, but each node is considered separately.

Building upon within-node balancing capabilities of the Accel framework, we are experimenting with new strategies for heterogeneous load balancing across nodes. A *vector load balancer* adds additional fields to the load database used in regular Charm++ load balancers. These additional fields can be used to store different load metrics and data to inform load balancer decisions. In particular, it can be used to store both CPU and GPU load for every object in the application. When combined with multidimensional load balancing algorithms, both CPU and GPU load can be balanced simultaneously. As the diversity and speed of high-performance computing (HPC) machines increase, this will likely be a very important technique in load balancing because it allows more data to be used when making load balancing decisions.

NAMD utilizes GPUs without using the Charm++ extensions because GPU support was added to NAMD before it existed in Charm++. The GPU related Charm++ additions have been driven by the needs of NAMD, however. For example, NAMD performance analysis concluded that CPUs were sitting idle when GPU enabled methods were being used and that GPU load balance was a significant disparity for certain types of problems and jobs. When control of GPU work is ceded to the RTS, it should be able to address these issues and better prepare NAMD and other applications to scale to future molecular systems and future HPC systems.

5.4.1.6 Parallel I/O

Charm++ provides a library for parallel I/O called CkIO. CkIO abstracts the parallel file system away from the application I/O code. By doing so, it is able to agglomerate I/O operations and decouple application file reading and writing operations from system configuration. Additionally, it simplifies and optimizes I/O operations made from migratable objects. CkIO allows the number of involved processors, arrangement of involved processors, and file stripe parameters to be specified to customize agglomeration. These parameters are used to construct and distribute objects that perform I/O. One key feature of allowing this customization is that it allows developers to properly size and align I/O blocks without having to change the access pattern of the application. Using the correct alignment is critical on parallel file systems, such as Lustre, because operations that are not well aligned can dramatically increase network and disk interference and load.

Because CkIO is implemented in Charm++, it utilizes asynchrony and overdecomposition to improve the effective performance of I/O. For a properly structured application with sufficient work, these features of the Charm++ RTS can hide the latency of messages used to agglomerate work and perform I/O operations to disk. Although other parallel I/O systems can accomplish similar feats, it is notable that the standard structure for Charm++ applications should fulfill these criteria without any changes.

NAMD writes out the positions of the particles in its molecular system periodically over the course of the simulation, and increasing molecular system size and a number of processors were creating performance woes. To address these, NAMD uses parallel I/O routines with rudimentary agglomeration. However, these do not align writes to file system or storage hardware boundaries and the implementation is somewhat NAMD specific. Thus, CkIO was written to solve a real application-driven problem while also being generic enough to be used in other Charm++ applications.

5.4.2 Portability

NAMD development has been specifically geared toward addressing both high-end and low-end platforms, with and without NVIDIA GPUs or Intel Xeon Phi coprocessors, and with x86_64, POWER, ARM, or self-hosted Xeon Phi CPUs. We do not see any future alternative for programming NVIDIA GPUs that will be as well supported as CUDA, but OpenMP 4.0 SIMD directives provide at last a standard cross-platform means of reliably accessing vector instructions. These single instruction multiple data (SIMD) directives will be most critical on non-GPU-accelerated machines, which will predominantly have x86_64 and/or Xeon Phi processors, both targets of the Intel compiler. Hence, we first write OpenMP 4.0 SIMD kernels vectorizable by the Intel compiler, and only later address ARM and POWER performance for those kernels that are not offloaded to the GPU via CUDA. We use identical data structures between Xeon Phi CPUs and offload coprocessors, and also other CPUs and GPUs whenever reasonable, using modern C++ features to keep as much code in common as possible.

Charm++ provides NAMD portability across many architectures and compilers. Charm++'s portable network communication layer is optimized for each vendor's network API. Unlike many other task-based parallel programming models, or partitioned global address space (PGAS) languages, Charm++ implements native communication API's of the vendors such as uGNI for Cray [15], PAMI for BlueGene [16], IB-Verbs for Infiniband [17], and TCP, UDP for Ethernet. Charm++ also has an MPI-based communication layer; however, native communication layers usually outperform MPI. For example, with a Gemini-based Charm++, NAMD performs up to 54% better than MPI-based version due to fine grained communication optimizations [13]. We are also working on an Open Fabrics Interfaces (OFI)-based communication layer for OmniPath-based architectures [17].

Charm++ supports many compilers, including Intel, IBM XL, GNU, Clang C/C++, Cray, Portland Group, and Microsoft compilers. Some of the supported operating systems are Linux, Mac OS X, Microsoft Windows (native and Cygwin), AIX, and various compute-node kernels.

5.4.3 External Libraries

5.4.3.1 FFTW

NAMD uses FFT calculations to approximate the long-range forces over the 3D grid. For this purpose, NAMD uses the popular FFT library Fastest Fourier Transform in the West (FFTW) for serial 1D or 2D FFTs, which are aggregated into 3D FFTs via NAMD-specific communication code written in Charm++. The MPI-parallelized 3D FFTs provided by the FFTW library are not used in production NAMD simulations, although they have been used as an example of how MPI and Charm++, two distinct approaches for parallelization, can interoperate and be used together [18]. MPI and Charm++ can share both resources and the data. In this case, at every iteration of NAMD, while

Charm++ tasks calculate the short-range forces, MPI tasks simultaneously calculate the long-range forces. Data are also shared between Charm++ objects and the MPI ranks through a data transfer repository. When Charm++ objects produce and deposit their data into the repository, the repository triggers the execution of a parallel FFT in MPI. Use of FFTW compared with a custom Charm++ based implementation provides similar performance, however, it has many advantages in terms of productivity [18]. It reduces the total source lines of code (SLOC), code development and maintenance. Moreover, vendors may provide highly optimized implementations in certain architectures that can be utilized without additional effort.

5.4.3.2   Tcl

NAMD leverages the Tcl scripting language to parse its configuration file, providing a complete and powerful programming language. Tcl is a mature, portable, and stable language; the interfaces have not changed during the history of NAMD over the past 20 years. The Tcl interpreter is self-contained as linked to the NAMD binary and does not require the presence of any external libraries. This lack of dependencies has enabled the NAMD Tcl interpreter to "just work" without issue on generations of supercomputers. In addition, most NAMD users are familiar with Tcl from its use in the popular codeveloped visualization program Visual Molecular Dynamic (VMD) [19]. It is a simple and easy to use scripting language that is ideal for nonprogrammer NAMD users.

5.4.3.3   Python

In response to user requests, it is possible to build NAMD with an optional Python interpreter, which can be called from and call back to the main Tcl interpreter. Unlike Tcl, the Python interpreter requires an extensive library, which complicates distribution, installation, and execution. It is, however, the extensive Python library that advanced users and method developers find appealing. The initial application of Python was in the development of constant pH simulation in NAMD, but with a small amount of work the dependencies on the Python library were replaced with Tcl equivalents such that the Python interpreter is no longer required.

5.5   SOFTWARE PRACTICES

5.5.1   NAMD

NAMD's software development practices have been in place for two decades and are largely shared with the codeveloped visualization program VMD. NAMD development is done primarily by a few key technical staff, with contributions of new features from scientists both within the National Institutes of Health (NIH) Center and externally. In order to retain full continuity of the source code revision history, NAMD uses Git version control system and uses Gerrit for code review and testing in same way that Charm++ is hosted, which will be described more later in this section. Branches are not used. NAMD is maintained as production-quality software at all times, with source code and read-only repository access publicly available and recent commits tracked publicly online.

NAMD Linux and Linux-CUDA binaries are automatically built nightly and posted for public download, whereas semiautomated NAMD builds across all supported local and remote supercomputer platforms are launched manually after major commits and installed as *latest* binaries available to all users of the machines, ensuring that features are regularly exercised by the researchers of the NIH Center prior to final release. New NAMD features are developed in collaboration with driving projects that provide eager test users, and the initial input sets are preserved for regression testing during refactoring. All NAMD executions include extensive internal checksums and stability tests that raise fatal errors on unexpected conditions.

A complete NAMD User's Guide is maintained, documenting all nonexperimental simulation features. Instructions for building and running the software are distributed as release notes both online and with each binary and source code download. More detailed information on running Charm++ programs, such as NAMD, is linked from the Charm++ website (http://charm.cs.illinois.edu). The NIH Center has since 2003 taught 46 hands-on computational biophysics workshops at locations across the US and the world. The hands-on components of the workshops are selected from 40 tutorials that teach both basic and high-end NAMD and VMD skills. The tutorials are publicly available via the Center/NAMD website at all times (http://www.ks.uiuc.edu/Research/namd/).

Although we enable and encourage a growing NAMD *user* community, with many independent projects requiring zero interaction with the development team, prospective outside *developers* are strongly encouraged to contact the development team, either privately or via the public e-mail list, before beginning any significant alteration or extension. In most cases, the desired functionality is either already available, inadvisable for some reason, or readily implemented via the NAMD Tcl scripting interfaces (documented in both the User's Guide and the tutorial "User-Defined Forces in NAMD"). For the remaining cases, the development team is often able to point to specific small modifications necessary, or to suggest analogous features that may be used as an exemplar without needing to understand the full parallel execution of the program. The NAMD source code is maintained in a legible state, with multiple-word variable and function names and brief explanatory comments as necessary to clarify algorithms and warn of potential hazards, and online NAMD source code documentation is generated nightly. We have experienced many times that graduate students will ask for guidance on adding some capability and, when asked several weeks later, report that they were successful without further assistance.

The primary barrier for new NAMD developers is learning Charm++ message-driven programming, which is not widely taught as is MPI, but is equally intuitive for beginners and presents a smaller set of concepts necessary for advanced usage. Complete documentation is available, and the Charm++ source distribution includes many tests and examples. The Charm++ performance visualization tool *Projections* allows a complete and interactive view into the message-driven execution of a program and is used extensively for parallel performance tuning by NAMD developers. Further, the Parallel Programming Laboratory has since 2002 hosted an annual "Workshop on Charm++ and its Applications" with archived presentation slides and videos, as well as a hands-on Charm++ tutorial.

The main NAMD and VMD software components are distributed under a custom open source license agreement developed by the NIH Center in 1999. The license is designed to minimize restrictions on scientific users of the software while ensuring that the center is able to collect use and impact statistics to justify continued funding. Thus, the primary restriction in the license is against redistribution of more than 10% of the software source code, or binaries created from it. The licensee has an obligation to acknowledge use of the software and to cite a specific reference paper in resulting publications. The license specifically grants the user the right to create and distribute complementary works such as scripts and patches and to redistribute without restriction works with up to half of their noncomment source code derived from at most a tenth of the noncomment source code of the software packages. The later clause allows unrestricted reuse of functionality, such as the core NAMD computational kernels. In contrast, under the GPU Public License a single line of copied code subjects the entire work to the license.

NAMD and VMD are distributed by web download. Registration for a username and password is required to enable tracking of download statistics. Charm++ and NAMD releases often coincide to support new platforms or NAMD features, and because NAMD beta releases provide the best test of a new Charm++ release. Matching Charm++ source code is included with the NAMD source distribution. The primary determinants of NAMD releases are the availability of new features of value to the wider user base and the observed stability of the code for internal users. We aim for one major release per year, and several beta releases over multiple months precede the final stable release.

5.5.2 Charm++

Charm++ uses a variety of software development practices to facilitate development. Git is used for version control. Git provides an ability to easily make lightweight branches, work in a decentralized manner, and encourages quick, iterative development, all of which are helpful for allowing the team to provide experimental research and stable software simultaneously. The Charm++ team makes use of precommit code review, requiring every commit to be reviewed by hand and automatically tested before being integrated into the mainline repository. This prevents breaking changes and regressions from being committed to the repository and helps the developers stay abreast of what changes are being made to the codebase because multiple people must look at every change by necessity. When code review was first adopted, there were some concerns that it would unnecessarily slow down the development process and cause patches to stagnate in code review limbo. These worries were not unfounded, as the development velocity did decrease in some cases, but the benefits of code review have vastly outweighed any of the negatives. Bugs are caught early, code style is maintained, implementations are well designed, and so on. Gerrit is used for code review and code hosting (https://charm.cs.illinois.edu/gerrit). Gerrit was chosen because it is easy and free to self-host, it offers more robust code review scoring than most competitors, it integrates well with testing systems, and it is fairly customizable to suit particular needs.

Charm++ uses two different services for testing and validation. The first is a home grown nightly build system called autobuild. Autobuild tests the bleeding edge version of Charm++ every night on various hardware and software platforms. Some of these tests are run on local machines with different configuration options (e.g., Windows vs. Linux vs. Darwin, GCC vs. Clang vs. other compilers), whereas others are run on remote clusters and HPC systems that cannot be tested locally due to lack of hardware. Because Charm++ supports a wide variety of different PC, cluster, and HPC configurations, a system like this is necessary to ensure that changes work well on all supported platforms. Autobuild is implemented as a Cron job, series of shell scripts, and an e-mail and web interface to view results. The second testing system is Jenkins. Jenkins is a widely used automation and build server. The Charm++ testing system integrates Jenkins with Gerrit, triggering a new Jenkins test with every commit made in Gerrit. When the test completes, the results are automatically posted to Gerrit so that code reviewers can use the results in making their reviews. A successful Jenkins result is required before a commit can be merged in Gerrit. Testing every platform for every commit would take too long and cause too much computational work, so commit-triggered Jenkins tests are only run on the default Linux configuration of Charm++. This is sufficient to detect most issues, and manual code review can identify situations when automated testing does not fully test a change, which can be resolved by manual testing. Occasionally, breaking changes make it through this process, but nearly all of these are caught by autobuild the following night.

Charm++ uses a custom build script that wraps configure and make. The custom build script parses various user specified options to customize the build for a particular system, selecting things like the desired network layer, compiler, and optional Charm++ modules to use. Many of the options have an associated shell script and header that are read when the option is selected; these files define the specific environment variables and compiler or linker flags for the selected option. These files are undocumented and are not intended for modification, but they allow advanced users to easily modify the configuration and even add new options.

Charm++ is implemented in C++. Charm++ programs are usually written in C++, but Fortran bindings also exist, as well as interoperation with any language that supports C++ bindings. Charm++ defines some language structures on top of C++. The most significant of these is called Structured Dagger (SDAG). SDAG can be used to express control flow dealing with sending, receiving, and ordering remote messages in Charm++. Charm++ also defines new keywords to deal with its object model, the most significant of which are *chare,* which indicates a class definition for an object exposed to the RTS, and *entry,* which indicates a method that can be remotely invoked with a message sent from a chare.

Charm++ is freely available for noncommercial usage under the Charm++/Converse license. The code is distributed from the website of the University of Illinois Parallel Programming Laboratory (http://charm.cs.illinois.edu). Commercial usage is allowed under a different licensing structure, as well.

Community contributions are allowed to Charm++. In order to contribute, a contributor should register with the code review system described earlier and submit a patch there. From there, the normal review and testing procedure occurs, eventually merging the patch into the mainline repository. Most community contributions come from collaborators of the Parallel Programming Laboratory and are shepherded along by a Charm++ developer, who often takes care of the logistical details and assists the submitter with any issues that may arise.

5.6 BENCHMARKING RESULTS

In order to benchmark NAMD performance for large simulations on petascale platforms we have developed freely distributable synthetic benchmarks by replicating the 1.06 M atom STMV benchmark (satellite tobacco mosaic virus—the first full virus simulation performed with NAMD in 2006). Two benchmarks were formed by tiling the 216.832 A cubic cell: a $5 \times 2 \times 2$ system of 21 million atoms and a $7 \times 6 \times 5$ system of 224 million atoms. The benchmarks use rigid water and bonds to hydrogen to enable a 2 fs time step and use a 12 A cutoff distance with PME every three steps using a 2A grid and eighth-order interpolation. Pressure control with a relaxed reduction-broadcast barrier is enabled to avoid global synchronization.

Below we compare existing petascale architectures in the United States (Figure 5.3). We note that the largest machines, Blue Waters and Titan, date from 2013 and use a Cray toroidal network and AMD Opteron CPUs. Edison and Theta both employ Intel processors, Edison having two older 8-core processors per node, whereas Theta uses a single newly released 64-core Xeon Phi KNL processor with AVX-512 vector units.

A clear performance advantage is provided by the NCSA Kepler GPUS on Titan, outperforming even the latest CPU on a per-node basis. However, as the non-Knights Landing (KNL) systems use two-socket nodes (for Titan one CPU is replaced with a GPU), whereas KNL is single-socket-only,

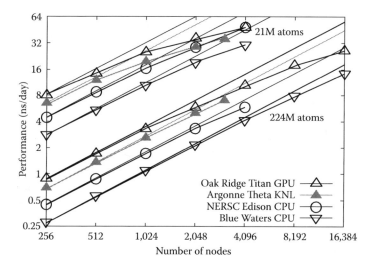

FIGURE 5.3 NAMD strong scaling performance on four systems with 21 and 224 million atoms up to 16 K nodes.

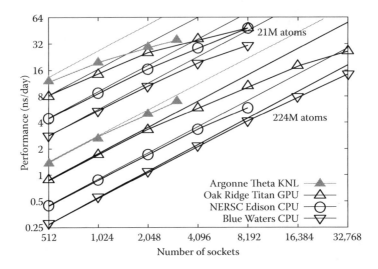

FIGURE 5.4   NAMD strong scaling performance on four systems with 21 and 224 million atoms up to 32 K sockets.

a fairer comparison may be on a per-socket basis (Figure 5.4), for which KNL is dominant (although over much older processors and GPUs).

A significant observation with regards to upcoming 200 petaflop and exaflop machines is that scaling is worse for systems using larger numbers of low-power cores (GPUs and Xeon Phi) than for traditional CPUs. In the case of Theta, this can be ascribed to serial bottlenecks, particularly in the Charm++ communication thread, even though seven processes are being used per-node. For GPU-accelerated Titan serial bottlenecks and communication are handled by the powerful CPU threads, and scalability is limited more by parallel decomposition and offloading inefficiencies in mapping work to the many cores of the GPU.

NAMD is part of the application readiness program for the Oak Ridge Summit machine based on IBM Power9 CPUs and NVIDIA Volta GPUs. We are able to make some performance predictions based on publicly available information. We assume that on the 3400 nodes of Summit, we will obtain a similar efficiency as we do currently on 4096 nodes of Titan, on which our 224 million atom benchmarks run at 9.0 ns/day, 19 ms/step. Scaling this performance by a factor of 4.5 to the full Titan machine, and then by another factor of five for the anticipated relative total performance of Summit, we arrive at an expected performance of roughly 200 ns/day, 0.9 ms/step. This corresponds to 150 16.4 A cubic domains of 440 atoms each per node, sending and receiving at each step positions and forces of 102 domains, a per-node network injection of 2.1 MB/step for short-range forces. PME on a 2 A grid would require $48 \times 48 \times 56$ grid points per node, plus transposes of $40 \times 40 \times 48$ grid points, yielding an additional 2.3 MB on PME steps, for a worst case of 4.4 MB/step. This is well within the 23 GB/s injection bandwidth of the internode network, which at 0.9 ms/step is 20.7 MB/step. Scaling smaller simulations across the full machine would require even higher injection rates and limited contention, so we are hesitant to predict beyond 200 ns/day for a multinode simulation. Multiple-copy algorithms with one or more replicas per node, however, should scale almost perfectly due to their limited interreplica communication.

Collective variables: Collective variables [20,21] (colvars in short) are reduced representations of degrees of freedom of the simulation, which can be either analyzed individually or manipulated in order to alter the dynamics in a controlled manner. Colvars can be sampled extensively to calculate certain statistical quantities accurately, unlike the far more numerous atomic positions. A single colvar can combine different functions of Cartesian coordinates, herein termed colvar components, which can be used to describe macroscopic phenomena.

Serial colvar calculations become a bottleneck as the complexity of user-defined variables increases (e.g., multiple root mean square displacement [RMSDs] [22]). The calculation of individual components of colvars are parallelized using CkLoop/OpenMP on multiple cores of a node, thus giving an even distribution of work across cores (on the same node) in the case of multiple colvars, each defined with many components. An individual component of a colvar is computed serially, thus the parallelization is beneficial only when there are multiple colvars or complex colvars which have multiple components. CkLoop parallelization of colvar in shared memory, as shown as SMP in Figure 5.5, improves colvar performance approximately by 2×.

5.6.1 EXTRAPOLATION TO EXASCALE

5.6.1.1 Science Goals

Looking forward to the next generation of 200 petaflop machines, in particular Oak Ridge Summit and Argonne Aurora, we foresee two classes of simulations. First, more routine and a greater variety of simulations in the 200-million-atom range, which encompasses a variety of small organelles, membrane budding processes, and even complete viruses. Second, highly accurate and detailed multicopy simulations of millisecond-timescale processes in 300,000-atom systems such as complex membrane transporters. Future exaflop machines will enable basic simulations of multibillion-atom assemblies including models of recently experimentally constructed *minimal cells,* as well as enhanced sampling of the full range of smaller simulations. What is yet to be seen is whether future technologies enhance internode communication latencies sufficiently to enable the further reduction of time per step rather than only extending simulation scale and sampling (Figure 5.6).

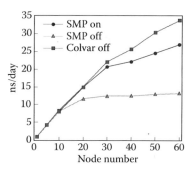

FIGURE 5.5 Colvar performance improvements.

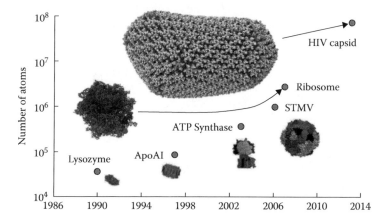

FIGURE 5.6 NAMD simulation sizes over the years.

5.6.1.2   Runtime System Enhancements Needed

What makes biomolecular simulation challenging to scale is the fact that scientists want to study relatively modest-size molecular systems but with huge number of time steps. Even a system with 100 million atoms, when running on tens of millions of cores, has only a few tens of classically represented atoms per core (with only about 24 bytes for each). Yet, because each time step simulates only a femtosecond, billions of time steps are needed for useful science. The challenge at exascale is to be able to complete a single time step in tens of microseconds! To continue to execute NAMD efficiently at exascale, the RTS of Charm++ will need to be improved in several essential ways.

5.6.1.3   Supporting Fine-Grain Computations

Grain size can be defined as the amount of computation per asynchronous (potentially remote) interaction. With current Charm++, average grain sizes as low as tens of microseconds can be handled without significant overhead. We need to reduce the overhead further in order to support the short time steps needed for NAMD. Other applications, including graph algorithms, are also driving this requirement. We have initiated a review of the Charm++ RTS to audit the overhead and to reduce it mainly by specializing implementations of individual chare collections so that only the necessary overheads are paid, and creating leaner representations of chare object handles.

5.6.1.4   Optimizations Related to Wide Nodes

One of the known characteristics of future machines is the large number of cores on each node. Several optimization challenges arise from such wide nodes, which we plan to address in the near future. Serialization due to locks is one such growing challenge, especially for many core chips such as Xeon Phi, but also on regular multicore chips. C11 atomics provide a potential way for avoiding or reducing serialization. One of the steps we are taking is to utilize atomics-based queues for within-node communication. For a widely used code, such as NAMD and Charm++, it is important to ensure that all machines and compilers support primitives that we use, which has made this transition challenging, in addition to the challenges arising from complex race conditions.

High bandwidth memory (HBM) is increasingly being used in next generation supercomputers, especially with accelerators such as general-purpose graphical processing units (GPGPUs) and Xeon Phi. Utilizing this effectively, via adaptive runtime support, is another planned objective. The ability to peak at the scheduler's queue gives the RTS an opportunity to effect prefetch and scheduling policies. At a lower level, scratchpad memories also provide potential for runtime optimization via prefetching.

Nonuniform memory access (NUMA) effects on large multicore chips present another challenge, which needs to be considered along with the issue of overload on communication threads. In the Charm++ RTS, in each process, a thread is typically dedicated to communication processing. Splitting a node's cores into multiple processes increases the ratio of communication threads to worker threads, reducing the potential bottleneck on communication threads. However, this leads to reduced opportunities for sharing resources such as HBM and caches. A potential solution we plan to explore is supporting multiple communication threads within a single process. The success and utility of this approach will depend on the ability of the hardware and the operating system to provide multiple concurrently usable ramps to the network FIFOs. Automatically controlling the number of SMT threads to use at runtime, based on runtime instrumentation, is another optimization we are considering in Charm++. Some of these optimizations will automatically benefit NAMD, whereas others will require some localized restructuring of NAMD code.

One insidious challenge we see to fine-grained performance is the increase in static and dynamic speed variability across cores within a chip, as well as across chips. We have shown the existence and magnitude of such variation in the presence of turbo-boost on supercomputers in use currently. As this variation increases in magnitude, stronger techniques that combine dynamic and static load balancing and reduce overhead without losing locality will be needed. We plan to develop such techniques for use in Charm++ and NAMD, but also to develop flexible mechanisms (such as integration of OpenMP or similar constructs) so that runtime research from other groups, such as the SOL-LVE project, OMPSS project from Barcelona and several others can be used to benefit Charm++ applications.

5.7 RELIABILITY AND ENERGY-RELATED CONCERNS

In the face of power, energy, and reliability challenges in exascale architectures, Charm++ offers multiple solutions to address these concerns for its applications. These features are implemented within the runtime and require minimal code change to be enabled in the applications. As specific needs for these features arise for NAMD users, it will be modified to utilize these features.

5.7.1 FAULT TOLERANCE

Charm++ provides various fault tolerance strategies to handle hard and soft errors: checkpoint/restart [23], message logging [24], proactive evacuation [25], and targeted protection for silent data corruption [26]. The checkpoint/restart mechanism is the most popular technique among them and is the one used in production Charm++ applications. Two schemes are provided:

1. On a shared file system with split execution where the application execution is interrupted and can be resumed later
2. With double local storage (in memory, local disk, or solid state drive) for online automatic fault tolerance

LeanMD is the mini-app version of NAMD and it is commonly used to prototype and test new features. Figure 5.7 shows in-memory checkpoint and restart time on a BlueGene/Q system with simulated failures with LeanMD application of two settings: 1.6- and 2.8-million-atom systems. The time it takes for both checkpoint and restart are in the few tens of milliseconds range. When the number of processors increases from 2 to 32 K, the checkpoint time for the 2.8-million-atom system decreases from 43 to 33 ms. During restart phase, there are several barriers to ensure consistency until full recovery of the crashed node. Because of these barriers, the restart time for the 2.8-million-atom system increases slightly from 66 to 139 ms when the number of processors is increased from 4 to 32 K.

With runtime support, failure detection and recovery is made automatic. Even though the application execution can take longer time to complete, the user does not need to be aware of the failure because the RTS will automatically put the application back on track and make forward

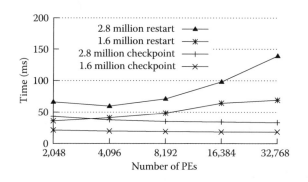

FIGURE 5.7 Checkpoint and restart time of LeanMD.

progress. NAMD uses its own disk-based checkpoint-restart mechanism. However, given this high-performance demonstration of online fault tolerance, we expect that NAMD will use the Charm++ mechanism in the future, especially when job schedulers start allowing applications to handle failures and continue in the presence of failures.

5.7.2 Energy, Power, and Variation

Not only the increasing power and energy consumption with scale, but also power and thermal variations among processors in the supercomputing platforms have become important concerns for HPC applications and data centers. Charm++ applications have a unique advantage: the object/task-based model and overdecomposition property enable applications to adapt to the variations in the hardware via dynamic load balancing and reconfigure the hardware resources transparently from the application while minimizing the performance overhead. Some of the strategies Charm++ offers are thermal-aware load balancing [27], variation-aware load balancing [28], power-aware job scheduling with malleable applications, and hardware reconfiguration [29,30].

5.7.2.1 Thermal-Aware Load Balancing

Thermal variations across components are common in large-scale data centers and may further increase in exascale. A thermal-aware load balancing strategy enables the runtime to remove the thermal hotspots among the processors by applying dynamic voltage and frequency scaling (DVFS) followed by load balancing. This technique can reduce the overhead of frequency reduction compared to naïve DVFS by 20%, while providing up to 57% cooling energy savings [27].

5.7.2.2 Speed-Aware Load Balancing

Future architectures may have heterogeneous components operating at different frequency levels. Even some of today's architectures having dynamic overclocking exhibit frequency variations across identical hardware components due to manufacturing-related reasons [28]. Charm++ offers speed-aware load balancing to obtain better performance under these variations. Any Charm++ load balancing strategies can be speed aware by taking into account the speed of the processors while moving the load from one to another. This way Charm++ can improve performance when the source of load imbalance is not the application but the hardware.

5.7.2.3 Power-Aware Job Scheduling with Malleable Applications

Future architectures may use power-aware resource managers where each application might have a limited power allocation. Charm++ applications can be malleable, that is, they can shrink or expand the number of processors that they are running on [31]. Malleability increases the data center utilization, reduces the mean response time, and achieve high job throughput under a strict power budget [32]. Demonstration of malleability with LeanMD application is shown in Figure 5.8.

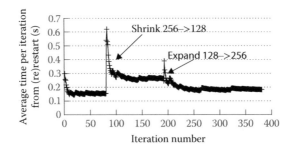

FIGURE 5.8   Runtime system (RTS) adapts the load as the number of processors shrinks and expand.

5.7.2.4 Hardware Reconfiguration

The RTS can adaptively turn on and off or reconfigure hardware components such as caches and network links to reduce power and save energy with minimal performance overhead [29,30]. Some applications, including NAMD, do not use the full caches. For example, with a configuration of 1000 atoms per processor and 80 B of data, per processor uses only 78 KB of memory, which is only a small fraction of a typical last level cache (LLC). Charm++'s runtime can predict the application's future usage and find the best cache hierarchy configuration. With this method 67% of the cache energy can be saved with only 2.4% performance degradation. A similar technique can be applied with network links if the application is not using the network links fully [30].

Many of the current HPC platforms do not give users the necessary rights to control frequency or power, or even to measure power and temperature data. This may change in future exascale architectures with the pressing needs for energy optimizations, and hence allowing the described optimizations to be put in production for applications including NAMD.

5.8 SUMMARY

NAMD is one of the best scaling, most used, and impactful parallel MD applications. Simulations of influenza virus (A/H1N1) and the HIV capsid are excellent demonstrations of NAMD's impact in the field of biomolecular simulations. NAMD is installed on many major supercomputers in the United States and supported on many architectures by means of its underlying robust Charm++ infrastructure. NAMD and Charm++ collaboration started years ago and has generated mutual benefits via its synergistic codesign process, illustrated in multiple features discussed in this paper. NAMD motivated implementation of new features, optimizations, and abstractions in Charm++, and these in turn have enabled implementation of new features and better performance for NAMD. NAMD was one of the first science applications using Charm++. Since then, many other Charm++ science/engineering applications have been developed, including OpenAtom [33], ChaNGA [34], and EpiSimdemics [35], with similar codesign and synergy. Implementation of the described features and optimizations within the runtime has not only benefited NAMD but also other Charm++ applications as well.

ACKNOWLEDGMENTS

First and foremost, we acknowledge the guidance and leadership of Prof. Klaus Schulten, who led the NIH center, the home of NAMD, from 1992 until his passing in late 2016. Many contributors to the development of the NAMD software have made this project a success, including: Bilge Acun, Ilya Balabin, Rafael Bernardi, Milind Bhandarkar, Abhinav Bhatele, Eric Bohm, Robert Brunner, Floris Buelens, Christophe Chipot, Jordi Cohen, Jeffrey Comer, Andrew Dalke, Surjit B. Dixit, Giacomo Fiorin, Peter Freddolino, Paul Grayson, Justin Gullingsrud, Attila Gürsoy, David Hardy, Chris Harrison, Jerome Henin, Bill Humphrey, David Hurwitz, Antti-Pekka Hynninen, Barry Isralewitz, Sergei Izrailev, Nikhil Jain, Neal Krawetz, Sameer Kumar, David Kunzman, Jonathan Lai, Chee Wai Lee, Charles Matthews, Ryan McGreevy, Chao Mei, Marcelo Melo, Esteban Meneses, Mark Nelson, Ferenc Ötvös, Jim Phillips, Brian Radak, Till Rudack, Osman Sarood, Ari Shinozaki, John Stone, Johan Strumpfer, Yanhua Sun, David Tanner, Kirby Vandivort, Krishnan Varadarajan, Yi Wang, David Wells, Gengbin Zheng, Fangqiang Zhu.

This work was supported in part by a grant from the National Institutes of Health NIH 9P41GM104601 "Center for Macromolecular Modeling and Bioinformatics." This research is part of the Blue Waters sustained-petascale computing project, which is supported by the National Science Foundation (award number OCI 07-25070) and the state of Illinois. Blue Waters is a joint effort of the University of Illinois at Urbana-Champaign and its National Center for Supercomputing Applications. This research used resources of the Argonne Leadership Computing Facility at

Argonne National Laboratory, which is supported by the Office of Science of the U.S. Department of Energy under contract DE-AC02-06CH11357. This research also used resources of the Oak Ridge Leadership Facility at the Oak Ridge National Laboratory, which is supported by the Office of Science of the US Department of Energy under Contract No. DE-AC05-00OR22725. This work used the Extreme Science and Engineering Discovery Environment (XSEDE), which is supported by National Science Foundation grant number OCI-1053575 (project allocations TG-ASC050039N and TG-ASC050040N).

REFERENCES

1. L. Kale and K. Schulten. Scalable molecular dynamics with NAMD. *J. Comput. Chem.*, 26, 1781–1802, 2003.
2. B. Acun et al. Parallel programming with migratable objects: Charm++ in practice. In *Proceedings of the International Conference for High Performance Computing, Networking, Storage and Analysis (SC '14)*, New Orleans, Louisiana. IEEE, 2014, 647–658.
3. L. V. Kale and A. Bhatele, Eds. *Parallel Science and Engineering Applications: The Charm++ Approach*. Boca Raton, FL: CRC Press, 2011.
4. J. C. Phillips, Y. Sun, N. Jain, E. J. Bohm, and L. V. Kalé. Mapping to irregular torus topologies and other techniques for petascale biomolecular simulation. *SC Conf. Proc.*, 2014, 81–91, 2012.
5. R. Das and J. Saltz. Parallelizing molecular dynamics codes using the parti software primitives. In *Proceedings of the Sixth SIAM Conference on Parallel Processing for Scientific Computing*, 1993, Norfolk, Virginia, March 22–24, 1993.
6. S. Plimpton. Fast parallel algorithms for shortrange molecular dynamics. *J. Comput. Phys.*, 117, 1–19, March 1995.
7. D. E. Shaw et al. Anton, a special-purpose machine for molecular dynamics simulation, *Commun. ACM*, 51, 91–97, July 2008.
8. G. S. Almasi et al. Demonstrating the scalability of a molecular dynamics application on a Petaflop computer. In *ICS '01: Proceedings of the 15th International Conference on Supercomputing*. New York, NY: ACM Press, 1998.
9. B. Fitch et al. Blue matter: Approaching the limits of concurrency for classical molecular dynamics. In *SC 2006 Conference, Proceedings of the ACM/IEEE*, Tampa, FL, 2006, 44–44.
10. G. Zheng, A. Bhatele, E. Meneses, and L. V. Kale. Periodic hierarchical load balancing for large supercomputers. *Int. J. High Perform. Comput. Appl.*, 25, 371–385, March 2011.
11. J. Comer, J. C. Phillips, K. Schulten, and C. Chipot. Multiple-replica strategies for free-energy calculations in NAMD: Multiple-walker adaptive biasing force and walker selection rules. *J. Chem. Theory Comput.*, 10, 5276–5285, December. 2014.
12. C. Mei et al. Enabling and scaling biomolecular simulations of 100 million atoms on petascale machines with a multicore-optimized message-driven runtime. In *Proceedings of 2011 International Conference for High Performance Computing, Networking, Storage and Analysis—SC '11*, Seattle, WA. IEEE, 2011, 1–11.
13. Y. Sun et al. Optimizing fine-grained communication in a biomolecular simulation application on Cray XK6. In *Proceedings of the International Conference on High Performance Computing, Networking, Storage and Analysis (SC '12)*, Salt Lake City, Utah. IEEE, 2012.
14. D. Kunzman. Runtime support for object-based message-driven parallel applications on heterogeneous clusters. Department of Computer Science, University of Illinois, 2012. http://hdl.handle.net/2142/34256.
15. Y. Sun, G. Zheng, L. V. Kale, T. R. Jones, and R. Olson. A uGNI-based asynchronous message-driven runtime system for cray supercomputers with Gemini Interconnect. In *2012 IEEE 26th International Parallel and Distributed Processing Symposium*, Shanghai, China. IEEE, 2012, 751–762.
16. S. Kumar, Y. Sun, and L. V. Kale. Acceleration of an asynchronous message driven programming paradigm on IBM Blue Gene/Q. In *2013 IEEE 27th International Symposium on Parallel and Distributed Processing*, Boston, MA. IEEE, 2013, 689–699.
17. Libfabric OpenFabrics. *Libfabric Programmer's Manual*. https://ofiwg.github.io/libfabric/.

18. N. Jain et al. Charm++ and MPI: Combining the best of both worlds. In *2015 IEEE International Parallel and Distributed Processing Symposium*, Hyderabad, India, 2015, 655–664.

19. Humphrey, W., Dalke, A., and Schulten, K. VMD: visual molecular dynamics. *Journal of molecular graphics*, 1996, 14.1, 33–38.

20. G. Fiorin, M. L. Klein, and J. Hénin. Using collective variables to drive molecular dynamics simulations. *Mol. Phys.*, 111, 3345–3362, December 2013.

21. NAMD User's Guide. Collective variable-based calculations. 2017. http://www.ks.uiuc.edu/Research/namd/2.12/ug/node53.html. Accessed: July 23, 2017.

22. NAMD User's Guide. Component rmsd: Root mean square displacement (RMSD) with respect to a reference structure. 2017. http://www.ks.uiuc.edu/Research/namd/2.9/ug/node55.html. Accessed: July 23, 2017.

23. L. V. Kale, G. Zheng, and X. Ni. A scalable double in-memory checkpoint and restart scheme towards exascale. In *IEEE/IFIP International Conference on Dependable Systems and Networks Workshops (DSN 2012)*, Boston, MA, 2012, 1–6.

24. J. Lifflander, E. Meneses, H. Menon, P. Miller, S. Krishnamoorthy, and L. V. Kale, Scalable replay with partial-order dependencies for message-logging fault tolerance, In *2014 IEEE International Conference on Cluster Computing (CLUSTER)*, Madrid, Spain, 2014, 19–28.

25. S. Chakravorty, C. L. Mendes, and L. V. Kalé. Proactive fault tolerance in MPI applications via task migration. *HiPC*, 4297, 485–496, Madrid, Spain, 2014, 19–28.

26. X. Ni and L. V. Kale. FlipBack: Automatic targeted protection against silent data corruption. In *Proceedings of the International Conference for High Performance Computing, Networking, Storage and Analysis (SC '16)*, Salt Lake City, Utah. IEEE, 2016.

27. O. Sarood and L. V. Kale. A 'cool' load balancer for parallel applications. In *Proceedings of 2011 International Conference for High Performance Computing, Networking, Storage and Analysis—SC '11*, Seattle, Washington: ACM, 2011.

28. B. Acun, P. Miller, and L. V. Kale. Variation among processors under turbo boost in HPC systems. In *Proceedings of the 2016 International Conference on Supercomputing*, ICS'16, Istanbul, Turkey: ACM, 2016.

29. E. Totoni, J. Torrellas, and L. V. Kale. Using an adaptive HPC runtime system to reconfigure the cache hierarchy. In *SC'14: International Conference for High Performance Computing, Networking, Storage and Analysis*, New Orleans, Louisiana. IEEE, 2014, 1047–1058.

30. E. Totoni, N. Jain, and L. V. Kale. Toward runtime power management of exascale networks by on/off control of links. In *2013 IEEE International Symposium on Parallel and Distributed Processing, Workshops and Phd Forum*, Cambridge, MA, 2013, 915–922.

31. A. Gupta, B. Acun, O. Sarood, and L. V. Kale, Towards realizing the potential of malleable jobs. In *2014 21st International Conference on High Performance Computing (HiPC)*, Dona Paula, India. IEEE, 2014, 1–10.

32. O. Sarood, A. Langer, A. Gupta, and L. Kale, Maximizing throughput of overprovisioned HPC data centers under a strict power budget. In *SC14: International Conference for High Performance Computing, Networking, Storage and Analysis*, New Orleans, Louisiana. IEEE, 2014.

33. N. Jain et al. OpenAtom: Scalable ab-initio molecular dynamics with diverse capabilities. In *High Performance Computing: 31st International Conference, ISC High Performance 2016*, Frankfurt, Germany, June 19–23, 2016, Proceedings. Vol. 9697. Springer, 2016.

34. H. Menon et al. Adaptive techniques for clustered N-body cosmological simulations. *Comput. Astrophys. Cosmol.*, 2, 1, March 2015.

35. M. V. Marathe, C. L. Barrett, K. R. Bisset, S. G. Eubank, and Xizhou Feng, EpiSimdemics: An efficient algorithm for simulating the spread of infectious disease over large realistic social networks. In *Proceedings of the 2008 ACM/IEEE conference on Supercomputing (SC '08)*, Austin, Texas. IEEE, 2008.

6 Developments in Computer Architecture and the Birth and Growth of Computational Chemistry

Wim Nieuwpoort and Ria Broer

CONTENTS

6.1 Introduction...145
6.2 Evolution of Computers and Their Use in Quantum Chemistry 145
6.3 Evolution of Quantum Chemistry Programs in the Early Years of Computational
   Chemistry..147
References 149

6.1 INTRODUCTION

It goes almost without saying that the impressive development of computational chemistry in the past 50–60 years is a direct consequence of the even more impressive developments in computer hardware and software in that same period. However, this development also required the vision and skills of pioneering scientists that saw the new possibilities early, such as Roothaan and students at the University of Chicago, Slater's group at MIT, and the Boys group at Cambridge, United Kingdom. Yet, we might recall in this context that it was a chemical physics problem, the calculation of the dielectric constant of Helium, that inspired John Atanasoff, working on his PhD thesis in Madison, Wisconsin in 1930, to think about designing an electronic calculating machine. About 10 years later, at Ames, Iowa, with the help of his student Clifford Berry, he succeeded in building the first electronic computer— the ABC, the Atanasoff–Berry computer. Because of war circumstances, this invention was never patented and only in October 1973 it was legally settled that the ABC and not the ENIAC was the first electronic computer ever built. A fairly complete account of this interesting episode in computer history can be found on the site www.columbia.edu/~td2177/JVAtanasoff/JVAtanasoff.html.

In the following, we limit ourselves to quantum chemistry, noting that similar developments took place in molecular dynamics. An interesting difference is that to the best of our knowledge dedicated machines have been built and used in the latter case but not in quantum chemistry.

6.2 EVOLUTION OF COMPUTERS AND THEIR USE IN QUANTUM CHEMISTRY

The use of computers to investigate the electronic structure of atoms, molecules, and solids, initiated by pioneering research groups as mentioned above, penetrated, rather slowly around 1960, into theoretical chemistry. The reasons for this are clear. Although by that time most first-generation machines such as IBMs 650 and 704 based on vacuum tubes had been replaced by more powerful (speed

100–1000 FLOPS, main memory 8–32 K words) and more reliable second-generation transistorized installations, they were still very expensive to buy and operate. The number of machines available for general research was therefore small and users had to reserve time and make travel arrangements to work on them. Moreover, they were difficult to use because of manufacturer-dependent differences in hardware (memory sizes, word lengths, peripheral memory) and system software. Input methods, such as paper tape or punched cards, were slow and error prone. Early versions of higher level languages, FORTRAN in particular, were emerging but in order to use a machine efficiently, programs had to be written in machine and/or assembler language and one had to be well aware of the machine architecture. In spite of this, the building blocks that form the foundation of present-day quantum chemical software were designed and implemented already at that time. The basic ingredients followed from Roothaan's groundbreaking work [1] on solving the Hartree–Fock equations for molecules by the linear combination of atomic orbitals-molecular orbital (LCAO-MO) method: one- and two-electron integral packages for atoms and molecules (at the time, based on Slater-type orbitals [STOs]) and methods to deal with nonlinear matrix eigenvalue equations.

Life became much easier when the next generation of machines, such as the International Business Machines (IBM) 7094 and 360 series and Control Data Corporation (CDC) 3200, 3600, and 6600 appeared in the period 1965–1975. Based on medium- to large-scale integrated circuit technology, these systems offered much more CPU-speed (up to a few MFLOPS), CPU-independent I/O channels or peripheral processors, larger main memories (100–1000 KB), magnetic tape units, and later magnetic disks. Punched card input remained. However, input decks could now be prepared and checked off-line, saving not only precious CPU time but also a lot of user frustration. System software generally became more transparent and user-friendly. FORTRAN was accepted as the programming language of choice for computationally intensive tasks, which improved the transferability of programs between machines of different companies considerably. This is the period that saw the birth of much that makes up the bulk of computational quantum chemistry packages today, for example, the general acceptance of Gaussian-type orbitals (GTOs) as a computationally far more efficient alternative for STOs and the assembling and integration of separate pieces of code into more or less complete program suites to calculate the electronic structure and properties of molecules. Important also was the fact that the growing number of molecular calculations brought to light quantitatively the limitations of the Hartree–Fock method leading to several approaches to include the effects of the electron–electron interaction beyond the mean-field theory, for example, second-order perturbation theory, configuration interaction (CI), and density-functional theory (DFT) *avant la lettre*, the Hartree–Fock–Slater (HFS) or X_α theory by Baerends, Ellis, and Ros.

In the period 1975–1990, two divergent lines in machine development can be recognized. At first, a continuation of the traditional mainframe line: very large system integration, specially designed processors culminating in the (multi)vector processor Cray systems, MFLOP performance, megabyte memories, and gigabyte disks. Supercomputer centers were instituted in several countries to make the most powerful facilities available to the scientific and engineering community. A successful endeavor considering the fact that data communication was still in its infancy. Simultaneously, there was a growing demand for smaller and cheaper machines or minis that could be used locally, such as the popular VAX-series of Digital Equipment Corporation (DEC). But the real game changer was the other line of development, the appearance of microprocessors. Having been around on a small scale in the 1970s, the development and use of these single-chip devices took off explosively in the 1980s by the unbelievably successful introduction of the personal computer. After 1990, it became increasingly clear that the era of sophisticated, specially designed powerful proprietary processors was over. Their cost became prohibitively high in comparison with the mass-produced commodity processing and memory chips. Together with breakthroughs in switching and network technology that occurred in the same period of time, these microdevices led to the design of a variety of computer configurations and architectures. The relatively short-lived efforts of building special machines for massive parallel processing around 1994 are worth mentioning, although their meaning for quantum chemistry was limited. The Connection Machine produced by Thinking Machines, in particular, the ingeniously designed CM-5, a multiple instruction, multiple data (MIMD) machine, and the Cray T3D and T3E. For quantum chemistry, the main-stream developments, from small to very large local clusters of powerful workstations, all the way to the gigantic multicore assemblies of today's petascale computers offered ample opportunities for method developments as well as for large-scale or high-precision applications. Quantum chemists have been quick to grab the opportunities by converting serial codes to parallel versions with the help of parallelizing compilers. Optimal use of the available hardware was in general a far-away goal, however. In a way, the situation resembled that of the 1960s. Writing a good parallel code is mainly a manual process that requires on the one hand detailed knowledge of the algorithms chosen to solve the physical problem and on the other hand knowledge of the machine architecture and available parallel programming models and tools. That is a lot to know! It is not surprising therefore that over the years, we have seen that the development of good quantum chemistry codes has become a group effort in which the expertise of specialists is combined. These groups can be a delocalized cooperation of workgroups such as the DIRAC group or a concentration of experts at one place working on a commercial or a publicly available product. Moving toward the exoscale era it seems that a continuation of this trend is a *sine qua non*.

6.3 EVOLUTION OF QUANTUM CHEMISTRY PROGRAMS IN THE EARLY YEARS OF COMPUTATIONAL CHEMISTRY

The first quantum chemical calculations were done as early as 1927 by Heitler and London [2]. Two decades later, in 1951, Roothaan showed that one could cast the Hartree–Fock equations into matrix form by introducing a basis set expansion for the MOs [1]. Only then, it became clear that Hartree–Fock calculations could become feasible for polyatomic molecules.

Although the first quantum chemical calculations were performed using single-purpose programs, scientists soon realized that it was more practical to write computer programs that, perhaps with some modifications, could be used for quantum chemical calculations on various atoms or molecules. In the 1960s, quantum chemists began to write scientific programs for quantum chemical calculations that could be used on more than one computer, primarily in FORTRAN II. At that time, most computers were IBM-made and words had 36 bits. In 1964–1965, CDC introduced computers with increased word lengths of 48 bits and later 60 bits. Between 1960 and 1970, in many research groups, PhD students and postdocs were producing computer codes for doing quantum chemical calculations for molecules based on valence bond theory or on the LCAO-MO approach. They needed to have knowledge not only of quantum chemistry but also of the operating system and of other technical details of the machines they were using. The programs were stored on punched cards or on magnetic tape. In parallel to programs for first principles Hartree–Fock calculations, methods and programs for semiempirical calculations were developed. Examples of successful semiempirical methods used at that time are the extended Hückel method of Roald Hoffmann [3], the complete neglect of differential overlap (CNDO) method of John Pople and coworkers [4], and, later, Mike Zerner's INDO/S semiempirical model (ZINDO) method for predicting UV/vis spectra [5]. In the early 1970s, Hehre and Pople had developed their Gaussian 70 package for *ab initio* quantum chemical calculations [6] and other efficient packages such as ATMOL and POLYATOM appeared, and also at IBM, a program suite IBMOL was released. These programs were able to perform not only Hartree–Fock but also CI and multiconfiguration (MC) calculations. Most of the programs mentioned here are no longer in use, but the Gaussian package has been updated many times and recently, a new version, Gaussian 16, has appeared.

In the 1970s and 1980s, a collection of quantum chemical packages was used, including among others HONDO, GAMESS(US), GAMES(UK), SYMOL, PCILO. The package for HFS calculations of the Free University of Amsterdam [7] deserves to be especially mentioned. It was based on the $X\alpha$ approximation introduced by Slater and Johnson and it was later renamed the Amsterdam density-functional (ADF) package because the HFS method could also be regarded as a flavor of the Kohn–Sham DFT. Spatial symmetry was exploited as much as possible to save computer time and space and to enable easier interpretation of the outcomes.

It was clear from the beginning that it was inefficient to have a program used by only the members of the research group in which it was generated. Moreover, in many groups sooner or later a continuity problem arose: once the student or postdoc who had written the program had left in the group often no one was present who knew the computational and technical details of the program. This was especially a problem if there was no documentation of the program (and its input) available in addition to the program. Often, the only documentation was in the form of so-called *comment cards* in the source code.

The idea of having an exchange of quantum chemistry programs among the various groups who were active in quantum chemistry has arisen, as early as 1962. In 1963, the Quantum Chemistry Program Exchange (QCPE) system was launched as a service to collect and distribute computer software in source form. The reasons for setting up the exchange system were threefold. First, it was inefficient for graduate students at one university to have to write a program that was similar to a program that had already been written elsewhere; second, if a student or postdoc left a university the results of their labor would not be lost, and, third, an intermediary between the code writers/owners and users could be created [8]. Moreover, depositing a program to the QCPE was a way to *publish* the program. The QCPE thus served as a system through which individual researchers could donate their programs to the quantum chemistry community. The availability of the programs was announced through QCPE's newsletters and a catalog. Programs were sent to subscribers who paid a modest distribution and handling fee. QCPE shipped not only the source code and documentation, but also a sample input and a printout of the corresponding output [8]. The idea turned out to be successful, the library quickly grew from 23 programs in 1963 to 71 programs in 1965. The QCPE system not only gathered and distributed program packages, it also organized workshops on applications of quantum chemistry, in particular discussing the latest programs and the advantages and limitations of each method. Around 1980, QCPE flourished, and about 2500 programs were distributed per year. The MOPAC program, that was able to carry out geometry optimizations in an efficient way, was the most popular program. The success of MOPAC coincided with the introduction of the hugely successful VAX 11/780 superminicomputers from DEC [8]. The cheap VAX machines with their easy-to-use operating system significantly changed the way computational chemistry was being done [8]. In the mid-1970s, vector machines came to the market, leading to the need to vectorize the existing quantum chemical software. From the mid-1980s, QCPE started to become less successful, partly due to the commercialization of computational chemistry software. The next decade showed a further commercialization of computational chemistry software. In the 1990s, the Internet came to life, making it easier for individuals to exchange software, another reason why fewer and fewer programs were distributed through QCPE.

In 1989, Enrico Clementi launched the Modern Techniques in Computational Chemistry (MOTECC) initiative advocating a *global simulation* methodology. The main idea was to "pull together researchers from different fields, sharing enthusiasm for computing, modeling, and for learning from each other the different ways the scientific method evolves today, stressing logic, mathematics, and physics, but also capable of dealing more and more with the complexity of biology, neurology and, eventually, sociology" [9]. The MOTECC initiative lead to a series of books, titled MOTECC and later Methods and Techniques in Computational Chemistry (METECC). The MOTECC [9] and METECC [10] book series collected chapters containing detailed descriptions of computational chemistry program packages with the underlying theories, mostly in the areas of molecular dynamics and quantum chemistry. A selection of quantum chemistry packages described in the series: KNG-MOL, HONDO, SIRIUS, ATOMCI, AMPAC, HYCOIN, MCHF, MCDF, HCI, LINMOL, SIRIUS,

MOLCAS, MELD, ALCHEMY II, SAPT, and IGLO. The last volume in the series, METECC95, contains descriptions of parallelized versions of SINDO1, MNDO94, MOLFDIR, IGLO, ADF, and TURBOMOLE.

The noncommercial quantum chemistry program packages listed in the above have in most cases been created in one single research group, or at most through collaboration between a few groups. In the past 20 years, we have seen a concentration of programming efforts: many scientists, from many different groups all contribute to one common package. As an example, we consider the package for four-component relativistic calculations MOLFDIR that was created in a long-lasting effort of only one research group [11]. Although the MOLFDIR package is still used for specific problems (especially when the use of non-Abelian symmetry is important), the majority of researchers in this field, including some authors of MOLFDIR, have moved to another four-component package for relativistic calculations, DIRAC [12]. DIRAC is a package with contributions from many authors working in many groups. The example given here reflects a general tendency toward using more general, more efficient but also more complex quantum chemistry program packages that are maintained and improved in a joint effort of many researchers. The adaptation of such complex packages to novel massively parallel computer architectures requires team efforts that are far from trivial. On the other hand, when successful, such adaptations yield extremely powerful programs that can benefit optimally from initiatives like the INCITE program.*

REFERENCES

1. Roothaan, C. C. New developments in molecular orbital theory. J. *Rev. Mod. Phys.* 23 (1951) 69.
2. Heitler, W. and London, F. Wechselwirkung neutraler Atome und homöopolare Bindung nach der Quantenmechanik. *Zeitschrift für Physik.* 44 (1927) 455–472.
3. Hoffmann, R. An extended Hückel theory. I. Hydrocarbons. *J. Chem. Phys.* 39 (1963) 1397–1412.
4. Pople, J. and Beveridge, D. *Approximate Molecular Orbital Theory.* New York, NY: McGraw-Hill, 1970.
5. Ridley, J. and Zerner, M. Intermediate neglect of differential overlap technique for spectroscopy— pyrrole and azines. *Theor. Chim. Acta.* 32 (1973) 111–134.
6. Hehre, W. J., Lathan, W. A., Ditchfield, R., Newton, M. D., and Pople, J. A. *Gaussian70*, Quantum Chemistry Program Exchange, Program No. 237, 1970.
7. Baerends, E. J., Ellis, D. E., and Ros, P. Self-consistent Hartree–Fock–Slater calculations I. The computational procedure. *Chem. Phys.* 2 (1973) 41–51.
8. Boyd, D. B. Quantum chemistry program exchange, facilitator of theoretical and computational chemistry in pre-Internet history. *In Pioneers of Quantum Chemistry.* Strom, E. et al. ACS Symposium Series. Washington, DC: American Chemical Society, 2013.
9. Clementi, E. (Ed.). *Modern Techniques in Computational Chemistry: MOTECC-89.* New York, NY: IBM Kingston, 1989.
10. Clementi, E. and Corongiu, G. (Eds.). *Methods and Techniques in Computational Chemistry: METECC-89.* Strasbourg, France: Université L. Pasteur, 1995.
11. Visscher, L., Visser, O., Arts, P. J. C., Merenga, H., and Nieuwpoort, W. C. Relativistic quantum chemistry—The MOLFDIR program package. *Comp. Phys. Comm.* 81 (1994) 120–144.
12. Visscher, L., Jensen, H. J. A., and Saue,T. with new contributions from Bast, R., Dubillard, S., Dyall, K. G., Ekström, U., Eliav, E., Fleig, T., Gomes, A. S. P., Helgaker, T. U., Henriksson, J., Ilias, M., Jacob, Ch. R., Knecht, S., Norman, P., Olsen, J., Pernpointner, M., Ruud, K., Salek, P., and Sikkema, J. DIRAC, a relativistic ab initio electronic structure program. http://dirac.chem.sdu.dk.

---

* The Innovative and Novel Computational Impact on Theory and Experiment (INCITE) program of the U.S. Department of Energy Office of Science provides access to some of the fastest supercomputers to researchers from academia, government labs, and industry.

7 On Preparing the Super Instruction Architecture and Aces4 for Future Computer Systems

Jason Byrd, Rodney Bartlett, and Beverly A. Sanders

CONTENTS

7.1 Scientific Methodology 151
7.2 Algorithmic Details 153
7.3 Programming Approach 155
    7.3.1 Aces4 and Domain Scientists 155
    7.3.2 Aces4 System Development 155
        7.3.2.1 Structure of the SIA 155
        7.3.2.2 Workers 157
        7.3.2.3 Servers 157
        7.3.2.4 Load Balancing 158
        7.3.2.5 Barriers 158
        7.3.2.6 Exploiting GPUs 159
7.4 Scalability 159
7.5 Performance 159
7.6 Portability 160
7.7 External Libraries 161
7.8 Software Practices 161
7.9 Benchmark Results 162
7.10 Other Considerations 162
    7.10.1 Fault Tolerance 162
7.11 Conclusion 163
Acknowledgments 163
References 164

7.1 SCIENTIFIC METHODOLOGY

The Super Instruction Architecture (SIA) [1–3] is an approach to parallel programming targeting scientific applications with extremely large arrays. Its original instantiation was ACES III [4], a software platform for computational chemistry focused on highly accurate *ab initio* electronic structure methods. Aces4 [5] is a from-the-ground up reimplementation of the SIA with extensions to the DSL, new capabilities including the ability to handle block-sparse arrays, and an improved software architecture. Although the SIA is used as a parallel programming platform for other suitable domains,* the

---

* For example, a meteorology application has been implemented using our SIA implementation.

majority of the effort in SIA/Aces4 development has been focused toward supporting quantum chemistry (QC) algorithm and theory development, in particular highly accurate many-body solutions to the Schrödinger equation. Table 7.1 gives a list of QC methods that have been implemented using Aces4 to date. It should be noted that the last two methods in Table 7.1 are made computationally feasible only by the new capabilities in the new implementation of the SIA.

The goal of many-body theory in QC is to be able to compute any desired intrinsic property of molecular systems ranging from small gas phase compounds to proteins and crystals, and to do so *ab initio* without any empirical corrections. Such molecular properties are obtained by solving the (time independent) Schrödinger equation

$$\hat{H}|\psi\rangle = E|\psi\rangle \tag{7.1}$$

for the wavefunction $|\psi\rangle$ and total energy E. The wavefunction $|\psi\rangle$ is an orthonormal basis that spans the Hilbert space of $\hat{H}$, such that the inner product $\langle\psi|\psi\rangle = 1$ and $Pdr = \psi(r)\psi(r)dr$ is a probability density.

Once the wavefunction has been obtained, any desired property can be (at least in principle) obtained using response theory using the Hellman–Feynman theorem

$$\frac{\partial}{\partial\lambda}E = \frac{\langle\psi|\frac{\partial}{\partial\lambda}\hat{H}|\psi\rangle}{\langle\psi|\psi\rangle} \tag{7.2}$$

Here the desired property is represented as the derivative (e.g., response) of the total energy E with respect to the independent variable λ. The Born–Oppenheimer* electronic Hamiltonian $\hat{H}$ has five components representing kinetic energy of its electrons and nuclei, the attraction of the electrons to the nuclei and the interelectronic and internuclear repulsions

$$\hat{H} = -\sum_i \frac{\hbar^2}{2m_e}\nabla_i^2 + \sum_{ik} \frac{e^2 Z_k}{r_{ik}} + \sum_{i<j} \frac{e^2}{r_{ij}} \tag{7.3}$$

TABLE 7.1
QC Methods Implemented in Aces4

Restricted and unrestricted Hartree–Fock and gradients

Restricted and unrestricted second-order Möller–Plesset (MP2) energy and gradients

Restricted coupled-cluster singles and doubles (CCSD) including linearized variations (LCC) and gradients

Restricted standard and lambda triples corrections to CC singles and doubles (CCSD(T))

Restricted configuration interaction singles (CIS) including the corrections from perturbative doubles (CIS(D)) or CC perturbation theory (CIS-CCPT)

Restricted equation-of-motion (EOM) CC theory for second-order Möller–Plesset, CCSD including linearized variations, and second-order CC perturbation theory

One-electron response properties for all CC-based methods

Restricted second-order molecular-cluster perturbation theory (MCPT) [6]

Restricted fragment-effective-field coupled-cluster perturbation theory (FEF-CCPT) and CIS corrected by FEF-CCPT [7–9]

Note: Those in bold font are novel methods not available in other quantum chemistry methods and which are not feasible in ACES III.

---

* Nuclei are assumed to be so massive compared to the mass of an electron that the nuclei are effectively frozen in space.

where i and j label electrons and k labels nuclei; $\hbar$ is Plank's constant divided by 2π, m_e is the mass of an electron, e is the charge of an electron, Z is the atomic number of nucleus k, and r_{ab} is the distance between particles a and b and ∇_i^2 is the Laplacian operator. In Cartesian coordinates, ∇_i^2 has the form

$$\nabla_i^2 = \frac{\partial^2}{\partial x_i^2} + \frac{\partial^2}{\partial y_i^2} + \frac{\partial^2}{\partial z_i^2} \tag{7.4}$$

which makes Equation 7.1 a second-order differential equation. There are a few customary ways to solve the Schrödinger equation: numerically solve the differential equation, path integrals, integral transformations, or by expanding in a basis set of polynomials, which transforms the differential equation into a system of nonlinear equations that are solved using linear algebra techniques. It is this latter method of expanding in a basis that Aces4 was designed to support, thus the driving focus of all theory and algorithm development fundamentally revolves around the efficient manipulation of very large matrices.

Current research in theoretical computational chemistry is aimed at developing new methods to deal with large molecular systems such as proteins, nucleic acids, and crystals, as well as handling excited states, the types of problems where exascale computers will find application. These calculations are typically infeasible with standard methods, and difficult or impossible to implement with most existing software platforms; they generate many terabytes of data and have scaling characteristics ($O(n^5)$ or worse) that prove challenging for developing efficient programs. Methodological developments in many-body theory to allow these systems to be studied involve reducing the size of the data and improving the overall scaling, for example, by replacing a tensor with a lower dimensional approximation, exploiting spacial symmetry, and/or taking advantage of locality to create sparsity. Sparsity reduces both the amount of computation and the amount of meaningful data, but taking advantage of this requires appropriate data structures. Aces4 was designed with this demand in mind.

7.2 ALGORITHMIC DETAILS

While the methods given in Table 7.1 vary, they share a computational structure based on iterative solutions of problems dominated by tensor algebra using very large multidimensional arrays that need to be partitioned and distributed among the nodes in the system. Data access patterns are irregular and the algorithms are complex. Calculations typically comprise several well-defined steps where the amount of data shared between steps, while significant, is much less than the amount of data used within a step.

To help deal with the complexity of managing the large arrays, most QC software packages use some sort of middleware. Perhaps the most well known such middleware for dealing with large distributed arrays is the Global Arrays (GA) toolkit [10] and its successors such as the DArray library [11]. GA was used, for example, to implement NWChem [12], one of the most prominent parallel QC suites. GA offers a one-sided communication abstraction that allows arbitrary sections of distributed, dense, arrays to be accessed with a shared memory-like style using a global index space. The basic programming model is to copy a section of an array from GA to local memory (get), update it locally, then copy or accumulate it back in global memory (put). The GA toolkit has been in development for more than 20 years and offers a variety of enhancements to the basic functionality described above.

The SIA/Aces4 programming model shares some aspects of GA, but is based on the abstraction of blocked arrays rather than GAs. SIA/Aces4 distributed arrays are statically partitioned into blocks, which need not have uniform size, and distributed among a set of server processes. Blocks are the object of interest rather than individual array elements and a key aspect of the SIA is that algorithms are expressed in terms of operations on blocks, not individual floating point numbers. A rich, *extensible* set of operations on blocks is provided. These operations, or computational kernels, are called

```
1 pardo M,N,I,J
2   tmpsum[M,N,I,J] = 0.0
3   do L
4     do S
5       get T[L,S,I,J]
6       get V[M,N,L,S]
7       tmp[M,N,I,J] =
8           V[M,N,L,S] * T[L,S,I,J]
9       tmpsum[M,N,I,J] += tmp[M,N,I,J]
10    enddo S
11   enddo L
12   put R[M,N,I,J] = tmpsum[M,N,I,J]
13 endpardo M,N,I,J
14 sip_barrier
```

FIGURE 7.1   SIAL implementation of tensor contraction $R_{ij}^{\mu\nu} = \sum_{\lambda\sigma} V_{\lambda\sigma}^{\mu\nu} T_{ij}^{\lambda\sigma}$.

super instructions and operate on super numbers (blocks), hence the name Super Instruction Architecture. Some super instructions are built in to the DSL; others are provided by domain scientists and are implemented using a general purpose programming language.

As an example that will serve to illustrate both flavor of the DSL and the notion of super instructions, consider the code fragment in Figure 7.1, which computes a four index tensor contraction. T, V, and R are distributed arrays.* The range of each dimension is partitioned into segments; M, N, I, J, L, and S are indices that count segments rather than individual elements. Thus, T[L,S,I,J], V[M,N, L,S] and R[M,N,I,J] are blocks of the distributed arrays, not (in general) individual floating point numbers. tmpsum[M,N,I,J] and tmp[M,N,I,J] are appropriately sized temporary blocks automatically allocated when referenced and deallocated when no longer in scope. The assignment in line 2 assigns a scalar value, here 0.0, to every element in the block tmpsum[M,N,I,J]. The calculation performed in lines 7 and 8 executes a built-in super instruction performing a contraction on the blocks. Line 9 shows a += operation performed elementwise on the two blocks.

Other aspects of the DSL seen in the example include a parallel worksharing loop construct (line 1), serial loops (lines 3 and 4), and one-sided communication constructs (lines 5, 6, and 12) that transmit the block to or from its home location. In the code fragment shown here, the complete calculation of a block of the result array R is performed by a single process while the work is distributed according to the M, N, I, J indices. Larger or smaller parallel tasks could be created by simply moving indices between the parallel **pardo** and serial **do** loops. The available one-sided **put** += (accumulate) command would replace the **put**= if multiple tasks participate in the computation of R.

As can be seen from the example and previous discussion, the SIAL program expresses the coarse-grained structure of the parallel computation in a high level-way. The details necessary for robust, high performance calculations are managed by the runtime system. Among these details include memory management, performing communication in a highly asynchronous way allowing overlapping communication and computation, load balancing, and checkpointing. These will be discussed in more detail below.

The SIA programming model is well suited for hierarchical parallelism. SIAL programs express coarse-grained parallelism between processes while the super instructions, where the vast majority of the computational effort takes place, are implemented in a general purpose programming language, and may exploit fine-grained parallelism in any convenient way. For example, the built-in

---

* The necessary declarations of arrays and indices are not shown.

super instruction, denoted $\star$ in line 8 of Figure 7.1 is a single-node tensor contraction that could be performed serially, vectorized, parallelized with OpenMP, or implemented in CUDA and executed on a graphical processing unit (GPU). In the latter case, the runtime system would also manage transferring data between the host and GPU device.

7.3 PROGRAMMING APPROACH

We will describe the programming approach for Aces4 developers from two points of view. The first is the viewpoint of domain scientists using the system to implement QC methods. The second is the viewpoint of developers of the SIA system, that is, the SIAL compiler and the runtime system.

7.3.1 ACES4 AND DOMAIN SCIENTISTS

Domain scientists use the DSL SIAL to express their algorithms, usually supplemented with new super instructions. A typical QC method is composed of several steps, with each step implemented with a separate SIAL programs. Distributed arrays that should be passed from one SIAL program to a subsequent one are marked as persistent and given a label that can be used to load them in subsequent steps.

A SIAL program is analyzed by the SIAL compiler, error checking and optimization are performed, and the program is translated into a more compact representation we'll call SIAL byte code.* When the program is executed, the byte code is interpreted by the SIA runtime system.

The domain scientist will probably also need to provide implementations for some user-defined super instructions. These super instructions simply take some blocks as input and update and/or generate some blocks as output. They do not engage in communication. The super instructions are compiled into a library that is linked with the Aces4 executable. To fully exploit an exascale computer, super instructions will need to utilize fine-grained parallelism. Although this will require attention to, and possibly reimplementation of each super instructions, this burden is mitigated by several factors. First, it can be done incrementally, one super instruction at a time. Second, super instructions may be implemented in almost any convenient general purpose programming language; the only requirement is that procedures in the language interoperate with C++. Super instructions need to be able to invoke C++ in order to request certain services, such as memory allocation and deallocation, from the runtime system. Special support is provided for super instructions that will be executed on accelerators. This is described in Section 7.3.2.6.

To execute an Aces4 program, the input data in the form of a ZMAT file is given to a preprocessor called AcesInit. This program parses the ZMAT file and generates a file with input data for Aces4. This includes the list of SIAL programs to execute, information about array sizes and segmentation, and initial values for predefined constants and arrays.

7.3.2 ACES4 SYSTEM DEVELOPMENT

The other point of view is of a developer of the SIAL compiler or runtime system. As most of the preparation to exascale will occur here (with the effect of benefiting all of the application programs), this will be discussed in more detail.

7.3.2.1 Structure of the SIA

The workhorse version of the SIA is implemented in C++ using message-passing interface (MPI). Each MPI process is either a server or a worker. The main task of a server process is to own and

---

* Technically, SIAL byte code is not a byte code, but we'll call it that due to its conceptual similarity to Java byte code, which is created by a compiler and then interpreted or further translated by the Java virtual machine.

manage blocks of distributed arrays. This includes responding to worker requests to send or update a block. The server keeps blocks in memory if possible, and writes them to disk if necessary. Servers can perform simple computations on single blocks (like initializing all the elements of a block to a given value) and atomically accumulate blocks arriving from workers into existing blocks. In computational chemistry, the shapes of all arrays used in the program can be determined statically once the input data is known. This information is replicated at both servers and workers.

Workers carry out computations using blocks of distributed arrays transmitted to and from servers as required. The SIAL program fragment above would be executed by a worker process with the `get` and `put` operations involving communication with the server owning the specific block.

This structure dedicates a number of processes to manage data rather than perform computation. In a more typical program structure, all processes perform computation and own part of the data. The design decision in Aces4 to incorporate dedicated servers was based on the realization that as the number of cores increase, dedicating some to tasks other than computation becomes more justifiable, and influenced by experience with ACES III. ACES III offered two types of distributed arrays, one type was managed by servers, the other was distributed among workers. As time went on, more and more ACES III codes were using only the server-based arrays, as these tended to perform better, even though they required more communication. In computational chemistry, access to data over the course of a computation tends to be irregular, so *owner-computes* type strategies are less valuable than in domains where the data access patterns are more regular and predictable. In ACES III, a worker could only respond to messages from other workers requesting blocks between execution of super instructions. This could lead workers to experience significant *block-wait* time while waiting for another worker to send a block. In Aces4, servers are designed to provide fast response to worker messages, and the decision that all distributed arrays would be managed by servers allowed significant simplification of the worker code.

The most important function of the Aces4 runtime system is managing the blocks. Some important aspects of the way blocks are represented and accessed include the following:

- Each block has a convenient identifier that serves as a key into specialized maps. In particular, blocks of an n dimensional array are identified by an $n + 1$ element vector: the first entry identifies the array, the remaining elements identify the particular block in the array. IDs are totally ordered, and given the shape of the array, can also be serialized to integer values and vice versa.
- No memory for data or metadata is allocated for a block (except for blocks of so-called static arrays, which are replicated at all processes and should be small) until it is used in the calculation.
- At each process, block-specific metadata only exists for blocks that are currently located in that process.*
- Block IDs are maintained and known to the runtime system wherever a version of a block is located. (This is in contrast to GA where once a region of data is copied into a local buffer, it is independent of GA.)
- The SIAL compiler deduces how blocks will be accessed (read, write, update, accumulate into, etc.) and encodes this information in the byte code.

The second and third item are key to exploiting sparse arrays. If the algorithm avoids creating blocks that are all zero, which is usually the case, then no resources are required for such blocks. The latter two items, along with a small amount of per/block metadata, make it possible for the runtime system to perform several valuable functions including ensuring consistency of multiple copies of

---

* This is a significant improvement over ACES III, which replicated metadata for every potential block in every array at every node.

a block, performing inexpensive runtime checks for data races at servers,* transparently caching copies of blocks until either they are no longer valid or the memory is required, and efficiently and automatically transferring of blocks between a host and a GPU. Since all of these actions occur at the granularity of a block, the overhead is acceptable.

7.3.2.2 Workers

Worker processes interpret SIAL byte code using a typical interpreter loop; at each iteration, an instruction is fetched then executed. At any point in time, the worker may hold blocks from distributed arrays. Each such block is either a copy of a block held at its owning server, or an updated version that will replace or be accumulated into a block at its owning server. The blocks are held in a specialized map along with blocks from (small) replicated arrays and local arrays. A block's metadata includes its shape and status, for example, whether the block is part of a pending asynchronous operation.

Although the runtime system for workers is (currently) single threaded, there is still significant opportunity for asynchrony. In particular, all communication with servers uses asynchronous computation. For example, the `get T[L,S,I,J]` implementation sends a message requesting the block to the appropriate server, then posts an asynchronous receive. Computation continues, and when the block is needed in line 8, receive status is checked and the worker waits if necessary. One can often arrange things so that more work is done between the get and use, thus completely overlapping communication with computation. (Having the worker automatically prefetch blocks is a possibility for future work.) Similarly, the implementation of `put R[M,N,I,J] = tmpsum[M,N,I,J]` in line 12 performs an asynchronous protocol to transfer the block to the server. The runtime system ensures that the block `tmpsum[M,N,I,J]` is not overwritten or deleted until it is safe to do so. In the meantime, the worker continues with subsequent commands.

SIA/Aces4 is under active development and at this writing several obvious optimizations related to memory management remain to be explored. For example, communication and memory usage could be reduced by using shared memory to allow processes located on the same node to share some data. The SIAL semantics and already available metadata make it possible to do this safely. Currently the system uses the normal C++ memory allocators without any attention to alignment and does not take advantage of opportunities to reuse blocks rather than freeing and reallocating them.

7.3.2.3 Servers

The most basic function of a server is to own blocks of a distributed array and send them to workers when requested. A server executes a loop that receives messages from workers and responds. If the message is a `get`, the server responds by sending the block. If the message is a `put`, the worker retrieves or allocates new memory for the block, posts an asynchronous receive, and sends an acknowledgment to the worker. On receipt of the acknowledgment, the worker transmits the block itself and posts an asynchronous receive for a second acknowledgment message from the server. The initial implementation of `put` simply sent the block. However, experience has shown that performance is much better using asynchronous receives that bypass the MPI buffers, thus the explicit two phase protocol is used. The second acknowledgment message is needed by the protocol implementing the `sip_barrier` to ensure that there are no in-transit messages. The server provides fast response time prioritizing incoming messages over handling asynchronous events. Both servers and workers use a lightweight framework specifically designed to manage asynchronous events.

Because Aces4 is specifically intended for calculations that involve very large amounts of data, it is quite possible for data generated during a calculation to exceed available memory. Thus, servers can write blocks to disk in order to free memory and restore blocks when needed. Except for the

---

* In contrast, one race detection tool for general MPI one-sided communication [13] reported an average of 45% overhead. Overhead in Aces4 is negligible and experience has shown these checks to be very valuable in revealing SIAL programming errors.

possibility of increased latency, this process is transparent to the workers (and the SIAL programmer). It is also necessary to write blocks to disk between SIAL programs, both to serve as a checkpoint and to transfer data between a series of SIAL programs in a job.

7.3.2.4 Load Balancing

Although not seen in the example SIAL code, SIAL **pardo** and **do** loops may contain a **where** clause that constrains the set of iterations that will be performed. For example

```
pardo I,J
   where I<=J
   ...
endpardo I,J
```

would indicate that iterations should be performed over the declared ranges of I and J, but only where $I \leq J$. Clearly, implementing work sharing by partitioning the declared ranges of I and J and assigning the resulting regions to workers is likely to lead to poor load balance. On the other hand, if the tasks to actually be executed are allocated with a knowledge of the where clause, then the load balance is much better. It is also possible that sparsity has arisen in a program because of locality issues. In this case, the iterations that should be executed are not known when the SIAL program is written, but are known before the loop begins. In either case, the obvious solution to have each worker run through all the possible values for I and J, selecting every nth one (with proper offset, where n is the number of workers) that should be executed is inefficient. Thus, Aces4 offers several optimized schemes that can be selected in the SIAL program by adding a pragma to the **pardo** statement.

7.3.2.5 Barriers

The **sip_barrier** instruction on line 14 is the only synchronization primitive in SIAL. Its semantics guarantee that all operations before the **sip_barrier** are complete before the operations after the **sip_barrier** are executed. The implementation of **sip_barrier** by the runtime system is a quiescence detection algorithm [14,15] that detects when (1) all workers have reached the **sip_barrier** and all servers are quiescent (or appear to be), (2) there are no application messages in-transit that could cause a recipient to become nonidle, and (3) the previous conditions are checked in the context of a consistent global state [16]. This is a classical problem in distributed algorithms for which a variety of solutions have been proposed. The algorithm implemented in Aces4 takes advantage of special characteristics of Ace4, in particular, that workers never receive unexpected messages. Servers reply to every message from a worker, either by sending a requested block or sending an acknowledgment message. Thus, workers can easily determine whether all of their own messages have been received by servers. On reaching a **sip_barrier** command, the worker waits for all of its outstanding asynchronous receives to be resolved, then invokes an MPI barrier using a communicator containing only the worker processes. Requiring the server to respond to all messages from workers allows the **sip_barrier** to execute with minimal latency since no message passing is required by the barrier implementation itself to determine state of sent messages, and no involvement is needed from servers. The cost is the extra acknowledgment messages for operations that send blocks to servers.

In contrast to, for example, OpenMP, there are no implicit barriers at the end of loops, rather they are all explicitly given in the SIAL program. This means that if a SIAL program contains two parallel loops with no barrier in between, a worker will immediately proceed to the subsequent loop after finishing the first loop. The ability to execute non conflicting loops concurrently provides a starting point for dealing with situations where the computation in the loop does not scale well when distributed among all the available workers. Instead, the loop could share its work over a subset, and the remainder could be applied to do other work. Successful experiments along these lines have been carried out with ACES III using simple performance models to determine how many workers to assign to a loop.

Barriers naturally divide the computation into a sequence of phases agreed upon by all workers, with the property that the sending phase of messages received at all servers is monotone. In other words, if the worker's phase number is included in messages that are sent, a server should never receive a message with a phase number less than any already received. If we record information about the last access to a block at the server, including access type, worker, and phase number, data races can be dynamically detected. A data race occurs if two accesses in the same phase conflict, that is, they originate from different workers and at least one modifies the block in a way that is not an accumulate.

7.3.2.6 Exploiting GPUs

GPUs can be exploited in the SIA by providing implementations of super instructions that utilize them and arranging for blocks to be transferred to and from the device memory. This was successfully done in ACES III [17]. Special super instructions were provided to allocate and delete blocks on the device, and copy blocks from host to device and vice versa. Instructions were also provided to mark regions of SIAL programs which should be executed on a GPU. Readers are referred to the paper for detailed performance studies, but to give an example, the time to perform CCSD(T) calculation of an RDX molecule with 1000 CPU cores on Blue Waters was reduced from nearly 10 hours using CPUs only to 3 hours with GPUs.

A similar capability is currently being added to Aces4 with a significant improvement, because the access mode of each operation is encoded in each super instruction, the runtime system has sufficient information to automatically, and efficiently, handle data movement between the GPU and host. A block that will be written on the GPU (writing, as opposed to update, requires that all elements of the block be written) is only allocated, a block that is only read on the GPU need not be copied back to the CPU. Once transferred to the GPU, blocks remain there as long as they are valid and the memory is not required for another purpose.

7.4   SCALABILITY

Efforts to scale the SIA/Aces4 have focused on new QC methods that handle very large molecular systems. This combines methodological advancements, such as the introduction of our molecular-cluster perturbation theory [6] approach to solving the fragmented Schrödinger equation is a linear-scaling method (w.r.t. number of fragments) that generates a highly structured and sparse representation of the entire physical system while retaining many of the desirable aspects of infinite-order perturbation theory together with software infrastructure designed to support such algorithms. Scaling results are shown in Section 7.9.

In addition to the features such as disk backing of distributed arrays and the support for sparsity discussed elsewhere, ensuring that Aces4 is scalable required attention to less exciting details within the runtime such as ensuring that values do not exceed the bound of the variables that hold them (e.g., using an int where a long is needed), and in general, always keeping scalability in mind when making design choices.

7.5   PERFORMANCE

An important prerequisite to improving performance is to understand the program behavior. Aces4 is instrumented to collect useful timing data. In particular, we obtain average time for each super instruction execution and loop. This information reveals problem areas where performance can be improved. The coarse-grained nature of SIAL programs is beneficial here; we can time every instruction without significant overhead. A course-grain profile of memory and CPU bottlenecks is particularly helpful when solving nonlinear systems of equations as there are usually several different ways to formulate the problem through the use of intermediate arrays and orders of operation as a way to balance communication, redundant calculation, and total memory footprint. By providing the domain programmer as much coarse-grain information as possible, it is much easier to see how to improve the performance of any given implementation.

There are a number of performance aspects to consider when developing a software package intended to be used on the largest supercomputers. Where possible, we have exploited SIAL semantics and other specifics of the SIA to design high performance features. For example, with the initial straightforward implementation of disk backing during the computation and checkpointing between SIAL programs, disk IO proved to be a significant bottleneck in large calculations. This led to the design of special purpose memory and disk file layouts specialized for the SIA/Aces4 along with protocols implemented directly using MPI-IO.

A first requirement of the new design was that the *virtual memory* facility and the checkpointing facility should be integrated so that blocks that have been written to disk during the calculation and are still valid do not need to be rewritten during a checkpoint. While fully exploiting the potential of this feature remains future work, it clearly opens up opportunities to optimize when data is written to disk.

A second requirement was that the design should exploit and support block sparse arrays; blocks that have not been created during the computation should not use resources of the file system or cause IO. To accomplish these goals with an efficient implementation, the notion of a chunk was introduced. Chunks are intended to hold several blocks and their size is determined using information about the shape of the array, including its maximum block size. Server memory for blocks is allocated in chunks, which are then further divided into blocks as required. A file is created for each disk at startup time. Chunk-sized regions are assigned in a round-robin way among the servers, so that each server can independently determine which parts of the file it has exclusive access to, and as many chunks as needed during the computation may be acquired. Each server keeps track of the location in the file of each block it has written to disk; when a checkpoint is complete, the per-server indices are combined into a single, complete, index for the checkpoint file. Disk IO is performed by reading and writing entire chunks. Disk backing during the computation uses independent IO operations; during a checkpoint, MPI-IO collective functions are performed. Because of the semantics of SIAL, blocks of an array tend to be accessed in approximately the same order at various places in the code; thus fetching a chunk on account of one block is likely also to prefetch additional blocks that will be needed soon. In practice, this design and its implementation have proven to be robust and efficient.

Another example concerns communication latency between servers and workers. Questions include "How quickly does data get from a server to a worker?," "Is there a lot of worker idle time during this process?," "When no data is being requested, how idle are the servers?." The first is a function of how fast the node interconnects are but even so, the developer does have some room to tune performance by adjusting the size of each block of data being transmitted (this is the segment size in Aces4). We denote worker idle time during a request to a server as block-wait time. As the number of workers goes up, any given server has a growing probability of becoming overloaded with requests for data, which was a serious limitation in the ACES III implementation. This remained an issue in the first prototype Aces4 implementation placed on Titan. By implementing asynchronous servers in Aces4, the block-wait time has been reduced to a hundredth of what it was in the original prototype.

7.6 PORTABILITY

Aces4 has been used on a variety of machines including HiPerGator @ University of Florida, DOD Garnet Cray XE6, DOD Haise IBM iDataPlex, DOD Armstrong Cray XC30, DOD Excalibur Cray XC40, DOD Topaz SGI Ice X, DOE Cray XK7 Titan. For convenience of developers, it also can be run on OSX 10.9 and greater and Windows64 under cygwin. To date, the sometimes onerous burden of porting a program between supercomputers has been absent when installing Aces4 on the various

Department of Defense (DOD) and Department of Energy (DOE) machines. This is in counter point to the growing pains seen in ACES III where installing the program on new clusters often required the direct help from one of the developers.

7.7   EXTERNAL LIBRARIES

Aces4 utilizes BLAS libraries and an electron repulsion integral library [18]. An ongoing project is to replace our current shared memory tensor library with the TAL_SH library [19]. The SIAL compiler is written with the assistance of the LPG compiler generator [20].

7.8   SOFTWARE PRACTICES

An important goal of a software platform such as SIA/Aces4 is to make it easier for domain scientists to develop correct, performant code. To a large extent, this is accomplished by isolating the parts of the code that perform communication and asynchronous operations in the runtime system. Debugging and tuning the runtime system then benefits all software developed using the platform.

Nevertheless, effort has been made to provide support for debugging both SIAL programs and the runtime system itself. The code of the runtime system contains liberal use of assertions to document and check invariants. Macros for several types of assertions have been defined: these are conditionally compiled so can be (separately) disabled if desired. In particular, assertions that are likely to be violated due to an error in a SIAL program are distinguished from those that are likely due to a bug in the runtime system itself. In both cases, an assertion failure should yield a useful error message. Aces4 SIAL byte code includes the names defined in the SIAL program, thus error messages can display the line number of the SIAL code and names of involved variables. Although we plan to implement a more useful tool in the future, the structure of the interpreter makes it reasonable to use a standard debugger to inspect the execution of SIAL programs using the source of the interpreter: one can put breakpoints either at the top of the interpreter loop to step through the byte-code instructions, or at the place where a particular instruction of interest is implemented.

The Aces4 runtime system is written in C++ and benefits from adherence to coding standards adopted from Google's coding standards for C++ and especially from a clean, modular design. In the future, we may want to experiment with a communication layer other than MPI; the code structure will allow this to be done with reasonable effort. This claim is supported by the fact that we already easily maintain a version based on MPI that requires at least two processes to run (one server and one worker) and another, single-node version with no dependency on MPI, which is useful when debugging SIAL programs with a debugger.

It is very important when modifying the runtime implementation to ensure that nothing was unintentionally broken by the changes. Aces4 includes several test suites providing sufficient coverage that changes can be made with confidence. The test suites utilize the googletest [21] framework. Googletest with MPI is not completely satisfactory but it is workable. Tests of the runtime system typically use a test_controller class that replaces the usual Aces4 main program. This class accepts synthetic input and allows access to the internal state of the runtime system after the interpretation of a SIAL program is finished. Each test executes at least one small SIAL program designed to exercise specific parts of the runtime system. This approach is easier and more effective than true unit tests with mock environments. The code generation part of the compiler is tested similarly; each feature of SIAL is tested by providing a test that executes a minimal SIAL program containing that feature. Tests are also provided for each quantum mechanics (QM) method. These execute the method for a small input and check the results of the calculation. Where possible, QM methods implemented in Aces4 have been validated by comparing results with results obtained from a different software package.

Certain aspects of our software practices are evolving. Initial development of Aces4 was done on a public github repository, under the GNU open source license. The original build system was based on autoconf tools, and later a CMake-based system was added. The autoconf system no longer seems necessary, so in the future only the CMake-based system will be supported. We are also changing future releases to an Apache open source license and plan to do active development on a private repository with pushes of the master branch to the public repository at sensible points. Due to limitations of the compilers that were available on certain supercomputers, Aces4 has avoided C++11 features. This no longer seems to be necessary, and in the future, C++11 features will be used where appropriate. A small but important future project to better support reproducibility will be to add the capability to output the git hash of the AcesInit, Aces4 and SIAL compiler versions that were used for a run.

7.9   BENCHMARK RESULTS

The coupled-cluster singles and doubles (CCSD) many-body theory is a gold standard method in the computational chemistry field. Despite an unparalleled track record of accuracy, the very high $O(n^6)$ computational scaling and large data requirements place great burden on any implementation. We have illustrated in Table 7.2 the strong scaling efficiency $(t_0/(NProc \times t_N)) * 100)$ of a single Aces4 CCSD iteration for a range of processors and molecular size.*

The molecular-cluster perturbation theory is a linearly scaling (with respect to the mean number of molecules included in a neighbor radius) novel computational approach to computing electronic properties of extremely large liquids and crystals. The method is entirely enabled by the sparse array management introduced in the Aces4/SIA parallel framework. The scaling slopes of the MCPT method for a large (1600 explicit water molecules) water ice is given in Table 7.3, where the slope for each mean number of waters is relative to the one water limit and should remain a constant[†] as the number of waters is increased. As a counter point, if implemented in the ACES III package, just tracking the zero matrix elements of this calculation would entail storing $\sim 100 \times 10^{12}$ metadata elements, a completely unrealistic prospect.

7.10   OTHER CONSIDERATIONS

7.10.1   FAULT TOLERANCE

Aces4 currently makes checkpoints transparently at well-defined points in a job, namely between SIAL programs where the only data that needs to be saved are the arrays that were marked persistent in the SIAL program. This takes advantage of the fact that in QM methods, the quantity of data used within a step of a method is significantly larger than the output. The mechanism is set up so that it is easy to rerun a computation starting with any SIAL program in a job whose predecessors have successfully finished. Although this simple and efficient checkpointing facility has thus far proved to be completely satisfactory for current supercomputers, it remains to be seen whether this will continue to be the case on larger machines. If not, then the goal would be to implement mechanisms for fault tolerance that would be implemented within the SIAL compiler and runtime system without requiring changes to SIAL programs or user provided super instructions. By exploiting knowledge of the SIAL semantics and utilizing the coarse-grained representation of the program (the SIAL byte-code) available at runtime, this should be possible.

---

* A lattice of neon atoms is an easy way to systematically add 10 electrons at a time to a calculation when performing benchmarks. The Dunning aug-cc-pVTZ basis was used for this benchmark.
[†] A constant until memory and disk I/O limitations become a factor.

TABLE 7.2

Strong Scaling Efficiency of a Single CCSD Iteration for an Increasingly Larger Molecular System Size

Number of Processors	Ne_{20}	Ne_{25}	Ne_{30}
1,280	100%		
2,560	89%	100%	
5,120	62%	96%	100%
10,240	46%	73%	79%
20,480	37%	56%	65%

Note: Calculations were performed on the DOD DSRC Excalibur supercomputer, a 3.7 PFLOP Cray XC40 using Cray-LibSci v. 13.1.0, Cray-MPICH v. 7.1.0, and Intel v. 15.0.1.133.

TABLE 7.3

Scaling Slope for an MCPT Calculation of (1600 Explicit Molecules) Water Ice Ih for a Range of Processors

Mean Number of Water	1200 proc.	2400 proc.	4800 proc.
7	0.61	0.43	0.31
13	0.55	0.36	0.25
20	0.51	0.34	0.23
29	0.47	0.31	0.22

Note: Calculations were performed on the DOD Shepard supercomputer, a 817 TFLOP Cray XC30 using Cray-LibSci v. 12.2.0, Cray-MPICH v. 6.3.1, Intel v. 14.0.2.144.

7.11 CONCLUSION

Aces4 is a software package under continuing development that has already enabled new, data-intensive QM methods to be implemented. Several of its features were designed with the ability to handle very large problems that require very large computer systems as a goal. As Aces4 is used to develop additional, data-intensive QM methods, which will be executed on ever larger computers, its design decisions will be tested and we hope validated.

ACKNOWLEDGMENTS

This project was supported by the U.S. Department of Defense HASI program. This research used resources of the Oak Ridge Leadership Computing Facility at the Oak Ridge National Laboratory, which is supported by the Office of Science of the U.S. Department of Energy under Contract No. DE-AC05-00OR22725. The authors acknowledge University of Florida Research Computing for providing computational resources and support.

REFERENCES

1. V Lotrich, N Flocke, M Ponton, AD Yau, A Perera, E Deumens, and RJ Bartlett. Parallel implementation of electronic structure energy, gradient and Hessian calculations. *J. Chem. Phys.*, 128:194104 (15 pages), 2008.

2. V Lotrich, JM Ponton, AS Perera, E Deumens, RJ Bartlett, BA Sanders. Super Instruction Architecture for Petascale Electronic Structure Software: The Story. *Mol. Phys.*, 108: 3323, 2010.

3. BA Sanders, R Bartlett, E Deumens, V Lotrich, and M Ponton. A block-oriented language and runtime system for tensor algebra with very large arrays. In *SC '10: Proceedings of the 2010 ACM/IEEE International Conference for High Performance Computing, Networking, Storage and Analysis,* New Orleans, LA, 2010.

4. Aces III. personal communication with T. Straatsma.

5. Aces4. personal communication with T. Straatsma.

6. JN Byrd, N Jindal, RW Molt, Jr., RJ Bartlett, BA Sanders, and VF Lotrich. Molecular cluster perturbation theory I: Formalism. *Mol. Phys.*, 113:1–12, 2015.

7. JN Byrd, RW Molt, Jr., RJ Bartlett, and BA Sanders. Computing solvated excited states using fragment-effective-field coupled-cluster perturbation theory with application to the electronic spectra of nucleobases in water. Unpublished manuscript, 2017.

8. JN Byrd, VF Lotrich, and RJ Bartlett. Correlation correction to configuration interaction singles from coupled cluster perturbation theory. *J. Chem. Phys.*, 140(23), 2014.

9. JN Byrd, V Rishi, A Perera, and RJ Bartlett. Approximating electronically excited states with equation-of-motion linear coupled-cluster theory. *J. Chem. Phys.*, 143(16), 2015.

10. RJ Harrison. Global arrays. *Theor. Chim. Acta*, 84:363–375, 1993.

11. D Ozog, A Kamil, Y Zheng, P Hargrove, JR Hammond, A Malony, WD Jong, and K Yelick. A Hartree–Fock application using UPC++ and the new darray library. In *2016 IEEE International Parallel and Distributed Processing Symposium (IPDPS)*, Chicago, IL, 453–462, May 2016.

12. RA Kendall, E Apra, DE Bernholdt, EJ Bylaska1, M Dupuis, GI Fann, RJ Harrison, J Ju, JA Nichols, J Nieplocha, TP Straatsma, TL Windus, and AT Wong. High performance computational chemistry: An overview of NWChem a distributed parallel application. *Comput. Phys. Comm.*, 128(1):268–283, June 2000.

13. Z Chen, J Dinan, Z Tang, P Balaji, H Zhong, J Wei, T Huang, and F Qin. Mc-checker: Detecting memory consistency errors in mpi one-sided applications. In *SC '14: Proceedings of the International Conference for High Performance Computing, Networking, Storage and Analysis*, New Orleans, LA, 499–510, 2014.

14. EW Dijkstra, WHJ Feijen, and AJM van Gasteren. Derivation of a termination detection algorithm for distributed computations. *Inf. Process. Lett.*, 16(5):217–219, 1983.

15. BA Sanders. The derivation of probe-based termination detection algorithms. In MH Barton, EL Dagless, and GL Reijns, editors, *Proceedings of the IFIP Working Conference on Distributed Processing*, North Holland, 249–258, 1988.

16. KM Chandy, and L Lamport. Distributed snapshots: Determining global states of distributed systems. *ACM Trans. Comput. Syst.*, 3(1):63–75, February 1985.

17. N Jindal, V Lotrich, E Deumens, and BA Sanders. Exploiting GPUs with the super instruction architecture. *Int. J. Parallel Prog.*, 44(2):309–324, 2016.

18. N Flocke, and V Lotrich. Efficient electronic integrals and their generalized derivatives for object oriented implementations of electronic structure calculations. *J. Comp. Chem.*, 29(16):2722–2736, 2008. doi:10.1002/jcc.21018.

19. D Lyakh. *Tal sh*. https://github.com/DmitryLyakh/TAL_SH.

20. LPG. *An LALR Parser Generator*. http://sourceforge.net/projects/lpg/.

21. Google Test. https://github.com/google/googletest.

8 Transitioning NWChem to the Next Generation of Manycore Machines

*Eric J. Bylaska, Edoardo Aprà, Karol Kowalski,
Mathias Jacquelin, Wibe A. de Jong, Abhinav Vishnu,
Bruce Palmer, Jeff Daily, Tjerk P. Straatsma,
Jeff R. Hammond, and Michael Klemm*

CONTENTS

8.1 Introduction.. 165
8.2 Plane-Wave DFT Methods.. 166
    8.2.1 FFT Algorithm... 168
    8.2.2 Nonlocal Pseudopotential and Lagrange Multiplier Algorithms........................ 170
    8.2.3 Overall Timings for AIMD on KNL... 172
8.3 High-Level Quantum Chemistry Methods 174
    8.3.1 Tensor Contraction Engine 174
    8.3.2 Implementation for the Intel Xeon Phi KNC Coprocessor................................ 177
    8.3.3 Benchmarks 178
8.4 Large-Scale MD Methods.. 179
    8.4.1 Domain Decomposition 180
    8.4.2 Synchronization and Global Reductions 180
    8.4.3 DSLs for Force and Energy Evaluation 181
    8.4.4 Hierarchical Ensemble Methods 181
8.5 GAs Parallel Toolkit 181
8.6 Conclusions... 183
Acknowledgments 183
References 184

8.1 INTRODUCTION

The NorthWest chemistry (NWChem) modeling software is a popular molecular chemistry simulation software that was designed from the start to work on massively parallel processing supercomputers [1–3]. It contains an umbrella of modules that today includes self-consistent field (SCF), second order Møller–Plesset perturbation theory (MP2), coupled cluster (CC), multiconfiguration self-consistent field (MCSCF), selected configuration interaction (CI), tensor contraction engine (TCE) many body methods, density functional theory (DFT), time-dependent density functional theory (TDDFT), real-time time-dependent density functional theory, pseudopotential plane-wave density functional theory (PSPW), band structure (BAND), ab initio molecular dynamics (AIMD), Car–Parrinello molecular dynamics (MD), classical MD, hybrid quantum mechanics molecular mechanics (QM/MM), hybrid ab initio molecular dynamics molecular mechanics (AIMD/MM), gauge independent atomic orbital nuclear magnetic resonance (GIAO NMR), conductor like screening solvation model (COSMO), conductor-like screening solvation model based on density

(COSMO-SMD), and reference interaction site model (RISM) solvation models, free energy simulations, reaction path optimization, parallel in time, among other capabilities [4]. Moreover, new capabilities continue to be added with each new release.

The development of the software began in 1992 as part of the Environmental Molecular Sciences Laboratory (EMSL) [5] construction project and it had its first major release in 1997 (version 3.3). The initial goal for the software was for it to "provide 10–100 times the capability of what was currently available on conventional supercomputers" [2], which at the time were vector-based computers such as the Cray T90. At that time, the parallel infrastructure of NWChem, which relied on the internally developed partitioned global address space (PGAS) software, global arrays (GAs) [2,6,7], as well as the standard message passing interface (MPI) [8], was cutting edge. However, today, approximately 20 years later, the size of the machines and underlying changes in computer architecture since that time are now requiring significant changes to the underlying parallel infrastructure used in the code. The current GA/MPI programming model needs to be superseded by multilevel parallelism where at the lowest level thread centric programming paradigms are used within a node while at a higher level the GA/MPI model is utilized.

Even though, the GA/MPI model can be used on many of today current architectures with large numbers of cores [9], e.g., the second generation Intel Xeon Phi hardware named Knights Landing (KNL), and in principle can take advantage of the fact that memory is shared, this programming model has several drawbacks. Using this programming model, performance hits can happen because of its lack of ability to control memory at the node resulting in a lack of memory coherency, higher latencies, and slower synchronizations. An alternate approach for providing higher performance on large core architectures is to use a hybrid execution model [10], where data movement between nodes is handled by GA/MPI and the data movement and execution within a node is handled by a multithreading model such as OpenMP [11–13]. The advantages of this model are that synchronizing between threads is faster, extra data movements can be avoided, and the memory footprint is potentially smaller since particular data structures may not need to be duplicated among threads. The current target platforms for our work on developing multilevel parallelism are the NERSC supercomputer Cori, the PNNL supercomputer Cascade, and the ORNL supercomputer Titan.

In this chapter, we present our recent efforts to add thread-level parallelism to three core modules in NWChem, including plane-wave DFT (NWPW), TCE, and the MD. For the NWPW and TCE modules, thread-level parallelism is being implemented using OpenMP constructs. The developments for the MD module are focused on implementing a domain specific library (DSL) with algorithms tuned to various processing elements, including CPU, graphics processing unit (GPU), and the Intel MIC architecture.

8.2   PLANE-WAVE DFT METHODS

In this section, our recent developments of adding thread-level parallelism to the plane-wave DFT methodology implemented in NWChem is presented [14–17]. The parallelism in the prior implementation of this method is achieved using only MPI, without explicit thead-level parallelism. In our current development, thread-level parallelism is integrated into the code using OpenMP constructs and threaded mathematical libraries where possible. Similar efforts are underway with other codes [18], however, to our knowledge our development is at present unique in that the focus is on having AIMD simulations with very fast iteration times (i.e., small times per AIMD step). All the timing presented in this section were carried out using the NERSC-8 supercomputer Cori, which is composed of 9000+ self-hosted Intel KNL CPUs connected with a Cray Aries high-speed interconnect with Dragonfly topology (0.25–3.7 μs MPI latency, 8 GB/s MPI bandwidth).

One of the main uses of plane-wave DFT methods is to perform AIMD simulations [16,19–22]. AIMD is a method where the motions of the atoms are simulated using Newton's laws in which the forces on the atoms are calculated directly from the electronic Schrödinger equation, or more specifically in this work the Kohn–Sham DFT equations [23]. These simulations are computationally expensive because the DFT equations are solved at every time-integration step in the simulation, and for an AIMD simulation to be practical each step in the simulation must be able to complete with seconds. The need for such a fast solution to the DFT equations is driven primarily by the fact that the time-integration step of a conventional AIMD simulation can be quite small (~ 0.2 femtosecond $= 2 \times 10^{-16}$ s) along with the fact that the length of the simulation will need be at least 10 picoseconds, and for many chemical processes of interest, the simulations would need to run on the order of nanoseconds (10^{-9} s and larger). Given that each DFT minimization step will require 10 or more evaluations of the Schrodinger equation, this results in simulations where Schrodinger equation is evaluated half a million times or more during simulation (10^{-11} s / (2×10^{-16} s/step) $\times 10$ evaluations/step $=$ 500,000 evaluations). Another type of AIMD simulation developed by Car and Parrinello in 1985 [19], which simultaneously propagates the electronic and ionic degrees of freedom, reduces the cost of the simulation to have only one evaluation of the DFT Schrodinger equation per time-step, instead of the 10 or more evaluations per time-step in a regular AIMD simulation. The proviso of these simulations is that the time-step of the simulation is dictated by the propagation of electronic degrees of freedom, which is even smaller than a traditional AIMD simulation, but rarely less than 0.1 femtoseconds $= 10^{-16}$ s. Regardless of the type of AIMD simulation, these simulations are considerably more expensive than merely optimizing a molecular or solid-state structure, which requires at most a few hundred evaluations of the DFT equations.

The bulk of the computational work in solving plane-wave DFT equations revolves around the solution of N_e nonlinear eigenvalue equations, $H\psi_i = \epsilon_i \psi_i$, for the electron orbitals ψ_i, subject to the orthogonality constraint

$$\int_\Omega \psi_i(r)\psi_j(r)dr = \delta_{i,j} \tag{8.1}$$

For hybrid-DFT, the Hamiltonian operator H may be written as

$$H\psi_i = \begin{pmatrix} -\frac{1}{2}\nabla^2 + V_l + V_{NL} + V_H[\rho] \\ +(1-\alpha)V_x[\rho] + V_c[\rho] \end{pmatrix} \psi_i$$
$$- \alpha \sum_j K_{i,j}\psi_j \tag{8.2}$$

where the one electron density is given by

$$\rho(r) = \sum |\psi_i(r)|^2 \tag{8.3}$$

The local and nonlocal pseudpotentials, V_l and V_{NL}, represent the election–ion interaction. The Hartree potential V_H is given by

$$\nabla^2 V_H = -4\pi\rho \tag{8.4}$$

The local exchange and correlation potentials are V_x and V_c and exact exchange kernels K_{ij} are given by

$$\nabla^2 K_{i,j} = -4\pi\psi_j^* \psi_i \tag{8.5}$$

The algorithmic cost to evaluate $H\psi$ and maintain orthogonality are shown in Figure 8.1.

In an AIMD simulation, the electron gradient $H\psi_i$, (Equation 8.2) and orthogonalization (Equation 8.1) need to be calculated as efficiently as possible. The main parameters that determine the cost of a calculation are N_g, N_e, N_a, and N_{proj}, where N_g is the size of the three-dimensional (3D) fast Fourier transform (FFT) grid, N_e is the number of occupied orbitals, N_a is the number of atoms, and N_{proj} is the number of projectors per atom. The evaluation of the electron gradient (and

$$(-1/2)\nabla^2 \Psi + V_{\text{ext}} \Psi + V_H \Psi + V_{xc} \Psi = E\Psi$$

$$\left\langle \Psi_i \mid \Psi_j \right\rangle = \delta_{ij}$$

$$(-1/2)\nabla^2 \Psi :: N_e N_{\text{pack}}$$

$$V_{\text{ext}} \Psi :: (N_a N_{\text{pack}} + N_g \text{Log} N_{\text{pack}} + N_e N_{\text{pack}}) + N_a N_e N_{\text{pack}}$$

$$V_H \Psi :: N_e N_g \text{Log} N_g + N_e N_g + 2N_g \text{Log} N_g + N_g + N_e N_g$$

$$V_{xc} \Psi :: N_e N_g \text{Log} N_g + N_e N_g$$

$$\left\langle \Psi_i \mid \Psi_j \right\rangle :: N_e N^2 N_{\text{pack}} + N_e{}^3$$

N_a—number of atoms N_e—number of electrons

N_g—size of FFT grid N_{pack}—size of reciprocal space

FIGURE 8.1   Operation count of $H\psi_i$ in a plane-wave DFT simulation.

orthogonality) contains three or four major computational pieces (note that conventional DFT does not compute the exact exchange term; $\alpha = 0$ in Equation 8.2):

- The Hartree potential V_H, including the local exchange and correlation potentials $V_x + V_c$. The main computational kernel in these computations is the calculation of N_e 3D FFTs
- The nonlocal pseudopotential, V_{NL}. The major computational kernel in this computation can be expressed by the following matrix multiplications: $W = P^T \times Y$, and $Y_2 = P \times W$, where P is an $N_g \times (N_{\text{proj}} \times N_a)$ matrix, Y and Y_2 are $N_g \times N_e$ matrices, and W is an $(N_{\text{proj}} \times N_a) \times N_e$ matrix. We note that for most pseudopotential plane-wave calculations $N_{\text{proj}} \times N_a \approx N_e$.
- Enforcing orthogonality. In AIMD orthogonality is typically maintained using a Lagrange multiplier algorithm. The major computational kernels in this computation are following matrix multiplications: $S = Y^T \times Y$ and $Y_2 = Y \times S$, where Y and Y_2 are $N_g \times N_e$ matrices, and S is an $N_e \times N_e$ matrix.
- and when exact exchange is included, the exact exchange operator $\sum_j K_{i,j} \psi_j$. The major computational kernel in this computation involves the calculation of $(N_e + 1) \times N_e$ 3D FFTs.

8.2.1   FFT Algorithm

For each evaluation of the electron gradient $H\psi_i$, the N_e orbitals, $\psi(G, 1:N_e)$, are converted from reciprocal space to real space and subsequently the N_e orbital gradients $H\psi(r, 1:N_e)$ are transformed from real space to reciprocal space. That is during each evaluation of $H\psi_i$, N_e reverse 3D FFTs ($\psi(r, 1:N_e)$ = reverse-3DFFT($\psi(G, 1:N_e)$)) and N_e forward 3D FFTs ($H\psi(G, 1:N_e)$ = forward-3DFFT($H\psi(r, 1:N_e)$)) are computed. The FFTs in a plane-wave DFT code are slightly different than in other FFT applications in that only a sphere of radius E_{cut} (or hemisphere for a Γ-point code) in reciprocal space and contained within the 3D FFT block, is needed and saved in the program. Each 3D FFT is carried out in six steps as illustrated in Figure 8.2.

For the forward 3D FFT the steps are as follows:

1. Unpack the reciprocal space sphere into a 3D cube, where the leading dimension of the cube is the z-direction, second dimension is the x-direction, and the third dimension is the y-direction, i.e., cube stored as z, x, y.
2. Perform $nx \times ny$ FFTs along the z-direction, where only the arrays that intersect the sphere need to be computed.
3. Rotate the cube to make the first dimension the y-direction, z, x, $y \rightarrow y$, z, x.

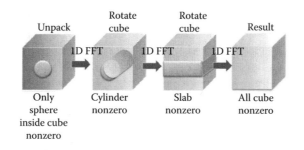

FIGURE 8.2   Illustration of computational steps in a specialized 3D FFT used in a AIMD program.

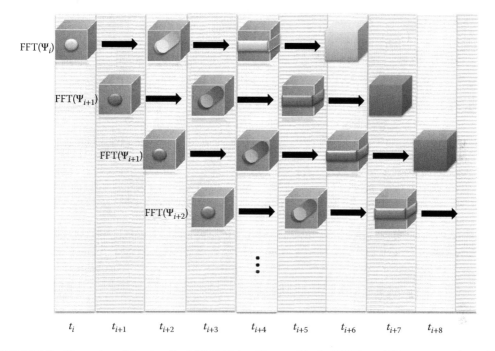

FIGURE 8.3   Illustration of the pipe-lined 3D FFT algorithm used in the NWChem AIMD program. Note that only the spheres inside the cube have nonzero values when in reciprocal space.

4. Perform $nz \times nx$ 1D FFTs along the y-direction.
5. Rotate the cube to make the first dimension the x-direction, y, z, $x \to x$, y, z.
6. Perform $ny \times nz$ 1D FFTs along the x-direction.

To carry out a reverse 3D FFT, the steps are simply carried out in reverse order. Since the forward and reverse 3D FFTs are called N_e times, N_e FFTs can be carried out several different ways. One way is to just call the N_e FFTs one after another. The N_e FFTs could also be done in tandem in a block-like fashion through each of the six steps. In this work, the algorithm computes the N_e FFTs in a pipe-lined way as illustrated in Figure 8.3.

The 3D FFTs used in this study were implemented by modifying the existing parallel 3D FFTs contained in the NWChem NWPW module. More details on the implementation of these FFTs can be found in prior work by Bylaska et al. [9,15,24]. In the existing parallel FFT code, the 3D cube is distributed along the second and third dimension. This distribution is block-mapped using a two-dimensional Hilbert curve spanning the grid of the second and third dimensions (see [24]).

This two-dimensional Hilbert parallel FFT was built using a 1D FFT and a parallel block rotation. The FFTPACK library [25] is used to perform the one-dimensional (1D) FFTs, and the parallel block rotation was implemented using MPI. We generalized the above FFTs to an MPI-OpenMP hybrid model by making the following changes. Each plane of 1D FFTs in steps 2, 4, and 6 were threaded by using an `!$OMPT DO` directive so that each 1D FFT is carried out on only one thread. The data rearrangement in steps 1, 3, and 5 were threaded by using an `!$OMPT DO` directive on the loops that perform the data-copying on the node.

8.2.2 NONLOCAL PSEUDOPOTENTIAL AND LAGRANGE MULTIPLIER ALGORITHMS

Computing the nonlocal pseudopotential and the Lagrange multiplier kernels are two key components of AIMD [16]. These two kernels share a lot of similarities in that they both can be thought of as a series of matrix multiplies and are critical to optimize. The three matrix multiplies, shown in Figure 8.4, are labeled as **FFM**, **MMM**, and **FMF**. The letter **F** refers to an N_{pack}-by-M matrix or its transpose, and the letter **M** refers to a M-by-M matrix (Lagrange multiplier algorithm: $M = N_e$, nonlocal pseudopotential algorithm: $M = N_a \times N_{proj}$). In general, $N_{pack} >> M$, thus, **F** matrices (or their transpose) are *tall-skinny* matrices. The three matrix–matrix products $C = A \times B$ can then be described as a combination of three of these two letters, the first referring to matrix A, the second to matrix B, and the last to matrix C.

The most straightforward way to implement these three matrix multiplies is to just use a multi-threaded BLAS (basic linear algebra subprograms) library, e.g., the Intel MKL library. Unfortunately, it was found from extensive benchmarks of the multithreaded DGEMM in the Intel MKL library that performance drops significantly when N_{pack} is significantly larger than M [26]. We have therefore developed our own multithreaded matrix multiplies using OpenMP. Details about their implementations can be found in Jacquelin et al. [26]. Here, we give a brief description of the three algorithms.

For the **FMF** matrix product, the parallelization is done block-wise along the N_{pack} rows of the **F** matrices A and C, and the **M** matrix, B is completely accessed by every thread. The pseudo-code for this algorithm, which makes use of a nonthreaded DGEMM subroutine, is as follows:

```
*
*  FFM matrix multiply
*
r   = mod(Npack,nthreads)
```

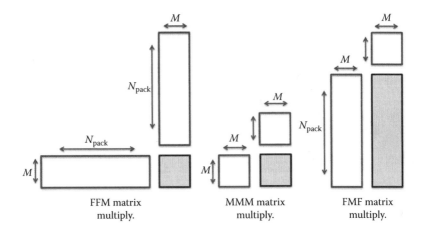

FFM matrix
multiply.

MMM matrix
multiply.

FMF matrix
multiply.

FIGURE 8.4   Matrix multiplies used in the NWChem AIMD program.

```
if (threadid.lt.r) then
   Itid = threadid*(Npack/nthreads+1) + 1
   Mtid = Npack/nthreads + 1
else
   Itid = r + threadid*(Npack/nthreads) + 1
   Mtid = Npack/nthreads
end if
call DGEMM('N','N',Mtid,M,M,ALPHA,
           A(Itid,1),LDA,B,LDB,
           BETA,C(Itid,1),LDC)
```

The **FFM** matrix product is also parallelized block-wise along N_{pack} rows of the **F** matrices. The partial result from each thread is stored in its own resultant **M** matrix, *Ctmp*, and then reduced to the final result. The pseudo-code for this algorithm is as follows:

```
*
* FFM matrix multiply
*
r   = mod(Npack,nthreads)
if (threadid.lt.r) then
   Itid = threadid*(Npack/nthreads+1) + 1
   Mtid = Npack/nthreads + 1
else
   Itid = r + threadid*(Npack/nthreads) + 1
   Mtid = Npack/nthreads
end if
call DGEMM('T','N',M,M,Mtid,ALPHA,
           A(Itid,1),LDA,B(Itid,1),LDB,
           BETA,Ctmp,LDC)
$OMP CRITICAL
do I=1,N
   do J=1,N
      C(I,J) = C(I,J) + Ctmp(I,J)
   end do
end do
$OMP CRITICAL
```

The last matrix product is the **MMM** matrix product. All three matrices are square and small compared to the **F** matrices, and as a result this operation does not dominate the computational cost except for very large problems. For many size problems, this operation can be done in serial. A pseudo-code for threading this algorithm is as follows:

```
subroutine Parallel_matrixblocks(nthr,m,n,mb,nb)
implicit none
integer nthr,m,n,mb,nb
integer ii,jj,mm,nn
real*8 ratio
ratio = dble(m)/dble(n)
mb = nthr
nb = 1
do nn =1,nthr
   ii = (nthr-1)/nn - 1
   jj = (nthr+1)/nn + 1
   if (ii.lt.1) ii = 1
```

```
      if (jj.gt.nthr) jj=nthr
      do mm=ii,jj
         if ((nn*mm).eq.nthr) then
            if (dabs(dble(mm)/dble(nn) - ratio) .lt.
               dabs(dble(mb)/dble(nb) - ratio)) then
                  mb = mm
                  nb = nn
            end if
         end if
      end do
end do
return
end

integer function Parallel_1dblocksize(m,mb,b)
implicit none
integer m,mb,b
integer i,bsz
bsz = 0
do i=1,m
        ib= mod(i-1,nthrnp) + 1
        if (ib.eq.b) bsz = bsz + 1
end do
Parallel_1dblocksize = bsz
return
end
*
* MMM matrix multiply
*
call Parallel_matrixblocking(nthreads,M,M,mblock,nblock)
i = mod(threadid,mblock
j = (threadid-i)/mblock
mstart = Parallel_index_1dblock(M,mblock,i)
mend   = Parallel_index_1dblock(M,mblock,i+1)
nstart = Parallel_index_1dblock(M,nblock,j)
nend   = Parallel_index_1dblock(M,nblock,j+1)
ishiftA = 1 + mstart
ishiftB = 1 + nstart*M
ishiftC = 1 + mstart + nstart*M
call DGEMM('N','N',mend-mstart,nend-nstart,M,
           alpha,
           A(ishiftA), M,
           B(ishiftB), M,
           beta,
           C(ishiftC), M)
```

8.2.3   OVERALL TIMINGS FOR AIMD ON KNL

In Figure 8.5, the timing results for a full AIMD simulation of 64 water molecules on 1, 2, 3, and 4 KNL nodes are shown. These simulations were taken from a Car–Parrinello simulation of 64 H_2O with an FFT grid of $N_g = 108^3$ ($N_e = 256$, restricted) using the plane-wave DFT module (PSPW) in NWChem. In these timings, the number of threads per node grows from 1 to 68. The size of

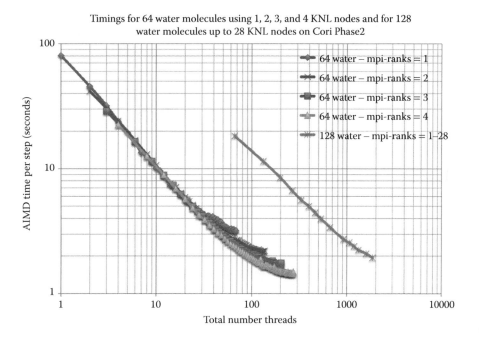

FIGURE 8.5 AIMD step timings for 64 water molecules versus number of total number of threads using 1, 2, 3, and 4 KNL nodes, and timings for 128 water molecules using 1,...,28 nodes with 66 threads per node on the NERSC Cori supercomputer.

this benchmark simulation is typical of many mid-size AIMD simulations, which have been carried out in recent years, e.g., in recent work by Bylaska et al. [27–32]. At least, some amount of scaling is seen up to all the threads on 4 nodes (272 threads), however, by 272 threads (68 threads × 4 nodes) the parallel efficiency has dropped to 20%. The lowest time per step was found to be 1.43 s when using 4 nodes with 66 threads per node. Also shown in Figure 8.5 are the timing results for an AIMD simulation of 128 water molecules on 1,...,28 KNL nodes using 66 threads per node are shown. This simulation had a grid of $N_g = 140^3$ ($N_e = 512$, restricted). The parallel scaling efficiency was found for this larger simulation was found to be similar to the 64 water simulation, with the parallel efficiency dropping down to 34% by 28 nodes (1848 threads). The lowest time per step for this larger simulation was found to be 1.94 s when using 28 nodes with 66 threads per node.

The timings of the major kernels, the pipe-lined 3D FFTs, nonlocal pseudopotential, and Lagrange multiplier kernels all display speedups the number of threads per nodes increases from 1 to 68. As seen in Figure 8.6, the pipelined 3D FFTs is the dominate kernel, followed by the nonlocal pseudopotential and Lagrange multiplier kernels. The 3D FFTs also display the worst parallel efficiency of the three kernels. This result is expected in that the 3D FFT algorithm is not expected to scale beyond $N_g^{1/3}$ [15,24] and demonstrates that the hybrid OpenMP-MPI program ultimately has the same bottlenecks of a standard MPI parallel program. The cause of the lack of performance in a 3D FFT is well known and is do to their reliance on global all to all operations that use all the threads in the calculation. A strategy for overcoming these scalability issues is to design the parallel algorithm to distribute the across orbitals as well as over space [15,16,33]. This decomposition reduces the cost of the global operations in the major parts of an AIMD step. We are currently adapting our hybrid OpenMP-MPI algorithms to use these advanced distributions.

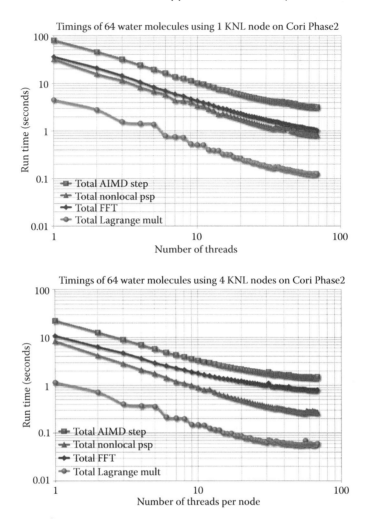

FIGURE 8.6 Illustration of computational steps in a specialized 3D FFT used in a AIMD program.

8.3 HIGH-LEVEL QUANTUM CHEMISTRY METHODS

8.3.1 TENSOR CONTRACTION ENGINE

The TCE is designed to automate the derivation and parallelization of quantum many-body methods utilizing the formalism of second quantization. Among various methodologies falling into this category one should mention many-body perturbation theory, CI methods, and CC formalisms. This is accomplished using in a three stages approach: first, operator expressions are transformed into multidimensional array expressions; second, array expressions are transformed into an optimized operation tree using performance heuristics; finally, the optimized operation tree is implemented in a parallel FORTRAN code using a source template.

Due to its accuracy and prominent role, it has assumed in high-accuracy modeling of chemical processes; the class of CC methods [34–39] has been and continues to be a major application of the TCE framework. In the CC formulation, the ground-state wavefunction $|\Psi\rangle$ of many-electron system is represented in the form of exponential Ansatz

$$|\Psi\rangle = e^T |\Phi\rangle \tag{8.6}$$

where the so-called reference function $|\Phi\rangle$ is typically represented by the Hartree–Fock (HF) determinant and the cluster operator T generates excited determinants when action onto the reference function. For widely used coupled-cluster singles and doubles (CCSD) approach [38] (CC with singles and doubles), the cluster operator includes singly (T_1) and doubly (T_2) excited clusters

$$T = T_1 + T_2 \tag{8.7}$$

where using second quantization T_1 and T_2 operators can be written as

$$T_1 = \sum_{a;i} t_a^i a_a^\dagger a_i \ , \ T_2 = \sum_{i<j;a<b} t_{ab}^{ij} a_a^\dagger a_b^\dagger a_j a_i \tag{8.8}$$

In Equation 8.8, t_a^i and t_{ab}^{ij} tensors correspond to the singly and double excited cluster amplitudes whereas $a_p^\dagger$ (a_p) operators are the creation (annihilation) operators for the electron in p-th spinorbital. We also assume that indices $i, j, k, \ldots$ and $a, b, c, \ldots$ correspond to occupied and unoccupied (in reference function $|\Phi\rangle$) spinorbital indices. The cluster operators are determined from energy independent CC equations that are expressed in terms of connected diagrams. For the CCSD formulation, we have

$$\langle \Phi_i^a | (He^{T_1+T_2})_C | \Phi \rangle = 0 \ \forall_{i;a} \tag{8.9}$$
$$\langle \Phi_{ij}^{ab} | (He^{T_1+T_2})_C | \Phi \rangle = 0 \ \forall_{i<j;a<b} \tag{8.10}$$

where subscript "C" designated connected part of a given operator expression and $\langle \Phi_i^a |$, $\langle \Phi_{ij}^{ab} |$, ... designates singly, doubly, etc. excited configuration with respect to the reference function. Once these equations are converged CC energy is calculated from the formula

$$E = \langle \Phi | (He^{T_1+T_2} | \Phi \rangle \tag{8.11}$$

The accuracy of the CCSD is not always sufficient to attain chemical accuracy. To this end, in order to obtain high-order spectrocopic constants and thermochemistry data, one needs to include effect of triple excitations. In the CCSD(T) formulation [40], this is achieved by adding noniterative correction to the CCSD energy:

$$E^{\text{CCSD(T)}} = E^{\text{CCSD}} + \sum_{i<j<k;a<b<c} \frac{\langle \Phi | (T_2^+ V_N) | \Phi_{ijk}^{abc} \rangle \langle \Phi_{ijk}^{abc} | V T_2 | \Phi \rangle}{\epsilon_i + \epsilon_j + \epsilon_k - \epsilon_a - \epsilon_b - \epsilon_c}$$
$$+ \sum_{i<j<k;a<b<c} \frac{\langle \Phi | T_1^+ V_N | \Phi_{ijk}^{abc} \rangle \langle \Phi_{ijk}^{abc} | V T_2 | \Phi \rangle}{\epsilon_i + \epsilon_j + \epsilon_k - \epsilon_a - \epsilon_b - \epsilon_c} \tag{8.12}$$

where ϵs are the orbital energies and V_N representing two-body part of the electronic Hamiltonian in normal product form. The numerical cost of the CCSD(T) is a sum of N^6 scaling (per iteration) of the iterative CCSD part and N^7 numerical cost of (T) perturbative corrections (N stands for system size). The TCE framework can be used to generate hierarchy of approximate methods utilizing various reference functions (based on the restricted, restricted open-shell, and unrestricted HF determinants).

The data representation in TCE generated codes is based on the partitioning of the entire spinorbital domain into smaller subsets (tiles), which contain subset of spinorbital indices corresponding to the same spatial and spin symmetries. This partitioning imposes a block structure on all tensors involved in the CC formalisms including one- and two-electron integrals, cluster amplitudes, and recursive intermediates used in factorization of the CC equations.

The occupied and unoccupied tiles are designated as $[i], [j], [k], \ldots$ and $[a], [b], [c], \ldots$, respectively. This division entails the partitioning of all tensors involved in the CC calculations including cluster amplitudes, recursive intermediate, and integrals (see Figure 8.7). For example, the tensor corresponding to doubly excited amplitudes is stored in the block form defined by smaller four dimensional tensors

$$t^{[i][j]}_{[a][b]} \tag{8.13}$$

representing a subset of doubly excited amplitudes defined by the indices belonging to the $[i], [j], [a]$, and $[b]$ tiles. This block structure of CC tensors also defines data granularity needed to parallelize the code. The `tilesize` also defines local memory requirements. While for the iterative CCSD approach, the local memory demand is proportional to $(\texttt{tilesize})^4$, the analogous demand for perturbative triples part amounts to $2(\texttt{tilesize})^6$ (see Figure 8.7).

The $\texttt{tilesize}^6$ local memory requirements of the (T) part are different than the ones characterizing the iterative CCSD approach (which is the first step in the CCSD(T) calculations). In contrast to the noniterative part, the local memory usage of the CCSD method is proportional to $\texttt{tilesize}^4$. While larger `tilesize` guarantees more efficient use of the `dgemm` procedures and kernels in the CCSD and (T) implementations, respectively, the increase in the `tilesize` value quickly leads to the memory bottleneck associated with storing $\texttt{tilesize}^6$ objects in the (T) part. For example, for `tilesize=40` (which is considered as an optimal choice for large CCSD calculations), the corresponding $2*\texttt{tilesize}^6$ memory is equivalent to 60 GB. In order to address this issue, the algorithm is introduced, where tensors are sliced with respect to two particles dimensions corresponding to tiles [a] and [b]. Tiles [a] and [b] can be dynamically partitioned into p and q intervals that leads to the reduced memory requirement associated with storing $\texttt{tilesize}^6$ slice, proportional to $\texttt{tilesize}^6/(\texttt{p} * \texttt{q})$. The above algorithm also introduces another level of parallelism that can be effectively utilized in our code. The sliced algorithm enables one to use large tile size without stumbling into the memory bottleneck.

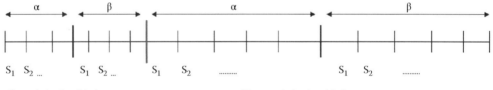

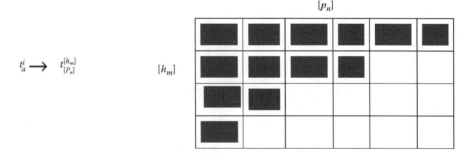

FIGURE 8.7 Schematic representation of tiling structure and induced block structure of tensors employed in the CC equations (in this picture block structure of t^i_a tensor is shown).

8.3.2 Implementation for the Intel Xeon Phi KNC Coprocessor

Due to the nature of the TCE algorithm, the Intel Xeon Phi coprocessor is a natural choice to accelerate the computation. The first generation Xeon Phi coprocessor (code name Knights Corner—KNC) can be utilized in two distinct ways [41]. As an autonomous computing device (also called *native mode*), each coprocessor acts as a full node of the compute cluster. In the *offload model*, the coprocessors receive requests and data for computation from the Xeon processors of their respective host systems. As some essential components used in preliminary steps of the CCSD(T) approach (e.g., the evaluation of two-electron repulsion integrals) require substantial tuning efforts if the native approach is used, the coprocessor implementation of the CCSD(T) method uses the offload model. The offload model provides the optimal balance between executing the compute-intensive parts with high-floating point requirements on the coprocessor and handling all other tasks such as I/O, communication, and less parallel algorithms on the host.

The compute demand for the (T) part of the CCSD(T) makes it an ideal target for offloading to a coprocessor device. We have used the directive-based Intel® Language Extensions for Offloading (LEO) [41] and OpenMP [13] for both kernel parallelization on the coprocessor and the CPU host.

Figure 8.8 shows the one typical kernel out of 18 kernels that implement the (T) part of the computation. This kernel corresponds to a summation of the occupied indices. As typical tile sizes are rather small (in the range of up to 24), simply outer-loop parallelization will not exhibit enough parallelism to satisfy the high number of threads needed on the coprocessor. We use the `collapse` clause of the OpenMP `do` construct to instruct the compiler to create a product loop for the p4 to p6 loops. The `OMPCOLLAPSE` parameter can be defined during compile time of NWChem and is set to three by default, effectively collapsing the loop nest of `p4`, `p5`, and `p6`. This results in a product loop with about one or two orders of magnitude more iterations than the number of threads (depending on `tilesize`), which then provides a sufficiently large loop iteration space.

We also add compiler directives to the code to inform the compiler about the expected trip count of the inner-most loops. This helps the compiler to determine the proper SIMD vectorization pattern and

```
!$omp parallel do &
!$omp private(p4,p5,p6,h2,h1,h3,h7) &
!$omp collapse(OMPCOLLAPSE)
do p4=1,p4d
do p5=1,p5d
do p6=1,p6d
do h1=1,h1d
do h7=1,h7d
!dec$ loop count max=1000, min=20
do h3h2=1,h2d*h3d
P3(h3h2,h1,p6,p5,p4) =
P3(h3h2,h1,p6,p5,p4)
- t2sub(h7,p4,p5,h1)*v2sub(h3h2,p6,h7)
enddo
enddo
enddo
enddo
enddo
enddo
!$omp end parallel do
```

FIGURE 8.8 Example kernel with OpenMP and loops collapsing.

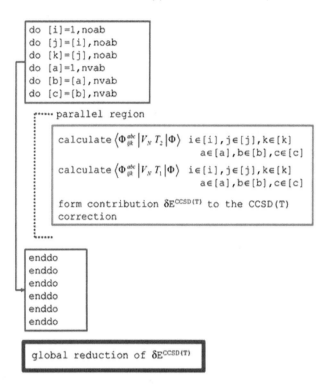

FIGURE 8.9 Schematic representation of the main kernel of the noniterative part of the CCSD(T) approach. The `noab` and `nvab` parameters refer to the total number of occupied and unoccupied tiles, respectively.

avoids more complex optimizations, such as loop blocking and tiling, which are counter productive to performance. Another micro-optimization applied to the kernels is to create product loops for the two inner-most loops (if applicable), as shown in Figure 8.8. This increases the trip count of the vectorized loop and thus reduces SIMD vectorization overhead while it increases the average utilization of the SIMD registers.

The final modification is to reorder the inner-most loops such that the P3 is accessed with a stride of one, so that the compiler can generate the most efficient load/store operations for it. In these cases, the t2sub and v2sub arrays become loop invariant scalar values that can be broadcast into a SIMD register by the compiler. For some of the kernel, this modification is a simple exchange of the loop order; for some kernels, it involves transposing one or both of the t2sub and v2sub arrays.

Key to offload performance is not only to obtain kernel performance on the coprocessor, but also to avoid excessive data transfers over the low-bandwidth, high-latency PCIe bus. The algorithm accumulates the contributions of t2sub and v2sub to the P3 array on the coprocessor during the offload phases, because no other host thread or GA process needs to make any updates to it, as shown in the schematic representation of the CCSD(T) workflow is shown of Figure 8.9. The main data movements required are the updates of the four-dimensional arrays t2sub and v2sub on the coprocessor after they have been computed and processed on the host. The final result $\delta E^{CCSD(T)}$ (see Figure 8.9) is computed on the coprocessors by combining the P3 arrays and then is sent back to the CPU host, therefore resulting in negligible data movement.

8.3.3 BENCHMARKS

The performance of the KNC Intel Xeon Phi implementation of the perturbative part of the CCSD(T) formalism is shown in Figure 8.10 on the example of pentacene molecule. The parallel performance

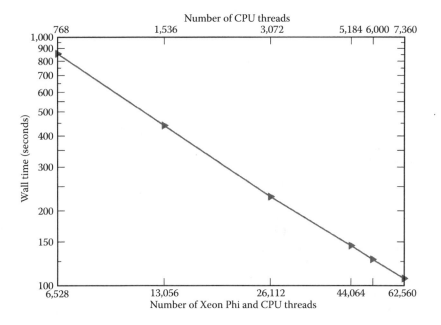

FIGURE 8.10 Wall time to solution (in seconds) for the perturbative triples correction to the CCSD(T) correlation energy of the pentacene molecule ($C_{22}H_{14}$). A logarithmic scale is used on all the axis.

of the (T) code was greatly improved by dynamics alignment of particular tasks in Figure 8.9 in the descending order with the numerical overhead corresponding to a given task. A very good scalability of our algorithm was achieved with concurrent utilization of processors and coprocessor (amounting to 62,560 cores in the largest calculation). Moreover, offloading algorithm was shown to be roughly three times faster compared to the CPU algorithm.

8.4 LARGE-SCALE MD METHODS

MD simulations using the classical and hybrid quantum-mechanical and classical force fields are important computational tools for the study of molecular condensed phase chemical and biomolecular systems and processes. The time-resolved trajectories generated by MD simulations provide atomic level detail of structural, dynamic, and thermodynamic features that can lead to important insights into the function of complex molecular processes such as protein association, enzymatic catalysis, protein–DNA interactions, and trans-membrane transport. MD provides an important complement to more accurate but computationally more demanding electronic structure calculations because it provides a method to study molecular systems at more realistic (i.e., longer) time-scales and in more realistic (i.e., larger) molecular environments. The primary challenge is the generation of time trajectories with millions of realizations of systems consisting of hundreds of thousands to millions of atoms. Even with the relatively simple representation of the interaction functions in classical MD, these simulations quickly become very complex. Many of the existing MD simulation codes, including NWChem, have gone through several cycles of redesign and reimplementation. Originally designed and implemented for single processor computers, refactoring was needed to take advantage of vector processors [42], then parallel clusters, and now for massively parallel hybrid systems with manycore and massively threaded accelerators. The current MD module in NWChem was designed for parallel systems with hundreds of nodes and includes a wide range of advanced features, such as free energy evaluation methods and special support of the setup and simulation of asymmetric biological membranes. A new prototype MD module is now being developed with a complete redesign of the data structure and communication scheme that will keep many of these features but provides much better scalability to the largest parallel computers of today. This prototype code is addressing the three main technical challenges for MD simulations on large-scale parallel architectures. These are the communication requirements for the calculation of many-body atomic interactions, the synchronization between force evaluations and coordinate updates, and the load balancing of simulations of heterogeneous and highly dynamic molecular systems.

8.4.1 DOMAIN DECOMPOSITION

Practically, all implementations of MD on massively parallel machines use domain decomposition to distribute the atomic data over the available processes. Earlier atom list or force decomposition schemes have been proposed [43], but the increased communication requirements make these less effective for large-scale parallelism. Domain decomposition in the current MD module in NWChem is based on a division of the physical space into a number of cells equal or larger than the number of available processes. Each cell is permanently assigned to a process, and interactions between atoms in different cells are evaluated by the process assigned to one of the two cells. The advantage here is that data for one cell in the cell pair does not have to be moved. This scheme allows for additional optimizations to reduce the amount of communication, such as the shift [44] and zonal [45] methods, and effective ways to carry out load balancing [46]. For simulation using large process counts, this approach results in relatively a small physical region per cell and, consequently, large numbers of small messages if cells beyond nearest neighbors need to be considered for the force and energy evaluations. This is a distinct disadvantage. The redesigned module also distributes the physical space into cells, but rather than assigning cells to processes, cell–cell pairs are assigned to processes. In addition, individual cell–cell pairs may appear more than once in the list to allow for a load balancing, for example, for face-sharing pairs of cells that have more computational load than edge- or corner-sharing cell pairs. This approach allows for more than an order of magnitude more processes to be used with the same number of cells in the domain decomposition. Advancing coordinates is done by the process assigned to evaluate the cell self-interactions.

8.4.2 SYNCHRONIZATION AND GLOBAL REDUCTIONS

The basic MD process consists of evaluation of atomic Newtonian forces and using these and the atomic velocities to advance atomic coordinates. These coordinates are periodically recorded, providing the atomic trajectories for further analysis. The easiest way to ensure that all coordinates are current before force and energy evaluations are performed, and that force evaluations are completed before coordinates are updated, is to put explicit synchronization before and after the force evaluations. This is done in the current MD module in NWChem. In addition, for constant temperature and constant pressure ensembles, the total system kinetic energy and virial are needed at every time step. These are global quantities that in the current MD module are evaluated using a global summation operation, which is essentially an additional synchronization. The wall clock time for a single global synchronization on large core counts can be an order of magnitude higher than the computational cost of evaluation of the forces. We have designed an MD implementation based on the GAs programming model and its asynchronous one-sided communication features, which is used to avoid the explicit synchronizations as well as the explicit global summations. The basic idea is to use asynchronous one-sided accumulation operations in a way that allows the receiving process to query the number of contributions and proceed when the expected number of contributions has been received. This process is essentially an implicit synchronization and is used for the force evaluation updates, the coordinate communication, as well as the global quantities needed for the time stepping. This procedure has been demonstrated to improve scalability by at least an order of magnitude [47].

8.4.3 DSLs for Force and Energy Evaluation

The evaluation of forces and energies, and other quantities that directly depend on atomic distances, constitute the computationally most demanding part of any MD simulation. The evaluation of these quantities will be implemented as a DSL with highly optimized algorithms for a variety of processing elements, including CPU, GPU, and Intel MIC. The objective is to develop this DSL with well-defined application programming interfaces (APIs), such that it can be used by other MD codes as well.

8.4.4 Hierarchical Ensemble Methods

The primary premise of MD simulations is the ability to appropriate sample a systems available phase space for the evaluation of properties as ensemble averages. Especially for large molecular systems with many local minima on the potential energy surface, single MD simulations need to be run for very long time spans. Methods like replica exchange, in which simulations at higher temperature or modified Hamiltonians are used, can significantly remedy the problem of slow sampling. This is commonly referred to as an ensemble method. Applications such as free energy evaluations, either through thermodynamic integration or thermodynamic perturbation, also involve a series of simulations providing free energy derivatives over which is integrated to obtain free energy differences between an initial and a final state of a system. If each of these derivatives is evaluated from a replica exchange simulation, the two combined approaches result in a hierarchical ensemble method that is highly parallel in nature and requires resources at the peta-scale or more. In our current MD prototype such a hierarchical ensemble sampling has been implemented.

8.5 GAS PARALLEL TOOLKIT

A critical component in NWChem is the GA toolkit [2,6,7]. This toolkit was co-developed with NWChem to facilitate the parallelism of many algorithms used in computational chemistry. The GA toolkit implements a shared-memory programming model in which data locality can be managed explicitly by the programmer. This management is achieved by explicit calls to functions that transfer data between a global address space (a distributed array) and local storage. Additional information on the distributed array is available that allows programmers to determine which data in the array is local and which is remote. This enables further performance gains by allowing programmers to optimize the amount of locally available data used in computations while minimizing the amount of remotely accessed data. It is important to stress that the GA programming model is designed to be complimentary to message passing and is not a total replacement. Both libraries can and do coexist in an application without any instability or incompatibility.

The GA programming model is fully compatible with message-passing, and in particular the programmer can use the full MPI functionality on both GA and non-GA data. The GA library can be used in C, C++, Fortran 77, Fortran 90 and Python programs and offers support for both task and data parallelism. The task parallelism is supported through one-sided (noncollective) get/put operations that transfer data between global memory (distributed/shared array) and local memory. Each process is able to directly access data held in a section of a GA that is logically assigned to that process. Atomic operations are provided that can be used to implement synchronization and assure correctness of an accumulate operation (floating-point sum reduction that combines local and remote data) executed concurrently by multiple processes and targeting overlapping array sections. The accumulate operation is a fundamental operation needed by distributed-data implementations of several electronic structure algorithms in computational chemistry. The one-sided semantics used to access data are also well-adapted to calculations with potentially large amounts of asynchronous work. When combined with global counters (implemented using atomic read—modify—write operations), one-sided communication can be used to organize calculations so that they are automatically load-balanced with little programming overhead.

The basic GA operations include

- Initialize: initialize the libary with or without limits on the size of global memory.
- Create: create an array with a regular or irregular distribution.
- Duplicate: create a new GA array that inherits the distribution. data type, and dimensions of an existing array.
- Destroy: remove a single GA.
- Terminate: shutdown the GA interface.
- One-sided or asynchronous operations that include
  - Remote block-wise read or write (get or put),
  - Remote atomic update (accumulate or read and increment of integers),
  - Remote element read or write (gather or scatter).
- Interprocess synchronization operations, e.g., lock with mutex to exclusively access a critical section of execution.
- Fence: guarantee that the GA operation(s) issued from the calling process are complete (the fence is local to the calling process).
- Sync: a barrier combined with fence that forces completion of all outstanding communication in the system. This operation can be used to guarantee that data in a GA is in a known state.
- Collective GA operations are basic operations on an entire array or patch of an array. They include
  - Zero, fill with a value, or scale
  - Copy or duplicate an array
- Linear algebra operations
  - Scale and add GAs and patches
  - Matrix multiplication of GAs and patches
  - Dot product of GAs and patches
  - symmetrize a GA
- Interfaces to third party packages
  - Scalapack linear equation solvers
  - PeIGS eigensystem solvers

A typical example of an algorithm that uses this common computational kernel is the evaluation of the Fock matrix elements in a HF calculation [2] using a local basis set (also known as a Fock build); a very similar algorithm is the evaluation of the Kohn–Sham matrix elements in DFT calculations. The data parallel computing model is supported through the set of collectively called functions that operate on either entire arrays or sections of GAs. The set includes BLAS-like operations (copy, additions, transpose, dot products, and matrix multiplication). Some of them are defined and supported for all array dimensions (e.g., addition). Other operations, such as matrix multiplication, are limited to two-dimensional arrays (however, multiplication is also offered on two-dimensional subsections of higher dimensional arrays). GA extends its capabilities in the area of linear algebra by offering interfaces to third party libraries, e.g., standard and generalized real symmetric eigensolvers (PeIGS) [2,48], and linear equation solvers (ScaLAPACK) [49].

There is currently a major effort underway to update the GA toolkit in several different ways. These include making the GA toolkit completely thread safe, eliminating all nonscaling features in the GA interface (e.g., data structures that are proportional to the number of processors) and eliminating internal data structures that scale with the number of processors. More support for sparse or unstructured data is also planned for GA. An initial release for this new development is expected in the

Spring of 2017. Even though GA is not currently thread safe, it can still be used with OpenMP by having the GA library calls contained with a $OMP MASTER block. This is equivalent to an MPI thread funneled model [8].

8.6   CONCLUSIONS

The very high amount of parallelism available on machines with manycore processors requires developers to carefully revisit the implementation of their programs in order to make use of this hardware efficiently. In this chapter, we have demonstrated that rewriting key kernels in the NWPW and TCE modules in NWChem to use a hybrid OpenMP-MPI and hybrid OpenMP-GA programming models, respectively, was able to provide good scalability on manycore machines. For the NWPW module, the parallelism within a node needed to be programmed at a very fine grain level, to achieve the performance needed for AIMD simulations. The unique implementations of key kernels used in AIMD such as sphere to cube 3D FFTs and the matrix multiplication of tall-skinny matrices did not perform well with the standard threaded Intel MKL library. However, by rewriting these kernels from scratch using the hybrid OpenMP-MPI model at a very fine grain level, we were able to obtain good performance. For the TCE module, adding parallelism within a node was able to achieve very good performance using OpenMP and offloading directives at a coarse grain parallel level along with vectorization of inner loops. For CCSD(T) calculations, very good scalability was able to be achieved with concurrent utilization of processors and coprocessor (62,560 cores in the largest calculation) using a hybrid OpenMP-GA model. Moreover, the offloading algorithm was shown to be roughly three times faster compared to the GA only algorithm.

Other key parts of NWChem are also being upgraded to use manycore processors. The current focus for the MD module is in redesigning the fundamental parallel algorithms to be able to effectively use processor counts that are an order of magnitude larger than the existing algorithms in NWChem. The low level kernels in an MD simulation are extremely challenging to parallelize, because they only take milliseconds or less to complete and represent a very hard Amdahl scaling problem. In addition, ensemble techniques that are used in conjunction with MD, which, in principle, are able to distribute large chunks of work across very large numbers of processors, suffer from extreme load imbalances. The GA parallel library is also undergoing a major rewrite to make it completely thread safe. In addition, all nonscaling features in the current GA interface and internal data structures are being eliminated.

ACKNOWLEDGMENTS

This work was supported by the NWChem project in the William R. Wiley Environmental Molecular Sciences Laboratory (EMSL), the U.S. Department of Energy, Office of Science, Advanced Scientific Computing Research ECP program (NWChemEx project), and E.J.B was also supported by the the U.S. Department of Energy, Office of Science, Office of Basic Energy Sciences, Chemical Sciences, Geosciences, and Biosciences Division at PNNL, DE-AC06-76RLO 1830. EMSL operations are supported by the DOE's Office of Biological and Environmental Research. M.J. and W.A.D. were partially supported by the Scientific Discovery through Advanced Computing (SciDAC) program funded by U.S. Department of Energy, Office of Science, Advanced Scientific Computing Research and Basic Energy Sciences. We wish to thank the Scientific Computing Staff, Office of Energy Research, and the U.S. Department of Energy for support through the NERSC NESAP program the National Energy Research Scientific Computing Center (Berkeley, CA). This work was also supported by Intel as part of its Parallel Computing Centers effort, and this work was supported in part through resources of the Oak Ridge Leadership Computing Facility, which is a DOE Office of Science User Facility supported under Contract DE-AC05-00OR22725.

REFERENCES

1. D. E. Bernholdt, E. Apra, H. A. Früchtl, M. F. Guest, R. J. Harrison, R. A. Kendall, R. A. Kutteh, et al. Parallel computational chemistry made easier: The development of NWChem. *International Journal of Quantum Chemistry*, 56(S29):475–483, 1995.

2. R. A. Kendall, E. Aprà, D. E. Bernholdt, E. J. Bylaska, M. Dupuis, G. I. Fann, R. J. Harrison, J. Ju, J. A. Nichols, J. Nieplocha, T. P. Straatsma, T. L. Windus, and A. T. Wong. High performance computational chemistry: An overview of NWChem a distributed parallel application. *Computer Physics Communications*, 128(1):260–283, 2000.

3. M. Valiev, E. J. Bylaska, N. Govind, K. Kowalski, T. P. Straatsma, H. J. J. van Dam, D. Wang, et al. NWChem: A comprehensive and scalable open-source solution for large scale molecular simulations. *Computer Physics Communications*, 181:1477–1489, 2010.

4. PNNL Environmental Molecular Sciences Laboratory. NWChem: Open Source High-Performance Computational Chemistry, http://www.nwchem-sw.org, 2017.

5. PNNL Environmental Molecular Sciences Laboratory. Pacific Northwest National Laboratory, Richland, WA, https://www.emsl.pnl.gov/emslweb/, 2017.

6. J. Nieplocha, B. Palmer, V. Tipparaju, M. Krishnan, H. Trease, and E. Aprà. Advances, applications and performance of the global 36 transitioning NWChem to many-cores arrays shared memory programming toolkit. *International Journal of High Performance Computing Applications*, 20(2):203–231, 2006.

7. M. Krishnan, B. Palmer, A. Vishnu, S. Krishnamoorthy, J. Daily, and D. Chavarria. *The Global Arrays User Manual*. February 12, 2012, Pacific Northwest National Laboratory Technical Report Number PNNL-13130.

8. W. Gropp and R. Thakur. Issues in developing a thread-safe MPI implementation. In *European Parallel Virtual Machine/Message Passing Interface Users Group Meeting*, pp. 12–21. Bonn, Germany: Springer, 2006.

9. W. A. de Jong, E. Bylaska, N. Govind, C. L. Janssen, K. Kowalski, T. Müller, I. M. Nielsen, et al. Utilizing high performance computing for chemistry: Parallel computational chemistry. *Physical Chemistry Chemical Physics*, 12(26):6896–6920, 2010.

10. R. Rabenseifner, G. Hager, and G. Jost. Hybrid MPI/OpenMP parallel programming on clusters of multi-core SMP nodes. In *Proceedings of the 17th Euromicro International Conference on Parallel, Distributed and Network-based Processing*, pp. 427–436. Weimar, Germany: IEEE, 2009.

11. L. Dagum and R. Menon. OpenMP: An industry standard API for shared-memory programming. *IEEE Computational Science and Engineering*, 5(1):46–55, 1998.

12. T. Cramer, D. Schmidl, M. Klemm, and D. an Mey. OpenMP Programming on Intel Xeon Phi Coprocessors: An Early Performance Comparison. In *Proceedings of the Many Core Applications Research Community (MARC) Symposium*, Aachen, Germany, pp. 38–44, 2012.

13. OpenMP Architecture Review Board. OpenMP Application Program Interface, Version 4.5, http://www.openmp.org/, November, 2015

14. E. Aprà, E. J. Bylaska, D. J. Dean, A. Fortunelli, F. Gao, P. S. Krstić, J. C. Wells, and T. L. Windus. NWChem for materials science. *Computational Materials Science*, 28(2):209–221, 2003.

15. E. J. Bylaska, K. Glass, D. Baxter, S. B. Baden, and J. H. Weare. Hard scaling challenges for ab initio molecular dynamics capabilities in NWChem: Using transitioning NWChem to many core 37 100,000 CPUs per second. *Journal of Physics: Conference Series*, 180(1):012028, 2009.

16. E. Bylaska, K. Tsemekhman, N. Govind, and M. Valiev. Large-scale plane-wave-based density functional theory: Formalism, parallelization, and applications. In *Computational Methods for Large Systems: Electronic Structure Approaches for Biotechnology and Nanotechnology*, J. R. Reimers (Ed.), Wiley, pp. 77–116. Hoboken, NJ: John Wiley & Sons, 2011.

17. Y. Chen, E. Bylaska, and J. Weare. First principles estimation of geochemically important transition metal oxide properties. *Molecular Modeling of Geochemical Reactions: An Introduction*, J. D. Kubicki (Ed.), Hoboken, NJ: Wiley-Blackwell, p. 107, 2016.

18. NERSC. Measuring arithmetic intensity. http://www.nersc.gov/users/application-performance/measuring-arithmetic-intensity/. Accessed on December 22, 2016.

19. R. Car and M. Parrinello. Unified approach for molecular dynamics and density-functional theory. *Physical Review Letters*, 55(22):2471, 1985.

20. D. K. Remler and P. A. Madden. Molecular dynamics without effective potentials via the Car-Parrinello approach. *Molecular Physics*, 70(6):921–966, 1990.

21. G. Kresse and J. Hafner. Ab initio molecular dynamics for liquid metals. *Physical Review B*, 47(1):558, 1993.
22. D. Marx and J. Hutter. Ab initio molecular dynamics: Theory and implementation. In *Modern Methods and Algorithms of Quantum Chemistry*, J. Grotendorst (Ed.), Jülich: John von Neumann Institute for Computing, pp. 301–449, 2000.
23. W. Kohn and L. J. Sham. Self-consistent equations including exchange and correlation effects. *Physical Review*, 140(4A):A1133, 1965.
24. E. J. Bylaska, M. Valiev, R. Kawai, and J. H. Weare. Parallel implementation of the projector augmented plane wave method for charged systems. *Computer Physics Communications*, 143(1):11–28, 2002.
25. P. N. Swarztrauber. Multiprocessor FFTs. *Parallel Computing*, 5(1):197–210, 1987.
26. M. Jacquelin, W. De Jong, and E. Bylaska. Towards highly scalable ab initio molecular dynamics (AIMD) simulations on the Intel Knights Landing many core processor. In *31st IEEE International Parallel & Distributed Processing Symposium*, Orlando, Florida, page accepted. IEEE Computer Society, 2017.
27. T. W. Swaddle, J. Rosenqvist, P. Yu, E. Bylaska, B. L. Phillips, and W. H. Casey. Kinetic evidence for five-coordination in AlOH(aq)2+ ion. *Science*, 308(5727):1450–1453, 2005.
28. J. R. Rustad and E. J. Bylaska. Ab initio calculation of isotopic fractionation in B(OH)3(aq) and BOH$_4^-$(aq). *Journal of the American Chemical Society*, 129(8):2222–2223, 2007. PMID: 17266314.
29. R. Atta-Fynn, D. F. Johnson, E. J. Bylaska, E. S. Ilton, G. K. Schenter, and W. A. De Jong. Structure and hydrolysis of the U (IV), U (V), and U (VI) aqua ions from ab initio molecular simulations. *Inorganic Chemistry*, 51(5):3016–3024, 2012.
30. J. L. Fulton, E. J. Bylaska, S. Bogatko, M. Balasubramanian, E. Cauet, G. K. Schenter, and J. H. Weare. Near-quantitative agreement of model-free DFT-MD predictions with XAFS observations of the hydration structure of highly charged transition-metal ions. *Journal of Physical Chemistry Letters*, 3(18):2588–2593, 2012.
31. S. O. Odoh, E. J. Bylaska, and W. A. de Jong. Coordination and hydrolysis of plutonium ions in aqueous solution using car-Parrinello molecular dynamics free energy simulations. *Journal of Physical Chemistry A*, 117:12256–12267, 2013.
32. R. Atta-Fynn, E. J. Bylaska, and W. A. de Jong. Importance of counteranions on the hydration structure of the curium ion. *Journal of Physical Chemistry Letters*, 4(13):2166–2170, 2013.
33. F. Gygi. Architecture of Qbox: A scalable first-principles molecular dynamics code. *IBM Journal of Research and Development*, 52(1.2):137–144, 2008.
34. F. Coester. Bound states of a many-particle aystem. *Nuclear Physics*, 7(4):421–424, 1958.
35. F. Coester and H. Kümmel. Short-range correlations in nuclear wave functions. *Nuclear Physics*, 17(3):477–485, 1960.
36. J. Čížek. On correlation problem in atomic and molecular systems. Calculation of wavefunction components in ursell-type expansion using quantum-field theoretical methods. *Journal of Chemical Physics*, 45(11):4256–4266, 1966.
37. J. Paldus, I. Shavitt, and J. Čížek. Correlation problems in atomic and molecular systems. 4. Extended coupled-pair many-electron theory and its application to BH3 molecule. *Physical Review A*, 5(1):50–67, 1972.
38. G. D. Purvis and R. J. Bartlett. A full coupled-cluster singles and doubles model: The inclusion of disconnected triples. *Journal of Chemical Physics*, 76(4):1910–1918, 1982.
39. R. J. Bartlett and M. Musial. Coupled-cluster theory in quantum chemistry. *Reviews of Modern Physics*, 79(1):291–352, January–March 2007.
40. K. Raghavachari, G. W. Trucks, J. A. Pople, and M. Head-Gordon. A 5th-order perturbation comparison of electron correlation theories. *Chemical Physics Letters*, 157(6):479–483, 1989.
41. J. Jeers and J. Reinders. *Intel Xeon Phi Coprocessor High Performance Programming*. Burlington, MA: Morgan Kaufmann, February 2013.
42. T. P. Straatsma and J. A. McCammon. ARGOS, a vectorized molecular dynamics program. *Journal of Computational Chemistry*, 11:943–951, 1990.
43. S. Plimpton. Fast parallel algorithms for short-range molecular dynamics. *Journal of Computational Chemistry*, 117:110, 1995.
44. K. J. Bowers, R. O. Dror, and D. E. Shaw. Zonal methods for the parallel execution of range-limited N-body simulations. *Journal of Computational Physics*, 221:303–329, 2007.

45. S. Kumar, C. Huang, G. Almasi, and L. V. Kale. Achieving strong scaling with NAMD on Blue Gene/L. In *Proceedings 20th IEEE International Parallel Distributed Processing Symposium*, Rhodes Island, Greece, p. 10, April 2006.

46. J. A. McCammon, T. P. Straatsma, M. Philippopoulos. NWChem: Exploiting parallelism in molecular simulations. *Computer Physics Communications*, 128:377385, 2000.

47. T. P. Straatsma and D. G. Chavarria-Miranda. On eliminating synchronous communication in molecular simulations to improve scalability. *Computer Physics Communications*, 184:2634–2640, 2013.

48. D. Elwood, G. Fann, and R. Littlefield. *Parallel Eigensolver System Users Manual*. Richland, WA: Pacific Northwest National Laboratory, 1993.

49. L. S. Blackford, J. Choi, A. Cleary, E. D'Azevedo, J. Demmel, I. Dhillon, J. Dongarra, et al. *ScaLAPACK Users' Guide*. Philadelphia, PA: SIAM Publications, 1997.

9 Exascale Programming Approaches for Accelerated Climate Modeling for Energy

Matthew R. Norman, Azamat Mametjanov, and Mark Taylor

CONTENTS

9.1 Overview and Scientific Impact of Accelerated Climate Modeling for Energy 188
9.2 GPU Refactoring of ACME Atmosphere 188
    9.2.1 Mathematical Considerations and Their Computational Impacts 189
        9.2.1.1 Mathematical Formulation 189
        9.2.1.2 Grid...189
        9.2.1.3 Element Boundary Averaging...190
        9.2.1.4 Limiting..190
        9.2.1.5 Time Discretization 190
    9.2.2 Runtime Characterization 191
        9.2.2.1 Throughput and Scaling...191
    9.2.3 Code Structure 191
        9.2.3.1 Data and Loops...191
        9.2.3.2 OpenMP...192
        9.2.3.3 Pack, Exchange, and Unpack 192
        9.2.3.4 Bandwidth and Latency in MPI Communication 192
    9.2.4 Previous CUDA FORTRAN Refactoring Effort......................................193
    9.2.5 OpenACC Refactoring...193
        9.2.5.1 Thread Master Regions...193
        9.2.5.2 Breaking Up Element Loops 194
        9.2.5.3 Flattening Arrays for Reusable Subroutines..........................194
        9.2.5.4 Loop Collapsing and Reducing Repeated Array Accesses 195
        9.2.5.5 Using Shared Memory and Local Memory 195
        9.2.5.6 Optimizing the Boundary Exchange for Bandwidth 197
        9.2.5.7 Optimizing the Boundary Exchange for Latency 197
        9.2.5.8 Use of CUDA MPS 197
    9.2.6 Optimizing for Pack, Exchange, and Unpack...197
    9.2.7 Testing for Correctness 198
9.3 Nested OpenMP for ACME Atmosphere 199
    9.3.1 Introduction...199
    9.3.2 Algorithmic Structure 199
    9.3.3 Programming Approach...201
    9.3.4 Software Practices 202
    9.3.5 Benchmarking Results...203

9.4   Portability Considerations 203
       9.4.1   Breaking Up Element Loops 204
       9.4.2   Collapsing and Pushing If-Statements Down the Callstack 204
       9.4.3   Manual Loop Fissioning and Pushing Looping Down the Callstack 205
       9.4.4   Kernels versus Parallel Loop 205
9.5   Ongoing Codebase Changes and Future Directions 206
Acknowledgments 206
References 206

9.1   OVERVIEW AND SCIENTIFIC IMPACT OF ACCELERATED CLIMATE MODELING FOR ENERGY

The Accelerated Climate Modeling for Energy (ACME) program is a leading-edge climate and Earth system model designed to address the U.S. Department of Energy (DOE) mission needs. The ACME project involves collaboration between seven National Laboratories, the National Center for Atmospheric Research, four academic institutions, and one private-sector company. The ACME project funds ongoing efforts to ensure the model continues to make efficient use of DOE Leadership Computing Facilities (LCFs) in order to best address the three driving grand challenge science questions:

1. Water cycle: How will more realistic portrayals of features important to the water cycle (resolution, clouds, aerosols, snowpack, river routing, land use) affect river flow and associated freshwater supplies at the watershed scale?
2. Cryosphere systems: Could a dynamical instability in the Antarctic ice sheet be triggered within the next 40 years?
3. Biogeochemistry: What are the contributions and feedbacks from natural and managed systems to current greenhouse gas fluxes, and how will those factors and associated fluxes evolve in the future?

For the water cycle, the goal is to simulate the changes in the hydrological cycle with a specific focus on precipitation and surface water in orographically complex regions such as the western United States and the headwaters of the Amazon. For the cryosphere, the objective is to examine the near-term risk of initiating the dynamic instability and onset of the collapse of the Antarctic ice sheet due to rapid melting by warming waters adjacent to the ice sheet grounding lines. For the biogeochemistry, the goal is to examine how coupled terrestrial and coastal ecosystems drive natural sources and sinks of carbon dioxide and methane in a warmer environment.

The ACME model is composed currently of five main components: atmosphere, ocean, land surface, sea ice, and land ice. All of these components are connected through a model coupler. The atmospheric component, the community atmosphere model-spectral element (CAM-SE) [1,2] is based on a local, time-explicit spectral element (SE) method on cubed-sphere topology with static adaptation capability. CAM-SE uses either a hydrostatic or non-hydrostatic assumption and it includes an evolving set of physics packages. The atmospheric component is the most expensive currently. The next most expensive component is the Ocean model, the Model for Prediction Across Scales-Ocean (MPAS-O) [3], which is based on fully unstructured lower ordered finite-volume approximations with split-explicit handling of the barotropic equations as opposed to a more traditional elliptic solve approach.

9.2   GPU REFACTORING OF ACME ATMOSPHERE

CAM-SE can be decomposed ideologically and practically into three separate parts: dynamics, tracers, and physics. The *dynamics* are gas-only stratified fluid dynamics on a rotating sphere with a hydrostatic approximation to eliminate vertically propagating acoustic waves. The *tracers* component transports quantities used by the dynamics and physics along wind trajectories, and examples of these quantities are the three forms of water, CO_2, methane, aerosols, etc. Finally, the *physics* are everything either below spatiotemporal truncation in the dynamics or not included in the equation set to begin with. This includes eddy viscosity, gravity wave drag, shallow and deep moist convection, radiation, microphysics, and other physical phenomena. The physics and transport are operator split in an advection-reaction manner.

9.2.1 MATHEMATICAL CONSIDERATIONS AND THEIR COMPUTATIONAL IMPACTS

9.2.1.1 Mathematical Formulation

CAM-SE, as the name suggests, is based upon the SE (Spectral Element) numerical discretization of the underlying partial differential equations (PDEs) for stratified, hydrostatic fluid dynamics on the rotating sphere. The SE belongs to the class of finite element methods, or more specifically, continuous Galerkin methods. It, therefore, uses a variational form of the PDEs, and the underlying basis in this case is 2-D tensored Lagrange interpolating polynomials on a grid of Gauss–Legendre–Lobatto (GLL) points. When integration by parts is performed, the subsequent Dirichlet element boundary terms drop out due to the fact that C^0 continuity is maintained at element boundaries by a straightforward linear averaging operation.

Body integrals are solved using GLL quadrature, which in coordination with the GLL Lagrange interpolating basis leads to a discrete orthogonality that renders the mass matrix diagonal, though it is inexactly integrated. This discrete orthogonality also simplifies the body integral calculations such that they are simply one-dimensional (1-D) sweeps of quadrature, and this is the core calculation of the SE method. What this ends up looking like computationally is 1-D sweeps of $N \times N$ matrix-vector multiplies (where N is the spatial order of accuracy), and indeed, the SE method can be cast in this way. Thus, the computational complexity of the SE spatial operator is DN^{D+1}, where D is the spatial dimensionality ($D = 2$ in this case), and this is generally the optimal computational complexity for a time-explicit, hyperbolic spatial operator.

The fact that the SE method at its core is essentially made up of 1-D sweeps of matrix-vector multiplies is advantageous for graphics processing units (GPUs) because this means that data fetched from dynamic random-access memory (DRAM) will be reused. The data access scales as N^2, while the computation scales as $2N^3$, making this a compute intensive operation—all the more so as N increases. On the ground level, matrix-vector multiplies always have an innermost loop that is dependent, and it accesses different indices of the same arrays multiple times. This means that for efficient execution on a GPU, these kernels must use CUDA Shared Memory, if only due to the fact that arrays are accessed noncontiguously in the innermost loop. Another advantage to SE methods is that they can cluster the majority of the computation into a single unbroken loop/kernel. It's good to have numerical methods that do not inherently require a significant amount of global dependence between global data structures.

9.2.1.2 Grid

CAM-SE realizes spherical geometry by use of the so-called cubed-sphere [4] grid, which in this case utilizes a nonorthogonal equal-angle gnomonic projection from a cube onto the sphere. This grid is advantageous in several ways: (1) it provides nearly uniform grid spacing; (2) it avoids the strong polar singularities experienced by a latitude–longitude grid and instead has eight weaker singularities at cube corners; (3) it is logically structured, which can simplify operations like element edge averaging; and (4) it avoids conservation and/or Courant–Friedrichs–Lewy (CFL) difficulties experienced by overset grids such as Yin-Yang [5]. The main disadvantageous aspect is the use of nonorthogonal coordinates, which complicates the numerics and also makes maintaining desirable properties from orthogonal schemes difficult. Each of the cubed-sphere's six panels is subdivided into $ne \times ne$ elements.

9.2.1.3 Element Boundary Averaging

Again, the element boundaries are kept continuous by a linear averaging of the element boundary basis function coefficients (the unknowns being evolved by SE). This is the only means of element-to-element communication in the SE method, as all spatial operators must have some mechanism of communication between DOFs on the grid. Though it is low in computation, it actually consumes more time than the body integral calculations because (1) it is heavy in terms of data movement, and (2) the data accesses are typically not regular. This averaging can easily induce a race condition in that one element's data must not yet be replaced with the linear average before the other element accesses it. To keep race conditions from occurring, the developers of CAM-SE chose to pack all element edge data into a process-wide buffer, and then during the unpacking procedure, the averaging is performed. The benefit of this is that it makes arbitrarily unstructured meshes easy to implement. The downsides are that (1) DRAM accesses are doubled by packing to and from yet another global buffer, (2) many DRAM accesses are neither contiguous nor ordered, and (3) this often leads to having to access data owned by another core on the CPU. In hindsight, it would be better to traverse the edges themselves and average the data in place, and there are efforts to investigate this.

In parallel implementations, SE is particularly advantageous because the only communication between adjacent compute nodes is the element edge data that overlay a domain decomposition boundary. This is an optimal communication requirement per time step.

9.2.1.4 Limiting

There are two mechanisms for limiting oscillations that are frequently paired with the SE method in the presence of nonsmooth data. The first option is a fourth-order *hyperviscosity*, which is expressed in PDE form as: $\partial q / \partial t = -\nu \nabla^4 q$, where ν is the coefficient of viscosity. This fourth-order derivative term is computed by applying a Laplacian operator twice, each application followed by an averaging of element boundary data. The hyperviscosity operator is included along with the standard advective update in the tracer transport, and therefore, the last stage of element edge data averaging of the hyperviscosity, being coupled with the advective update, is hidden. Also, it is only applied once for tracers. For the dynamics, the hyperviscosity operator is used essentially as an energy cascade closure scheme, and it must be applied multiple times (i.e., *subcycled*) per time step in order to achieve a proper energy cascade. For the dynamics, hyperviscosity takes up more time than the fluid update step.

The other mechanism for limiting oscillations is a monotone limiter that is only applied to tracer transport. The limiter begins by gathering the maximum and minimum basis coefficient values among all neighboring elements and the element in question. Then, after the advective step takes place, a limiter routine is called that begins by replacing any data that exceeds the maximum (minimum) bounds with the maximum (minimum) value. It then proceeds to redistribute the changed mass in an optimal manner that best maintains similarity to the original data shape, and this limiter is applied in multiple iterations. There must also be included a hyperviscosity tendency along with the advective step to avoid artifacts known as *terracing* that also are common in flux corrective transport methods when a viscous process is excluded [6].

9.2.1.5 Time Discretization

The SE method is cast into semidiscrete form, meaning the temporal derivative is left in place as the last remaining continuously differentiated term in the PDE. Then, an ordinary differential equation (ODE) solver is applied in time to close the scheme in time, and in this case, Runge–Kutta (RK) integrators are used. The RK integrator currently being used for the dynamics is a five-stage, third-order accurate integrator from [7] that provides a very large maximum stable CFL (MSCFL) value for a large time step. The tracers use a three-stage, second-order accurate strong stability preserving (SSP) RK method [8], and the reason for using an SSP method is that it guarantees that if the spatial operator is monotonic, then the temporal integrator will not introduce monotonicity violations of its own.

9.2.2 Runtime Characterization

9.2.2.1 Throughput and Scaling

Climate simulation has a unique and unfortunate constraint not experienced by many scientific fields in that it must be run for very long lengths of simulation time. Many climate runs must complete on the order of 100 years to give useful results and statistics, and multiple scenarios with different anthropogenic forcings must be completed. The current simulations that use an average distance of about 28 km between grid points require a dynamics time step on the order of a minute. This means that we must take roughly 50 million dynamics time steps before the simulation is completed. The physics are evaluated on a time step of half an hour, meaning roughly two million physics time steps must be taken. In order for the results to be obtained in a reasonable amount of time, the climate community has converged on a widely accepted throughput requirement of about five simulated years per wall clock day (SYPD). This is about 2,000× realtime.

There are significant problems with having such a constraint. If one holds the throughput as a constant, which we generally must, then each 2× refinement in horizontal grid spacing results in an 8× increase in the amount of work to be performed, 4× of which comes from spatial refinement in two spatial dimensions, and another multiple of 2× that comes from being forced to refine the time step with the grid spacing, since CAM-SE uses an explicit SE method. However, there is only 4× more data/parallelism available because the time dimension is dependent. This means that in order to achieve the same throughput, the refined simulation will have to run on 8× more compute nodes with only 4× more data. Thus, the amount of data and the amount of work per node cuts in half with each 2× spatial refinement. And this is in a perfect world. In reality, since scaling is not perfect, more than 8× nodes will be required in order to keep the same throughput, and we are left with *less than* half the work per node for each 2× refinement.

What this means, practically, is that the CAM-SE code, when run in production mode, is always strong scaled to a significant extent. This means that message passing interface (MPI) data exchanges are always consuming a significant amount of the runtime. Coming exascale platforms will interact with this reality in an interesting way. First, it appears that exascale computing will, on the whole, prefer *fatter* nodes, which we take here to mean that aggregate memory bandwidth on a node will grow faster than the number of nodes (a node being a collection of processing elements separated from other nodes by network interconnect). The only benefit this really buys us is that we can spend comparatively more money on interconnect for improved bandwidth. Latency, however, appears to be remaining stagnant unless a transformative technology comes along.

Also of significant concern on current architectures is the regrettable dependence on a high-latency and low-bandwidth peripheral component interconnect (PCI) express bus. First, the PCI-e latency is an order of magnitude higher than the latency of nearest-neighbor MPI communication. Second, the bandwidth of PCI-e is roughly the same as the interconnect used by MPI, meaning the MPI bandwidth costs are doubled. It appears that while future technologies will significantly increase the bandwidth, the latency, again, will remain mostly the same. Thus, smaller transfers are increasingly penalized.

9.2.3 Code Structure

9.2.3.1 Data and Loops

CAM-SE uses spectral elements only in the horizontal direction, so a node contains `ntimelevels` time levels of `nelemd` columns of elements with `nlev` vertical levels, each of which has `np×np` basis functions coefficients. These form the fundamental dimensions of the code and also the loop bounds. For tracer transport, there are also `qsize` tracers that need to be transported, creating an additional dimension for that portion of the code. For most of the code, loops over each of these dimensions provide data-independent work that is relatively easy to vectorize, especially on the GPU.

The data is laid out such that `np×np` is the fastest varying dimension, followed by `nlev`, then `qsize` (for tracers only), followed by `timelevels`, followed by `nelemd`. A single column of elements for all time levels is stored in a FORTRAN derived type called `element_t`, which also contains the geometric, basis function, and differentiation data that is used in the SE method for that column of elements. As expected, the looping structure mirrors the data layout with looping over elements as the outermost, followed by tracers, then vertical levels, and finally the basis coefficients.

The practice in the original CAM-SE code is to structure the code largely into two groups of tightly nested loops. The outer tightly nested loops are those over elements, tracers, and vertical levels. The next tightly nested loops are those over the basis functions, and these loops are often inside routines that calculate various derivative-based quantities. This is advantageous in terms of cache locality since often times, the same data is being accessed as the code progresses from one routine to another.

9.2.3.2   OpenMP

For the sake of efficiency, the choice was made at the creation of CAM-SE to structure the OpenMP implementation using parallel regions instead of loop-level OpenMP. The reasoning for this was to ensure that thread pools weren't needlessly destroyed and recreated. So, in the initialization of the code, each thread is assigned a range of element columns: `[nets,nete]`, and that thread only operates on those element columns. Generally this works efficiently, the main exception being the packing and unpacking of element edge data to and from process-wide buffers.

There are difficulties with this when performing a GPU refactoring of the code, however. At least at the time the code was refactored using CUDA FORTRAN and then later in OpenACC, poor performance was experienced when multiple threads were launching kernels on the same process and device. Therefore, a thread master region had to be generated for every section of GPU code between MPI boundary exchanges, which themselves cannot be inside master regions.

There is also the option of using all MPI, and simply using the CUDA MPS server (previously called *proxy*), and this works well for many codes. However, since CAM-SE is significantly strong scaled, there ends up being too little work per kernel call for this to be efficient in practice.

9.2.3.3   Pack, Exchange, and Unpack

Averaging of the element edge data ends up being by far the most expensive operation in the CAM-SE tracers and dynamics because this is where the majority of the data movement occurs, both to and from on-node DRAM and over network interconnect.

Packing essentially just places element edge data into predefined locations in a process-wide buffer. Data is placed into this buffer in such a manner that data moving from one process to another is contiguous in memory so as to minimize the number of MPI data transfers and reduce latency costs. Next, the boundary exchange takes data from the process-wide buffer that must go to a neighbor and performs an `MPI_Isend` on that data. It then performs an `MPI_Irecv` on data it is receiving and places that data into a receive buffer. Next, the boundary exchange routine copies internal data from the edge buffer to the receive buffer so that all data is in the receive buffer. Then, finally, the unpack routine takes data from predefined locations, sums them up, and replaces its value with that sum. Later, a multiplier will be applied to scale this sum down into an average.

9.2.3.4   Bandwidth and Latency in MPI Communication

For production runs, MPI messages in the dynamics between adjacent elements will generally at most be on the order of 10–20 KB, while messages between corner elements will be at most 1–2 KB. If we assume, for instance, that nearest-neighbor interconnect latency is roughly a microsecond, and the bandwidth is roughly 6 GB/sec (which is true for OLCF's Titan), the bandwidth costs for adjacent element messages is only three times larger than the latency costs, and for corner elements, latency clearly dominates. When running larger problem sizes, often times, the MPI communication is not the nearest neighbor and latency costs increase substantially. For this reason, the dynamics are extremely sensitive to interconnect latency costs. For the tracer transport routines, however, the message sizes are increased by a factor of about 40 because there are currently 40 tracers being evolved, which can all be sent together. Therefore, the tracer transport is bandwidth bound in the MPI rather than latency bound. This means one must take different optimization strategies when refactoring the dynamics versus the tracer transport.

9.2.4 Previous Cuda Fortran Refactoring Effort

Previously, CUDA FORTRAN was used to refactor only the atmospheric tracer transport to GPUs [9–11]. This, generally speaking, was a poor design choice, but compiler implementations of the OpenACC standard were nowhere near ready for use in a real code due to bugs and performance problems. Even at the time of writing this, the CUDA FORTRAN implementation significantly outperforms the OpenACC implementation. There were, however, software maintainability problems with using CUDA in a FORTRAN context:

- The syntax of CUDA FORTRAN is so foreign to the existing FORTRAN standards that the resulting code is usually unrecognizable. The chevron syntax is truly bizarre.
- It requires separating out each kernel into unnecessary subroutines that will not be reused and are now distant from their previous context in the code.
- It changes every kernel from a simple in-file loop structure to a kernel call to a subroutine.
- At the time, it required breaking all variables used in kernels out of their derived types.
- There are no longer any actual loops, which is quite foreign to any domain science programmer and makes the subroutine much more difficult to read.
- CUDA shared memory had to be used in order to gain efficiency in the kernels, and being ideologically a local temporary variable, it is very confusing to see a local variable used in a context that looks global with CUDA FORTRAN's "`__syncthreads()`" used in between.
- There is no incremental refactoring to CUDA FORTRAN once the code is placed into a kernel, which makes debugging more difficult.
- The same code will not run on CPU and GPU like it does in OpenACC, and this also makes debugging significantly more difficult.

The original CUDA FORTRAN refactoring gave good performance, but it turned out that the codebase was still somewhat in flux. This meant that changes needed to be propagated into the CUDA code, which was quite difficult to support. The refactoring effort had to be reperformed once because of structural changes that didn't agree well with the previous CUDA structure. For this reason, the GPU code was transferred into OpenACC code for better software maintainability.

9.2.5 OpenACC Refactoring

9.2.5.1 Thread Master Regions

As mentioned before, we made a choice to use a master thread for all GPU kernel activity instead of calling kernels within threaded regions for several reasons. First, the amount of work per kernel call was typically too low to gain efficiency on the GPU when running from threads, and therefore aggregating all threads to a single kernel call was preferred. Also, when running threaded, even with separate asynchronous ids for each CPU thread, we found that the kernels were being serialized on the device and not running concurrently.

This process created no shortage of difficulties in the refactoring process, as creating master regions in an OpenMP parallel region code can be very prone to producing bugs. As an example, many array temporaries are passed wholly into subroutines, where it is assumed they only have memory for the range [nets,nete]. Once inside a master region, it is easy to forget to add this array slice to the subroutine input. Also, the thread's element bounds are often passed into routines, and they must be changed to [1,nelemd]. This process begins on a small routine bracketed by barrier statements and master / end master statements inside them. Then, the next small section of code is added into the master region, and so forth until the entire section of code being ported was included in the master region.

9.2.5.2   Breaking Up Element Loops

For several reasons, the long loop bodies inside the element and vertical level loops were broken up. First, OpenACC was having a lot of difficulty with the use of the routine directive at the time, especially for functions rather than subroutines, and the compiler also was unable to inline the routines. Second, it's important not to let the body of a kernel get too long because eventually, register pressure sets in, and the occupancy of the kernel becomes poor, as does the performance. Therefore, routines that previously only iterated over the basis functions within one element and one vertical level were transformed so that they iterate over all loop indices in the problem.

This, in turn, required several other transformations. First, small array temporaries that previously only dimensioned as (np,np) had to be changed into globally sized variables. This increased the memory requirements of the code, but it isn't a problem for ACME, which runs strong scaled to such an extent that most of the on-node memory still remains unused. This is another opportunity for bugs to be introduced because many FORTRAN compilers allow (I believe erroneously) one to use fewer subscripts than those that actually exist in the declared array. Therefore, failing to add the extra subscripts to something that used to be a small array temporary will produce a bug.

Also, this requires pushing loops down the callstack, which is a common operation when refactoring codes to run on GPUs. Instead of having each of the derivative-type routines operate only on the basis functions of a single element and vertical level, they are now transformed to work on all levels of parallelism up to the element level. This requires adding more dummy parameters to be passed in because the routine needs to know, for instance, what time level to work on or what range of elements to loop over. This, in turn, leads to another transformation requirement in order to allow the subroutines to be reusable, and this is discussed in the next subsection.

9.2.5.3   Flattening Arrays for Reusable Subroutines

There is an issue that arises when pushing loops down the callstack while using a derived type to hold each element's data. When the subroutine must now loop over elements, it must know the actual name of the variable within the element derived type being operated upon. Previously, one could simply pass, for instance, elem(ie)%state%T(:,:,k,n0) to the routine, where elem is an array of the element derived type, k is the vertical level index, and n0 is the time level being operated upon. This is because the routine doesn't operate over the element loop. However, once the element loop is pushed down into the routine in question, one must pass all of elem(:), and inside the routine, elem (ie)%state%T would have to be explicitly named. At this point, that routine is no longer reusable for other variables besides T.

It is poor software engineering practice to create multiple routines that do identical work, and therefore it is desirable to make these routines reusable for any particular data. In order to accomplish this, we created a flat, global array, called state_T(np,np,nlev,timelevels,nelemd) for this example. Then, in order to make sure the rest of the code doesn't need to know about this new flattened array, we use a pointer inside the element data structure, and we point into this array so that either syntax may be used for the CPU code at least. Then, state_T can be passed into the routine in question, or any other flattened array. We only flattened the arrays that needed to be passed into a reusable subroutine.

This, in turn led to yet another potential for bugs, and this was difficult to find at times. Because the form of elem(ie)%state%T is now a pointer, and at least for the PGI compiler, only the flattened array syntax, state_T(...) is allowed in any of the OpenACC code, be it a data statement or a kernel. This means one must remember to add the flattened array to the present clause of the kernel along with elem to ensure correctness, and one must replace all kernel references to that array with the flattened expression, including previously ported kernels (which is where the problem usually arose). Probably, the worst part of it all was that one does not always get an invalid address error in the kernel when using the derived type syntax inside a kernel. Sometimes, it can give the right answer, and then later, it will give an invalid memory address error. The intermittency of this error led to difficulty in finding bugs in the refactored code.

9.2.5.4 Loop Collapsing and Reducing Repeated Array Accesses

For the majority of the kernels, some relatively simple optimizations gave good enough performance. First of all, because there are always four to five levels of tightly nested loops in each of the kernels, loop collapsing is absolutely essential. This is mainly because one cannot nest multiple "`!$acc loop gang`" or "`!$acc loop vector`" directives when using the "`parallel loop`" construct in OpenACC. We hope this restriction will be removed in later versions of OpenACC.

Also, the "`kernels`" construct is not relied upon in the ACME code because it generally gave very poor performance. The `kernels` construct is one in which most of the modifiers are simply suggestions to the compiler about what is parallelizable. It leaves a lot of room for the compiler itself to make choices about where to put the parallelism for various loops. For CAM-SE, the compiler usually made very poor decisions. The `parallel loop` construct is much more like the OpenMP parallel do construct, in that it is highly prescriptive rather than descriptive, giving the user a larger amount of control. Therefore, the `kernels` construct has been abandoned, and the `parallel loop` construction is relied upon in all of the kernels.

The last general optimization strategy is to replace repeated array accesses with a temporary scalar variable. Presumably, this makes the compiler more likely to place the scalar into register rather than repeatedly accessing the array over and over again from DRAM. It is our opinion that this is a performance bug in the compiler because the compiler should already be placing repeated array accesses into register and keeping it there. But for now, this is an optimization technique that provides a lot of benefit.

9.2.5.5 Using Shared Memory and Local Memory

As mentioned earlier, because the Spectral Element method is essentially 1-D sweeps of matrix-vector products, at least at its core, these kernels stand a lot to gain from the use of CUDA shared memory. The inner parts of these routines' loops often look somewhat like the following:

```
1   do ie = 1 , nelemd
2     do k = 1 , nlev
3       do j = 1 , np
4         do i = 1 , np
5           dsdx00=0.0d0
6           dsdy00=0.0d0
7           do s = 1 , np
8             dsdx00 = dsdx00 + deriv(s,i)*glob_1(s,j,k,ie)
9             dsdy00 = dsdy00 + deriv(s,j)*glob_1(i,s,k,ie)
10          enddo
11          glob_2(i,j,1,k,ie) = c1 * dsdx00 + c2 * dsdy00
12          glob_2(i,j,2,k,ie) = c3 * dsdx00 + c4 * dsdy00
13        enddo
14      enddo
15    enddo
16  enddo
```

We have found that the most efficient way to run this on the GPU is to thread the outer four loops on the GPU and to leave the inner loop sequential. What this means is that (1) the kernel is accessing `glob_1` and `deriv` multiple times in each thread and (2) neither `glob_1` nor `deriv` are being accessed contiguously in the GPUs DRAM memory. Since the GPU is actually threading the i and j indices, the presence of the s index in those arrays means they are not being accessed contiguously. In order to alleviate this problem, these arrays are both put into CUDA shared memory.

However, in order to do this, the vertical levels loop must be tiled. When run in production mode, np is always set to four, meaning the i and j loops have 16 iterations together. Also, 72 vertical levels are currently being used. For most cases, the best kernel performance is obtained by using 128–256 threads within an SM. Therefore, the i and j loops do not have enough parallelism to fill an SM, and the i, j, and k loops together have too much parallelism. Therefore, with kernels that need to use shared memory, the vertical level loop is tiled to produce something that looks like the following:

```
1   do ie = 1 , nelemd
2     do kc = 1 , nlev/kchunk + 1
3       do kk = 1 , kchunk
4         do j = 1 , np
5           do i = 1 , np
6             k = (kc-1)*kchunk + kk
7             if (k <= nlev) then
8               dsdx00=0.0d0
9               dsdy00=0.0d0
10              do s = 1 , np
11                dsdx00 = dsdx00 + deriv(s,i)*glob_1(s,j,k,ie)
12                dsdy00 = dsdy00 + deriv(s,j)*glob_1(i,s,k,ie)
13              enddo
14              glob_2(i,j,1,k,ie) = c1 * dsdx00 + c2 * dsdy00
15              glob_2(i,j,2,k,ie) = c3 * dsdx00 + c4 * dsdy00
16            endif
17          enddo
18        enddo
19      enddo
20    enddo
21  enddo
```

With this loop, one can now place glob_1(:,:,(kc-1)*kchunk+1:kc*kchunk,ie) and deriv(:,:) into shared memory on the GPU, and the loops over i, j, and kk can be collapsed down and placed within an SM with the "vector" loop clause modifier, and the kc and ie loops can be collapsed down and distributed over SMs with the "gang" loop clause modifier. Routines of this form will typically use kchunk=8 because it gives a vector length of 128, which often performs best on the K20x and K40 GPUs at least. The cache clause is used to place the above variables into CUDA shared memory. Also note that on the CPU, this code will not vectorize because of the if-statement inside the inner loop. On the GPU, this is not a problem because if k gets larger than nlev at any point, the cores will simply no-op. During runtime, you do not even notice the presence of the if-statement on the GPU.

We did encounter a problem when trying to cache contiguous chunks of a global variable like mentioned in the previous paragraph. The PGI compiler has a performance bug wherein the correct amount of shared memory is allocated, and it is used correctly, but it is loaded multiple times from DRAM, causing significant performance degradation. This issue is being addressed, but currently, we are having to create our own array temporary of the size (np,np,kchunk), declare it private at the gang level, and then use the cache clause on that variable above the collapsed vector loops. It is a tedious process, and it requires knowing about the array temporary at the gang loop levels where the code doesn't actually need to know about the array. Once this performance bug is fixed, the code that uses shared memory will certainly be simpler.

9.2.5.6 Optimizing the Boundary Exchange for Bandwidth

The tracer transport section of CAM-SE has much larger MPI transfers than the dynamics, and it turns out that these MPI transfers experience more runtime cost due to bandwidth than due to latency. Also, given that on OLCF's Titan, the PCI-e bandwidth is roughly the same as that of MPI, it is important to hide as much of that as possible so that the bandwidth costs are not fully tripled because of GPU usage.

- Prepost the `MPI_Irecv` calls
- Call "`!$acc update host(...)`" of each stage's data in its own asynchronous ID ("`asyncid`")
- Begin a polling loop
  - Once a stage's data is on the CPU (tested by "`acc_async_test`")
    - Call the `MPI_Isend` for that stage
  - Once a stage's `MPI_Irecv` is completed (tested by "`MPI_Test`")
    - Call "`!$acc update device(...)`" for that stage's received data
- Call "`!$acc wait`" to ensure all data is finished being copied

In this way, PCI-e copies from device to host, PCI-e copies from host to device, and MPI transfers are all able to overlap with one another.

9.2.5.7 Optimizing the Boundary Exchange for Latency

For the dynamics, however, the issue is not bandwidth but latency. The aforementioned strategy would basically end up with a ton of PCI-e traffic, each call incurring a measure of latency. Therefore, the strategy for minimizing latency costs is to simply: (1) pre-post the `MPI_Irecv` calls; (2) call "`!$acc update host(...)`" for each stage asynchronously followed by an "`!$acc wait`"; (3) call `MPI_Isend` for each stage; and (4) call "`!$acc update device(...)`" for each stage asynchronously followed by an "`!$acc wait`." The extra calls to "`acc_async_test`" cost more in terms of latency than they are worth, since there isn't really any bandwidth to overlap in the first place. Also, calling the PCI-e copies via "`!$acc update`" asynchronously and then waiting after the fact helps avoid noticing the PCI-e latency for each call.

9.2.5.8 Use of CUDA MPS

It turns out that it is more efficient to use a combination of MPI tasks and OpenMP tasks on a node rather than simply using all OpenMP. We aren't entirely sure of the reason, but it appears to be the case that running multiple MPI tasks on the same GPU gives a natural overlap between PCI-e and MPI costs that isn't being effectively exploited by the technique in Section 9.2.5.6. The optimal way to run on Oak Ridge Leadership Computing Facility's (OLCF) Titan is to use four OpenMP threads per MPI task (so that nonported sections of the code like the physics still run on all of the CPU cores) and four MPI tasks per node. Since there are multiple MPI tasks running per GPU, this also requires the use of CUDA MPS.

9.2.6 OPTIMIZING FOR PACK, EXCHANGE, AND UNPACK

In the past, there was an optimization over all of the pack, exchange, and unpack calls. First, during initialization, it would be determined which elements on the node are internal (meaning they have no dependency on MPI) and which elements are external. Then, when pack, exchange, and unpack are called, the external pack is first called asynchronously in one thread, and then the internal pack and unpack are called asynchronously in another thread (since there is no exchange for internal elements). Then, after waiting on the first thread, the boundary exchange is performed as normal. In this way, the internal pack and unpack routines, which are costly, are overlapped with the PCI-e copies and the MPI transfers.

However, in practice, this is of little use for production runs because they are strong scaled to the point where there aren't usually any internal elements in the first place. Therefore, this practice is not currently used. In the future, when more work per node is expected, this technique will likely be reinstituted.

9.2.7 TESTING FOR CORRECTNESS

Testing for correctness is one of the most crucial parts of the refactoring process, and it is also one of the most difficult to perform in an efficient way. As is true with most production science codes, CAM-SE can be run in many different ways with different runtime options. This can certainly make things difficult when refactoring the code, and for this reason, a single configuration was chosen at a time, and any other configuration is caused to abort with an error message until that code can be properly tested during the refactoring effort. Once a given configuration is ported, then more options can be added one at a time while exercising that specific code in a test between development cycles.

It's significantly easier to find and fix a bug when only a small amount of code could have caused it. For this reason, it's crucial to have a fast running test that can be performed frequently. This was an easy situation for the tracer transport routines, as they are linear in nature, thus giving machine precision similar results as the CPU code. For the tracers, a simple recompile, test run, and diff on the output would only require a couple of minutes at most. There was a small amount difficulty getting the code to work for a generic number of vertical levels, but this was not too cumbersome.

The dynamics were another story entirely for three main reasons: (1) there are more options to exercise at runtime than for the tracers; (2) the dynamics are nonlinear in nature, which causes small differences to grow exponentially; and (3) a lot of bugs in the dynamics do not show up after a single time step.

Regarding the first issue, there are options chosen in the full CAM-SE code that are not available or not default in the standalone *dynamical core* (dynamics + tracer) code. Also, the CAM-SE standalone code is fairly deprecated, meaning that to exercise CAM-SE as a whole, it really needs to be run inside of a full ACME compilation. Data models are used for the ocean, land ice, and sea ice, but the land model has to be active when the atmosphere is active when running inside ACME. Compiling the land model and CAM-SE takes much more time than simply compiling the standalone dynamical core, which lengthens the time spent testing between development cycles.

Regarding the second issue, climate is inherently a statistical science and not a deterministic one, and this owes itself to the fact that small differences in a fluid state will amplify exponentially in simulation time due to the nonlinearity of the PDEs that describe fluid flow. This is precisely why there are limits to our weather forecasts, in fact. What it means for correctness testing is that there isn't really a truly robust way to test for correctness in a quick manner. Therefore, we developed as robust of a smoke test as possible that could be run in a reasonable amount of time. First, the CPU code is run with different compiler flags (-O0 and -O2 in our case) for a few time steps, outputting every time step. Next, the global norm of the absolute difference is computed at each of the time steps. This establishes the expected baseline of answer differences over several time steps that are solely due to machine precision changes, and it only needs to be done once. Finally, since the GPU runs should only cause machine precision differences, the global norm of the absolute difference between GPU and CPU must be similar to that of the difference between -O0 and -O2.

Finally, the reason we do this over several time steps (eight ended up being good enough) is that some bugs do not manifest in this smoke test after only one time step. What this means, in total, is that the smoke testing between development cycles to most easily find and fix bugs during the refactoring process took on the order of 15 minutes. It meant that the porting of the dynamics was a very slow process compared to the tracer transport. However, in the end, there was a large amount of confidence that the solution was indeed correct.

The only truly robust way to test for correctness with machine precision changes without having to re-validate the code against observations is to run a suite of ensembles to establish a baseline, and then to run a suite of ensembles with the new version. Then, statistical tests can be performed to ensure that the two sets of ensembles were, in fact, sampled from the same population, not just for globally and annually averaged quantities but also with spatial and intra-annual variations. This is a capability under current development in ACME.

9.3   NESTED OPENMP FOR ACME ATMOSPHERE

9.3.1   INTRODUCTION

ACME is a global climate model with coupled components for atmosphere, land, ocean, sea ice, land ice, and river runoff. Model Coupling Toolkit combines individual components into various component sets that are focused on atmosphere, ocean and sea ice, land, or on fully coupled global climate issues such as water cycle and biogeochemistry. Each compset together with the discretization grid for its components contains a different computational workload. Time integration of the workloads on the 80-year time scale of 40-year hindcast and 40-year forecast can answer various climate simulation questions. Computational performance of a workload is measured with a throughput rate of model Simulated Years Per Day (SYPD). Our goal is the optimization of the throughput rate of these workloads on peta- and pre-exascale computing machines.

Currently, the highest resolution grid used in ACME production runs provides an average grid spacing of 28km between gridpoints. At this resolution, the primary throughput rate limiting component is atmosphere and in particular the time integration of atmospheric dynamics by the High-Order Methods Modeling Environment (HOMME) atmospheric dynamical core. Improving scalability and performance of HOMME can lead to the overall throughput acceleration of ACME's water cycle and biogeochemistry climate simulations.

In this work, we describe improvements to OpenMP-based threading and vectorization in HOMME. We use nested threading to increase the current scaling limit of one spectral element column per core. Nested threads iterate over subelemental data structures and lead to performance improvements. In addition, we discuss OpenMP-based vectorization of iterations over innermost data structures. Together with other code refactoring to help compilers generate efficient code, vectorization leads to more efficient use of on-chip hardware resources and better performance. Overall, combination of the OpenMP-based parallelization techniques exhibits up to 40% overall improvement in the throughput rate of high-resolution atmosphere component set on ALCF's Mira.

The outline of our discussion is as follows. In Section 9.3.2, we overview the existing algorithmic structure of HOMME's dynamics and summarize the hybrid MPI+OpenMP parallelization approach. In Section 9.3.3, we present the programming approach of nested threading. Section 9.3.4 discusses software engineering issues affecting performance engineering. In Section 9.3.5, we conclude with performance benchmarking results.

9.3.2   ALGORITHMIC STRUCTURE

The discretization grid in HOMME is a cubed sphere with six cube faces and ne number of elements per cube edge for a total of 6 ne^2 elements. The high-resolution 28 km grid has ne $= 120$ elements on a cube edge and a total of 86,400 elements columns. Each element column contains a number of vertical levels to represent the vertical dimension. The hierarchical decomposition of an element's state lends itself naturally to the structure-of-arrays data storage. Thus, the layout of element data along with the typical parameters used in ACME have the following form in Fortran:

```fortran
1   !Vertical Levels
2   integer, parameter, public :: nlev = 72
3   !Number of Tracers
4   integer, parameter, public :: qsize = 35
5   !Number of basis functions in 1-D per element
6   integer, parameter, public :: np = 4
7   !Number of time levels
8   integer, parameter, public :: timelevels = 3
9   type, public :: elem_state_t
10    !Temperature
11    real (kind=real_kind) :: T(np,np,nlev,timelevels)
12    !Tracer Concentration
13    real (kind=real_kind) :: Q(np,np,nlev,qsize)
14    ...
15  end type elem_state_t
16  type, public :: element_t
17    type (elem_state_t) :: state
18    ...
19  end type element_t
```

For parallel efficiency, only basis function points that are collocated between two elements (that is, ones that share an edge or corner) are transferred in parallel. Most of the computation is done within an element and takes advantage of low intraprocess shared memory latencies.

For flexibility, elements can also be mapped onto OpenMP threads making MPI ranks and OpenMP threads interchangeable in element processing. During initialization, each rank and thread is assigned the starting and ending global indices of owned elements.

HOMME provides procedures to initialize its state and then drive the integration forward in time with subcycling to account for tracer advection, vertical remapping, and fluid flow substeps within a coarse dynamics step, which is interleaved with physics timestepping. All subroutines operating on an element's state are driven by the dyn_run procedure described below with relevant aspects.

```fortran
1   subroutine dyn_run( dyn_state, rc )
2   #ifdef HORIZ_OPENMP
3     !$OMP  PARALLEL NUM_THREADS(nthreads), DEFAULT(SHARED), &
4     !$OMP& PRIVATE(ithr,nets,nete,hybrid,n)
5   #endif
6   #ifdef COLUMN_OPENMP
7     call omp_set_num_threads(vthreads)
8   #endif
9     ithr=omp_get_thread_num()
10    nets=dom_mt(ithr)%start
11    nete=dom_mt(ithr)%end
12    hybrid = hybrid_create(par,ithr,nthreads)
13    do n=1,nsplit ! nsplit vertical remapping sub steps
14      ! forward_in_time Runge_Kutta, with subcycling
15      call prim_run_subcycle(dyn_state%elem, dyn_state%fvm, hybrid, &
16                           nets, nete, tstep, TimeLevel, hvcoord, n)
17    end do
18  #ifdef HORIZ_OPENMP
19    !$OMP END PARALLEL
20  #endif
```

```
21  end subroutine dyn_run
22  subroutine prim_run_subcycle(elem, nets, nete, nsubstep)
23    do r=1,rsplit ! rsplit tracer advection sub steps
24      call prim_step(elem, fvm, hybrid, nets, nete, dt, tl, hvcoord, &
25                       cdiags, r)
26    enddo
27    call vertical_remap(...)
28  end subroutine prim_run_subcycle
29  subroutine prim_step(elem, fvm, hybrid, nets, nete, dt, tl, &
30                       hvcoord, compute_diagnostics, rstep)
31    do n=1,qsplit ! qsplit fluid dynamics steps
32      call prim_advance_exp(elem, deriv(hybrid%ithr), hvcoord,   &
33                 hybrid, dt, tl, nets, nete, compute_diagnostics)
34    enddo
35    call euler_step(... step1 ...)
36    call euler_step(... step2 ...)
37    call euler_step(... step3 ...)
38  end subroutine prim_step
```

The essential feature of this algorithmic structure is the OpenMP parallel region at the beginning of the dynamics time stepping, which can be disabled with CPP preprocessing directives for pure-MPI simulation runs. Parallelization over elements is programmed explicitly with first and last element indices corresponding to the thread's ID. Inside the subroutines, each element-level loop is ranged as **do** `ie=nets,nete` ... **enddo** ensuring that the workload of iterating over elements is shared by `nthreads` number of OpenMP threads.

Within the `euler_step` subroutine, there are calls to pack (and then unpack) an element's boundary data and exchange it with neighboring elements using a sequence of `MPI_Isend`, `MPI_Irecv`, and `MPI_Waitall` calls. If multiple elements are owned by an MPI rank, then only the external boundary data is exchanged and intraprocess internal element boundaries are handled within the process's memory. To reduce the overall number of messages and avoid potential network congestion, only the master thread communicates boundaries owned by all other threads within an MPI rank.

9.3.3 PROGRAMMING APPROACH

Having described the hybrid MPI+OpenMP parallelization over elements, we now turn to performance improvements in computations over subelemental data structures. Since most of the computation time is spent in loops, we focus on two forms of parallelization of loops: nested threading over subelemental loops and vectorization of innermost loops.

As a coarse summary of loops in HOMME, we note that there are approximately 3390 loops in ≈56 thousand lines of code (including comments) of 63 Fortran90 source files. Performance profiling results indicated a relatively flat profile with no hot spots of computation. Instead, there were several *warm* loops that could be targeted by nested OpenMP parallel regions. We used General Purpose Timing Library (GPTL) timers [12] to measure the computation time taken by a loop and added parallel regions to compute-intensive loops. The following listing shows an example extracted from euler_step subroutine:

```
1  real(kind=real_kind) :: Qtens_biharmonic(np,np,nlev,qsize,nets:nete)
2  do ie = nets, nete
3    !$omp simd
4    do k = 1, nlev
```

```
5       dp(:,:,k) = elem(ie)%derived%dp(:,:,k) - rhs_multiplier*dt*   &
6                   elem(ie)%derived%divdp_proj(:,:,k)
7    enddo
8  call t_startf('eul_nested')
9  #if (defined COLUMN_OPENMP)
10   !$omp parallel do private(q,k) collapse(2)
11 #endif
12   do q = 1, qsize
13     do k=1, nlev
14       Qtens_biharmonic(:,:,k,q,ie) = &
15               elem(ie)%state%Qdp(:,:,k,q,n0_qdp)/dp(:,:,k)
16       qmin(k,q,ie)=minval(Qtens_biharmonic(:,:,k,q,ie))
17       qmax(k,q,ie)=maxval(Qtens_biharmonic(:,:,k,q,ie))
18     enddo
19   enddo
20   call t_stopf('eul_nested')
21 enddo
```

This example demonstrates three parallelization levels: (1) at the horizontal element level with threads iterating over (nete − nets + 1) elements, (2) at the vertical column level with nested threads iterating on the collapsed loop over tracer q and level k indices, and (3) at the vertical and subcolumn levels with data-parallel SIMD vectorization. Nested threading directive is protected with a C preprocessor (CPP) macro to disable it for a specific system and compiler combinations: for example, nested threading is not as beneficial on Titan when compiled with PGI compiler compared to OpenACC-based approach. To ensure that nested threads have a sufficiently large workload to justify the overhead of a nested parallel region, the two q and k loops are collapsed into one combined loop using OpenMP collapse clause. Finally, if a compiler does not autovectorize innermost loops over npxnp subcolumns that are denoted by Fortran array notation, we provide an !$omp simd hint to the compiler that the annotated loop should be vectorized.

9.3.4 SOFTWARE PRACTICES

ACME uses Git versioning system to track software development in a shared repository hosted at http://github.com. Developers track their work in git branches that are submitted to component (ATM, LND, ICE, OCN) integrators in pull requests (PRs). Component integrators review PRs and merge them into an integration branch called next. Nightly integration tests run on major system and compiler combinations (e.g., Edison+Intel, Mira+IBM, Titan+PGI) to ensure that the new branch does not break existing and tested features. After all tests are passed, the new branch is merged into the master branch.

Besides the expected checks for absence of build- and run-time errors, ACME uses baseline comparison testing. Simulation tests produce NetCDF output files that capture the state of a simulation at the end of a run. A baseline output is first analyzed for correctness: for example, results of a hindcast run match observational data. Then, every subsequent run of the same simulation test must produce outputs that are bit-for-bit (BFB) identical with the established baseline. If the results of the test on a developer's branch are not BFB, then the branch needs to be reworked to become BFB. If the developer provides evidence that the new results should be accepted as a new baseline, then the nightly integration testing framework's set of baselines is updated with the new results.

Initial work on nested threading required substantial efforts into ensuring that outputs after adding nested parallel regions were BFB identical to outputs produced with pure-MPI simulation runs. The most common source of errors that were producing non-BFB results was missed identification of variables that should be private within a nested parallel region. Compilers and OpenMP implementations do not autodetect that a variable is overwritten by multiple threads and such hazards had to be manually inspected and removed.

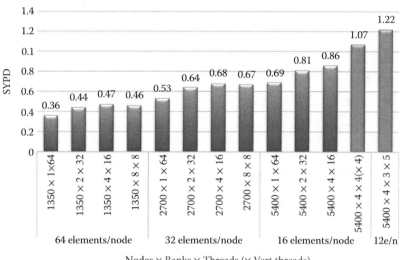

FIGURE 9.1 Hybrid MPI+OpenMP scaling and nested threading on Mira. A workload of 86,400 elements is scaled from 1350 nodes to 8192 nodes at various *rank × thread* configurations. Results indicate that MPI is more efficient than horizontal OpenMP. Also, nested threading in the rightmost two configurations provides an ≈ 40% speedup at the highest scales.

9.3.5 BENCHMARKING RESULTS

To measure the benefits of nested threading, we have benchmarked an atmosphere-focused component set on Mira. Figure 9.1 shows the results of hybrid MPI+OpenMP benchmarking at various rank and thread combinations including nested threads. We strong scale the workload of the high-resolution component set, which contains $6 \times 120^2 = 86,400$ elements, from 1350 to 5400 Mira nodes. This corresponds to allocating 64, 32, and 16 element workloads on a 16-core four-way hyperthreaded node. Within each workload, we also benchmark various combinations of MPI ranks and OpenMP threads to determine the optimal throughput rate of simulation years per day (higher is better). Based on the results, we can observe that MPI is more efficient than horizontal OpenMP threads: that is, 4×16 clearly outperforms all 1×64 configurations with 1 MPI rank and 64 OpenMP threads.

In addition to MPI versus OpenMP efficiency comparison, we scale the workload to 8192 nodes. In this mode, there are fewer elements than there are available cores (and hardware threads) and we can use them for nested threading. For example, the $5400 \times 4 \times 4(\times 4)$ configuration uses four nested threads to iterate over vertical column and tracer loops and this configuration achieves ≈ 25% speedup over the traditional hybrid MPI+OpenMP approach. Further, job scheduling on Mira enforces power-of-two partition sizes, which implies that a 5400-node job occupies an 8192-node partition. To fully occupy the partition, we use five nested threads in the $8192 \times 4 \times 3 \times 5$ configuration. This achieves the highest throughput rate of 1.22 SYPD, which is a speedup of ≈ 40%.

9.4 PORTABILITY CONSIDERATIONS

Portability is one of the most important considerations for our efforts in ensuring ACME continues to efficiently utilize LCF architectures both now and in the future. Portability is something that must be balanced with performance, as one cannot obtain ideal performance on multiple architectures with the exact same codebase. However, it may be that one can eventually obtain suitably good performance with the same codebase.

The real goal is probably better labeled *maintainability* than strict portability. For now, one must use a separate codebase for the CPU and GPU. More than likely, Intel's MICs will be roughly similar enough to a CPU that one could unify the CPU and MIC code base with a likely large amount of refactoring beforehand. However, there are several issues precluding a unified CPU/GPU codebase at present. We hope some of these issues are abated in the future by standards changes to OpenACC/OpenMP as well as improved compiler implementation and hardware features.

For the atmospheric code, we have implemented a CMake target-based handling of the multiple architectures. We have a target for the CPU, and then we have that same target name with "_acc" appended for the OpenACC version of the code. Both codes look similar enough that it's easy to move changes from CPU development OpenACC code when necessary. All common code shared between the CPU and GPU versions are kept in a shared directory. Then, when something needs to be specialized for the GPU, the base module is used, and only those routines that need altering are reimplemented.

Ideally, this functionality would be implemented using Fortran 2003 classes. However, we and others have found that compiler implementation of recent Fortran standards is quite poor when used inside OpenACC regions (and often even when implemented only for the CPU). We wish to avoid running into those current limitations. Improved compiler implementation coupled with the coming Unified Memory between CPU and GPU (which will greatly improve pointer functionality) should enable the use of modern Fortran inside OpenACC portions of the code.

9.4.1 BREAKING UP ELEMENT LOOPS

One of the main reasons that CPU and GPU code cannot currently share a single source code is that as the amount of code in a GPU kernel gets larger, the register pressure increases substantially. This means that the kernel will have poor occupancy on the GPU and will not execute efficiently. The only remedy for this at present is to break the kernel up into multiple kernels. Using function calls will also ameliorate this, but it doesn't solve the problem, as successive function calls also increase register pressure, albeit more slowly. Also function calls on the GPU are quite expensive because the register page and cache are flushed.

On the CPU, one would want to keep an element loop fused together for caching reasons, since the CPU is heavily utilized keeping cached data resident for avoiding the high latency and low bandwidth of DRAM accesses. The GPU, however, is mainly a vector machine and relies mostly on quick thread switching to hide costs of DRAM accesses, so this is less of a concern on the GPU. Thus, one would not want the broken-up loops to run on the CPU, since it would be less efficient. In our GPU port, we have to branch at the element loop level into separate GPU code in order to break up the element loop for nearly every function call to avoid register pressure problems.

We believe that improvements to compiler implementation of OpenACC compilers that improve register handling would remove this block to a unified codebase.

9.4.2 COLLAPSING AND PUSHING IF-STATEMENTS DOWN THE CALLSTACK

In the atmospheric code, we have up to five levels of looping: element (ie), tracer index (q), vertical level (k), *y*-basis function (j), and *x*-basis function (i). The problem with having so many levels of looping is that with the parallel loop construct in OpenACC, one cannot nest multiple "acc loop gang" directives or multiple "acc loop vector" directives. One can do this with "acc kernel" rather than "acc parallel loop," but we will give reasons in the next subsection as to why this is not a good idea for portable or performant code.

When there are more than three levels of looping, then one must use the OpenACC collapse clause in order to access all of them for threading. In our case, we tightly nest the loops into two groups: a collapsed vector loop and a collapsed gang loop. One of the issues that comes from collapsing loops is that if-statements that were in that loop nesting but not on the innermost level will have to be pushed down to the innermost level, since collapsing requires tightly nested looping. We have a number of instances where this happens.

There is no penalty on the GPU for doing this, as long as there is no branching within a half-warp. However, on the CPU, the code will not land on the vector units if there are if-statements inside the innermost loop that is being vectorized. Thus, this is another reason why CPU and GPU code must be separate currently.

We believe that allowing nested "acc loop gang" and "acc loop vector" directives would remove this block to a unified codebase.

9.4.3 MANUAL LOOP FISSIONING AND PUSHING LOOPING DOWN THE CALLSTACK

It doesn't always look that great to manually fission (or block or tile) a loop. However, we believe that for some codes, it is a wise decision whether or not one is running on an accelerator. The reason is twofold: (1) caching is an important consideration on all architectures, but they are quite different in size between CPU, GPU, and MIC and (2) effective vector lengths will differ substantially between different architectures. In order to provide flexibility to respect varying cache sizes and vector lengths, it is often helpful to manually fission a loop in the code base.

If there is no real callstack, then this isn't necessary, but rarely is that the case for a scientific code. When there is a callstack present, then it is important to be able to flexibly push and pull looping through the callstack. This is most easily done if the developer chooses a loop (possibly two) to manually fission. Then the routines can loop over any range of that fissioned loop, and the amount of looping assigned to the callee as opposed to the caller can be determined at either compile time or run time.

In our case, the atmospheric code originally was coded for the CPU code, and thus, most of the routines only operated on a single vertical level for a single element and for a single variable, looping only over the horizontal basis functions, which in production runs translates to 16 loop iterations. That isn't nearly enough for the OpenACC vector loop on the GPU. The next loop is the vertical level loop, and there are 72 vertical levels. We cannot push the entire vertical level loop into those routines because this would violate CPU cache and GPU vector loop length requirements. Thus, we chose to manually fission the vertical loop, and this gave two improvements.

First, it allows the same routines to keep the CPU code performing as it currently does while also allowing a much larger amount of looping inside the routine to be used for the GPU. We typically push a block of 8 loop indices of the vertical level into the routines for the GPU, which provides 128 loop indices for the OpenACC vector loop, which is typically ideal or close to it. It would also allow some medium between these two to be specific for the MIC. Second, it allows us to size the vector loop correctly for the right amount of data to be cached in L1 cache using shared memory. Again, a block of 8 from the vertical loop seems about right for most kernels. Everything outside the vector loop is collapsed down into a gang loop.

9.4.4 KERNELS VERSUS PARALLEL LOOP

There are two main ways of porting a section of code to an accelerator using OpenACC: (1) the "acc kernels" construct and (2) the "acc parallel loop" construct. The advantage of using kernels is that it theoretically offloads many of the decisions to the compiler, since it is more descriptive in nature than prescriptive. However, we have two problems with the kernels construct. First, we've found that for our code, the compiler rarely makes good decisions in terms of vector and gang loop placement. In fact, it often made such poor decisions that the code would perform over 100× worse than the parallel loop construct. Also, with the compiler making more decisions, our performance would be more exposed to the compiler version, making it less robust overall.

Second, we eventually are going to transition to using OpenMP 4.x when it becomes available and is relatively mature in compiler implementation. OpenMP has more similarities with parallel loop than with kernels, since it is also prescriptive in nature. Thus, if we eventually wish to transition to OpenMP, it is a better route to go ahead and use the parallel loop construct.

9.5 ONGOING CODEBASE CHANGES AND FUTURE DIRECTIONS

Aside from the atmospheric dynamics and tracers, efforts are underway to port other parts of the code, including the MPAS-Ocean model, a so-called *super-parametrization* for the atmospheric physics, and other parts of the atmospheric physics. These combined would account for the vast majority of the runtime cost of ACME, and they are each planned to be ported using OpenACC and migrated later to OpenMP 4.x.

ACKNOWLEDGMENTS

This research used resources of the Oak Ridge Leadership Computing Facility at the Oak Ridge National Laboratory, which is supported by the Office of Science of the U.S. Department of Energy under Contract No. DE-AC05-00OR22725.

REFERENCES

1. MA Taylor, J Edwards, S Thomas, and R Nair. A mass and energy conserving spectral element atmospheric dynamical core on the cubed-sphere grid. *Journal of Physics: Conference Series*, 78:012074. IOP Publishing, 2007.

2. MA Taylor, J Edwards, and A St Cyr. Petascale atmospheric models for the Community Climate System Model: New developments and evaluation of scalable dynamical cores. *Journal of Physics: Conference Series*, 125:012023. IOP Publishing, 2008.

3. T Ringler, M Petersen, RL Higdon, D Jacobsen, PW Jones, and M Maltrud. A multi-resolution approach to global ocean modeling. *Ocean Modelling*, 69:211–232, 2013.

4. C Ronchi, R Iacono, and PS Paolucci. The "cubed sphere": A new method for the solution of partial differential equations in spherical geometry. *Journal of Computational Physics*, 124(1):93–114, 1996.

5. A Kageyama and T Sato. "Yin-Yang grid": An overset grid in spherical geometry. *Geochemistry, Geophysics, Geosystems*, 5(9), 2004.

6. ST Zalesak. Fully multidimensional flux-corrected transport algorithms for fluids. *Journal of Computational Physics*, 31(3):335–362, 1979.

7. IPE Kinnmark and WG Gray. One step integration methods of third-fourth order accuracy with large hyperbolic stability limits. *Mathematics and Computers in Simulation*, 26(3):181–188, 1984.

8. S Gottlieb. On high order strong stability preserving Runge–Kutta and multi step time discretizations. *Journal of Scientific Computing*, 25(1):105–128, 2005.

9. M Norman, J Larkin, A Vose, and K Evans. A case study of CUDA FORTRAN and OpenAcc for an atmospheric climate kernel. *Journal of Computational Science*, 9:1–6, 2015.

10. I Carpenter, R Archibald, KJ Evans, J Larkin, P Micikevicius, M Norman, J Rosinski, J Schwarzmeier, and MA Taylor. Progress towards accelerating HOMME on hybrid multi-core systems. *International Journal of High Performance Computing Applications*, 27(3):335–347, 2012.

11. M Norman, L Larkin, R Archibald, I Carpenter, V Anantharaj, and P Micikevicius. Porting the community atmosphere model—Spectral element code to utilize GPU accelerators. *Cray User Group*, Stuttgart, Germany, 2012.

12. JM Rosinski. GPTL—General Purpose Timing Library. http://jmrosinski.github.io/GPTL/, 2016.

10 Preparing the Community Earth System Model for Exascale Computing

John M. Dennis, Christopher Kerr, Allison H. Baker,
Brian Dobbins, Kevin Paul, Richard Mills, Sheri Mickelson,
Youngsung Kim, and Raghu Kumar

CONTENTS

10.1 Introduction...207
10.2 Background..209
    10.2.1 CESM 209
    10.2.2 Exascale Challenges and Expectations...210
10.3 Scientific Methodology...211
    10.3.1 Performance Analysis...211
    10.3.2 Kernel Extraction..211
    10.3.3 Folding Analysis...212
    10.3.4 Ensemble Verification...212
    10.3.5 Platforms...213
10.4 Case Study: Dynamical Core..213
    10.4.1 Algorithm Details 214
    10.4.2 Parallelization Improvements 216
    10.4.3 Single-Core Optimization...218
    10.4.4 Benchmarking Results...218
10.5 Case Study: Data Analytics 225
10.6 Conclusions and Future Work...227
Acknowledgments 227
References 227

10.1 INTRODUCTION

Understanding the complex interactions present in the Earth's physical and biological systems within the ocean, atmosphere, land, and ice has long been of scientific interest. Recently, with the advent of accelerated climate change, the need to understand the societal and policy implications in such areas as water resource planning, agricultural practices, and human activities (e.g., deforestation, CO_2 emissions) [1] has made the need to understand the earth system even more critical. Fortunately, advances in computing capabilities over the last few decades have enabled the development of climate simulation codes, or Earth-system models (ESMs) that have significantly advanced climate science and improved understanding of the complexities of the climate system [2].

ESMs are a particularly important vehicle in geoscience research as they serve as a virtual laboratory for investigating the effects of external influences, such as increasing CO_2 levels in the atmosphere, or more generally answering "What if ...?" questions, such as "What if the Greenland ice sheet melts?" (e.g., see [3–5]). In fact, ESMs are increasingly being utilized to better prepare for future climate scenarios and a prime example is the Coupled Model Intercomparison Projects (CMIPs) that occur every few years. The CMIPs are international efforts to coordinate and compare climate model experiments from as many as 20 modeling groups (e.g., CMIP Phase 5 [6]), and the results are widely utilized by scientists and policy makers alike to inform and prepare for future climate states. Indeed, climate science has a strong computational component that has prominently contributed to our current knowledge of the Earth's system.

To be both effective and accurate, ESMs must well represent the physical processes of the major components in the climate system (atmospheres, oceans, sea ice, land ice, land, and rivers) and capture the complicated interactions between them. Current ESMs typically consist of models for each major earth system component that are coupled together such that feedback occurs between components [1]. The model equations for the dynamics of the atmosphere, ocean, and sea ice components are derived from basic conservation laws for energy, mass, and momentum, which are tailored for each component. In addition, an equation of state (i.e., gas law) is needed by the ESM to relate the temperature, pressure, and density quantities. It is not surprising that the complexity of the Earth's climate system has led to ESMs that are similarly complex, large in size, and typically the result of many years or even decades of development [7,8].

ESMs continue to grow even larger and more complex as they evolve to include new scientific features and take advantage of advances in high-performance computing technologies. In particular, advances in computing over the last few decades have allowed for finer resolutions, the inclusion of more physical processes, and improved boundary interactions (e.g., air–sea) [2]. Yet despite steadily increasing high-performance computing capabilities, the ESM community's requests for computing resources continue to outpace available resources. High-resolution (spatial and temporal) models are increasingly in demand as scientists seek to produce more accurate and realistic simulations, the importance of which should not be underestimated in light of potential societal implications. High-resolution simulations are often required to resolve particular phenomena; for example, ocean eddies, narrow ocean currents, tropical cyclones, and fronts all require finer resolutions than can be reasonably run for multicentury simulations and ensemble studies [2,9]. Therefore, in practice, scientists must carefully balance their scientific needs (e.g., resolution, simulation length, physics complexity) with the amount of available computing power and compromises are frequently required [2]. Given current limitations, if climate model codes are adapted to successfully run on future exascale machines, the potential gains for climate science knowledge may very well be transformative.

In this chapter, we discuss the preparation for exascale of a widely used and well-respected climate model: the Community Earth System Model (CESM) [10]. CESM is a fully coupled ESM with a large complex code base (currently over one and a half million lines) that has been developed over the last 20 years. Although its active development is led by the National Center for Atmospheric Research (NCAR), CESM has a long history of contributions from the university and Department of Energy research communities. It has become increasingly popular with scientists around the globe and is a key contributor to the aforementioned CMIPs. The path to exascale for CESM contains many challenges, and our goal in this chapter is to provide an overview of our current efforts and plans to address some of these challenges, focusing specifically on the CESM software itself. Clearly, we cannot detail all modifications and decisions made thus far, so instead we describe our overall optimization methodology and strategy, sharing details via two illustrative case studies.

This chapter is organized as follows. In Section 10.2, we provide more details on the CESM software as well as challenges and expectations for exascale computing. Then in Section 10.3, we outline our incremental approach to adapting CESM for exascale. Two optimization case studies are presented in Sections 10.4 and 10.5, the first of which provides an in-depth look at the atmospheric dynamical core. We provide concluding remarks in Section 10.6.

10.2 BACKGROUND

Before addressing the specifics of CESM's path to exascale, we provide additional details on CESM, including its current performance and expected challenges and gains at exascale.

10.2.1 CESM

CESM consists of multiple geophysical component models of the atmosphere, ocean, land, sea ice, land ice, and rivers. These components are primarily written in Fortran 90 (with some usage of Fortran 95 and 2003 features), utilize a hybrid (MPI-OpenMP) programming model, and can all run on different grid resolutions with different time steps, exchanging boundary data with each other (via Message Passing Inter-face [MPI]) through a central coupler. Because CESM supports a variety of spatial resolutions and time-scales, simulations can be run on both state-of-the-art supercomputers as well as on laptops. Both the myriad model configurations available and the sheer number of physical processes represented contribute to the size and complexity of the CESM code. Note that periodic releases of CESM to the community ensure that new science developments are accessible by all, and CESM's open-source availability encourages collaboration in the climate science community.

For CESM, and its atmospheric model component (the Community Atmosphere Model [CAM]), in particular, the term *high resolution* is commonly used to describe grids with resolutions that are less than 1° latitude/longitude ($\approx$100 km). High-resolution climate simulations are computationally expensive but are important to climate scientists as they improve small- and large-scale interactions and can reduce bias in some large-scale features [9]. As an example, a recent high-resolution CESM simulation at NCAR by Small et al. [9] completed 100 years of "present-day" simulation, with the atmospheric component at 0.25° grid spacing and the ocean component at 0.1°, which was the longest run to date with CESM with 0.25° atmosphere. This computationally expensive simulation consumed nearly 45 million CPU hours on NCAR's Yellowstone supercomputer [11] (machine details given in Section 10.3.5), with a throughput of two simulated model years per day [9]. Further, we note that 0.25° resolution simulations are effectively limited to approximately 10–20 K cores due to simulation costs and scalability limitations, which does not bode well for effective use of future exascale systems. Certainly, a single simulation of multiple centuries or an ensemble of many simulations at the 0.25° resolution is well beyond the computing capacity available to most climate scientists and achieving 0.125° (considered *ultra-high-resolution*), which requires 8× the computer resources of a 0.25° simulation, is daunting at best.

In fact, computing and data limitations result in most production CESM studies using a 1° resolution for CAM. For example, the recent and well-regarded CESM Large Ensemble (CESM-LE) Project for studying internal climate variability [12], which currently includes a 40-member ensemble of 180-year fully coupled CESM climate simulations (in addition to several additional multicentury control simulations), ran all simulations at approximately 1° horizontal resolution. The first 30 members required 17 million CPU hours on NCAR's Yellowstone machine, indicating that while performing such a large ensemble study at a higher resolution (e.g., 0.25°) may be desirable, such a task is simply not feasible at this time. Further, model runs of CESM (and other models) for CMIP Phase 5 (discussed in Section 10.1) used a 1° horizontal resolution or coarser due to computing costs, even though higher resolutions would clearly be more useful for many studies. In addition to CPU hour requirements, the amount of data generated by a climate simulation is an important consideration when determining the simulation resolution. For example, CESM simulations (at 1° and coarser) produced nearly 2.5 PB (petabytes) of raw output data that were subsequently postprocessed, resulting in a similar volume of processed data, of which 170 TB of data were extracted for submission to CMIP5 [13]. Likewise, the CESM-LE project produced 300 TB of raw data from 1° simulations that were then postprocessed to time-series format (roughly doubling the data volume). Ultimately, only 200 TB of the raw and processed data combined could be stored long term due to storage limitations [14]. High-resolution simulations generate even larger data volumes: the high-resolution 0.25° simulation done by Small et al. [9] generated more than half a terabyte of data per simulation year, despite rather conservatively chosen data output frequencies.

10.2.2 EXASCALE CHALLENGES AND EXPECTATIONS

The basic issues faced in achieving exascale computations have been articulated in several reports [15–18] and there is a broad consensus on the major bottlenecks that are encountered in both the hardware and software. Power management is the driving factor in the design of exascale systems and provides a first-order constraint for computations, as well as memory and communication traffic. One of the fundamental issues arising from the power constraints is the need for greater concurrency on the system, as the end of Dennard scaling has meant that practical processor clock speeds have not increased over the past decade, and performance gains must, therefore, come from increased parallelism. This constraint leads to a total concurrency of several billions of processes on exascale systems. At the highest-level, these systems implement a message passing paradigm. The key issue is how to deal with the on-node concurrency at a level of order (100–10,000) processes. On the node, power constraints and technology scaling are pushing toward hybrid memory architectures. Current architectural trends also indicate continued pressure on the amount of memory available per node. Potentially, a significant fraction of the storage is in nonvolatile memory. Finally, the mean time between failures on the systems is expected to be on the order of days.

Exascale computing is used to enable climate simulations to be performed at finer spatial and temporal resolutions, contains more complex physical process models, and enables longer timescales—all of which are desirable for more accurate and realistic climate experiments. Accomplishing these goals at exascale requires significantly increasing the concurrency exposed in climate models and exploiting that concurrency on many levels/granularities when performing climate simulations. The first level results from the scientific need to perform multiple climate scenarios, which enables several copies of the experiments to be run across the entire system. The second level is to separate the component models (atmosphere, ocean, land, sea ice, land ice, and rivers) to enable them to run concurrently on separate MPI communicators. Within each of these component models, further concurrency can be exposed by separating the components and executing the subcomponents concurrently within a separate pool of threads. The third level would be to expose concurrency within the component models to maximize the number of threads utilized. Finally, parallelism can be exploited at the loop level in the code through vectorization.

The high-fidelity climate simulations that are run at exascale produce massive output data sets and processing these is challenging at best. The current methodology used to manage the work flow in climate models uses a weakly coupled approach. Here, the climate experiments are performed and the data are written to a storage device from which the climate diagnostics are computed. The designs outlined for exascale systems, which integrate advances in system interconnects and memory technologies, enable a more tightly coupled work flow in climate models.

The current climate analysis tools also need to be mapped to these developing technologies to provide a solution that can handle the volume of data expected from these exascale systems. We note that data storage concerns have prompted recent investigations into utilizing more aggressive data compression for climate simulation data (e.g., [14,19–21]), but the high data production rates that accompany exascale computing also require a reexamination of the CESM postprocessing and data analytics as well [13]. In particular, existing analysis tools are largely serial and parallel technologies will be critical at exascale. Note that we discuss our preparation for postprocessing of exascale data volumes in the case study in Section 10.5.

10.3 SCIENTIFIC METHODOLOGY

There has been a long history of refactoring efforts of CESM and CESM component models to enhance parallelism and improve computational efficiency [22–27]. Although adapting the CESM code base to exploit exascale technology is necessary, this task is nontrivial at best; a complete rewrite of a large, complex, and actively developed code like CESM is simply not possible with the resources at hand. Because CESM undergoes rapid evolution, the implementation of optimization and parallelization changes in CAM needs to coexist with other scientific improvements to the model. Development for different architectures [22] would require significant changes and produce multiple code versions, which would be difficult to maintain. We have, therefore, chosen to develop a single unified version of CAM for the Intel x86 architecture with 64 byte reals and 32 byte integers.

Given these constraints, we, therefore, take an incremental approach toward preparing CESM for exascale computing. Our methodology, which we detail in subsequent subsections, can be summarized as follows. We first profile the CESM code and look for expensive and inefficient kernels whose improvement would impact the overall runtime. We then extract the kernel in question to more easily experiment with optimizations. Next, we apply both generic and specialized optimizations. Finally, the optimized kernel is incorporated back into the CESM code base, and we verify that the resulting simulation output has not had a climate-changing impact. We note that some optimizations have a significant impact in terms of speedup, whereas other modifications are minor in comparison. We incorporate optimizations at both ends of the spectrum as the computing resources required to run climate simulations are so significant that even a seemingly minor improvement can translate into nontrivially lower computing resources.

10.3.1 PERFORMANCE ANALYSIS

The CESM timing profile is acquired with the General Purpose Timing Library (GPTL) [28], which is present in each of the CESM components. Once these performance-critical regions have been identified, computational kernels are extracted from CESM with KGEN [29,30], a Python-based tool for automated Fortran kernel generation and verification. A function-level and line-by-line timing of the computational kernel is generated with the GNU profiler (gprof). These timings are coupled with the compiler optimization and vectorization reports generated by the Intel compiler. These combined reports serve as a starting point to indicate how the code could be improved by employing one of several techniques, which include exposing thread-level parallelism at the highest level in the code, restructuring loops to improve vectorization, rearranging loops to improve data locality and optimizing cache reuse, using compile time specifications for loop and array bounds, and replacing division with inversion of multiplication. Further optimizations are possible with studying the results from the profile-guided optimizations generated by the Intel compiler. This report assists in organizing the code to reduce instruction-cache misses and branch mispredictions. To further understand the CPU utilization in the computational kernel, hardware performance countermonitors are used. This is accomplished with the folding analysis tool developed at the Barcelona Supercomputing Center (BSC).

10.3.2 KERNEL EXTRACTION

To address the challenges of optimizing a large complex application that requires expert knowledge to build and execute, we utilize Kernel Generator (KGEN), a Python tool that automatically extracts a kernel from Fortran applications [29,30]. These smaller kernels allow pieces of a much larger Fortran application to be built and run outside the complexity of the main application, significantly reducing the time it takes to modify, build, and execute code. The KGEN generated kernel, which is typically from 1 to 30 K lines long, may take minutes to compile and only seconds to execute, and is easily portable to different compute platforms.

KGEN has been used extensively in our CESM optimization work. It has simplified and enhanced collaborations between application developers and compiler and hardware vendors. It has allowed the extraction of a number of kernels for different physical processes including the Morrison Gettelman Microphysics version 2 (MG2) [31], the Rapid Radiation Transport Model (RRTMG) [32], and the Mozart implicit chemistry solver [33]. The general optimization techniques for each of these modules are similar to those used on High-Order Method Modeling Environment (HOMME) that are described later in Section 10.4.3. The MG2 optimization effort is described in [34,35].

Because kernel extraction is essential to our optimization work we maintain a repository of all extracted kernels in Git [36]. Each kernel directory contains several versions of the kernel code. We store the original kernel created by KGEN into a directory called orig and each optimization milestone gets its own numbered directory. This enables us to test each optimization step on new architectures.

10.3.3 FOLDING ANALYSIS

Folding analysis is one of the performance analysis techniques developed at the BSC [37]. Using this technique, a detailed evolution of performance-related events can be obtained for an arbitrary block of software source code. The strength of this tool is two-fold. First, the fine-grain performance evolution collected using this tool helps to show how the program runs on the target processor, which would be very difficult with coarse-grain performance event collection. Second, it is possible to map specific regions of the performance evolution to a set of source code lines. This feature is particularly important because it helps identify a small number of source code lines for optimization and also to decide what types of optimization techniques to apply. To perform folding analysis, a user first needs to collect performance event samples using Extrae, which is another tool from BSC for instrumentation and sampling [38]. Through interpolating the collected samples, the Folding tool generates fine-grain evolution plots of collected performance events. For further investigation of the Folding outputs, a user can use Paraver, which is an offline trace analyzer from BSC [39].

10.3.4 ENSEMBLE VERIFICATION

Given the ongoing state of CESM development, quality assurance via software verification is particularly critical to both detect and reduce errors, particularly in the context of modifying and optimizing code to prepare for exascale machines. Verifying the correctness of a change or update to the CESM hardware/software stack (e.g., a new compiler flag, a different compiler, a different machine, a code modification) is trivial when the simulation results after the update are bit-for-bit (BFB) identical with the original results. However, when the new results are no longer BFB with the original results, determining whether the difference is significant (i.e., *climate changing*) is key. In the course of optimizing the code, we frequently find ourselves in this latter situation, as the chaotic nature of climate model simulations means that tiny code modifications or a compiler change can easily result in non-BFB solutions [40].

We consider our code modifications to be admissible if their results are statistically indistinguishable from the original results, and this determination is made via the recently developed CESM Ensemble Consistency Test (ECT) tool [41]. CESM-ECT is regularly used by the CESM software engineers when porting to new machines and releasing code updates, and the objective tool is useful in detecting errors and providing tangible feedback. In particular, the capability to determine consistency between simulation results that are not BFB provides the much-needed flexibility, for example, to pursue more aggressive code optimizations and utilize heterogeneous execution environments, both of which are critical in the path to exascale.

As its name implies, the CESM-ECT tool uses an ensemble-based approach to determine climate consistency (i.e., statistical indistinguishability). In particular, an ensemble of *accepted* simulations that represent the same ESM allows us to gauge the natural variability of the models climate by providing a quantitative measurement of variability with which to compare future simulations. The CESM-ECT issues a pass or fail for new non-BFB CESM output (e.g., from one of our code optimizations) by comparing it to the variability represented by the accepted ensemble (e.g., the original nonoptimized code) and determining whether it is statistically distinguishable. The CESM-ECT is a suite of tools tailored for an individual CESM component, and, at present, ECT modules have been developed and released for the CAM component (CAM-ECT) [41] and ocean component (CESM-POP) (POP-ECT) [42]. (Developing modules for the other components is work in progress.) Note that although all CESM-ECT modules rely on ensembles, the underlying testing algorithms may be surprisingly different due to characteristics of the respective models. In some situations, ECT modules can even detect errors from other components [41].

Although CESM-ECT has greatly contributed to quality assurance for CESM, at present its scope is limited to coarse-grain verification. In other words, although CESM-ECT can readily indicate that a problem exists by issuing a *fail*, an easy means of identifying the root cause of the failure is not provided, and identifying the error in a complex code with more than one million lines can be daunting and nontrivial (e.g., see Section 6 in [40]). Therefore, to more easily trace a CESM-ECT failure to its root cause, we have begun working on a utility to identify potential problematic CESM kernels based on information from the CESM-ECT failure. Then, once the kernel(s) has been identified, we use KGEN (Section 10.3.2) to extract the kernel(s) from CESM to render the debugging task reasonable. Finally, we note that our latest efforts have been focused on the development of a so-called *ultrafast* version of CESM-ECT, which requires only nine model steps (equivalent to 4.5 hours in model time), has shown promising preliminary results, and will greatly benefit software developers.

10.3.5 PLATFORMS

For the analysis of our optimization work, we utilize several platforms. Our primary machines are Yellowstone [11], a Sandy Bridge-based Infiniband Linux cluster installed at the NCAR Wyoming Supercomputing Center (NWSC), and Carl, a Knights Landing (KNL)-based test platform deployed at the National Energy Research Supercomputing Center (NERSC) [43]. We also provide timing information for several kernels on the HPCFL machine, a 4-socket Haswell-based system in NCAR's HPC Futures Laboratory. The specific configuration of each platform is provided in Table 10.1. On these platforms we use the Intel 16 and 17 compilers and the Intel 5.1.3.210 MPI libraries. We use the compiler flag "-O3 -fp-model fast -qno-opt-dynamic-align" for all runs along with the correct architecture flag.

Our postprocessing and data analysis results were obtained on numerous platforms as well. Our primary test platform is Geyser, an Intel Xeon (Westmere EX) Infiniband Linux cluster installed at the NWSC with NCAR's Yellowstone system [11]. Additionally, we have tested our analysis and workflow utilities on the flash-based XSEDE system, Wrangler. Wrangler, located at the Texas Advanced Computing Center (TACC), is an Intel Haswell-based cluster with a 500 TB shared flash storage system capable of 1 TB/s I/O bandwidth [44].

10.4 CASE STUDY: DYNAMICAL CORE

Our first case study focuses on CAM's spectral element dynamical core, known as HOMME [45,46]. The HOMME dynamical core is used within CAM for very high 0.25° resolution simulations. We focus on HOMME because the dynamical core can consume 30%–88% of the total runtime of CAM and is particularly influential in its overall scalability, and therefore the performance of CESM. We first provide more information on the HOMME dynamical core and then detail several modifications aimed at both enhancing its scalability and reducing its cost.

TABLE 10.1
Computational Platforms

Property	Yellowstone	HPCFL	Carl
# of nodes	4,356	1	8
Processor type	Intel Xeon E5-2670, Sandy Bridge (SNB)	Intel Xeon E7-8890-v3, Haswell (HSW)	Intel Xeon Phi 7250 Knights Landing (KNL)
Sockets per node	2	4	1
Cores per socket	8	18	68
Cores per node	16	72	68
Total # of cores	72,576	72	68
Base frequency (GHz)	2.6	2.5	1.4
L1 D+I cache	32 + 32 KB per core	32 + 32 KB per core	32 + 32 KB per core
L2 cache	256 KB per core	256 KB per core	1,024 KB shared by 2 cores
L3 cache	20 MB per socket	45 MB per socket	None
DRAM per node	32 GB	256 GB	96 GB
MCDRAM	None	None	16 GB flat
RAM type	DDR3-1600	DDR4-2133	DDR4-2133
Vector Length (bits)	256	256	512
fp Vector units	1	1	2

10.4.1 Algorithm Details

HOMME uses a cubed-sphere topology where each face of the cube is projected onto a sphere to generate a global computational grid. HOMME uses a spectral element method to discretize in the horizontal and a finite difference approximation [47] in the vertical. A continuous Galerkin finite-element method [48] is used for the spectral element method. The integrals used in the Galerkin formulation are computed from a Gauss–Lobatto quadrature rule within each element. HOMME decomposes each time step into different components. The equation, for a compressible fluid with hydrostatic and a shallow water approximation, can be written in terms of a vector U containing the prognostic state variables (velocity, temperature, and surface pressure) as $dU/dt = F + D + A + T + R$, where F represents the forcing from physics, D the dissipation, A the dynamics from the primitive equations, T the tracer advection, and R the vertical remapping of the mass and momentum variables. A diagram of the overall computational flow of the HOMME algorithm is provided in Figure 10.1.

HOMME solves these equations in a time split fully explicit form. For time steps involving the forcing (F) and dissipation (D) terms, a forward Euler time scheme is used. The dynamics (A) and tracer advection (T) are computed using an N-stage Runge–Kutta time–scheme. The dynamics (A) computes the primitive equations' prognostic variables. The tracer advection (T) is based on a finite-volume algorithm and advances the specific humidity, liquid water, ice variables, and additional tracer constituents. For advection, a vertical Lagrangian approach is used [49] where the horizontal advection on Lagrangian vertical levels is followed by remapping (R) the mass and momentum variables back to the reference vertical levels at the end of the time step.

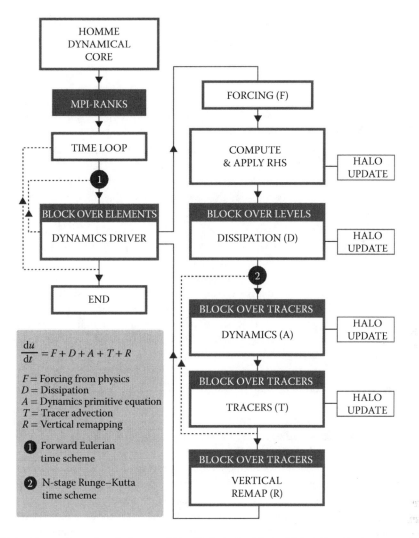

FIGURE 10.1 The computational phases within High-Order Method Modeling Environment (HOMME). Note that OpenMP threading is now presented in both the *Dynamics Driver* as well as several submodules.

As with the rest of CESM, the HOMME dynamical core is written in Fortran 90 and utilizes a hybrid (MPI-OpenMP) programming model. The resolution of HOMME's cubed-sphere computational grid is indicated by the number of spectral elements on a side of a cube (ne). In particular, each side of the cubed-sphere computational grid is tiled with $ne \times ne$ spectral elements, resulting in a total number of spectral elements of $nelem = 6 \times ne \times ne$, where $nelem$ is the total number of elements. For reference, CAM is typically run in production at $ne = 30$, which roughly corresponds to 1° resolution with grid points approximately every 100 km. High-resolution simulations at quarter-degree resolution (i.e., $ne = 120$) with grid points approximately every 25 km, such as the simulation in [9], typically require very large core counts for extended periods of time. The spectral elements are statically partitioned [50] during initialization across either MPI tasks or OpenMP threads. Note that the 25 km resolution where $nelem = 86,600$ used a total of 21,600 hardware cores, which resulted in the assignment of four elements per hardware core. The performance of HOMME is strongly influenced by the number of spectral elements allocated to each hardware core, and historically, although it was technically possible to scale HOMME to core counts equal to the number of spectral elements *nelem*, in practice, it was cost prohibitive in CESM version 1 (CESM1).

10.4.2 PARALLELIZATION IMPROVEMENTS

Although HOMME has historically exhibited excellent scalability on large core counts, which was demonstrated on 173 K hardware cores [45] for ultrahigh 0.125° resolution, significant opportunities to increase efficiency and reduce simulation cost are still present. In particular, several scalability issues were identified that were associated with the boundary exchange and the way in which a particular calculation was performed in the Eulerian advection algorithm. In the case of the boundary exchange, a piece of user code was present that copied part of the send message buffer into the receive message buffer in a nonthreaded fashion. This deficiency resulted in boundary exchanges that were more expensive than necessary when the model was executed in hybrid OpenMP/MPI mode. We modified the user boundary exchange code so as to use threads not otherwise involved in MPI message passing code to perform the memory copy.

The opportunity for an additional parallel optimization was identified using the Extrae folding described Section 10.3.1. In particular, large L3 total cache misses rates were observed in the Eulerian advection module, which indicated excessive data movement through the memory hierarchy. Further analysis revealed not only excessive data movement in the memory hierarchy but also through the message passing network as well. In order to address this excessive message passing traffic, specialized communication operators were written that reduced the total message traffic by 4× for certain boundary exchanges within the Eulerian advection module. Additionally, an asynchronous version of these specialized communication operators was provided to allow the hiding of the network latency for one of the two invocations of this operator.

Both of these parallel optimizations improved the scalability of the HOMME dynamical core by reducing the cost of MPI message traffic. However, they did not increase the amount of parallelism available to HOMME. Recall that as mentioned in the previous subsection, the original HOMME parallelization scheme partitioned over elements at a high level and loops at lower levels. Although it was possible to use the loop-level threading to use more than a single hardware core per spectral element, in practice it was never used due to the inefficiencies in the loop-level threading implementation. In effect, HOMME was limited to only using a single hardware core per spectral element.

A revised scheme that uses the same high-level parallelization over elements replaces an inefficient lower-loop level parallelization approach with a high-level parallelization over the vertical level and tracers dimensions. Note that the dimension chosen to parallelize is dependent on the region (dynamics, tracer advection, dissipation, or vertical remapping) of the code. This revised scheme has several improvements over the original approach. These improvements include a significant reduction in the number of threaded regions created, allowing a greater number of threads to be assigned and the replacement of MPI ranks with OpenMP threads, which reduces the total MPI communication, thereby providing greater flexibility in determining where the threads are assigned in the regions of the code; simplification of the scoping of variables in the parallel region as variables local to the routines, inside the parallel region, are private; a reduction in wastage of thread resources as the parallelization over the vertical level and tracer blocks do not overlap so threads can be shared between these regions.

The revised parallelization structure of HOMME is shown in Figure 10.1. Elements within the global domain are initially assigned over MPI ranks and horizontal threads at the highest level. The dynamics driver calls the dynamics, tracers, dissipation, and vertical remapping modules. These are parallelized over tracers or vertical levels; the blocking scheme chosen was dependent on the parallelism available in the module. In the dynamics, tracer, and vertical remap modules, the code is blocked over tracers and in the dissipation module the code is blocked over the vertical levels.

The only component module where we have maintained the original loop-level parallelism is in the "compute_and_apply_rhs" module as dependencies in the vertical prevent blocking over levels. The code changes needed to implement the revised parallelization strategy were extensive. However, they could be made systematically, simplifying the implementation process.

At the highest level in the code, we have the following structure:

```
!$OMP PARALLEL num_threads(horz_num_threads) &
!$OMP DEFAULT(SHARED), PRIVATE(ie,nets,nete,hybrid)
hybrid = config_thread_region(par,'horizontal')
call get_loop_ranges(hybrid,ibeg=nets,iend=nete)
do ie=nets,nete
    call dynamics_driver(hybrid)
enddo
!$OMP END PARALLEL
```

Where the string *horizontal* indicates the type of threading parallelism, which in this case is over spectral element; *hybrid* is a derived type that contains the mapping from the thread number to the starting and ending element indices *nets* and *nete*, respectively.

Invocation of the blocked regions (dynamics, tracers, dissipation, and vertical remap) can be understood with a simple example:

```
subroutine dynamics_driver()
...
call omp_nested(.true.)
!$OMP PARALLEL NUM_THREADS (num_threads_region) &
!$OMP& DEFAULT(SHARED), PRIVATE(hybrid_region)
  hybridreg = config_thread_region(hybrid, regtype)
  call dissipation_driver (hybridreg, ...)
!$OMP END PARALLEL
call omp_nested(.false.)
```

Where the variable *regtype* is the name of the parallelism implemented in the dissipation_driver; *hybridreg* is a derived type that contains the mapping from the thread number to the starting and ending indices for each *regtype*; "num_threads_region" is the number of threads assigned in the *regtype*. Although the extent of the level and tracer dimension is set at compile time, the actual determination of the number of MPI ranks and OpenMP threads for the elements, tracer, and vertical is performed at runtime. Prior to the blocked implementation, the loop bounds in a lower level code like dissipation_driver were written:

```
subroutine dissipation_driver()
real(r8) :: var(np,np,plev,qsize,nelemd)
do q=1,qsize
  do k=1,plev
    variable_name(1:np,1:np,k,q,ie) = ...
  enddo
enddo
```

Note that the inner indices (np, np) that correspond to the points in the quadrature grid are known at compile time, which allows the compiler to determine variable alignment and extent and generate more efficient vectorized code. The variable "plev" is the number of vertical layers, "qsize" is the number of tracer variables, "nelemd" is the number of elements per MPI-Rank, and "ie" is the element index. The specification of the number of levels and tracers is made at runtime without impacting performance.

The loop structure within the dissipation, dynamics, tracers, and vertical remapping modules after blocking on threads is

```
subroutine dissipation_driver()
real(r8) ::  var(np,np,nlevel,ntracer,nelem)
call get_loop_ranges (hybridreg, qbeg=qbeg, &
           qend=qend, kbeg=kbeg,kend=kend)
do q=qbeg,qend
 do k=kbeg,kend
   var(1:np,1:np,k,q,ie) = ...
  enddo
 enddo
```

where "kbeg, kend, qbeg, and qend" loop ranges are defined from "get_loop_ranges" for the vertical levels and tracers, respectively, and are evaluated for each thread number within the parallel region. The algorithm used to create these loop bounds is designed to minimize the load imbalance in the work that each thread performs. For the dimensions that are not blocked within the parallel region, they are simply defined as "1:plev" and "1:qsize".

10.4.3 SINGLE-CORE OPTIMIZATION

Although increasing the scalability of HOMME is critical to efficiently utilize the additional on-node parallelism that exascale architectures provide, improving the performance of the code on a single core is also critical. The single-core performance optimizations implemented included improving vectorization by redesigning data structures, improving data locality by rearranging calculations for cache reuse, minimizing the number of expensive divides, and enhancing compile-time specifications for loop and array bounds [51]. Each of these techniques, although modest, have a powerful cumulative effect and have contributed to the overall single-core performance improvements in the code. The overall results of these optimizations are discussed in Section 10.4.4.

Unfortunately, several features used in HOMME have inhibited certain compiler optimizations. These features include the use of small inner-loop extents and the extensive use of function statements. Although it is possible to improve serial code performance by the use of vendor-specific alignment directives, we have not actively pursued this capability at this time. In particular, we find vendor-specific compiler directives to be excessively burdensome and create a long-term maintainability issue for only modest performance gain. We next describe the impact of small inner-loop extents and our use of function statements.

As described in the previous section, the first two dimensions are of size (np, np) where $np = 4$. Ideally, a compiler would collapse and vectorize the inner-loop and generate unit-stride instructions. Unfortunately, the current Intel and Cray compilers do not collapse and linearize but instead generate nonunit stride gather and scatters. We are working with compiler developers to ensure that more efficient vectorized code is generated in the near future.

The use of function statements throughout the code has an impact performance. When inlining was forced with a compiler directive, the compiler did inline the function; however, the code generated was not as efficient as when the function was manually inlined. This was the result not of the data copy across the function boundary but from the optimizer performing more aggressive loop optimizations for the manually inlined code.

10.4.4 BENCHMARKING RESULTS

We now evaluate the modifications described in the previous subsections using two different configurations of the HOMME dynamical core. The *CAM-like* configuration uses 26 vertical levels ($plev = 26$) and 25 tracers ($qsize = 25$) and is similar to the default configuration in CAM.

The *Whole Atmosphere Community Climate Model (WACCM)-like* configuration uses 70 vertical levels (*plev* = 70) and 135 tracers (*qsize* = 135) and is comparable to the default configuration used by the WACCM [52]. For each configuration, we also use several different horizontal resolution configurations (*ne* = 4, *ne* = 8, and *ne* = 30). The *ne* = 4 resolution has a total of 96 total spectral elements, whereas the *ne* = 8 and *ne* = 30 have 384 and 5,400, respectively. We use the *ne* = 4 and *ne* = 8 configurations to measure the impact of our modifications on either a single node or small number of nodes. The *ne* = 30 configuration is to measure the impact of our modifications at much larger node counts.

The execution time in sec/model-day on Yellowstone and the computational cost for both the original (*orig*) and optimized (*opt*) code for the *CAM-like* configuration at *ne* = 4 are provided in Figure 10.2. In Figure 10.3, a similar plot of execution time and computational cost is provided for the *ne* = 30 configuration. Note that the number of nodes is on the *x*-axis in Figures 10.2 and 10.3. The left *y*-axis corresponds to the execution time in sec/model day, whereas the right *y*-axis corresponds to the computational cost. The lines in Figures 10.2 and 10.3 that are solid and dashed with diamond data points correspond to execution time, whereas the dashed lines correspond to computational cost. Figures 10.2 and 10.3 clearly illustrate that the *opt* code significantly reduces execution time versus the *orig* code. For example, for the *ne* = 4 resolution on a single node, the execution time is reduced from 4.3 to 3.3 seconds a reduction of 23%. An even larger reduction in execution time of 37% is seen on 57 nodes for the *ne* = 30 resolution. It should be noted that the choice of 57 nodes was not arbitrary but rather to match the per hardware core problem size between the *ne* = 4 and *ne* = 30 resolutions. Both *ne* = 4 on a single node and *ne* = 30 on 57 nodes of Yellowstone assign six spectral elements per hardware core. Both Figures 10.2 and 10.3 also illustrate that *ne* = 4 and *ne* = 30 resolutions show similar scaling characteristics and speedup of the *opt* versus *orig* codes. It is for this reason that we believe a similar speedup will be apparent at a target production resolution of *ne* = 120 as well. The *orig* version of HOMME, indicated by the solid line in Figures 10.2 and 10.3, which was used in previous versions of the CESM code base, was limited to six nodes for *ne* = 4 and 338 nodes for *ne* = 30 the resolutions. Figures 10.2 and 10.3 also illustrate that it is possible to scale HOMME to larger core counts. In particular, speedup for configurations with a much smaller one-fourth element per hardware core is now possible.

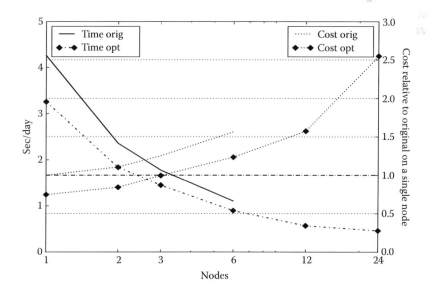

FIGURE 10.2 Strong scaling for HOMME in *Community Atmosphere Model (CAM)-like* (plev = 26, qsize = 25) configuration on Yellowstone. Note that each panel illustrates the execution time and computational cost of HOMME on core counts that range from 6 to 1/4 spectral element per core. Resolution *ne* = 4 on 1–24 nodes.

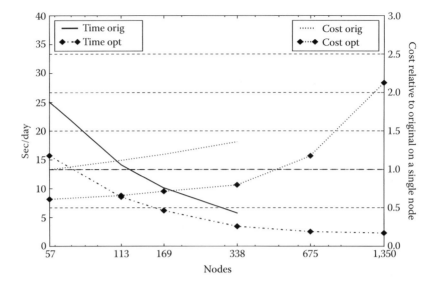

FIGURE 10.3   Strong scaling for HOMME in *CAM-like* (plev = 26, qsize = 25) configuration on Yellowstone. Note that each panel illustrates the execution time and computational cost of HOMME on core counts that range from 6 to 1/4 spectral element per core. Resolution *ne* = 30 on 57–1,350 nodes.

To illustrate the impact our optimizations may have on how HOMME is run in production, the computational costs relative to the *orig* code are included in Figures 10.2 and 10.3. The plain dotted line is the computational cost of the *orig* code normalized to *orig* code with a per problem size of six elements per hardware core, whereas the dotted line with diamond points is for the *opt* version of the code. Although it is clear that use of the *opt* version of the code reduces computational cost on a fixed number of cores, it also introduces the possibility of using more hardware cores for a fixed computational cost. For example, in Figure 10.3, if we draw a horizontal line from the cost to run HOMME using six elements per hardware core from the original code (plain dotted line) to the optimized code (dotted line with diamonds) it is apparent that is now possible to increase the number of nodes used from 57 node to 675 with only a modest 19% increase in the computational cost. Use of the additional nodes would allow a factor of 12 decrease in the time-to-solution for a 19% increase in computational cost.

Figures 10.2 and 10.3 also illustrate HOMME's excellent weak scaling characteristics when increasing the resolution from *ne* = 4 to *ne* = 30. Although we do not have measured timing data for our target resolution of 25 km, it is reasonable to believe that it will be possible to run HOMME in a *CAM-like* configuration at a small 1/2 element per hardware core. Recall that the 25 km simulation [9], which utilized 21,600 cores of Yellowstone, has 86,400 spectral elements. We anticipate that it will be possible to utilize 172,800 cores of a Yellowstone-like system, reduce the time-to-solution by a factor of 8 while only increasing the computational cost of the dynamical core by approximately 20%.

The *WACCM-like* configuration has 70 vertical levels and 135 tracers (*plev* = 70, *qsize* = 125) as is similar to a standard configuration of the WACCM. Figures 10.4 and 10.5 illustrate both the execution time and computational cost of the *orig* and *opt* versions of the HOMME dynamical core. The advantage of the *opt* versus *orig* code is even greater in the *WACCM-like* configuration. For the six element per hardware core size, the *opt* code reduces execution time by 38% for the *ne* = 4 resolution and 43% for the *ne* = 30 resolution. Due to the increased number of tracers and vertical levels, which represents additional threading parallelism, it is possible to scale to 1/8 element per hardware core. An evaluation of computational cost indicates that it is possible to reduce the time-to-solution by a

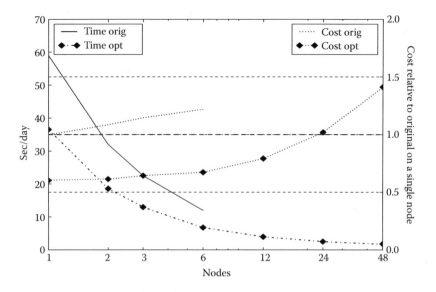

FIGURE 10.4 Strong scaling for HOMME in *WACCM-like* (plev = 70, qsize = 135) configuration on Yellowstone. Note that each panel illustrates the execution time and computational cost of HOMME on core counts that range from 6 to 1/4 spectral element per core. Resolution *ne* = 4 on 1–48 nodes.

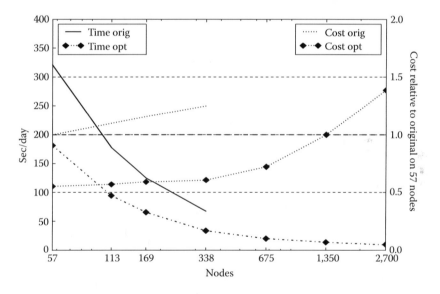

FIGURE 10.5 Strong scaling for HOMME in *WACCM-like* (plev = 70, qsize = 135) configuration on Yellowstone. Note that each panel illustrates the execution time and computational cost of HOMME on core counts that range from 6 to 1/4 spectral element per core. Resolution *ne* = 30 on 57–2,700 nodes.

factor of 24 times when using the *opt* version of HOMME in a *WACCM-like* configuration on 1,350 nodes versus the *orig* code on 57 nodes for the same computational cost.

It is possible to examine the reason for the significant reduction in execution time for the *opt* versus *orig* code by looking at the way in which each code utilizes the underlying hardware. In particular, we use the Extrae-based folding technique and Performance Application Programming Interface (PAPI) hardware counters described in Section 10.3.1 to examine the way in which each code interacts with the hardware. We use the *WACCM-like* configuration at *ne* = 4 resolution on a single node for this evaluation and focus in particular on the Eulerian advection code. We sample a single MPI rank and plot the rates of L2 data cache misses, L3 total cache misses, and 8-byte vector instructions for both the *orig* and *opt* codes. Figure 10.6 illustrates the L3 total cache miss rates for both codes. The *y*-axis is in millions of events/sec, whereas the *x*-axis is normalized time. The very large L3 miss rates for the *orig* that are apparent in Figure 10.6 were a primary motivation for much of the optimization work done in the Eulerian advection module. Our optimizations reduced the average L3 cache miss rate by 42% from 11.1 to 6.5 million events/sec. Figure 10.7 illustrates the L2 data cache miss rates for both codes. The reduction in L2 cache miss rates for the *opt* versus *orig* code is even more significant, dropping from an average 70–17 million events/sec, a reduction of 76%. The code modifications clearly significantly reduced the L2 and L3 cache miss rates. Figure 10.8 also illustrates that a significantly larger amount of the *opt* code is vectorized. The amount of vectorized 8-byte floating-point operations increased from 253 million events/sec to 1,116 million events/sec, an increase of 342%

Although Yellowstone is an excellent system to test the impact of our optimizations at large core counts, its rather small per node core count of 16 is not a good analog for the likely number of cores per node on an exascale system. Instead, we utilize a single KNL socket to test the impact of our optimizations for large on-node parallelism. We use a 68-core single socket system to compare the performance of HOMME for both a CAM and *WACCM-like* configuration with an $ne = 8$ resolution. In Figure 10.9, the execution time for a number of different MPI and OpenMP configurations is provided for both the original *orig* and *opt* code bases. Because the amount of parallelism in this problem is not divisible by 68 we choose to idle 4 hardware cores and configure HOMME

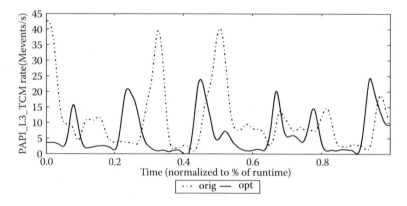

FIGURE 10.6   L3 data cache miss rate on Sandy Bridge for a *WACCM-like* configuration at $ne = 4$ resolution.

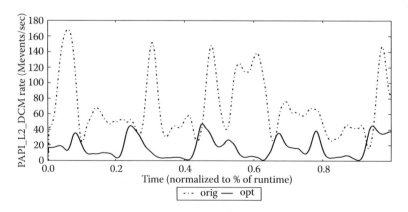

FIGURE 10.7   L2 data cache miss rate on Sandy Bridge for a *WACCM-like* configuration at $ne = 4$ resolution.

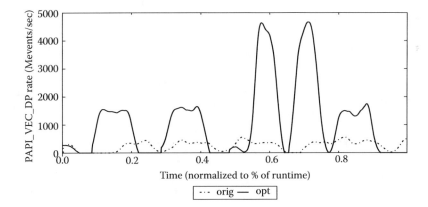

FIGURE 10.8   8-byte real vector instruction rate on Sandy Bridge for a *WACCM-like* configuration at *ne* = 4.

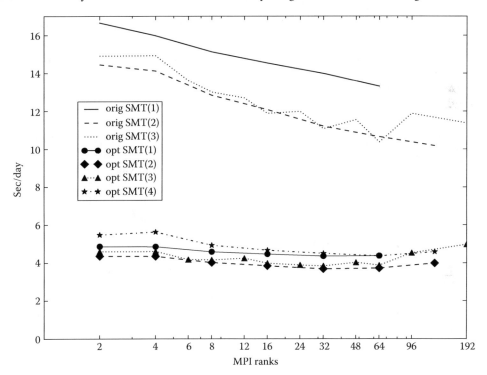

FIGURE 10.9   Execution time for HOMME using 64 hardware cores of a KNL node for a *CAM-like* configuration at *ne* = 8 (*plev* = 26, *qsize* = 25).

to use 64, 128, 192, or 256 threads of execution. These configurations that use 1, 2, 3, and 4 execution threads per hardware core are indicated by SMT1, SMT(2), SMT(3), and SMT(4) in Figure 10.9. Recall that the new version of HOMME now supports threading over both the horizontal dimension as well as the vertical and tracer dimensions. We define the particular parallel configuration of HOMME by the tuple

$$nRanks, horz\_nthreads, max(vert\_nthreads, tracer\_nthreads)$$

where *nRanks* is the number of MPI ranks, *horz_nthreads* is the number of OpenMP threads in the horizontal dimension, *vert_nthreads* is the number of threads in the vertical dimension, and *tracer_nthreads* is the number of threads in the tracer dimension. For example, a parallel config-uration of *8,16,2* corresponds to a configuration that uses 8 MPI ranks, 16 OpenMP threads over the horizontal, and 2 over the vertical and tracer dimension, and has a total of 256 execution threads. Despite the fact that we consider only those configurations that have a balanced number of spectral elements per MPI rank, a total of 45 configurations are possible for the *opt* code base and 25 for the *orig* code. In the cases where there are multiple configurations for a given number of MPI ranks we only plot the configurations with the lowest execution time. As expected, Figure 10.9 reveals a significant advantage for the *opt* code versus the *orig* code base. The best execution time for the *opt* code for the *CAM-like* configuration at *ne* = 8 is 3.69 seconds, a reduction of nearly 64% from the *orig* code. For the *opt* code the best execution time is for the *32,4,1* configuration with 128 execution threads, whereas for the *orig* code it is for a *128,1,1* configuration. Both versions of code appear to benefit from the use of multiple execution threads per hardware core and prefer a larger number of MPI ranks and a modest number of OpenMP threads. Interestingly, there appears to be a moderate penalty to using a total of 256 execution threads per node for the *opt* code. It is also noteworthy that the *opt* version of the code appears to be more tolerant of moving parallelism from MPI ranks to OpenMP threads. In particular, the performance impact of varying the number of on-node MPI ranks from 8 to 128 is less than 10%. The ability to move from one type of parallelism to another with modest impact on performance may provide a significant advantage when running HOMME on a system with multiple KNL nodes.

The execution time for the *WACCM-like* configuration at *ne* = 8 on a single KNL node is provided in Figure 10.10. In the *WACCM-like* configuration, the minimum execution time for HOMME of 31.1 seconds is obtained on a *64,2,2* configuration and represents a nearly 75% reduction in the execution time. The *WACCM-like* configuration has similar characteristics to the *CAM-like* configuration. In particular, best performance is achieved using two or more execution threads per hardware core.

Unfortunately, at the current time we do not have access to a large multi-node KNL-based sys-tem to evaluate execution time and computational cost of the HOMME dynamical core. However, based on our results on Yellowstone, a single node KNL-based system, and preliminary results on a small number of KNL nodes, we believe that HOMME will exhibit similar scaling and performance

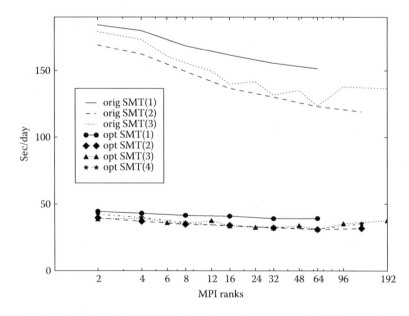

FIGURE 10.10 Execution time for HOMME using 64 hardware cores of a KNL node for a *WACCM-like* configuration at *ne* = 8 (*plev* = 70, *qsize* = 135).

characteristics. Recall that our target resolution of 0.25° resolution has a total of 86,400 spectral elements. If we assume that a large multinode KNL-based system will have similar scaling characteristics to Yellowstone, we can expect to be able to run the spectral element dynamical core in production at 1/2 spectral element per hardware core, utilizing 172,800 hardware cores per run.

10.5   CASE STUDY: DATA ANALYTICS

Our last case study pertains to the postprocessing and analysis of the large data sets that result from the higher resolution and frequency that is achievable at exascale. Data production rates from climate simulations are expected to grow significantly and the previously existing technologies for postprocessing and analysis are mostly serial. We choose to take an approach that introduces the fewest changes to the workflow with which scientists are already familiar, namely parallelizing individual steps in the workflow by swapping out the serial scripts with parallel alternatives. This approach requires less user reeducation and should improve the adoption rate of the newly developed technologies. This should also lead to fewer problems encountered during and after deployment of these new technologies. Note that preliminary efforts for our new approach are detailed further in [13].

We have chosen to develop our new tools with parallel Python, using MPI4PY for parallelism. Python is excellent for rapid development, reducing the time to deployment. It also allows us to develop tools with a very small code-base, making it easier to maintain. Python seamlessly integrates into the existing script-driven CESM workflow and is cross-platform, reducing the need to address build issues on multiple exascale platforms.

The general CESM postprocessing workflow proceeds from the generation of the raw simulation data in the following way. First, because postprocessing and analysis tend to be performed on time-series data of only a handful of variables, the data must be transformed from files storing synoptic data (time slice) to files storing a single variable over a long period of simulated time (time series). This reduces the amount of data moved when scientists copy data sets for local analysis, and it makes time-series data contiguous within a single file, improving I/O performance with conventional spinning disk storage. The serial tool for performing this transformation is a script written using the NetCDF Operators (NCO), and this serial transformation step has been observed to take as long as the data generation with CESM itself. The parallel Python tool that we have written to replace this NCO script is called the PYRESHAPER.

The PYRESHAPER implements a task-parallel approach to speeding up this data transformation. Each output file is defined and written by a single MPI process. In particular, for the time slice to time series transformation, this means one output file for each time series variable in the data set. This, obviously, has limited scalability, but the immediate benefit is obvious for large data sets, which quite often have more than a few dozen time-series variables.

The next step in the workflow is an analysis step, from the time-series files generate with the PYRESHAPER. In this step, temporal and spatial averages are computed for variables in the time-series data set. These averages, called climatologies, are primarily used to create plots that help scientists quickly assess the correctness of various aspects of the CESM component models. The original script used to perform this task was written using NCO and the new parallel Python tool that we have written to replace it is called the PYAVERAGER.

The PYAVERAGER implements a parallel approach based on the PYRESHAPER, namely task parallel with parallelization over the output files. However, because the computation of the averages is a data reduction, it is possible to add an additional level of parallelization over the reduction steps, themselves. Therefore, unlike the PYRESHAPER that dedicates a single MPI process to each output file, the PYAVERAGER dedicates an MPI subcommunicator to each output file, allowing separate MPI processes to be independently responsible for reading and operating on the data. This allows for greater scalability than the PYRESHAPER can achieve, but is still limited by the data set and the kind and number of averages requested.

Both tools were tested in parallel on NCAR's Yellowstone supercomputing platform, located at the NWSC [11]. Two complete CESM data sets were used in these tests: a low-resolution (1°, 232 GB) data set and a high-resolution (0.25° atmosphere/land, 0.1° ocean/ice, 4.4 TB) data set, both spanning 10 simulated years of data. At maximum parallelism, the PYRESHAPER reduced the timeslice to timeseries transformation step for the low-resolution data sets from 3.7 hours to 30 minutes, a 7.6× speedup. For the high-resolution data set, PYRESHAPER achieves a 11.7× speedup, reducing execution time from 23 to 2 hours. The PYAVERAGER reduced the computation of climatologies for the low-resolution data sets from 9 hours to 4 minutes, a speedup of 130×. For the high-resolution data sets, execution times are reduced from 40 hours to 16 minutes a 150× speedup, respectively. See Table 10.2 for a summary of these speedup results.

Current and future development has been directed toward the application of these parallel Python techniques to more general and sophisticated problems, including the problem of data standardization and publication. During model intercomparison projects, such as the CMIP, data generated by many modeling codes and institutions must be collected together for direct comparison with each other. This necessitates data standardization, including conventions for variable naming, units, and additional metadata. The process of standardization involves many of the same procedures as those involved in the PYRESHAPER and PYAVERAGER utilities, and these standardizations can even be seen as a generalization of the same tools. This tool, in development at the time of this writing, is called PYCONFORM.

PYCONFORM is a general tool for performing dependent tasks, in a prescribed sequence, on large data sets, to produce second data sets of a different format. Such a sequence of tasks comprises a directed acyclic graph (DAG), the flow of which describes the flow of data from an input data set to an output data set. Obvious parallelism can be achieved across independent components of the DAG, whereas greater parallelism can be achieved by further splitting components of the DAG and connecting the components with MPI communication. Further development and testing is planned and a release is expected before the end of 2016.

The kinds of operations described above should see significant further improvement with hardware advances such as nonvolatile memory (NVM)-based storage devices, which are expected to be commonplace in exascale machines. Some hint of the performance improvements this technology can bring can be found through testing on existing NVM-based platforms, though the use of NVM in existing machines varies greatly from platform to platform. On Wrangler, we have seen the PYRESHAPER tests improve by 1.5× to 2.5× for both low- and high-resolution data sets, with the bulk of the speedup coming directly from the improved read speeds of NVM. The PYAVERAGER, which has a much more asymmetric data flow (large read and small write), sees an even greater improve-

TABLE 10.2

Summary of Performance Speedup of Lightweight Parallel Python Tools over Their Equivalent Serial NetCDF Operators (NCO)-Based Scripts

Python Utility	Low Resolution	High Resolution
PyReshaper	7.6×	11.7×
PyAverager	130×	150×

Note: The low-resolution test-case data set contained 1° data from CESM atmosphere, land, ocean, and ice components, comprising 232 GB. The high-resolution test-case data set contained 0.25° data from Community Earth System Model (CESM)'s atmosphere and land components and 0.1° data from CESM's ocean and ice components, comprising 4.4 TB.

ment of 4× to 15×. These speedups are already on top of the performance improvements seen by parallelizing the operations. We expect solid state drive (SSD) and I/O hardware improvements to dramatically improve the performance of these kinds of operations in future machines and make previously impossible operations commonplace.

10.6 CONCLUSIONS AND FUTURE WORK

We have described the ongoing work to prepare CESM to efficiently utilize the upcoming exascale computing resources that will be deployed in the next 4–6 years. We describe a two-pronged approach that includes refactoring the dynamical core used by CESM for high-resolution simulations to enable more parallelism and accelerating the postprocessing of output data using a parallel Python approach. Early results have demonstrated that such an approach is having a significant impact. In particular, refactoring of the spectral element dynamical core has illustrated that it is possible to use 6–12 times the number of cores for the same computational cost on existing hardware. We have also illustrated that these code modifications have sped up the execution of HOMME on a single KNL node by 4×. Finally, we have achieved speedups that range from 25× to 300× for the postprocessing.

Although our current results are promising, there remains much work to done. Future work will center on both further optimizations of the HOMME dynamical core and the postprocessing work-flow as well as optimizations of other pieces of the CESM model. For the HOMME dynamical core, we expect that improvements to compiler optimizations along with further refinement of the thread-ing will reduce execution time of HOMME at large core counts by an additional 25%. We have MPI-3 neighborhood collective support in HOMME and intend to evaluate its impact at very large core count. Preliminary results to systems that use nonvolatile memory as a part of the file-systems have been promising and we continue to evaluate its potential impact on our workflow. We also continue to optimize other performance critical pieces of CESM.

ACKNOWLEDGMENTS

This effort was supported through funding through the National Science Foundation grant NSF01 as well as an Intel Parallel Computing Center for Weather and Climate Simulation. Additional support has been made through the NERSC Exascale Science Application Program.

REFERENCES

1. W.M. Washington and C.L. Parkinson. *An Introduction to Three-Dimensional Climate Modeling*. Sausal-ito, CA: University Science Books, 2005.
2. W.M Washington, L. Buja, and A. Craig. The computational future for climate and earth system mod-els: On the path to petaflop and beyond. *Philosophical Transactions of the Royal Society of London A: Mathematical, Physical and Engineering Sciences*, 367(1890):833–846, 2009.
3. N.G. Heavens, D.S. Ward, and M.M. Natalie. Studying and projecting climate change with earth system models. *Nature Education Knowledge*, 4(5):4, 2013.
4. R.E. Zeebe. Where are you heading earth? *Nature Geoscience*, 4(7):416–417, 2011.
5. S.M. Easterbrook and T.C. Johns. Engineering the software for understanding climate change. *Comput-ing in Science and Engineering*, 11(6):65–74, 2009.
6. Coupled Model Intercomparison Project phase 5. CMIP5: Coupled Model Comparison Project Phase 5. http://cmip-pcmdi.llnl.gov/cmip5/ [Accessed: 19 July 2017], 2013. [Accessed 1 June 2016].
7. S.M. Easterbrook, P.N. Edwards, V. Balaji, and R. Budich. Guest editors' introduction: Climate change—Science and software. *IEEE Software*, 28(6):32–35, 2011.

8. J. Pipitone and S. Easterbrook. Assessing climate model software quality: A defect density analysis of three models. *Geoscientific Model Development*, 5(4):1009–1022, 2012.

9. R. Justin Small, J. Bacmeister, D. Bailey, A. Baker, S. Bishop, F. Bryan, J. Caron, J. Dennis, P. Gent, H.-M. Hsu, et al. A new synoptic scale resolving global climate simulation using the Community Earth System Model. *Journal of Advances in Modeling Earth Systems*, 6(4):1065–1094, 2014.

10. J. Hurrell, M. Holland, P. Gent, S. Ghan, J. Kay, P. Kushner, J.-F. Lamarque, W. Large, D. Lawrence, K. Lindsay, et al. The community earth system model: A framework for collaborative research. *Bulletin of the American Meteorological Society*, 94:1339–1360, 2013.

11. R. Loft, A. Andersen, F. Bryan, J.M. Dennis, T. Engel, P. Gillman, D. Hart, I. Elahi, S. Ghosh, R. Kelly, et al. Yellowstone: A dedicated resource for earth system science. In J.S. Vetter, editor, *Contemporary High Performance Computing: From Petascale Toward Exascale*, Vol. 2, Chap. 8, pp. 185–224. Boca Raton, FL: Chapman and Hall/CRC, 2015.

12. J.E. Kay, C. Deser, A. Phillips, A. Mai, C. Hannay, G. Strand, J. Arblaster, S. Bates, G. Danabasoglu, J. Edwards, et al. The Community Earth System Model (CESM) large ensemble project: A community resource for studying climate change in the presence of internal climate variability. *Bulletin of the American Meteorological Society*, 96: 1333–1349, 2015.

13. K. Paul, S. Mickelson, H. Xu, J.M. Dennis, and D. Brown. Light-weight parallel Python tools for earth system modeling workflows. In *IEEE International Conference on Big Data*, Santa Clara, CA: IEEE, pp. 1985–1994, October 2015.

14. A.H. Baker, D.M. Hammerling, S.A. Mickleson, H. Xu, M.B. Stolpe, B. Naveau, B. Sanderson, I. Ebert-Uphoff, S. Samarasinghe, F. De Simone, et al. Evaluating lossy data compression on climate simulation data within a large ensemble. *Geoscientific Model Development Discussions*, 9, 4381–4403, 2016.

15. S. Ahern, S.R. Alam, M.R. Fahey, R.J. Hartman-Baker, R.F. Barrett, R.A. Kendall, D.B. Kothe, R.T. Mills, R. Sankaran, A.N. Tharrington, et al. Scientific application requirements for leadership computing at the exascale. Technical Report, Oak Ridge National Laboratory (ORNL); Center for Computational Sciences, 2007.

16. S. Ashby, P. Beckman, J. Chen, P. Colella, B. Collins, D. Crawford, J. Dongarra, D. Kothe, R. Lusk, P. Messina, et al. The opportunities and challenges of exascale computing. Summary Report of the Advanced Scientific Computing Advisory Committee (ASCAC) Subcommittee, pp. 1–77, 2010.

17. J. Shalf, S. Dosanjh, and J. Morrison. Exascale computing technology challenges. In *Proceedings of the 9th International Conference on High Performance Computing for Computational Science, VECPAR'10*, Berkeley, CA, pp. 1–25, 2011.

18. J. Dongarra, P. Beckman, T. Moore, P. Aerts, G. Aloisio, J.-C. Andre, D. Barkai, J.-Y. Berthou, T. Boku, B. Braunschweig, et al. The international exascale software project roadmap. *International Journal of High Performance Computing Applications*, 25(1):3–60, February 2011.

19. A.H. Baker, H. Xu, J.M. Dennis, M.N. Levy, D. Nychka, S.A. Mickelson, J. Edwards, M. Vertenstein, and A. Wegener. A methodology for evaluating the impact of data compression on climate simulation data. In *Proceedings of the 23rd International Symposium on High-performance Parallel and Distributed Computing, HPDC '14*, Vancouver, Canada, pp. 203–214, 2014.

20. N. Hübbe and J. Kunkel. Reducing the HPC-datastorage footprint with MAFISC—multidimensional adaptive filtering improved scientific data compression. In *Computer Science–Research and Development,* Hamburg/Berlin/Heidelberg: Springer, 2012.

21. J. Woodring, S.M. Mniszewski, C.M. Brislawn, D.E. DeMarle, and J.P. Ahrens. Revisiting wavelet compression for large-scale climate data using JPEG2000 and ensuring data precision. In D. Rogers and C.T. Silva, editors, *IEEE Symposium on Large Data Analysis and Visualization (LDAV)*, pp. 31–38. Providence, RI: IEEE, 2011.

22. I. Carpenter, R.K. Archibald, K.J. Evans, J. Larkin, P. Micikevicius, M. Norman, J. Rosinski, J. Schwarzmeier, and M.A. Taylor. Progress towards accelerating homme on hybrid multi-core systems. *International Journal of High Performance Computing Applications*, 27(3):335–347, August 2013.

23. J.M. Dennis and H.M. Tufo. Scaling climate simulation applications on IBM Blue Gene. *IBM Journal of Research and Development: Applications for Massively Parallel Systems*, 52(1.2):117–126, 2008.

24. J.M. Dennis. Inverse space-filling curve partitioning of a global ocean model. In *IEEE International Parallel & Distributed Processing Symposium*, Long Beach, CA, 2007.

25. J.M. Dennis, M. Vertenstein, P.H. Worley, A.A. Mirin, A.P. Craig, R. Jacob, and S. Mickelson. Computational performance of ultra-high-resolution capability in the Community Earth System Model. *International Journal of High Performance Computing Applications*, 26(1):5–16, February 2012.

26. P.W. Jones, P.H. Worley, Y. Yoshida, J.B. White III, and J. Levesque. Practical performance portability in the Parallel Ocean Program (POP). *Concurrency and Computation Practice and Experience*, 17:1317–1327, 2005.

27. P.H. Worley, A.P. Craig, J.M. Dennis, A.A. Mirin, M.A. Taylor, and M. Vertenstein. Performance of the community earth system model. In *Proceedings of the 2011 International High Performance Computing, Networking, Storage and Analysis (SC)*, Seattle, WA, November 2011.

28. GPTL-General Purpose Timing Library, 2016. http://jmrosinski.github.io/GPTL/. [Accessed 19 July 2017].

29. Y. Kim, J. Dennis, C. Kerr, R.R.P. Kumar, A. Simha, A. Baker, and S. Mickelson. KGEN: A Python tool for automated Fortran kernel generation and verification. In *Procedia Computer Science, Volume 80 of ICCS 2016. The International Conference on Computational Science*, San Diego, CA, pp. 1450–1460, 2016.

30. KGEN: Fortran Kernel Generator. https://github.com/NCAR/KGen, 2016.

31. A. Gettelman and H. Morrison. Advanced two-moment bulk microphysics for global models part I: Off-line tests and comparison with other schemes. *Journal of Climate*, 28(3):1268–1287, 2015.

32. M.J. Iacono, J.S. Delamere, E.J. Mlawer, M.W. Shepard, S.A. Clough, and W.D. Collins. Radiative forcing by long-lived greenhouse gases: Calculations with the AER radiative transfer models. *Journal of Geophysical Research*, 113(D13103): 1–8, 2008.

33. L.K. Emmons, S. Walters, P.G. Hess, J.-F. Lamarque, G.G. Pfister, D. Fillmore, C. Granier, A. Guenther, D. Kinnison, T. Laepple, et al. Description and evaluation of the model for ozone and related chemical tracers, version 4 (MOZART-4). *Geoscientific Model Development*, 3:43–67, 2010.

34. CESM case study. http://www.nersc.gov/users/computationalsystems/cori/application-porting-and-performance/application-casestudies/cesm-case-study/, 2016. [Accessed 19 July 2017].

35. T. Barnes, B. Cook, J. Deslippe, D. Doerfler, B. Friesen, Y. He, T. Kurth, T. Koskela, M. Lobet, T. Malas, et al. Evaluating and optimizing the NERSC workload on knight's landing. In *7th International Workshop in Performance Modeling, Benchmarking and Simulation of High Performance Computing Systems (PMBS16)*, ACM/IEEE Supercomputing 2016 (SC16), Salt Lake City, UT, November 2016.

36. NCAR climate/weather kernels. https://github.com/NCAR/kernelOptimization, 2016. [Accessed July 19, 2017].

37. H. Servat, G. Llort, J. Germanand, K. Huck, and J. Labarta. Folding: Detailed analysis with coarse sampling. In H. Brunst, M.S. Mueller, W.E. Nagel, and M.M. Resch, editors, *Tools for High Performance Computing*, pp. 105–118. Berlin/Heidelberg: Springer-Verlag, 2011.

38. Extrae instrumentation package. https://tools.bsc.es/extrae, 2016. [Accessed 20 July 2017].

39. Paraver: A tool to visualize and analyze parallel code, https://tools.bsc.es/paraver, 2016. [Accessed 20 July 2017].

40. D.J. Milroy, A.H. Baker, D.M. Hammerling, J.M. Dennis, S.A. Mickelson, and E.R. Jessup. Towards characterizing the variability of statistically consistent community earth system model simulations. In *Procedia Computer Science: Volume 80 of ICCS 2016. The International Conference on Computational Science*, San Diego, CA, pp. 1589–1600, 2016.

41. A.H. Baker, D.M. Hammerling, M.N. Levy, H. Xu, J.M. Dennis, B.E. Eaton, J. Edwards, C. Hannay, S.A. Mickelson, R.B. Neale, et al. A new ensemble-based consistency test for the community earth system model. *Geoscientific Model Development*, 8:2829–2840, 2015.

42. A.H. Baker, Y. Hu, D.M. Hammerling, Y. Tseng, H. Xu, X. Huang, F.O. Bryan, and G. Yang. Evaluating statistical consistency in the ocean model component of the community earth system model (pyCECT v2.0). *Geoscientific Model Development*, 9:2391–2406, 2016.

43. Cori. http://www.nersc.gov/users/computational-systems/cori/, 2016. [Accessed 19 July 2017].

44. Wrangler: Ground breaking data intensive computing. https://www.tacc.utexas.edu/systems/wrangler, 2016. [Accessed 19 July 2017].

45. J.M. Dennis, J. Edwards, K.J. Evans, O. Guba, P.H. Lauritzen, A.A. Mirin, A. St-Cyr, M.A. Taylor, and P.H. Worley. CAM-SE: A scalable spectral element dynamical core for the community atmosphere model. *International Journal High Performance Computing Applications*, 26(1):74–89, January 2012.

46. R.D. Loft, S.J. Thomas, and J.M. Dennis. Terascale spectral element dynamical core for atmospheric general circulation models. *In Proceedings of SC2001 ACM/IEEE Conference on Supercomputing*, Denver, CO: IEEE, 2001.

47. A.J. Simmons and D.M. Burridge. An energy and angular-momentum conserving vertical finite-difference scheme and hybrid vertical coordinates. *Monthly Weather Review*, 109:758–766, 1981.

48. M. Taylor, J. Tribbia, and M. Iskandarani. The spectral element method for the shallow water equations on the sphere. *Journal of Computational Physics*, 130(1):92–108, 1997.

49. S.-J. Lin. A "vertically Lagrangian" finite-volume dynamical core for global models. *Monthly Weather Review*, 132:2293–2397, 2004.

50. J.M. Dennis. Partitioning with space-filling curves on the cubed-sphere. In *Proceedings of Workshop on Massively Parallel Processing at IPDPS'03*, Nice, France, April 2003.

51. T. Henderson, J. Michalakes, I. Gokhale, and A. Jha. Numerical weather prediction optimization. In J. Reinders and J. Jeffers, editors, *High Performance Parallelism Pearls Multicore and Many-Core Programming Approaches*, Vol. 2, Chap. 2, pp. 7–23. New York, NY: Elsevier, 2015.

52. D. Marsh, M. Mills, D.E. Kinnison, and J.F. Lamarque. Climate change from 1850 to 2005 simulated in CESM1 (WACCM). *Journal of Climate*, 26:7372–7391, 2013.

11 Large Eddy Simulation of Reacting Flow Physics and Combustion

Joseph C. Oefelein and Ramanan Sankaran

CONTENTS

11.1 Scientific Methodology 231
11.2 Algorithmic Details 233
11.3 Programming Approach 236
    11.3.1 Scalability 236
    11.3.2 Performance 239
    11.3.3 Portability 240
    11.3.4 External Libraries 240
11.4 Software Practices 240
11.5 Benchmarking Results 241
11.6 Hybrid Shared Memory Parallelism in RAPTOR 246
11.7 Other Considerations 251
Acknowledgments 252
References 252

11.1 SCIENTIFIC METHODOLOGY

The multiphysics, multiscale nature of turbulent chemically reacting flows makes these systems one of the most challenging to understand and control. A multitude of strongly coupled fluid dynamic, thermodynamic, transport, chemical, multiphase, and heat transfer processes are intrinsically coupled and must be considered simultaneously in complex domains to satisfy the governing conservation equations. The problem is compounded by the broad range of time and length scales over which interactions occur due to turbulence and differences in chemical reaction rates. The nonlinear nature of the system significantly limits the number of simplifying assumptions that can be made. Conversely, some form of modeling is always required and significant sets of assumptions must be made to derive multiscale closures that are both accurate and affordable. This combination of challenges significantly complicates the process of scientific discovery, model development, and model validation. Model development is challenging because it is nearly impossible to decouple various processes without introducing potentially significant sources of error. Model validation is problematic because detailed data over relevant parameter spaces are limited and comparisons using available data are difficult to interpret. In many cases, deviations between measured and modeled results cannot be directly linked to a particular modeling approach and the related assumptions because many other factors (e.g., boundary conditions, grid resolution, numerical errors, etc.) can have dominating effects on the results. To mitigate these challenges, there is a significant need to carefully assess the physical characteristics of the governing system and the related first-principles constitutive relations. This assessment must be done within the formalism of a generalized multiscale closure that exposes the effects of turbulence on molecular processes, within the fully coupled governing system, over clearly defined ranges of scales. The analysis must be performed using numerical methods that are well suited for the treatment of broadband multiscale processes and do not introduce additional extraneous errors that compete with the submodels and further complicate the analysis. The large eddy simulation (LES) approach described here is focused on addressing these needs.

There is no experimental or simulation technique is capable of providing a complete description of the multiscale processes described above. The highest quality experiments only provide partial information due to limitations associated with various diagnostic techniques and/or harsh operating environments. Likewise, simulations of these processes will always be limited by computational power. Even with exascale computing (and beyond), direct numerical simulation (DNS) of the fully coupled equations of fluid motion, transport, and chemical reaction can only be applied over a limited range of turbulence scales (e.g., in highly confined domains), in the high wavenumber, low-Reynolds-number regime of turbulence. Results from DNS are useful to better select underlying physical assumptions used in modeling and to understand fundamentals related to small-scale phenomena. However, such results must be interpreted with caution and cannot be considered fully conclusive due to the limited dynamic range present in the simulations. This limited range, coupled with the indirect relation to the physical conditions present in practical systems of interest (e.g., high-Reynolds number, anisotropic, wall-dominated turbulence) introduces limitations for using DNS directly for model development. Similar to experiments and DNS, simulation techniques such as LES aimed at treating the full range of time and length scales present in a given system introduce other forms of uncertainties. Treating the full range of time and length scales in practical systems of interest must always begin with some form of formal filtering of the governing conservation equations. The Reynolds-averaged Navier–Stokes (RANS) approximation, for example, employs filtering in time to derive the governing equations for the mean state. For this approach, thermo-physical interactions across the full range of scales are modeled to represent the net effect on the largest energy containing scales (e.g., turbulent eddies). All dynamic degrees of freedom smaller than these largest energy containing eddies are represented by bulk averages and no information exists that describes phenomenological interactions between the small scales. RANS is currently the primary method used for industry relevant engineering calculations since it is the least costly. However, the models employed are the least universal and provide the lowest level of accuracy because they inherently involve many tuning constants that must be calibrated on a case by case basis.

In contrast to RANS, LES has historically employed spatial filtering techniques to split field variables into time-dependent resolved-scale and subfilter-scale components. It is important to note, however, that the mathematical formalism associated with LES can be generalized to also include temporal filtering techniques, which allows one to account for subfilter-scale fluctuations in time also. This is an extremely important consideration for problems involving the reacting flow physics described above. It represents the most general form of filtering with mathematically exact definitions of the resultant space–time subfilter variance and covariance fields that must be modeled. This generalization is a key focal point of the LES approach described here. Using this formalism, LES can be viewed as a tool that operates between the limits of DNS and RANS. The large energetic scales are resolved directly. The small subfilter scales must be modeled. However, just like one chooses the resolution at which a photographic image is resolved, one can conceptually choose the resolution at which pertinent broadband structures of a flow are resolved. As the spatial and temporal resolution is increased (for a fixed problem size, geometry, and set of operating conditions), the cost of a calculation increases, but detailed localized information from the resolved scales also increases and becomes available as additional input for selected subfilter closures. Similarly, the range of scales over which the system of subfilter-scale models must work becomes proportionately less and they tend to be more universal in character. Using these characteristics, LES is applied with mathematically rigorous definitions of the space–time subfilter variance and covariance fields to investigate the full range of multidimensional time and length scales that exist, in a given flow system of interest, at a given set of operating conditions. Subfilter models are formulated using first-principles derivations that, in the limit as the spatial and temporal resolution of a calculation approach the smallest relevant system scales, the subfilter contributions approach zero and the solution converges to a DNS. This is in contrast to typical engineering-based models (e.g., the flamelet approximation), which by design do not relax in this limit. The mathematical formalism of LES also facilitates use of mathematical identities that eliminate the need for tuning constants. Thus, the only adjustable parameters in our simulations are resolution and boundary conditions.

Research performed using LES addresses priority research directions identified in recent Department of Energy (DOE) workshop reports; namely, the Office of Science (SC), Basic Energy Sciences (BES), *Workshop on Clean and Efficient Combustion of 21st Century Transportation Fuels*, and the jointly sponsored SC–BES and Energy Efficiency and Renewable Energy (EERE), Vehicle Technologies Office (VTO), *Workshop to Identify Research Needs and Impacts in Predictive Simulations for Internal Combustion Engines*. Thus, it is well aligned with DOE missions in both basic and applied research. The scale of simulations facilitate new insights that are directly aligned with companion target experiments and that provide new quantitative insights into the key processes that must be accounted for in engineering models. The ultimate impact of this work is in enabling predictive simulations of advanced concepts that have the potential to accelerate the design cycle of devices such as reciprocating internal combustion engines and gas turbines. Note that the parallel model and approach are general and applicable to a wide variety of computational fluid dynamics (CFD) application codes.

11.2 ALGORITHMIC DETAILS

Using the approach described above, LES provides the formal ability to treat the full range of multidimensional time and length scales that exist, in a given flow system of interest, at a given set of operating conditions, with emphasis on fundamental inquiries into the structure and dynamics of turbulent combustion processes that are dominated by high pressure, high-Reynolds number, geometrically complex flows. LES calculations are performed using a single unified code framework called RAPTOR. Unlike conventional LES codes, RAPTOR is a DNS solver that has been optimized to meet the strict algorithmic requirements imposed by the LES formalism. The theoretical framework solves the fully coupled conservation equations of mass, momentum, total energy, and species for a chemically reacting flow. It is designed to handle high-Reynolds number, high pressure, real gas, and/or liquid conditions over a wide Mach operating range. It also accounts for detailed thermodynamics and transport processes at the molecular level. A noteworthy aspect of RAPTOR was designed specifically for LES using nondissipative, discretely conservative, staggered, finite-volume differencing. This eliminates numerical contamination of the subfilter models due to artificial dissipation and provides discrete conservation of mass, momentum, energy, and species, which is an imperative requirement for high quality LES. Details related to the baseline formulation and subfilter models are given by Oefelein [1]. Representative results and case studies are given by Oefelein et al. [2–17].

For LES, the filtered version of the Navier–Stokes equations are solved. These equations are given as

$$\frac{\partial \overline{\rho}}{\partial t} + \nabla \cdot (\overline{\rho}\tilde{\mathbf{u}}) = 0 \tag{11.1}$$

$$\frac{\partial}{\partial t}(\overline{\rho}\tilde{\mathbf{u}}) + \nabla \cdot \left[\left(\overline{\rho}\tilde{\mathbf{u}} \otimes \tilde{\mathbf{u}} + \frac{P}{M^2}\mathbf{I} \right) \right] = \nabla \cdot \vec{\overline{\mathcal{T}}} \tag{11.2}$$

$$\frac{\partial}{\partial t}(\overline{\rho}\tilde{e}_t) + \nabla \cdot \left[(\overline{\rho}\tilde{e}_t + P)\tilde{\mathbf{u}} \right] = \nabla \cdot \left[\left(\vec{\overline{\mathcal{Q}}}_e + M^2(\vec{\overline{\mathcal{T}}} \cdot \tilde{\mathbf{u}}) \right) \right] + \overline{\mathcal{Q}}_e \tag{11.3}$$

$$\frac{\partial}{\partial t}(\overline{\rho}\tilde{Y}_i) + \nabla \cdot (\overline{\rho}\tilde{Y}_i\tilde{\mathbf{u}}) = \nabla \cdot \vec{\overline{S}}_i + \overline{\dot{\omega}}_i \tag{11.4}$$

In these equations,

$$e_t = e + \frac{M^2}{2} \mathbf{u} \cdot \mathbf{u}$$

$$e = \sum_{i=1}^{N} h_i Y_i - \frac{p}{\rho}$$

$$h_i = h_{f_i}^\circ + \int_{p^\circ}^{p} \int_{T^\circ}^{T} C_{p_i}(T,p) T p$$

represent the total internal energy, internal energy, and enthalpy of the ith species, respectively. The terms $\vec{\mathcal{P}}$, $\vec{\mathcal{T}}$, $\vec{Q}_e$, and $\vec{S}_i$ represent respective composite (i.e., molecular plus subfilter) stresses and fluxes. Quantities $\dot{Q}_e$ and $\bar{\omega}_i$ represent the filtered energy and chemical source terms, respectively.

The viscous stress tensor is assumed to follow Stokes' hypothesis, and the heat release due to chemical reaction is accounted for in the description of the specific enthalpies, h_i, as given by the enthalpy of formation, $h_{f_i}^\circ$. The heat release rate can be represented equivalently as the product of the enthalpy of formation and the local rate of production of all the species considered in the system. Using this representation, the source term and specific enthalpies are defined as

$$\dot{Q}_e = -\sum_{i=1}^{N} \dot{\omega}_i h_{f_i}^\circ \tag{11.5}$$

$$h_i = \int_{p^\circ}^{p} \int_{T^\circ}^{T} C_{p_i}(T,p) dT dp \tag{11.6}$$

The subfilter closure is obtained using the *mixed* dynamic Smagorinsky model by combining the models proposed by Erlebacher, Hussaini, Speziale, and Zang [18] and Speziale [19] with the dynamic modeling procedure [20–24]. The composite stresses and fluxes in Equations 11.1 through 11.4 are then given as

$$\vec{\mathcal{T}} = (\mu_t + \mu) \frac{1}{Re} \left[-\frac{2}{3} (\nabla \cdot \tilde{\mathbf{u}}) \mathbf{I} + (\nabla \tilde{\mathbf{u}} + \nabla \tilde{\mathbf{u}}^T) \right] - \bar{\rho} \left(\widetilde{\mathbf{u} \otimes \mathbf{u}} - \tilde{\mathbf{u}} \otimes \tilde{\mathbf{u}} \right) \tag{11.7}$$

$$\vec{Q}_e = \left(\frac{\mu_t}{Pr_t} + \frac{\mu}{Pr} \right) \frac{1}{Re} \nabla \tilde{h} + \sum_{i=1}^{N} \tilde{h}_i \vec{S}_i - \bar{\rho} \left(\widetilde{h \mathbf{u}} - \tilde{h} \tilde{\mathbf{u}} \right) \tag{11.8}$$

$$\vec{S}_i = \left(\frac{\mu_t}{Sc_{t_i}} + \frac{\mu}{Sc_i} \right) \frac{1}{Re} \nabla \tilde{Y}_i - \bar{\rho} \left(\widetilde{Y_i \mathbf{u}} - \tilde{Y}_i \tilde{\mathbf{u}} \right) \tag{11.9}$$

The term μ_t is the subfilter eddy viscosity, given by

$$\mu_t = \bar{\rho} C_R \Delta^2 \Pi_{\tilde{\mathbf{S}}}^{\frac{1}{2}} \tag{11.10}$$

where

$$\Pi_{\tilde{\mathbf{S}}} = \tilde{\mathbf{S}} : \tilde{\mathbf{S}}, \text{ and } \tilde{\mathbf{S}} = \frac{1}{2} \left(\nabla \tilde{\mathbf{u}} + \nabla \tilde{\mathbf{u}}^T \right) \tag{11.11}$$

The terms C_R, Pr_t, and Sc_{t_i} are the Smagorinsky, subfilter Prandtl, and subfilter Schmidt numbers and are evaluated dynamically as functions of space and time. The overall model includes the Leonard and cross-term stresses and provides a Favre averaged generalization of the Smagorinsky eddy viscosity model [25] coupled with gradient diffusion models to account for subfilter mass and energy transport processes.

In recent work [26–30], we have demonstrated that when operating pressures exceed the critical pressure of the injected fuel, interfacial diffusion layers can develop due to thickening of the gas–liquid interfaces combined with a significant reduction in the mean free molecular path and surface tension forces. These interfaces eventually enter the continuum length scale regime and disappear as interfacial fluid temperatures rise above the critical temperature of the local mixture. The lack of intermolecular forces, coupled with broadening interfaces, promote diffusion dominated mixing processes prior to atomization. As a consequence, injected jets evolve in the presence of exceedingly large but continuous thermo-physical gradients in a manner markedly different from classical sprays. Given these observations, the supercritical jet considered here (and related scalar-mixing processes) are modeled using a property evaluation scheme that is designed to account for real-fluid thermodynamic nonidealities and transport anomalies for multicomponent mixtures.

The property evaluation scheme is designed to operate over a wide range of pressures and temperatures. The scheme is comprehensive and intricate, thus only a skeletal description can be given here. The extended corresponding states model [31,32] is employed with a cubic equation of state. In past studies, both the Benedict–Webb–Rubin (BWR) equation of state and cubic equations of state have been used to evaluate the pressure–volume–temperature (PVT) behavior of the inherent dense multicomponent mixtures. Use of modified BWR equations of state in conjunction with the extended corresponding states principle has been shown to provide consistently accurate results over the widest range of pressures, temperatures, and mixture states, especially at near-critical conditions. A major disadvantage of BWR equations, however, is that they are not computationally efficient. Cubic equations of state can be less accurate, especially for mixtures at near-critical or saturated conditions, but are computationally efficient. Experience has shown that both the Soave–Redlich–Kwong (SRK) and Peng–Robinson (PR) equations, when used in conjunction with the corresponding states principle, can give accurate results over the range of pressures, temperatures, and mixture states of interest in this study. The SRK coefficients are fit to vapor pressure data and are, thus, more suitable for conditions when the reduced temperature is less than one. The PR coefficients, on the other hand, are more suitable for conditions when the reduced temperature is greater than one. Here the PR equation of state is used exclusively. A summary of the cubic equations of state and recommended constants is given by Reid et al. [33, Chapter 3].

Having established an analytical representation for real mixture PVT behavior, the thermodynamics properties are obtained in two steps. First, respective component properties are combined at a fixed temperature using the extended corresponding states methodology to obtain the mixture state at a given reference pressure. A pressure correction is then applied using departure functions of the form given by Reid et al. [33, Chapter 5]. These functions are exact relations derived using the Maxwell relations (e.g., see VanWylen and Sonntag [34, Chapter 10]) and make full use of the real mixture PVT path dependencies dictated by the equation of state. Standard state properties are obtained using the databases developed by Gordon and McBride [35] and Kee et al. [36]. Chemical potentials and fugacity coefficients are obtained in a manner similar to that outlined above. Molecular transport properties are evaluated in a manner analogous to the thermodynamics properties. Viscosity and thermal conductivity are obtained using the extended corresponding states methodologies developed by Ely and Hanley [37,38]. The mass diffusion coefficients and thermal diffusion coefficients are obtained using the methodologies outlined by Bird et al. [39] and Hirschfelder et al. [40] in conjunction with the corresponding states methodology proposed by Takahashi [41].

The temporal integration scheme in RAPTOR employs an all-Mach-number formulation using dual-time stepping with generalized preconditioning. The approach is up to fourth-order accurate in time and provides a fully implicit solution using a fully explicit (highly scalable) multistage scheme. Preconditioning is applied on an inner pseudo-time loop and coupled to local time-stepping techniques to minimize convective, diffusive, geometric, and source term anomalies (i.e., stiffness) in an optimal manner. The formulation is A-stable, which allows one to set the physical time-step based solely on accuracy considerations. This attribute typically provides a two to three order of magnitude increase in the allowable integration time-step compared to conventional compressible flow solvers, especially in low Mach number regimes.

The spatial scheme in RAPTOR is designed specifically for LES using discretely conservative, staggered, finite-volume differencing. The second-order accurate staggered-grid formulation (where we store scalar values at cell centers and velocity components at respective cell faces) fulfills two key accuracy requirements. First, it is spatially nondissipative, which eliminates numerical contamination of the subfilter models due to artificial dissipation. Second, it provides discrete conservation of mass, momentum, energy, and species, which eliminates the artificial build up of energy at the high wavenumbers. This is an imperative requirement for LES. The algorithm includes appropriate switches to handle shocks, detonations, flame-fronts, and contact discontinuities typically present in a variety of transportation, propulsion, and power systems. The code uses structured curvilinear (i.e., body-fitted) grids with a generalized multiblock domain decomposition technique that uses fully unstructured connectivity at the block level.

11.3   PROGRAMMING APPROACH

RAPTOR is designed using a hybrid parallel model for use on distributed manycore systems with accelerators. Distributed-memory message-passing is performed across blocks using the messaging passing interface (MPI) and the single-program, multiple-data (SPMD) model. Within blocks, shared-memory symmetric multiprocessing (SMP) is employed using a directive-based approach. The generic structure of the code is illustrated in Figure 11.1. RAPTOR contains over 100,000 lines of source code and over 250 subroutines. Interdependencies between various routines closely follows the physical coupling of the governing conservation equations and related operators. Data structures used to evaluate respective operators in the governing equations (e.g., the multistage integrator and spatial differencing operators shown on the left in Figure 11.1) are strongly coupled. The workflow begins with a set of preprocessing operations, including generation and decomposition of the grid and related connectivity files, initial, and boundary conditions. Detailed output involves *in situ* processing to minimize the amount of raw data generated coupled with the calculation of a wide array of compound quantities derived both locally and statistically in the domain as a function of the baseline primitive variables. The code has built in, self-contained, checkpoint, and restart capability. Explicit fault tolerance handling (e.g., using fault-tolerant [FT]-MPI to detect and correct failures) can be easily adapted.

Associated with the evaluation of the governing conservation equations are various libraries used to perform required system, math, and submodel-specific operations. Submodels in RAPTOR required for treatment of the turbulent reacting flow physics describe above are listed on the right side of Figure 11.1. An inherent characteristic of many mutiphysics codes is the fact that evaluation of these routines can consume a significant portion of the runtime (e.g., 50% is not uncommon). The isolated functional form of these routines (i.e., kernels) and floating point intensive characteristics makes them ideal candidates for application of both multicore and multithreaded parallelism in the outermost layers of the algorithm. A second set of kernels are those associated with the construction of the right-hand-side spatial operators in the code (e.g., convection, diffusion, and chemical source terms). These involve implicit dependencies on globally shared data structures. Ultimately, these dependencies need to be removed to enable hybrid shared memory parallelism.

11.3.1   SCALABILITY

The distributed memory model using MPI in RAPTOR has been highly optimized over time. The message-passing layer makes significant use of interleaving and overlap of computation and

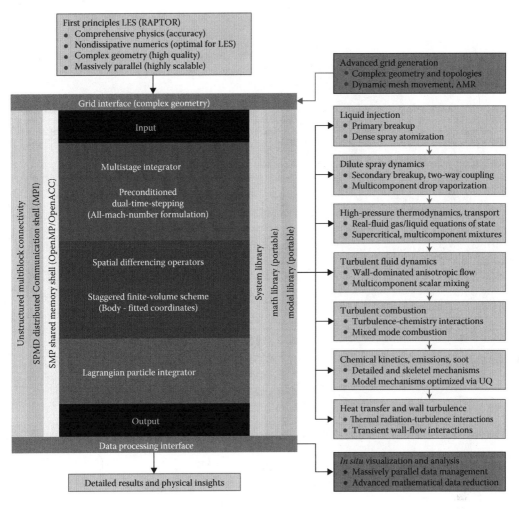

FIGURE 11.1 Generic structure of RAPTOR.

communication. Using this model, RAPTOR has consistently scaled very well on platforms such as the Oak Ridge Leadership Computing Facility (OLCF) JAGUAR and TITAN systems beyond 100,000 cores. For example, scaling results on TITAN have demonstrated the favorable strong and weak scaling attributes of the solver. The test case used is the jet-in-cross-flow configuration shown in Figure 11.2, which is a representative production level high-resolution LES result that was enabled by previous INCITE awards at OLCF. A grid size of 500-million cells was used in the domain shown. Detailed visualizations show the dynamics of a high-Reynolds-number liquid n-decane jet (Re_{Jet} = 93,000) injected into a high-Reynolds-number cross flow of air ($Re_{Channel}$ = 620,000) at high pressure supercritical conditions encountered in an actual gas turbine engine. The strong scaling attributes using MPI only are shown in Figure 11.3. Holding the total grid size fixed, and scaling from 24,000 cores to 120,000 cores, demonstrates near linear scalability while maintaining over 90% parallel efficiency. Weak scaling tests were also conducted using the same test case. For these studies, the grid was increased from 500-million cells using 24,000 cores to 2-billion cells using 120,000 cores. Using the characteristic computational time (CCT):

$$CCT = \frac{(\text{Wall Clock Time}) \times (\text{Number of Cores})}{(\text{Number of Grid Cells}) \times (\text{Number of Time-Steps})} \quad (11.12)$$

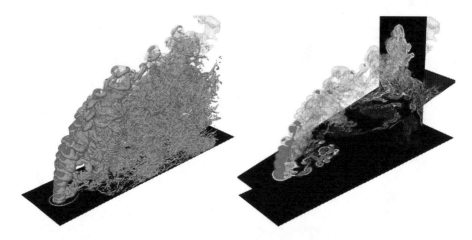

FIGURE 11.2 (See color insert.)   High fidelity LES of a high-Reynolds-number liquid n-decane jet (Re_{Jet} = 93,000) injected into a high-Reynolds-number cross flow of air ($Re_{Channel}$ = 620,000) at high-pressure super-critical conditions encountered in gas turbine engines. The left image shows the structure of the vorticity field. The right image shows iso-contours of mixture fraction, Z, that mark the location of the compressed liquid core, $Z = 0.88$ (red), and where $Z = 0.5$ (yellow). Also shown are cross-sections of the scalar dissipation rate.

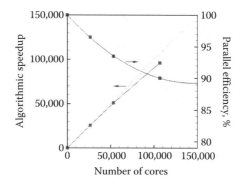

FIGURE 11.3   Baseline strong scaling attributes exhibited by RAPTOR on TITAN using MPI only.

the performance varied from CCT = 125 μ*s* to CCT = 129 μ*s*, which is less than a 4% increase. These results represent an improvement over a similar set of studies conducted as part of the 2009 DOE, Office of Advanced Scientific Computing Research (ASCR), annual *Joule Metric on Computational Effectiveness* [42] and provide a baseline to establish the enhanced performance attributes produced by the hybridized MPI + X (X = OpenMP, OpenACC) version of the code.

Having a well-optimized starting point, the strategy for refactoring the code to manycore systems with accelerators using the MPI + X model was first to optimize the kernels associated with the model library listed on the right side of Figure 11.1. One approach to systematically refactor these kernels is using Kokkos, a performance portable C++ programming model [43]. Equation level parallelism within the kernels is expressed as function operators that are executed using Kokkos.

Several groups have published their efforts for accelerating the flow solvers used for multicomponent reacting flow calculations using graphics processing units (GPUs). The most common approach is to select the most time-consuming and compute-intensive kernels and offload them to the GPUs for accelerated solution [44]. In the case of combustion solvers, the chemical reaction kinetics is a typically predominant kernel that is offloaded to the GPUs. Spafford et al. [45] was one of the first to use this approach to offload the chemical kinetics evaluation to a GPU using the CUDA programming model and thereby accelerate a DNS solver for turbulent combustion. Their approach was to use grid-level parallelism for acceleration by computing the reaction kinetics across the large number of grid points in parallel on the GPU. In contrast, Shi et al. [46] utilized the parallelism available in the reaction network itself to simultaneously calculate all the reaction rates for a single kinetic system. They observed a considerable speedup for large chemical reaction mechanisms greater than 1000 species, but could not obtain any acceleration for smaller mechanisms with fewer than 100 species. Stone and Davis [47] measured the performance of implicit integration schemes for stiff chemical kinetics and compared against the speedups obtained for explicit integration. More recently, Niemeyer and Sung [48] used fully explicit and stabilized explicit methods for accelerating chemical kinetics with low to moderate levels of stiffness. These studies have shown that the potential for acceleration of chemical kinetics through GPUs are large when explicit integration is feasible. But in the presence of stiffness, the performance on GPUs was highly susceptible to thread divergence due to varying levels of stiffness across the multiple states. Sankaran [49] has proposed exploiting a third level of parallelism beyond the reaction network and grid-level parallelisms used in most of the current works. In this approach, locally one-dimensional subfilter models used for the reaction rate closure, such as the linear eddy model (LEM) or interactive unsteady flamelets provide an additional level of parallelism to fully saturate the potential of the accelerators.

In RAPTOR, GPUs are used to accelerate the computation of the thermophysical and molecular transport properties of multicomponent mixtures. The evaluation of the closure equations and constitutive properties at the supercritical pressures is highly computationally intensive, as will be demonstrated later. Furthermore, they have a high level of parallelism because the evaluation across multiple grid points are independent. These physics models have not been accelerated on GPUs before. The results presented here are the first demonstration of their potential for acceleration and the impact on high fidelity simulations of flows at supercritical pressures.

11.3.2 Performance

The RAPTOR code framework has been systematically developed for over two decades with significant effort invested on verification, validation, and algorithmic tuning. A complete rewrite of the code is not required or justified because the software has been consistently morphed over time using modern programming approaches with emphasis on quality and correctness. The main challenge is to refactor the existing data structures to expose hybrid shared memory parallelism using either OpenMP or OpenACC directives to achieve scalability on the new target architectures. Earlier experience [50] has provided proven confidence that the RAPTOR framework can be ported to target architectures using this approach. The baseline data structures are systematically modified so that they perform optimally on heterogeneous nodes. Extensive changes are not required, but the process is labor intensive in terms of applying uniform changes across the entire code framework due to its size. Hybrid parallelism is targeted to be scalable on $\mathcal{O}(100)$ threads in the long term. This requires that each of the parallel OpenMP regions have enough computation to offset the fixed cost of spawning an OpenMP region. This becomes even more important when a hybrid parallel region has to be launched on an accelerator. Thus, data structures are reconstructed with this in mind to obtain much larger granularity than is typically available at the loop levels.

While pursuing the strategy outlined above, a foundational focal point is to devote a significant level of effort toward ensuring that RAPTOR continues to remain scalable across present and future leadership class systems. The underlying dual-time stepping integration scheme uses an explicit inner-loop multistage scheme that only requires nearest neighbor point-to-point communication. Global communication is only required for synchronization and monitoring and has not been a major barrier to scalability. As more on-node parallelism is exposed and acceleration is achieved, off-node or scalable message passing parallelism will become more and more important. As the compute performance is accelerated, the extent of time available to hide communication latencies becomes lower. This can expose previously hidden communication latencies that have a negative impact on parallel scalability. To circumvent this, a task mapping strategy is applied that facilitates optimal task placement both on-node and off-node [51]. Using the Gampi task placement library, developed by Sankaran et al. in 2015 [51], allows one to reduce the communication time needed for message passing, improve scalability, and reduce performance variability due to network congestion. This approach has been extended in RAPTOR by adopting Gampi to its unstructured multiblock communication topology. This provides control of on-node and off-node MPI task placement.

11.3.3 PORTABILITY

Throughout its development, the RAPTOR code framework has been ported to a wide variety of computer architectures. Ease in porting the code to different architectures is attributed to its self-contained design and use of standard ANSI coding practices. In addition to a variety of commodity clusters, the code has been ported to all of the leading architectures available at DOE computing facilities, including (a) JAGUAR and TITAN at the OLCF, (b) MIRA at the Argonne Leadership Computing Facility, and (c) HOPPER and EDISON at the Lawrence Berkeley National Energy Research Scientific Computing Center. RAPTOR has been used to perform LES of combustion as part of several previous INCITE awards from 2007–2014, and it is also been used as the application code for a 2014–2015 ALCC award entitled *Development of High-Fidelity Multiphase Combustion Models for LES of Advanced Engine Systems*. Most recently, it has been selected as one of the application codes under the 2015–2018 Oak Ridge National Laboratory (ORNL) *Center for Accelerated Application Readiness* program in preparation for the early science campaign on the next generation SUMMIT platform. Results from these awards have led to several publications (e.g., most recently [11,12,14]). The code has also been used as a benchmark application for several studies related to massively parallel computing (e.g., [42,52–55]).

11.3.4 EXTERNAL LIBRARIES

The RAPTOR code framework is a completely self-contained software package. It does not use or depend on the application of third-party libraries. The input/output (I/O) modules use the HDF5 data model/library as a default option, but are not dependent on this and easily converted, as necessary. It is important to note, however, that RAPTOR has the capability to be interfaced with third-party libraries as deemed prudent.

11.4 SOFTWARE PRACTICES

Codes such as RAPTOR take decades in time and millions of dollars to develop from scratch due to complexity in the numerics, models, and strongly coupled nature of the governing physical system. As such, development of these codes involves many delicacies. An optimal strategy is to continually refactor the software in a manner that adopts the latest computational architectures and parallel programming models, including version control and testing. At the same time, rigorous verification studies are performed to insure modifications are systematically checked and compatible (e.g., bug free) when combined with the master version of the code. From this perspective, the coupled theoretical–numerical framework (i.e., governing equations, physical subfilter models, numerics, and parallel model) has been extensively developed over more than two decades by the code author (Oefelein), with carefully monitored and integrated contributions from various staff, postdoctoral researchers, and students. A positive consequence of this approach is that the code is continually well positioned to achieve aggressive technical goals related to combustion science and the development of predictive models.

As of this writing, distributed version control managed using Git. The setup involves a hierarchical model using a central "`master`" repository coupled to a set of "`developer`" repositories.

Respective `developer` branches are established as a function of projects, teaming, and the distribution of the code. Between the `master` and `developer` repositories are a set of supporting branches to aid simultaneous development of different kernels between team members, ease tracking of features, and stage new releases for final testing before they are merged into the `master` repository. Supporting branches include "`feature`," "`release`," and "`hotfix`" branches, as needed during various stages of the verification and validation process.

Given the complex and delicate nature of CFD codes like RAPTOR, the code author (Oefelein) manages the `master` repository to insure modifications interface cleanly and accurately with the master version of the code without injecting bugs. Developers have the ability to push and pull changes from other `developer` repositories, and other branches have strict rules as to which may be an originating branch. For example, `feature` branches are only allowed to exist between developer repositories and have no linkage to the `master` repository. These are used to develop and test new and/or experimental features for distant future releases and exist only as long as the feature is in development. They are eventually merged back into the `developer` branches. Some might also be simply discarded if the idea fails. Similarly, `release` branches, only exist between `developer` branches and the `master`. These are designed to support the preparation of a new production release. Developers have the ability to push/pull code to/from `release` branches; however, they are only able to pull releases from the `master`. Conceptually, `Hotfix` branches can also be made available to provide intermediate modifications between releases when it becomes necessary to distribute modified or corrected versions of the master code prior to the next release. Developers are able to access these modifications immediately by pulling them from the `hotfix` repository.

11.5 BENCHMARKING RESULTS

Operating regimes of many present and future propulsion systems are characterized by high pressures and low temperatures suited for optimal fuel injection and combustion. The efficiency of these systems is heavily dependent on mixing of fuel at elevated pressures and has been the focus of research for many years. With the advent of novel investigative methods and increase in computational resources, significant progress has been achieved in understanding multiphase flows for rocket propulsion, gas turbines, and internal combustion engines. Most of these past studies, however, dealt with low pressure conditions. Evidence from experimental investigations suggests that the process of fuel injection is markedly different in systems at high pressures. Unlike the low pressure injection where the fuel undergoes primary and secondary atomization leading to droplet formation and fuel evaporation, high pressure fuel injection is characterized by diminished surface tension and dense fluid mixing [30,56]. As the pressure in the system exceeds the thermodynamic critical value, mixing and the combustion become diffusion dominated with reduced interfacial structure. Hence, it is vital to understand turbulent mixing and the supercritical combustion regime to develop efficient propulsive devices. Although some studies have been performed [14] to analyze the supercritical mixing and combustion processes, the turbulent mixing of dense fuel jet with ambient gases is not yet well understood. Here, this is addressed by simulating a fuel jet-in-cross-flow (JICF) where the fuel (n-Decane) is injected into a high pressure turbulent cross flow and the properties of the flow are evaluated considering the nonideal nature of the fluids at elevated pressures.

Investigation of JICF configurations have multiple uses due to the prevalence in many applications such as plume and pollutant dispersion, fuel injection, and cooling in gas turbine combustors. Extensive knowledge of the flow physics is essential to optimize mixing in all these applications and many past investigations have helped to develop key insights into the processes involved in the evolution of a jet in a cross flow [15,57]. Although developments, so far, have been valuable, the knowledge of the interaction of the several instabilities and structures in the flow is lacking and needs further investigation. Developing a complete understanding of relevant scales in the flow provides valuable information and enables optimal ways to resolve key features. This can be achieved only in a limited way using experiments because available techniques, like particle image velocimetry (PIV) [58], are not capable of the characterizing the small scales of turbulence. DNSs are also limited and, in most cases, prohibitively expensive due to the required resolution, domain dimensions, and long integration times involved. LESs show promise in this aspect and can be successfully employed to investigate complex flow physics of the JICF [15]. However, here again, the knowledge of resolution requirements is still lacking, especially for cases with supercritical flow. Additionally, supercritical flow adds the complexity of the real-fluid properties. Here, simulations are performed that provide insights into the resolution requirements for a JICF in high pressure flows.

Achieving a detailed understanding of turbulent mixing in a JICF is canonically relevant to many practical devices. The objective is to control and optimize transport and mixing, which depends on having a detailed understanding of the turbulence characteristics of the flow. For this reason, JICF configurations have been extensively studied experimentally, theoretically, and numerically. Comprehensive reviews can be found in papers by Margason et al. [59], Karagozian [60], and Mahesh [57]. Details of turbulence and mixing have been investigated over a wide range of Reynolds numbers, momentum-flux ratios, density ratios, and Mach numbers; LES has been applied by a number of researchers (like for predicting film-cooling in gas turbines [61–66]). These studies have improved our understanding of turbulent mixing processes in a JICF.

In a recent study [15], we have applied the theoretical–numerical framework, described in Section 11.2 to characterize the nature of the turbulence dynamics and scales in the canonical JICF studied experimentally by Su and Mungal [67]. The major focal points were first to validate the LES calculations over a range of different spatial and temporal resolutions, then to perform a comprehensive characterization of the turbulence field, the related turbulence scales and scalar-mixing processes, and their evolution. Results were compared with experimental measurements of mean velocity, mean concentration, the Reynolds stress tensor, and the turbulent scalar flux to quantify the accuracy of the calculations and establish a good understanding of how the system of subfilter models used in the calculations performed at different resolutions. Then, after establishing the *optimal* resolution, data from the LES were used to study the turbulence characteristics. Four quantities were mapped within the computational domain: (1) the integral scales, (2) the Taylor microscales, (3) the turbulent energy spectra, and (4) the scalar gradient thicknesses. For the dynamic quantities (i.e., integral scales, Taylor microscales, and turbulent energy spectra), a similar spatial evolution was observed. Scales were observed to first grow at various rates before breakdown of the jet turbulence, then a one-third power law was observed along the jet trajectory and consistently within the wake. For the scalar fluctuations and gradients, the trend was different. A sharp linear evolution of the gradient thickness was observed before the complete destabilization of the jet followed by a secondary, almost constant, evolution further downstream. As one example, this demonstrated how overly aggressive stretching of the mesh from the injection point to the far-field can significantly alter far-field predictions by filtering out key spatially evolving turbulent flow structures in the wake of the jet. Details related to the approach and insights gained from these calculations are given by Ruiz, Lacaze, and Oefelein [15].

To facilitate the objectives related to code performance and scalability, we have extend the approach taken in Reference [15] by now considering a liquid fuel jet (n-decane) in a hot cross flow of air at supercritical pressure. This provides a platform that emulates liquid injection processes at high pressure conditions typically encountered in advanced propulsion and power systems (e.g., gas turbines, and liquid rockets), starting from a good baseline understanding related to the physics and its representation using LES. Most importantly, it exercises the relevant subfilter models (starting with the kernels that calculate high pressure real-fluid thermodynamics and transport of multicomponent gas–liquid mixtures) over the precise set of conditions that the GPU equipped computer nodes must scale over. Performing the simulations at high pressure adds significant computational complexity since real-fluid gas-liquid phenomena departs substantially from the ideal gas equation of state (see

Refs. [26–30]). Evaluation of the thermophysical and transport properties at these pressures is thus performed via a major computational kernel that is ideal for acceleration on GPUs.

The computational domain with key dimensions is shown in Figure 11.2. Liquid n-decane is injected as a model fuel for Jet A in a direction perpendicular to an ambient cross flow of air. The ambient air pressure is 40 bar. The streamwise cross flow is parallel to the cartesian x direction, the jet axis is aligned with the cartesian y direction, and the spanwise direction is in the z direction. The jet diameter is assumed to be $d_j = 1$ mm, and respective coordinates are nondimensionalized using this as the reference length scale. The computational domain is $32d_j$ in the streamwise direction $24d_j$ in the wall normal direction, and $16d_j$ in the spanwise direction. The fuel jet is centered at a location of $8d_j$ from the cross flow inlet in the streamwise direction, and is centered in the spanwise direction. The bulk injection velocity of the fuel jet is $\bar{V}_j = 40$ m/s, and the of the cross-stream is $\bar{U}_a = 80$ m/s. The corresponding inflow temperatures are 450 K and 1000 K, respectively. This produces a momentum flux ratio of approximately $J = 9.9$, a Reynolds number based on the fuel jet diameter of $Re_f = 125,000$, and a Reynolds number based of the cross flow channel half height of $Re_a = 277,000$. These and other relevant operating conditions are summarized in Table 11.1.

The influence of the inlet boundary conditions on the penetration of a jet in a cross flow has been observed in the literature [15,67,68]. Thus, the time-evolving turbulent boundary layer dynamics are imposed at both the fuel and channel inlets using the synthetic eddy method developed by Jarrin et al. [69,70]. This method uses the integral scale distribution, Reynolds stress tensor, and mean velocity profiles calculated from the detailed treatment of the geometry to reconstruct the correct time evolving turbulent boundary layer dynamics. Fully developed turbulence with smooth wall boundary layers is assumed. No-slip boundary conditions are imposed along the channel walls. The domain was discretized using an equi-spaced grid at four resolutions. The sizes of the four grids generated, named as G1, G2, G3, and G4, are shown in Table 11.2. The grids have progressively higher resolutions such

TABLE 11.1

Flow Conditions for Jet-in-Cross-Flow Case

Ambient Pressure: 40 bar	
n-Decane Jet:	
T	450 K
$(Y_{C_{10}H_{22}}, Y_{O_2}, Y_{N_2})$	(1.0, 0.0, 0.0)
ρ	590 kg/m^3 (PR EOS)
μ	1.88×10^{-4} Pa · s
Speed of sound (c)	618 m/s
$\bar{V}_a$	40 m/s (1/7th profile)
Reynolds number (Re_f)	125,000
Mach number (M)	0.0647
Air Cross Flow:	
T	1000 K
$(Y_{C_{10}H_{22}}, Y_{O_2}, Y_{N_2})$	(0.00, 0.77, 0.23)
ρ	14.9 kg/m^3 (PR EOS)
μ	5.15×10^{-5} Pa · s
Speed of sound (c)	599 m/s
$\bar{U}_a$	80 m/s (1/7th profile)
Reynolds number (Re_a)	277,000
Mach number (M)	0.134

TABLE 11.2

Grid Sizes

Grid	Cells Across d_j	Total Size
G1	8	6.29×10^6
G2	16	50.33×10^6
G3	32	402.65×10^6
G4	64	3221.23×10^6

TABLE 11.3

Grid Topologies and Respective Block Sizes

Topology	Number of Blocks	Number of Cells per Block			
		G1	G2	G3	G4
T1	24	64^3	128^3	256^3	512^3
T2	192	32^3	64^3	128^3	256^3
T3	1,536	16^3	32^3	64^3	128^3
T4	12,288	8^3	16^3	32^3	64^3
T5	98,304		8^3	16^3	32^3

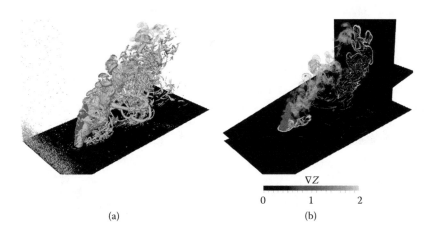

(a)   (b)

FIGURE 11.4 (See color insert.) Instantaneous three-dimensional visualizations of the baseline flow field using the G3 mesh: (a) the blue iso-contour is $Q = 5$, the red iso-contour marks where the mixture fraction is $Z = 0.88$ (the compressed liquid core) and the yellow iso-contour marks where $Z = 0.5$ (gaseous jet), (b) cutting planes that highlight the scalar dissipation rate with red and yellow iso-contours that are the same as (a).

that there were 8, 16, 32, and 64 cells across the jet diameter. A multiblock domain decomposition was performed to achieve distributed memory parallelism using MPI. Table 11.3 shows the five sets of multiblock domain topologies that are produced and the corresponding block sizes for the grids listed in Table 11.3. The topologies have been generated to span a wide range of block sizes starting from 8^3 to 512^3.

Figure 11.4 shows two instantaneous three-dimensional visualizations of the flow field using the G3 mesh with 32 cells across the jet nozzle exit. On the left (Figure 11.4a), the blue iso-contour represents the second invariant of the velocity gradient tensor, Q, which is also known as the

Q-criterion. The expression for Q in dimensionless form is

$$Q = -\frac{1}{2}\left(\left(\frac{\partial u}{\partial x}\right)^2 + \left(\frac{\partial v}{\partial y}\right)^2 + \left(\frac{\partial w}{\partial z}\right)^2 + 2\frac{\partial u}{\partial y}\frac{\partial v}{\partial x} + 2\frac{\partial u}{\partial z}\frac{\partial w}{\partial x} + 2\frac{\partial v}{\partial z}\frac{\partial w}{\partial y}\right)/(\bar{V}_j/d_j) \qquad (11.13)$$

The value of $Q = 5$ shown here enables one to identify the coherent turbulence structures in the flow field. Also shown are two iso-contours in mixture fraction. The red iso-contour marks a mixture fraction of $Z = 0.88$, which illuminates the structure of the compressed liquid core. The yellow iso-contour marks where $Z = 0.5$ (gaseous jet). Figure 11.4b) shows two cutting planes that illuminate the structure of the scalar dissipation rate along with the same red and yellow iso-contours shown in Figure 11.4a). The scalar dissipation fields show qualitatively how the turbulent structures captured with a given mesh drive the mixing of fuel and air. The two cuts shown show how the turbulence dynamics induces a combination of broadband and coherent vortex shedding in the wake of the jet. This can have a significant impact on the local fuel distribution and its spatial evolution as the flow evolves downstream.

The impact of resolution on the flow structures described above is shown in Figure 11.5. Here, the G3 mesh is coarsened by a factor of two in each direction, at two successive levels (G2 and G1). The images on top show Q-criterion values of $Q = 5$ with the blue iso-contours (see Equation 11.13), and mixture fraction values of $Z = 0.88$ and 0.5 with the red and yellow iso-contours, respectively. Images on the bottom show cutting planes colored by scalar dissipation rate along with the red and yellow iso-contours shown on the top figures. The resolved velocity gradients depend on grid spacing so that a comparison of a particular Q iso-value for different resolutions is only qualitative. In addition, because of the chaotic nature of a turbulent flow, any comparison of instantaneous features is merely

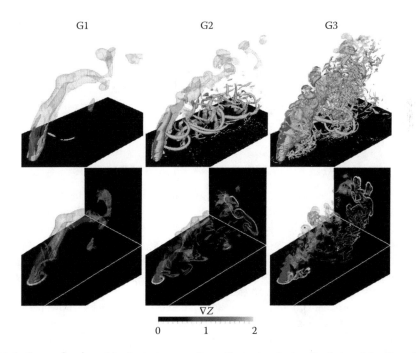

FIGURE 11.5 (See color insert.) Instantaneous three-dimensional visualizations of the flow field using meshes G1, G2, and G3. Images on top show Q-criterion values of $Q = 5$ with the blue iso-contours, and mixture fraction values of $Z = 0.88$ and 0.5 with the red and yellow iso-contours, respectively. Images on the bottom show cutting planes colored by scalar dissipation rate along with the red and yellow iso-contours shown on the top figures.

qualitative. Given these observations, using $Q = 5$ for all three resolutions leads to the following observations. The small horseshoe vortex upstream of the jet has disappeared in both the G2 and G1 meshes. No vortex rings are observed along the jet edges. For the G2 mesh, the presence of a $Q = 5$ iso-surface on the edges of the jet tends to indicate a region of formation of coherent structure, but with this particular threshold of Q, no clear vortex ring is observed. Although smaller coherent structures appear to be developing near the tip of the fuel in the G2 calculation, the spectrum of resolved turbulent structures is clearly diminished in comparison to the G3 calculation. Vortex shedding in the wake of the jet is observed in the G2 calculations. This is not the case for the G1 calculation. The development of wall turbulence appears to be very limited in the G2 calculation and wall turbulence is absent from the G1 calculation. The impact of coarsening the mesh on the resolved wrinkling of the mixture fraction field is also clearly observed on the three-dimensional mixture fraction iso-surfaces. In the G1 calculation, the $Z = 0.88$ (red) iso-surface is very smooth whereas some wrinkling is observed in the G2 calculation, which is qualitatively similar to the highly wrinkled iso-surface in the reference G3 calculation. The same observation can be made on the $Z = 0.5$ (yellow) isosurface. One of the main challenges in LES is, thus, to determine when the predictions for turbulent mixing start to fail as the mesh is coarsened.

Images shown on the bottom of Figure 11.5 also show qualitatively how the turbulent structures, captured with a given mesh, drive the mixing of fuel and air. Here, the blue iso-contour of Q is replaced by a representation of the scalar dissipation rate. In a DNS, the scalar dissipation field represents the intensity of molecular mixing. In LES, it shows the exchange surface between fuel and air, where both molecular viscosity and the related subfilter models mix species and energy together. These visualizations show that convective transport of fuel actually occurs in the wake of the jet, even in the G2 and G1 calculations. This was not observable using solely the $Q = 5$ iso-contour. However, in the G1 calculation, the transport of fuel appears to be linked to a deterministic vortex shedding mechanism, as in a low-Reynolds-number flow that is slightly above the laminar-to-turbulent transition. In the G2 calculation, a more chaotic vortex shedding is observed, which clearly indicates that the effective turbulent Reynolds number of the flow has increased. Even though vortex shedding can also be identified in the G3 calculation, its chaotic nature makes it hard to distinguish. This chaotic nature of the flow in the G3 calculation results from nonlinear interactions of the coherent structures resolved on the fine mesh, where the filtering induced by coarser resolution induces relatively more coalescence of coherent structures. This coalescence eliminates some of the nonlinear coupling associated with the true broadband turbulence field.

11.6 HYBRID SHARED MEMORY PARALLELISM IN RAPTOR

The goal in the previous section was to highlight the dominant physics in the JICF configuration and demonstrate the effects of resolution on its representation using LES. Important coherent structures include vortex rings along the jet edges, which eventually breakdown into turbulence; vortex shedding in the wake of the jet, which creates alleys of long vortices; and wall turbulence and the related hairpin vortices, which results from the large shear in the vicinity of the wall. Detailed visualizations enabled the observation of these coherent structures how they are coupled and how they drive turbulent mixing of fuel and air. In past work [15], extensive validation was performed using precisely the same approach. Quantities such as the mean penetration were shown to agree well with available experimental results, and the reader is referred to this publication for a detailed discussion on the quantitative nature of these types of calculations. Having demonstrated the the relevance of these calculations to "real-world" problems, in this section, the matrix of test cases are now used to establish a good baseline for improving the parallel performance.

The primary system used to test the performance of RAPTOR has most recently been the Titan supercomputer at the OLCF. Titan is a hybrid architecture Cray XK7 system. The system contains 18,688 compute nodes (*nodes* for short) that are interconnected by Cray's Gemini network. The

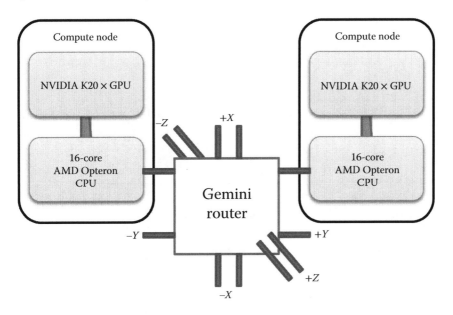

FIGURE 11.6   Titan's heterogeneous compute node architecture.

node architecture of the system is shown in Figure 11.6. Each node is composed of a 16-core AMD Opteron processor and NVIDIA Tesla K20 × GPU as an accelerator. Also, each node has 32 GB of memory on the host Opeteron processor and 6 GB of memory on the GPU accelerator.

As demonstrated in Section 11.3.1, the distributed memory model using MPI in RAPTOR has been highly optimized over time and consistently scales well beyond 100,000 cores. Thus, the major goal related to porting the code is focused on optimization of the shared-memory model. To demonstrate net performance gains, a series of strong and weak scaling studies are performed using the grids and topologies listed in Tables 11.2 and 11.3. Studies are focused on block sizes ranging from 16^3 to 64^3. Block sizes 8^3 and smaller are not investigated since they are too fine grained for the scale of simulations considered here. Similarly, block sizes larger than 64^3 are not practical due the memory and resource constraints on the node.

The time taken to advance the solution over a known number of time steps is measured in the performance tests. The computational performance is quantified using a metric similar to Equation 11.12; that is, the time per grid cell per iteration. However, in the case of multicore systems, it is customary to measure the computational time in terms of core-hours. Such a definition is not suitable for a heterogeneous node architecture with GPU accelerators that contain both a multicore processor and also a shared GPU accelerator. Therefore, here, we replace the number of cores in Equation 11.12 with the number of nodes.

Figure 11.7 shows the initial baseline performance of the code using MPI only while scaling from 12 to 768 nodes on the Titan system. The performance for a given block size with increasing number of nodes shows little degradation and exhibits near-ideal weak scaling behavior. We can also study the strong scaling performance of the code by comparing the simulation cost for multiple block sizes for the same node count. It is seen that the cost is lowest for the largest block size, 64^3, and then progressive increases for smaller block sizes. In a solely distributed memory parallel code, a large number of blocks are required to engage the large number of processor cores available in the computer system. The current effort to hybridize the software to a mixed distributed and shared memory parallel model would allow larger block sizes per task and hence lead to a lower cost.

After measuring the baseline performance and scalability, a deeper profiling study was conducted to analyze and identify the percentage of time spent in individual kernels. Cray performance

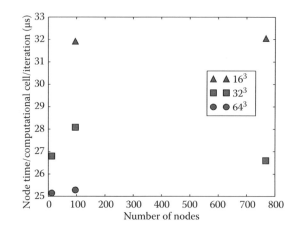

FIGURE 11.7   Scalability of RAPTOR prior to acceleration. The block size per MPI task for each case is indicated.

measurement and analysis tools are used to instrument the code for tracing. This tracking operation provides the time spent in the various routines along with the hardware performance counters that show the volume of computation being performed in these routines. The results are shown as percentages in Figure 11.8. Figure 11.8a shows the fraction of time spent in the different kernels as a fraction of the total computational time. The time for computation of the thermodynamics and transport closures is the largest comprising of 51.2% of the total time followed by the time for turbulent closures at 33.8% of the total time. The computation of time derivatives requires 5.1% of total time with rest of the routines occupying 9.9% of the total time. This analysis clearly demonstrates the dominant cost associated with the calculation of the real-fluid properties. The analysis of the floating point operations (see Figure 11.8b) performed within in each kernel further demonstrates the dominance of the thermodynamic, transport, and turbulence closure routines. To reduce the total computational time through hybridization, routines for the thermodynamic, transport, and turbulence closures are targeted for porting to the accelerators.

Software written for conventional processor architectures, except in rare occasions, fails to obtain good performance on the modern processor architectures. GPUs and manycore processors require particular attention and assessment of the critical aspects of the software such as data locality, layout, and how well the parallelism is exposed. At times, a complete rewrite of the software in a newer programming model is the only viable path forward. In such cases, application developers can choose to program in an architecture specific programming model, such as the CUDA for nVidia GPUs and Pthreads for Intel chips. However, choosing a hardware-specific programming model limits the portability of the software and requires multiple vendor specific implementations of the software to be maintained. Programming models, like OpenCL, that promise to be portable across multiple vendors and architectures, have not been adopted widely and also suffer from lack of performance portability. In contrast to programming models that require a rewrite of the software, directive-based approaches, such as OpenACC and OpenMP, can be adopted. Directives are used to expose the underlying data structures and levels of parallelism, thereby allowing a compiler to generate efficient code for the target architecture. Levesque et al. [50] converted a legacy fortran DNS code to a hybrid distributed and shared memory parallel code and then used the OpenACC directives to offload almost all of the computations to a GPU accelerator. They found that efficient hybridization of the code for multicore processors and accelerators requires refactoring of the data structures and rearrangement of loops to achieve higher granularity in the loop regions. The experiences of Levesque et al. and others [71] have shown that, while the directives approach emphasizes reuse of legacy code, a considerable

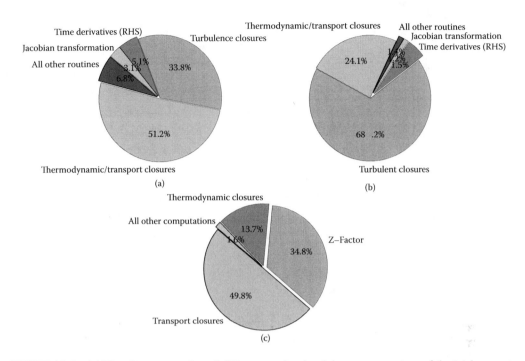

FIGURE 11.8 (a) Time for computation of different routines/modules as a percentage of the total computational time. (b) Floating point operations as a percent of total flops for all routines. (c) Floating point operations for different thermodynamics and transport routines.

refactoring of the original code, in some cases is, necessary to get the right data layout and loop granularity.

Since the thermophysical properties kernel is shown to consume a significant fraction of the computational time and can also be treated as an independent software module, we choose to rewrite the kernel in a form suitable for the accelerated architectures. Instead of an accelerator-specific programming model, we prefer a performance portable programming model for rewriting the kernel. The programming model provided by the Kokkos C++ library [43] is selected for abstracting the data location, layout, and parallel execution. Kernels programmed in the Kokkos programming model provide an additional layer of parallelism separate from the distributed memory parallelism. This provides an additional on-node parallelism through threads and vectorization. The computational kernel is written to express the work at the finest granularity; in this case, the calculations that are needed to evaluate the thermophysical properties in one computational cell. A parallel work dispatch pattern is used to generate parallel code over all the computational cells that belong to a single shared memory node. The parallel work dispatch method used by Kokkos is not very different from the corresponding approaches used by other C++ based programming libraries such as the standard template library (STL), thread building blocks (TBB), or thrust. However, Kokkos also provides an abstraction for the data location and layout through its memory spaces and arrays, which allows the computations to be expressed using multidimensional arrays that are similar to the fortran array syntax.

We build on the multidimensional array class provided by Kokkos to form the data structures needed for our code development. We define a class, named Yarn, consisting of the datatypes required to store scalar, vector, and tensor quantities across all the computational cells. We also define utilities to convert back and forth with the traditional fortran layout arrays to allow systematic porting of the code while ensuring interoperability with the initial baseline version of the code. Below, we illustrate a simple kernel that converts species mass fractions to mole fractions and then show its parallel launch.

```
struct convertYtoX {
  VectorFieldType ys;
  VectorFieldType xs;
  ScalarFieldType ws_bar;

  // constructor
  convertYtoX(
      VectorFieldType ys_, VectorFieldType xs_, ScalarFieldType ws_bar_)
      : ys(ys_), xs(xs_), ws_bar(ws_bar_) {};

  KOKKOS_INLINE_FUNCTION
  void operator()(const size_type i) const {
    for(int m=0; m<ns; m++){
      xs(i,m)=ys(i,m)/ws(m)/ws_bar(i);
    }
  }
};

Kokkos::parallel_for(nCells, convertYtoX(ys, xs, ws_bar));
```

The sample kernel, called convertYtoX, is initialized with the data structures that are in the Kokkos multidimensional array formats. The kernel operator performs the operation on the data for the smallest unit of work, which, in this case, is the conversion from mass fraction to mole fraction for one cell only. While the kernel has been programmed in the familiar fortran array style indexing notation, the actual layout of data in memory is abstracted and can be controlled at compile time to target multicore processors or GPUs. The kernel is then launched in parallel using parallel_for over all the computational cells in the domain with a simple integer argument that gives the number of cells. The parallel launch can also be controlled further using more sophisticated range arguments that execute the kernel over a subset of cells or exclude a subset.

The example above illustrates the approach used for developing the thermophysics kernels using the Kokkos programming model. The complete software library is around 1000 lines long and no larger in complexity than the fortran implementation of the same kernel. Expressing the data structures and parallel launch using the Kokkos syntax allow us to utilize the C++ language features for targeting multiple hardware and storage architectures at compile time and generate optimized code. For example, the array layout is changed from row-major to column-major depending on whether the code is being compiled for multicore processors or GPUs without requiring any changes to the kernel implementations. It also allows the parallel kernel launch to be portable across multithreaded and vector architectures.

The performance of the accelerated code is shown in Figure 11.9. In comparison to the original code, the computational time for all grid sizes and nodes used is reduced by three times. The thermodynamics routines, which accounted for 51.2% of the total time in the original code, now occupy only 11.7% in the accelerated version whereas the time for computing the turbulent closures has increased from 33.8% of the total time to 44.3% of the total time. In absolute wall clock seconds, the time taken by the the thermodynamics, transport, and turbulence closures were all reduced. However, the GPU acceleration was much more effective on the thermodynamics and transport kernels, in comparison to the turbulence closures. The thermodynamic and transport properties kernel was accelerated by a factor of 12.8×, while the turbulence closure kernel was accelerated by a factor of 2.3× with respect to their performance on a central processing unit (CPU) core. The disparity in speedup factor between these two kernels has caused the turbulence closure kernel's profile as a percentage of time to increase with respect to the original code. There are some important differences between these two kernels that could have caused the disparity in speedup between these two kernels. For instance, while the thermo and transport kernel compute pointwise properties, the dynamic subfilter-scale turbulence

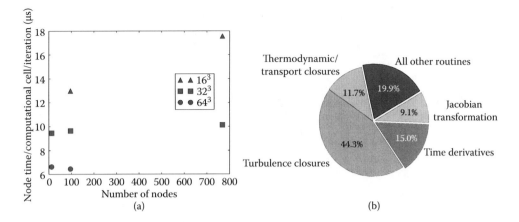

FIGURE 11.9 (a) Performance of RAPTOR with GPU acceleration. The block size per MPI task for each case is indicated. (b) Time for computation of different routines/modules as a percentage of the total computational time after acceleration.

closure model performs spatial filtering. The pointwise kernel leads to well-coalesced memory access on the GPU, while spatial filtering operator has strided accesses, considering the three spatial dimensions involved in filtering. This suggests that there is room for improvement in the turbulence closure kernel through improvement of the data layout and optimal coalesced memory accesses. Overall, the acceleration of these two kernels gives an application level speedup of 3× with respect to the multicore performance. This clearly demonstrates the effectiveness of code acceleration in reducing the computational cost associated with the thermodynamics and transport routines. The results from the accelerated code provide confidence that the methods used here are applicable to investigate large-scale numerical simulations with considerable reduction in computational cost.

11.7 OTHER CONSIDERATIONS

The dramatic increase in the amount of data generated in future simulations will necessitate a relative decrease in the amount of I/O possible. Currently, the size of respective datasets from DNS and LES is doubling every year, which is not sustainable. This will necessitate a large movement toward *in situ* data analysis and visualization at runtime. Thus, the role of scientific data management middleware will be key in providing the software infrastructure that exploits current petascale and future exascale architectures ability to overlap data analysis with I/O. The computational power of processors will be used to reduce the need for data output by building I/O systems that enhance massively parallel methods for data movement with extensive processing resources to manipulate and reduce data before it is moved. Since I/O will become a very constrained resource, it is expected that the adoption of I/O middleware libraries will significantly accelerate over the next 5 years, with the majority of high-performance computing (HPC) applications using some middleware system after this point. This is especially true for combustion science applications. Scientific workflow tools are also likely to increase in prominence, especially for scientific analyses that involve parameter sweeps or manage ensemble runs. Workflows that extend beyond the individual computer system are also likely to see greater adoption, including management of archival storage, off-site data transfer, exploitation of multiple computing systems, and distribution to web-based portals.

In addition to optimization of the hybrid parallel model in terms of FLOPS, treatment of I/O must also be integrated efficiently with the selected parallel model. In this regard, RAPTOR has a built in, self-contained, parallel I/O capability for checkpoint, restart, and output of case-specific analysis data. The I/O portion has been optimized over time to improve its performance and generality. Developmental research performed as part of previous INCITE grants (e.g., Sankaran and Oefelein [55]) have now been completely generalized so that the I/O routines in RAPTOR achieve good scalability on over 100,000 processors while also improving the productivity of analyzing data from large and complex geometries. Using HDF5, a new I/O layer has been developed to decompose the MPI domain into subcommunicators using a spokesperson model for parallel I/O. In this model, one MPI rank from every subgroup aggregates the data from respective groups and writes to a file. The number of MPI subcommunicators defined determines both the number of files generated and the I/O throughput. This effectively decouples the number of data files from the number of MPI ranks and allows one to optimize the I/O performance as a function of the problem size, number of MPI cores, and the type of filesystem employed. Restart solutions are stored in double precision, while processed data and snapshots are generally stored in single precision to reduce the use of hard drive space. Note that storing data in double or single precision is a runtime option. Since RAPTOR solves the governing equations in dimensionless and normalized form, the error induced by single precision output is typically negligible at the postprocessing stage. The same output solution (restart or snapshots) can be used both in RAPTOR or in postprocessing software (e.g., Paraview, MATLAB®, etc.), which eliminates the need to convert output solutions and thus unnecessary data duplication. Current I/O costs are between 1% and 10% of the total run time. The cost, performance, and portability associated with I/O in RAPTOR will be given full consideration as part of this effort.

ACKNOWLEDGMENTS

This work was supported by the U.S. Department of Energy, Office of Basic Energy Sciences, Division of Chemical Sciences, Geosciences, and Biosciences. This research also used resources of the Oak Ridge Leadership Computing Facility at the Oak Ridge National Laboratory, which is supported by the Office of Science of the U.S. Department of Energy under contract DE-AC05-00OR22725. Sandia National Laboratories is a multiprogram laboratory operated by Sandia Corporation, a Lockheed Martin Company, for the U.S. Department of Energy under contract DE-AC04-94-AL85000.

REFERENCES

1. Oefelein, J. C. (2006) Large eddy simulation of turbulent combustion processes in propulsion and power systems. *Progress in Aerospace Sciences* 42, 2–37.
2. Oefelein, J. C., Schefer, R. W., and Barlow, R. W. (2006) Toward validation of LES for turbulent combustion. *American Institute of Aeronautics and Astronautics Journal* 44, 418–433.
3. Oefelein, J. C. (2006) Mixing and combustion of cryogenic oxygen-hydrogen shear-coaxial jet flames at supercritical pressure. *Combustion Science and Technology* 178, 229–252.
4. Oefelein, J. C., Sankaran, V., and Drozda, T. G. (2007) Large eddy simulation of swirling particle-laden flow in a model axisymmetric combustor. *Proceedings of the Combustion Institute* 31, 2291–2299, Distinguished Paper Award.
5. Williams, T. C., Schefer, R. W., Oefelein, J. C., and Shaddix, C. R. (2007) Idealized gas turbine combustor for performance research and validation of large eddy simulations. *Review of Scientific Instruments* 78, 035114–1–9.
6. Frank, J. H., Kaiser, S. A., and Oefelein, J. C. (2011) Analysis of scalar mixing dynamics in LES using high-resolution imaging of laser Rayleigh scattering in turbulent non-reacting jets and non-premixed jet flames. *Proceedings of the Combustion Institute* 33, 1373–1381.
7. Kempf, A. M., Geurts, B. J., and Oefelein, J. C. (2011) Error analysis of large eddy simulation of the turbulent non-premixed sydney bluff-body flame. *Combustion and Flame* 158, 2408–2419.
8. Hu, B., Musculus, M. P., and Oefelein, J. C. (2012) The influence of large-scale structures on entrainment in a decelerating transient turbulent jet revealed by large eddy simulation. *Physics of Fluids* 24, 1–17.

9. Lacaze, G. and Oefelein, J. C. (2012) A non-premixed combustion model based on flame structure analysis at supercritical pressures. *Combustion and Flame* 159, 2087–2103.

10. Oefelein, J. C., Dahms, R. N., and Lacaze, G. (2012) Detailed modeling and simulation of high-pressure fuel injection processes in diesel engines. *SAE International Journal of Engines* 5, 1–10.

11. Oefelein, J. C. (2013) Large eddy simulation of complex thermophysics in advanced propulsion and power systems. *Proceedings of the 8th U.S. National Combustion Meeting*, Park City, UT: Invited Plenary Presentation and Paper.

12. Oefelein, J., Lacaze, G., Dahms, R., Ruiz, A., and Misdariis, A. (2014) Effects of real-fluid thermodynamics on high-pressure fuel injection processes. *SAE International Journal of Engines* 7, 1–12.

13. Khalil, M., Lacaze, G., Oefelein, J. C., and Najm, H. N. (2015) Uncertainty quantification in LES of a turbulent bluff-body stabilized flame. *Proceedings of the Combustion Institute* 35, 1147–1156.

14. Lacaze, G., Misdariis, A., Ruiz, A., and Oefelein, J. C. (2015) Analysis of high-pressure diesel fuel injection processes using LES with real-fluid thermodynamics and transport. *Proceedings of the Combustion Institute* 35, 1603–1611.

15. Ruiz, A. M., Lacaze, G., and Oefelein, J. C. (2015) Flow topologies and turbulence scales in a jet-in-cross-flow. *Physics of Fluids* 27, 1–35.

16. Hakim, L., Lacaze, G., and Oefelein, J. C. (2016) Large eddy simulation of autoignition transients in a model diesel injector configuration. *SAE International Journal of Fuels and Lubricants* 9, 165–176.

17. Ruiz, A. M., Lacaze, G., Oefelein, J. C., Mari, R., Cuenot, B., Selle, L., and Poinsot, T. (2016) A numerical benchmark for high-reynolds number supercritical flows with large density gradients. *Institute of Aeronautics and Astronautics Journal* 54, 1445–1460.

18. Erlebacher, G., Hussaini, M. Y., Speziale, C. G., and Zang, T. A. (1992) Toward the large eddy simulation of compressible turbulent flows. *Journal of Fluid Mechanics* 238, 155–185.

19. Speziale, C. G. (1985) Galilean invariance of subgrid-scale stress models in the large eddy simulation of turbulence. *Journal of Fluid Mechanics* 156, 55–62.

20. Germano, M., Piomelli, U., Moin, P., and Cabot, W. H. (1991) A dynamic subgrid-scale eddy viscosity model. *Physics of Fluids* 3, 1760–1765.

21. Moin, P., Squires, K., Cabot, W., and Lee, S. (1991) A dynamic dubgrid-scale model for compressible turbulence and scalar transport. *Physics of Fluids* 3, 2746–2757.

22. Lilly, D. K. (1992) A proposed modification of the Germano subgrid-scale closure method. *Physics of Fluids* 3, 633–635.

23. Zang, Y., Street, R. L., and Koseff, J. R. (1993) A dynamic mixed subgrid-scale model and its application to turbulent recirculating flows. *Physics of Fluids* 5, 3186–3195.

24. Vreman, B., Geurts, B., and Kuerten, H. (1994) On the formulation of the dynamic mixed subgrid-scale model. *Physics of Fluids* 6, 4057–4059.

25. Smagorinsky, J. (1963) General circulation experiments with the primitive equations. I. The basic experiment. *Monthly Weather Review* 91, 99–164.

26. Dahms, R. N. and Oefelein, J. C. (2013) On the transition between two-phase and single-phase interface dynamics in multicomponent fluids at supercritical pressures. *Physics of Fluids* 25, 1–24.

27. Manin, J., Bardi, M., Pickett, L. M., Dahms, R. N., and Oefelein, J. C. (2014) Microscopic investigation of the atomization and mixing processes of diesel sprays injected into high pressure and temperature environments. *Fuel* 134, 531–543.

28. Dahms, R. N. and Oefelein, J. C. (2015) Non-equilibrium gas-liquid interface dynamics in high-pressure liquid injection systems. *Proceedings of the Combustion Institute* 35, 1587–1594.

29. Dahms, R. N. and Oefelein, J. C. (2015) Atomization and dense-fluid breakup regimes in liquid rocket engines. *Journal of Propulsion and Power* 31, 1221–1231.

30. Dahms, R. N. and Oefelein, J. C. (2015) Liquid jet breakup regimes at supercritical pressures. *Combustion and Flame* 162, 3648–3657.

31. Leland, T. W. and Chappelear, P. S. (1968) The corresponding states Principle. A review of current theory and practice. *Industrial and Engineering Chemistry Fundamentals* 60, 15–43.

32. Rowlinson, J. S. and Watson, I. D. (1969) The prediction of the thermodynamic properties of fluids and fluid mixtures–I. The principle of corresponding states and its extensions. *Chemical Engineering Science* 24, 1565–1574.

33. Reid, R. C., Prausnitz, J. M., and Polling, B. E. *The Properties of Liquids and Gases*, 4th ed., New York, NY: McGraw-Hill, 1987.

34. VanWylen, G. J. and Sonntag, R. E. *Fundamentals of Classical Thermodynamics*, 3rd ed., New York, NY: John Wiley & Sons, Incorporated, 1986.

35. Gordon, S. and McBride, B. J. *Computer Program for Calculation of Complex Chemical Equilibrium Compositions, Rocket Performance, Incident and Reflected Shocks and Chapman-Jouguet Detonations*, Washington, DC: National Aeronautics and Space Administration, 1971.

36. Kee, R. J., Rupley, F. M., and Miller, J. A. *Chemkin Thermodynamic Data Base*, Livermore, CA: Sandia National Laboratories, 1990; Supersedes SAND87-8215 dated April 1987.

37. Ely, J. F. and Hanley, H. J. M. (1981) Prediction of transport properties. 1. Viscosity of fluids and mixtures. *Industrial and Engineering Chemistry Fundamentals* 20, 323–332.

38. Ely, J. F. and Hanley, H. J. M. (1981) Prediction of transport properties. 2. Thermal conductivity of pure fluids and mixtures. *Industrial and Engineering Chemistry Fundamentals* 22, 90–97.

39. Bird, R. B., Stewart, W. E., and Lightfoot, E. N. *Transport Phenomena*, New York, NY: John Wiley & Sons, Incorporated, 1960.

40. Hirschfelder, J. O., Curtiss, C. F., and Bird, R. B. *Molecular Theory of Gases and Liquids*, 2nd ed., New York, NY: John Wiley & Sons, Incorporated, 1964.

41. Takahashi, S. (1974) Preparation of a generalized chart for the diffusion coefficients of gases at high pressures. *Journal of Chemical Engineering of Japan* 7, 417–420.

42. Kothe, D., Roche, K., Kendall, R., Adams, M., Ahern, S., Chang, C. -S., Childs, H., et al. FY 2009 Annual Report of Joule Software Metric SC GG 3.1/2.5.2, Improve Computational Science Capabilities, Rept. ORNL/TM-2009/322, Oak Ridge National Laboratory, December 2009.

43. Carter Edwards, H., Trott, C. R., and Sunderland, D. (2014) Kokkos: Enabling manycore performance portability through polymorphic memory access patterns. *Journal of Parallel and Distributed Computing* 74, 3202–3216.

44. Niemeyer, K. E. and Sung, C. -J. (2014) Recent progress and challenges in exploiting graphics processors in computational fluid dynamics. *Journal of Supercomputing* 67, 528–564.

45. Spafford, K., Meredith, J., Vetter, J., Chen, J., Grout, R., and Sankaran, R. In *Euro-Par 2009 Parallel Processing Workshops. Lecture Notes in Computer Science*, Lin, H. -X., Alexander, M., Forsell, M., Knüpfer, A., Prodan, R., Sousa, L., and Streit, A., Eds., Berlin/Heidelberg: Springer, 2010; pp 122–131.

46. Shi, Y., Green Jr., W. H., Wong, H. -W., and Oluwole, O. O. (2011) Redesigning combustion modeling algorithms for the graphics processing unit (GPU): Chemical kinetic rate evaluation and ordinary differential equation integration. *Combustion and Flame* 158, 836–847.

47. Stone, C. P. and Davis, R. L. (2013) Techniques for solving stiff chemical kinetics on graphical processing units. *Journal of Propulsion and Power* 29, 764–773.

48. Niemeyer, K. E. and Sung, C. -J. (2014) Accelerating moderately stiff chemical kinetics in reactive-flow simulations using GPUs. *Journal of Computational Physics* 256, 854–871.

49. Sankaran, R. GPU-accelerated software library for unsteady flamelet modeling of turbulent combustion with complex chemical kinetics. *51st AIAA Aerospace Sciences Meeting including the New Horizons Forum and Aerospace Exposition*, Grapevine, Texas. 2013.

50. Levesque, J., Sankaran, R., and Grout, R. Hybridizing S3D into an exascale application using OpenACC: An approach for moving to multi-petaflops and beyond. *International Conference for High Performance Computing, Networking, Storage and Analysis (SC12)*, Salt Lake City, Utah. 2012; pp 1–11.

51. Sankaran, R., Angel, J., and Brown, W. M. (2015) Genetic algorithm based task reordering to improve the performance of batch scheduled massively parallel scientific applications. *Concurrency and Computation: Practice and Experience*, 27, 1532–0634.

52. Oefelein, J. C. and Sankaran, R. (2009) Large eddy simulation of turbulence-chemistry interactions in reacting flows: Experiences on the ORNL NCCS Cray-XT platforms (Jaguar). *Proceedings of the 21st International Conference on Parallel Computational Fluid Dynamics,* Moffett Field, CA.

53. Oefelein, J. C., Chen, J. H., and Sankaran, R. (2009) High-fidelity simulations for clean and efficient combustion of alternative fuels. *Journal of Physics* 180, 1–5.

54. Oefelein, J. C. and Sankaran, R. (2011) High-fidelity large eddy simulation of combustion for propulsion and power. *Proceedings of the SciDAC 2011 Meeting, Denver*, CO. http://press.mcs.anl.gov/scidac2011/.

55. Sankaran, R. and Oefelein, J. C. (2011) Efficient data management and analysis for high-fidelity combustion simulations on petascale supercomputers. *Proceedings of the 13th International Conference on Numerical Combustion*, Corfu, Greece.

56. Oschwald, M., Smith, J., Branam, R., Hussong, J., Schik, A., Chehroudi, B., and Talley, D. (2006) Injection of fluids into supercritical environments. *Combustion Science and Technology* 178, 49–100.

57. Mahesh, K. (2013) The interaction of jets with crossflow. *Annual Review of Fluid Mechanics* 45, 379–407.

58. Scarano, F. (2013) Tomographic PIV: Principles and practice. *Measurement Science and Technology* 24, 1–28.

59. Margason, R. J. Fifty years of jet in cross flow research. In *AGARD, Computational and Experimental Assessment of Jets in Cross Flow,* CP-534, Neuilly Sur Seine, France. 1993.

60. Karagozian, A. R. (2010) Transverse jets and their control. *Progress in Energy and Combustion Science* 36, 531–553.

61. Yuan, L. L. and Street, R. L. (1998) Trajectory and entrainment of a round jet in crossflow. *Physics of Fluids* 10, 2323.

62. Schlüter, J. and Schönfeld, T. (2000) LES of jets in cross flow and its application to a gas turbine burner. *Flow, Turbulence and Combustion* 65, 177–203.

63. Guo, X., Schröder, W., and Meinke, M. (2006) Large-eddy simulations of film cooling flows. *Computers and Fluids* 35, 587–606.

64. Renze, P., Schröder, W., and Meinke, M. (2008) Large-eddy simulation of film cooling flows at density gradients. *International Journal of Heat and Fluid Flow* 29, 18–34.

65. Rozati, A. and Tafti, D. K. (2008) Large-eddy simulations of leading edge film cooling: Analysis of flow structures, effectiveness, and heat transfer coefficient. *International Journal of Heat and Fluid Flow* 29, 1–17.

66. Mendez, S. and Nicoud, F. (2008) Large-eddy simulation of a bi-periodic turbulent flow with effusion. Journal of Fluid Mechanics 598, 27–66.

67. Su, L. and Mungal, M. (2004) Simultaneous measurements of scalar and velocity field evolution in turbulent crossflowing jets. *Journal of Fluid Mechanics* 513, 1–45.

68. Muppidi, S. and Mahesh, K. (2005) Study of trajectories of jets in crossflow using direct numerical simulations. *Journal of Fluid Mechanics* 530, 81–100.

69. Jarrin, N., Benhamadouche, S., Laurence, D., and Prosser, R. (2006) A synthetic-eddy-method for generating inflow conditions for large-eddy simulations. *International Journal of Heat and Fluid Flow* 27, 585–593.

70. Jarrin, N. Synthetic inflow boundary conditions for the numerical simulation of turbulence. Ph.D. thesis, The University of Manchester, Manchester, 2008.

71. Hernandez, O., Ding, W., Chapman, B., Kartsaklis, C., Sankaran, R., and Graham, R. In *Facing the Multicore—Challenge II. Lecture Notes in Computer Science,* Vol. 7174, Keller, R., Kramer, D., and Weiss, J. -P., Eds., Berlin/Heidelberg: Springer, 2012; pp 96–107.

12 S3D-Legion

An Exascale Software for Direct Numerical Simulation of Turbulent Combustion with Complex Multicomponent Chemistry

Sean Treichler, Michael Bauer, Ankit Bhagatwala, Giulio Borghesi, Ramanan Sankaran, Hemanth Kolla, Patrick S. McCormick, Elliott Slaughter, Wonchan Lee, Alex Aiken, and Jacqueline Chen

CONTENTS

12.1 Introduction..258
    12.1.1 Direct Numerical Simulation 258
    12.1.2 Programming for HPC and Exascale 259
    12.1.3 Contributions 259
12.2 S3D 260
    12.2.1 Fortran+MPI Implementation 261
    12.2.2 Hybrid OpenACC+MPI Implementation..261
    12.2.3 Legion Implementation 262
12.3 Legion 263
12.4 Singe 263
12.5 Simulation Setup..264
    12.5.1 RCCI Simulation 264
    12.5.2 Temporal Jet Simulation 265
12.6 Combustion Results 266
    12.6.1 RCCI Simulation Results...266
    12.6.2 Temporal Jet Simulation Results 270
12.7 Performance..271
    12.7.1 Performance Bottlenecks..272
    12.7.2 Weak Scaling 274
    12.7.3 Strong Scaling...274
12.8 Conclusions..275
Acknowledgments 276
References 276

12.1   INTRODUCTION

New methods are needed to explore novel fuels and design better engines that can substantially increase combustion efficiency, extend the longevity of finite fossil fuel reserves, and reduce carbon dioxide (CO_2) and other emissions. Government mandates to reduce petroleum use by 25% by 2020 and greenhouse gas emissions by 80% by 2050 are also exerting pressure on industry and will require significant retooling of all aspects of energy use in the United States. Achieving these aggressive goals requires the automotive industry to significantly shorten its design cycle. The transportation sector alone accounts for two-thirds of the nation's petroleum use and one-quarter of the nation's greenhouse gas emissions. Compounding these challenges, fuels are also evolving, adding another layer of complexity and further highlighting the need for rapid development cycles. We believe the optimal path to fast design cycles is through predictive modeling and simulation, enabled by recent advances in supercomputing.

In the past quarter century, the continual growth of high-performance computing (HPC) has had a major impact on the progress of science and engineering. It has accelerated the pace of research and development in many important fields including combustion science. For example, it has enabled new fundamental insights into turbulent combustion, a key subprocess that requires accurate modeling for the design of practical combustion devices. Turbulent combustion poses daunting modeling and computational challenges. The challenges arise due to (1) the large number of chemical species needed to describe a practical fuel (more than 100 species), (2) the large dynamic range of length and time scales (10^{-7}–10^{-2} m and 10^{-9}–10^{-2} seconds in an internal combustion engine), and (3) the coupling of highly nonlinear Arrhenius chemistry and large turbulent fluctuations of species concentrations with pressure and temperature.

12.1.1   Direct Numerical Simulation

To cope with these challenges, combustion scientists rely on a diverse set of simulation infrastructures and modeling techniques. The multiscale nature of combustion demands a hierarchy of computational fluid dynamics (CFD) methods ranging from fine-grain first principles direct numerical simulation (DNS) to coarse-grain large-eddy simulations (LES) and Reynolds averaged simulations (RANS). In particular, statistics from DNS simulations are used to bootstrap coarse-grain simulations and therefore DNS requires the most accuracy. In DNS, the instantaneous governing reactive Navier–Stokes, energy, and species continuity equations are solved without averaging or filtering; all relevant flow and combustion scales are resolved on the grid with accurate numerical methods.

Although DNS yields very high-fidelity results, it also makes simulations computationally expensive, especially because the Reynolds number scaling between the largest and smallest turbulent scales is proportional to $N^{9/4}$, where N is the number of grid points and practical devices operate at high Reynolds numbers for efficiency. A single simulation often takes longer than a month to complete, can consume millions of node hours on today's petascale supercomputers, and generates upward of a petabyte of raw data. To date, these costs have constrained scientists to perform simulations that are reduced in one or more ways:

- A reduction in the size of the physical volume being simulated decreases the number of grid points required, but limits the Reynolds number and the size of the turbulent structures that can be studied. It also limits the statistical sample size needed for model development and validation.
- Existing studies have focused on simpler fuel chemistries such as dimethyl ether (DME) [1]. Future fuels of interest for internal combustion engines are generally much more complicated. For example, the reduced primary reference fuel (PRF) mixture [2] we consider in this work requires over 4× more computation and storage resources than DME, even after reduction techniques have been applied [3].

- In studies that unavoidably require examining large structures with complicated chemistry, experimental diagnostics have been performed with reduced spatial dimensionality, using at most two-dimensional (2D) planar information in a probe volume [4,5]. Unfortunately, turbulence is intrinsically three-dimensional (3D), so these studies are unable to accurately capture the mixing effects of combustion in a turbulent environment.

12.1.2 Programming for HPC and Exascale

Simulation codes themselves are not steady-state either. A new DNS experiment often requires new functionality to be incorporated into an existing code while maintaining good performance characteristics. Within existing programming models, this process requires a good understanding of the underlying science, the intricacies of the existing source code, and the performance characteristics of the machine. Introducing new features mandates the optimization of both the new code as well any old code impacted by the change. Under such conditions, the programming effort and cognitive load placed on programmers explodes as the number of interacting features grows. The resulting complexity coupled with the time necessary to implement, test, and optimize code changes for new features can easily exceed the running time of the simulation itself.

The evolving architectures of modern supercomputers compound the problem of programmer productivity. As machines integrate heterogeneous processors such as graphics processing units (GPUs) and deep memory hierarchies, the programmer becomes responsible for deciding how to effectively *map* applications onto a target architecture. The process of mapping requires programmers to decide how computations are assigned to processors and how data are placed in caches and memories. Existing programming models force the programmer to implement these mapping decisions directly in the source code. This entangles the functionality of the code with its mapping in a way that inhibits either aspect of the code from being modified without a complete understanding of the other. Even worse, achieving performance portability across different architectures requires implementing two or more different mappings in the same source code, which is commonly done with a preponderance of conditional compilation directives and/or complete forks of the source tree.

12.1.3 Contributions

In this work, we report a dramatic improvement on the state of the art by presenting the first 3D DNS of the PRF mechanism that is able to resolve micrometer-scale turbulent flame structures. The simulation was run on two heterogeneous supercomputers, Piz Daint at CSCS [6] and Titan at ORNL [7], obtaining over 80% of the maximum achievable performance. Our code is based on a port of S3D [8] to the Legion programming model [9]. The port is significantly simplified by the ability of Legion to interoperate with existing message passing interface (MPI) applications—everything outside of the main simulation loop is left in its original MPI Fortran form. We leverage the novel mapping interface provided by the Legion runtime to easily tune our code for the PRF mechanism and map our code onto different architectures. The addition of the new PRF chemistry to our code was made effortless by the singe domain-specific language (DSL) compiler [10], which generates optimized Legion tasks for the main computational kernels.

We also present the DNS of a temporally evolving n-dodecane fuel jet surrounded on either side by air. For this simulation, the stencil tasks of the S3D-Legion code were enhanced to support stretched grids and to allow one-sided stencils at the domain boundaries. In addition, nonreflecting outflow boundary conditions (BCs) [11] were added as additional tasks. The BC implementation is based on the algorithm proposed by Sutherland and Kennedy [12] and can be readily extended to BCs for inflows and walls. Although this algorithm favors programmability over performance, most of the BC-related tasks operate on a small subset of the grid and their impact on the simulation time is negligible.

We believe that this work demonstrates a general approach and framework in which exascale computers can be productively programmed via the construction of performance-portable programs. By decoupling the specification of programs from their mapping using Legion, we make it easier to modify, tune, and port applications. The use of domain-specific compilers such as Singe, in conjunction with Legion, enables fast task implementations to be generated without requiring programmer input. These innovations greatly reduce both the time and effort required by human programmers to implement, maintain, and augment simulations. Furthermore, this approach often yields higher performance than hand-tuned codes. In our case, the use of Legion with a DSL compiler has made the 3D turbulent simulation of a complex PRF chemical mechanism, which is representative of other large hydrocarbon fuels, feasible for the first time.

12.2   S3D

S3D [8] is a massively parallel direct numerical solver to simulate turbulent combustion in canonical configurations and thereby gain fundamental insights into the physical and chemical interactions in turbulent reacting flows.

S3D solves the governing equations for reacting flows, namely the conservation equations for mass, momentum, total energy, and species written as

$$\frac{\partial}{\partial t}\rho = -\nabla_\beta \cdot (\rho u_\beta) \tag{12.1}$$

$$\frac{\partial}{\partial t}(\rho u_\alpha) = -\nabla_\beta \cdot (\rho u_\alpha u_\beta) + \nabla_\beta \cdot \boldsymbol{\tau}_{\beta\alpha} - \nabla_\alpha p \tag{12.2}$$

$$\frac{\partial}{\partial t}(\rho e) = -\nabla_\beta \cdot \left[u_\beta(\rho e_0 + p) \right] + \nabla_\beta \cdot (\boldsymbol{\tau}_{\beta\alpha} \cdot u_\alpha)$$
$$- \nabla_\beta \cdot \mathbf{q}_\beta \tag{12.3}$$

$$\frac{\partial}{\partial t}(\rho Y_i) = -\nabla_\beta \cdot (\rho Y_i u_\beta) - \nabla_\beta \cdot (\rho Y_i V_{\beta i}) + W_i \dot{\omega}_i \tag{12.4}$$

where the indices α and β indicate spatial dimensions (with repeated indices denoting summation) and i indicates the species index. Additional terms include the mass density (ρ), the mass fraction and molecular weights of each species (Y_i, W_i), the fluid velocity (u), the total energy (e), and pressure (p).

Species-specific heat capacities and enthalpies are needed to compute the thermodynamic relations between the internal energy and temperature. S3D uses NASA thermodynamic polynomials with seven coefficients and two temperature ranges for computing the detailed species thermodynamic properties.

$\dot{\omega}_i$ is the molar production rate of species due to chemical reaction and is computed using a detailed chemical kinetics model of the chemistry being simulated [13]. The complexity of these models depends on the number of species and reactions, which varies considerably between mechanisms, as shown in Table 12.1. For all but the simplest mechanisms, the computation of these production rates is the most important and computationally intensive kernel in S3D.

The next major computational kernels are in the evaluation of the diffusive transport terms: the stress tensor ($\boldsymbol{\tau}$), thermal diffusive flux (q), and species mass diffusion velocities (V_i). These quantities are computed from the known conserved variables using constitutive relations that require continuum models for molecular transport properties such as viscosity, thermal conductivity, multicomponent species diffusivity, and thermal diffusivity. S3D uses detailed multi-component and mixture-averaged models for the transport properties [14] that are also dependent on density, temperature, and local mixture composition.

TABLE 12.1

Summary of S3D Chemical Mechanisms

Mechanism	Reactions	Species	Unique	QSSA	Stiff
H2	15	9	9	–	–
DME	175	39	30	9	22
Dodecane	268	53	35	18	22
Heptane	283	68	52	16	27
PRF	861	171	116	55	93

*DME, dimethyl ether; PRF, primary reference fuel; QSSA, quasi-steady-state approximation.

The final significant computation kernel is used in the gradient operators (∇_β), which use a 9-point centered stencil to compute an explicit 8th order finite difference approximation of the partial derivative in direction β. Although not captured in the equations above, the solution is filtered periodically using an explicit 10th order finite difference filter on an 11-point stencil for numerical stability [15]. Navier-Stokes characteristic boundary conditions (NSCBC) [11,12] were used to prescribe the outflow BCs for the temporal jet simulation.

The solution is advanced through time integration using an explicit fourth order six-stage Runge–Kutta scheme [16] with built-in error estimates. A single time step, therefore, consists of six evaluations of the right-hand-side function (rhsf), each followed by three scaled accumulations for integration.

12.2.1 FORTRAN+MPI IMPLEMENTATION

S3D is a complex piece of software that has evolved over the course of 30 years, during which time it has been worked on by a number of different scientists and engineers. The original version of S3D consists of approximately 200 K lines of Fortran and uses MPI [17] for communication between threads.

Parallelism is achieved by running a separate MPI rank on each core and trusting the Fortran compiler to vectorize loops to take advantage of the wider datapaths in today's central processing units (CPUs). This approach works well at smaller node counts and the simplicity of the source code is appreciated by domain scientists when they wish to add features to the code base.

However, the limitations of this initial approach become evident at scale (see Figure 12.1), as well as on machines with heterogeneous computational resources where there is no way for the pure Fortran version to target the accelerators.

12.2.2 HYBRID OPENACC+MPI IMPLEMENTATION

A second version of S3D targets heterogeneous systems by combining MPI with OpenACC [18] directives in a hybrid implementation [19]. This version was ported from the Fortran+MPI version by a team of scientists and engineers from Cray, ORNL, NREL, and NVIDIA. By leveraging both CPUs and GPUs, the MPI/OpenACC version of S3D is roughly two times faster than the Fortran/MPI version for the heptane mechanism with 52 species, between a third and a half of the complexity of the PRF mechanism used in this study. This implementation runs with a single MPI rank per node and uses OpenMP to recover the parallelism available from the multiple CPU cores on each node.

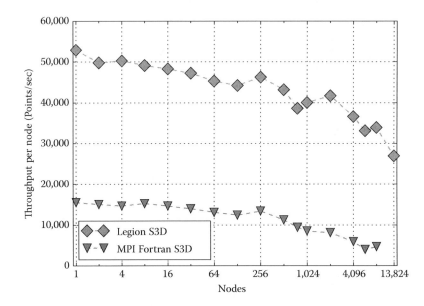

FIGURE 12.1 Weak scaling of PRF on Titan.

The effort to port the original Fortran version to the hybrid OpenACC model was significant. Over 2500 lines of directives were added to the code, each representing a decision by a person familiar with both the application and the then-current hardware of whether a given loop body should be offloaded to the GPU and which arrays must be copied from CPU memory to GPU memory or vice-versa. Because the directives apply to lexical constructs (specifically Fortran `do` loops), a large number of loops were manually fused (or occasionally split) and entire function bodies were inlined into loops when the compiler could not (or would not) perform the necessary inlining automatically.

The OpenACC implementation of S3D assumes a system with a single GPU per node. Systems with multiple GPUs per node are already common (e.g., Keeneland [20]), and although the OpenACC 2.0 standard includes support for multiple accelerators, additional directives would be required for all offloaded loops. Furthermore, when newer GPUs are introduced with different performance characteristics, the original decisions of which loops to fuse or split would need to be revisited. Any changes would be structural modifications to the source code, requiring conditional compilation directives to maintain support for versions targeted at different machines.

12.2.3 LEGION IMPLEMENTATION

The most recent version of S3D is an implementation in the Legion programming model [21]. The Legion version runs inside of the original Fortran+MPI version, offloading the main simulation loop but using the existing Fortran code for everything else. The changes to the Fortran source consist of a few calls to initialize the Legion runtime and the replacement of the Fortran `integrates` call with one call into Legion code, all controlled by a single build option.

The Legion code consists of approximately 25,000 lines of C++ and CUDA source, representing several person–months of effort. All of the mechanism-specific codes are generated by the Singe DSL compiler (discussed in detail in Section 12.4). Approximately 1000 lines of the C++ source are in a custom *mapper* class that allows the same Legion application code to run efficiently on systems with different types and ratios of CPUs and GPUs, different memory hierarchies, and different interconnect networks.

12.3   LEGION

Legion is a task-based parallel programming model designed for high-performance computing on systems with heterogeneous computational resources and deep memory hierarchies [21]. Like other dynamically scheduled task-based execution models [22–24], Legion maintains a *task graph* in which the nodes are *tasks* to be executed and edges are ordering constraints between the tasks. The Legion runtime executes the tasks in the graph, guaranteeing that a given task does not start until all of its predecessors in the graph have been completed. The Legion programming model differs from other task-based models in the combination of two significant features that are visible to the programmer.

First, the programmer does not directly construct the task graph. The programmer instead declares what data each task uses and how each task uses it (e.g., reading or writing). The Legion runtime uses this information to infer the necessary ordering constraints. Legion provides a *logical region* abstraction that allows precise specification of the data being used by a task. Logical regions may be partitioned into *subregions*, and the runtime tracks subregion relationships, allowing it to correctly and efficiently detect when two tasks may access the same data. Importantly, logical regions provide a data model that decouples the description of data from how it is placed and laid out in the memory hierarchy.

Second, the Legion application code does not specify on what processors tasks run or where in the memory hierarchy logical regions are placed, neither does the runtime attempt to automatically make these decisions. Instead, these decisions are made in a separate *mapper* object that encapsulates all machine-specific decisions about program execution. The mapper is part of the application code and implements an interface that responds to queries from the Legion runtime. When a task t is ready to be mapped, the runtime provides information about the current locations of *physical instances* of logical regions needed by t. Using this information, the mapper responds with the processor on which t should run and where t's physical instances should be placed. If the requested locations of physical instances differ from the current locations, the runtime automatically adds the necessary data movement tasks and dependencies to the task graph to ensure that t executes only after its data have arrived.

A key property of Legion programs is that changing the mapping cannot change the program's input–output behavior; the Legion runtime understands the semantics of task dependencies and guarantees that every task executes on the right data, regardless of the mapping. Thus, Legion programs truly constitute a machine-independent specification of an application. This property of Legion is crucial to achieving high productivity: unlike other task-based models and bulk synchronous models such as MPI, task and data placement decisions are not baked directly into a Legion program. Mapping decisions that impact performance are completely isolated within mapper objects. In a Legion application, it is easy to change the mapping to tune or port for new machines without compromising correctness.

Like other dynamic runtime systems, Legion incurs runtime overhead to compute and manage the task graph. The Legion runtime is highly optimized to keep this overhead off of the critical path of execution by performing much of the dynamic analysis in parallel with the rest of the application [25]. However, any dynamic runtime has a minimum granularity of useful work that it can efficiently support. For Legion, we find that the analysis overhead can be hidden if tasks take at least 500 μsec to execute on average. Many of S3D's operations are below this threshold, and exploiting fine-grain vector parallelism within Legion tasks requires an alternative static scheduling approach.

12.4   SINGE

As mentioned in Section 12.2.3, the mechanism-specific kernels (the reaction rate and transport coefficient calculations) are both the most complicated and the most performance-critical kernels. Tuning these kernels is essential for improving time to solution, but their size and complexity exceed the optimization capabilities of general purpose compilers. Hand optimization can yield significant benefits for an individual mechanism, but the immense time investment must be repeated for each new mechanism.

Our solution is Singe, a domain-specific language compiler for combustion chemistry kernels [10]. Singe takes as input a standard description of a chemical mechanism along with tables describing the transport and thermodynamic properties of chemical species. From these specifications, Singe synthesizes high-performance kernels for use in combustion simulations. Singe leverages domain-specific optimizations, such as the quasi-steady-state approximation (QSSA) [26], which groups similar reactions into a single computation, and special handling of *stiff* species [27], permitting larger time steps.* It also applies advanced compiler optimizations such as warp specialization for GPUs [28]. Singe then emits highly specialized variants of kernels for different architectures. Code is specialized differently for the Fermi and Kepler GPU architectures, as well as for CPUs with different cache sizes and vector instruction sets (e.g., streaming SIMD extensions [SSE] vs. advanced vector extensions [AVX]. SIMD is single instruction, multiple data [SIMD]). All of these kernel variants are registered with the Legion runtime and made available to the mapper, allowing it the runtime option of mapping tasks to a CPU or a GPU.

The performance of kernels generated by Singe is excellent. In cases where the OpenACC compiler generated GPU code for these kernels, the Singe versions are two to five times faster, with the largest speedups coming from sophisticated mechanisms, such as PRF, that have complicated computations and huge working sets. However, the productivity improvement is even more dramatic. Singe comes equipped with its own autotuning framework [10], which can optimize all the required kernels for a new chemical mechanism in less than 24 hours without any human intervention or direction. Instead of spending days or weeks hand-optimizing individual kernels that are thousands of lines long, adding support for a new mechanism is now an automated process that frees combustion scientists to engage in other useful work.

12.5 SIMULATION SETUP

The S3D-Legion code was used to simulate two separate scientific problems that are described below. The first scientific problem was the simulation of reactivity controlled compression ignition (RCCI) that was performed using all periodic boundaries and therefore did not require any special BC. The second scientific problem simulated using the S3D-Legion is the temporal evolution of a nonpremixed jet mixing layer, which required the application of outflow BCs and nonuniform mesh spacing through algebraic stretching. The setup of the simulation experiments for these two problems is presented below.

12.5.1 RCCI SIMULATION

The first simulation case is part of a larger parametric study of the impact of turbulence on combustion under RCCI conditions [29]. In RCCI, a mixture of fuels is used—iso-octane (i-C_8H_{18}) and n-heptane (n-C_7H_{16}) in the case of PRF—to obtain reliable ignition while operating in a regime that minimizes both particulate and NO_x emissions and still provides very high (i.e., diesel-like) thermal efficiency. Combustion phasing (ignition timing with respect to piston motion) and subsequent heat release rate are strongly sensitive to turbulence—the uneven mixing due to turbulence causes behavior to deviate significantly from that of a perfectly homogeneous mixture. Thus, it is essential to be able to couple turbulent simulations with detailed chemistry.

The physical volume of the DNS was chosen to be a cube 3 mm on a side, with the expectation that a grid spacing of 2.6 μm would be sufficient to resolve all the flame structures that were generated.

---

* The number of QSSA and stiff species in PRF and the other chemical mechanisms we study are listed in Table 12.1.

This results in a grid size of 1152^3 (just over 1.5 billion grid points) with 123 conserved variables per grid point (including two additional transported variables for model assessment discussed later in this section). With guidance from 1D flame simulations on the size of PRF flame structures under these conditions, this choice was conservative; we gave ourselves enough margin to refine the grid spacing to as small as 2.1 μm (using nearly 3 billion grid points) if necessary. Our run needed to simulate 3 ms of physical time using 2.5 ns time steps, requiring 1.2 million time steps to observe interesting combustion effects.

As described earlier, the flame structures that demand such high spatial resolution do not appear until well into the simulation at a time when the first ignition kernel undergoes thermal explosion and forms a flame, so the simulation was initially started with a grid size of 576^3. The factor of 2 increase in grid spacing also permitted a factor of 2 increase to 5 ns per time step for the first 2.7 ms of the run, improving the time to solution. The computational resource usage is improved by the 2× reduction in grid points in each dimensions and the reduced number of time steps required for an overall reduction factor of 16 for the majority of our run.

S3D is formulated as a constant-volume simulation, and models compression of the cylinder with an additional source term in each of the governing equations that adds mass in a way that does not cause spurious pressure fluctuations [30,31]. To maximize the number of turbulent eddies in the simulation domain, the simulation was set up with periodic BCs, which can cause the fraction of volume associated with heat release to be higher than under realistic engine conditions. To address this issue, the mass injection strategy was extended to not only account for volume changes due to piston motion, but to also replicate the pressure history of a fired engine. This required the introduction of a global reduction (to compute the average rate of change of pressure over the entire volume) to be incorporated into an application whose communication pattern previously consisted entirely of nearest-neighbor exchanges.

These changes were initially implemented in the Fortran+MPI version of S3D by domain scientists. As they all impact the main simulation loop, they had to be ported to the Legion application code. The existing Fortran code served as an excellent reference, as it succinctly described the necessary functionality without being obfuscated by directives and/or manual code transformations for optimization reasons. Because Legion tasks are also unencumbered by mapping considerations, the porting was a simple matter of translating Fortran code into the corresponding C++. The porting (and verification) effort required less than a programmer-week of work. Once the desired functionality was achieved, adding mapping policies for the new tasks required less than 10 lines of code in the Legion S3D mapper and was tuned in less than an hour.

For this simulation, we also explored the use of the OpenACC version of S3D for performance comparisons. This required the OpenACC code to be updated for both the PRF mechanisms and the additional features described above. We estimated that at least another month of implementation work as well as an unknown amount of debugging and tuning would be required to reach a production version of the OpenACC code. Our previous performance results demonstrated that the Legion version outperforms the OpenACC version by 2–3× on the smaller DME and heptane mechanisms [9]. This led us to abandon the effort and collect performance data for the MPI and Legion version of S3D only. We performed the RCCI production simulation entirely using the Legion version of S3D.

12.5.2 Temporal Jet Simulation

The second simulation case addressed ignition and flame propagation in a temporally evolving air/ n-dodecane jet at pressure (25 bar) and temperature ($T_{air} = 960$ K) conditions that are relevant to low-temperature diesel engine combustion. The scope of this simulation was to investigate the effect of low-temperature chemistry on the ignition of the jet and to better understand how burning kernels develop into flames propagating in a non-homogeneous mixture.

The physical volume of the DNS was a cuboid of 3.6 mm × 14 mm × 3 mm in the x, y, and z directions. The computational grid was uniform and periodic in the x and z directions and stretched along the inhomogeneous y direction, with a fine mesh region that spanned a distance equal to 4.0 mm at the beginning of the simulation to 6.6 mm at the end of the simulation. The grid spacing in the periodic directions and in the fine mesh region was 2.9 micron. The grid size at the end of the simulation was 1216 × 2400 × 1024 (3 billion grid nodes). The chemical mechanism used in the simulation was the 35-species reduced n-dodecane one listed in Table 12.1 and the number of conserved variables per grid point was 42. The variable count includes two auxiliary transported variables, a mixture fraction and a cumulative scalar dissipation rate, which were used to better understand the physics of the simulated problem. The run needed to simulate 1 ms of physical time using 4 ns time steps to observe mixture ignition and the propagation of burning flames throughout the simulation domain.

The code used for the simulation described in 12.5.1 supported neither stretched grids nor open BCs. To overcome these limitations, we rewrote the existing stencil tasks to support stretched grids and one-sided stencils at the domain boundaries. We also implemented the necessary tasks to simulate problems in which nonreflecting outflow BCs [11] are used. Our BCs implementation is based on the algorithm proposed by Sutherland and Kennedy [12] and can be readily extended to other types of BCs, such as walls and inlets. Most of the boundary-related tasks operate on a small subsets of the grid associated with each computational node and their impact on the simulation time is negligible (less than 1%).

The jet was initialized as a laminar jet of half-width equal to 0.25 mm with superposed turbulent fluctuations. Chemical reactions were not activated until $t = 0.28$ ms, when the jet had become fully turbulent. Early during the flow development, the jet width remained small and a computational grid with a fine mesh region of 4.0 mm was used to save computational resources. We used the full three billion nodes grid only after $t = 0.65$ ms, shortly before the jet extended beyond the refined region of the starting computational grid.

Initially, the simulation was run on Titan using the MPI version of S3D because at that time support for BCs and nonuniform grids was not implemented yet within the Legion version of the code. We eventually adopted the modified Legion version when the simulation reached $t = 0.45$ ms. The per node problem size was 64^3 grid points for the smallest computational grid used during the simulation and $64 \times 96 \times 64$ grid points for the largest one. Both these problem sizes allowed an all-GPU mapping of the simulation tasks, which turned out to be the optimal mapping strategy on Titan for problems using the n-dodecane chemistry. Due to the lack of a ready-to-use n-dodecane mechanism for the OpenACC version of S3D, no attempt was made in running this DNS simulation with that code.

12.6 COMBUSTION RESULTS

12.6.1 RCCI Simulation Results

Figure 12.2 shows the temporal evolution of normalized domain average species mass fractions of several key species along with temperature and heat release for the 3D simulation. It can be seen that n-heptane (n-C_7H_{16}) is consumed significantly earlier in the cycle than iso-octane. After the low-temperature heat release stage marked by the first peak in the heat release rate (HRR in Figure 12.2) profile, the original n-heptane is almost entirely consumed, most of it having decomposed to CH_2O, C_2H_4, and other smaller molecules. Consumption of CH_2O and production of OH appear to coincide with the consumption of all remaining iso-octane (i-C_8H_{18}). This period also coincides with the oxidation of CO. The generation of intermediate species such as CH_2O and CO occurs primar-

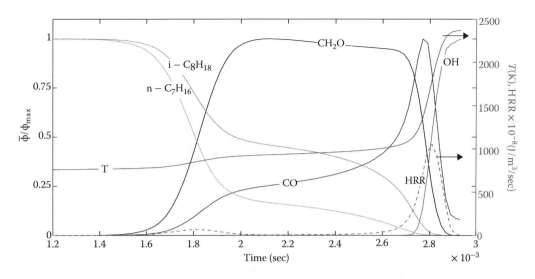

FIGURE 12.2 Temporal evolution of normalized species concentrations (left axis) and temperature and heat release rate (right axis) for the three-dimensional reactivity controlled compression ignition simulation. The overbar (¯.) denotes an average over the domain. The dashed line represents heat release rate.

ily through the breakdown of n-heptane. Because most of the heat released later in the simulation (i.e., the high-temperature heat release) is driven by the oxidation of these intermediates, it follows that combustion is driven by the staged consumption and oxidation of n-heptane and its intermediates followed by a rapid decomposition and oxidation of iso-octane.

Compression ignition configurations generally ignite by generating a series of ignition fronts. One of the most interesting aspects of the RCCI configuration is the appearance of flame fronts in conjunction with spontaneous ignition fronts. Ignition fronts are completely reaction driven: individual locations in the domain react and spontaneously ignite independently of their neighbors, down a gradient of ignition delay imposed by spatial variations in reactivity, temperature, or composition. In contrast, a flame front is diffusion driven: individual locations in the domain react and propagate only when heat and reactants diffuse into them from neighboring locations. Visually, flame fronts appear as thin, spindly structures, whereas ignition fronts appear as thick, blob-like structures. Figure 12.3 shows the overall heat release rate in the simulation volume at the time corresponding to approximately 50% of the total heat release. Figure 12.4 shows slices of the heat release rate in the simulation domain taken at one of the midplanes. The three images correspond to time instances at which 30%, 50%, and 80% of the total heat release has taken place. Most of the heat release occurs through the thin flame fronts.

One of the methods for quantifying whether the mode of combustion in a given location is through ignition or flames is to compute a reaction-diffusion balance across the burning surface. Figure 12.5 shows a volume rendering of these surfaces at a time instance corresponding to approximately 50% of the total combustion heat release. It can be seen that diffusion and reaction track each other quite closely in space, though they are not collocated. Moreover, their magnitudes are roughly equal throughout the domain. This suggests that most of the combustion in the simulation occurs through the flame propagation mode. Figure 12.6 shows profiles of the diffusion rate and reaction rate of one of the key intermediate species, OH, at the same time instant. The profile on the left shows that the peak magnitudes of both reaction and diffusion are comparable, which is a key characteristic of premixed flames. The profile on the right shows a *spontaneous ignition front*, which is dominated by reaction and hence the magnitude of the diffusion rate is negligible compared to that of the reaction rate.

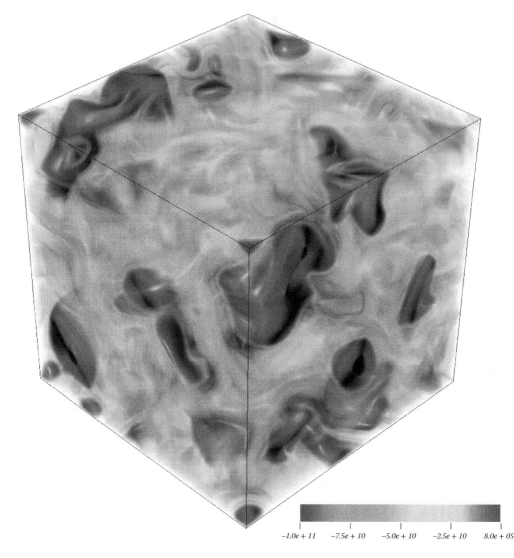

FIGURE 12.3 (See color insert.)   Volume rendering of the heat release rate at the time corresponding to 50% of total heat release. Values are in $J/m^3/sec$.

The PRF mixture used in this study had a large fraction of n-heptane. Although both flames and ignition fronts are found to coexist in this simulation, more than 80% of the combustion heat release occurs through the flame propagation mode. This observation appears to be unique to fuel blends that contain high reactivity fuels, such as n-heptane. This has broad implications for modeling of RCCI combustion and this 3D dataset provides a unique benchmark for the development and validation of models applicable to an engine operating in the mixed combustion modes under RCCI conditions. Optical engine experiments provide complementary statistics but have insufficient resolution to discern flames from spontaneous ignition fronts. This is the first study of the details of the underlying flame and ignition structure with fuel blending. The results suggest that by conducting future DNS simulations, in conjunction with experiments and LES, it will be possible to find the optimal fuel blend for different engine operating conditions.

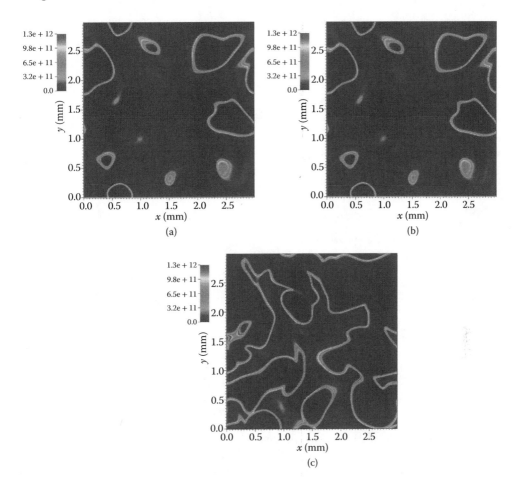

FIGURE 12.4 Slices taken at the midplane of the simulation volume showing contours of heat release rate at time instances corresponding to 30% (a), 50% (b), and 80% (c) of the total heat release. Values are in J/m³/sec.

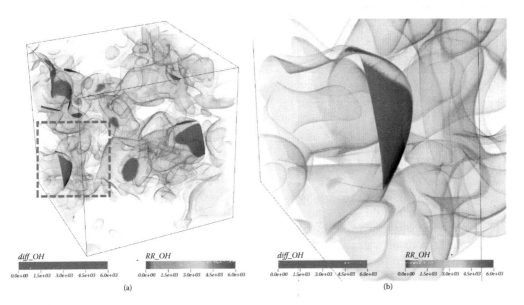

FIGURE 12.5 (See color insert.) Contours of diffusion rate and reaction rate of the OH radical for the full simulation volume (a) and a zoomed in view of the same in the boxed region near a flame kernel (b).

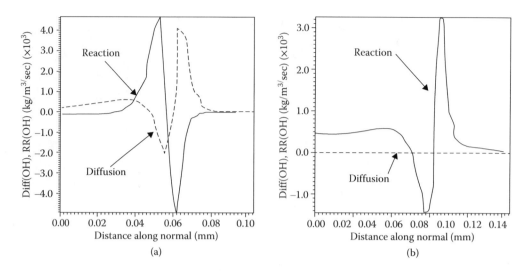

FIGURE 12.6 Comparison of the diffusion rate and reaction rate of the OH radical along a flame surface (a) and an ignition front (b).

12.6.2 TEMPORAL JET SIMULATION RESULTS

Understanding the physics of ignition is very important for achieving high efficiency and low pollutant emissions in modern diesel engines. Generally speaking, ignition in a nonpremixed flow is the result of a competition over time between heat production due to chemical reactions and heat dissipation due to transport processes. Under turbulent conditions, the probability is small that two flow elements will be subjected to the same reacting and mixing histories and, as such, we should expect that different flow regions with the same composition will ignite at different times. Our simulation reveals this fact to hold true for both low- and high-temperature ignition.

Characterizing the topology of the low- and high-temperature ignition kernels is very important to optimize the mixture preparation phase in compression-ignition engines. Figure 12.7 shows the scatterplots of temperature versus mixture fraction (ξ) at selected instants of time during the ignition transient. It can be seen that the turbulent jet ignites first in those regions where $\xi = 0.06$ and that ignition is followed by the rapid propagation of the burning kernels toward regions with fuel-rich compositions. During this phase, most of the heat release rate is due to low-temperature reactions and therefore the maximum increase in gas temperature remains below 400 K. This increase in temperature, albeit modest, is sufficient to initiate the second phase of the ignition transient by triggering the high-temperature reactions pathway of n-dodecane, which in turn initiates the hot flame ignition process. High-temperature regions are seen developing first around $\xi = 0.16$: their subsequent ignition results in several reaction fronts that propagate toward the flow regions with leaner mixture compositions, leading eventually to the establishment of fully burning conditions throughout the entire flow.

In this study, low-temperature reactions play a key role in triggering autoignition by raising the gas temperature just enough to activate the high-temperature reactions pathways. As their name suggests, low-temperature reactions are non-negligible only when the ambient gas temperature is below a fuel-dependent threshold, which is approximately $T_g = 1000$ K for n-dodecane. Diesel engines used to be operated above this threshold value to maximize their efficiencies: as such, low-temperature combustion phenomena received little or no attention in most of the existing DNS studies on turbulent autoignition. The situation has changed in recent years due to the emergence of the low-temperature diesel combustion concept as a promising engine technology for abating pollutants emissions while maintaining the high efficiency typical of traditional diesel engines. Unfortunately, investigating tur-

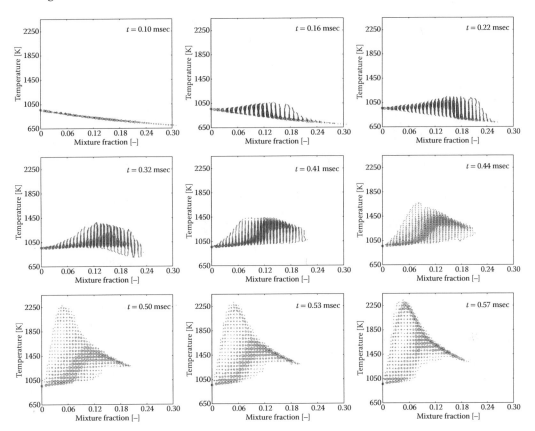

FIGURE 12.7 Scatterplots of temperature versus mixture fraction at selected instants of time during the ignition and transient. Shading and size of the symbols are indicative of the number of samples in the computational domain with that composition and temperature.

bulent flow autoignition with DNS at conditions representative of low-temperature diesel combustion has proven to be an extreme challenge due to the combination of long induction times, small length scales, and large sizes of the chemical mechanisms required to capture the low-temperature reaction pathways. This study constitutes a pioneering work that will shed light on the combustion physics behind the next generations of clean, efficient engines.

12.7 PERFORMANCE

This section describes the performance characteristics of the RCCI science case described in Section 12.5.1. Portions of the RCCI simulations were run on two different Cray machines: Titan at the Oak Ridge Leadership Computing Facility (currently second on the Top500 list [32]) and Piz Daint at the Swiss National Supercomputing Center (sixth in the Top500). Both systems rely on a Kepler K20X GPU per node for the bulk of their computational power. Titan [7] has 18,688 nodes, pairing the GPU with a 16-core Opteron CPU and a Gemini interconnect. Piz Daint [6] is a newer, but smaller, machine, consisting of 5272 nodes each with an 8-core Sandy Bridge-EP CPU and a Cray Aries interconnect.

12.7.1 PERFORMANCE BOTTLENECKS

With roughly 90% of the available floating point operations per second (FLOPS) on the GPU, the first priority is to make sure the GPU is fully utilized. Figure 12.8 breaks down the GPU utilization on Titan by kernel, showing that the GPU is kept busy over 95% of the time. Due to all the internode communication for the gradient operations, every time step requires the movement of a large amount of data over the PCI-Express bus that links the CPU and GPU: 1.6 GB is moved from CPU to GPU and 4.0 GB is moved back. The bus is bidirectional and can transfer approximately 6 GB/sec in each direction; the CPU to GPU channel operates at 15% capacity, whereas the return link operates at 38% capacity. The Legion runtime automatically overlaps data transfers with kernels running on the GPU, requiring no effort from the programmer.

Although ensuring that the GPU is nearly always busy is important, it is equally important that the individual tasks make efficient use of the GPU. The roofline analysis used by Rossinelli et al. [33] remains an excellent tool for this purpose. We measured the achievable memory bandwidth of the K20X to be 159.6 GB/sec and the peak computational throughput to be 1121.9 GFLOPS. The computational intensity of the top kernels from Figure 12.8 and their performance are shown in comparison to the roofline in Figure 12.9. The two key kernels (getrates and stencil) are both memory bound. The getrates kernel is benefiting slightly from L1 cache hits for its register spills, which is why it appears slightly above the roofline. The Singe-generated diffusion and viscosity kernels are able to move their working sets into CUDA shared memory, eliminating the external memory bottleneck (their arithmetic intensities would be reduced to 1.4 and 2.8 FLOPS/B without this optimization), but are still unable to fully utilize the arithmetic potential of the GPU because they saturate the available shared memory bandwidth. Finally, the temp and ydiffflux kernels must iterate over all 116 species of the PRF mechanism, resulting in less efficient global memory access patterns and reduced memory system performance. By considering each kernel's

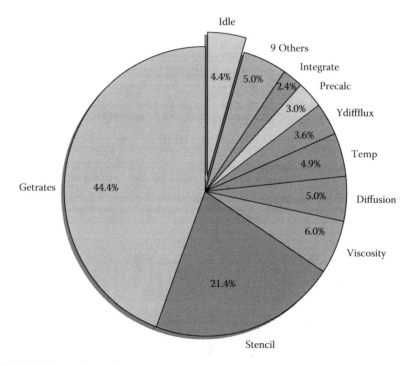

FIGURE 12.8   GPU usage by task on Titan.

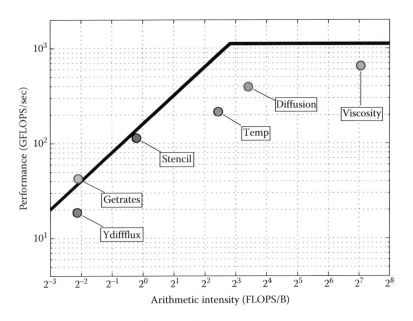

FIGURE 12.9 Roofline analysis of key GPU kernels.

distance from the roofline and weighting by the average fraction of a time step spent in a kernel, we determined that we are obtaining 80.4% of the achievable performance of the GPU.

Prior work on CFD [33] has shown that cache blocking on CPUs can be used to significantly improve the computational intensities of important kernels. Unfortunately, these techniques are not available to us for two reasons. First, although the caches and shared memories on the Kepler K20X are similar in capacity to those of modern CPUs, they service an order of magnitude more concurrent threads that considerably reduces the space available to each thread. Second, the introduction of complex chemistry to the system dramatically increases the working set of each thread. The computation of the right hand side for a single grid point in the PRF mechanism requires approximately 2000 double-precision intermediate temporaries. The L1 cache on a modern CPU would only be able to fit a handful of grid points, and even all the combined L2 caches in a CPU socket could support only a few hundred—not nearly enough to compensate for the edge effects (e.g., warming up prefetch engines and fetching stencil neighbors) in a cache-blocked formulation. Additionally, the chemistry, transport, and enthalpy calculations make heavy use of constants for polynomial coefficients. For PRF, there are over 34,000 of these constants, increasing the working set by a further 272 KB.

The viscosity and diffusion kernels achieve the best computational intensities because Singe is able to fit their working sets into on-chip memories. To handle large working sets, Singe uses a technique called warp specialization to extract task parallelism from a computation and distribute it across the warps of a thread block [28]. For smaller mechanisms such as DME and heptane, Singe has been able to use the same technique for the chemistry kernel [10]. However, for the PRF mechanism, the chemistry kernel working set consists of all the forward and backward reaction rates (1722 doubles per grid point), which exceeds the available on-chip memory and the resulting code is limited by external memory bandwidth due to register spills.

Our discussion has focused on the Kepler GPU because nearly all tasks related to the main computation are placed there by the Legion S3D mapper. Only a few tasks related to inter-node communication or interfacing with the Fortran application are performed on the CPU. By changing only a few lines of code in the mapper, we explored many alternative mappings that use the computational resources of the CPUs as well. Although we have found mapping strategies that use the CPUs to be effective on other machines with other chemical mechanisms [9], the increased size of the PRF mechanism results in the need to move much more intermediate data between the CPU and GPU. In all of the proposed mappings for the PRF mechanism, the PCI-Express bus became the bottleneck and GPU utilization dropped. Although the CPU utilization increased, overall performance was reduced.

12.7.2 Weak Scaling

We next measured weak scaling performance on both Titan and Piz Daint. We held the per node problem size constant at 48^3 grid points, the size on which we had based our simulation plan. The original Fortran+MPI version of S3D was run as a reference. The throughput achieved per node (higher is better) is shown for Titan and Piz Daint in Figures 12.1 and 12.10 respectively. Perfect weak scaling would be a flat line.

In addition to being significantly faster than the Fortran baseline (3.95× faster on Piz Daint and 3.93× faster on Titan for a single node), the Legion version demonstrates greatly improved scalability. Whereas the Fortran version drops below 50% efficiency beyond 2048 nodes on Titan,* the Legion version is above 64% even out to 8192 nodes, where the performance difference has grown to a factor of 7.2×. The asynchronous, task-based nature of the Legion programming model allows the Legion runtime to dynamically discover considerable task parallelism and use it to hide communication latency better than the MPI version that relies on manual overlap of communication and computation. The improvement from the Gemini to the Aries interconnect is evident in the Piz Daint data, with both the Fortran and Legion versions achieving parallel efficiencies of 82% at 4096 nodes. Note that the Legion version must reduce network overhead by a factor of 4 compared to the Fortran version to achieve the equivalent parallel efficiency.

12.7.3 Strong Scaling

In an effort to further improve our time to solution, we explored the strong scaling properties of the Legion implementation. Using the smaller 576^3 grid size that we hoped to use for the majority of

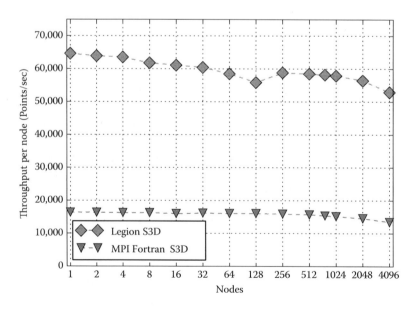

FIGURE 12.10   Weak scaling of PRF on Piz Daint.

---

*   All runs on Titan made use of a *reranking* script that attempts to optimize the assignment of MPI ranks to nodes to match S3D's specific communication pattern to the underlying network topology [34].

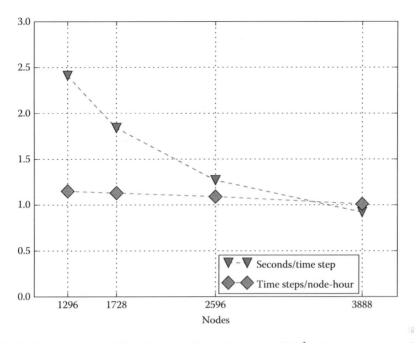

FIGURE 12.11   Strong scaling: PRF on Piz Daint for problem size of 576^3 points.

the simulation, we ran at node counts of 1296, 1728, 2596, and 3888. (The S3D code requires that the grid size divide evenly among the nodes, limiting the number of allowable node counts.) The results were excellent. Figure 12.11 shows the wall clock time per time step as well as the throughput per node (in timesteps/node-hour) as a measure of efficiency. Based on these results, we chose to run nearly all of the simulation of the 576^3 grid with 3888 nodes instead of the originally planned 1728 nodes, improving our simulation speed by a factor of 2 for only an 11% loss in efficiency.

12.8 CONCLUSIONS

We have presented the first 3-D DNS simulation of a PRF chemical mechanism at scale. Our results represent a substantial improvement in both the kind of combustion chemistry that can be studied as well as the time required to perform simulations. We reduced overall time to solution (including both development time and actual run time) over the previous state of the art, making it feasible to perform these computations in a reasonable amount of time and within existing computational budgets. Our approach also demonstrates that it is computationally tractable to conduct simulations of realistic chemical mechanisms such as PRF that had previously been possible only in reduced capacities, thus advancing the state of the art for computational combustion science both quantitatively and qualitatively.

Although our results have a significant impact within the domain of turbulent combustion with complex chemistry, our approach is more generally applicable. We have shown that by raising the level of programming abstraction using a task-based programming model, we can both reduce programming time and improve performance. Specifically, we have demonstrated the value of decoupling the specification of an application from how it is mapped onto a target architecture. By isolating correctness concerns from performance issues, we can develop performance-portable codes capable of being easily modified, tuned, and ported with a minimum amount of programmer effort.

In this work we have shown how the Legion programming model provides one approach to decoupling the specification of a program from how it is mapped. In particular, Legion isolates the specification of a program in the form of tasks and logical regions from how mapping decisions are made through the mapping interface. This property allowed us to easily port our version of S3D to several different architectures with minimal programmer effort and overhead. We further built on top of the Legion abstractions by creating the even higher-level Singe DSL compiler capable of emitting high-performance Legion tasks for a myriad array of chemical mechanisms and target architectures. The result is a cohesive framework for combustion chemistry coupled with compressible reacting fluid dynamics that addresses currently pressing problems while ensuring our code remains adaptable to upcoming machine architectures, and as yet unknown domain-specific variations, all the while achieving much improved performance over existing techniques.

The challenges we encountered are not unique to computational combustion science. For many current applications, time to solution is already dominated by programming effort, even for conceptually small extensions to existing codes. This situation will only become worse as both applications and machines continue to grow in complexity. Consequently, there will be an increasing need to develop performance-portable codes that can be easily reconfigured for new experiments and quickly adapted to new machine architectures. Under these circumstances, providing high-productivity computing environments such as Legion will be imperative to ensure that machines are efficiently utilized and codes achieve high performance. Our work demonstrates that this goal is not just a dream, but an achievable reality for actual production codes.

ACKNOWLEDGMENTS

This research used resources of the Oak Ridge Leadership Computing Facility at the Oak Ridge National Laboratory, which is supported by the Office of Science of the U.S. Department of Energy (DOE) under Contract No. DE-AC05-00OR22725, as well as computing resources at the Swiss National Supercomputing Centre (CSCS). We thank the operations and user support staff at OLCF and CSCS for their tireless assistance. We thank Dr. Jack C. Wells of ORNL and Dr. Thomas C. Schulthess of CSCS for their support. This research was supported in part by the Director, Office of Advanced Scientific Computing Research, Office of Science, of the US DOE through the ExaCT Combustion Co-Design Center.

REFERENCES

1. G. Bansal, A. Mascarenhas, and J. H. Chen. Direct numerical simulation of autoignition in stratified dimethyl-ether (dme)/air turbulent mixtures. *Combustion and Flame*, 162:688–702, 2015.
2. P. Gaffuri et al. A comprehensive modeling study of iso-octane oxidation. *Combustion and Flame*, 129:253–280, 2002.
3. T. Lu and C. K. Law. A directed relation graph method for mechanism reduction. *Proceedings of the Combustion Institute*, 30(1):1333–1341, 2005.
4. M. B. Loung et al. A dns study of the ignition of lean prf/air mixtures with temperature inhomogeneities under high pressure and intermediate temperature. *Combustion Flame*, 162:717–726, 2015.
5. M. B. Luong et al. Direct numerical simulations of the ignition of lean primary reference fuel/air mixtures under hcci condition. *Combustion and Flame*, 160:2038–2047, 2013.
6. Piz Daint & Piz Dora - CSCS. http://www.cscs.ch/computers/piz_daint_piz_dora/, 2013.
7. Introducing Titan—The World's #1 Open Science Supercomputer. https://www.olcf.ornl.gov/titan/, 2012.
8. J. H. Chen et al. Terascale direct numerical simulations of turbulent combustion using S3D. *Computational Science and Discovery*, 2:015001, 2009.

9. M. Bauer, S. Treichler, E. Slaughter, and A. Aiken. Structure slicing: Extending logical regions with fields. In *Proceedings of the International Conference for High Performance Computing, Networking, Storage and Analysis*, New Orleans, LA, 845–856. Piscataway, NJ: IEEE Press, 2014.

10. M. Bauer et al. Singe: Leveraging warp specialization for high performance on GPUs. In *Symposium on Principles and Practice of Parallel Programming*, February 2014.

11. T. J. Poinsot and S. K. Lele. Boundary-conditions for direct simulations of compressible viscous flows. *Journal of Computational Physics*, 101:104–129, 1 July 1992.

12. J. C. Sutherland and C. A. Kennedy. Improved boundary conditions for viscous, reacting, compressible ows. *Journal of Computational Physics*, 191:502–524, 2 November 2003. A.

13. R. Kee et al. *CHEMKIN-III: A Fortran Chemical Kinetics Package for the Analysis of Gas Phase Chemical and Plasma Kinetics*. Report number UC-405 SAND96-8216, Livermore, CA: Sandia National Laboratories. 1996.

14. R. B. Bird et al. *Transport Phenomena*. Hoboken, NJ: John Wiley & Sons, 1960.

15. C. A. Kennedy and M. H. Carpenter. Several new numerical methods for compressible shear- layer simulations. *Applied Numerical Mathematics*, 14(4):397–433, June 1994.

16. M. H. Carpenter et al. Fourth-order Runge-Kutta schemes for fluid mechanics applications. *Journal of Scientific Computing*, 25:157–194, October 2005.

17. M. Snir et al. *MPI-The Complete Reference*. Cambridge, MA: MIT Presss, 1998.

18. OpenACC Standard. http://www.openacc-standard.org.

19. J. Levesque et al. Hybridizing S3D into an exascale application using OpenACC: An approach for moving to multi-petaops and beyond. In *SC'12 Proceedings of the International Conference on High Performance Computing, Networking, Storage and Analysis*, pp. 15:1–15:11, Salt Lake City, UT: IEEE, 2012.

20. J. Vetter et al. Keeneland: Bringing heterogeneous GPU computing to the computational science community. *Computing in Science Engineering*, 13:90–95, 2011.

21. M. Bauer et al. Legion: Expressing locality and independence with logical regions. In *Supercomputing Conference (SC)*, 2012.

22. The Open Community Runtime Interface. https://xstackwiki.modelado.org/images/1/13/Ocr-v0.9-spec.pdf, 2014.

23. C. Augonnet et al. StarPU: A unified platform for task scheduling on heterogeneous multicore architectures. *Concurrency and Computation: Practice and Experience*, 23:187–198, February 2011.

24. Q. Meng et al. Investigating applications portability with the uintah dag-based runtime system on petascale supercomputers. In *Supercomputing Conference (SC)*, pp. 96:1–96:12, New York, NY: ACM, 2013.

25. S. Treichler et al. Realm: An event-based low-level runtime for distributed memory architectures. In Parallel Architectures and Compilation Techniques (PACT), 2014.

26. T. Lu and C. K. Law. Toward accommodating realistic fuel chemistry in large-scale computations. *Progress in Energy and Combustion Science,* 35:192–215.

27. T. Lu et al. Dynamic stiffness removal for direct numerical simulations. *Combustion and Flame*, 156(8):1542–1551, 2009.

28. M. Bauer et al. CudaDMA: optimizing GPU memory bandwidth via warp specialization. In *SC '11 Proceedings of 2011 International Conference for High Performance Computing, Networking, Storage and Analysis*, Seattle, WA, 2011.

29. S. L. Kokjohn et al. Fuel reactivity controlled compression ignition (rcci): A pathway to controlled high-efficiency clean combustion. *International Journal of Engine Research*, 12:209–226, 2011.

30. Bhagatwala et al. Direct numerical simulations of hcci/saci with ethanol. *Combustion Flame*, 161:1826–1841, 2014.

31. P. Domingo and L. Vervisch. Triple flames and partially premixed combustion in autoignition of non-premixed turbulent mixtures. In *Symposium (International) on Combustion*, Vol. 26, pp. 233–240. Elsevier, 1996.

32. November 2014—top500 supercomputer sites. http://top500.org/lists/2014/11/, 2014.

33. D. Rossinelli et al. 11 pop/s simulations of cloud cavitation collapse. In *Supercomputing Conference (SC)*, pp. 3:1–3:13, New York, NY: ACM, 2013.

34. R. Sankaran et al. Genetic algorithm based task reordering to improve the performance of batch scheduled massively parallel scientific applications. *Concurrency and Computation: Practice and Experience*, 27(17):4763–4783, December 2015.

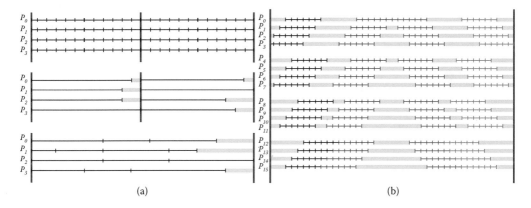

FIGURE 1.2 Example application communication templates showing periods of idleness (green) suitable for energy savings. Synchronous patterns are shown in (a), while asynchronous patterns requiring application knowledge for energy optimization are shown in (b).

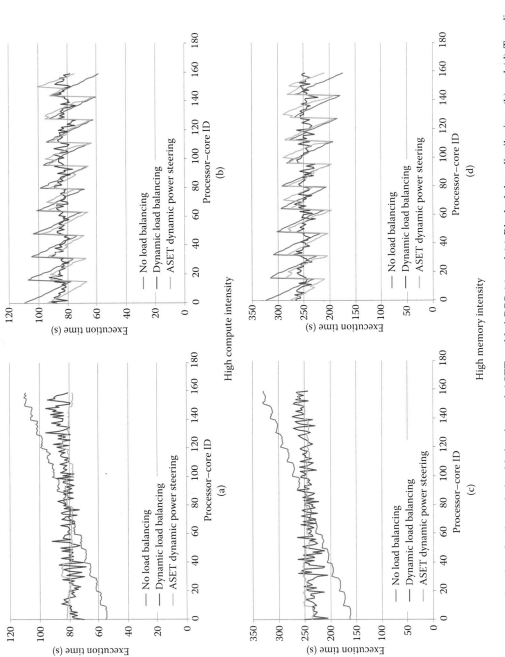

FIGURE 1.8 Per-core execution time for dynamic load balancing and ASET-enabled DPS (a) and (c) Blocked data distribution. (b) and (d) Two-dimensional data distribution.

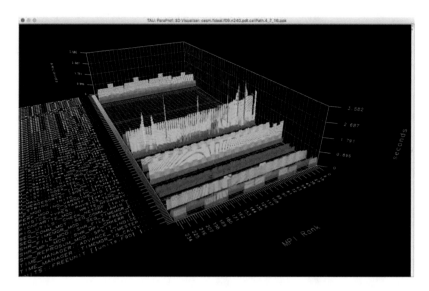

FIGURE 2.2 ParaProf.

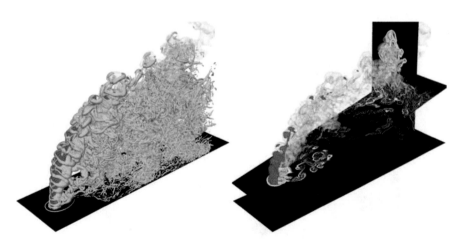

FIGURE 11.2 High fidelity LES of a high-Reynolds-number liquid n-decane jet (Re_{Jet} = 93,000) injected into a high-Reynolds-number cross flow of air ($Re_{Channel}$ = 620,000) at high-pressure supercritical conditions encountered in gas turbine engines. The left image shows the structure of the vorticity field. The right image shows iso-contours of mixture fraction, Z, that mark the location of the compressed liquid core, $Z = 0.88$ (red), and where $Z = 0.5$ (yellow). Also shown are cross-sections of the scalar dissipation rate

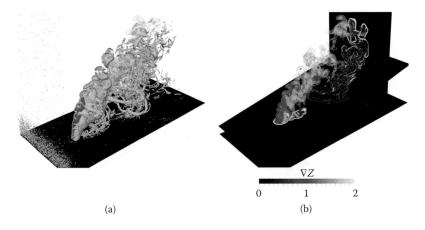

FIGURE 11.4 Instantaneous three-dimensional visualizations of the baseline flow field using the G3 mesh: (a) the blue iso-contour is $Q = 5$, the red iso-contour marks where the mixture fraction is $Z = 0.88$ (the compressed liquid core) and the yellow iso-contour marks where $Z = 0.5$ (gaseous jet), (b) cutting planes that highlight the scalar dissipation rate with red and yellow iso-contours that are the same as (a).

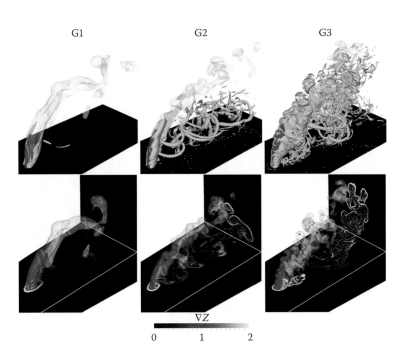

FIGURE 11.5 Instantaneous three-dimensional visualizations of the flow field using meshes G1, G2, and G3. Images on top show Q-criterion values of $Q = 5$ with the blue iso-contours, and mixture fraction values of $Z = 0.88$ and 0.5 with the red and yellow iso-contours, respectively. Images on the bottom show cutting planes colored by scalar dissipation rate along with the red and yellow iso-contours shown on the top figures.

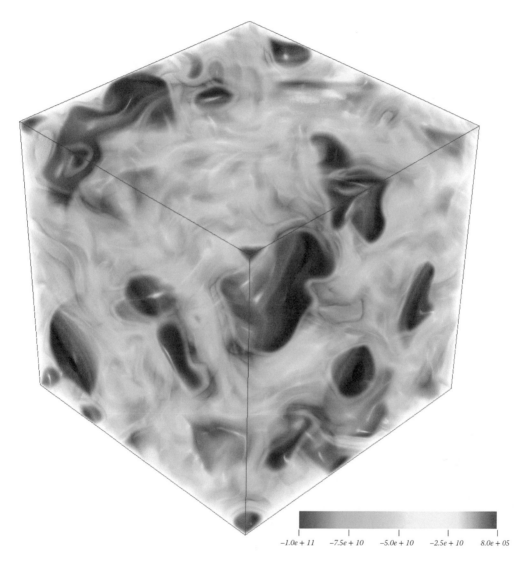

-1.0e + 11   -7.5e + 10   -5.0e + 10   -2.5e + 10   8.0e + 05

FIGURE 12.3 Volume rendering of the heat release rate at the time corresponding to 50% of total heat release. Values are in J/m^3/sec.

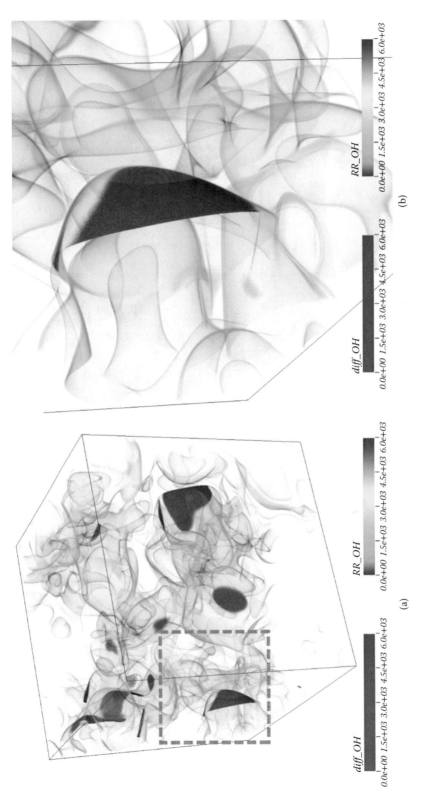

FIGURE 12.5 Contours of diffusion rate and reaction rate of the OH radical for the full simulation volume (a) and a zoomed in view of the same in the boxed region near a flame kernel (b).

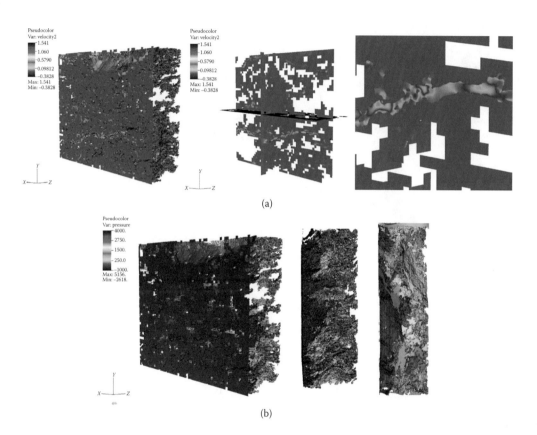

FIGURE 14.2 Direct numerical simulation from image data of fractured shale with pruning (From D. Trebotich and D. Graves. *Comm. App. Math. Comp. Sci.*, 10, 43–82, 2015.) (a) Velocity data mapped to surface and data slice. Pruned boxes are shown as white space. (b) Pressure mapped onto the mineral surface with side views. The simulation has been performed on NERSC Edison using 40,000 cores.

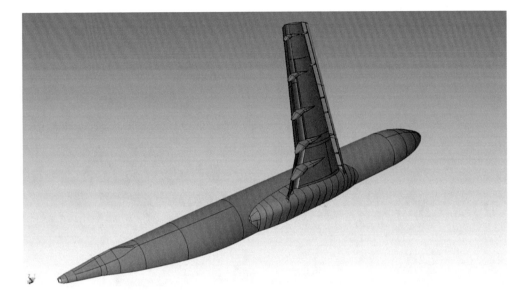

FIGURE 15.1 Geometric model of a high lift wing/fuselage configuration.

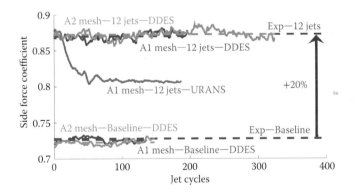

FIGURE 15.2 Plot of side force with and without flow control. Experiment shown in dashed lines. Detached eddy simulation (CFD) predictions with two levels of adaptivity show excellent agreement with the experiment and grid independence.

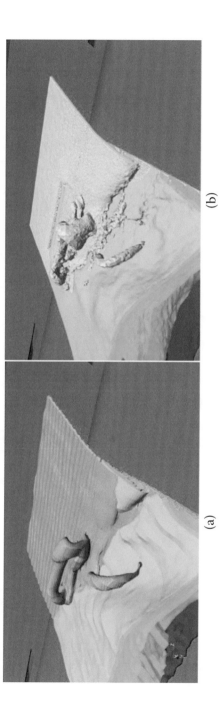

(a)

(b)

FIGURE 15.3 Synthetic jets apply flow periodic flow. By averaging like phases (phase-averaging) the coherent structures produced by the jets are revealed and excellent agreement between the CFD (a) and the experiment (b) are observed.

FIGURE 15.4 Isosurface of instantaneous value of Q (measure of vorticity or rotational features of the flow) colored by speed.

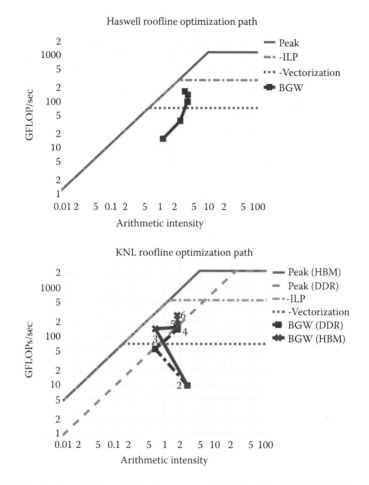

FIGURE 22.11 Visualization of the optimization process of GW bottleneck C within the roofline [15,22,23] model on Haswell and Knights Landing (KNL). The two separate diagonal lines on the KNL roofline represent the different ceilings when running out of DDR and MCDRAM with the KNL booted into flat memory mode. The three ceilings correspond to peak performance, peak minus the use of fused multiply-add instructions and instruction saturation on dual VPUs, and the additional loss of vector parallelism (a factor of 4 for Haswell and a factor of 8 for KNL). See text for more details.

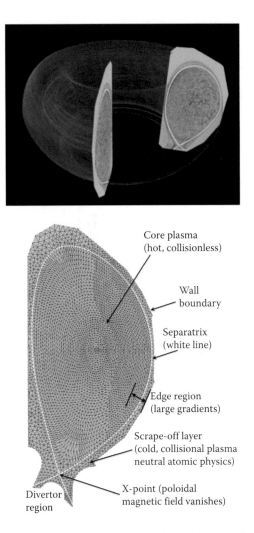

Core plasma
(hot, collisionless)

Wall
boundary

Separatrix
(white line)

Edge region
(large gradients)

Scrape-off layer
(cold, collisional plasma
neutral atomic physics)

X-point (poloidal
magnetic field vanishes)

Divertor
region

FIGURE 24.1 Unstructured triangular mesh in XGC maps the entire tokamak cross-section, including the divertor separatrix surface and the irregular wall structure. Figure used ITER geometry with artificially coarsed mesh for a better visualization.

13 Data and Workflow Management for Exascale Global Adjoint Tomography

Matthieu Lefebvre, Yangkang Chen, Wenjie Lei,
David Luet, Youyi Ruan, Ebru Bozdağ, Judith Hill,
Dimitri Komatitsch, Lion Krischer, Daniel Peter,
Norbert Podhorszki, James Smith, and Jeroen Tromp

CONTENTS

13.1 Introduction 279
13.2 Scientific Methodology 280
13.3 Solver: SPECFEM3D_GLOBE 282
    13.3.1 Overview and Programming Approach 282
    13.3.2 Scalability and Benchmarking Results 283
13.4 Optimizing I/O for Computational Data 289
    13.4.1 Parallel File Formats and I/O Libraries for Scientific Computing 290
    13.4.2 Integrating Parallel I/O in Adjoint-Based Seismic Workflows 292
13.5 A Modern Approach to Seismic Data: Efficiency and Reproducibility 293
    13.5.1 Legacy 293
    13.5.2 The Adaptable Seismic Data Format 294
    13.5.3 Data Processing 295
13.6 Bringing the Pieces Together: Workflow Management 297
    13.6.1 Existing Solutions 297
    13.6.2 Adjoint Tomography Workflow Management 298
    13.6.3 Moving Toward Fully Managed Workflows 299
    13.6.4 Additional Challenges 300
13.7 Software Practices 301
13.8 Conclusion 303
Acknowledgments 304
References 304

13.1 INTRODUCTION

Striving to comprehend Earth's interior has been a longstanding pursuit for humankind, and has been fantasized by many, from Dante Alighieri [1] to Jules Vernes [2]. Seismologists see Earth's interior through seismic waves generated by seismic sources such as earthquakes, oceanic noise, or man-made explosions and recorded by seismic instruments deployed at the surface. The information inherent to these seismic waves, which are sensitive to physical parameters of the medium they propagate through, is used to construct three-dimensional (3D) images of the Earth based on *seismic tomography*. Advances in the theory of wave propagation and 3D numerical solvers, supported by dramatic increases in the amount and quality of seismic data as well as the unprecedented amount of computational power provided by large-scale high-performance computing centers, enables us to greatly improve our understanding of the physics and chemistry of Earth's interior.

Adjoint methods are very efficient at incorporating 3D numerical wave simulations in seismic tomography. They have, for instance, successfully been applied to regional- and global-scale earthquake tomography [3–6] and, to some extent, in exploration seismology studies [7,8].

Adjoint tomography workflows consist of a series of iterations. Each iteration is composed of a few shared operations (e.g., mesh generation, model updates) and a large number of embarrassingly parallel operations (e.g., forward and adjoint simulations for each seismic event, pre- and postprocessing of seismic data). One of the main computational challenges is to increase the quality of seismic models while keeping the time to solution as short as possible. Having fast and efficient solvers still remains an important concern [9], but new obstacles have emerged: large-scale experiments and big data sets create bottlenecks in workflows causing significant I/O problems on high performance computing (HPC) systems.

In this chapter, we devise and investigate strategies to scale global adjoint tomography to unprecedented levels by assimilating data from thousands of earthquakes. We elaborate on improvements targeting not only current supercomputers, but also next generation systems, such as the Oak Ridge Leadership Computing Facility's (OLCF) *Summit*. The following remarks and developments stem from lessons learned while performing simulations on OLCF's *Titan* for the first-generation global adjoint tomography model, which is the result of 15 iterations and a limited data set of 253 earthquakes [6].

We begin in Section 13.2 by laying down the scientific methodology of the global adjoint tomography problem and providing explanations of the scientific workflow and its components. We then follow a reductionist approach, considering each component individually. Section 13.3 examines the computational aspects of SPECFEM3D_GLOBE [10,11], our 3D seismic wave equation solver and the most computationally demanding part of our workflow. We provide a brief overview and discuss the programming approach it follows. We also present scalability analyses. Section 13.4 is closely related to the solver and describes the approach we chose to optimize I/O for computational data, that is, data related to meshes and models. We then describe a modern seismic data format in Section 13.5. Assimilating a large number of seismic time series from numerous earthquakes requires improvement of legacy seismic data formats, including provenance information and the ability to seamlessly integrate in a complex data processing chain. With the previous points in mind, Section 13.6 returns to our initial holistic approach and discusses how to bring the global adjoint tomography workflow bits and pieces under the control of a scientific workflow management system. Finally, Section 13.7 explains our approach to software practices, an often overlooked but crucial part of scientific software development.

13.2 SCIENTIFIC METHODOLOGY

Seismic tomography involves the construction of images of Earth's interior by minimizing a predefined misfit function involving observed and synthetic data. Typically, seismic signals are generated by a wide variety of sources and recorded by a set of receivers as time series of a physical quantity. Sources can be either passive (e.g., earthquakes, ambient noises) or active (e.g., nuclear explosions, air guns). In the most common cases, the receivers record a quantity, such as displacement, on vertical and/or horizontal components. In the past decades, seismic data sets have grown very fast with accumulating sources and a dramatic increase in the number of receivers (e.g., cross-continental array deployments in passive seismology, 3D surveys in active source seismology). Such growth provides more information to constrain the model in greater details but also poses a challenge for processing massive seismic data. In addition to the data deluge, a numerical solver capable of accurately simulating seismic wave propagation is very important for tomography to achieve higher resolution. To solve the (an)elastic or acoustic wave equation in realistic 3D models, a spectral-element method [10–12] is used to achieve high accuracy for realistic Earth models in complex geometries. Because the most computationally expensive part of a tomographic inversion involves wavefield simulations, excellent performance of the solver is crucial, as discussed in Section 13.3.

Using data from 253 earthquakes and adjoint tomography technique, we generated model GLAD-M15 [6], the first Earth model where forward and Fréchet derivative calculations were performed numerically in 3D background models. To further improve the resolution of this model, we have expanded the data set to 1000 earthquake and will eventually include all available earthquakes (Mw 5.0–7.0) on the global scale (more than 6,000 as of 2016). Inverting such a large data set requires optimizing I/O performance within data processing and simulations as well as efficient management of the entire workflow. To prepare for exascale computing in seismic tomography, we address the bottlenecks encountered in our current adjoint tomography studies and discuss possible solutions. A typical adjoint tomography workflow is shown in Figure 13.1. It involves three major stages: (1) calculating

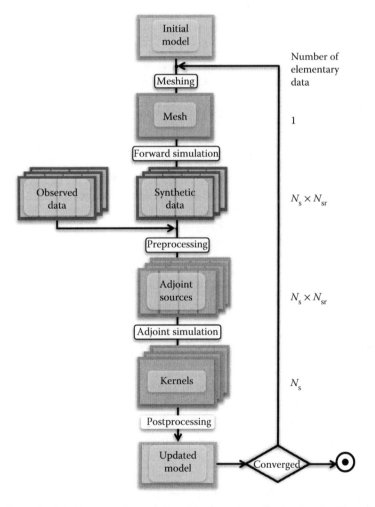

FIGURE 13.1 General adjoint tomography workflow. The focus is on the data involved in each step. Seismic data are depicted by dark-framed boxes and for each of the N_s seismic events they are recorded by N_{sr} receivers. Computational data are represented by shaded boxes. The amount of elementary data varies depending on the workflow stage and can eventually be grouped into a smaller number of files.

synthetic seismograms and preprocessing observed and synthetic data, (2) calculating the gradient of the misfit function (the Fréchet derivatives), and (3) postprocessing and updating the model.

The preprocessing stage is dedicated to assimilating data: (1) signal processing (i.e., tapering, resampling, filtering, deconvolving the instrument response, etc.), (2) window selection to determine the usable parts of seismograms according to certain criteria by comparing observed and simulated seismograms, and (3) making measurements in each window and computing the associated adjoint sources [3,7,8,13]. The preprocessing stage can easily involve dealing with millions of seismograms and has become a bottleneck in tomographic inversions. Although some groups have their own data formats, the Seismic Analysis Code (SAC) format has been the main seismic data format for earthquake seismology. Seismic data in SAC files are single time series stored independently for each component of each receiver, along with limited metadata as header. For large-scale studies, this format produces severe I/O traffic in data processing and numerical simulation due to the large numbers of files to be dealt with. An Adaptable Seismic Data Format (ASDF) has been developed as an alternative to SAC. It enables flexible data types and allows storing large data in a single file through its hierarchical organization. We illustrate the advantages of this new data format in Section 13.5.

Computing the gradient of the misfit or objective function is accomplished through the interaction of a forward seismic wavefield with its adjoint wavefield; the latter is generated by back-projecting seismic measurements simultaneously from all receivers [14]. This procedure requires two numerical simulations for each source: one for the forward wavefield from the source to the receivers and another for the adjoint wavefield from all the receivers to the source. The simulations are run for each source, resulting in *event kernels*, and the gradient of the misfit function is simply the sum of all event kernels. Between forward and adjoint simulations, we need to store a vast amount of wavefield data to account for the full attenuation of seismic waves in an anelastic Earth model. For thousands of events, the I/O approach we used to efficiently calculate the kernels is discussed in Section 13.4. In the postprocessing stage, the gradient is preconditioned and smoothed. Based upon the gradient, a model update is obtained using a conjugate gradient or Limited-memory Broyden-Fletcher-Goldfarb-Shanno (L-BFGS) [15] optimization scheme. Usually, tens of iterations have to be performed to obtain a stable model, and in each iteration we have to process data, and run forward and adjoint simulations for thousands of events. At each stage, an additional complication is the necessity to check results and relaunch jobs when failure occurs. Manually handling the procedures described above is difficult because of the size of the data set and also because of the distributed resource environment. Therefore, an efficient scientific workflow management system is needed and various options are discussed in Section 13.6.

13.3 SOLVER: SPECFEM3D_GLOBE

13.3.1 OVERVIEW AND PROGRAMMING APPROACH

Our seismic inversion workflow includes many parts, each of them implemented based on a different software package. The most computationally expensive part involves the solver SPECFEM3D_GLOBE, a spectral-element code capable of modeling seismic wave propagation using fully unstructured hexahedral meshes of 3D Earth models of essentially arbitrary complexity. Capabilities for its application in adjoint tomography are extensively illustrated in [16]. Table 13.1 outlines the importance of the solver relatively to other parts of the workflow for a shortest period of 17 seconds. As the number of earthquakes grows, the importance of having a computationally efficient solver to perform both forward and adjoint simulations becomes clear.

At the top level, SPECFEM3D_GLOBE is parallelized using message passing interface (MPI). The parallelization extensively relies on the analytical decomposition of a cubed-sphere mesh, splitting the mesh into partitions with close-to-perfect load balancing and assigning a single partition to a

TABLE 13.1

Summary of Computational Requirements for a Maximum Resolution of 17 sec Using 180 min Seismograms on Oak Ridge Leadership Computing Facility (OLCF) Infrastructure

	1 Earthquake	253 Earthquakes	6,000 Earthquakes
Shortest period: 17 sec		1 Iteration	1 Iteration
Duration: 180 m			
Solver (core hours) (forward + adjoint)	~11,520	~2,764,560	~69,120,000
Preprocessing (core hours)	~15	~3,795	~90,000
Postprocessing (core hours)	~5,760	~5,760	~5,760

single MPI process. This coarse domain decomposition parallelism may even include a finer level of parallelism inside each domain.

To take advantage of architectures of modern processors, in particular multicore CPUs and hardware accelerators such as graphics processing units (GPUs), a finer level of parallelization is added to each MPI process. For computations on multicore CPU clusters, partial OpenMP support and vectorization has recently been added by Tsuboi et al. [17].

The SPECFEM software suite has been running on CUDA-enabled GPUs since 2009 [18]. It has continuously been adapted to take advantage of advances in NVIDIA GPUs [9,19].

With the advent of heterogeneous supercomputers, our code base, as for many other applications, has become increasingly complex. Due to a relatively large user base and the variety of targeted systems, several code paths need to be maintained. The most important issue is to ensure that similar capabilities are available, regardless of the system. Another matter is to provide the user with optimized software.

Our solution has been to use the BOAST [20] transpiler to generate both CUDA and OpenCL kernels. The performance of these BOAST-generated kernels is very close to those implemented directly in CUDA and provides code optimized for various NVIDIA GPU generations.

On the one hand, because SPECFEM3D_GLOBE strives to be a tool capable of solving unexplored geophysical problems, most of it is written by geophysicists. On the other hand, BOAST provides the programmer with a Ruby-based domain-specific language (DSL), an unfamiliar language for most of our developers. This implies that although BOAST is equally capable of generating Fortran or C code, we maintain the original Fortran code as the reference source for CPU execution.

In the future, and in particular for the next generation heterogeneous supercomputers, it will be interesting to investigate directive-based approaches. This is increasingly true with accelerator support becoming more mature in OpenMP 4. If performance is close to hand written code, it will certainly be the solution of choice toward a portable, geophysicist-friendly code base.

13.3.2 SCALABILITY AND BENCHMARKING RESULTS

During a large-scale inversion, the proportion of computational time spend to simulate wave propagation mandates the solver to be performant and to scale well. Figure 13.2 shows SPECFEM3D_GLOBE strong scaling results for simulations running on multiple GPUs with a shortest period of 9 seconds. It demonstrates that very close to ideal efficiency can be achieved with a minimum size of about 100 MB of data per GPU. Thus, the code exhibits excellent scalability and can be run with almost ideal code performance in part because communications are almost entirely overlapped with calculations.

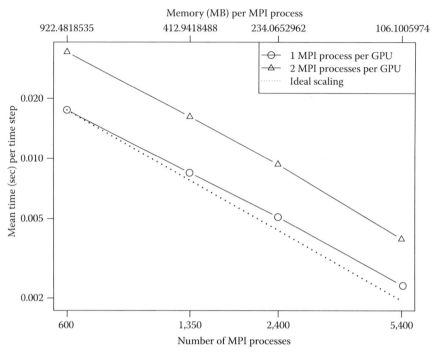

Strong scaling for the PREM, NEX_XI = 480

FIGURE 13.2   Strong scaling for the spherically symmetric PREM on Titan for a minimum seismic period of 9 sec. The mesh is comprised of ~ 25 million elements for this resolution. Values are plotted against the number of MPI processes. PREM, preliminary reference earth model.

Also note that the average elapsed calculation time per time-step is inversely proportional to the number of GPUs whether one uses a single MPI process per GPU or two MPI processes. This indicates that we can run multiple MPI processes on a single GPU without loss of performance. Depending on the number of simultaneous seismic events that we want to simulate and the expected maximum time-to-solution, these benchmarks help us select both the optimal number of GPUs and the organization of the MPI tasks across them. Scalability improves as the scale of the simulation becomes larger and larger. Although scalability performance at the lowest resolution (27 seconds) is not optimal (Figures 13.3 and 13.4), as the resolution increases to 9 seconds scalability performance is very close to ideal (Figures 13.2 and 13.5). It should also be admitted that as the number of MPI tasks used in the simulation increases, though scalability performance is not affected, the overall performance of each GPU node will be slightly decreased. This is because communications between MPI processes on each node will slightly slow down the computation.

Figure 13.6 shows SPECFEM3D_GLOBE weak scaling test results for an increasing problem size while keeping the work load for each MPI process almost the same. Weak scaling tests are a bit unusual for SPECFEM3D_GLOBE. What we do is find a setup where slices have more or less the same number of elements, then correct the timing by dividing the actual number of elements per slice and multiply with the setup, which we take as a reference. It can be observed from the weak scaling plot that parallel efficiency for GPU simulations is excellent and scales within 95% of the ideal runtimes across all benchmark sizes. Comparing weak scaling performances between GPU and CPU-only simulations on a Titan node, that is, a K20x GPU (Kepler) versus 16 CPU-cores (AMD Interlagos), we see a speedup factor of the GPU simulations of ~18x for the chosen mesh sizes. For high-end Intel Xeons microprocessor or IBM Power microprocessor, the speedup factor

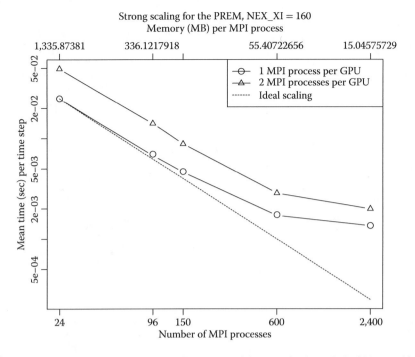

FIGURE 13.3 Strong scaling for the PREM on Titan for a minimum seismic period of 27 sec with full attenuation. Values are plotted against the number of MPI processes. PREM, preliminary reference earth model.

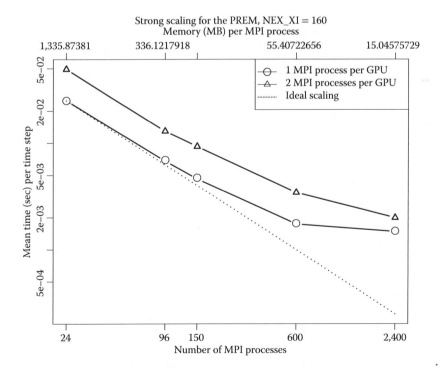

FIGURE 13.4 Strong scaling for the PREM on Titan for a minimum seismic period of 27 sec in the case of
without *full attenuation*. Values are plotted against the number of MPI processes. PREM, preliminary reference
earth model.

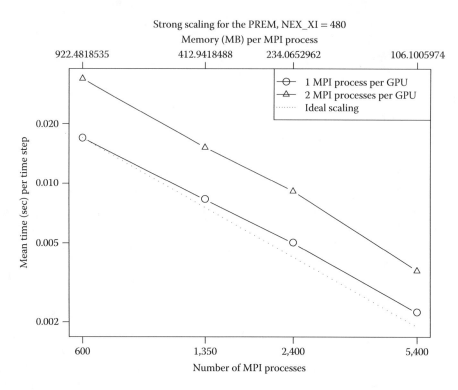

FIGURE 13.5 Strong scaling for the PREM on Titan for a minimum seismic period of 9 sec in the case of without *full attenuation*. Values are plotted against the number of MPI processes. PREM, preliminary reference earth model.

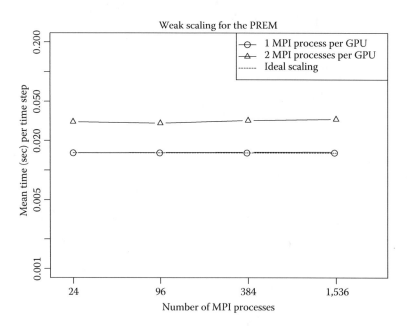

FIGURE 13.6 Weak scaling for the PREM on Titan. Performance is measured as the averaged mean time for each time step. The same work load is applied to each MPI process in different cases. PREM, preliminary reference earth model.

of the GPU simulation may not be as high as 18x. To make SPECFEM3D_GLOBE ready for next-generation supercomputers, e.g., GPU-based machines with high-end microprocessors, continuous work should be carried out in order to maintain a high speedup factor for GPU simulations. Our scaling results indicate that asynchronous message passing is nearly perfect in hiding network latency on Titan. Note that this becomes more of a challenge for smaller mesh sizes, where the percentage of outer elements increases compared to inner elements, but such small meshes are never used in practice.

Next, we discuss scaling results for low- and high-resolution simulations. We anticipate that the scalability of low-resolution simulations will be worse because the computing resources of each node will not be effectively used and MPI communication costs will take up a larger percentage of the total computation time. An additional issue is that accurate calculation of the gradient of the misfit function requires full attenuation. This involves storage of snapshots of the forward wavefield, which are read back during the adjoint simulation, leading to increased I/O. We expect this additional I/O to adversely affect performance.

Figures 13.3 and 13.4 show strong scaling tests for a low-resolution model (minimum period of 27 seconds) with and without full attenuation, respectively. In each figure, the two diagrams show the scaling results for 1 and 2 MPI tasks per GPU node. The scalability does not change as the number of MPI tasks increases. However, when the number of GPU tasks increases from 1 to 2, the processing time per time step doubles. The scalability does not change wether or not we are considering full attenuation. As anticipated, when considering full attenuation the processing time per time step is a bit longer than when not considering full attenuation. Table 13.2 shows a comparison of processing time per time step in two different cases for a different number of MPI processes. Processing time with full attenuation is a bit longer than without.

Figures 13.7 and 13.8 show strong scaling tests for a medium resolution model (minimum period of 17 seconds) in two different situations: with and without full attenuation. The scalability in the two cases is similar, except that when considering full attenuation the scalability is a bit better than that of the opposite case. The number of MPI processes still does not affect scalability but is inversely proportional to computational performance. Compared to the 27 seconds case, the 17 seconds simulation shows scalability. Processing efficiency with full attenuation is also a bit lower than that of without full attenuation, as shown in Table 13.3.

Figure 13.2 shows strong scaling tests for 9 seconds resolution with full attenuation. Figure 13.5 is the same test without full attenuation. It is obvious that scalability is close to ideal in both cases. Comparing the actual processing time for the two cases, as shown in Table 13.4, we conclude again that full attenuation causes slightly lower efficiency.

We compare the processing time per time step for different output data formats, including ASDF, our newly design modern seismic data file format (see Section 13.5). Whether or not we use ASDF, the processing time per time step is almost unaffected. This observation is encouraging, because when we enjoy the benefits of ASDF we do not lose the superb scalability performance of the code. We also compare scalability performance with respect to the number of receivers. We find that the number of receivers does not influence scalability of SPECFEM3D_GLOBE. This

TABLE 13.2

Comparison of Processing Time per Time Step With and Without Full Attenuation (27 sec Resolution)

No. of MPI processes	24	96	150	600	2,400
With full attenuation	0.025	0.00679	0.00462	0.00174	0.00148
Without full attenuation	0.025	0.00678	0.00462	0.00171	0.00135

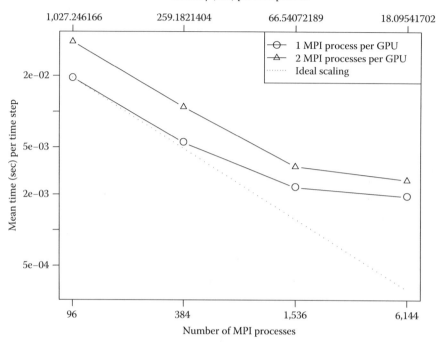

FIGURE 13.7   Strong scaling for the PREM on Titan for a minimum seismic period of 17 sec in the case of with *full attenuation*. Values are plotted against the number of MPI processes. PREM, preliminary reference earth model.

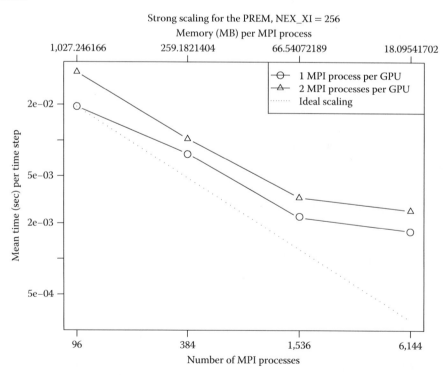

FIGURE 13.8   Strong scaling for the PREM on Titan for a minimum seismic period of 17 sec in the case of without *full attenuation*. Values are plotted against the number of MPI processes. PREM, preliminary reference earth model.

TABLE 13.3

Comparison of Processing Time per Time Step With and Without Full Attenuation (17 sec resolution)

No. of MPI processes	96	384	1,536	6,144
With full attenuation	0.0194	0.00542	0.0226	0.0189
Without full attenuation	0.0193	0.00490	0.0023	0.00171

TABLE 13.4

Comparison of Processing Time per Time Step With and Without Full Attenuation (9 sec Resolution)

No. of MPI processes	600	1,350	2,400	5,400
With full attenuation	0.0175	0.00848	0.0051	0.00233
Without full attenuation	0.017	0.00827	0.005	0.00229

indicates that as the global seismographic network becomes denser and denser, scalability does not degrade.

In order to prepare SPECFEM3D_GLOBE for the next generation supercomputers, for example, *Summit*, code improvements should be done to ensure a linear speedup of GPU simulations to at least 20 % of current Titan nodes (~3,600 MPI processes) for both high-resolution and medium-resolution simulations. High-resolution simulations (~9 s) have already demonstrated perfect scalability to at least 5,400 nodes, which ensures a stable application for the intermediate future.

13.4 OPTIMIZING I/O FOR COMPUTATIONAL DATA

The workflow sketched in the previous sections deals with two main types of data, namely, seismic and computational data. In this section, we focus on computational data.

Computational data are, in general, characterized by discretization and representation of the scientific problem. In our case, these are mesh and model files, and data sensitivity kernels, which are the output of SPECFEM3D and SPECFEM3D_GLOBE used in the postprocessing stage. They are shown shaded in the workflow chart in Figure 13.1. The size of these files depends on the spatial and temporal resolutions. For instance, a transversely isotropic global adjoint simulation (100 minutes long seismograms at a resolution going down to 27 seconds with 1,300 receivers) reads 49 GB of computational data and writes out 8 GB of data for adjoint data sensitivity kernels. When increasing the resolution of the simulations by going down to a shortest period of 9 seconds, all these numbers should be multiplied by 3^3, yielding about 1.3 TB of computational data. In practice, the number of events drastically reduces the simulation size we are able to reach, even on the latest supercomputers. This problem becomes even more prevailing when more realistic physics is used. To compute anelastic sensitivity kernels with full attenuation, a parsimonious disk storage technique is introduced [21]. It avoids instabilities that occur during time-reversing dissipative wave propagation simulations. In practice, this leads to a dramatic increase for I/O, as outlined in Figure 13.9.

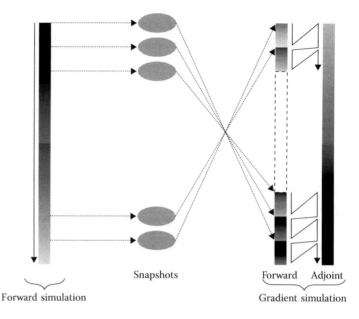

Snapshots Forward Adjoint

Forward simulation Gradient simulation

FIGURE 13.9 I/O pattern with full attenuation. The forward simulation (left bar) produces snapshots at regular intervals. During the kernel simulation (right bars) these snapshots are read in reverse order to piecewise reconstruct the forward wavefield for a number of time steps. The interaction of this reconstructed wavefield with the adjoint wavefield yields to the so-called event kernels, the sum of which is the misfit gradient. Solid arrows depict computation order; dashed arrows represent I/O. Bars shade from black to light gray jointly with forward time.

13.4.1 PARALLEL FILE FORMATS AND I/O LIBRARIES FOR SCIENTIFIC COMPUTING

Although I/O libraries and file systems must be considered together to improve I/O techniques, only I/O libraries are exposed to the scientific programmer. In what follows, we try to focus our efforts on a more library-oriented approach.

The POSIX standard defines an I/O interface designed for sequential processing. Its single *stream of bytes* approach is well known to poorly scale on distributed memory machines. Extensive studies tried to extend this standard [22], most of the time in combination with research on parallel file systems.

When considering parallel software, developer choices have a great impact on how I/O calls perform. There are two simple ways to use the POSIX I/O interface in such parallel software:

1. The most straightforward method is having separate files for each task. Eventually, subsequent processing can be applied on subsets of files to gather data. For a large number of tasks, this approach leads to a large number of files, potentially causing contention on the file system metadata servers.
2. Each process sends its data to a set of master MPI tasks in charge of writing data to disk. Conversely, data can be read from disk by a subset of MPI tasks and then distributed among all MPI tasks. This approach has a few drawbacks. In particular, nodes running master tasks might run out of memory if the number of scattered/gathered data is high. Moreover, network traffic is likely to reach high levels, slowing down the execution.

Over the past decades, a number of techniques have been developed and incorporated into libraries to ease writing, sometimes large amounts of data, to disk. Our goal is not to provide a full description of parallel I/O techniques and libraries evolution. Fur this purpose, the reader may consult Schikuta and Vanek [23] or Liu et al. [24]. However, we do believe that understanding where the need for simple and generic application program interfaces (APIs) comes from helps determine a solution satisfying our needs. In what follows, we consider distributed memory systems and software programmed using MPI to date the most common paradigm to address large scientific simulations on modern HPC systems.

The first version of MPI did not define a dedicated I/O API. Parallel I/O libraries were often developed to match particular architectural or applicative requirements. For instance, ChemIo [25] was developed as an interface for chemistry applications and SOLAR [26] to accelerate writing dense out-of-core matrices. Thakur et al. developed ADIO (Abstract-Device interface for parallel I/O) [27] to fill the gap between various parallel file systems and parallel I/O libraries. This work ultimately led to ROMIO [28], one of the most popular MPI-IO libraries that was later integrated as a component into the well-known MPICH2 library. Although MPI-2 introduced support for parallel I/O with the MPI-I/O approach, using it in large scientific code is not always straightforward: it is a low-level API writing raw data to files and demands a concerted effort from the scientific programmer.

Scientific software can benefit from libraries wrapping complexity into a higher level function set. Hence, libraries accommodating metadata were designed to ease further exploitation of produced data. Two of the most popular parallel IO libraries embedding metadata are netCDF [29] and HDF5 [30]. Both of them provide a parallel interface. Although netCDF is principally oriented toward the storage of arrays, the most common scientific data structure, HDF5 is more versatile and based on user-defined data structures. The distinction between these two libraries became blurrier as netCDF, starting from version 4, is implemented through HDF5. Metadata allow further analysis on potentially large data sets in providing the necessary information to fetch values of interest. A significant number of well-established tools are based on this format, showing their durability.

An alternative is the ADIOS library [24] released by ORNL. Compared to netCDF and HDF5, it works on simpler data structures because its main focus is parallel performance. Besides metadata availability similar to the other formats previously mentioned, it also lets users change the transport method to target the most efficient I/O method for a particular system or platform. A set of optimizations is embedded in so-called *transport methods*. I/O experts have the option to develop new transport methods while scientific developers have to pick one matching the platform, their software and their simulation case. In particular, transport methods describe the file pattern, that is to say, if a number of MPI tasks write data to the same number of files, to a smaller number of files or to a single file. The underlying optimizations contain methods such as aggregation, buffering, and chunking and are transparent to the user.

Two APIs are available, one POSIX-like API with a reduced number of functions and one XML API that is not as flexible as the regular API but allows one to keep I/O separated from the main program.

From a user perspective, reading data is very similar to writing them. It must be noted that it is possible to read and write data from a *staging* memory area, thus limiting disk access when produced data are consumed right away.

Although the ADIOS library needs to be improved by providing optimizations tuned for a larger number of file systems (IBM General Parallel File System [GPFS] for instance), its architecture allows domain scientists to focus on the actual problem.

Liu et al. [24] demonstrated excellent improvements in terms of I/O speed. For instance, both S3D, software simulating reactive turbulent flows, and PMCL3D, a finite-difference-based software package simulating wave propagation, show a 10-fold improvement when switching from MPI-IO to ADIOS. They managed to write at a 30 GB/sec rate when using 96 K MPI tasks for S3D and 30 K MPI tasks for PMCL3D.

13.4.2 INTEGRATING PARALLEL I/O IN ADJOINT-BASED SEISMIC WORKFLOWS

Computational data, in general, do not require complex metadata because they are well structured within numerical solvers. In legacy SPECFEM3D_GLOBE, the way computational data were written to disk was not problematic on local clusters for smaller size scientific problems (e.g., regional- or continental-scale wave propagation, small seismic data sets). However, to run simulations more efficiently on HPC systems for more challenging problems, such as global adjoint tomography or increased resolution regional- and exploration-scale tomography, we needed to revise the way the solver handles computational data. In the previous version of the SPECFEM3D packages, for each variable or set of closely related variables, a file was created for each MPI process. The number of files, for a single seismic event, was proportional to the number of MPI processes P. For a full iteration of the workflow, the number of files was $\mathcal{O}(P.N_s)$. Accessing these files during large-scale simulations did not only have an impact on performance but also on the file system due to heavy I/O traffic. This is because of the difficulty for the file system metadata server to handle all requests to create the whole set of files. The new implementation uses ADIOS to limit the number of files accessed during reading and writing of computational data, independent of the number of processes, that is $\mathcal{O}(N_s)$. Because writing a simple file is also a potential bottleneck due to lock contention, we are sometimes asked to change the transport method to output a limited number of files. This is mostly invisible to the code user as these files are output as subfiles that are part of a single file. As an additional benefit, using ADIOS, HDF5, or netCDF let us define readers for popular visualization tools, such as Paraview and VisIt.

Tests have been run to assess the I/O speed to write models in SPECFEM3D_GLOBE on the Titan supercomputer. The test case has been chosen to match the number of processes and to result in the same amount of I/O as the complete simulation for our 253 earthquake database. This results in more than 6 million ($6 \times 1,007^2$) spectral elements on the Earth's surface, processed through 24,576 MPI tasks. ADIOS experts at ORNL indicated that the preferred way to get I/O performance is to use the MPI_AGGREGATE transport method. This method is carefully tuned for large runs on Lustre file systems. Suitable parameters for this transport method were given by ORNL ADIOS experts in order to match both OLCF Spider file system characteristics and the test case parameters. A single ADIOS writer process was associated with 256 Object Server Targets (OST), 32 MPI tasks running on 2 Titan nodes. The test was later reproduced on the new OLCF Atlas file system. Results in Table 13.5 show that switching from Spider to Atlas brings a significant improvement in terms of I/O bandwidth. Moreover, in this case, the peak bandwidth on the Spider file system is 16.9 GB/sec, while on the Atlas file system the peak bandwidth is 51.5 GB/sec. This is likely to be beneficial for our research, especially when the spatial resolution is increased, yielding large data volumes on a high number of nodes.

TABLE 13.5

Bandwidth for SPECFEM3D_GLOBE Output Using the ADIOS MPI_AGGREGATE Transport Method for Mesh Using 24,576 MPI Tasks

Mesh Region	Output Size (GB)	Spider (GB/sec)	Atlas (GB/sec)
Crust–Mantle	2,548	14.3	40.6
Outer Core	317	7.4	8.47
Inner Core	177	4.8	7.6

Note: Results are presented both for the old (Spider) and new (Atlas) OLCF file systems. Numbers for different regions outline that large files benefit the most from use of the ADIOS library.

13.5   A MODERN APPROACH TO SEISMIC DATA: EFFICIENCY AND REPRODUCIBILITY

Seismology is a science driven by observing, modeling, and understanding data. In recent years, large volumes of high-quality seismic data are becoming more and more easily accessible through the dramatic growth of seismic networks. Together with the development of modern compute platforms, both large-scale seismic simulations and data processing can be done very efficiently. However, the seismological community is not yet ready to embrace the era of big data. Most seismic data formats and processing tools were developed more than a decade ago and are becoming obsolete in many aspects. In this chapter, we present our thoughts and efforts in bringing modernity into the realm.

The very basic unit of seismic data is the seismogram, which is a time series of ground motion recorded by a seismometer during an earthquake. Most seismometers on land are able to record three-component data: one vertical component and two horizontal components perpendicular to each other (usually east and north), whereas seismometers in the ocean vary, some are equipped with a water pressure sensor and others record three-component displacement of the seabed (an Ocean Bottom Seismometers). Seismometers are recording 7/24 and data are archived at data centers. The data we are primarily interested in are 3-hour time windows after an earthquake, during which time seismic waves propagate inside the earth and gradually damp out.

13.5.1   LEGACY

Most earthquake seismologists are familiar with SAC (see [31]), a general-purpose program designed for the study of time series. It provides basic analysis capabilities, including general arithmetic operations, Fourier transforms, filtering, signal stacking, interpolation, etc., which fit the general requirements of researchers.

Alongside the software package, SAC also defines a data format, which has been widely used by seismologists over the last few decades. In the SAC data format, each waveform (time series) is stored as a separate data file, containing a fixed length header section to store metadata, including time and location information), followed by the data section (the actual time series). Thus, for one earthquake recorded by 2,000 seismometers, one would expect 6,000 independent SAC files. The main reason SAC is popular within the seismological community is its ease of use and its interactive command line tools. Even though functionality is limited, SAC covers the most frequent needs of seismologists. For example, visualizing seismograms is very easy in SAC and is frequently used to check the effect of operations applied to data.

However, things are evolving quickly as more and more seismometers are installed. For a single network, the number of seismometers can go easily beyond 1,000, leading to sizable data sets. Thus, SAC and its associated format is no longer a good choice for the following reasons.

- It has limited nonprogrammable functionalities. SAC tools have to be invoked by system calls (shell scripts) and the lack of APIs for programming languages, such as C and Fortran, makes it difficult to customize workflows.
- The SAC data format only stores one waveform per file. Given 5,000 earthquakes and 2,000 stations for each earthquake, $3 * 10^7$ SAC files have to be generated and stored. Reading or writing such a large number of files is highly inefficient.
- The header in the SAC data format is very limited, with only a fixed number of predefined slots to store metadata. However, a modern data format should be flexible enough for users to define metadata relevant to the problem they are solving. Imposing predefined offsets in bytes to access information is a recipe for disaster.
- Station information, which contains instrument response information, is stored in separate files. This approach increases the number of files to deal with and the possibility of making errors. Having the ability to store station information along with the waveform data greatly reduce the chances of mistakes.

13.5.2 THE ADAPTABLE SEISMIC DATA FORMAT

We looked for existing solutions lacking the drawbacks listed in the previous section. Because introducing a new data format should ideally be avoided, the seismological community has been postponing the definition of more modern approaches. We believe that the advantage of a new data format is significant enough to quickly outweigh the initial difficulties of switching to a new format. We identify five key issues that the new data format must resolve:

1. **Robustness and stability**: The data format should be stable enough to be used on large data sets while ensuring data integrity and the correctness of scientific results.
2. **Efficiency**: The data format should be exploitable by efficient, parallel tools.
3. **Data organization**: Different types of data (waveform, source and station information, derived data types) should be grouped at certain levels to perform a variety of tasks. The data should be self-describing so no extra effort is needed to understand the data.
4. **Reproducibility**: A critical aspect of science is the ability to reproduce results. A modern data format should facilitate and encourage this.
5. **Mining and visualization of data**: Data could be queried and visualized anytime in an easy manner.

The ASDF [32] was introduced to solve these issues. Using HDF5 at its most basic level, it organizes its data in a hierarchical structure inside a container; in a simplified manner a container can be pictured as a file system within a file.

HDF5 was chosen as the underlying data format for a variety of reasons. First, HDF5 has been used in a wide variety of scientific projects and has a rich and active ecosystem of libraries and tools. It has a number of built-in data compression algorithms and data corruption tests in the form of check summing. Second, it also fulfills our hard requirement of being capable of parallel I/O with MPI [30,33]. Besides, there is no need to worry about the endianness of data, which historically has been a big issue in seismology.

An ASDF file is roughly arranged in four sections, as follows:

1. Details about seismic events of any kind (earthquakes, mine blasts, rock falls, etc.) are stored in a QuakeML document.
2. Seismic waveforms are grouped seismic station along with meta information describing the station properties (a StationXML document).
3. Arbitrary data that cannot be understood as a seismic waveform are stored in the auxiliary data section.
4. Data history (provenance) is kept as a number of SEIS-PROV documents (an extension to W3C PROV).

Existing and established data formats and conventions are utilized wherever possible. This keeps large parts of ASDF conceptually simple and delegates pieces of the development burden to existing efforts. The ASDF structure is summarized in Figure 13.10. With such a layout, every seismogram of a given earthquake can be naturally grouped into one ASDF file. Also, event information and station information are incorporated so no extra files have to be retrieved during processing.

Reproducibility is frequently discussed and widely recognized as a critical requirement of scientific results. In practice, it is a cumbersome goal to achieve and is frequently ignored. Provenance is the process of keeping track of and storing all constituents of information that were used to arrive at a certain result or at a particular piece of data. The main goal of storing provenance data directly in ASDF is that scientists looking at data described by it should be able to tell what steps were taken to generate that particular piece of data. Each piece of waveform and auxiliary data within ASDF can optionally store provenance information in the form of a W3C PROV or SEIS-PROV document.

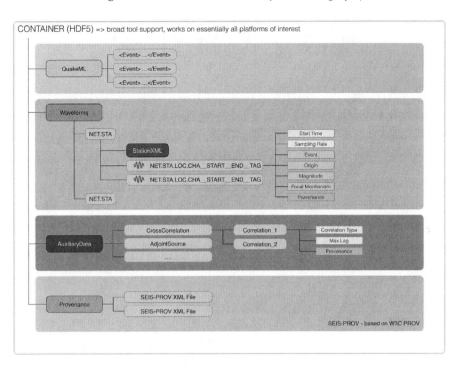

FIGURE 13.10   Layout of the Adaptable Seismic Data Format (ASDF), including earthquake event information, waveforms and station meta information, auxiliary data, and provenance. In such layout, different types of data are grouped into one file and ready for later retrieval.

Thus, such a file can be safely archived and exchanged with others and information that led to a certain piece of it is readily available.

More details about ASDF may be found in [32].

13.5.3   DATA PROCESSING

Global adjoint tomography is ideal for the ASDF data format. First, the data volumes involved are massive, easily containing millions of seismograms. Second, it necessitates sophisticated processing to turn raw data into meaningful results. Here, we present a typical data processing workflow occurring in full seismic waveform inversions with adjoint techniques [5,13,34,35]. The general idea also translates to other types of tomography (see [36] for a recent review).

To enable a physically meaningful comparison between observed and synthetic waveforms, time series need to be converted to the same units and filtered in a way that ensures a comparable spectral content. This includes standard processing steps like detrending, tapering, filtering, interpolating, deconvolving the instrument response, and others. Subsequently, time windows in which observed and simulated waveforms are sufficiently similar are selected and adjoint sources are constructed from the data within these windows, see Figure 13.11 for a graphical overview.

The following is an account of our experiences and compares a legacy workflow to one utilizing the ASDF format, demonstrating the latter's clear advantages. Existing processing tools oftentimes work on pairs of SAC files, observed and synthetic seismic data for the same component and station, and loop over all seismic records associated with any given earthquake. Given the large number of seismic receivers and earthquakes, the frequent read and write operations on a very large number of single files create severe I/O bottlenecks on modern compute platforms. The implementation centered around ASDF shows superior scalability for applications on high-performance computers: observed

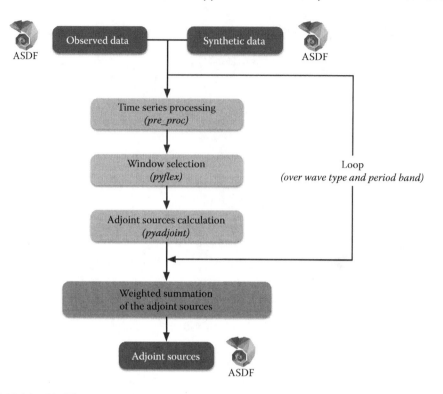

FIGURE 13.11 Workflow of seismic data processing using ASDF. First, time series analysis is applied to raw observed and synthetic data to ensure a comparable spectral content at a later stage. Then, time windows are selected for pairs of processed observed and synthetic data, inside which measurement are made. Finally, adjoint sources are generated as the final preprocessing output. ASDF speeds up these tasks by using parallel processing (MPI). For each step, processing information is added to the provenance.

and synthetic seismograms of a single event are stored in only two ASDF files, resulting in a significantly reduced I/O load. What is more, it is beneficial to keep meta information in the same file. For example, one does not need to reach out for separate files that keep track of the stations' instrument information or files containing earthquake information, which greatly reduces the complexity of operations and the possibility of making mistakes. Last but not least, provenance information is kept to increase reproducibility and for future reference.

Other than the data format itself, the data processing workflow benefits from the extensive APIs provided by ASDF. ASDF is supported in the SPECFEM3D_GLOBE package [11]. Synthetic ASDF files are directly generated, meaning synthetic data can seamlessly be fed as an input into the workflow. To maximize performance, we rewrote our existing processing tools. A big drawback in the old versions was that codes were written in different languages and unable to communicate with each other easily. For example, the SAC package was used for signal processing and the Fortran-based FLEXWIN program [37] for window selection. In the new version, we treat tasks as individual components in a single cohesive workflow. Relying on the seismic analysis package ObsPy [38], we redeveloped all workflow components in Python. Therefore, all components integrate with each other and stream data from one unit to the next. I/O only happens at the very beginning, when we read the seismogram into memory, and at the very end, when we write out the adjoint sources. All in all these changes empower us to increase the scale of our inversions, in terms of frequency content, number of earthquakes, and number of stations, and fully exploit modern computational platforms.

13.6 BRINGING THE PIECES TOGETHER: WORKFLOW MANAGEMENT

The importance of a performant solver to simulate forward and adjoint wavefields is well understood and accepted. In our case, sustained efforts are being made to adapt and tune SPECFEM3D_GLOBE to newer architectures. One of the most significant benefits of this work is the ability to use GPU acceleration.

The ever increasing performance level of the wave simulation software goes along with rapidly growing data volumes. This offers new opportunities for improving our understanding of the physics and chemistry of Earth's interior, but also brings new data management and workflow organization challenges.

Existing geoscience inversion workflows were designed for smaller scientific problems and simpler computational environments and suffer from I/O inefficiency, lack of fault tolerance, and inability to work in distributed resource environments. Workflow management challenges are by no means limited to earthquake seismology, let alone to global adjoint tomography. In exploration seismology, streamers can contain 60,000 hydrophones, and the number of shots can reach 50,000, requiring petabytes of storage. Even then, a crying lack of scalable seismic inversion software (outside of proprietary, closely guarded oil industry codes) poses a continuing obstacle to robust and routine inversions. Given data volumes in the petabytes and compute time requirements in tens to hundreds of millions of core-hours, new workflow management strategies are needed.

This section starts with a discussion of some of the most widely used scientific workflow management system and solutions that have been brought in order to manage seismic workflow. We then expose the requirements for large-scale seismic inversions and the design of a solution. Finally, additional challenges are outlined.

13.6.1 Existing Solutions

When researching which workflow engine would be the most appropriate for global adjoint tomography, the need to restrict the signification of the word *workflow* emerges. Indeed, workflow and workflow management have very different meanings depending on the application domain. Although there is some degree of similarity between business and scientific workflows [39], we exclusively consider tools focusing on the latter.

Even then, the number of tools available to manage scientific workflows is large. What follows discusses the main options and is by no means exhaustive. For more in-depth reviews, the reader should consult [40–42].

Focusing on usability by domain scientists, we also restrict ourselves to tools providing a higher level of abstraction and forbid ourselves to directly work with powerful but complex tools such as HTCondor [43].

From an user experience point of view, and forgetting about technical details, two competing approaches are available. The first one relies on graphical user interfaces (GUI). Examples include Kepler [44], Taverna [45], and Triana [46]. The second approach involves scripting and is implemented by a number of workflow management systems, such as Swift [47], Pegasus [48], or Radical-Ensemble [49]. Scripting is particularly well suited to scientists familiar with both HPC systems and software development. It allows for fast prototyping and flexible definition of workflows. As such, it provide users with powerful exploratory tools.

In the field of geophysics, fully managed workflows seem to be the exception rather than the norm. Of course, proprietary software geared toward the oil industry exists, but their closed nature forbids us to adapt and use them to perform global adjoint tomography. Most of the daily research and production computational work rely on a mixture of hand-written scripts steering more computationally expensive software such as solvers and data processing packages. Each scientist, or group of scientists, has their own set of scripts embedding a fair amount of specific knowledge about the system they are running on. Needless to say, such an approach is nonportable and error prone. Attempts to provide a more streamlined way of running these hand-written scripts have been made. Starting from the ever increasing importance of reproducible research [50], Fomel et al. developed Madagascar [51], where dependencies between tasks are managed with SCons, a software build tool similar in essence to GNU Make.

As science workflows, computer systems, and workflow engines grow more mature and complex, interdisciplinary collaboration is mandatory to bring seismic simulation and processing to the next level. One major exception to the lack of fully managed seismic workflows is CyberShake [52], which aims to compute probabilistic seismic hazard maps. CyberShake developers have been experimenting and using a number of workflow managers to schedule computations on a wide range of HPC centers. Among the workflow managers CyberShake has been run under the control of are Condor, Globus, Pegasus [53], and Radical-Pilot.

The Hadoop ecosystem, a popular paradigm to perform distributed computations, is worth mentioning. It has been used in production environments for many years by the industry and is gaining traction in scientific computing, especially to solve data-driven problems. For many scientific problems relying on HPC systems, involving large, multinodes simulations, it has so far remained an exotic approach. The frontier tends to blur, thanks to approaches such as Yarn. A noncritical, but interesting feature for a suitable workflow engine, is to be able to address both Hadoop and HPC systems. This is the case for some of Radical-Pilot [54] most recent developments.

13.6.2 Adjoint Tomography Workflow Management

As each problem and domain has widely different requirements, we fill focus on ad hoc solutions suited to large-scale seismic inversions on leadership-class resources, such has the ones provided by the U.S. Department of Energy (DOE) computing centers.

The first requirement for large-scale seismic inversions is performance along with efficiency. Indeed, the number of core-hours required to perform a global inversion being in the hundreds of millions, a suitable workflow management system needs to ensure that a minimum amount of compute cycles are wasted. Large compute centers have requirements on the size of jobs that are allowed to run; as they are primarily designed to accommodate computations that would not fit any other place. Although elementary computations of a seismic inversion do not fulfill this condition, the large number of simulations involved does fulfill this condition. This means that in order to match the queueing requirements, smaller-scale jobs have to be bundled in batches. Ideally, the workflow management system should be in charge of such accounting matters. This is, for instance, one of the features offered by the Pilot approach.

A second condition is the ability to execute in a relatively heterogeneous environment. Here, the concept of heterogenous environment is understood differently than for its more traditional grid computing definition. Each elementary part of the workflow is run on an homogenous machine, while different parts are not. For instance, for our current global inversions, Titan is used to run simulations while Rhea, an Intel-based commodity cluster, is used to process data. Appropriate resources are also used for visualizing data and data transfer.

Another reason to run seismic inversion under a workflow management system is reliability. On large systems, the mean time between failures [55] is reduced compared to smaller-scale systems. This is specially true for systems, such as the ones provided by DOE facilities, which are on the edge of what is technically feasible. Job failures due to hardware and software errors as well has corrupted data do happen. Hand-tracking causes of such failures when dealing with large data sets and numerous simulations is time consuming and error prone. The ability for a workflow to account for this failures and eventually relaunch jobs has become even more critical as we are the number of earthquake we assimilate data from raise.

It is equally important to keep the user in mind and follow the science problem logic. The typical user is a domain scientist with experience running simulations on large-scale supercomputers. Although the computer science details must be hidden from such users, they are usually fluent in developing scripts, allowing some level of technical details to be exposed to them. For this scenario, scripting is particularly well adapted as it provides a dynamic environment to define and iteratively improve workflows. This flexibility is a very desirable feature for our global tomographic inversion, where the numerical algorithmic strategy needs to be adapted according to the decrease in the misfit function and the lessons learned performing previous iterations. Domain logic is also better accommodated by a flexible ad hoc approach. From experience, concepts such as direct acyclic graphs (DAG) are, surprisingly enough, difficult to convey to domain scientists. It is important to note that the previous remarks are specific to large-scale exploratory computations by power users on leadership systems. A better approach for the broader community might very well be such that it includes a graphical interface and does not require any knowledge of the underlying system.

Additionally, a desirable feature is workflow portability. As newer distributed paradigms, such has Hadoop, are gaining traction across the scientific community, being able to run part of the workflow, most likely data processing, on such infrastructures would undoubtedly benefit us.

13.6.3 Moving Toward Fully Managed Workflows

From this panorama of existing workflow management systems and requirements, we can see what a suitable solution for large-scale seismic inversion is.

Past experience on defining scripts ranging from simple bash scripts to sophisticated modular python scripts taught us the need to separate the application domain from the engine running the workflow. This has several benefits, the most immediate being able to take advantages of the most recent advances in the application domain and in workflow science. Decoupling is also a good software engineering practice, as each part can be implemented, tested, and maintained independently. For instance, the application domain pieces might be used as standalone applications or plugged in different workflow engines. Similarly, the workflow engine can evolve to exhibit common patterns for a class of problems and thus be reused. However, this does not mean that each side should not be designed with the other one in mind. Indeed, complex inverse workflows impose significantly more complex and sophisticated resource management and coordination requirements than simple parameter sweeps in that they support varying degrees of coupling between the ensemble members, such as global or partial synchronization. In addition, all parts of the workflow must successfully complete to yield a meaningful scientific result.

From the previous requirements, and after surveying a few workflow management systems, it appears that two of them are particularly well suited to our needs: Pegasus and Radical-Pilot.

The first step to be able to take advantages of such tools is to ensure the most desired separation between the science software and the workflow management system. Due to the number of steps involved in processing seismic data (as explained in Section 13.5) and to create adjoint sources, we picked this stage of the inversion workflow as the first subworkflow to implement.

An important preparation has been to define clear interfaces for each of the executables. That is, each executable must clearly define its inputs and outputs without assumptions such as their relative location. Different parts can then be assembled, either as a DAG (in the case of Pegasus) or as an adhoc dependency description (in the case of Radical-Pilot). We experimented with both and consideration over the end-user experience oriented our choice toward the latter.

To operate, workflow engines store information about job statuses along with data useful to their internal machinery. Information of interest to the scientist and to the science workflow, regardless of the engine, also need to be tracked. We have chosen to have information relevant to our preprocessing workflow stored in an SQLite database. This database is regularly polled by the workflow engine to dynamically create and launch jobs along with relevant parameters. Its purpose is two-fold: to feed the workflow engine and to keep track of the assimilated data. The process is described in Figure 13.12.

For now, the workflow engine is rudimentary and relies on Radical-SAGA to launch jobs. Adapting it to a more complete solution, similar to Ensemble-MD, is ongoing. Relying on Radical-Pilot, patterns common to seismic inversions would be exhibited and released to the public.

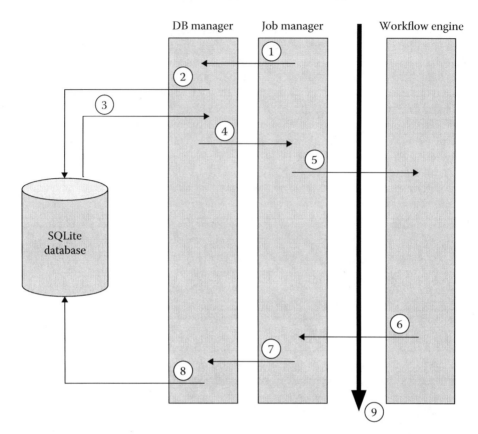

FIGURE 13.12 Steering process of the workflow management system. Two objects (DB manager and job manager) serve as an interface between the database and the workflow engine. The job manager requests data from the DB manager (1), which polls an SQLite database (2). Once the request is served (3), executables, parameters, and inputs are formatted (4) to feed the workflow engine (5). The workflow engine transparently launches the job and monitors its status, which is then returned (6). The status is used to update the database (7, 8) in order to keep track of the inversion status. This process is repeated (9), until everything has been processed.

13.6.4 ADDITIONAL CHALLENGES

As we progress through the implementation and the understanding of automated seismic inversion workflows, several challenges worth mentioning need to be taken care of. Taking full advantage of large-scale resources requires tight software integration. For instance, some next generation super-computers have burst buffers allowing staging of data between computing steps. Although this is a promise of greater performance, this is problematic from a workflow management perspective. Indeed, such techniques disrupt the control flow and defy the purpose of a workflow engine. How to solve this is an open question. A second challenge comes from the desire of scientists to visualize intermediate data. This is motivated by the necessity to take informed decisions to steer the inversion process in the best direction possible. This calls for a level of interactivity that interfaces well with an automated approach.

It is equally important to start thinking from the beginning about the general geophysicist popula-tion and how they can benefit from developments made for large-scale inversions. We are confident that the pattern defining approach of the Ensemble toolkits, along with the system-agnostic Radical-SAGA backend, is a step toward dissemination.

13.7   SOFTWARE PRACTICES

The number and complexity of the software packages that we have developed, in order to be able to perform exascale seismic inversions, have required us to adopt more rigorous software practices.

Compared to professional software development teams, scientific software developers face particular issues. First, they are often a group of independent researchers that are in different physical locations. Second, there is often a large range in the level of programming experience. To address these issues, we implemented a simple collaborative workflow based on modern software development techniques, such as Agile development and Continuous Integration/Continuous Development.

The two main goals of this workflow are to facilitate communication between the developers and to ensure that new software developments meet some agreed upon quality criteria before being added to the common code base. In practice, we have implemented this workflow using the tools provided by the GitHub platform, and the automatic testing frameworks Buildbot, Jenkins, and Travis CI.

As illustrated in Figure 13.13, our collaborative software development practice is organized around three Git repositories.

The central repository: where the code is shared among developers, and where users can download releases of the code.
Forks of the central repository: where each developer can post their changes to be tested by Buildbot, before being committed into the central repository.
Local clones: where each developer builds his/her changes.

The first two repositories are hosted on the Computational Infrastructure for Geodynamics (CIG) GitHub organization. The third repository is on the developers desktop or laptop.

An essential part of our workflow is the differentiation between production and development code: the production code is in a Git branch called `master`, and the development code is in a branch called `devel`. The code in the `master` branch is intended for users that are only interested in running the code, whereas the `devel` branch is intended for code developers. The changes to the code are first committed to the `devel` branch. Development code is transferred to the `master` branch only after extensive testing.

A fundamental rule of our workflow is that code changes can only be committed to the `devel` branch of the central repository through a pull-request. This provides us with two important features: first, it allows us to test the changes before they are committed to the central repository, and, second, it sends an e-mail notification to the group of developers. The notification to the developers is important because it lets them review the changes before they are committed to the shared repository.

We have two distinct roles within our developer's community: *code maintainer* and *code developer*. The code maintainer role consists of accepting the code changes proposed by the developers. The code maintainers have push/pull (or read/write) permissions to the central repository, whereas the developers only have pull (or read only) permissions. In addition, code maintainers cannot accept their own source code changes.

Assuming that a developer already has a clone of the central repository on his or her local machine, a typical workflow for committing new code developments to the central repository is as follows (see Figure 13.13):

1. The developer pushes his or her changes to his or her fork on GitHub.
2. He or she opens a pull-request.
3. Opening a pull-request triggers automatic testing of the changes.
4. The maintainers and developers are notified of the results of the tests. If the changes failed the tests, then the developer needs to fix the problems and follow the steps 1 through 4, if they pass the test, go to step 5.

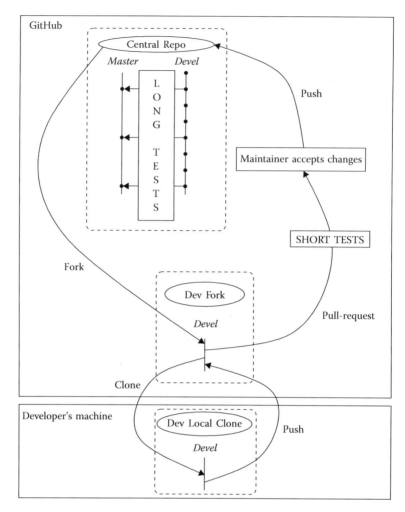

FIGURE 13.13   There are three types of Git repositories: the central repository on GitHub, the developer's fork on GitHub, and the developer's clone on their own machine. The only way for developers to commit their changes to the central repository is through a pull-request, which the code maintainers must accept. Short tests are run before the maintainers accept the changes. Longer, more extensive tests are run on daily and weekly bases, as well as when new developments are transferred from the `devel` to the `master` branches of the central repository.

5. Before they can be committed in the central repository, the code needs to be reviewed by other developers.
6. The maintainers accept the changes and the new code is committed to the `devel` branch of the central repository.

For this workflow to be successful, it is crucial to have a carefully designed test suite. We use three types of tests for our codes.

Compilation tests: they consist in looping through all the available compiling options (e.g., OpenMP, MPI, CUDA) and using different compilers (GCC and Intel compilers).
Unit tests: they test individual functions by checking the output for some predefined set of input parameters.

Functional tests: in our case, functional testing refers to the testing of the a set of features of the code. Concretely, we run full examples for which we compare the computed seismograms with some precomputed reference seismograms.

The compilation, unit, and some functional tests can be all done within 15 minutes of opening the pull-request. These quick tests are the only ones that are done before a pull-request is accepted. Other tests, that take longer to execute, are run on daily and weekly bases. If these tests fail, then some changes need to be reverted, but at least the changes are recent and there are few of them, so it is easy to find what needs to be fixed; we failed but we failed early.

In conclusion, our experience over the past 3 years has shown this software development workflow to balance simplicity and effectiveness. Its simplicity has made it easy to adopt both experienced and new developers, without hindering new developments. Its effectiveness at detecting problems early has ensured the stability of our central repository. In addition, by making the changes to the code more visible, this workflow has improved the communication within our developer's community and enabled the release of increasingly more sophisticated software.

13.8   CONCLUSION

We have outlined some of the difficulties arising in modern computational seismology. They stem from the need to simultaneously handle large data sets and increased Earth model resolution. This is even more true when performing large-scale inversions at leadership supercomputer centers. Even though the data volumes might not be comparable to what is commonly referred as *big data*, data and workflow management are creating performance and file system issues on supercomputers.

In order to be able to pursue our scientific goals on the next-generation supercomputers, we have devised several strategies. For heavy computational I/O we now rely on ADIOS. The developers of ADIOS have either tight links with US computing centers or are part of them. We rely on the improvements they bring to the so-called transport method to continue getting a satisfying level of performance. To accommodate the attenuation snapshots required for anelastic simulations, several additional strategies might need to be developed as the specificities of next-generation machine are unveiled. We can think about overlapping I/O calls with computations, or using on-node nonvolatile memory (NVMe) as a burst buffer.

Interestingly enough, the focus of seismic inversion is shifting from pure computations to a more balanced approach, where data are a first-class citizen, seen as equally important as computations. Using a modern file format, such as ASDF, including comprehensive metadata, not only helps increase computational performance, but also ensures reproducibility, and in the long term brings a standard to seismic and computational data that ultimately increases collaboration within the seismological community.

The shear number of data and simulations is becoming increasing difficult to manage. Workflow management has been sparsely used within the seismological community and, to the best of our knowledge, not in production-scale inversions. This last sentence might be controversial, but, in our opinion, to be considered as managed, a workflow must provide the user with automation going beyond a simple dependency description. Workflow management is an exciting challenge, particularly with the present effort of infrastructure designers (both hardware and software) to bring *HPC* and *Big Data* systems closers.

Many other challenges remain and keep arising during our journey to perform global adjoint tomography problems on exascale systems. Some of the more thrilling include exploring deep-learning methods to assimilate data in a more sensible fashion, as well as newer visualization techniques allowing scientists to discover features in global Earth models with unprecedented levels of detail.

ACKNOWLEDGMENTS

This research used resources of the Oak Ridge Leadership Computing Facility at Oak Ridge National Laboratory, which is supported by the Office of Science of the U.S. Department of Energy under Contract No. DE-AC05-00OR22725.

Additional computational resources were provided by the Princeton Institute for Computational Science & Engineering and the Princeton University Office of Information Technology Research Computing Department.

REFERENCES

1. Dante. *Divine Comedy (Italian: Divina Commedia)*. 1320.
2. J. Vernes. *The Journey to the Center of the Earth (French: Voyage au centre de la Terre)*. Paris, France: Pierre-Jules Hetzel, 1864.
3. A. Tape, Q. Liu, A. Maggi, and J. Tromp. Adjoint tomography of the southern California crust. *Science*, 325(5943):988–992, 2009.
4. A. Fichtner, B. L. N. Kennett, H. Igel, and H.-P. Bunge. Full seismic waveform tomography for upper-mantle structure in the Australasian region using adjoint methods. *Geophysical Journal International*, 179(3):1703–1725, 2009.
5. H. Zhu, E. Bozdag, D. Peter, and J. Tromp. Structure of the European upper mantle revealed by adjoint tomography. *Nature Geoscience*, 5(7):493–498, 2012.
6. E. Bozdag, D. Peter, M. Lefebvre, D. Komatitsch, J. Tromp, J. Hill, N. Podhorszki, and D. Pugmire. Global adjoint tomography: Firstgeneration model. *Geophysical Journal International*, 207(3):1739, 2016.
7. H. Zhu, Y. Luo, T. Nissen-Meyer, C. Morency, and J. Tromp. Elastic imaging and time-lapse migration based on adjoint methods. *Geophysics*, 74(6):WCA167, 2009.
8. Y. Luo, J. Tromp, B. Denel, and H. Calandra. 3D coupled acoustic-elastic migration with topography and bathymetry based on spectral-element and adjoint methods. *Geophysics*, 78(4):S193–S202, 2013.
9. M. Rietmann, P. Messmer, T. Nissen-Meyer, D. Peter, P. Basini, D. Komatitsch, O. Schenk, J. Tromp, L. Boschi, and D. Giardini. Forward and adjoint simulations of seismic wave propagation on emerging largescale GPU architectures. In *Proceedings of the International Conference on High Performance Computing, Networking, Storage and Analysis, SC '12*, pp. 38:1–38:11, IEEE Computer Society Press: Los Alamitos, CA, 2012.
10. D. Komatitsch and J. Tromp. Spectral-element simulations of global seismic wave propagation—II. Three-dimensional models, oceans, rotation and self-gravitation. *Geophysical Journal International*, 150(1):303–318, 2002.
11. D. Komatitsch and J. Tromp. Spectral-element simulations of global seismic wave propagation-I. Validation. *Geophysical Journal International*, 149(2):390–412, 2002.
12. D. Komatitsch and J. Tromp. Introduction to the spectral element method for three-dimensional seismic wave propagation. *Geophysical Journal International*, 139(3):806–822, 1999.
13. J. Tromp, C. Tape, and Q. Liu. Seismic tomography, adjoint methods, time reversal and banana-doughnut kernels. *Geophysical Journal International*, 160(1):195–216, January 2005.
14. A. Tarantola. Inversion of seismic reflection data in the acoustic approximation. *Geophysics*, 49(8):1259–1266, 1984.
15. J. Nocedal. Updating quasi-Newton matrices with limited storage. *Mathematics of Computation*, 35(151):773–782, 1980.
16. D. Peter, D. Komatitsch, Y. Luo, R. Martin, N. Le Goff, E. Casarotti, P. Le Loher, F. Magnoni, Q. Liu, C. Blitz, T. Nissen-Meyer, P. Basini, and J. Tromp. Forward and adjoint simulations of seismic wave propagation on fully unstructured hexahedral meshes. *Geophysical Journal International*, 186(2):721–739, 2011.
17. S. Tsuboi, K. Ando, T. Miyoshi, D. Peter, D. Komatitsch, and J. Tromp. A 1.8 trillion degrees-of-freedom, 1.24 petaflops global seismic wave simulation on the s/k/K computer. *International Journal of High Performance Computing Applications*, 30(4):411–422, 2016.

18. D. Komatitsch, D. Michéa, and G. Erlebacher. Porting a high-order finite-element earthquake modeling application to NVIDIA graphics cards using CUDA. *Journal of Parallel and Distributed Computing*, 69(5):451–460, 2009.

19. D. Komatitsch, G. Erlebacher, D. Göddeke, and D. Michéa. High-order finite-element seismic wave propagation modeling with MPI on a large GPU cluster. *Journal of Computational Physics*, 229(20):7692–7714, 2010.

20. J. Cronsioe, B. Videau, and V. Marangozova-Martin. Boast: Bringing optimization through automatic source-to-source transformations. In *IEEE 7th International Symposium on Embedded Multicore Socs (MCSoC)*, Tokyo, Japan, pp. 129–134, September 2013.

21. D. Komatitsch, Z. Xie, E. Bozdag, E. S. De Andrade, D. B. Peter, Q. Liu, and J. Tromp. Anelastic sensitivity kernels with parsimonious storage for adjoint tomography and full waveform inversion. *Geophysical Journal International*, 206(3):1467, 2016.

22. M. Vilayannur, S. Lang, R. Ross, R. Klundt, and L. Ward. Extending the POSIX I/O interface: A parallel file system perspective. Technical report, Argonne National Laboratory, 2008.

23. E. Schikuta and H. Vanek. Parallel I/O. *International Journal of High Performance Computing Applications*, 15(2):162–168, May 2001.

24. Q. Liu, J. Logan, Y. Tian, H. Abbasi, N. Podhorszki, J. Y. Choi, S. Klasky, R. Tchoua, J. Lofstead, R. Oldfield, M. Parashar, N. Samatova, K. Schwan, A. Shoshani, M. Wolf, K. Wu, and W. Yu. Hello ADIOS: The challenges and lessons of developing leadership class I/O frameworks. *Concurrency and Computation: Practice and Experience*, 26(7):1453–1473, August 2013.

25. J. Nieplocha, I. Foster, and R. A. Kendall. ChemIo: High performance parallel I/O for computational chemistry applications. *International Journal of High Performance Computing Applications*, 12(3):345–363, September 1998.

26. S. Toledo and F. G. Gustavson. The design and implementation of SOLAR, a portable library for scalable out-of-core linear algebra computations. In *Workshop on I/O in Parallel and Distributed Systems*, Philadelphia, PA, pp. 28–40, New York, NY: ACM, 1996.

27. R. Thakur, W. Gropp, and E. Lusk. An abstract-device interface for implementing portable parallel-I/O interfaces. In *Proceedings of 6th Symposium on the Frontiers of Massively Parallel Computation (Frontiers '96)*, Annapolis, MD. Piscataway, NJ: IEEE Press, 1996.

28. R. Thakur, W. Gropp, and E. Lusk. Data sieving and collective I/O in ROMIO. In *Proceedings. Frontiers'99. 7th Symposium on the Frontiers of Massively Parallel Computation*, Annapolis, MD. Piscataway, NJ: IEEE Press, 1999.

29. J. Li, M. Zingale, W.-k. Liao, A. Choudhary, R. Ross, R. Thakur, W. Gropp, R. Latham, A. Siegel, and B. Gallagher. Parallel netCDF: A high-performance scientific I/O interface. In *Proceedings of the 2003 ACM/IEEE conference on Supercomputing—SC '03*, pp. 39, New York, NY: ACM Press, November 2003.

30. M. Howison, Q. Koziol, D. Knaak, J. Mainzer, and J. Shalf. Tuning HDF5 for lustre file systems. In *Proceedings of Workshop on Interfaces and Abstractions for Scientific Data*, Heraklion, Crete. Piscataway, NJ: IEEE Press, 2010.

31. G. Helffrich, J. Wookey, and I. Bastow. *The Seismic Analysis Code: A Primer and User's Guide*. New York, NY: Cambridge University Press, 2013.

32. L. Krischer, J. Smith, W. Lei, M. Lefebvre, Y. Ruan, E. Sales de Andrade, N. Podhorszki, E. Bozdag, and J. Tromp. An adaptable seismic data format. *Geophysical Journal International*, 207(2):1003, 2016.

33. MPI Forum. Message Passing Interface (MPI) Forum Home Page. http://www.mpi-forum.org/, 2009.

34. A. Fichtner, H.-P. Bunge, and H. Igel. The adjoint method in seismology: I. Theory. *Physics of the Earth and Planetary Interiors*, 157(1–2):86–104, 2006.

35. A. Tape, Q. Liu, A. Maggi, and J. Tromp. Seismic tomography of the southern California crust based on spectral-element and adjoint methods. *Geophysical Journal International*, 180(1):433–462, 2010.

36. Q. Liu and Y. Gu. Seismic imaging: From classical to adjoint tomography. *Tectonophysics*, 566–567:31–66, 2012.

37. A. Maggi, C. Tape, M. Chen, D. Chao, and J. Tromp. An automated time-window selection algorithm for seismic tomography. *Geophysical Journal International*, 178(1):257–281, 2009.

38. M. Beyreuther, R. Barsch, L. Krischer, T. Megies, Y. Behr, and J. Wassermann. ObsPy: A python toolbox for seismology. *Seismological Research Letters*, 81(3):530–533, 2010.

39. A. Ludascher, M. Weske, T. McPhillips, and S. Bowers. Scientific workflows: Business as usual? *Business Process Management*, 5701:31–47, 2009.

40. J. Yu and R. Buyya. A taxonomy of scientific workflow systems for grid computing. *ACM SIGMOD Record*, 34(3):44–49, 2005.

41. A. Barker and A. Hemert. Scientific workflow: A survey and research directions. In *International Conference on Parallel Processing and Applied Mathematics*, Gdansk, Poland, pp. 746–753. Berlin-Heidelberg, Germany: Springer, 2008.

42. V. Curcin and M. Ghanem. Scientific workflow systems-can one size fit all? *Cairo International Biomedical Engineering Conference*, Cairo, Egypt, pp. 1–9. Piscataway, NJ: IEEE Press, 2008.

43. A. Thain, T. Tannenbaum, and M. Livny. Distributed computing in practice: The condor experience. *Concurrency–Practice and Experience*, 17(2–4):323–356, 2005.

44. I. Altintas, C. Berkley, E. Jaeger, M. Jones, B. Ludascher, and S. Mock. Kepler: An extensible system for design and execution of scientific workflows. In *Proceedings of the 16th International Conference on Scientific and Statistical Database Management, SSDBM '04*, pp. 423–424, Washington, DC: IEEE Computer Society, 2004.

45. K. Wolstencroft, R. Haines, D. Fellows, A. Williams, D. Withers, S. Owen, S. Soiland-Reyes, I. Dunlop, A. Nenadic, P. Fisher, J. Bhagat, K. Belhajjame, F. Bacall, A. Hardisty, A. Nieva de la Hidalga, M. P. Balcazar Vargas, S. Sufi, and C. Goble. The Taverna workflow suite: Designing and executing workflows of Web Services on the desktop, web or in the cloud. *Nucleic Acids Research*, 41(Web Server issue):W328–W561, May 2013.

46. I. Taylor, M. Shields, I. Wang, and A. Harrison. *The Triana Workflow Environment: Architecture and Applications*, pp. 320–339. London: Springer, 2007.

47. M. Wilde, I. Foster, K. Iskra, P. Beckman, Z. Zhang, A. Espinosa, M. Hategan, B. Clifford, and I. Raicu. Parallel scripting for applications at the petascale and beyond. *Computer*, 42(11):50–60, 2009.

48. E. Deelman, K. Vahi, G. Juve, M. Rynge, S. Callaghan, P. J. Maechling, R. Mayani, W. Chen, R. Ferreira da Silva, M. Livny, and K. Wenger. Pegasus, a workflow management system for science automation. *Future Generation Computer Systems*, 46:17–35, 2015.

49. M. Turilli, M. Santcroos, and S. Jha. A Comprehensive Perspective on Pilot-Jobs, (under review) 2016.

50. S. Fomel and G. Hennenfent. Reproducible computational experiments using scons. In *IEEE International Conference on Acoustics, Speech and Signal Processing (ICASSP '07)*, Honolulu, HI, pp. 1520–6149. Piscataway, NJ: IEEE Press, 2007.

51. S. Fomel, P. Sava, I. Vlad, Y. Liu, and V. Bashkardin. Madagascar: Open-source software project for multidimensional data analysis and reproducible computational experiments. *Journal of Open Research Software*, 1(1):e8, 2013.

52. R. Graves, T. H. Jordan, S. Callaghan, E. Deelman, E. Field, G. Juve, C. Kesselman, P. Maechling, G. Mehta, K. Milner, D. Okaya, P. Small, and K. Vahi. CyberShake: A physics-based seismic hazard model for Southern California. *Pure and Applied Geophysics*, 168(3–4):367–381, 2011.

53. S. Callaghan, E. Deelman, D. Gunter, G. Juve, P. Maechling, C. Brooks, K. Vahi, K. Milner, R. Graves, E. Field, D. Okaya, and T. Jordan. Scaling up workflow-based applications. *Journal of Computer and System Sciences*, 76(6):428–446, 2010.

54. A. Luckow, P. Mantha, and S. Jha. Pilot-abstraction: A valid abstraction for data-intensive applications on HPC, Hadoop and Cloud Infrastructures? *The 24th International ACM Symposium on High Performance Distributed Computing, Portland, OR, (under review)*, 2015.

55. F. Cappello, A. Geist, B. Gropp, L. Kale, B. Kramer, and M. Snir. Toward exascale resilience. *International Journal of High Performance Computing Applications*, 23(4):374–388, 2009.

14 Scalable Structured Adaptive Mesh Refinement with Complex Geometry

Brian Van Straalen, David Trebotich,
Andrey Ovsyannikov, and Daniel T. Graves

CONTENTS

14.1	Introduction	307
14.2	Distributed Geometry	308
	14.2.1 EBIndexSpace Generation	308
	14.2.2 Local Geometry Caching	309
14.3	Feedback-Based Load Balancing	310
14.4	Aggregated Stencils	310
14.5	Sparse Plot File	311
14.6	Covered Box Pruning	312
14.7	Hybrid MPI+OpenMP Refactoring	312
14.8	Evaluated Platforms	313
14.9	Chombo-Crunch: Production Supercomputing	313
14.10	Burst Buffer and In-Transit Data Processing Workflow	315
14.11	Conclusions and Future Work	317
	Acknowledgments	317
	References	317

14.1 INTRODUCTION

Cartesian structured adaptive mesh refinement (AMR) methods have become an increasingly popular modeling approach to solving partial differential equations (PDEs) in complex or irregular geometries. Though there are several Cartesian grid approaches (e.g., immersed boundary, immersed interface, and ghost fluid), we focus on the cut-cell, embedded boundary (EB) approach of [1], and, specifically, as it pertains to unsteady flow problems where the arrangement of mesh refinement is not expressible *a priori*. In the EB cut cell approach, finite volume approximations are used to discretize the solution in the cut cells that result from intersecting the irregular boundary with a structured Cartesian grid. Conservative numerical approximations to the solution can be found from discrete integration over the nonrectangular control volumes with fluxes located at centroids of the edges or faces of a control volume [1].

Block structured AMR is a technique to add grid resolution efficiently and dynamically in areas of interest while leaving the rest of the domain at a coarser resolution. Effectively, grid resolution is added where and when it is needed in a simulation depending on a refinement criterion such as gradients in the data. AMR was originally applied to finite difference methods for inviscid shock hydrodynamics [2] and has been extended to inviscid, incompressible flow [3], and viscous flow [4,5] in rectangular domains.

The goal of the *EB-AMR* approach is to maintain the high-performance computing (HPC) advantages of Cartesian structured AMR methods while permitting a wider range of applications requiring complex boundary representation. Building on top of the Chombo package (`chombo.lbl.gov`), we have extended the Single Program Multiple Data (SPMD) parallelism model to include alteration of the finite-volume stencils in the region adjacent to an irregular boundary.

In this chapter, we describe the primary framework optimizations that enable scalable simulations (Sections 14.2 through 14.6). The application code Chombo-Crunch is presented as an example of production scale computing with the EB-AMR approach (Section 14.9).

14.2 DISTRIBUTED GEOMETRY

The regions of the domain requiring highly refined finite volumes will be changing over the course of the simulation. For regular structured AMR, this leads to the familiar tag-regrid-advance modeling loop. For the EB-AMR approach, this basic cycle needs to be augmented with geometric moments from the constructive solid geometry implicit function [6].

The alteration to control volumes that interact with the EB does not interact directly with the implicit function. Instead, the implicit function is first preprocessed into the EB index space object, referred to as the *EBIndexSpace*. The EBIndexSpace contains the preprocessed geometric moments of an implicit function restricted to the intersection with a Cartesian grid. The formalism for this construct is described in [6]. For any implicit function geometric representation and a specified minimal grid spacing, the EBIndexSpace is a preprocessing step. It can be done in line with the fluid simulation calculation, or it can be done off line as a preprocessing calculation and read in from a file. While the algorithm is automatic and robust, it is a nontrivial amount of computation and is too large to be replicated across processing elements. Hence, EBIndexSpace generation must be computed in parallel and stored in a distributed fashion.

14.2.1 EBIndexSpace Generation

EBIndexSpace generation takes the implicit function as input. The active simulation domain is defined as all the space where the implicit function is greater than zero. While the EBIndexSpace might be a larger distributed data structure, the implicit data itself can be extremely compact. We first cover the case where the implicit function has compact representation and can be replicated across all processing elements.

EBIndexSpace generation is parallelized with domain decomposition. We employ the simple *domainsplit* or the more involved *recursive bisection*:

- *Domainsplit* segments the Cartesian box global domain that encloses the simulation volume into equal-sized subdomains. Within each subdomain, a processing element executes the moment generation algorithm and stores on its local memory its portion of the EBIndexSpace. While domainsplit is simple to understand, explain, and implement, it only executes in a load-balanced fashion for problems that have equal distribution of geometric complexity throughout the domain.

   A number of applications fall into this category (e.g., subsurface flow and transport at the pore scale). The geometric information for the micron scale pore geometry demonstrates self-similarity and produces geometric information that evenly spreads throughout the domain. Other examples from both nature and engineering also have these properties (e.g., coral, battery electrodes, paper manufacturing).

- *Recursive bisection* is an alternative domain decomposition technique. Starting with base domain and minimal grid spacing, the implicit function is queried as to the amount of geometric material that is contained within that volume. If larger than a desired threshold, the domain is split in its longest dimension and each subdomain is then queried in a similar fashion. The recursive algorithm is repeated redundantly across each processing element until an acceptable threshold for geometric complexity within a subvolume is reached. The threshold is determined experimentally.

This recursive bisection tiling of the simulation domain is then partitioned across the processing elements using a space-filling curve. The EBIndexSpace is computed in parallel as was done in the domainsplit algorithm. While having a higher upfront cost, the recursive bisection algorithm can result in a better distribution of the EBIndexSpace across the compute platform.

14.2.2 Local Geometry Caching

PDE simulations on a structure-adaptive mesh have their own mapping of processing elements to computational regions. An adaptive mesh computation with subcycling represents a different computational pattern than the computational geometry problem. This requires a separate parallelization strategy and load-balancing technique.

In the standard subcycled AMR compute pattern, the advance step advances the solution one time step at a level. If a regrid interval is reached, then a new collection of Boxs called a BoxLayout are generated and then initialized. New stencils for the EB cut cells need to be generated. Although on a specific Box, there might be more than a dozen different stencils generated for any given finite volume based on the type of operator being discretized, they will all need to access the same underlying geometric moments.

EBISLayout EBIndexSpace::fillEBISLayout(const ProblemDomain\& p, const BoxLayout\& b, int ghostCells) is the Chombo operation that creates a local view of the EBIndexSpace for each specific patch. In this way, processing elements never need to know about the entire simulation process domain or about the EBIndexSpace. In this model, the pair [p,b, ghostCells] serves two roles: (1) indicates where geometric moments need to be communicated (what level of the refinement hierarchy and which patches are local to this processing element) and (2) provides a key into a caching map. EBISLayout is a refcounted object. In this way, we minimize how often the EBIndexSpace needs to be interrogated. The effects of local geometry caching for a high-fidelity incompressible viscous flow computation in a pore-scale geometry of packed spheres are shown in Table 14.1.

TABLE 14.1
Local Geometry Caching Effects

	Total Runtime (s)	Peak Memory Usage per Rank (MB)
Single-level, no caching	65	105
Single-level, caching	57	89
Adaptive mesh 2-levels, no caching	653	760
Adaptive mesh 2-levels, caching	348	413

14.3   FEEDBACK-BASED LOAD BALANCING

AMR modeling techniques pose a significant challenge to load balancing on distribute memory compute platforms. These algorithms have succeeded to the degree in which the compute load experienced by a processing element over a single patch of data is predictable. One can typically use the number of cells within a patch as a proxy for the expected compute load on a patch [7]. While not perfect, this approach has served well for AMR applications. Given a predictable compute load, various methods for distributing a load across processing elements like Knapsack and space-filling curve have been widely employed [8,9].

However, with the presence of the EB the number of cells on a patch is a poor proxy for compute load. Scaling tends to top out at 10X speedup. The first improvement to the cell count balancing is to apply a heuristic where irregular cells incur a larger penalty. This is meant to reflect the increased computation for irregular computation and the nonunit stride data access patterns. This heuristic model can become increasingly sophisticated, including the effects of cut cells crossing course-fine AMR interfaces, domain boundaries, and localized nonlinear physics. Though promising, this approach has not produced desired results. A performance model combining compile and run-time information with a tool like ExaSat [10,11] may illuminate better parameterizations than can be managed manually.

An alternative is to use a feedback-based load-balancing scheme. Each patch can construct a simple EB operator (e.g., Poisson). One step of a multigrid relaxation is computed and for each patch the actual compute time is measured. While the computational result is discarded, the compute time serves as the stand-in for the load-balance cost for that patch. This can be input into the existing load-balance algorithms, the operators are then reconstructed with the new distribution, and the simulation is allowed to advance (Table 14.2).

14.4   AGGREGATED STENCILS

The formalism described in [6] provides a powerful generalization of high-order finite volume methods for PDEs in arbitrary geometry. This expressive syntax is reflected in the EB-AMR class library in Chombo (EBChombo). While this powerful extraction layer is desirable in a framework that must target many application domains, it does come at a cost. The C++ abstractions for expressing arbitrary control volume discretizations imposes an unreasonable computational burden. While there are known techniques for faster execution of C++ abstractions, like template metaprogramming [12], these require knowledge of the stencil shape at compile-time. High-order EB stencils are produced from a least-squares algorithm at runtime based on the geometric moments and current mesh hierarchy. Therefore, EB stencil classes are passed through a run-time trace when they are bound to their target data structures. This process involves iterating through the abstract representation of a

---

TABLE 14.2
Different Load Balancers

	Total runtime (s)	Compute (%)/Commun. (%)
Standard, no ordering	41.60	45.83/54.17
Standard, Morton ordering	46.08	41.32/58.68
Feedback-based, no ordering	39.51	48.37/51.63
Feedback-based, Morton ordering	43.12	45.62/54.38

Note: Statistics are averaged over 10 iterations and five separate runs.

TABLE 14.3

Total Runtime Spent in AMR V-Cycle of Elliptic Solver: AggStencil versus EBStencil

	EBStencil (s)	AggStencil (s)	Speedup (s)
$128 \times 128 \times 128$ mesh			
Dual-socket *Ivy Bridge* node	3.00	1.99	1.50
Dual-socket *Haswell* node	2.22	1.62	1.37
KNL node (DDR)	5.71	3.54	1.61
KNL node (HBM)	5.27	2.88	1.83

Note: Runtime is averaged over eight iterations and five separate runs.

complex stencil one time and recording the final floating-point data deference locations and caching these offsets in the *AggStencil* (Table 14.3).

These AggStencil objects can be reused throughout multiple PDE operators across multiple time steps and only need to be reconstructed at the same time that the AMR level is regridded. The AggStencil represents a compromise between compilation efficiency and the need to have runtime data dependency integrated. AggStencil construction cannot be done as an offline processing step since it represents the combination of the preprocessed EBIndexSpace and the run-time processing of the AMR grid hierarchy.

14.5 SPARSE PLOT FILE

In the same way that the AggStencil object represents the object of a preprocessed geometry database and a runtime adaptive mesh hierarchy, the plot file has to capture this combination of static and dynamic data structures. In our initial implementation of plot file generation, the geometric moments were turned into self-centered state variables and output with the rest of the simulation data. We referred to this as the dense output. The problem with this approach is that for a relatively simple PDE like compressible Navier–Stokes, you might have five state variables (rho, X-momentum, Y-momentum, Z-momentum, and energy.) You can easily have twice that number of geometric moment information (e.g., volume fraction, area fractions, area centroids, and volume centroids) for even a second-order algorithm. For most cells, this data is trivial because they are not participating in a cut cell.

The benefit is that the visualization and data analytics pipeline in the visit analytics program could utilize moment geometric information for more complex postprocessing. In particular, the visualization postprocessor could reconstruct the geometric information and projections of state data onto the boundary manifold. The drawback for this approach is a tremendously large file that takes a long time to write and store and move. The alternative is to only write out nontrivial geometric moments. We call this the sparse format.

The sparse format creates an additional dataset next to the state data dataset in the HDF5 file that represents an unstructured graph embedded in the structured AMR grid hierarchy. This is very much like a vertex-edge data structure you find in traditional unstructured grid methods except we exploit the fact that the nodes in this graph are embedded in a Cartesian index space. Finite volume faces are the analog to edges. The edges will only pass along one cardinal access. Volumes correspond to vertices. The graph still must be written out to the data file in a parallel write operation into a collective HDF5 dataset and thus requires at least one invocation of a parallel prefix sum to compute unique processing element offsets.

In practice, there is not one clear choice between these two approaches. For geometrically dense problem definitions, the sparse file format has little savings in terms of final storage and it incurs the collective communication burden of creating the prefix sum. The two options are available in Chombo, and users can experiment to suit their application (Table 14.4).

14.6   COVERED BOX PRUNING

The runtime mesh generation algorithm is free to construct patches anywhere within the computational domain. It is oblivious to the geometric implicit function or the EBIndexSpace itself. After a regrid phase, it is not impossible to have patches that have fallen completely outside of the region of positive implicit function. While EBChombo will not expend computational work executing the finite volume method within these boxes and hence no payment of a runtime penalty, Chombo will still generate attendant data structures and storage for these patches as if the state variable extended into these regions.

It might seem like a rare case for this to happen, but by the quirk of how domain boundary conditions are specified in most legacy Chombo application examples, it is assumed that Level 0 extends throughout the entire extent of the Cartesian computational domain. In fact, the original Berger–Rigoutsos mesh segmentation algorithm takes that as an axiom. For applications that use shallow grid hierarchies, a significant amount of the computational finite volumes can fall outside of the intersection of the active domain and the Cartesian computational domain. A simple improvement is called pruning where boxes are visited prior to load-balancing, and if they have fallen completely outside of the active simulation domain, then they are removed from that level's patches (Table 14.5).

14.7   HYBRID MPI+OPENMP REFACTORING

Our previously published work for scalable EB calculations has relied upon a pure message-passing interface (MPI) 1.1 Application program interface (API) (e.g., [1,13]). The result of this reliance is that redundant meta-data used to optimize the communication phase of our calculations stressed the limited memory availability per processing element. On Knights Landing (KNL), there are 16 GB

TABLE 14.4

Plot File Size and Averaged Write Time to Lustre PFS: Dense Plot Format versus Sparse Plot Format

# Cores	File Size (GB)		Write time (s)	
	Dense	**Sparse**	**Dense**	**Sparse**
512	7.37	1.85	6.00	15.99
1,024	14.75	3.81	6.45	18.55
2,048	29.5	7.72	12.75	21.47
4,096	59	10.53	16.41	30.38
8,192	118	21.54	25.63	63.02
16,384	236	43.5	32.52	76.53
32,768	472	67.1	46.74	82.27

TABLE 14.5

Effect of Pruning of Covered Boxes on Plot File Size, Checkpoint File Size, and Total System Memory Usage

# Boxes	# Covered Boxes	Plotfile (GB)		Checkpoint (GB)		Memory (GB)	
		Original	Pruning	Original	Pruning	Original	Pruning
512	32 (6.25%)	7.37	6.91	6.01	5.63	216.5	210.1
4,096	512 (12.5%)	59	51.62	48.09	42.08	1,459	1,369
32,768	6,862 (20.9%)	472	373.8	384.75	304.7	11,211	9,699

of high bandwidth memory available, which is approximately five times less than traditional DRAM capacities on current X86 processors such as Ivy Bridge or Haswell. In order to take advantage of this high bandwidth memory, which is five times faster than traditional memory, and keep the same memory footprint, we have moved to a hybrid MPI+OpenMP model. The logical change to the software would be for compute elements to share common metadata. Our options are to adopt an MPI-3 remote memory access API, or a hybrid MPI+OpenMP programming model. For complex dynamic memory-intensive algorithms written in C++, it can be difficult to utilize the shared memory semantics of MPI-3. The metadata in Chombo AMR makes significant use of the operating system's free store mechanisms. To use MPI-3, we would need to replicate a heap manager within our user-space MPI windows.

We have chosen to use MPI+OpenMP. It is a programming model that is recommended from vendors and National Energy Research Scientific Computing Center (NERSC). In course-grained hybrid OpenMP, threads take the place of MPI ranks but maintain the SPMD programming model. While we don't anticipate this to result in speed improvements in the code, we observe significant improvements in the total memory usage in our simulations. As an example, we solve the incompressible Navier–Stokes flow equations. On Edison, the memory usage for flat MPI (24 ranks) is a Resident Set Size of 2,880 MB/node. With Hybrid MPI+OpenMP this total drops to 1,672 MB/node.

14.8 EVALUATED PLATFORMS

Edison is a Cray XC30 MPP at NERSC. Each node contains two 2.6 GHz 12-core Xeon Ivy Bridge chips each with four DDR3-1600 memory controllers and a 30 MB cache. Each core implements the four-way AVX SIMD instruction set and includes both a 32 KB L1 and a 256B L2 cache.

Cori Phase 1 is a Cray XC40 MPP at NERSC. Each node contains two 2.3 GHz 16-core Xeon Haswell chips and four DDR4-2133 memory controllers and 80 MB of L3 cache (Table 14.6) [14].

14.9 CHOMBO-CRUNCH: PRODUCTION SUPERCOMPUTING

Chombo-Crunch is a high-performance simulation code used to model pore scale reactive transport processes associated with subsurface problems including carbon sequestration. It is based on adaptive, finite volume methods developed in the Chombo framework [1] and, thus, relies heavily on the EB infrastructure for treatment of very complex geometries. In particular, the cut cell approach allows for explicit resolution of reactive surface area between mineral and pore. Reactions are treated with the geochemistry module of CrunchFlow [15], which performs point-by-point computations, and thus scales ideally with Chombo solvers. The code makes use of a novel interface between Chombo

TABLE 14.6

Overview of Evaluated Platforms

Core Architecture	Intel Ivy Bridge	Intel Haswell
Clock (GHz)	2.40	2.3
DP (GFlop/s)	19.2 (17.1)	36.8 (26.2)
Data cache (KB)	32+256	64+512
Chip Architecture	**Intel Xeon E5-2695v2**	**AMD Opteron 6172**
Cores	12	6
Last-level cache	30 MB	5 MB
DP (GFlop/s)	230.4	50.4
STREAM bandwidth	45 GB/s	12 GB/s
Memory capacity	32 GB	8 GB
System	**Cray XC30 (Edison)**	**Cray XC40 (Cori Phase 1)**
CPUs/node	2	4
Compiler	icc 14.0.0	icc 13.1.3

Source: Y. J. Lo et al., *International Workshop on Performance Modeling, Benchmarking and Simulation of High Performance Computer Systems*, Springer, 2014.

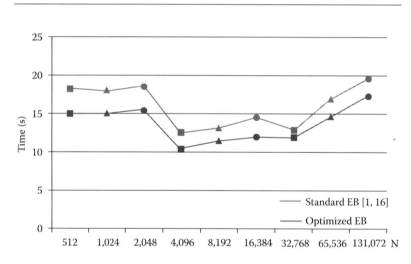

FIGURE 14.1   Weak scaling study of Chombo-Crunch on Cray XC30 NERSC Edison system.

and PETSc for access to algebraic multigrid solvers that do not possess the inherent shortcomings of geometric coarsening for very complex geometries [16]. Altogether, Chombo-Crunch is able to perform direct numerical simulation of real rock samples from image data at full-machine scale on DOE supercomputers. It has been validated by reactive transport experiments involving CO_2 injection [17,18].

Chombo-Crunch scales to full machine capability on NERSC supercomputers Edison [1], Cori Haswell, and now Cori KNL. The optimization work described in this chapter has improved that performance by 10%–15% and reduced the memory footprint of the application code by approximately 75% in order to fit into the high bandwidth memory on Cori KNL. Figure 14.1 shows the results of

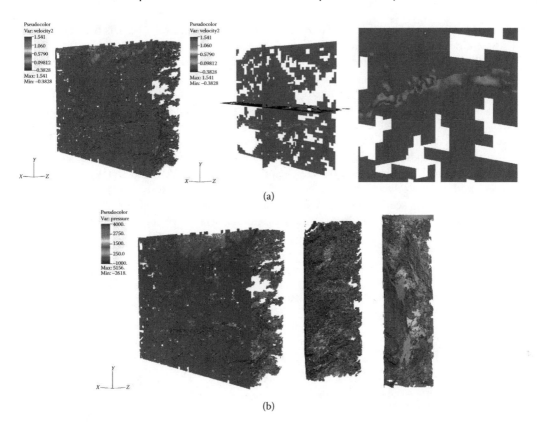

FIGURE 14.2 (See color insert.) Direct numerical simulation from image data of fractured shale with pruning. (From D. Trebotich and D. Graves. *Comm. App. Math. Comp. Sci.*, 10, 43–82, 2015.) (a) Velocity data mapped to surface and data slice. Pruned boxes are shown as white space. (b) Pressure mapped onto the mineral surface with side views. The simulation has been performed on NERSC Edison using 40,000 cores.

weak scaling on Edison. Comparison of the performance of standard EB from [1] and [16] versus optimized EB is provided.

As an example of Chombo-Crunch production capability, we perform direct numerical simulation of flow in the fractured shale example in [1] using the aforementioned optimizations. The computational domain has been discretized using nearly 2 billion grid points ($1,920 \times 1,600 \times 640$) and a resolution of 48 nanometers. Figure 14.2 shows velocity and pressure plots. Pruning of covered boxes, shown as white space, has resulted in a reduction of at least one third of both the total boxes and computational cores in [1]. (The memory bandwidth sweet spot for Chombo-Crunch due to domain decomposition and load balancing is 32^3 grid cells per box, one box per core on current Intel architectures Ivy Bridge and Haswell.)

14.10 BURST BUFFER AND IN-TRANSIT DATA PROCESSING WORKFLOW

Emerging exascale supercomputers have the ability to accelerate the time-to-discovery for scientific workflows. However, the I/O systems have not kept the same pace as computational systems: technology trends show a growing gap in performance of computational systems versus I/O systems. Moving forward exascale systems and the anticipated increase in the size of scientific datasets, the I/O constraint will become more critical. To address this problem, advanced memory hierarchies, such as burst buffers, have been recently proposed as intermediate layers between the compute nodes and the parallel file system. Recently, the Cray DataWarp Burst Buffer [19] has been deployed on NERSC's Cori Phase 1 system. It consists of 144 I/O nodes with overall more than 900 TB of non-volatile RAM (NVRAM). Keeping data in burst buffer, close to a processing element, allows a simulation to accelerate the checkpoint/restart process. We utilize Cray DataWarp burst buffer coupled with in-transit processing mechanisms, to demonstrate the advantages of advanced memory hierarchies in preserving traditional coupled scientific workflows. Figure 14.3 shows the schematic of in-transit workflow [20], which couples simulation (Chombo-Crunch) with on-the-fly visualization (VisIt). The burst buffer is used as a file-based coordination mechanism between workflow components. Figure 14.4 shows the amount of I/O in total simulation-visualization workflow runtime. The comparison of total I/O time in two cases when the Lustre file system and DataWarp burst buffer are used to store and exchange data between workflow components. Results are provided for different I/O intensities. As seen in Figure 14.4, the burst buffer provides a definite I/O improvement to the Chombo-Crunch+VisIt workflow at high-I/O intensity when a plotfile is written and processed

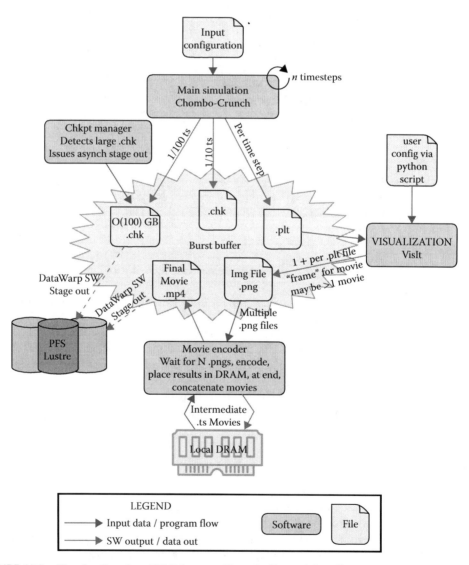

FIGURE 14.3 Chombo-Crunch and VisIt integrated burst buffer workflow diagram.

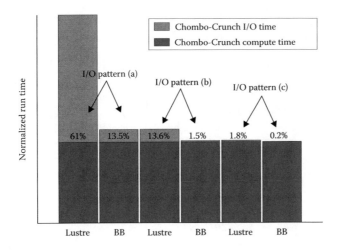

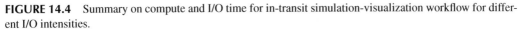

FIGURE 14.4   Summary on compute and I/O time for in-transit simulation-visualization workflow for different I/O intensities.

at every time step: 61% of I/O time for Lustre file system versus 13.5% of I/O time for for burst buffer; and medium I/O intensity when plot file is written and processed every 10 timesteps: 13.6% for Lustre file system and 1.5% of I/O time for burst buffer.

14.11   CONCLUSIONS AND FUTURE WORK

The combination of AMR and EB cut cell methods provides a powerful tool for multiscale, multiphysics simulation of PDEs in complex geometries. Mesh generation from arbitrary image data is not only tractable but even flexible in this approach. In order to scale to larger simulation domains and experimental time scales, however, we will need to alter our underlying software to accommodate new architectures on the path to exascale computing. A number of optimizations have been discussed in this chapter toward this end of distributed memory computing. Optimal mixture of these optimizations will be platform-dependent and will require further codesign with computer center experts and vendor support. We will be able to make use of a number of optimizations and features of these new machines. For example, off-chip NVRAM can be used for multicode coupling and more advanced workflows.

ACKNOWLEDGMENTS

This chapter is based upon work supported by the U.S. Department of Energy, Office of Science, Office of Advanced Scientific Computing Research, and in part by the Office of Basic Energy Sciences Energy Frontier Research Centers, and used resources of the National Energy Research Scientific Computing Center, all under contract number DE-AC02-05CH11231.

REFERENCES

1. D. Trebotich and D. Graves. An adaptive finite volume method for the incompressible Navier–Stokes equations in complex geometries. *Communications in Applied Mathematics and Computational Science*, 10(1):43–82, 2015.

2. M. J. Berger and P. Colella. Local adaptive mesh refinement for shock hydrodynamics. *Journal of Computational Physics*, 82(1):64–84, 23 May 1989.

3. D. Martin and P. Colella. A cell-centered adaptive projection method for the incompressible Euler equations. *Journal of Computational Physics*, 163(2):271–312, 2000.

4. A. S. Almgren, J. B. Bell, P. Colella, L. H. Howell, and M. J. Welcome. A conservative adaptive projection method for the variable density incompressible Navier-Stokes equations. *Journal of Computational Physics*, 142(1):1–46, May 1998.

5. D. F. Martin, P. Colella, and D. T. Graves. A cell-centered adaptive projection method for the incompressible Navier-Stokes equations in three dimensions. *Journal of Computational Physics*, 227:1863–1886, 2008.

6. P. Schwartz, J. Percelay, T. J. Ligocki, H. Johansen, D. T. Graves, D. Devendran, P. Colella, and E. Ateljevich. High-accuracy embedded boundary grid generation using the divergence theorem. *Communications in Applied Mathematics and Computational Science*, 10(1):83–96, 2015.

7. B. Van Straalen, P. Colella, D. Graves, and N. Keen. Petascale block-structured AMR applications without distributed meta-data. *Euro-Par 2011 Parallel Processing*, pp. 377–386, Bordeaux, France: Springer, 2011.

8. A. Dubey and B. Van Straalen. Experiences from software engineering of large scale AMR multiphysics code frameworks. *arXiv preprint arXiv:1309.1781*, 2013.

9. A. Dubey, A. Almgren, J. Bell, M. Berzins, S. Brandt, G. Bryan, P. Colella, D. Graves, M. Lijewski, F. Löffler et al. A survey of high level frameworks in block-structured adaptive mesh refinement packages. *Journal of Parallel and Distributed Computing*, 74(12):3217–3227, 2014.

10. C. Chan, D. Unat, M. Lijewski, W. Zhang, J. Bell, and J. Shalf. Software design space exploration for exascale combustion co-design. In J. M Kunkel, T. Ludwig, and H. W. Meuer (eds.), *Supercomputing*, pp. 196–212. New York, NY: Springer, 2013.

11. D. Unat, C. Chan,W. Zhang, S. Williams, J. Bachan, J. Bell, and J. Shalf. Exasat: An exascale co-design tool for performance modeling. *International Journal of High Performance Computing Applications*, 29(2):209–232, 2015.

12. S. W Haney, P. F. Dubois. Beating the abstraction penalty in c++ using expression templates. *Computers in Physics*, 10(6):552–557, 1996.

13. D. Trebotich, B.Van Straalen, D. T. Graves, and P. Colella. Performance of embedded boundary methods for CFD with complex geometry. *Journal of Physics: Conference Series*, 125:012083, 2008.

14. Y. J. Lo, S. Williams, B. Van Straalen, T. J. Ligocki, M. J. Cordery, N. J. Wright, M. W. Hall, and L. Oliker. Roofline model toolkit: A practical tool for architectural and program analysis. In *International Workshop on Performance Modeling, Benchmarking and Simulation of High Performance Computer Systems*, pp. 129–148. Springer, 2014.

15. C. I. Steefel. New directions in hydrogeochemical transport modeling: Incorporating multiple kinetic and equilibrium reaction pathways. In L.R. Bentley, J. F. Sykes, C. A. Brebbia, W. G. Gray, and G.F. Pinder (eds.), *Computational Methods in Water Resources XIII*, Rotterdam: A. A. Balkema, 2000.

16. D. Trebotich, M. F. Adams, S. Molins, C. I. Steefel, and C. Shen. High-resolution simulation of pore-scale reactive transport processes associated with carbon sequestration. *Computing in Science and Engineering*, 16(6):22–31, 2014.

17. S. Molins, D. Trebotich, C. I. Steefel, and C. Shen. An investigation of the effect of pore scale flow on average geochemical reaction rates using direct numerical simulation. *Water Resources Research*, 48:W03527, 2012.

18. S. Molins, D. Trebotich, L. Yang, J. B. Ajo-Franklin, T. J. Ligocki, C. Shen, and C. I. Steefel. Pore-scale controls on calcite dissolution rates from flow-through laboratory and numerical experiments. *Environmental Science and Technology*, 48(13):7453–7460, 2014. PMID: 24865463.

19. W. Bhimji, D. Bard, D. Paul, M. Romanus, A. Ovsyannikov, B. Friesen, M. Bryson, J. Correa, G. Lockwood, V. Tsulaia et al. Accelerating science with the NERSC burst buffer early user program. *Annual Cray User Group Meeting*, 2016.

20. A. Ovsyannikov, M. Romanus, B. Van Straalen, G. Weber, and D. Trebotich. Scientific Workflows at DataWarp-Speed: Accelerated Data-Intensive Science Using NERSC's Burst Buffer. *PDSW-DISCS*, to appear, 2016.

15 Extreme Scale Unstructured Adaptive CFD for Aerodynamic Flow Control

Kenneth E. Jansen, Michel Rasquin, Jed Brown,
Cameron Smith, Mark S. Shephard, and Chris Carothers

CONTENTS

15.1 Introduction...319
15.2 Scientific Methodology..320
    15.2.1 CFD and Turbulence Modeling 320
    15.2.2 Discretization Choices...321
    15.2.3 Aerodynamic Flow Control 321
        15.2.3.1 Economic Impact of Flow Control on a Vertical Tail/Rudder Assembly ..322
    15.2.4 Vertical Tail/Rudder Assembly Simulations: Past and Future.....................324
        15.2.4.1 Past Simulations at Re = 0.35 Million 324
        15.2.4.2 Estimating Resources for Flight Reynolds Number Flow Control...........326
        15.2.4.3 INCITE 2017: Five-Fold Rise in Reynolds Number Flow Control...........326
        15.2.4.4 Aurora ESP: Flight Reynolds Number Flow Control..............................327
    15.2.5 Computational Approach..328
        15.2.5.1 Parallel Flow Solver...328
        15.2.5.2 Adaptive Mesh Control...329
15.3 Algorithmic Details 330
    15.3.1 Flow Solver..330
    15.3.2 Adaptive Mesh Control and Partitioning 332
    15.3.3 Algorithm Summary 333
15.4 Programming Approach..334
    15.4.1 Scalability 335
    15.4.2 Performance 335
    15.4.3 Portability 335
    15.4.4 External Libraries 336
15.5 Software Practices 337
15.6 *In Situ* Visualization and Computational Steering...337
15.7 Summary..340
References 340

15.1 INTRODUCTION

Understanding the flow of fluid, either liquid or gas, through and around solid bodies has challenged man since the dawn of scientific inquiry. Many of the great minds of science and mathematics have progressively built up a hierarchy of fluid models since fluid flow impacts our lives in so many fundamental ways—from the early days of flow of water in viaducts to cities, to today's flight of planes.

This chapter is concerned with the computational modeling of turbulent flow around aerodynamic bodies such as planes and wind turbines. In this case, viscous effects near the solid bodies create very thin boundary layers that yield highly anisotropic (gradients normal to the surface may be 10^6 larger than gradients along the surface) solutions to the governing nonlinear partial differential equations (PDEs): the Navier–Stokes equations. Furthermore, turbulent flows develop extremely broad ranges of length and timescales motivating the use of discretization methods capable of employing adaptivity and implicit time integration. The combination of these features—nonlinear, anisotropy, adaptivity, and implicit time integration—dramatically raise the complexity of the discretization posing large challenges to efficient scalable parallel implementation. However, through careful design, the more complex algorithms can provide great reductions in computational costs relative to simpler methods (e.g., Cartesian grids with explicit time integration) that are easier to mate efficiently to hardware. In this chapter, we not only describe our approach, but we also address the fact that while complex algorithms may never be as efficient flop-for-flop as simple methods, in the important measure of science-per-core-hour, they can still win big by making complex features such as adaptivity and implicit methods as efficient and scalable as possible.

15.2 SCIENTIFIC METHODOLOGY

15.2.1 CFD AND TURBULENCE MODELING

Computational fluid dynamics (CFD) is a general field wherein a given fluid model is discretized for a computer solution by some technique. We further restrict our attention to continuum-based PDEs, viscous fluid models based on the Navier–Stokes equations. Viscous aerodynamic flows are characterized by Mach and Reynolds numbers. The Mach number is the ratio of flight speed, V, to *local* sound speed ranging from 0.18 at landing ($V = 200$ km/hr) to 0.8 at high-altitude cruise ($V = 880$ km/hr) for typical commercial airlines. The Reynolds number is the ratio of inertial forces to viscous forces and is given by $Re = Vc/\nu$, where V is the flow or plane speed, c is the dimension of the wing in the flow direction (e.g., mean chord of 6 m), and ν is the kinematic viscosity (e.g., for air $1.8 \times 10^{-5} \text{m}^2/\text{s}$). Simulation methods that resolve all of the continuum-level spatial and temporal scales are referred to as direct numerical simulation (DNS). For geometries encountered in aerodynamics, DNS produces too broad a range of scales in length and time to be fully resolved. Indeed, turbulence theory (Pope 2000; Wilcox 1998) provides that the ratio of the largest to smallest length scale increases proportional to the 3/4 power of Reynolds number. As turbulent scales interact in all three dimensions, this leads to cell count growth proportional to the 9/4 power of Reynolds number. Furthermore, time accuracy of the dynamics of these small scales requires the number of time steps to increase with the 3/4 power making the overall cost of the simulation grow cubically with Reynolds number. Since flight Reynolds numbers for commercial airlines range from $18 - 80 \times 10^6$ even for smaller components such as the vertical tail and rudder over the flight speed range, DNS will remain out of reach for aerodynamic modeling for several decades (Spalart et al. 1997). Wings have longer chords yielding higher Reynolds numbers. Full plane simulations must consider boundary layers growing over the fuselage that add another order of magnitude to the Reynolds number. This brings about the need to pursue a secondary model for some or all of the turbulent scales, typically referred to as a turbulence model. Models that average the equations in time to remove all of the turbulent scales are referred to as Reynolds-averaged Navier–Stokes (RANS) models. Intermediate to RANS and DNS are large-eddy simulation (LES) models, which, as the name implies, resolve the large eddies and model the smaller eddies. For aerodynamic problems at flight scale, LES, is typically still too costly and is expected to remain so (Spalart et al. 1997, 2001) for the next several decades. As the references explain, the impractical cost comes primarily from the near-wall regions where the flow is attached which have three-dimensional resolution needs that, while having improved constants, still scale spatially with the 9/4 power of Reynolds number. The modeling short-falls of RANS and the unacceptable cost of LES have motivated the development of hybrid models. One particularly promising model family is referred to as detached eddy simulation (DES) (Spalart et al. 1997, 2009; Shur et al. 2008). DES was developed specifically to behave as a RANS model in attached flows (where its predictive capacity is often acceptable) and behave as an LES model in separated or detached flow regions (where LES is often superior to RANS and where its cost is not excessive owing to the fact that separated flows are not strongly dependent on near-wall turbulence structures). DES escapes the often quoted 9/4 power cell growth with Reynolds number. DES cost is usually dictated by the volume the LES zone and the large- to medium-size eddies within it that must be resolved to provide its improved predictive capacity.

15.2.2 DISCRETIZATION CHOICES

While many groups are now making progress in scale-resolving methods such as LES and DES, our mature finite-element flow solver and anisotropic adaptive meshing procedures are ideally suited to address the challenging fluid flow problems involving complicated geometries with complex anisotropic solution features. The generality of our current algorithms, together with their ability to scale with an implicit solve on massively parallel systems (e.g., 786,432 cores of IBM BG/Q with more the 3 million processes [Rasquin et al. 2014], and 192 Ki Knights Landing [KNL] cores), make our tools a strong candidate for this class of problem. While there do exist alternatives that may be appropriate for other classes of problems, each has at least one limitation that makes their application to the complex problems considered here problematic. For example, structured grid approaches such as spectral methods (Mansour and Wray 1994; Moser et al. 1999; Kim et al. 1987) and finite difference methods (Morinishi et al. 1998) are outstanding choices for simple geometries but not complex ones like those considered here. Unstructured grid approaches such as hexahedral-based spectral element methods (Fischer et al. 2007) and finite volume methods (Mahesh et al. 2004; Muppidi and Mahesh 2005) are also good choices when all of the grid length scales in every region are known *a priori*, but these and other similar methods have not been successfully mated with parallel, anisotropic mesh adaptivity to the level that our solver has. Finally, block-structured adapted mesh refinement (SAMR) solvers (Dubey et al. 2014) address both the complex geometry and adaptivity issues but are inefficient for resolving the highly anisotropic solutions over curved geometries since the mesh scale is forced to match the smallest physical scale in all three dimensions near boundaries. For DES of aerodynamic boundary layer flows considered here, boundary layer meshes have normal resolution needs that range from 10^3 to 10^6 finer than the other two directions. In these cases, the near-wall region would require 10^6–10^{12} more points with isotropic SAMR than anisotropic AMR since the very fine wall resolution is needed in only one of the SAMR Cartesian axes.

In our approach, which is described in detail in subsequent sections, we employ anisotropic adaptivity for general unstructured meshes as well as boundary layer hybrid meshes. This anisotropic capability, together with the desire to automatically and efficiently discretize very complex geometries such as shown in Figure 15.1 are two of the reasons we focus our research on an anisotropic, adaptive unstructured grid solver.

15.2.3 AERODYNAMIC FLOW CONTROL

The goal of aerodynamics is to improve the vehicle performance over a wide range of operating conditions. Traditionally, this has been done solely with mechanical flaps but a new, alternative approach is to use active flow control, such as fluidic modification of the apparent aerodynamic shape of lifting surfaces using synthetic jet actuators (Amitay et al. 1998; Glezer and Amitay 2002). Synthetic jet actuators oscillate a piezoelectric membrane to alternately expel and ingest fluid (zero-net-mass-flux

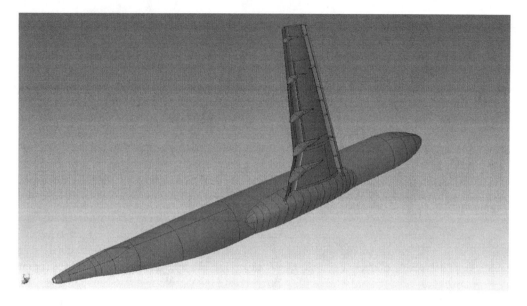

FIGURE 15.1 (See color insert.) Geometric model of a high lift wing/fuselage configuration.

driven electrically). Experimentally, synthetic jets have been shown to produce large-scale flow changes (e.g., reattachment of separated flow or virtual aerodynamic shaping of lifting surfaces) from microscale input (e.g., a 0.1 W piezoelectric disk resonating in a cavity); a process that has yet to be understood fundamentally.

These actuators are driven at frequencies that are much larger than the characteristic frequency of the flow and, thus, can be applied over a broad range of operating conditions. Synthetic jet actuation offers the prospect of not only augmenting lift but also other forces and moments in a dynamic and controlled fashion. This makes them an attractive solution for a wide variety of challenging flows. Some recent synthetic jet applications involve dynamic *aero-shaping* of wings and blades to improve the performance around design conditions and to alleviate unsteady aerodynamic loading. For example, previous and continuing experimental studies (Farnsworth et al. 2008; Maldonado et al. 2010) conducted at Rensselaer Polytechnic Institute have successfully demonstrated the ability of synthetic jets to restore and maintain flow attachment. Applications of this technology include flight control of aerial vehicles (including commercial airplanes) where conventional control surfaces (e.g., flaps, rudder, etc.) can be augmented or even replaced with active flow control, thus improving their lift-to-drag ratio and/or control power.

15.2.3.1 Economic Impact of Flow Control on a Vertical Tail/Rudder Assembly

The selection of a vertical tail/rudder assembly for past and future study is based on a number of factors. First, there is a clear case for dramatic energy savings with a flow control-based redesign of aeronautical control surfaces. With active flow control, greater control force can be obtained at lower deflection angles for a given control surface size. Or, more importantly, a smaller and lighter control surface can create the required force. We speak generically about control surfaces but these can include flaps, elevators, or even rudders on a vertical tail as was the subject of past Department of Energy 's (DOE) Innovative and Novel Computational Impact on Theory and Experiment (INCITE) (Argonne Leadership Computing Facility 2012, 2014, 2016) and Early Science Program (ESP) (Argonne Leadership Computing Facility 2010, 2015) studies. In the case chosen (vertical tail/rudder assembly), the energy impact is directly related to the size of the stabilizer since it is a significant contributor to drag in the cruise condition where much of the fuel is expended. A recent study

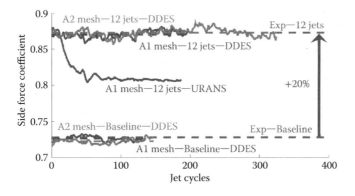

FIGURE 15.2 (See color insert.) Plot of side force with and without flow control. Experiment shown in dashed lines. Detached eddy simulation (CFD) predictions with two levels of adaptivity show excellent agreement with the experiment and grid independence.

at Boeing estimated that a 777-class airplane could reduce its fuel consumption by 0.75%–1.0% on a 3000 Nautical mile trip if its vertical tail size could be reduced by 25%. Our joint experimental/computational studies suggest that active flow control can achieve this size reduction (see Figure 15.2). Reducing the size of the vertical tail by 25% is not an arbitrary number. Current designs are in fact 25% larger than needed for all but one certification condition; when a plane needs to be capable of landing after losing one engine (so called *engine out* landing). Current vertical tail/rudder assemblies are sized 25% larger so that they can produce the required side force to compensate for unequal thrust via the rudder. Using active flow control, the smaller vertical tail can extend the rudder deflection angle past the point it would otherwise separate, thus producing the needed additional side force to satisfy this certification condition while reducing drag for all other parts of the flight envelope. To get an idea of the economic/energy impact of this improvement, consider that commercial airlines consumed over 20 billion (B) gallons of jet fuel a year. At current prices, this is $60 B and projections are for higher future prices.

Indeed, airline manufacturers have already optimized the wings and fuselage to a level that resizing the vertical tail is viewed as one of the new *lowest hanging fruits* for further energy efficiency. While not all of that 20 B gallons of fuel was expended on long flights where the cruise component of fuel consumption is such a high percentage, even a fairly conservative estimate of 0.5% savings would result in $0.3 B per year. It is also worth noting that the 25% reduction in the vertical tail size will also reduce the weight creating even more savings not only in cruise but also in the takeoff phase and in manufacturing costs. A similar analysis of wind turbines suggests that if flow control could reduce the unsteady loads that gusts create on blades then substantial savings would be reaped from reduced maintenance and replacement costs on the gear boxes that currently absorb those loads.

The other large fuel consumption phase of the flight is naturally takeoff which becomes even more significant for short and medium haul flights. A similar analysis as for the vertical tail can be made for high-lift wings that include a slat, a main body, and a flat. These high-lift devices are designed to provide the adequate additional lift during the takeoff and landing phases of the flight but the slats and flats are then closed in cruise configuration. Flow control offers the capacity to produce the same required lift during takeoff and landing phases but with smaller slats and flaps so that reduction in size of these devices will also reduce the total weight, creating even more savings not only in cruise but also in the takeoff phase. Marketing of planes today involves many factors but safety and energy efficiency are high on the list.

15.2.4 Vertical Tail/Rudder Assembly Simulations: Past and Future

While we have performed similar simulation of lab scale for wings (Wood et al. 2009; Sahni et al. 2011) and half planes (Chitale et al. 2014a,b; Rasquin 2014; Balin et al. 2017), for this chapter we will focus our discussion on our simulation of a vertical tail/rudder assembly. This choice is not only because it has complex geometry and complex flow features that can demonstrate tremendous cost reduction through the use of the anisotropic adaptive methods, but also because there are many practical engineering applications that could benefit from a more fundamental understanding of flow control applied to such a configuration. While lower fidelity models (such as RANS) can give some insight, the highfidelity models discussed here (such as DES) can provide a much deeper insight into the underlying flow physics at a broader range of length and timescales and much higher confidence in the resulting predictions. It is well established that RANS predictions of separated flow are less accurate than LES but, resolution of the eddies in the region where the flow control interacts with the separated regions is essential for accuracy and for obtaining fundamental insight into how the flow control works and how it might be further improved/altered to maintain effectiveness with increasing Reynolds number.

15.2.4.1 Past Simulations at Re = 0.35 Million

Numerical simulations provide a complementary and detailed view of the flow interactions and in turn give the insight required to understand and exploit the underlying physical mechanisms related to active flow control. The modeling approach that is the focus of this chapter is called parallel hierarchic adaptive stabilized transient analysis (PHASTA). PHASTA has already been validated with a closely coordinated experiment on a prismatic wing (Sahni et al. 2011), and more recently on a full 3D vertical tail/rudder assembly geometry (Jansen et al. 2017) at wind tunnel lab scale. The lab-scale numerical predictions were found to be in excellent agreement with the experimental measurements. Specifically, Figure 15.2 demonstrates that our simulations were able to accurately predict the side force (the engineering quantity of interest) while also showing excellent agreement with the phase-averaged structures (key to understanding the fundamental physics of how flow control achieves this benefit) as shown in Figure 15.3. Furthermore, numerical calculations were also able to provide instantaneous flow structures that are inaccessible via experimental measurements and thus, complement the joint study, see Figure 15.4. Our past success with DES models, as shown in these figures and in other studies (Sahni et al. 2011; Jansen et al. 2017; Rasquin et al. 2014), confirm that this is the right method for this study.

A second important aspect of the past vertical tail simulations was the relatively short time they required. Here, we summarize the process described in Jansen et al. (2017). The simulation was

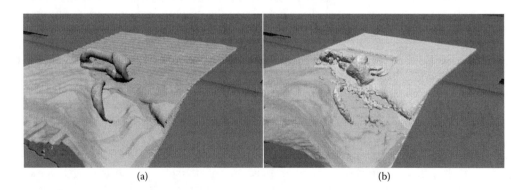

(a)   (b)

FIGURE 15.3 (See color insert.) Synthetic jets apply periodic flow. By averaging like phases (phase-averaging) the coherent structures produced by the jets are revealed and excellent agreement between the CFD (a) and the experiment (b) are observed.

FIGURE 15.4 (See color insert.)   Isosurface of instantaneous value of Q (measure of vorticity or rotational features of the flow) colored by speed.

started on a relatively coarse mesh (M0 mesh with 500 million tetrahedral elements) of the baseline flow configuration (e.g., all 12 of the jets were discretized but zero velocity applied). To get through the start-up transient, a RANS model and a relatively large time step (1.25e-3 seconds) was used to reach a steady state. The mesh was adapted (details on adaptive methods in subsequent section) to produce a mesh with 780 million tetrahedra, M1 mesh. Our adaptive tools transfer the solution to the new mesh which shortens the transient. On the M1 mesh, a series of DES with zero, 1, 6, and 12 jets activated at 1600 Hz were performed at a time step of 120 steps per jet period. Larger (60 steps per period) and smaller (180 steps per period) time steps were tested and shown to be almost indistinguishable and thus, 120 steps per period was deemed time-step independent. Each case was run for 50 jet periods to clear the transient (e.g., the flow changes both due to better modeling capacity of the DES model and due to the flow control's modification of the flow—typically a reduction in the separation). Once a statistically stationary limit cycle was observed, the flow was integrated another 200 jet periods. It was observed that statistics from the first 100 jet periods were almost indistinguishable from those from the second 100 jet periods. To prove grid independence, a second adaptation (M2 mesh) was performed, resulting in 1.25 B elements. Results from the M2 mesh and the M1 mesh were compared and also found to be almost indistinguishable. This series of simulations represents a complete and unprecedented demonstration of spatial, temporal, and statistical convergence that was only possible with Mira-scale resources for such a complicated aerodynamic simulation.

It is our plan to use Aurora to do the same at flight scale. Before we can do that, we need to discuss the computational resources that were required to complete the lab-scale simulation on Mira. The M2 mesh was run on 256k cores, four processes per core, 64k cores. Completing 150 jet cycles consumed 3.75 million Mira core-hours—1.25 million for the transient 50 jet cycles and 2.5 million for collecting statistics over 100 jet cycles. This corresponds to 19 and 38 wall-clock hours respectively on 1/12 of Mira. Since the solution on 0.75 billion elements and 1.25 B elements is almost indistinguishable, we will keep the math somewhat simpler (and introducing a modest, conservative resource adjustment factor) by the using the following numbers in subsequent resource estimates: 1 B elements, integrated 150 jet cycles consuming 3.75 M Mira core-hours on 64k cores in less than 2 days.

15.2.4.2 Estimating Resources for Flight Reynolds Number Flow Control

PHASTA has met the challenge of validating against experiments with a vertical tail at $Re = 3.5e10^5$. However, the target designs must perform at flight Reynolds numbers that are 53 times higher compared with the lab model. There are two fundamental challenges associated with the higher Reynolds numbers. The first challenge comes with estimating the resolution needs of DES, which, as noted earlier is not the simple 9/4 power of Reynolds number. Looking first at the attached flow regions, they are the same as that needed by lower fidelity models like RANS, since in fact DES does apply RANS models in those regions. Therefore, the mesh resolution needs are dominated by the separated flow regions and the active flow control regions. First, even without the jets on, the separated region for the vertical tail involves most of the rudder region which grows both in length, span, and height with rise in Reynolds number. This factor is mitigated substantially by the fact that what constitutes a large eddy for these flows also grows owing to the fact that it scales with the height of the boundary layer at separation, DES switches from RANS mode to LES mode through the Kelvin–Helmholtz shear layer roll-up process (Kundu and Cohen 2012; White 2000) of the separating shear layer and thus this rollup length scale sets the size of the largest turbulent eddy for the separated flow zone that is resolved via LES. From this largest eddy length scale, the LES zones of the DES model must then resolve a few (3–4) wave number doublings to resolve sufficiently far into the inertial range for the LES model to represent the effects of the remaining turbulence scales on the resolved field, and thereby provide a better model than RANS which represented only the average of all of the turbulent scales. To be clear, with the rise of the Reynolds numbers by increasing of the characteristic length scale of the geometry while holding velocity fixed, the chord and span increases and thus the thickness of the boundary layer at separation also increases. Thus, the turbulence energy spectrum broadens into lower wave numbers and this is what makes the largest eddy grow larger. This large eddy growth partially offsets the growth in resolution needs arising from the growth in the LES volume with the increased size of the aerodynamic control surfaces (vertical tail/rudder assembly in this case).

The second challenge is associated with accurately simulating the active flow control. The optimal size of the actuator is still under research. In fact, it is one of the parameters we will vary in our Aurora simulations to determine what size is most effective. That said, our preliminary estimates suggest that the actuators will grow somewhat but not linearly with chord (and other length scales such as span and thickness). At flight scales and speeds, it is expected that the number of jets will likely grow from our current O(10) configurations to O(100). In the pending INCITE 2017, we will consider a Re = 1.75 M. This five-fold increase in Reynolds number appears to be within reach with Mira. In the pending Aurora ESP we have proposed the first flight scale DES of active flow-control on a vertical tail/rudder assembly We estimate these meshes will reach 160 B elements, a size that can only be efficiently simulated with Aurora's 50k KNH nodes. Furthermore, based on our Theta scaling, we are confident that this size mesh will scale well to the full 50k nodes of Aurora. More detail on both of these simulations is provided next.

15.2.4.3 INCITE 2017: Five-Fold Rise in Reynolds Number Flow Control

With the concept of our flow-control research now introduced, we can now turn our attention to the specific DES simulations planned for the INCITE 2017. The Mira five-fold increase in Reynolds number will be achieved in two ways. First, the geometric model will be scaled up by a factor of 2 (going from a 1/19th scale model up to a 2/19th scale). Since we saw great success with 12 jets on the lab-scale model, we plan to use the same size jets but use twice as many of them. This leaves the second factor of 2.5 to be achieved by increasing the flow speed from 20 m/s to 50 m/s. This is very close to the true landing speed of 56 m/s. More details about the impact of these choices on the demand for computational resources will be provided in later sections but, suffice to say that this five-fold jump in Reynolds number, through an increased number of jets at a higher flow speed (and thus smaller grid spacing) will consume 60 million Mira core-hours. That is to say, it will not be possible to sweep over a parameter space of jet speeds, rudder deflection angles, jet sizes, or other parameters. This plan has relatively low risk, since our experience with the lab-scale experimental validation gives us high confidence that keeping these parameters at the best choice observed at the lab-scale experiment will provide the most insight owing to the ability to understand directly the influence due to only changing the Reynolds number. Specifically, we plan to simulate a 20°degree rudder deflection and maintain the same jet geometry (1 mm by 19 mm rectangle oriented at 20°to the stabilizer surface) and position (5% of chord upstream of the hinge line). The detailed data obtained (e.g., the phaseaveraged structures shown in Figure 15.3) compared for two otherwise identical cases with a five-fold Reynolds number separation will enable several important advances in the field of aerodynamic flow control. These include but are not limited to (1) developing scaling laws for jet structures, (2) collecting statistics from the scale-resolving simulations to better understand why RANS models are failing (two Reynolds numbers with sufficient separation are needed to do this properly), and (3) visualizing the vortex dynamics interaction of both instantaneous and phase-averaged jet structures with the separated boundary layer flow that they are intended to alter. A second simulation of the baseline flow (no jets present) will also be performed so that the action of the flow control at the fivefold higher Reynolds number can be compared similar to that shown in Figure 15.2. The simulations will be adapted until they have shown grid independence. Our experience with validating at the lab scale gives us confidence that these early adapted runs will consume only a small portion of the allocation, leaving more than 90% of the resources to be carried out on the final, grid independent mesh. Statistics will be collected and analyzed to provide insight into improvements that can be made to RANS models of active flow control. By collecting this information at Re = 1.75 M, we expect to be able to make quality inferences to the application of these active flow-control approaches at flight Reynolds numbers as required for final designs to achieve the same level of safety with substantial fuel savings. In this way, the 2017 INCITE campaign will gather key data to enable our Aurora ESP to confidently tackle a flight Reynolds number DES of a vertical tail/rudder assembly.

15.2.4.4   Aurora ESP: Flight Reynolds Number Flow Control

Using Aurora, we have proposed extending these experiment scale and five-fold higher Reynolds number simulations to a full-flight Reynolds number. Specifically, we proposed performing 12 simulations at 1/4 flight scale (varying jet width, spacing, and aspect-ratio), down selecting the most promising four cases for a 1/2 flight-scale simulation, that will then down select to two flight-matched cases. This suite of 18 simulations will help us understand how the flow-control structures and the jets that create them must be adjusted for Reynolds number.

Using our past DES experience, boundary layer theory, and the best available guides (Spalart 2001, 2009; Shur et al. 2008), we estimate that matching flight conditions require 182.4 B elements while 1/2 and 1/4 flight scale requires 48 B and 20 B respectively. Note, the element count on the full flight case is about a factor of two higher than what we anticipate, being required because it is for this most economically and physically relevant case that we have planned the most extensive grid independence checks. Similarly, the number of steps required remains the same due to flow control scaling. More specifically, in all flow control simulations to date, the unsteadiness of the jet has been substantially more rapid than that of the resolved turbulence (e.g., lab scale jet is at 1600 Hz). The jet frequency is chosen to have a fixed relationship to the timescale related to the chord flight (e.g., c/V the chord over the plane speed). Relative to the lab scale, the flight speed will grow by 2.38 while the chord length grows by 19. To generate sufficient momentum at these larger length scales, larger jets operating at lower frequencies are expected to be more effective, thus maintain a relatively constant chord flight time and total number of time steps required.

Factoring in the mesh growth and the expected speedup of Aurora, we project that we will be able to integrate the full-scale vertical tail/rudder assembly for 150 jet periods in 68.4 M Aurora core-hours (ACH) or 94.4% of a full machine compute day) per flight-scale simulation, 18 M ACH per 1/2-scale, and 7.5 M ACH per 1/4-scale. Thus, the 12 simulations at 1/4-scale (on 4.5k nodes), four

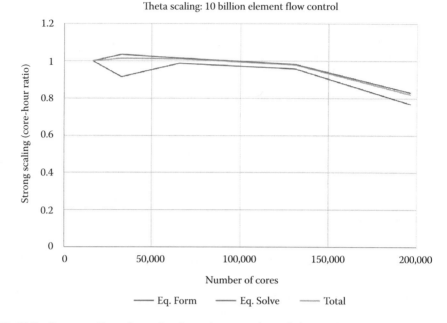

FIGURE 15.5 Strong scaling of equation formation, equation solution, and total solver on Theta using a 10 billion element mesh. Scaling perfect to 2 Ki nodes, 128 Ki cores with 76k elements per core. Only slight degradation (0.82) at 3 Ki nodes, 192 Ki cores, and 51k elements per core.

simulations at 1/2-scale (on 10k nodes), and two flight-scale simulations (on full machine) require 350 M ACH (node counts chosen according to strong scaling from Figures 15.5 and 15.6. Even our current KNL efficiency would enable us to perform half these simulations which will be a dramatic step forward for the state of the art in aerodynamic simulation; the first ever flight Reynolds number DES simulation of flow control on an aerodynamic component.

15.2.5 COMPUTATIONAL APPROACH

A mature finite-element flow solver (PHASTA) (Whiting and Jansen 2001; Whiting et al. 2003) is paired with anisotropic adaptive meshing procedures (Ibanez, Dunn, and Shephard 2016; Ibanez and Shephard 2016; Ibanez et al. 2016; Smith et al. 2015) (which we have developed within the Scientific Discovery through Advanced Computing (SciDAC) Interoperable Technologies for Advanced Petascale Simulations (ITAPS) and now FASTMath project) to provide a powerful tool for attacking fluid flow problems where boundary and shear layers develop highly anisotropic solutions that can only be located and resolved by applying adaptive methods (Chitale et al. 2014b, 2015). The solver and adaptive mesh control are described in the next two sections.

15.2.5.1 Parallel Flow Solver

PHASTA is a parallel, hierarchic (second- to fifth-order accurate), adaptive, stabilized (finite element) transient analysis tool for the solution of compressible or incompressible flows. It falls under the realm of computational/numerical methods for solving PDEs which have matured for a wide range of physical problems including ones in fluid mechanics, electromagnetics, biomechanics, to name a few. PHASTA (and its predecessor ENSA) was the first massively parallel unstructured grid LES/DNS code (Jansen 1993, 1994, 1999) and it has been applied to flows ranging from validation benchmarks (Sahni et al. 2011; Araya et al. 2011; Doosttalab et al. 2016) to cases of practical interest (Chitale et al. 2014a,b, 2015; Vaccaro et al. 2009, 2014, 2015; Rasquin 2014; Balin 2017; Jansen

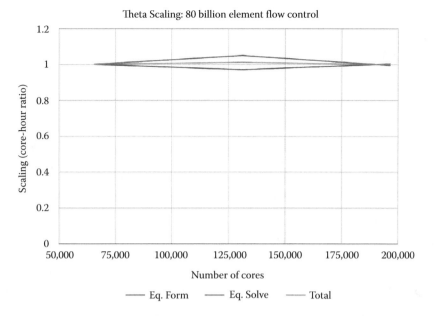

FIGURE 15.6 Strong scaling of equation formation, equation solution, and total solver on Theta using a 80 billion element mesh. Scaling perfect to full machine 3 Ki nodes, 92 Ki cores. Still performing well in MCDRAM with 1.2 million elements per core due to efficient element blocking.

2017). The practical cases of interest not only involve complicated geometries (such as detailed aerospace configurations or human arterial system) but also complex physics (such as fluid turbulence or multiphase interactions).

PHASTA has been shown (Jansen 1999; Karanam et al. 2008; Whiting and Jansen 2001; Whiting et al. 2003) to be an effective tool using implicit techniques for bridging a broad range of time and length scales in various flows including turbulent ones (based on URANSS, DES, LES, and DNS). It has also effectively applied recent anisotropic adaptive algorithms (Mueller et al. 2005; Sahni et al. 2006, 2008) along with advanced numerical models of fluid turbulence. Note that DES, LES, and DNS are computationally intensive even for single-phase flows. PHASTA has extended this capability to two phase flows using the level set method (Nagrath et al. 2005, 2006; Rodriguez et al. 2013; Bolotnov et al. 2011; Bolotnov 2013; Mishra and Bolotnov 2015; Fang et al. 2016) to implicitly track the boundary between two immiscible fluids. PHASTA is also the flow simulator for SimVascular (Zhou et al. 2010; Kim et al. 2009; Sahni et al. 2009; Figueroa et al. 2006; Vignon-Clementel et al. 2006) supported by National Science Foundation (NSF) and National Institutes of Health (NIH). Furthermore, many of its application cases have been sufficiently complex that grid independent results could only be obtained by efficient use of anisotropically adapted unstructured grids on meshes capable of maintaining highquality boundary layer elements (Sahni et al. 2008) and through scalable performance on massively parallel computers (Shephard et al. 2007). Further details on the numerical methods employed in PHASTA are given in Section 15.3; but, it is worth mentioning that both of the primary work components of the flow solver, that is, equation formation and equation solution, have been carefully constructed for parallel performance and scaling to 786,432 cores (on 1, 2, and 4 processes per core which exceeded 3M processes) on Mira as shown in Sections 15.4.1 and 15.4.2 (Rasquin et al. 2014).

15.2.5.2 Adaptive Mesh Control

The application of reliable numerical simulations requires them to be executed in an automated manner with explicit control of the approximations made. Since there are no reliable *a priori* methods to efficiently control the approximation errors, adaptive methods must be applied where the mesh resolution is determined in a local fashion based on the spatial distribution of the solution and errors associated with its numerical approximation. For instance, regions of the mesh that require more resolution can be identified by the high root mean square value of the solution and/or high value of the nonlinear residual of the discretized Navier–Stokes equations. Furthermore, the reliability and accuracy of the simulations are also strong functions of the mesh quality and configuration. Many physical problems of interest, especially in the field of fluid mechanics solution features are most effectively resolved using mesh elements, which are oriented and configured in a certain manner (Sahni et al. 2006, 2008). For example, in the case of viscous flows use of boundary layer meshes is central to the ability to effectively perform the flow simulations due to their favorable attributes, that is, high aspect ratio, orthogonal, layered, and graded elements near the viscous walls.

The parallel unstructured mesh infrastructure (PUMI) (Ibanez et al. 2016), adaptive meshing tools, and the ParMA partitioning tool (Smith et al. 2016) have already been ported over to Mira and Theta and allowed the generation and the partitioning of a 92 B element mesh which was then used as a scaling benchmark of our flow solver PHASTA to >3M processes (Rasquin et al. 2014). Unstructured parallel mesh adaptation procedures based on local modification operators are used to adaptively construct the meshes required for the target applications. PUMI supports these operations through the use of a component based design. At PUMI's core is an array-based mesh representation component that provides efficient mechanisms to query and modify the mesh while maintaining a small memory footprint (Ibanez and Shephard 2016; Ibanez et al. 2016). Parallel mesh operations such as the definition of the partition graph, the migration of elements, and synchronization of off-process boundary data, is provided by the APF component. These parallel mesh operations provide the supporting functionality to implement mesh adaptation and fast dynamic load balancing components, MeshAdapt (Alauzet et al. 2006; Ovcharenko et al. 2013) and ParMA (Smith et al. 2016; Zhou et al. n.d.) respectively.

ParMA APIs are used to predictively balance mesh elements during mesh adaptation to avoid memory exhaustion, after adaptation operations are completed to ensure that the applications mesh entity balance requirements are met. For a PHASTA analysis, ParMA first targets the reduction of mesh vertex imbalance to ensure the scalability of the dominant equation solution analysis stage and then balances elements, without disturbing the vertex imbalance, to scale the equation formation stage (forming the LHS A and the RHS b). PHASTA's strong scalability on Mira was improved by over 35% using ParMA meshes relative to meshes prepared with only graph and geometric-based partitioning methods (Smith et al. 2015). All tools scale well on Mira and Theta.

15.3 ALGORITHMIC DETAILS

While the previous section gave the background on the science of CFD, flow control, and our parallel adaptive approach, efficient execution of these ideas requires many algorithmic details to be explained so as to ultimately understand how best to execute them with great efficiency and high scalability. This section revisits the flow solver and its supporting adaptive control software to describe those algorithmic details.

15.3.1 FLOW SOLVER

Flow computations are performed using a stabilized, semidiscrete finite element method for the transient, incompressible Navier–Stokes PDE governing fluid flows. In particular, we employ the streamline upwind/Petrov–Galerkin (SUPG) stabilization method introduced in Brooks and Hughes (1982) to discretize the governing equations. The stabilized finite element formulation currently utilized has been shown to be robust, accurate, and stable on a variety of flow problems (see, e.g., Taylor et al. 1998; Whiting and Jansen 2001).

In our flow solver (PHASTA), the Navier–Stokes equations (conservation of mass, momentum and energy) plus any auxiliary equations (as needed for turbulence models or level sets in two-phase flow) are discretized in space and time. Since Galerkin's method has been shown for equal-order basis to be unstable for advection dominated flows, we carry out the discretization in space with a stabilized finite element method (Brooks and Hughes 1982), thus allowing the effective use of equal-order basis functions for all variables.

The stabilized finite element method leads to a so-called weak form of the governing equations which is then discretized. In PHASTA, we employ interpolating linear shape functions for the base element and then employ hierarchic, piecewise polynomials (Whiting and Jansen 2001; Whiting et al. 2003) for higher order discretizations. The resulting element integrals are computed using Gauss quadrature. Implicit integration in time is then performed using a generalized-α method (Jansen et al. 1999) which is second-order accurate and provides precise control of the temporal damping to reproduce Gear's Method, Midpoint Rule, or any blend in between. On a given time step, the resulting nonlinear algebraic equations are linearized to yield a system of equations which are solved using iterative procedures, for example, generalized minimal residual method (GMRES) (Saad and Schultz 1986; Shakib et al. 1989) is applied to the linear system of equations $Ax = b$ (where b is the right-hand side or residual-vector of the weak form and A is the left-hand side or linearized tangent-matrix of b with respect to unknown solution coefficients x).

Next, we focus our attention on the parallel paradigm to make clear our approach to developing a exaflop flow solver for a diverse class of flow phenomena. Finite element methods are very well suited for use on parallel computers as substantial computational effort is in the calculation of element level integrals and in the solution of the resulting system of algebraic equations using iterative methods (which employ matrix-vector $A\,p$ products).

Both of these work types can be equally divided among the processors by partitioning the aggregate mesh into equal load parts. One important point to consider during partitioning is that the computational load (in any part) during the system formation stage (i.e., during formation of A and b) depends on the elements present in the part whereas in the system solution stage, it depends on the degrees-of-freedom-holders dof_h and the number of degree-of-freedom-variables dof_v whose product yields the total unknowns in the system of equations on that part.

For example, in the case of linear finite elements of all the same topology, work involved in equation formation is proportional to the number of mesh elements in the part while during equation solution the work is proportional to the number of mesh nodes (dof_h) in the part since the unknowns are associated with the nodes.

Though many approaches elect to redistribute data for load balancing (or change the partition) between two stages of implicit solvers, for example, after formation the matrix (A) is (re) distributed based on rows, in our native approach the same partition is maintained throughout both stages and no redistribution of data is performed in between. PHASTA also supports using the Portable, Extensible Toolkit for Scientific Computation (PETSc) (Balay et al. 1997, 2016a,b) solver library, which has been recently modified to perform this data redistribution with dramatic improvement to the scalability.

Element-based partitioning is currently used as it is natural for element-integration/equation-formation stage making it highly scalable. So long as the dof_h balance is also preserved, this partitioning also maintains the scalability of the iterative linear solve. In element-based partitioning, each element is uniquely assigned to a single part but in turn leads to shared dof_h at interpart boundaries.

Typically, element balance (with sufficient load per part) and minimization of amount of communications during partitioning results in a reasonable dof_h balance as well. For a mesh with fixed element topology and order, balanced parts within a partition implies that each part contains as close to the average number of mesh entities as possible. For other cases, such as ones with mixed element topology or order, weights reflecting the work load for every individual element are assigned to create parts with a balanced load.

Each processor core first performs interpolation and numerical integration over the (interior and boundary) elements on its local part to form the linearized equations, that is, the tangent matrix (A) and the residual vector (b). Subsequently, Krylov iterative solution techniques are used to find x. These techniques employ repeated products of A with a series of vectors (say, p) to construct an orthonormal basis of vectors to approximate x. After each local $q = Ap$ product we apply communications to obtain complete values in q. To describe the interactions and communications among parts within a partition, we employ the concept of a *partition graph*. Each partition-graph vertex represents a part whereas each partition-graph edge represents interaction between a pair of parts sharing dof_h. This is done in a distributed way such that a part contains information only in terms of its portion of the work (or submesh) along with its interaction with neighboring parts. The interaction between neighboring parts is defined based on shared dof_h, where every shared dof_h resides as an image on each part sharing it. Only one image among all images of a shared dof_h is assigned to be the *owner*, thereby making all other images explicitly declared to be nonowners.

This process insures that the sum total of dof_h based on owner images over all the parts within a partition is independent of the partitioning and is equal to the number of (unique) dof_h in the aggregate mesh. Such a control relationship among images of shared dof_h allows the owner image of each shared dof_h to be *in-charge* for data accumulation and update and in turn limits communications to exist only between owner and nonowners (i.e., nonowner images do not communicate to each other). Furthermore, data exchange is done only for vector entries (e.g., in b) as this is sufficient to advance the computations. Thus, partition-graph edge connects only those pairs of parts that involve owner image(s). Typically, for the three-dimensional unstructured meshes, each part contains on the order of 40 partition-graph edges connecting it with neighboring parts.

Before proceeding to the next matrix-vector product in the series, it is important to note that computation of norms is required to perform orthonormalization. In this step, the norm of vector q, and its dot-product with the previous vectors in the series, are computed. To compute a norm or dot-product, first a local dot-product is computed (requiring no communication) but then, to obtain a complete dot-product, a sum across all cores is needed. A collective communication (of all reduce type) is used to carry out the global summation. It is important to point out that while computing a local dot-product value, only the owner image of each shared dof_h takes active part to correctly account for its contribution in the complete (or global) dot-product. Successive Ap products are carried out along with communications to obtain complete values and to carry out the orthonormalization. This leads to an orthonormal basis of vectors which is used to find an approximate update vector x and marks the end of a nonlinear iteration step.

PHASTA has two forms of I/O; one file per message passing interface (MPI) process and MPI-IO (Liu et al. 2010), which allows multiple parts to be written and read from a given file. The latter has proven scalable to >3 M processes on Mira and 256Ki on Theta. PHASTA has been coded for pure MPI and MPI+X where for KNL, X is currently OpenMP but other options are being developed. While MPI+X has been shown to scale at better than 75% efficiency on a variety of architectures, on Mira and Theta, pure MPI has scaled at >90% efficiency (Figures 15.5 through 15.7).

15.3.2   ADAPTIVE MESH CONTROL AND PARTITIONING

Mesh adaptation procedures based on local mesh modification operators are used in this project to adaptively construct significantly large meshes (in the order of 1–10 B or more elements) required for the target applications. This requires effective execution of adaptive meshing techniques on meshes that are distributed over massively parallel systems. The current process of adapting a distributed mesh in parallel relies on a flexible and distributed mesh representation (Seol and Shephard 2006; Seol 2005) that is designed to fulfill its needs such as allowing local mesh migration to move groups of mesh entities.

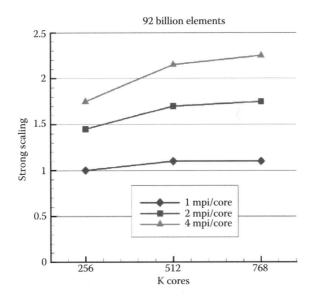

FIGURE 15.7   Scaling of PHASTA on Mira BG/Q with 1, 2, and 4 processes per core on a 92-billion element mesh.

The parallel implementation of such a representation scheme builds on a formal partition model that describes the distribution of parts in the partition in terms of adjacency relations of mesh entities (Seol et al. 2012; Zhou et al. 2012). Conceptually, the partition model lies between the geometry and mesh, and maintains the relationships across the interpart boundaries.

Moreover, since mesh adaptation selectively refines and/or coarsens a mesh to control the mesh discretization error, it is clear that a mesh partitioning that was balanced is likely to be imbalanced after the mesh is adapted. Thus, we have developed the capacity to dynamically alter the mesh partitioning on a distributed mesh as the adaptive simulation proceeds in order to maintain a good load balance, both in terms of elements per part and *dofs* per part.

Both graph- (adjacency) and geometry-based procedures are available to support this process. Based on past experience the graph-based procedures tend to do a better job on connected meshes and can be extended to account for geometrically based interactions like near contact.

We are currently also developing and using iterative load balancing algorithms that operate by doing iterations of load migration between neighboring processors. Scalable methods of this type have been developed (Ozturan et al. 1994). Although scalable, these methods were found to be more expensive than effective graph partitioners when the mesh is repartitioned in a general way. However, there are steps in the parallel adaptive process where the current partitions are known to be nearly balanced and only some minor movement of mesh entities is needed to yield load balance and/or reduce interprocessor communications. Iterative load balancing should prove effective for this. Moreover, we have used these methods to improve balance in degrees of freedom while maintaining the balance in elements, especially for partitions with lightly loaded parts.

15.3.3   ALGORITHM SUMMARY

Summarizing, PHASTA addresses the challenges of obtaining solutions to nonlinear PDE's, generating highly anisotropic solutions with an enormous range of spatial and temporal scales that require implicit time integration with complex algorithms that pose major challenges to solution efficiency including: indirect addressing (irregular memory access) for unstructured grids and for sparse linear algebra, collective communications (dot products for GMRES), simultaneous balancing of multiple mesh entities, and finally, the difficulty of balancing work for adaptation. While these aspects prevent PHASTA from achieving large fractions of theoretical peak performance, retreating to simpler methods would result in the following: (1) Cartesian grid methods would require 4e17 cells (3e8 savings), (2) an octree/AMR mesh would require 3.35e13 cells (2.8e4 savings). These savings factors refer only to the spatial discretization. Implicit time integration, while complicated algorithmically, provides a factor of 875 savings in time step. From these factors, which are multiplicative, it is clear that the efficiency reduction as measured in FLOPS or fractions of peak performance of the more complicated algorithms is recovered many times over in the more important metric of science-progress-per-core-hour.

15.4   PROGRAMMING APPROACH

The Theta ESP workshop runs also gave us an opportunity to study performance, scalability, and memory limits across multiple nodes. For performance, there is a strong interaction between vectorization and the use of cache. This further ties into the parallel performance/scalability through changes to the communication fabric and through the on-package memory. We defer discussions of performance and scalability to the following subsections until after we have explained our parallel programming approach.

Our parallel programming approach can be summarized as MPI+X. Specifically, MPI across nodes (allowing the possibility of multiple MPI processes per node) combined with some other communication mechanism, X, within nodes. We have pursued this general approach for 7 years. As we have tuned to each new platform (e.g., BGP, BGQ) we evaluate the available choices for X, but we also compare it to X=null, that is pure MPI on all of the processes within and across the nodes and even multiple processes per core. On the entire BlueGene family, pure MPI was always the clear winner and thus, all the simulations discussed above were done with four MPI processes per core (64 MPI processes on 16 cores of each node for BGQ).

On Theta, the gap was sufficiently less as, during the workshop, we demonstrated that we could use X = OpenMP for distributing our block level loop with reduced MPI processes with acceptable scaling. We say acceptable because the performance was still a bit below that of pure MPI (80% for OpenMP for problems that pure MPI was still at 100%) but this at least demonstrates that we have a viable strategy should MPI across all processes of Aurora degrade, as some have predicted. While we plan to continue to develop and improve our OpenMP code, we also have already started efforts on alternatives. MPI endpoints (Sridharan, Dinan, and Kalamkar 2014) are other possible options for X, and relative to OpenMP, would operate on larger part-level constructs. We also intend to explore other on-node shared memory models such as MPI 3.0 shared memory windows (Hoefler et al. 2013; Zhu et al. 2015) and XSI shmem (Group, n.d.). Note that PHASTA only requires O(10) MPI functions and thus places low demands on +X options. Even more fine-grain thread parallelism has already been developed and demonstrated in PHASTA and this will also be pursued (more details in Section 15.4.3).

To guide the choices and improvement, Co-PI, Carothers' developments within the DOE CODES project (Various, n.d.) will be used to model dragonfly topology communication patterns. During the development phase, we will collect full-scale MPI trace data from PHASTA runs on Mira and then predict how P2P and collective operations scale on a simulated Aurora-scale system using our massively parallel dragonfly network model in CODES. We will further use Intel SDE (Intel Software Development Emulator 2015) to understand the work that takes place between each MPI communication completing the performance analysis model for PHASTA. We also plan to extend our iterative partition improvement code partition improvement code (ParMA; Smith et al. 2016) to alter the element and node balance to improve overall performance based on the performance analysis model.

In summary, we propose to leverage our past success with MPI across all cores to >3 M processes and compare this to the MPI+X variants. All three will be continuously analyzed in our Aurora performance analysis model for potential performance gains. The best versions of all three will be evaluated on Theta to confirm emulated projections and then the best performing option will be used for the Aurora science production runs.

15.4.1 SCALABILITY

The Theta ESP workshop runs gave us an opportunity to study scalability and memory limits at large node counts. Despite common concerns about 64 cores sharing 16 GB of fast (MCDRAM) memory, we found that even with 1.2 M elements per core, the code stayed within MCDRAM (e.g., 80 B element mesh run on 1 Ki nodes Figure 15.7. When the same mesh was run on node counts up to 3 Ki, strong scaling was maintained in the equation formation. Strong scaling was demonstrated in a 10 B element case (Figure 15.6). Both equation formation and solution scaled equally well to 2 Ki nodes and dropped off only slightly at 3 Ki. While these results suggest that, at least for PHASTA, our MPI-only approach may remain viable, we understand that it is prudent to have alternatives in place and thus, we have already developed and seen promising results from other options.

15.4.2 PERFORMANCE

Achieving the highest possible portable performance on new architectures has been a major focus of the PHASTA development since its inception 15 years ago and this in fact built upon the same objectives of its predecessor (ENSA) developed at Stanford 15 years prior to that. Throughout this 30-year development period, considerable flexibility has been built into the code to make it highly adaptable to hardware and software advances. For example, the element equation formation phase which involves intensive loads, stores, multiplies, and adds was originally developed for the Cray vector architecture but it has been generalized over the years to improve cache performance and we find it is again able to strongly exploit vectorization in the KNX hardware. Looking at the hotspots identified by VTUNE runs on KNL, we have confirmed that a very high percentage of our computationally intensive kernels are already highly vectorized. While tuning for single-core performance is critical, we have also focused intensively over the years on maintaining parallel scaling.

Recent runs on Theta suggest that our per-core performance is roughly five times that of Mira. In the short time that KNL has been available, we have used VTUNE and Advisor on both the full code and representative computational kernels to identify ways to achieve even greater vectorization and stronger acceleration. Under the development period, further specialization for KNH will be the primary focus. Based on our representative kernel tests, we anticipate a further factor of two acceleration for a total of ten-fold acceleration per core relative to Mira.

15.4.3 PORTABILITY

Portability across high-performance computing (HPC) platforms has been a major objective for the PHASTA project; the code has been used on workstations and supercomputers dating back to the Cray X-MP shared-memory vector systems. Portability between manycore track systems (Theta/Aurora) and CPU–GPU track systems (Titan/Summit) presents a significant challenge. The most important difference for PHASTA (and many other codes) is the available high bandwidth memory (HBM) per computational *core* (Single Instruction Multiple Data [SIMD] unit). Theta has 260 MB of MCDRAM per core and Aurora is projected to have a similar amount, but Summit, like most GPU systems, will likely have much less. While PHASTA is shown to have sufficient HBM for pure MPI on Theta, and MPI+X alternatives can further reduce that usage as needed, these options are likely nonviable for CPU–GPU track systems.

To maintain a truly portable option and to provide another alternative fine-grained parallelism we will also develop a parallel paradigm where the MPI process count is substantially smaller than the total number of computational cores (including GPU cores). Work for parts assigned to these processes is distributed to threads. This approach has already been developed and scaled well (greater than 75% efficiency) on several previous platforms. The basic idea for equation formation is to distribute the blocks of elements across the processing units since this is embarrassingly parallel work. Theta should perform well under this approach. Portability to CPU–GPU systems, where HBM per core is much smaller, will likely require even finer grained parallelism (e.g., down to interior loops of the integral quadrature operations using OpenMP or similar). Regarding equation solution, we have also threaded the matrix-vector product of our native solver.

This, plus our recent integrated development with the PETSc team as part of the FASTMath project, suggests that other than the usual tuning to improve performance, equation solution will continue to scale well on Theta and Aurora and be portable to other architectures.

15.4.4 EXTERNAL LIBRARIES

Programming Languages:

- C, C++, and Fortan >=90

Libraries:

- Meshing tools for adaptivity and partitioning:
  - PUMI: Unstructured mesh tools
  - Zoltan: Partitioning interface to high-quality ParMETIS multilevel graph methods, and faster, lower quality, recursive geometric sectioning methods
  - ParMETIS: Multilevel graph-based partitioning methods.
- File I/O:
  - MPI-IO: PHASTA parallel IO using SyncIO (Liu et al. 2010. Massively parallel I/O for partitioned solver systems. *Parallel Processing Letters* 20 (4): 377–95 World Scientific Publishing.)

Parallel Methods:

- Pure-MPI: MPI 1.0
- MPI+OpenMP: OpenMP 3.0
- MPI+MPI: Opportunities to reduce communication costs are provided by the proposed MPI endpoints interface, MPI 3's shared memory window functionality (MPI_Win_allocate_shared), and by the XSI shared memory interface (cited above).

Optional Libraries:

- Linear equation solver:
  - PETSc: Open source algebraic equation solver
  - svLS: Open source incompressible solver
- Meshing tools for adaptivity and partitioning:
  - Simmetrix: Geometric model, parallel in-memory mesh generation and adaptation
- Visualization and data analysis:
  - ParaView: *In situ* visualization (Fabian et al. 2011; Rasquin et al. 2011; Ayachit et al. 2016) postprocessing.

File I/O:

- Bzip: Compressed mesh file format
  - hbw_malloc/hbw_free: HBM *flat* mode to support planned developments for quickly staging checkpoint/restart files to dynamic random-access memory (DRAM) (when capacity permits).

15.5 SOFTWARE PRACTICES

PHASTA is open-sourced through GitHub and maintained through Git. It is built using CMAKE. The code is a mixture of Fortran >=90, C, and C++. The choice of language for a given routine roughly follows the following guidelines. The most computationally intensive routines are written in fortran for, possibly now historic, computational efficiency and familiarity of the engineering-based developers. The I/O routines, both large data for geometry and small data for problem parameter definitions are written in C and C++, again, due to a possibly now historic perception that these languages provide greater flexibility. As these languages have evolved, the relative advantages have become less obvious and thus the continuation of these guidelines has at least as much to do with inertia/legacy. Obtaining financial support for a complete rewrite is unlikely, and, despite the challenges presented by working with three languages, PHASTA has proven agile with regard to rapid implementation and testing of new physics models, new math discretizations, and new solver technologies.

Where practical, portions of the code have been made into libraries so that portions of the code that are used by multiple executables are isolated (e.g., shape functions and parallel IO). The developer community spans more than 10 universities and several small and large companies. A growing set of regression tests are being developed to ensure that new developments do not break existing capability.

15.6 *IN SITU* VISUALIZATION AND COMPUTATIONAL STEERING

For visualization of field data on full 3D domain, PHASTA relies on the ParaView and Catalyst libraries developed by Kitware Inc. For that purpose, two visualization strategies have been implemented in close collaboration with Kitware: classical *a posteriori* visualization of checkpoint data and live, *in situ* visualization.

Classical *a posteriori* visualization illustrated in Figure 15.8 relies on the ParaView library used in client-server mode. A dedicated reader has been implemented in order to load checkpoint data saved to file systems under the PHASTA format. Two PHASTA formats are available: the first format corresponds to one file per mesh part and the second more flexible format is based on MPI-IO and allows any arbitrary number of parts per file. Due to the growing size of the data generated by simulations in general, it is becoming impractical to transfer or move data. Consequently, most computing centers operate their own visualization cluster, which share the same file system as the compute resource. The current visualization cluster at Argonne Leadership Computing Facility (ALCF) is named Cooley and includes a total of 126 compute nodes; each node has 12 CPU cores and one NVIDIA Tesla K80 dual-GPU card. We typically use from 2 to 16 nodes to visualize the flow resulting from our simulations. We also exercise remote visualization on meshes prior to final adapted ones to understand the behavior of flow structures, recognize regions of critical interest, and extract reduced data in those regions for detailed quantitative analysis such as flow variables in homogeneous lines or planes to collect turbulence statistics.

However, for the high-petascale machines of the near future and the exascale machines currently being codesigned, the amount of solution data that must be stored for later retrieval and

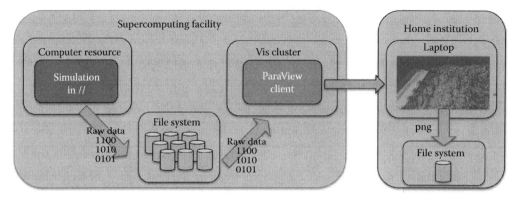

FIGURE 15.8   Classical *a posteriori* visualization workflow.

postprocessing will become prohibitive due to the widening gap between computational and I/O rates. Clearly, the classical paradigm of data creation, storage, and retrieval later for subsequent analysis (e.g., generation of animations) must be reconsidered in order to perform any assessment of the insight in a reasonable time frame. This situation strongly motivates *in situ* processing of the data, where visualization processing is performed while data is still resident in memory.

We started to collaborate with Kitware Inc. and ALCF in 2011 on the development of the ParaView Coprocessing Library, which is Catalyst's predecessor. We applied successfully *in situ* and interactive visualization techniques at large scale on BG/P compute resource Intrepid and Linux visualization resource Eureka (Fabian et al. 2011; Rasquin et al. 2011).

The term *in situ* is rather generic and several *in situ* configurations are available. The opposite extreme from the classical run/store/read/analyze is to embed the entire data analytics process into the solver. Here images of a predefined data analytics filter chain are processed within the primary simulation and exported either to files or directly via sockets to a coprocessing resource whose only requirement/capability is to display the output. While this approach has proven productive in some application areas, it typically limits the extent to which the data analytics can be reconfigured.

In many situations, it is highly desirable to be able to set up an initial definition of the filter chain, view several live frames from an ongoing simulation, and then redefine the filter chain to provide a more insightful window into the ongoing simulation. Indeed, if these views can be provided at a live frame rate (display completed before the next data set is delivered to the visualization compute resource), computational steering becomes possible wherein not only can the data analytics be redefined in a way that maintains temporal continuity of the insight but also key parameters of the solve can be adjusted and their influence on the simulation observed.

The best coprocessing approach to realize this vision of experiential simulation and/or less aggressive visions of near time but not necessarily live covisualization will likely be best accomplished by something between the two extremes described above. In this work, we consider two covisualization models which are both illustrated in Figure 15.9. The first will be referred to as classical covisualization (CCV) wherein the entire data set is exported from the ongoing simulation to a smaller visualization compute resource. No data reduction is performed on the solver compute resource, which has the advantage of not burdening it with the computational load of filtering the data. However, it can burden it with the time to ship the data to the visualization compute resource, which typically blocks the solution process to some extent.

Once the full data is shipped to the covisualization resource, any desired filter chain can be executed there since the full data is resident. The second approach, which we will refer to as *in situ* data extracts (ISDE) performs the currently defined filter chain on the solver compute resource and then ships only the data extract to the visualization compute resource. While ISDE consumes time

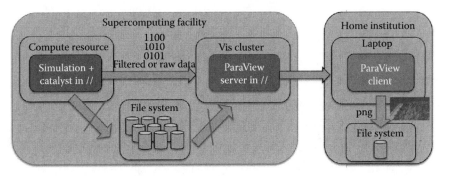

FIGURE 15.9 *In situ* visualization workflow with transfer of raw or filtered data from the compute resource to the visualization cluster.

from the solver compute resource to do the data extraction, for many filter chains, it can dramatically reduce the amount of data that must be transported to the visualization compute resource. It is important to note that the filter chain performed by ISDE can be dynamically reconfigured without stopping the run and thus, both are suitable candidates for interactive monitoring of ongoing jobs and/or computational steering.

In this work, four main components have therefore been carefully combined in order to demonstrate our covisualization capabilities, namely (1) a massively parallel CFD solver PHASTA, (2) the visualization tool ParaView (Moreland 2011; ParaView 2009) and its Coprocessing library (Fabian et al. 2011), (3) a fast I/O forwarding tool called GLEAN (Vishwanath et al. 2011), and (4) the full ALCF architecture.

Many configurations of our covisualization stack were considered in this study, including the covisualization approach (CCV or ISDE), the data transport between the solver resource and the visualization cluster (GLEAN or VTK sockets available in the Coprocessing library) and the number of pvserver running on the visualization cluster to name a few.

In summary, a live data analysis was demonstrated to provide continuous and reconfigurable insight into massively parallel simulations. Specifically, the full Intrepid ALCF resource (with 163,840 cores of a BG/P machine, tightly linked through a high-speed network to an 800 cores and 200 GPUs visualization cluster [Eureka]), was engaged to evaluate the current software and hardware's ability to deliver visualizations from an ongoing simulation. CCV was explored using two data transport mechanisms. While both data transport approaches were able to deliver data from the compute to the visualization nodes at a high rate (O(50 GB/s)), the visualization cluster needed intensive resource (especially in terms of CPU) to filter and, to a lesser extent, render the data at a rate that kept up with the solver on a step-by-step basis.

Consequently, these results suggest that ISDE can be more suitable for live simulations that use relatively simple filters that parallelize well, provided that these filters substantially reduce the data that must be transported to a size that is however not (yet) penalized by the latency of the network.

Clearly, much more complex analysis of the data could be performed on the visualization server with the full data resident and in cases where that is needed, the lack of live interactivity could be a fair trade. A similar conclusion could be made for simulations running at a small-enough time step that a significant number of steps could be skipped with an acceptable loss of interactivity. It was also noted that ISDE was successful at 10% of the full ALCF visualization facility which bodes well for exascale hardware which is expected to not be as data analytics rich as current generation machines.

In 2014, we ported this technology to Mira BG/Q (Fang, Rasquin, and Bolotnov 2016) with a successful live demonstration of a PHASTA simulation running with 256k MPI processes and instrumented with the Catalyst *in situ* library at the NUFO (National User Facility Organization) conference in June 2014. Although similar *in situ* configurations as the one shown in Figure 15.9 are also

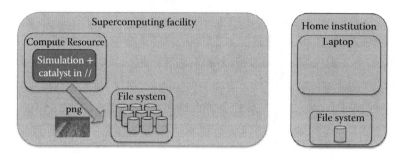

FIGURE 15.10   *In situ* visualization workflow with generation of flow pictures in batch mode.

possible with Catalyst, this demonstration implies a more straightforward *in situ* visualization work-flow illustrated in Figure 15.10. In this workflow, the filters and views for *in situ* visualization were chosen prior to the simulation. Consequently, flow pictures were created at each time step of the simulation in batch mode in order to generate an animation of turbulent flow structures in regions of interest.

More recently, this workflow has been updated for 512k cores and more than 1 M MPI processes in an accepted paper at SC16 (Ayachit et al. 2016) and to Theta. This paper examines several key design and performance issues related to the idea of *in situ* processing at extreme scale on modern platforms: scalability, overhead, performance measurement and analysis, comparison, and contrast with a traditional post hoc approach, and interfacing with simulation codes. These principles were illustrated in practice with studies, conducted on large-scale HPC platforms, which include a miniapplication and multiple science application codes. In this context, PHASTA demonstrated *in situ* methods in use at greater than 1 M-way concurrency.

15.7   SUMMARY

A massively parallel computational approach to modeling turbulent flow over aerodynamic bodies with active flow control was described. The science related to the fluid flow physics as well as the computational science were discussed in detail. The latter includes not only the efforts made to achieve near perfect strong scaling of the solver to more than 3M processes of IBM BG/Q and more recent scaling and performance efforts on XEON Phi KNL, but also efforts to extend *in situ* data analytics to the same scale to more effectively extract insight from ever-growing computational simulation data streams.

REFERENCES

Alauzet, F, X Li, E S Seol, and M S Shephard. 2006. Parallel anisotropic 3D mesh adaptation by mesh modification. *Engineering with Computers* 21 (3): 247–58. doi:10.1007/s00366-005-0009-3.
Amitay, M, B L Smith, and A Glezer. 1998. Aerodynamic flow control using synthetic jet technology. In *36th Aerospace Sciences Meeting and Exhibit, AIAA Paper* 208, *Reno, NV*.
Araya, G, L Castillo, C Meneveau, and K Jansen. 2011. A dynamic multi-scale approach for turbulent inflow boundary conditions in spatially developing flows. *Journal of Fluid Mechanics* 670: 581–605.
Argonne Leadership Computing Facility, Early Science Program. 2010. Https://www.alcf.anl.gov/projects/petascale-Adaptive-Cfd. Misc.
Argonne Leadership Computing Facility, Early Science Program. 2015. Https://www.alcf.anl.gov/projects/extreme-Scale-Unstructured-Adaptive-Cfd-Multiphase-Flow-Aerodynamic-Flow-Control. Misc.

Argonne Leadership Computing Facility, INCITE Program. 2012. Https://www.alcf.anl.gov/projects/adaptive-Detached-Eddy-Simulation-Vertical-Tail-Active-Flow-Control. Misc.

Argonne Leadership Computing Facility, INCITE Program 2014. Https://www.alcf.anl.gov/projects/adaptive-Detached-Eddy-Simulation-High-Lift-Wing-Active-Flow-Control. Misc.

Argonne Leadership Computing Facility, INCITE Program 2016. Https://www.alcf.anl.gov/projects/adaptive-Detached-Eddy-Simulation-High-Lift-Wing-Active-Flow-Control-1. Misc.

Ayachit, U, A Bauer, E P N Duque, G Eisenhauer, N Ferrierz, J Gu, K E Jansen, et al. 2016. Performance analysis, design considerations, and applications of extreme-scale in situ infrastructures. In *Proceedings of the 2016 ACM/IEEE Conference on Supercomputing*, Salt Lake City, UT, pp. 921–932.

Balay, S, S Abhyankar, M F Adams, J Brown, P Brune, K Buschelman, L Dalcin et al. 2016a. PETSc. Http://www.mcs.anl.gov/petsc.

Balay, S, S Abhyankar, M F Adams, J Brown, P Brune, K Buschelman, L Dalcin, V et al. 2016b. *PETSc Users Manual*. Technical report ANL-95/11 - Revision 3.7.

Balay, S, W D Gropp, L Curfman McInnes, and B F Smith. Efficient management of parallelism in object oriented numerical software libraries. In E. Arge, A. M. Bruaset and H. P. Langtangen, (eds.) *Modern Software Tools in Scientific Computing*, pp. 163–202. Basel, Switzerland: Birkhauser Press.

Balin, R, K E Jansen, M Rasquin, and K C Chitale. 2017. Investigation into the performance of turbulence models for the computation of high-lift flows at large angles of attack. 35th AIAA Applied Aerodynamics Conference, AIAA AVIATION Forum (AIAA 2017-3563), Atlanta, GA. doi:10.2514/6.2017-3563.

Bolotnov, I A. 2013. Influence of bubbles on the turbulence anisotropy. *Journal of Fluids Engineering* 135 (51301): 1–9.

Bolotnov, I A, K E Jansen, D A Drew, A A Oberai, R T Lahey Jr., and M Z Podowski. 2011. Detached direct numerical simulations of two-phase bubbly channel flow. *International Journal of Multiphase Flows* 37 (6): 647–59.

Brooks, A N, and T J R Hughes. 1982. Streamline Upwind / Petrov-Galerkin formulations for convection dominated flows with particular emphasis on the incompressible Navier-Stokes equations. *Computer Methods in Applied Mechanics and Engneering* 32: 199–259.

Chitale, K C, M Rasquin, J Martin, and K E Jansen. 2014a. Finite element flow simulations of the EUROLIFT DLR-F11 high lift configuration. In *52nd Aerospace Sciences Meeting, AIAA SciTech Forum (AIAA 2014-0749)*. National Harbor, MD.

Chitale, K C, M Rasquin, O Sahni, M S Shephard, and K E Jansen. 2014b. Anisotropic boundary layer adaptivity of multi-element wings. In *52nd Aerospace Sciences Meeting, AIAA SciTech Forum (AIAA 2014-0117)*. National Harbor, MD.

Chitale, K C, O Sahni, M S Shephard, S Tendulkar, and K E Jansen. 2015. Anisotropic adaptation for transonic flows with turbulent boundary layers. *American Institute of Aeronautics and Astronautics Journal* 53(2): 367–378.

Doosttalab, A, J G Araya, J Newman, R J Adrian, K E Jansen, and L Castillo. 2016. Effect of small roughness elements on thermal statistics of a turbulent boundary layer at moderate reynolds number. *Journal of Fluid Mechanics* 787: 84–115. doi:10.1017/jfm.2015.676.

Dubey, A, A Almgren, J Bell, M Berzins, S Brandt, G Bryan, P Colella, et al. 2014. A survey of high level frameworks in block-structured adaptive mesh refinement packages. *Journal of Parallel and Distributed Computing* 74 (12): 3217–27. Elsevier. doi:10.1016/j.jpdc.2014.07.001.

Fabian, N, K Moreland, D Thompson, A C Bauer, P Marion, B Geveci, M Rasquin, and K E Jansen. 2011. The ParaView Coprocessing Library: A scalable, general purpose in situ visualization library. In *Proceedings of the IEEE Symposium on Large-Scale Data Analysis (LDAV2011)*, Providence, RI, pp. 89–96.

Fang, J, M Rasquin, and I A Bolotnov. 2016. Interface tracking simulations of bubbly flows in PWR relevant geometries. *Nuclear Engineering and Design* 312: 205–13.

Farnsworth, J A N, J C Vaccaro, and M Amitay. 2008. Active flow control at low angles of attack: Stingray unmanned aerial vehicle. *American Institute of Aeronautics and Astronautics Journal* 46 (10): 2530–44.

Figueroa, A, I Vignon-Clementel, K E Jansen, T J R Hughes, and C A Taylor. 2006. Efficient anisotropic adaptive discretization of cardiovascular system. *Computer Methods in Applied Mechanics and Engineering* 195 (41–43): 5685–5706.

Fischer, P F, F Loth, S E Lee, Sang-Wook Lee, D Smith, and H Bassiouny. 2007. Simulation of high-reynolds number vascular flows. *Computer Methods in Applied Mechanics and Engineering* 196: 3049–60.

Glezer, A, and M Amitay. 2002. Synthetic jets. *Annual Review of Fluid Mechanics* 34 (1): 503–29.

Group, The Open. n.d. XSI Shared Memory Facility. Manual.

Hoefler, T, J Dinan, D Buntinas, P Balaji, B Barrett, R Brightwell, W Gropp, V Kale, and R Thakur. 2013. MPI + MPI: A new hybrid approach to parallel programming with MPI plus shared memory. *Computing* 95 (12): 1121–36.

Ibanez, D, I Dunn, and M S Shephard. 2016. Hybrid MPI-thread parallelization of adaptive mesh operations. *Parallel Computing* 52 (2): 133–43. Elsevier.

Ibanez, D, and M S Shephard. 2016. Modifiable array data structures for mesh topology. *SIAM Journal on Scientific Computing* 39: C144–61.

Ibanez, D A, E S Seol, C W Smith, and M S Shephard. 2016. Pumi: Parallel unstructured mesh infrastructure. *ACM Transactions on Mathematical Software (TOMS)* 42 (3): 17. ACM.

Intel Software Development Emulator. 2015. Https://software.intel.com/en-Us/articles/intel-Software-Development-Emulator/. Misc.

Jansen, K E. 1993. Unstructured grid large eddy simulation of wall bounded flow. In *Proceedings of Annual Research Briefs*, 151–56. Stanford, CA: NASA Ames/Stanford University.

Jansen, K E. 1994. Unstructured grid large eddy simulation of flow over an airfoil. In *Proceedings of Annual Research Briefs*, 161–73. NASA Ames/Stanford University.

Jansen, K E. 1999. A stabilized finite element method for computing turbulence. *Computer Methods in Applied Mechanics and Engineering* 174: 299–317.

Jansen, K E, M Rasquin, J A Farnsworth, N Rathay, M Monastero, and M Amitay. 2017. Interaction of a synthetic jet actuator on separated flow over a vertical tail. *35th AIAA Applied Aerodynamics Conference, AIAA AVIATION Forum, (AIAA 2017-3243)*. doi:10.2514/6.2017-3243.

Jansen, K E, C H Whiting, and G M Hulbert. 1999. A generalized-α method for integrating the filtered Navier-Stokes equations with a stabilized finite element method. *Computer Methods in Applied Mechanics and Engineering* 190: 305–19.

Karanam, A K, K E Jansen, and C H Whiting. 2008. Geometry based pre-processor for parallel fluid dynamic simulations using a hierarchical basis. *Engineering with Computers* 24 (1): 17–26. Springer.

Kim, H J, C A Figueroa, T J R Hughes, K E Jansen, and C A Taylor. 2009. Augmented lagrangian method for constraining the shape of velocity profiles at outlet boundaries for three-dimensional finite element simulations of blood flow. *Computer Methods in Applied Mechanics and Engineering* 198 (45–46): 3551–66.

Kim, J, P Moin, and R Moser. 1987. Turbulence statistics in fully developed channel flow at low Reynolds number. *Journal of Fluid Mechanics* 177: 133.

Kundu, P K, and I M Cohen. 2012. *Fluid Mechanics*, 5th Ed., New York, NY: Academic Press. doi:10.1016/B978-0-12-405935-1.18001-3.

Liu, N, J Fu, C D Carothers, O Sahni, K E Jansen, and M S Shephard. 2010. Massively parallel I/O for partitioned solver systems. *Parallel Processing Letters* 20 (4): 377–95. World Scientific Publishing.

Mahesh, K, G Constantinescu, and P Moin. 2004. A numerical method for large-eddy simulation in complex geometries. *Journal of Computational Physics* 197 (1). Elsevier: 215–40.

Maldonado, V, J Farnsworth, W Gressick, and M Amitay. 2010. Active control of flow separation and structural vibrations of wind turbine blades. *Wind Energy* 13 (2–3). Wiley Online Library: 221–37.

Mansour, N, and A Wray. 1994. Decay of isotropic turbuelnce and low reynolds number. *Physics of Fluids* 6(2): 808–14.

Mishra, A V, and I A Bolotnov. 2015. DNS of turbulent flow with hemispherical wall roughness. *Journal of Turbulence* 16 (3): 225–49.

Moreland, K. 2011. *The ParaView Tutorial*. Livermore, CA: Sandia National Laboratories.

Morinishi, Y, T S Lund, O V Vasilyev, and P Moin. 1998. Fully conservative higher order finite difference schemes for incompressible flow. *Journal of Computational Physics* 143 (1): 90–124.

Moser, R D, J Kim, and N N Mansour. 1999. Direct numerical simulation of turbulent channel flow up to Re_τ =590. *Physics of Fluids* 11: 943–45.

Mueller, J, O Sahni, X Li, K E Jansen, M S Shephard, and C A Taylor. 2005. Anisotropic adaptive finite element method for modeling blood flow. *Computer Methods in Biomechanics and Biomedical Engineering* 8 (5): 295–305.

Muppidi, S, and K Mahesh. 2005. Study of trajectories of jets in crossflow using direct numerical simulations. *Journal of Fluid Mechanics* 530: 81–100. Cambridge University Press.

Nagrath, S, K E Jansen, and R T Lahey. 2005. Three dimensional simulation of incompressible two phase flows using a stabilized finite element method and the level set approach. *Computer Methods in Applied Mechanics and Engineering* 194 (42–44): 4565–87.

Nagrath, S, K Jansen, R T. Lahey, and I Akhatov. 2006. Hydrodynamic simulation of air bubble implosion using a level set approach. *Journal of Computational Physics* 215 (1): 98–132.

Ovcharenko, A, K C Chitale, O Sahni, K E Jansen, and M S Shephard. 2013. Parallel adaptive boundary layer meshing for CFD analysis. In *Proceedings of the 21st International Meshing Roundtable*, edited by Xiangmin Jiao and Jean-Christophe Weill, 437–55. Berlin/Heidelberg: Springer. doi:10.1007/978-3-642-33573-0_26.

Ozturan, C, H L de Cougny, M S Shephard, and J E Flaherty. 1994. Parallel adaptive mesh refinement and redistribution on distributed memory machines. *Computer Methods in Applied Mechanics and Engineering* 119: 123–27.

ParaView. 2009. Http://www.paraview.org. Misc.

Pope, S B. 2000. *Turbulent Flows*. Cambridge, UK: Cambridge University Press.

Rasquin, M, K C Chitale, M Ali and K E Jansen. 2014. Parallel adaptive detached eddy simulations of the EUROLIFT DLR-F11 high lift configuration. *32nd AIAA Applied Aerodynamics Conference, AIAA AVIATION Forum (AIAA 2014-2570)*. doi:10.2514/6.2014-2570.

Rasquin, M, P Marion, V Vishwanath, B Matthews, M Hereld, K Jansen, R Loy, et al. 2011. Electronic poster: Co-Visualization of full data and in situ data extracts from unstructured grid Cfd at 160k cores. In *Proceedings of the 2011 Companion on High Performance Computing Networking, Storage and Analysis Companion* (SC '11), 103–4. New York, NY: ACM. doi:10.1145/2148600.2148653.

Rasquin, M, C Smith, K Chitale, S Seol, B A Matthews, J L Martin, O Sahni, R M Loy, M S Shephard, and K E Jansen. 2014. Scalable fully implicit finite element flow solver with application to high-fidelity flow control simulations on a realistic wing design. *Computing in Science and Engineering* 16 (6): 13–21.

Rodriguez, J M, O Sahni, R T Lahey Jr., and K E Jansen. 2013. A parallel adaptive mesh method for the numerical simulation of multiphase flows. *Computers and Fluids* 87: 115–31.

Saad, Y, and M H Schultz. 1986. GMRES: A generalized minimal residual algorithm for solving nonsymmetric linear systems. *SIAM Journal of Scientific and Statistical Computing* 7: 856–69.

Sahni, O, K E Jansen, M S Shephard, C A Taylor, and M W Beall. 2008. Adaptive boundary layer meshing for viscous flow simulations. *Engineering with Computers* 24 (3): 267–85.

Sahni, O, K E Jansen, C A Taylor, and M S Shephard. 2009. Automated Adaptive Cardiovascular Flow Simulations. *Engineering with Computers* 25 (1): 25–36.

Sahni, O, J Mueller, K E Jansen, M S Shephard, and C A Taylor. 2006. Efficient anisotropic adaptive discretization of cardiovascular system. *Computer Methods in Applied Mechanics and Engineering* 195 (41–43): 5634–55.

Sahni, O, J Wood, K E Jansen, and M Amitay. 2011. Three-dimensional interactions between a finite-span synthetic jet and a crossflow. *Journal of Fluid Mechanics* 671: 254–87. Cambridge University Press.

Seol, E S. 2005. FMDB: Flexible Distributed Mesh Database for Parallel Automated Adaptive Analysis. PhD Thesis, Troy, NY: Rensselaer Polytechnic Institute.

Seol, E S, and M S Shephard. 2006. Efficient distributed mesh data structure for parallel automated adaptive analysis. *Engineering with Computers* 22 (3–4): 197–213.

Seol, S, C W Smith, D A Ibanez, and M S Shephard. 2012. A parallel unstructured mesh infrastructure. In *SC Companion: High Performance Computing, Networking, Storage and Analysis (SCC)*, Salt Lake City, UT, 1124–32. doi:10.1109/SC.Companion.2012.135.

Shakib, F, T J R Hughes, and Z Johan. 1989. A multi-element group preconditioned GMRES algorithm for nonsymmetric systems arising in finite element analysis. *Computer Methods in Applied Mechanics and Engineering* 75: 415–56.

Shephard, M S, K E Jansen, O Sahni, and L A Diachin. 2007. Parallel adaptive simulations on unstructured meshes. *Journal of Physics: Conference Series* 78–012053: 12053.

Shur, M L, P R Spalart, M Kh Strelets, and A K Travin. 2008. A hybrid RANS-LES approach with delayed-DES and wall-modelled LES capabilities. *International Journal of Heat and Fluid Flow* 29 (6). Elsevier: 1638–49.

Smith, C W, M Rasquin, D Ibanez, K E Jansen, and M S Shephard. 2016. Improving unstructured mesh partitions for multiple criteria using mesh adjacencies. *SIAM Journal on Scientific Computing*. In Review.

Smith, C W, S Tran, O Sahni, F Behafarid, M S Shephard, and R Singh. 2015. Enabling HPC simulation workflows for complex industrial flow problems. In *Proceedings of the 2015 XSEDE Conference: Scientific Advancements Enabled by Enhanced Cyberinfrastructure*, (XSEDE '15, St. Louis, MO,). New York, NY: ACM, Article 41.

Spalart, P R. 2001. Young Person's Guide to Detached-Eddy Simulation Grids. Technical report, NASA Langley Technical Report Server.

Spalart, P R. 2009. Detached-eddy simulation. *Annual Review of Fluid Mechanics* 41. Annual Reviews: 181–202.

Spalart, P R, W H Jou, M Stretlets, and S R Allmaras. 1997. Comments on the feasibility of LES for wings and on a hybrid RANS/LES approach. In *Proceedings of Advances in DNS/LES*, 137–47. Columbus, OH.

Sridharan, S, J Dinan, and D D Kalamkar. 2014. Enabling efficient multithreaded MPI communication through a library-based implementation of MPI endpoints. In *Proceedings of SC14: International Conference for High Performance Computing, Networking, Storage and Analysis,* 487–98.

Taylor, C A, T J R Hughes, and C K Zarins. 1998. Finite element modeling of blood flow in arteries. *Computer Methods in Applied Mechanics and Engineering* 158: 155–96.

Vaccaro, J C, Y Elimelech, Y Chen, O Sahni, K E Jansen, and M Amitay. 2014. Experimental and numerical investigation on the flow field within a compact inlet duct. *International Journal of Heat and Fluid Flow* 44: 478–88.

Vaccaro, J C, Y Elimelech, Y Chen, O Sahni, K E Jansen, and M Amitay. 2015. Experimental and numerical investigation on steady blowing flow control within a compact inlet duct. *International Journal of Heat and Fluid Flow* 54: 143–52.

Vaccaro, J C, J D Vasile, J Olles, O Sahni, K E Jansen, and M Amitay. 2009. Active control of inlet ducts. *International Journal of Flow Control* 1: 133–54.

Various. n.d. Http://www.mcs.anl.gov/research/projects/codes/. Unpublished.

Vignon-Clementel, I, A Figueroa, K E Jansen, and C A Taylor. 2006. Outflow boundary conditions for three-dimensional finite element modeling of blood flow and pressure in arteries. *Computer Methods in Applied Mechanics and Engineering* 195: 3776–96.

Vishwanath, V, M Hereld, and M E. Papka. 2011. Toward simulation-time data analysis and I/O acceleration on leadership-class systems. In *Proceedings of the 1st IEEE Symposium on Large-Scale Data Analysis and Visualization (LDAV)* Providence, RI, 9–14. doi:10.1109/LDAV.2011.6092178.

White, F M. 2000. *Viscous Fluid Flow*. New York, NY: McGraw-Hill

Whiting, C H, and K E Jansen. 2001. A stabilized finite element method for the incompressible Navier-Stokes equations using a hierarchical basis. *International Journal of Numerical Methods in Fluids* 35: 93–116.

Whiting, C H, K E Jansen, and S Dey. 2003. Hierarchical basis in stabilized finite element methods for compressible flows. *Computer Methods in Applied Mechanics and Engineering* 192 (47–48): 5167–85.

Wilcox, D C. 1998. *Turbulence Modeling for CFD*. La Canada, CA: DCW Industries.

Wood, J,Please provide conference details for Wood et al. (2009). O Sahni, K Jansen, and M Amitay. 2009. Experimental and numerical investigation of active control of 3-D Flows. *39th AIAA Fluid Dynamics Conference, Fluid Dynamics and Co-located Conferences*. doi:10.2514/6.2009-4279.

Zhou, M, O Sahni, H J Kim, C A Figueroa, C A Taylor, M S Shephard, and K E Jansen. 2010. Cardiovascular flow simulation at extreme scale. *Computational Mechanics* 46: 71–82.

Zhou, M, O Sahni, T Xie, M S Shephard, and K E Jansen. n.d. Unstructured mesh partition improvement for implicit finite element at extreme scale. *Journal of Supercomputing* 59 (3). Netherlands: 1218–28. http://dx.doi.org/10.1007/s11227-010-0521-0.

Zhou, M, T Xie, S Seol, M S Shephard, O Sahni, and K E Jansen. 2012. Tools to support mesh adaptation on massively parallel computers. *Engineering with Computers* 28 (3): 287–301.

Zhu, X, J Zhang, K Yoshii, S Li, Y Zhang, and P Balaji. 2015. Analyzing MPI-3.0 process-level shared memory: A case study with stencil computations. In *15th IEEE/ACM International Symposium on Cluster, Cloud & Grid Computing*, Shenzhen, China, pp. 1099–1106. doi: 10.1109/CCGrid.2015.131.

16 Lattice Quantum Chromodynamics and Chroma

Bálint Joó, Robert G. Edwards, and Frank T. Winter

CONTENTS

16.1 Introduction...345
16.2 Lattice Quantum Chromodynamics..347
16.3 LQCD Simulations 349
    16.3.1 Gauge Generation 349
    16.3.2 Observable Calculation...352
    16.3.3 Schur Even–Odd Preconditioning 353
    16.3.4 Continuum Extrapolation..354
    16.3.5 Summary of the Methodology and Main Computational Challenges 354
16.4 The Chroma Code..355
    16.4.1 QDP++...356
    16.4.2 Chroma 358
        16.4.2.1 Generic Code through Base Class Defaults.............................358
        16.4.2.2 Polymorphism through Functors and Object Factory-Based
                  Construction...359
        16.4.2.3 Scripting via XML and the Command Pattern 360
    16.4.3 Performance Libraries 360
        16.4.3.1 The QPhiX Library for Xeon and Xeon Phi Architecture.........361
        16.4.3.2 The QUDA Library for GPUs 363
16.5 QDP-JIT and Performance Portability..364
    16.5.1 QDP-JIT/PTX..364
    16.5.2 QDP-JIT/LLVM...366
16.6 Summary and Conclusions 369
16.7 Acknowledgments 370
References 371

16.1 INTRODUCTION

Quantum chromodynamics (QCD) is the theory of the *strong nuclear* force of nature that is responsible for binding protons and neutrons together into atomic nuclei. It is one of the four fundamental forces of the Standard Model of particle interactions that consist of QCD, the nuclear weak force responsible for radioactive decays, electromagnetic forces that power our daily lives, and the force of gravity. The key constituents of the QCD are *matter particles* known as *quarks* and *force carrying particles* known as *gluons*. QCD is believed to provide mechanisms to address many key questions in nuclear and high-energy physics and calculations in QCD are needed to interpret the output of several major experimental programs in the United States and worldwide. The flagship Glue X experiment of the 12 GeV upgrade at Jefferson lab seeks to produce so called *exotic particles* in which gluons play a special role, which are predicted by QCD but have not yet been observed. QCD calculations are also needed to understand the distributions of quarks and gluons in protons and neutrons to improve our understanding of how hadronic matter is formed. At the same time, QCD should predict the properties of light nuclei, such as Helium and Tritium, and allow the computation of *effective nuclear forces* to bridge to higher level nuclear models which can then predict the properties of the rest of the nuclei in the periodic table of the elements. An understanding is needed of the behavior of QCD under conditions of extreme temperature and pressure, which can provide important information in astrophysics and cosmology, such as the behavior of quark–gluon plasma in the early stages of the universe. Such calculations are also needed to understand and interpret the experimental results of the heavy ion collision experiments such as Relativistic Heavy Ion Collider (RHIC) at Brookhaven National Laboratory and at CERN. High-energy physics also requires QCD calculations, for example, to understand the asymmetry of matter and antimatter in our universe. Current calculations in this direction provide a tantalizing tension with the predictions of the Standard Model and experimental results. High-precision calculations in this area are therefore strong candidates to point researchers to physics beyond the Standard Model.

The Chroma code is a toolbox of components needed for computational calculations of QCD in a theory called lattice quantum chromodynamics (LQCD). Chroma was developed during successive cycles of U.S. Department of Energy's Scientific Discovery through Advanced Computing (SciDAC) programs as part for the computational infrastructure for the USQCD collaboration. The primary use of Chroma within USQCD has been to perform calculations mostly in nuclear physics, however, the software is freely available and is used worldwide in many different calculations. The initial conference paper [1] describing Chroma has at the time of writing this chapter reached over 500 citations. During its lifetime, Chroma not only had to meet algorithmic challenges, but had to be ported to a wide variety of architectures. Perhaps, the most challenging of these was the port to systems containing graphical processing units (GPUs), which we discuss later in this chapter.

Heading toward the exascale, it has become clear that many of the underlying infrastructures which currently support Chroma may need to be redesigned and reimplemented. First, issues of performance of the underlying software layers need to be addressed, which have become sharpened in recent years on architectures such as graphics processing units (GPUs) and multicore CPUs with longer "short" vector processing units. Second, the algorithmic space used in QCD has developed substantially in the last decade or so, requiring software support that was not designed for in the original code-stack. USQCD is meeting these challenges under the auspices of the U.S. Department of Energy, Exascale Computing Project with the goal in the first four years to both push forward algorithmic research for QCD calculation, and to provide a software framework for use on the future exascale systems. One thing which is certain is that experiences, both successes as well as partial successes gained over the past 10 years of SciDAC software development will be fundamental in the design of any new systems toward the exascale. In this chapter, we hope to provide some of those experiences, both with regards to recent advanced architectures as well as performance portability.

This chapter is organized as follows. We start with a brief introduction to LQCD in Section 16.2, including the basic formulation and discretization. Section 16.3 is devoted to describing how an LQCD calculation is implemented as a computational campaign, highlighting the *phases of the computation* and the key algorithms used. These two sections are quite generic and apply to most LQCD calculations, irrespective of the software system used. Section 16.4 discusses the Chroma software stack in detail where we also discuss the layered USQCD SciDAC infrastructure. In Section 16.5 we discuss performance portability, and we summarize the material and look forward to the exascale challenges in Section 16.6.

16.2 LATTICE QUANTUM CHROMODYNAMICS

There are several textbooks with excellent descriptions of the LQCD [2–5] and we restrict ourselves here solely to the material that is needed to provide an understanding of the computational details that follow.

The fundamental fields of continuum QCD are the quarks, denoted $\psi(x)$ and the gluons denoted $A_\mu(x)$. The quarks are anti-commuting *Grassman* numbers, and the gluons (also known as the *gauge fields*) are members of the Lie Group $SU(3)$. The theory is defined by the QCD Lagrangian $\mathcal{L}_{QCD}$ written as

$$\mathcal{L}_{QCD}(x) = \sum_{f=1}^{N_f} \bar{\psi}_f(x) \left(\gamma_\mu D^\mu + m_f\right) \psi_f(x) - \frac{1}{2g^2} \text{Tr } G_{\mu\nu}(x) G^{\mu\nu}(x) \tag{16.1}$$

where f denotes the particular quark flavor (up, down, strange, charm, top, and bottom), m_f is the mass of the quark flavor in question, and γ_μ, $\mu = 0, 1, 2, 3$ are the basis of an anticommuting Dirac algebra, $D^\mu = \partial^\mu - iA_\mu(x)$ is the *gauge covariant derivative* constructed using the *minimal coupling prescription* which defines the interaction between quarks and gluons. The fields A_μ include a scaling with g, the interaction coupling constant. Finally, the term involving the field strength tensor $G_{\mu\nu}(x)$ denotes the *propagation and interactions of the gluons with themselves*. This notation implies a summation over repeated Greek indices.

Observables in QCD can then be defined in terms of *path integrals* introduced by Feynman:

$$\langle \mathcal{O} \rangle = \frac{1}{\mathcal{Z}} \int D\bar{\psi} \, D\psi \, DA_\mu \, \mathcal{O}\left(\psi, \bar{\psi}, A_\mu\right) \, e^{iS_{QCD}} \tag{16.2}$$

where the integrals are *functional integrals* over the quark, antiquark, and gauge fields,

$$S_{QCD} = \int dx \, \mathcal{L}_{QCD}(x) \tag{16.3}$$

is known as the *action*, and the normalization constant

$$\mathcal{Z} = \int D\bar{\psi} D\psi DA_\mu \, e^{-iS_{QCD}} \tag{16.4}$$

is the so-called *partition function*.

In order to make the computations tractable, we regularize the functional integrals by moving to a finite, hypercubic lattice with L_x, L_y, L_z, and L_t sites in the X, Y, Z, and T space-time directions, respectively. The lattice volume is $V = L_x L_y L_z L_t$ sites corresponding to a physical volume of $V_{phys} = a^4 V$ fm^4 where a is the lattice spacing between the sites, in this case in femtometers (fm). A site can then be written in terms of integer coordinates as $x = (n_x, n_y, n_z, n_t)$ where a particular coordinate n takes values between 0 and $L-1$ with L being the lattice extent corresponding to the dimension of the coordinate (e.g., L_x for n_x, etc.) Typically, the spatial dimensions are of equal length L_s and one discusses the lattice as having a volume of $V = L_s^3 L_t$ sites. Further, to counter the oscillatory behavior of the integrals due to the factor i appearing in front of the action, the system is formally analytically continued to imaginary time, making the replacement $\tau \to it$.

Moving from continuum to the lattice, the gauge fields A_μ are replaced by *parallel transport operators* $U_\mu(x)$:

$$U_\mu(x) = e^{iaA_\mu(x)} \approx e^{i \int_x^{x+\hat{\mu}} dx A_\mu(x)} \tag{16.5}$$

approximating the line integral from a point x along the link in the *forward* direction $\hat{\mu}$ ($\mu = 0, 1, 2, 3$) to the neighboring point $x + \hat{\mu}$ with a simple piecewise linear approximation along the link. The variables $U_\mu(x)$ are known as *lattice gauge fields* or *link fields* or *link variables*. Further, while $A_\mu(x)$

were members of the Lie Algebra $SU(3)$, their lattice counterparts $U_\mu(x)$ are members of the *Lie Group SU(3)*. In their fundamental representation, they are 3×3 *complex, unitary matrices with unit determinant*. In particular, $U_\mu^\dagger(x) = U_\mu^{-1}(x)$, and $U_\mu^\dagger(x)$ can be thought of as the link variable pointing *backward* from toward site x from site $x + \hat{\mu}$. The collection of $U_\mu(x)$ variables on all the links of a given lattice is referred to as a *gauge configuration* or *configuration* for short. One can then define a lattice *gauge action*, such as the plaquette action:

$$S_g = -\frac{\beta}{6} \sum_P \left(\text{Tr } U_P + \text{Tr } U_P^\dagger - 1 \right) = -\frac{1}{2g^2} \sum_x a^4 \text{Tr } G_{\mu\nu}(x) G^{\mu\nu}(x) + O(a^5) \qquad (16.6)$$

with $\beta = 6/g^2$. One can see that the Gauge propagation and self-interaction part of the action is recovered when comparing with Equation 16.1. Here U_P is the so called *plaquette variable*. On a given site x and a given plane $\mu - \nu$, the plaquette is defined as the product of the links around an elementary side one rectangular loop (plaquette) on the lattice

$$U_P(x, \mu, \nu) = U_\mu(x) U_\nu(x + \hat{\mu}) U_\mu^\dagger(x + \hat{\nu}) U_\nu^\dagger(x) \qquad (16.7)$$

and the sum in S_g is over all the distinct μ, ν planes on all the lattice sites. We note that the gauge action, and indeed the product of links around any closed loop on the lattice, remains *invariant under an arbitrary local transformation*: $U_\mu(x) = \Lambda^{-1}(x) U_\mu(x) \Lambda(x + \mu)$ *with* $\Lambda \in SU(3)$, and hence, the local gauge invariance property defining QCD is preserved exactly. Finally, we note that this is the simplest gauge action one can construct, and that there are other gauge actions in use featuring higher order loops than the plaquette, for example, the Lüscher–Weisz action [6] is in common use in USQCD lattice campaigns.

Discretizing the fermions is less straightforward. The naive central difference discretization of the Dirac equation is

$$\left(\gamma_\mu D^\mu + m \right) \psi(x) = \frac{\gamma_\mu}{2a} \left(U_\mu(x) \psi(x + \hat{\mu}) + U_\mu^\dagger(x - \hat{\mu}) \psi(x - \hat{\mu}) \right) + m\psi(x) \qquad (16.8)$$

with the U fields added in to affect parallel transport to the neighboring sites from site x, to make the derivative *gauge covariant*. However, this discretization admits extra species of fermions to exist at the corners of the Brillouin zone, leading to the infamous *fermion doubling problem*. There is a "no-go" theorem by Nielsen and Ninomiya [7] stating that it is not possible to find a discretization that at the same times allows for chiral symmetry, while being free of doublers, maintaining ultralocality, and still behaving like a fermion discretization. This gave rise to several formulation of fermions on the lattice, including *Wilson fermions*, where a second derivative term is added to the Dirac operator. This gives masses to the doublers on the order of $1/a$ that in the continuum limit, at the expense of explicitly violating chiral symmetry. The Wilson fermion action has discretization errors of $O(a)$. Following the Symanzik improvement program, Sheikoleslami and Wohlert (SW) [8] have added the so-called *clover term* to the Wilson fermion kernel, which can cancel the $O(a)$ errors, resulting in the so-called Wilson–Clover (or just "Clover") fermions with discretization errors of $O(a^2)$. The *Staggered* fermion formulations maintain a remnant chiral symmetry at the expense of breaking *flavor symmetry*. More advanced staggered fermions such ask AsqTAD [9] and HISQ [10] add terms to the fermion operator to reduce flavor (taste) symmetry breaking effects. Finally, a lattice version of chiral symmetry has been devised [11] that circumvents the ultralocality clause of the no-go theorem. These result in four-dimensional operators [12] that are local (but not ultralocal) and also five-dimensional formulations such as domain wall fermions (DWF) and their variants [13–17].

For the sake of concreteness, in this chapter, we focus on Wilson and Wilson Clover fermions as they are the ones that are mostly used in large-scale simulations using Chroma, although some initial research work on five-dimensional chiral formulations was carried out with Chroma in the past [18]. The workhorse code for Staggered Fermion calculation is the MIMD Lattice Collaboration (MILC) code [19], whereas production calculations using DWF within USQCD typiacally are done with the *Columbia Physics System (CPS)* code [20]. In the case of clover fermions, the fermion kernel in operator form is

$$M(U) = (\gamma_\mu D^\mu + m)_{x,y} = D(U)_{x,y} + A(m, c_{SW}, x)\delta_{x,y} \tag{16.9}$$

where the diagonal term

$$A(m, c_{SW}, x) = (4 + m) - \frac{ic_{SW}}{4}\sigma_{\mu\nu}F^{\mu\nu}(x) \tag{16.10}$$

is the mass term coupled to the so called *clover term** with coupling constant c_{SW}, and $D(U)$ is a nearest neighbor stencil like term to known as the Wilson "Dslash" term

$$D(U)_{y,x} = +\frac{1}{2}\sum_\mu \left(1 - \gamma_\mu\right) U_\mu(x)\delta_{y,x+\hat{\mu}} + \left(1 + \gamma_\mu\right) U_\mu^\dagger(x - \hat{\mu})\delta_{y,x-\hat{\mu}} \tag{16.11}$$

As we see later, solving linear systems with M features prominently in our program, so it is worth noting that M is complex, and sparse (nearest-neighbor). Further, M is *J-Hermitian* with $J = \gamma_5 \otimes I_{c,s,x}$ in the sense that $\gamma_5 M = M^\dagger \gamma_5$, where $\gamma_5 = \gamma_1\gamma_2\gamma_3\gamma_4$, $\gamma_5^2 = 1$, and γ_5 is maximally indefinite.

With this definition of M, one can construct a fermion action for two degenerate flavors of quarks as $S_f(U) = \bar{\psi}\left(M^\dagger(U)M(U)\right)\psi$. With both gauge and fermionic components of the action defined over the lattice, the path integral itself becomes a finite but high-dimensional integral over the gauge fields and the fermion fields:

$$\langle \mathcal{O} \rangle = \frac{1}{\mathcal{Z}} \int \prod_{x,\mu} dU_\mu(x) \prod_x d\bar{\psi}(x) \prod_x d\psi(x)\, \mathcal{O}(U, \bar{\psi}, \psi)e^{-S_g - S_f} \tag{16.12}$$

where the integrals over the U links are carried out using the $SU(3)$ Haar measure. For notational convenience, we will continue to use the functional integral notation going forward ($\mathcal{D}U$, etc.) but understand it to mean taking an integral over the finite degrees of freedom (e.g., links, sites, etc.). Finally, since $\int \mathcal{D}\bar{\psi}\, \mathcal{D}\psi\, e^{-S_f}$ is a Gaussian integral we can formally carry out the integration and arrive at

$$\langle \mathcal{O} \rangle = \frac{1}{\mathcal{Z}} \int \mathcal{D}U\, \mathcal{O}(U) \det\left(M^\dagger(U)M(U)\right)e^{-S_g(U)} \qquad \mathcal{Z} = \int \mathcal{D}U\, \det\left(M^\dagger(U)M(U)\right)e^{-S_g(U)} \tag{16.13}$$

for a theory with *two degenerate quark flavors*. We note that at this point, the integrals are well defined, and fermion fields comprised of Grassmann numbers have been integrated out. The remaining system is very similar to a statistical mechanical system such as a gas or a spin-glass system.

16.3   LQCD SIMULATIONS

16.3.1   GAUGE GENERATION

The path integral in Equation 16.13 can be evaluated numerically using Monte Carlo methods, by writing

$$\langle \mathcal{O} \rangle = \int \mathcal{D}U\, \mathcal{O}(U)P_{eq}(U), \quad P_{eq}(U) = \frac{1}{\mathcal{Z}}\det\left(M^\dagger M\right)e^{-S(U)} \tag{16.14}$$

---

* We will no longer refer to the continuum gauge field $A_\mu(x)$ and will use $A(m, c_{SW}, x)$ for denoting the clover term from here on.

treating the weight from the action as a Boltzmann like probabilistic weight. Then, one can approximate the answer as $\langle \mathcal{O} \rangle \approx \sum_U \mathcal{O}(U)P_{eq}(U)$. In this method, one needs to generate a large *ensemble* of gauge configurations U, evaluate the observable $\mathcal{O}(U)$ on every one, and the integral will be approximated by sum of the $\mathcal{O}$ weighted by their probability. If one can employ *importance sampling* so that the probability of U occurring in the ensemble is proportional to $P_{eq}(U)$ one has

$$\langle \mathcal{O} \rangle \approx \bar{\mathcal{O}} = \frac{1}{N} \sum_U \mathcal{O}(U). \tag{16.15}$$

The importance of sampling is typically carried out using hybrid molecular dynamics Markov chain Monte Carlo methods, the most common ones in use being hybrid Monte Carlo (HMC) [21], and Rational Hybrid Monte Carlo (RHMC) [22]. These methods select a trial configuration U' from a preceding configuration U, by performing Hamiltonian molecular dynamics (MD). The individual links $U_\mu(x)$ are treated as canonical coordinates, for which one defines *canonically conjugate momenta* $\pi_\mu(x)$. The action S is treated as a potential energy and kinetic energy is defined as $T(\pi) = \sum_{x,\mu} |\pi_\mu(x)|^2$. One can then define a Hamiltonian $H(\pi, U) = T(\pi) + S(U)$ and the simulation is carried out in the phase space $\{\pi, U\}$ rather than configuration space. The momentum integral can be carried out by refreshing the momenta regularly (typically, at the start of every trajectory) from a Gaussian Heatbath. Since the momenta do not enter the observables, the integral one carries out can be written as

$$\langle \mathcal{O} \rangle = \frac{1}{\mathcal{Z}'} \int d\pi^\dagger \, d\pi \, e^{-\sum_{\mu,x} |\pi_\mu(x)|^2} \int DU \mathcal{O}(U) e^{-S(U)} \tag{16.16}$$

and the momentum integral contributes an irrelevant constant that is canceled out by the extended partition function $\mathcal{Z}'$, which now also includes the exact same integral over the momenta in its definition. The MD needs to be carried out using *reversible and area preserving* integration schemes in order to guarantee *detailed balance* of the process that is a sufficient condition for the resulting Markov chain to converge to the desired fixed-point equilibrium. This is usually accomplished using an explicitly reversible combination of the symplectic updates:

$$U_\mu(x; \delta\tau) \quad \leftarrow \quad e^{i\delta\tau\pi_\mu(x)} U_\mu(x) \tag{16.17}$$

$$\pi_\mu(x; \delta\tau) \quad \leftarrow \quad \pi_\mu(x) + \delta\tau F_\mu(x) \qquad F_\mu(x) = \left[\frac{\partial S(U)}{\partial U_\mu(x)} \right]_{TA} \tag{16.18}$$

where the $\delta\tau$ is a time step in a *fictitious MD time*, and F is the *MD force* that is the traceless anti-Hermitian projection of the derivative of the action with respect to the link. Typical integrator schemes used are the second-order *leapfrog* method or recently, the second-order minimum error norm method suggested by Omelyan [23].

At the end of the MD, one arrives at a new configuration U' and momenta π', which are then subjected to a Metropolis [24] Accept/Reject step with the probability of acceptance being

$$P(\{\pi', U'\} \leftarrow \{\pi, U\}) = \min\left(1, \frac{P_{eq}(\{\pi', U'\})}{P_{eq}(\{\pi, U\})} \right) = \min\left(1, e^{-\delta H} \right) \tag{16.19}$$

with $\delta H = H(\pi', U') - H(\pi, U)$ being the change in the Hamiltonian along the MD trajectory. If $\{\pi', U'\}$ is rejected the original $\{\pi, U\}$ fields becomes the next configuration in the chain. Since the Hamiltonian system is conservative, δH is only nonzero because of *discretization errors* in the time integration. Correspondingly, the acceptance rate can be controlled with the MD step-size. Higher order integrators can be used, providing they are reversible and area preserving, such as are discussed in [23,25] and references therein, to allow one to take larger time steps. Further, one can split parts of the action onto different integration time scales following the techniques of [26]. A recent advancement in this area has been the construction of *force-gradient* integrators and the tuning of the MD integration using Shadow Hamiltonian Methods [27–29].

Simulating fermions poses an additional challenge. In this case, the determinant weight $\det(M^\dagger M)$ is (with the dependence of M on the gauge fields being implicit) rewritten as an integral over *non-Grassmann*, so-called *pseudofermion* fields ϕ and $\phi^\dagger$ as

$$\det\left(M^\dagger M\right) = \int d\phi^\dagger \, d\phi \, e^{-\phi^\dagger \left(M^\dagger M\right)^{-1}\phi} = \int d\phi^\dagger \, d\phi \, e^{-S_f(U,\phi^\dagger,\phi)} \qquad (16.20)$$

where we identify the *two-flavor fermion action as*

$$S_f(U,\phi^\dagger,\phi) = \phi^\dagger \left(M^\dagger M\right)^{-1} \phi \qquad (16.21)$$

The ϕ fields are kept fixed during the MD dynamics, and the Monte Carlo integral is carried out by refreshing them from a heat bath at the start of each trajectory as $\phi = M^\dagger \eta$ where η is a field of Gaussian random numbers. The fermionic contribution to the action can be evaluated as the inner product $S_f = \langle \phi, X \rangle$ where X is the solution of the linear system of equations

$$\left(M^\dagger M\right) X = \phi \qquad (16.22)$$

Further, to compute the (two-flavor) MD force one needs to evaluate

$$F = -X^\dagger \left[\frac{\partial M^\dagger}{\partial U} M + M^\dagger \frac{\partial M}{\partial U}\right] X \qquad (16.23)$$

and one again needs to solve the linear system in Equation 16.22 to compute X. Due to the sparsity and high dimension of M, universally X is found using sparse iterative solvers. Since $M^\dagger M$ is manifestly Hermitian positive definite, the standard workhorse has been the conjugate gradients algorithm [30]. However, for some fermion formulations such as Wilson–Clover, one can perform the solve in two steps as (1) $M^\dagger Y = \phi$, followed by (2) $MX = Y$ using a non-Hermitian solver such as BiCGStab (Bi-Conjugate Gradients Stabilized) [31,32], or others [33–38]. Since the condition number of $M^\dagger M$ is the square of the condition number of M, this can result in a considerable saving in computational cost.

The above discussion was for two degenerate flavors of quark (in the Wilson Formulation), and for a single flavor such as the strange quark one often employs the *square root trick* writing

$$\det(M) = \det\left(\sqrt{M^\dagger M}\right) = \int d\phi^\dagger \, d\phi \, e^{-\phi^\dagger \left(M^\dagger M\right)^{-1/2}\phi} \approx \int d\phi^\dagger \, d\phi \, e^{-\phi^\dagger R[M^\dagger M]\phi} \qquad (16.24)$$

where

$$R[M^\dagger M] = W \sum_j p_j \left[M^\dagger M + q_j\right]^{-1} \approx \left(M^\dagger M\right)^{-1/2} \qquad (16.25)$$

is a rational function approximation to the inverse square root expressed in a partial-fraction representation, and W, p_j, and q_j are coefficients that specify the particular approximation. Rapidly convergent rational fraction approximations for the square root can be computed using the Remez algorithm. It should be apparent, that to evaluate the action and force one now needs the quantities X_j (c.f. X in the two-flavor case) where

$$\left(M^\dagger M + q_j\right) X_j = \phi \qquad (16.26)$$

for every j. Since the q_j are just coefficients, the X_j are the solutions of a *shifted system* and are typically computed using a *shifted conjugate gradients* solver [39,40], where one solves the system with the smallest shift and in each iteration, one can compute updates to every shifted solution, without extra multiplications with M.

To finish off our current discussion of gauge generation, we should note that up to 95% of a gauge generation program is spent in computing forces, and a majority of that time is spent in solving the Dirac Equations 16.22 and 16.26. As the masses of the quarks approache their physical value, the condition number of M diverges, causing a *critical slowing down of the solver*. Algorithmic improvements over the last decade have focused on and are likely to continue to focus on more efficient solvers (especially for the light quark flavors), such as domain decomposed, deflated, and multigrid preconditioned solvers. Typically, the symplectic integrators that have been utilized have been *second order* requiring one or two evaluations of the force for an integrator time step. Higher order integrators have been constructed [23,25], however, these need four to five force evaluations per update step, which has outweighed their benefits until reasonably large lattices start to be used. Schemes that allow integrating parts of the action with different stiffness with different MD time steps have also been devised [26]. The recent introduction of *force-gradient* integrators [27], promises step-size discretization errors of $O(\delta\tau^4)$ over a unit length trajectory, but with only the cost of a second-order integrator (plus the evaluation of the second derivatives of the force). As lattice spacings are decreased to $a \approx 0.045$ fm and below, it has been noticed that the autocorrelation times of the ensembles generated have grown very long. In practice, this means that at finer lattice spacing, one has to perform more trajectories worth of computation than at coarse lattice spacing. This presents yet another *critical slowing down* of gauge generation algorithms, overcoming that will need new algorithmic approaches.

16.3.2 OBSERVABLE CALCULATION

For most calculations of interest in hadronic physics, the observables involve the construction of *quark line diagrams* representing the physical system under investigation (mesons, baryons, multiparticle systems), such as shown in Figure 16.1 for a meson.

The lines in Figure 16.1 represent quarks (Q) and antiquarks ($\bar{Q}$) being created at some time t_0, by a *color source* and destroyed at some final time t_f, by a *a color sink*. The object describing a quark created and destroyed in such a way is called a *quark propagator* $Q(t_0; t_f)$. In the simplest case, where the source and sink are point-like with the sink at point $y = (\vec{y}, t_f)$ and the source at point $x = (\vec{x}, t_0)$ we have that

$$M_{z,y}^{\alpha,\beta,a,b} Q_{y,x}^{\beta,\delta,b,c} = \delta_{z,x}\delta^{\alpha,\delta}\delta^{a,c} \qquad (16.27)$$

in other words, the *quark propagator* is the inverse of the *quark–gluon interaction matrix* from the action. However, computing this is not feasible due to the high dimensionality of M. By using translation invariance, we can always just consider the source to be at the origin of our lattice and then, we essentially compute just a single column (or a few columns) of Q.

As we can see the quark propagator, Q still has both spatial (source, sink) as well as spin and color indices. To make a colorless object such as one in nature, one needs to perform some contractions

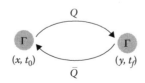

FIGURE 16.1 A quark line diagram depicting a meson created at point (x, t_0) and destroyed at (y, t_f).

over color and also one must fold in interpolating operators that give the final object the correct spin-parity quantum numbers. As an example

$$C(\vec{p}, t_f, t_0, \Gamma) = \sum_{\vec{x}} e^{i\vec{p}\cdot\vec{x}} \mathrm{Tr}_{s,c} \gamma_5 Q(0, x) \gamma_5 \Gamma Q(0, x) \Gamma \qquad (16.28)$$

is the *correlation function* wherein a meson is created at the origin and destroyed it some final space–time point x. The Fourier transform over the spatial indices forces the final state to have momentum $\vec{p}$. In the above, we have used the relation that $\bar{Q}(x, 0) = \gamma_5 Q(0, x) \gamma_5$, which results from the J-Hermiticity of M. The spin-parity quantum numbers of the meson come from the quantum numbers of the interpolating operators, denoted Γ, which are constructed out of the Dirac algebra γ matrices. The correlation function is actually composed of a tower of states:

$$C(\vec{p}, t_f, t_0, \Gamma) = \sum_{n}^{\text{all states}} F_n(\Gamma) e^{-E_n(t_f - t_0)} \qquad (16.29)$$

where E_n is the energy of the nth state, and $F_n(\Gamma)$ is a matrix element representing the overlap of the state with the interpolating operator. In a typical calculation, one looks at the correlation function at large time separations, when only the few lowest energy states survive (often only the lowest).

With newer analysis techniques, many propagators must be solved for. In the case of the *distillation* technique [41], used in studies of excited state mesons, one has to solve the Dirac equation for $O(100,000)$–$O(1M)$ propagators per lattice configuration. Typically, one aims for ensemble sizes of 200–500 configurations, although in the future, even larger datasets need to be considered. Contractions occur wherever interpolating operators are needed. In the case of distillation, every single contraction is comparable in effort to a dense complex matrix multiplication with matrices of dimension $O(700) - O(1000)$. As more complex processes are tackled, the number of quark line diagrams naively grows *combinatorially* with the number of quark lines, although many diagrams are related by symmetries. Hence, as more complex phenomena are studied, the computational cost of the contractions will become an increasingly large part of the calculations.

16.3.3 SCHUR EVEN–ODD PRECONDITIONING

Having to solve the various forms of the Dirac equation numerous times in LQCD simulations, an Even–Odd (Red–Black) preconditioning technique is frequently employed. The sites of the lattice are labeled as even or odd depending on whether the sum of their four-dimensional coordinates is even or odd. This is equivalent to a four-dimensional red–black checkerboarding of the sites. If one groups together all the even sites (e) and odd sites (o), pseudofermion fields can be represented by the vector $\phi = (\phi_e, \phi_o)^T$. The matrix M can then be Schur decomposed as

$$M = \begin{pmatrix} M_{ee} & M_{eo} \\ M_{oe} & M_{oo} \end{pmatrix} = L \begin{pmatrix} M_{ee} & 0 \\ 0 & M_{oo} - M_{oe}M_{ee}^{-1}M_{eo} \end{pmatrix} U \qquad (16.30)$$

with

$$L = \begin{pmatrix} 1 & 0 \\ M_{oe}M_{ee}^{-1} & 1 \end{pmatrix} \quad \text{and} \quad U = \begin{pmatrix} 1 & M_{ee}^{-1}M_{eo} \\ 0 & 1 \end{pmatrix} \qquad (16.31)$$

In these formulations, M_{ee}^{-1} is typically easy to evaluate, for Wilson fermions $M_{ee} = 1$ and for the Clover term one needs to apply the inverse of $A(m, c_{sw}, x)$, which is relatively inexpensive and local to each site. When computing quark propagators, one premultiplies both sides by L^{-1}, and sets $(\phi'_e, \phi'_o)^T = U(\phi_e, \phi_o)^T$ and turns to solve

$$\begin{pmatrix} M_{ee} & 0 \\ 0 & M_S \end{pmatrix} \begin{pmatrix} \phi'_e \\ \phi'_o \end{pmatrix} = \begin{pmatrix} b_e \\ b_o - M_{oe}M_{ee}^{-1}b_e \end{pmatrix} \qquad (16.32)$$

Now, one needs to solve two decoupled sets equations. One is trivial: $\phi'_e = M_{ee}^{-1} b_e$, and a sparse solver is only needed to compute ϕ'_o using M_S, after which the solution ϕ must be reconstructed from ϕ' by acting on it with U^{-1}.

Finally, the preconditioning can also affect the how the fermion determinant is included in the simulation. With the decomposition above, one has

$$\det (M) = \det \left(M_{ee} \right) \det \left(M_S \right) \tag{16.33}$$

In the case of pure Wilson fermions, $M_{ee} = 1$ and so one needs to simulate only with M_S in the fermion action, and keep pseudofermions on only a single checkerboard. In the case of the Wilson–Clover fermion action, M_{ee} is not unity, but one can write

$$\det \left(M_{ee} \right) = e^{- \sum_{x \in e} \log \det A(m, c_{SW}, x)} = e^{-S_{cl}} \tag{16.34}$$

and one can treat this determinant as a separate piece of the action, with its own force term, and so on.

16.3.4 Continuum Extrapolation

Finally, the lattice results need to be extrapolated to give continuum physics results. Strictly speaking, simulations need to be carried out with at least three values of the lattice spacing, to allow a basic linear extrapolation fit, though in practice results "along the way" are often published. Getting 3 lattice spacings is a major endeavor that frequently takes several years of computation.

16.3.5 Summary of the Methodology and Main Computational Challenges

Collecting together all that has been said, the high-level view of the computational task of modern LQCD calculations in nuclear physics proceeds as follows:

- *Gauge generation*: One generates ensembles of gauge configurations on which to measure observables. The majority of parallelism available in this phase is data parallelism over the sites of the lattice, and the problem is a strong scaling problem: the global system volume is fixed, and one would like to generate the ensemble as quickly as possible by adding extra parallelism in the form of compute nodes. The major topics of research relating to this phase are primarily to do with scalable sparse linear solvers, improved reversible, area preserving MD integrators, and new algorithms for controlling critical slowing down of the gauge generation algorithms at fine lattice spacing.
- *Propagator computation*: One computes quark propagators, treating each configuration in the ensemble as "identically independent." This phase of the computation is throughput oriented. Typically, relatively few nodes are needed to perform the necessary solutions of the Dirac Equation on a single configuration, and one can perform the work on several configurations in parallel. The costs of this phase are driven by the analysis methods that utilize the propagators, and research areas in this phase are to do with throughput oriented "multiple right hand side" solvers and with analysis methods that can extract better physical signal with fewer solves. This step also has the potential (again depending on analysis method employed) of generating very large amounts of data (solution vectors from propagator calculations, eigenvectors used in deflation-based schemes, etc.)
- *Contraction/correlation function construction*: The quark propagators computed are contracted into the necessary quark line diagrams. This step is usually carried out with a trivial parallelism. Beyond the configurations additional parallelism can be obtained by working concurrently on different ranges of Euclidean time. Again optimizations in this stage are driven by the particular analysis scheme. As mentioned earlier, a particular challenge is

the large number of quark line diagrams that need to be evaluated and optimizations focus on trying to reduce the number of graphs using symmetries and finding sets of common subgraphs that can be reused. This step can also be bound heavily by I/O.

16.4 THE CHROMA CODE

The Chroma code itself follows the USQCD software structure [42] developed under a series of USQCD SciDAC projects. This layered structure, described in [43], defined the following levels:

1. *Level 1* provides the QCD message passing (QMP) application programming interface (API) for communications needed in LQCD calculations and QCD Linear Algebra (QLA) API for site-local operations. These APIs are implemented as libraries. Chroma uses QMP for its message passing. QMP has been implemented for several network fabrics including a custom Torus Gigabit Ethernet fabric utilized in clusters at Jefferson Lab, over the high speed serial link interfaces of the QCDOC (QCD on a chip) supercomputer and over the Blue Gene and over the BlueGene SPI layer among others. However, these days, its reference MPI version is used on most systems. QMP provides node partitioning, a logical grid node topology, nearest neighbor and point-to-point messages, and several global reductions (sum, max, min). In the message-passing interface (MPI) reference implementation, QMP communicators map to MPI communicators, QMP message descriptors currently map onto MPI persistent communication handles, and global reductions map onto *MPI_Allreduce* calls. QMP has been the bedrock of portability of the code between a variety of different cluster and supercomputing systems.

2. *Level 2* provides a *QCD data parallel productivity layer* known as QDP. There are currently two types of implementation: QDP/C written in the C-language and QDP++ written in C++. The QDP++ system provides for LQCD specific types, and operations between them from a global data parallel viewpoint, similar to High Performance Fortran (HPF) and Connection Machine Fortran. In this view, nearest neighbor communications are encoded with *circular shifts*, QDP++ also provides for mathematical expressions, to allow writing code that is similar to the underlying mathematical structure. Finally, QDP++ interfaces with a variety of I/O packages including capabilities to read and write XML documents (for parameter and configuration files) using the *libxml2* library, and to provide a variety of binary I/O capabilities including reading and writing local databases (*FileDB* library), binary files for lattice data types using parallel I/O (USQCD *QIO* library), Hierarchical Data Format (HDF-5) files (contributed by the CalLat collaboration), and generic key-value files, where the keys can be arbitrary strings, and the values can be binary objects. In this way, QDP++ essentially provides *a domain specific language, embedded in C++* for QCD calculations or viewed alternatively a *data parallel machine abstraction* over which QCD programs can be expressed. The Chroma code is written almost completely in terms of QDP++ and we take a more indepth look at the QDP++ type structure and specific QDP++ implementations later on.

3. *Level 3* is a level designed to hold optimized libraries, which are tuned sufficiently close to specific hardware that they cannot be expressed with equivalent levels of code efficiency in Level 2. These components typically include highly tuned versions of the "Dslash" operator or the matrix M (and its Schur complement for use in preconditioned solvers), as well as custom implementations of particular solvers for the systems $Mx = b$, $M^\dagger Mx = b$ and the shifted solvers $(M^\dagger M + q_i)x = b$, which are needed for propagator, hybrid Monte Carlo energy and force term calculations, respectively. In some cases, complicated MD *force term calculations* could also be considered to be part of Level 3. The main numerical cost of the solvers is the evaluation of the matrix-vector operation $M(U)\psi$, and the relevant linear algebra for the

solvers could be carried out in Level 2 leading to *native solvers* in Chroma and other codes. However, the advent of graphical processing units (GPUs), and advances in solver technologies utilizing *multiple precision* [44], *domain decomposition* [45], *deflation* [35,46,47], and *multigrid methods* [37,38,48,49] have placed more and more optimized functionality in this layer as implementers hit the limitations of code frameworks designed in the early 2000s.

4. *The Application Level* is the level designed for concrete applications such as Chroma, MILC, the CPS and others. These application suites typically embody the higher level algorithms such as implementations of the *gauge generation* algorithms such as HMC or RHMC, as well as observable measurement including the creation of sources, implementation and/or wrapping of solvers from Level 3, basic correlation function construction.

When this structure was developed, *Chroma* was a brand new code written entirely in terms of packages and structures defined in these levels, while existing codes such as MILC and CPS moved to adapt specific features/libraries such as QMP as needed. The Chroma stack consisted of: QMP, QDP++, and QIO, several Level 3 packages for Dslash and solvers, and Chroma itself at the highest level. The basic philosophy was that since Chroma was written almost 100% in terms of QDP++, the work in porting Chroma to a new architecture would involve porting QDP++, and possibly writing a new Dslash Operator or Solvers package. QDP++ was written in standard C++, with some piecemeal template specializations to accelerate certain crucial operations, but not all QDP++ operations ran at the highest possible level of efficiency. This was considered to be reasonable, due to the majority of time being spent in solvers, which were heavily optimized. The arrival of GPUs and multicore systems with relatively long vector units changed this calculus somewhat for three reasons: First and foremost, QDP++ in its most production-ready branch, utilized array-of-structures (AOS) data layouts that were not well suited to vectorization or GPUs. Second, with increased thread and vector level parallelism available in hardware, it was less and less acceptable to have the operations in the data parallel layer operate inefficiently as they would produce a noticeable Amdahl's Law drag on the rest of the code. Finally, there was even a question of how to mount the expression template techniques of QDP++ on GPUs in the first place. Before discussing how this was achieved, we proceed to give an overview of the QDP++ and Chroma packages below.

16.4.1   QDP++

The key aim of the QDP++ library was to provide productivity to users in terms of simple programming constructs; types and expressions, which allowed programmers to focus on the underlying mathematics. In general, LQCD-specific types are composed of a base hypercubic lattice, where at each site one can have a tensor structure of internal indices. Specific indices in the case of nonstaggered formulations are *color* and *spin*. Since the numbers themselves are complex, one can also add an index referred to as *reality*. QDP++ implemented this structure as templated containers. The lattice structure was represented as the type OLattice<T> with T being the inner type on the lattice sites. One than had a variety of tensors for the sites: scalars (PScalar<T1>), vectors (PVector< T1, N>, where N is the dimension of the vector), and matrices (PMatrix<T1, N>) with dimension N. The types could be further composed by adding additional structure in place of T1, and so forth. This structure follows the mold originally devised in the preceding SZIN software system [50]. Some specific QDP++ types are given in Table 16.1.

The types are composed in C++ code, with nested templating. As an example consider the type

$$\text{LatticeColorMatrix} = \text{OLattice}< \text{PScalar}< \text{PColorMatrix}<$$
$$\text{RComplex}< \text{REAL} >, \text{ Nc} > > > \qquad (16.35)$$

where Nc, the number of colors is a compile time constant and one can see nested templated types for the tensor structure in action. This is a type that is commonly used to represent the gauge link

TABLE 16.1

Some Lattice and Tensor Structures Provided by QDP++

Name	Outer Lattice	Spin	Color	Reality
Real	*Scalar*	*Scalar*	*Scalar*	*Real*
Lattice Dirac fermion	*Lattice*	N_s *vector*	N_c *vector*	*Complex*
Lattice color matrix	*Lattice*	*scalar*	$N_c \times N_c$ *matrix*	*Complex*
Lattice Dirac propagator	*Lattice*	$N_s \times N_s$ *matrix*	$N_c \times N_c$ *matrix*	*Complex*

```
template< class T, class T1, class Op, class RHS>
void evaluate(OLattice<T>& dest, const Op& op,
                    const QDPExpr< RHS, OLattice<T1> >& rhs,
                    const Subset& s)
{
   const int *tab = s.siteTable().slice();            // Grab the
       site table
   const int numSiteTable = s.numSiteTable();   //  and the number of
       sites

#pragma omp parallel for
   for(int j=0; j < numSiteTable; ++j ) {             // visit each
       site
     int i=tab[j];

     // the ForEach functor traverses the expression tree
     op(dest.elem(i), forEach( rhs, EvalLeaf1(i), OpCombine() ) );
   }
}
```

FIGURE 16.2   The evaluate() function for a QDP++ expression template assignment.

variables *U*. In order to represent all four forward pointing links, one would define a one-dimensional array of LatticeColorMatrix types, provided by the QDP++ type multi1d.

QDP++ provided *expressions* between the types using the expression template mechanisms, implemented through a modified version of the *Portable Expression Template Engine (PETE)* [51], which was part of the Parallel Object-Oriented Methods and Applications. *POOMA* [52] framework. Expressions types could be built up using C++ overloading of the operators +,-, *, / as well as through specific functions such as sum or trace. Further communications could be specified using the shift operation. Eventually, the expression would be assigned to a result with some form of assignment operator, for example, operator=. The implementation of the assignment could then call a specific evaluate() function, templated on the expression, which would traverse the built up expression tree as shown in Figure 16.2. The operator= function in the expression calls the evaluate() function shown with Op being the type (OpAssign), the right hand side of the expression being held in the recursive template QDPExpr, which is templated by the actual expression type RHS and the type of the output (OLattice<T1>). The body of the evaluate function loops through the table of sites (s.siteTable()) of the subset s of sites from the lattice. On each site, the op is evaluated, and the ForEach functor traverses the Expression tree at ***compile time generating code within the C++ compiler to evaluate the expression***. This code will be a set of nested function calls to deal with the various tensor indices at lower levels (e.g., at the spin and color level). The code

```
Real plaquette( const multi1d<LatticeColorMatrix>& u)
{
    Double ret_val=0;
    for(int mu=0; mu < Nd; ++mu) {
        for(int nu=mu; nu < Nd; ++nu) {
            ret_val + = sum( real( trace(u[mu]*shift(u[nu], FORWARD, mu)
                * adj(shift(u[mu], FORWARD, nu))*adj(u[nu])))));
        }
    }
    return ret_val;
}
```

FIGURE 16.3   QDP++ listing to compute $\sum \mathrm{Re\ Tr}\ U_p$.

generated will not be executed until the evaluate function with these particular templates is *called at runtime* at which point there will be actual data to operate on.

Using these facilities, a user can generate a function to evaluate the *plaquette* operator of Equation 16.7 summed over the lattice as shown in Figure 16.3. The expression `shift(u[nu],` `FORWARD,mu)` performs communication to access $U_\nu(x+\hat{\mu})$, and the operator `*` is overloaded to perform matrix multiplication on every site of the lattice. The function `adj()` performs Hermitian conjugation. One can notice several things here: first, the user does not specify loops over the sites, or the internal indices. All that work is taken care of by the expression templates. Second, looking back to Figure 16.2 the one-line expression for `ret_val` will generate one single-site loop, which will be parallelized with OpenMP. Looking back at the type-structure in Equation 16.35, one can see that once everything unwinds, at the bottom level, one will invoke `operator*` between types `PColorMatrix<RComplex<REAL>,Nc>` and one will multiply 3×3 complex matrices in this instance. Since the types are using a user-defined complex type *RComplex* and the matrix dimensions do not immediately imply alignment, there is little information that could give the compiler hints about vectorization. However, it is possible to *specialize* (overload) the `operator*` for this type so that its internals could be coded with compiler intrinsics that can exploit the vector registers. Finally, had the code been written instead as in Figure 16.4, with three separate lines involving no shifts, one could specialize the entire `evaluate()` function over all the sites, giving perhaps more scope for optimizations. On the other hand, this would now introduce three lighter weight OpenMP regions and because three expressions are used instead of one, multiple memory traversals and potentially worse cache usage, compared to the one line case. Expression fusion could help here, but is difficult to express in a straightforward way with constructs that operate at the level of a single assignment, such as expression templates.

16.4.2 CHROMA

We turn our attention now to the Chroma code. While QDP++ gave us a domain-specific language to work with Chroma in turn uses this language to provide physics capability. Chroma is a *swiss army knife package* containing implementations of several fermion formulations, measurements, and algorithms related to gauge generation, propagator calculations, and correlator function construction. We neither can, nor wish to go into detail here of all the functionality, however, we highlight three broad features of Chroma that we think are noteworthy.

16.4.2.1 Generic Code through Base Class Defaults

A fundamental quantity in LQCD is the action S. As discussed earlier, the action can be split into pieces, for example, S_g for the gauge action or S_f for a fermion action. The fermion actions can be

```
Real plaquette2( const multild<LatticeColorMatrix>& u)
{
    Double ret_val=0;
    for(int mu=0; mu < Nd; ++mu) {
        for(int nu=mu; nu < Nd; ++nu) {
            LatticeColorMatrix u_nu_forw_mu, u_mu_forw_nu,u1,u2;
        u_nu_forw_mu = shift(u[nu], FORWARD, mu);
        u_mu_forw_nu = shift(u[mu], FORWARD, nu);

        u1 = u[mu] * u_nu_forw_mu; // specialized evaluate() hooks into
            optimized expression
        u2 = u1*adj(u_mu_forw_nu); // specialized evaluate() hooks into
            optimized expression
        u1 = u2*adj(u[nu]);                         // .. and here also

        ret_val + = sum( real( trace(u1) ) );
        }
    }
    return ret_val;
}
```

FIGURE 16.4   Alternative QDP++ listing to compute $\sum \mathrm{Re\ Tr}\ U_P$.

classified both in terms of the fermion formulations they implement, for example four-dimensional *Wilson*-like (Wilson, Clover, Twisted Mass), five-dimensional Wilson-like (domain wall, Möbius domain wall, extended overlap, five-dimensional continued fraction, etc.), and staggered-like. They may or may not be subjected to Schur style red–black (even–odd) preconditioning as discussed in Section 16.3.3. Chroma attempts to capture this structure in the inheritance hierarchy. Associated with the fermion action are linear operators: M and $M^\dagger M$ for a given gauge field. The linear operators are *functors* or function objects. Their operator() method applies the linear operator to a vector. The linear operators also provide their own force terms with the deriv() method that implement the operation: $X\frac{\partial M}{\partial U}Y$ for fermion fields X and Y, allowing the construction of more complicated force terms.

Schur decomposed fermion action classes inherit from a base class that specifies virtual methods for M_{ee}, M_{eo}, M_{oe}, M_{oo}, and M_{ee}^{-1}. The operator M_S can be constructed in the operator() method in the *base class* using these virtual methods to provide a *default implementation* for all Schur preconditioned fermion actions. A default implementation for the MD force term for the operator in the deriv() method of the base class can also be composed by utilizing calls to virtual functions that implement derivatives for these submatrices (derivEvenEven(), derivEvenOdd(), etc.). Of course, implementations classes can override these defaults for efficiency reasons if necessary.

Likewise, pieces of fermion action can also be categorized, for example, as a two-flavor piece (c.f. Equation 16.21), or a rational function piece (c.f. Equation 16.24). The force term for each such category can also be implemented as default code, making use of virtual functions. An example is the case for two-flavor pieces, which implement the term in Equation 16.23 as a default.

Thus, Chroma has parallel inheritance trees, based primarily on fermion type and preconditioning to allow as much code to be written in terms of default functions in the base classes as possible.

16.4.2.2   Polymorphism through Functors and Object Factory-Based Construction

The second noteworthy aspect of Chroma design is its *heavy use of function objects*, rather than functions. This is a way to implement polymorphism. Algorithms are defined in terms of function objects (often through their `operator()`). Examples of this are linear solvers and MD update steps. The goal is to implement an algorithm in terms of function objects satisfying constrained interfaces and then allowing, the user to select specific implementations from a configuration file. As an example, consider a solver to be used in a force calculation. This is an object of type `MdagMSysSolver`. The object has an `operator()` with signature

```
// T is a LatticeFermion type, psi is the result, chi is the source
void operator()(T& psi, const T& rhs);
```

and calculations can make use of this interface in default code. However, whether the solver itself uses conjugate gradients, or BiCGStab, or a specific solver for GPUs does not need to be specified to the force calculation. At some point, a concrete object will need to be created. This is done using the *ObjectFactory* design pattern. A piece of input XML from a configuration file specifying a solver will have a tag called, `<invType>` the value of which will act as a key to a construction method in an object factory. Object creation involves searching the XML snippet for this tag, looking up the creation function by the name specified in the tag, and then passing the XML snippet and other data (e.g., the gauge field) to the creation function, which will return a pointer to a newly created solver object. This *constructional design pattern* [53] *allows a diverse set of implementations of a base class to be created using a uniform method.* This approach of using factories gives Chroma a high degree of extensibility, and provides a plug-in architecture. This pattern is replicated throughout Chroma, from the creation of solver objects, boundary condition objects, and even to classes to perform I/O of different file formats. Anywhere where the user has a choice of implementation, algorithm, or one needs to make an important selection, this pattern is likely to be found. In particular, this approach allowed Chroma to absorb new solver technologies in third-party libraries without needing to change its look or feel.

16.4.2.3 Scripting via XML and the Command Pattern

The final aspect of Chroma we mention is related to the previous one. Observable measurement is also implemented in terms of function objects. Users of Chroma are free to add new measurements to their own version of Chroma. The Chroma executable itself, in fact, is quite simple. It reads in an XML input file containing a specification of a list of measurement objects. Using the factory pattern, it creates an array of measurement objects and then traverses this array one by one, providing the input gauge configuration to each measurement object. A set of basic measurements (sources, propagators, basic spectroscopy, etc.) soon grew. Advanced users of course are free to extend the system as they see fit, by writing new measurements. Pretty soon, it was discovered that tasks may need to exchange data, and so a basic "in memory filesystem" called the *NamedObject* store was provided. Once a piece of data has been produced, it can be associated with a name and persisted between measurements until a special measurement finally deletes it. This essentially turned the XML input file for Chroma into a kind of mini-language (*Command Pattern* [53]), that had variables (the persisted `NamedObjects`), but was a linear stream with no conditionals or loops. Users very quickly filled this gap by writing scripts to produce the XML, using the scripting language facilities for looping and conditionals. In our opinion, it was the availability of these basic measurements and maintenance and bundling of high-performance third-party components that made Chroma so popular. Many Chroma users never have to write any high-performance code in C++ explicitly, they only need to write input files, which they can do from the comfort of their scripting language.

16.4.3 PERFORMANCE LIBRARIES

There have been advances both in terms of solver algorithms, and architectures since the original design of QDP++ and Chroma. While some new solver algorithms have been explored directly in

Chroma, it was typical for the community to capture algorithmic advances and the code for maximal performance on specific hardware architectures, by creating libraries in "Level 3" of the USQCD infrastructure. Examples of such optimized libraries include *BAGEL/BFM* [54] for BlueGene/Q, *QUDA* [44] for Graphics Processing Units, *QPhiX* [55,56] for Xeon Phi, the *QOPQDP* [57], and others. These libraries could be wrapped by Chroma Solver objects as discussed previously. We do not go into a great level of detail regarding the optimizations here, leaving the reader to follow-up citations in the literature. However, we discuss briefly some performance examples of the libraries below.

16.4.3.1 The QPhiX Library for Xeon and Xeon Phi Architecture

The QPhiX Library [55,58] was originally developed to optimize the Wilson Dslash operator for the Intel Xeon Phi Knights Corner architecture. It utilized an *Array of Structure of Arrays* (AoSoA) data layout to effectively exploit the vector units of the Knights Corner processor. The fastest running (inner) dimension of the arrays split the logical X-direction into chunks of length S (known as the SOA length), limiting the X-dimension to be a multiple of $2S$ because of red–black coloring (checkerboarding). On Xeon Phi, the vector length is 16 single-precision (or eight double-precision) floating-point numbers. While the processor had a general gather instruction, in practice *load-unpack* and *pack-store* instructions proved faster, giving values of $S = 4, 8, 16$ in single precision, and $S = 4, 8$ in double. When S was less than the vector length V, chunks of length S were taken from $n_y = V/S$ rows of the Y-dimension. Hence, the vectorization strategy vectorized over two-dimensional tiles of dimension $n_y S$ sites. The lattice itself was blocked into blocks of size B_y and B_z along the Y- and Z-dimensions, respectively and streamed over the T-dimension implementing the so called 3.5D blocking strategy [59]. Finally, the temporal length of the $B_y \times B_z \times T$ blocks could be further subdivided to increase the number of blocks for better load balancing. Blocks were assigned to cores, and the work within a block was performed by SMT threads. Internode communication of the faces was carried out via MPI. The Dslash kernel applied to a single vector's worth of data was generated by a *code generator* that generated compiler intrinsics for Knights Corner, AVX, AVX2, and AVX512 instruction sets. QPhiX was ported to Intel Xeon Phi Knight's Landing (KNL) processor with support from the NERSC NESAP Program and initial performance on KNL was reported in [60]. Intrinsics for *QPX* for BlueGene/Q and legacy *SSE* were also added. Finally, by adding basic linear algebra routines, QPhiX implemented conjugate gradients, BiCGStab, and multiple-shift conjugate gradients needed for gauge generation and propagator calculations. An iterative refinement mixed-precision solver was also implemented which used BiCGStab as its inner solver.

We show in Figure 16.5 the weak scaling of QPhiX on the Theta system at Argonne Leadership Computing Facility (ALCF) featuring 64-core Intel Xeon Phi 2730 KNL processors, during its early user phase. Each individual line corresponds to a fixed local volume scaled out to 2048 nodes. The various symbols indicate different local volume sizes: 32^4, $24^3 \times 32$, and 16^4 sites per node, respectively. We also vary the S parameter of QPhiX, showing the $S = 8$ results for $V = 32$ and $V = 16$ sites per node, and the $S = 4$ sites per node for each volume. At any given node size, sliding from the larger volume symbols down to the smaller volume gives an indication of losses one would expect strong scaling. Figure 16.5a shows a term used in the Schur preconditioned clover operator: $A_{ee}^{-1} D_{eo}$, while on Figure 16.5b shows the scaling of the full even–odd preconditioned BiCGStab solver. Finally, on each plot, we show in maroon, the performance of the $V = 32^4$ sites/node, $S = 8$ result from the recently installed SciPhi Cluster at Jefferson Lab also featuring 64-core, Intel Xeon Phi 2730 processors. This cluster uses Intel's OmniPath fabric, whereas the Theta system features the Aires Dragonfly interconnect from Cray. Some clear structure is visible in the results below 32 nodes. This is due to switching on additional communications directions in the Dslash operator. Each additional direction involves packing buffers and exchanging them among nearest neighbors, and then reaccumulating the results on the surfaces of the lattice and the changes in performance

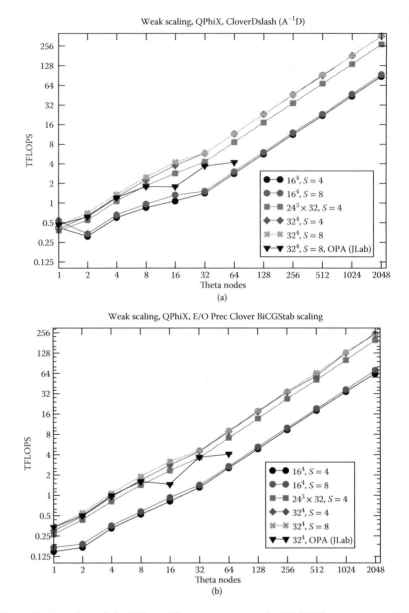

FIGURE 16.5 Weak scaling of the Wilson–Clover operator on the ALCF Theta system and the Jefferson Lab SciPhi Cluster: (a) the Clover Dslash operator $A^{-1}D$ and (b) the even–odd preconditioned BiCGStab solver.

are a result of this additional overhead (in the $V = 16^4$ case, going from one node to two actually seems to cause a decrease in performance). However, beyond 32 nodes, the scalings are nice and linear. The JLab Cluster results follow the Theta results well until around 8 nodes, whereafter we see dips going from 8 to 16 and from 32 to 64 nodes. We have not yet investigated the source of these dips at the current time. The pattern of behavior is consistent between the Dslash and the solver results.

16.4.3.2 The QUDA Library for GPUs

The QUDA [44,61] library for NVIDIA GPUs was developed at Boston University, shortly after the CUDA programming model for NVIDIA GPUs was publicly released. There have been previous attempts at using GPUs for LQCD computation, which used explicit graphics APIs [62]. QUDA was the source of now several standard techniques in optimizing LQCD Dirac solvers for performance, including emphasizing the importance of data layouts for performant memory access, the use of compression (to reduce bandwidth needs), and the role of reduced precision and mixed precision in solvers, for example, coupling reduced-precision preconditioners to accelerate full-precision solvers, or to use smarter restarting schemes than simple iterative refinement; such as reliable updates [32]. Two of the original developers from Boston University have since moved to NVIDIA, and QUDA continues to be developed and used both in production and as a toolkit in which novel algorithmic techniques are being explored. Various parts of the QUDA library have been interfaced with several major code packages, including Chroma, MILC, and others, and QUDA enjoys a high level of popularity in the world LQCD community. Some of the highlights from QUDA's development include its initial parallelization to multiple nodes [63], and eventually, by using a reduced communication, domain decomposed preconditioner, to over 100 nodes [45]. Most recently, a new adaptive multigrid solver has been added to QUDA [49] increasing its performance by a factor of 5× to 10× compared to its existing BiCGStab implementation for Wilson–Clover fermions. QUDA provides the main solvers used by Chroma in current gauge generation and propagator computation campaigns on GPU-based systems such as Titan at Oak Ridge Leadership Computing Facility (OLCF).

We show in Figure 16.6 the scaling of the QUDA library, as called from Chroma on the OLCF Titan System. On the left hand plot, we show the strong scaling in GFLOPS of the Generalised Conjugate Residual (GCR) algorithm preconditioned with an additive Schwarz domain decomposed preconditioner (DD+GCR) as discussed in [45] on synthetic data. It can clearly be seen that the DD+GCR solver has superior strong scaling to the BiCGStab solvers, which are also shown. However, we note that we are comparing FLOP rates on synthetic data here, rather than wall-clock times. Experience with equilibrated configurations reported in [45] shows that a wall-clock time improvement is also achieved. In Figure 16.6b we plot the strong scaling behavior of both BiCGStab and the recently developed adaptive multigrid algorithm from QUDA, running on an equilibrated lattice with $V = 64^3 \times 128$ sites, and light sea quarks. The mass of the resulting π-meson particle in this system would be around $m_\pi \approx 172$ MeV. The plot shows the same data (although replotted) as Ref. [49], however, the estimate of m_π has been updated since that publication where it was quoted as $m_\pi \approx 200$ MeV. Since the BiCGStab and multigrid algorithms are very different and the data are realistic (actually from production), the best figure of merit to compare is the wall-clock time for the solvers. It can be seen that the multigrid solver outperforms the BiCGStab solver by factors of between 5× and 10× depending on the partition size. For these results, the multigrid generated its coarse grids by two levels of blocking one with 4^4 sites per block and the other with 2^4 sites. Thus, the fine grid had $V = 64^3 \times 128$ sites, while the intermediate coarse grid had $V_1 = 16^3 \times 32$ sites, and the coarsest grid had $V_3 = 8^3 \times 16$ sites. A current limitation of the implementation is that a minimum of $V = 2^4$ sites are needed per node, giving the maximum partition size for these jobs to be 512 nodes, mapped as a virtual $4^3 \times 8$ grid of processors, although little improvement in wall-clock times is observed going from 256 to 512 nodes. The most cost-effective partition size for this problem appears to be 64 nodes per solve. In a distillation quark propagator computation scenario, with 384 K solves per configuration, with ≈ 5 seconds per solve and an ensemble size of 250 configurations, this could fill up 16,000 nodes of the Titan system in throughput mode for approximately 23 wall-clock days, neglecting other, for example, I/O costs, at a cost of approximately 265 M Titan Core-hours. Given that even with this most sophisticated of solvers, the cost of a calculation is so high, any new algorithmic and performance improvements could still have significant benefits. Current work in QUDA and Chroma is focused on integrating the solver into the gauge generation process.

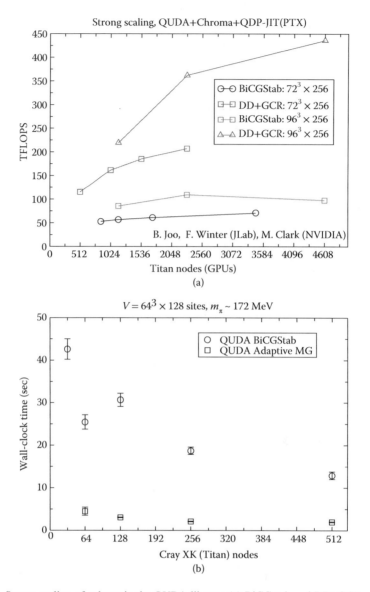

FIGURE 16.6 Strong scaling of solvers in the QUDA library: (a) BiCGstab and DD+GCR solvers on Titan and (b) adaptive multigrid preconditioned GCR versus BiCGStab in QUDA. (Data replotted from M. A. Clark, et al., Accelerating lattice QCD multigrid on GPUs using fine grained parallelization, in *Proceedings of SC16, The IEEE/ACM International Conference on High Performance Computing, Networking Storage and Analysis*, Salt Lake City, UT, 2016.)

16.5 QDP-JIT AND PERFORMANCE PORTABILITY

16.5.1 QDP-JIT/PTX

As discussed in Section 16.4.1 the bedrock of the Chroma stack is the QDP++ productivity layer, providing both a data parallel machine abstraction and in essence a domain specific language for LQCD. Further, it is relatively compact at approximately 106,000 lines of code without its support libraries (140,000 or so with support libraries), compared to Chroma that is approximately 250,000

(332,500) lines without (with) its associated submodules. Hence, when porting to a new architecture, apart from crafting/porting performant solver libraries a port of QDP++ is also required.

This came to light particularly with the arrival of GPUs. Three primary questions needed to be answered: (1) Given that expression templates generate code at compile time inside the "host" C++ compiler, how can one generate the necessary kernels to be offloaded? (2) Experience from the QUDA library and elsewhere raised the importance of *data layout flexibility* including different layouts between host and device, whereas the scheme in Section 16.4.1 essentially fixed the layout, and in the most supported branch fixed it to an array of structures (AoS), and (3) how to deal with the relatively small amount of memory available on GPU systems, compared to the host memory, and orchestrating data motion between the two spaces.

The solution constructed to solve these questions was the QDP-JIT [64] implementation of QDP++. The idea was that instead of generating executable kernels in the compiler, the expression templates would generate code generators. The code generators could then generate the actual kernels Just-In-Time (JIT) when first encountered at runtime. Initial versions of QDP-JIT, generated CUDA C kernels that could be compiled into dynamically linkable objects. The version described in [64] in turn generated GPU kernels directly in PTX (Parallel Thread eXecution) [65], which is an assembly-like language for programming NVIDIA GPUs. The PTX code, could then be turned into executable code directly by the GPU driver (through a secondary driver level JIT compilation). The JIT-ed kernels could be cached and reused, and some degree of performance autotuning (through grid and block-size variation) could be employed to tune performance.

To address the issue of movement and limited memory, a memory pool-cache was implemented. Data would be allocated into the pool on the device. As more objects were allocated to the pool, long unused objects could be evicted to the larger memory space of the host. Data transfers between host and device from the pool could also be made for purposes of I/O and communicating between nodes. In this sense the management of memory became automatic and hidden from the user.

The code generators, which were generated by the expression templates, looped over all the internal (fixed short trip-count) indices of spin and color while generating code, essentially completely unrolling these loops. The memory references they generated could be expressed via loop dimensions and offsets compared to a base pointer, by working with *views* of the data. Accesses via the views could then generate the appropriate offsets allowing for flexibility in terms of data layout, including working with different host and device data layouts. Using this technique and automatically implementing layout transformations as fields moved between the host and device via the pool-cache mechanism allowed efficient memory accesses to the data, both on the host and the device.

Finally, QDP-JIT improved on several aspects of the original QDP++ implementation, including, for example, the caching and reuse of communication resources (buffers, etc.) and overlapping computation with communication, in expressions containing circular shifts.

Figure 16.7 shows some of the key performance aspects of the QDP-JIT/PTX system. In Figure 16.7a, we show the memory bandwidth use of several kernels that are generated from QDP++ expression. This plot, showing performance on NVIDIA K20X GPUs with ECC enabled, is similar to one in [64] which shows the ECC disabled version. All the kernels are memory bandwidth bound. As particular examples, we can consider *lcm*, *matvec*, and *clover*. At each lattice site, respectively, the *lcm* operation multiplies two 3×3 complex matrices, and *matvec* multiplies a 3×3 matrix with a 3-vector. The *clover* operation applies the clover term $A(m, c_{\mathrm{sw}})$ to a *fermion field* at each site. This is relevant, since efficient implementation of the clover term is usually done outside of the usual tensor space enabled by QDP++ expressions and shows that the JIT approach can be successfully applied not just to QDP++ codes. All of the kernels depicted saturate memory bandwidth at a lattice size of about $V = 14^4 - 16^4$ sites. Below these volumes, one is exposed to the effects of kernel startup and termination. We refer to the region where performance starts to saturate as the *shoulder* region, and we anticipate it to be dependent on the device, the driver, and the underlying host, which drives the GPU. The shoulder also sets a strong scaling bound; for node-local volumes smaller than the lower

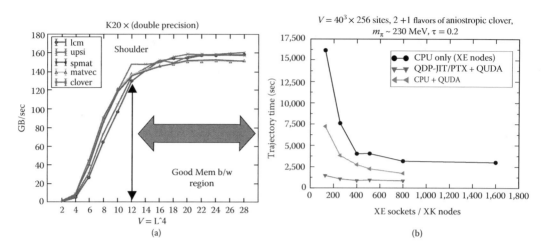

FIGURE 16.7 (a) Device memory bandwidth utilized by QDP-JIT/PTX on some key LQCD kernels on NVIDIA K20X GPUs with ECC enabled. (b) Short trajectory benchmark showing the strong scaling performance of Chroma with the QUDA library and QDP-JIT/PTX. (Data replotted from NVIDIA Corporation, Parallel Thread Execution ISA, v3.2.1, July 2013 Application Guide. http://docs.nvidia.com/cuda/pdf/ptx_isa_3.2.pdf, 2013.)

end of the shoulder region, we expect strong scaling to degrade irrespective of the underlying fabric in a multinode setting.

In Figure 16.7b, we show the wall-clock times from a short trajectory benchmark, again, having re-plotted the data from [64]. We show the strong scaling of the CPU only run, carried out on NCSA BlueWaters, on the XE partitions (which feature two CPU sockets per node). We then ran exactly the same problem, but accelerated the solvers only using the QUDA library. Because these were carried out on an XK7 system, which has one CPU and one GPU, we chose to compare socket to GPU. This resulted (because of Amdahl's law diluting the solver gains from acceleration) in an approximate speedup factor of 2× per socket, compared to the CPU results. Finally, we plot with red diamonds, the wall-clock time, for replacing QDP++ by the QDP-JIT/PTX implementation. We see substantial gains at 128 GPUs, and even in the extreme strong scaling limit, a rough speedup of 2× is seen over the results of running Chroma with only the solvers accelerated. This shows clearly that the QDP-JIT implementation reduces the effects of Amdahl's law, by moving all the computationally intensive calculations onto the GPU. Strong scaling is poor, but we point out that in this example one starts to enter the shoulder region for QDP-JIT at around 400 GPUs, and even the CPU shows strong scaling degradations beyond 400 sockets.

16.5.2 QDP-JIT/LLVM

QDP-JIT was not only successful on getting Chroma onto GPU-based systems, but also promised to remedy some of the shortcomings of QDP++ on regular CPU-based system, including the potential for vectorization of friendly layouts through the JIT-view mechanism. Further, if one could generate different code from the same source code tree, that could efficiently exploit the various different hardware architectures, that could reduce the burden of maintaining separate QDP++ implementations for the various architectures. With this in mind, QDP-JIT/(LLVM) was developed. While the original QDP-JIT generated PTX code directly, QDP-JIT/(LLVM) generates code in the internal representation (IR) of LLVM [66]. It can then be JIT compiled into executable code by the *LLVM-JIT* module and could target all the architectures LLVM can target, including ×86, BlueGene/Q, and power architectures, as well as NVIDIA GPUs through the *libnvvm* backend. This then provided for the following *performance portability strategy* for the Chroma stack

- Portable and performant QDP++ layer through the QDP-JIT/LLVM
- Custom solvers through Level 3 libraries tuned to architecture

We illustrate the efficacy of the QDP-JIT/LLVM approach on the KNL architecture in Figure 16.8a, we show wall clock times for a gauge generation short MD trajectory benchmark, featuring two-flavors, and rational approximation-based forces, having run on up to eight nodes of the Jefferson Lab SciPhi cluster with some GPU comparisons. The GPU runs were run on a single node containing 4 NVIDIA K80 GPUs (which contain two devices each, so 8 GPUs altogether back-aged as four PCIe card units) as well as on nodes of OLCF Titan and used QDP-JIT/PTX and the QUDA library. The specifics of the KNL runs are shown in Table 16.2.

On the KNL, we ran the code in several modes: To explain the notation regarding precision and gauge field compression in Figure 16.8 and Table 16.2 we note that QUDA and QPhiX admit the use of some gauge compression optimizations that were pioneered initially in the QUDA GPU library. These relate to the fact that instead of storing the full 18 real numbers that make up a 3×3 complex matrix, elements of the $SU(3)$ group can be represented as either the coefficients of the eight generators of $SU(3)$, or as 12 real numbers (six complex numbers) from which the full 18 can be reconstructed. QUDA therefore allows 8-, 12-, and 18-number representations for $SU(3)$ links while QPhiX allows 12- and 18-number representations. The compressed representations require less memory bandwidth, and this leads to a performance improvement when there are plenty of nearly flee floating-point operations to reconstruct the full matrices. Further, some solvers can use multiple precisions. For BiCGStab up to two precisions are typically used a so-called inner and an outer precision. The floating-point precisions can be 32-bit (single or "S") or 64-bit (double or "D"), and in the case of QUDA: 16-bit (half or "H"). Hence denoting a solver as S12-D12 (c.f. result E in Table 16.2) indicates that it is a mixed-precision solver, using 12 real number representations of gauge fields on both the inner and outer solves, with the inner solve being in 32-bit precision (single) and the outer solve being 64-bit precision. We only used mixed precision in the two-flavor solves; hence, the rational solves consist only of a single code: for example, D12, meaning they utilized double precision with 12-number representation of the gauge links. In the case of result C where QPhiX is used as a native Dslash, it is possible that non-$SU(3)$ links are passed to Dslash outside the solvers, for example, due to phase factors making up boundary conditions. For this reason it was not considered safe to enable the 12-number representation in the QPhiX Dslash for result C. However, in results D and E that used full QPhiX solvers, it was possible to guarantee on entry to the solvers that the fields are $SU(3)$ allowing the use of 12-number representations. The reference GPU results utilized either full double precision (D12–D12) for the two-flavor solve and D12 for the rational solve (+ symbols in Figure 16.8) or half single precision (H8-D12) for the two-flavor solve and D12 for the rational solver (* symbols in Figure 16.8).

Figure 16.8a shows that (with the exception of the 4-node case) result B proved faster than result A meaning that using a highly threaded solver with only one MPI process performed better than the out-of-the-box of the code but running as 64 MPI/node. However, the gain is not very big. We believe this is because the benchmark uses relatively heavy quarks, and so one cannot expect a very large gain from just improving the solvers alone. Result C used QDP-JIT/LLVM and QPhiX for the Chroma Dslash but keeping the solver implementations "native" in Chroma. This proved to outperform result B, showing gains from using QDP-JIT/LLVM and being able to interface with a "Level 3" (QPhiX) Dslash operator. Result D performs better than result indicating that having the full solver implementation in QPhiX can bring extra benefits, but the difference between C and D may just be due to enabling 12-compression in the solvers for D. Finally, result E outperforms result D showing improvement from the multiple-precision solver and result E is also competitive with the GPU mixed-precision solver results in terms of wall clock time.

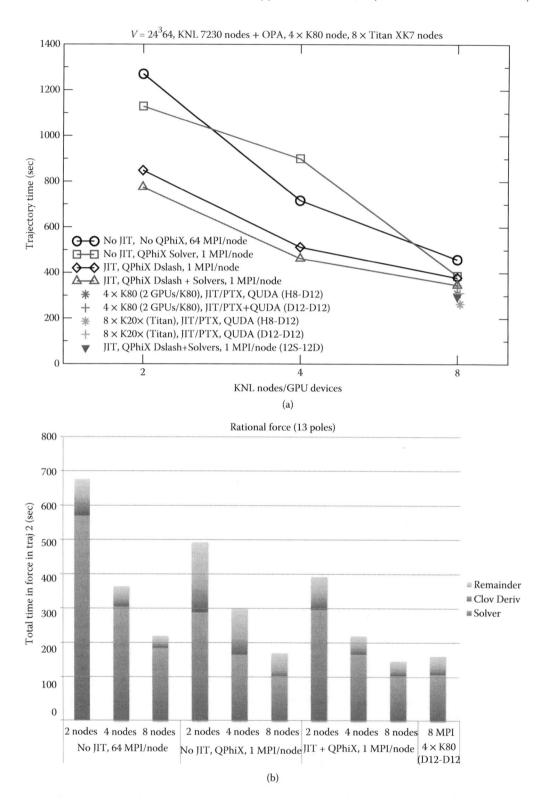

FIGURE 16.8   (a) Short trajectory benchmark times with QDP-JIT/LLVM and QPhiX, as well as QDP-JIT/PTX and QUDA. (b) Breakdown of components for the rational fermion force using various combinations of QDP++, QDP-JIT/LLVM, QDP-JIT/PTX, QPhiX, and QUDA.

TABLE 16.2

Parameters for the KNL Runs with and without QDP-JIT/LLVM

Result	JIT	MPI/Node	Threads/MPI	QPhiX	QPhiX	Precision	Precision	Symbol
				Dslash	Solvers	Two-flavor solver	Rational solver	
A	No	64	2	No	No	D18	D18	Circles
B	No	1	128	No	Yes	D12	D18	Squares
C	Yes	1	128	Yes	No	D18	D18	Diamonds
D	Yes	1	128	Yes	Yes	D12	D12	Up triangles
E	Yes	1	128	Yes	Yes	S12-D12	D12	Down triangles

In Figure 16.8b we show the breakdown of time spent in the calculation of the RHMC force term with 13 poles. There are three sets of three bars, and a final bar from a GPU comparison. The leftmost three bars are from run A (No-JIT, 64 MPI/node) and show the traditional trend that the largest proportion of the time is spent in the solvers and comparatively small amount of time is spent in the rest of the code. Comparing this with the results from run C (No-JIT, QPhiX, 1 MPI/node), we can see that the solver has been accelerated compared to the previous case (the three blue bars for this result are smaller than for the leftmost three for result A) but the portions of time spent outside the solver have increased considerably. Looking at the results from run D (JIT + QPhiX, 1 MPI/node), the primary gain is from making those nonsolver components smaller again, indeed the bars for the solver are unchanged compared with run C. Finally, we show the breakdown of the GPU force term and see that it is very similar to the run D case for eight nodes.

16.6 SUMMARY AND CONCLUSIONS

We have presented a basic introduction to the theory of LQCD, and have described in detail the *Chroma* code stack, highlighting the important roles played by high-performance domain-specific libraries such as *QUDA* and *QPhiX*; how the data parallel QDP++ provides for programmer productivity; and how Chroma at the highest level encodes important physics concepts such as the structure of the lattice fermion action, solvers used in propagator calculation, or force terms in gauge generation. Most users only ever have to deal with Chroma at the top level, and often only through input parameter files, to reap the benefits of the infrastructure.

Organizing our stack in this way gives us a performance portability strategy, as long as libraries of comparable capability are available on all target platforms, and the data parallel layer can be ported among systems. In particular, we demonstrated how the QDP-JIT approach solved the problem of porting QDP++ to GPUs and how its recent evolution using the LLVM JIT library allowed us to reach good performance on modern CPU architectures such as KNL. We anticipate that this structure will allow us to exploit the current and forthcoming CORAL generation of supercomputers.

As we head toward the exascale era, precision LQCD calculations in nuclear physics will require lattices with physical dimensions greater than 6 fm, quark mass parameters set to their physical values, and increasing resolution with lattice spacings approaching $a \approx 0.04$ fm. Further, in the case of some processes, the effects of electromagnetic interactions must also be included in the calculations. A rough scaling formula for the cost of generating gauge configurations using the hybrid Monte Carlo algorithm is Cost $\propto V^{9/8} a^{-2}$. In this cost estimate, we have assumed we are working at the physical quark masses and that the use of *multigrid* algorithms will have removed critical slowing down in terms of the quark mass. The $V^{9/8}$ term is the product of linear scaling with the volume (for solvers, link updates, etc.) and of a $V^{1/8}$ term, from having to reduce the step-size of a fourth-order

MD integrator, while increasing the volume, to keep the Monte Carlo acceptance rate constant. The a^{-2} is an estimate of the growth in Monte Carlo *autocorrelation time* as the lattice spacing is made finer. It is clear that the cost growth is substantially worse with decreasing lattice spacing, than with increasing volume. This argues both for research into algorithms that can tame the growth of auto-correlation times, and for solvers that have improved strong scaling properties, while maintaining the efficiency of current multigrid techniques. Further, as discussed earlier, advanced analysis techniques such as distillation have a high cost, both in terms of quark propagator solves and also in terms of the contractions needed. The cost of propagator computations and contractions is expected to get worse as lattice sizes are increased, and as one explores more complicated quark line diagrams with more particles. Algorithmic research is needed to tame this growing cost.

Finally, there is still some uncertainty as to the best software technologies to be used for most productively programming exascale systems and addressing performance portability across the expected architectures. In terms of software structures, especially to support research into multigrid and domain decomposed solvers, the existing data parallel layer needs to be extended with more primitive features that were not in the original design of QDP++, for example, to allow several lattice instances to exist concurrently with different geometries, and to easily define restriction and prolongation operators. Our discussion here explored only the first part of this question, which was to attain performance portability through a layered application infrastructure, using architecture specific high-performance libraries and a productivity oriented data parallel layer that has been made performance portable through the use of the JIT compilation technology.

The second part of the question, regarding the best technologies for productive and efficient software on future architectures and the primitive features required to support LQCD research on exascale systems, has not been answered here. Answers to these questions will come from future work, such as from USQCD participation in the US Department of Energy, Exascale Computing Project, participation in future application readiness programs, and through continued partnership with U.S. Department of Energy Leadership Computing Facilities, system vendors, and other strategic partners.

16.7   ACKNOWLEDGMENTS

The authors gratefully acknowledge funding under the U.S. DOE Scientific Discovery through Advanced Computing programs—SciDAC, SciDAC-2, and SciDAC-3—funded by the U.S. Department of Energy, Office Of Science, Office of Nuclear Physics, Office of High Energy Physics, and Office of Advanced Scientific Computing Research. We are grateful to our partners at large-scale U.S. Scientific Computing Centers for their support through the NERSC Exascale Applications Program, the Argonne Leadership Computing Facility Theta Early Science Program, and past Director's Discretionary Awards at Oak Ridge Leadership Computing Facility (OLCF) to assist in code porting and benchmarking both in terms of staff time and also in terms of organizing the NESAP dungeon sessions, the Theta early science program (ESP) hands-on workshop and OLCFHack GPU Hackathon, all of which proved invaluable in our recent work. In addition, some porting and benchmarking activities utilized portions of INCITE and ALCC awards (LGT003, NPH110) at OLCF, as well as parts of an NSF-PRAC award on the NCSA BlueWaters Petascale Computing Facility. We would like to thank and acknowledge the valuable contributions and collaboration of Dr. Thorsten Kurth, from NERSC, for all his assistance of the QPhiX library on KNL, and for his continued maintenance of the HDF-5 I/O capability in QDP++. Some of the QPhiX AVX512 Development for Knight's Landing was carried out by Aaron Walden, a Masters Degree Student at the time in the department of Computer Science at Old Dominion University. Finally, last, but not least, we are grateful to the enormous amount of support we have received and continue to receive from vendor partners, in particular, development time and collaboration from Kate Clark at NVIDIA, and Mikhail Smelyanskiy, Dhiraj D. Kalamkar and Karthikeyan Vaidyanathan at Intel. Notice: This manuscript has been authored by Jefferson Science Associates, LLC under Contract No. DE-AC05-06OR23177 with the U.S. Department of Energy. The U.S. Government retains and the publisher, by accepting the article for publication, acknowledges that the U.S. Government retains a nonexclusive, paid-up, irrevocable, world-wide license to publish or reproduce the published form of this manuscript, or allow others to do so, for U.S. Government purposes.

REFERENCES

1. R. G. Edwards and B. Joó. The Chroma software system for lattice QCD, *Nucl. Phys. Proc. Suppl.*, 140, 832, 2005.
2. H. J. Rothe. Lattice gauge theories: An introduction, *World Sci. Lect. Notes Phys.*, 74, 1–605, 2005.
3. M. Creutz. *Quarks, Gluons and Lattices*. Cambridge Monographs on Mathematical Physics, Cambridge: Cambridge University Press, 169 pp., 1983.
4. I. Montvay and G. Munster. *Quantum Fields on a Lattice*, Cambridge: Cambridge University Press, 491 p, 1994 (Cambridge Monographs on Mathematical Physics).
5. J. Smit. *Introduction to Quantum Fields on a Lattice*. Cambridge Lecture Notes in Physics, Cambridge, UK: Cambridge University Press, 2002.
6. M. Luscher and P. Weisz. On-shell improved lattice gauge theories, *Commun. Math. Phys.*, 97, 59, 1985.
7. H. Nielsen and M. Ninomiya. A no-go theorem for regularizing chiral fermions, *Phys. Lett. B*, 105(2), 219–223, 1981.
8. B. Sheikholeslami and R. Wohlert. Improved continuum limit lattice action for QCD with Wilson Fermions, *Nucl. Phys.*, B259, 572, 1985.
9. G. P. Lepage. Flavor symmetry restoration and Symanzik improvement for staggered quarks, *Phys. Rev.*, D59, 074502, 1999.
10. E. Follana et al. Highly improved staggered quarks on the lattice, with applications to charm physics, *Phys. Rev.*, D75, 054502, 2007.
11. M. Luscher. Exact chiral symmetry on the lattice and the Ginsparg–Wilson relation, *Phys. Lett.*, B428, 342–345, 1998.
12. H. Neuberger. Exactly massless quarks on the lattice, *Phys. Lett.*, B417, 141–144, 1998.
13. D. B. Kaplan. Chiral symmetry and lattice fermions, in *Modern Perspectives in Lattice QCD: Quantum Field Theory and High Performance Computing. Proceedings, International School, 93rd Session*, Les Houches, France, 223–272, August 3–28, 2009.
14. H. Neuberger. Alternative to domain wall fermions, *Nucl. Phys. Proc. Suppl.*, 109A, 63–69, 2002. [63 (2001)].
15. A. Borici, A. D. Kennedy, B. J. Pendleton, and U. Wenger. The overlap operator as a continued fraction, *Nucl. Phys. Proc. Suppl.*, 106, 757–759, 2002.
16. T.-W. Chiu. Optimal domain wall fermions, *Phys. Rev. Lett.*, 90, 071601, 2003.
17. R. C. Brower. H. Neff, and K. Orginos. *The Möbius Domain Wall Fermion Algorithm*, 2012.
18. R. G. Edwards, B. Joó, A. D. Kennedy, K. Orginos, and U. Wenger. Comparison of chiral fermion methods, *PoS*, LAT2005, 146, 2006.
19. C. Bernard et al. The MILC Code. http://www.physics.utah.edu/~detar/milc/milcv7.pdf, 2010.
20. RBRC Collaboration. The Columbia Physics System (CPS). http://qcdoc.phys.columbia.edu/cps.html, 2010.
21. S. Duane, A. D. Kennedy, B. J. Pendleton, and D. Roweth. Hybrid Monte Carlo. *Phys. Lett. B*, 195(2), 216–222, 1987.
22. M. A. Clark and A. D. Kennedy. Accelerating dynamical-fermion computations using the rational hybrid Monte Carlo algorithm with multiple pseudofermion fields, *Phys. Rev. Lett.*, 98, 051601, January 2007.
23. P. Omelyan, I. M. Mryglod, and R. Folk. Optimized Forest-Ruth- and Suzuki-like algorithms for integration of motion in many-body systems, *Comput. Phys. Commun.*, 146, 188–202, July 2002.
24. N. Metropolis, A. W. Rosenbluth, M. N. Rosenbluth, A. H. Teller, and E. Teller. Equation of state calculations by fast computing machines, *J. Chem. Phys.*, 21(6), 1087–1092, 1953.
25. T. Takaishi and P. de Forcrand. Testing and tuning new symplectic integrators for hybrid Monte Carlo algorithm in lattice QCD, *Phys. Rev.*, E73, 036706, 2006.

26. J. Sexton and D. Weingarten. Hamiltonian evolution for the hybrid Monte Carlo algorithm, *Nucl. Phys. B: Proc. Suppl.*, 26, 613–616, 1992.

27. D. Kennedy, M. A. Clark, and P. J. Silva. Force gradient integrators, *PoS*, LAT2009, 021, 2009.

28. M. A. Clark, B. Joó, A. D. Kennedy, and P. J. Silva. Improving dynamical lattice QCD simulations through integrator tuning using Poisson brackets and a force-gradient integrator, *Phys. Rev.*, D84, 071502, 2011.

29. H. Yin and R. D. Mawhinney. Improving DWF simulations: The force gradient integrator and the Móbius accelerated DWF solver," *PoS*, LATTICE2011, 051, 2011.

30. M. R. Hestenes and E. Stiefel. Methods of conjugate gradients for solving linear systems, *J. Res. Natl. Bur. Stand.*, 49, 409–436, December 1952.

31. H. A. van der Vorst. BI-CGSTAB: A fast and smoothly converging variant of bi-cg for the solution of nonsymmetric linear systems, *SIAM J. Sci. Stat. Comput.*, 13, 631–644, March 1992.

32. G. L. Sleijpen and H. A. van der Vorst. Reliable updated residuals in hybrid Bi-CG methods, *Computing*, 56(2), 141–163, 1996.

33. Y. Saad. A flexible inner-outer preconditioned GMRES algorithm, *SIAM J. Sci. Comput.*, 14, 461–469, March 1993.

34. A. Frommer, A. Nobile, and P. Zingler. Deflation and flexible SAP-preconditioning of GMRES in lattice QCD simulation, ArXiv e-prints, April 2012.

35. M. Luscher. Local coherence and deflation of the low quark modes in lattice QCD, *J. High Energy Phys.*, 0707, 081, 2007.

36. J. Brannick, R. C. Brower, M. A. Clark, J. C. Osborn, and C. Rebbi. Adaptive multigrid algorithm for lattice QCD, *Phys. Rev. Lett.*, 100, 041601, 2008.

37. A. Frommer, K. Kahl, S. Krieg, B. Leder, and M. Rottmann. Adaptive aggregation based domain decomposition multigrid for the lattice Wilson Dirac Operator, *SIAM J. Sci. Comput.*, 36, A1581–A1608, 2014.

38. J. Osborn et al. Multigrid solver for clover fermions, *PoS*, LATTICE2010, 037, 2010.

39. A. Frommer, B. Nöckel, S. Güsken, T. Lippert, and K. Schilling. Many masses on one stroke: Economic computation of quark propagators, *Int. J. Modern Phys. C*, 06(05), 627–638, 1995.

40. B. Jegerlehner. Krylov space solvers for shifted linear systems, arXiv:hep-lat/9612014, 1996.

41. M. Peardon et al. A novel quark-field creation operator construction for hadronic physics in lattice QCD, *Phys. Rev.*, D80, 054506, 2009.

42. US Lattice Quantum Chromodynamics Collaboration (USQCD), USQCD Software on GitHub. http://usqcd-software.github.io.

43. B. Joó. SciDAC-2 software infrastructure for lattice QCD, *J. Phys. Conf. Ser.*, 78, 012034, 2007.

44. M. Clark, R. Babich, K. Barros, R. Brower, and C. Rebbi. Solving lattice QCD systems of equations using mixed precision solvers on GPUs, *Comput. Phys. Commun.*, 181, 1517–1528, 2010.

45. R. Babich, M. A. Clark, B. Joó, G. Shi, R. C. Brower, and S. Gottlieb. Scaling lattice QCD beyond 100 GPUs, in *Proceedings of 2011 International Conference for High Performance Computing, Networking, Storage and Analysis, SC '11*, New York, NY: ACM, pp. 70:1–70:11, 2011.

46. A. Stathopoulos and K. Orginos. Computing and deflating eigenvalues while solving multiple right-hand side linear systems with an application to quantum chromodynamics., *SIAM J. Sci. Comput.*, 32, 439–462, 2010.

47. S. Heybrock et al. Lattice QCD with domain decomposition on intel® xeon phi™ co-processors, in *Proceedings of the International Conference for High Performance Computing, Networking, Storage and Analysis, SC '14*, Piscataway, NJ: IEEE Press, pp. 69–80, 2014.

48. R. Babich et al. Adaptive multigrid algorithm for the lattice Wilson-Dirac operator, *Phys. Rev. Lett.*, 105, 201602, 2010.

49. M. A. Clark, B. Joó, A. Strelchenko, M. Cheng, A. Gambhir, and R. C. Brower. Accelerating lattice QCD multigrid on GPUs using fine grained parallelization, in *Proceedings of SC16, The IEEE/ACM International Conference on High Performance Computing, Networking Storage and Analysis*, Salt Lake City, UT, November 2016.

50. R. G. Edwards, A. D. Kennedy, and C. Vohwinkel. SZIN: An Object-Oriented Macro-Based System for Lattice Field Theory. http://www.jlab.org/~edwards/szin/macros_v2.ps, August 1996.

51. S. Haney, J. Crotinger, S. Karmesin, and S Smith. PETE: The portable expression template engine, *Dr. Dobbs Journal*, online http://www.drdobbs.com/cpp/pete-the-portableexpression-template-en/184411075#, October 1999.

52. S. Karmesin et al. POOMA. NERSC ACTS Collection. http://acts.nersc.gov/formertools/pooma/.

53. E. Gamma. R. Helm, R. Johnson, and J. Vlissides. *Design Patterns: Elements of Reusable Object-Oriented Software*, 1st ed., Upper Saddle River, NJ: Addison-Wesley Professional, November 1994.

54. P. A. Boyle. The BAGEL assembler generation library, *Comput. Phys. Commun.*, 180(12), 2739–2748, 2009. 40 YEARS OF CPC: A celebratory issue focused on quality software for high performance, grid and novel computing architectures.

55. B. Joó et al. *Lattice QCD on Intel Xeon Phi Coprocessors, in Supercomputing* (J. Kunkel, T. Ludwig, and H. Meuer, eds.), vol. 7905 of Lecture Notes in Computer Science, Berlin/Heidelberg: Springer, pp. 40–54, 2013.

56. J. L. G. Projects. QPhiX Library. https://github.com/jeffersonlab/qphix.git.

57. J. Osborn. QOPQDP Software Library. http://usqcd-software.github.io/qopqdp/.

58. B. Joó, M. Smelyanskiy, D. D. Kalamkar, and K. Vaidyanathan. "Wilson Dslash kernel from lattice QCD optimization, in *High Performance Parallelism Pearls Volume Two: Multicore and Many-Core Programming Approaches* (J. Reinders and J. Jeffers, eds.), Chap. 9, Boston, MA: Morgan Kaufmann, pp. 139–170, 2015.

59. A. Nguyen, N. Satish, J. Chhugani, C. Kim, and P. Dubey. 3.5dd blocking optimization for stencil computations on modern CPUs and GPUs, in *Proceedings of the 2010 ACM/IEEE International Conference for High Performance Computing, Networking, Storage and Analysis, SC'10*, Washington, DC: IEEE Computer Society, pp. 1–13, 2010.

60. B. Joó, D. D. Kalamkar, T. Kurth, K. Vaidyanathan, and A. Walden. *Optimizing Wilson-Dirac operator and linear solvers for KNL*. In M. Taufer, B. Mohr, and J. M. Kunkel, (eds.), *High Performance Computing: ISC High Performance 2016, International Workshops, ExaComm, E-MuCoCoS, HPC-IODC, IXPUG, IWOPH, P^3MA, VHPC, WOPSSS*. Cham, Switzerland: Springer International Publishing, pp. 415–427, 2016.

61. M. Clark and R. Babich. QUDA: A Library for QCD on GPUs. http://lattice.github.io/quda/.

62. G. I. Egri, Z. Fodor, C. Hoelbling, S. D. Katz, D. Nógràdi, and K. K. Szabó. Lattice QCD as a video game, *Comput. Phys. Commun.*, 177(8), 631–639, 2007.

63. R. Babich, M. A. Clark, and B. Joó. Parallelizing the QUDA library for multi-GPU calculations in lattice quantum chromodynamics, in *ACM/IEEE International Conference on High Performance Computing, Networking, Storage and Analysis*, New Orleans, LA, 2010.

64. F. T. Winter, M. A. Clark, R. G. Edwards, and B. Joó. A framework for lattice QCD calculations on GPUs, in *Proceedings of the 2014 IEEE 28th International Parallel and Distributed Processing Symposium, IPDPS '14*, Washington, DC: IEEE Computer Society, pp. 1073–1082, 2014.

65. NVIDIA Corporation, Parallel Thread Execution ISA, v3.2.1, July 2013 Application Guide. http://docs.nvidia.com/cuda/pdf/ptx_isa_3.2.pdf, 2013.

66. C. Lattner and V. Adve. LLVM: A compilation framework for lifelong program analysis & transformation, in *Proceedings of the International Symposium on Code Generation and Optimization: Feedback-directed and Runtime Optimization, CGO '04*, Washington, DC:IEEE Computer Society, p. 75, 2004.

17 PIC Codes on the Road to Exascale Architectures

Henri Vincenti, Mathieu Lobet, Remi Lehe, Jean-Luc Vay, and Jack Deslippe

CONTENTS

17.1 Introduction...376
    17.1.1 Context and Goal of This Study 376
    17.1.2 Outline of This Study..376
17.2 Challenges to Port PIC Codes on Exascale Computers................................377
    17.2.1 The HPC Landscape: Present and Future Evolutions........................377
        17.2.1.1 The Energy Problem..377
        17.2.1.2 Impacts at Software and Hardware Levels 378
    17.2.2 The PIC Method 378
    17.2.3 Future Needs for Exascale Supercomputers 379
    17.2.4 The Classical Particle-In-Cell Algorithm 379
        17.2.4.1 The PIC Loop 379
        17.2.4.2 Data Structures Used for Particles and Fields 381
        17.2.4.3 Distributed Parallelization of the PIC Algorithm 382
    17.2.5 Performance Analysis of the Different PIC Steps 383
        17.2.5.1 The WARP and PXR PIC Codes 383
        17.2.5.2 Performance Analysis of Particle and Field Routines 383
17.3 Parallelism and Algorithm Optimizations 386
    17.3.1 Improving Memory Locality in PIC Codes Using Tiling..................386
    17.3.2 Efficient Intranode OpenMP Implementation...................................387
    17.3.3 Highly Scalable Pseudo-Spectral Maxwell Solvers 390
        17.3.3.1 Limits of Current Low Order FDTD Solver......................390
        17.3.3.2 High Order and Pseudo-Spectral Maxwell's Solvers for Much Higher Accuracy 391
        17.3.3.3 Limits to the Scaling of Pseudo-Spectral Solvers to an Arbitrary Large Number of Cores 392
        17.3.3.4 A New Method to Scale Pseudo-Spectral Solvers on up to a Million Cores 392
    17.3.4 An Optimized SIMD Charge and Current Deposition Algoritm 393
    17.3.5 Field Gathering Optimization 396
    17.3.6 Particle Sorting 397
17.4 Analysis of the Global Performance and Benchmarks on Different Architectures.............398
    17.4.1 Single Node Performances 398
    17.4.2 Performance Profiling Tools 399
    17.4.3 Application of the Roofline Performance Model...............................401
    17.4.4 Performances on Large-Scale Supercomputers 403
17.5 Conclusion and Propsects 404
Acknowledgments 404
Glossary 404
References 405

17.1 INTRODUCTION

17.1.1 CONTEXT AND GOAL OF THIS STUDY

In many fields of plasma physics where the particle-in-cell (PIC) method is used (e.g., laser–plasma physics, magnetohydrodynamics, and plasma astrophysics); the progress of experimental technology has paved the way for new physical processes and discoveries, which requires considerably more computing power and memory.

For plasmas created by lasers, the advent of petawatt (PW) lasers has enabled relativistic laser intensities, for which plasma particles (electrons and ions) can be accelerated to relativistic velocities and radiate extreme ultraviolet (X-UV) harmonic radiations of the driving laser. This gave access to a new branch of physics called "ultra-high-intensity" physics from which we expect very compact sources of light and particles. The understanding and control of these sources will largely rely on the strong coupling between experiments and PIC simulations. However, the three-dimensional (3D) accurate modeling of relativistic particles and short-wavelengths harmonic radiations spanning hundreds of harmonic orders will require very high spatio-temporal resolutions in simulations, therefore pushing for a tremendous increase in computing memory and power to make these new regimes accessible.

In magnetohydrodynamics, the next generation of fusion tokamaks being developed (ITER [International Thermonuclear Experimental Reactor] and DEMO [DEMOnstration Power Plant]) will be significantly larger than current machines and will require an unprecedented amount of computing resources. In particular, there is a growing requirement for full tokamak modeling (modeling all the processes within the plasma) rather than the current practice that is generally to model one part of the plasma (i.e., the core plasma or the edge plasma) by itself. Both of these requirements mean that the computational resources required for fusion modeling in the future will be significant.

If, on the one hand, the influence of technological breakthroughs is certainly one factor that pushes toward the developments of future exascale supercomputing architectures, it is also true that, on the other hand, the constant increase in computing power also drove scientific investigations that could not be envisaged before due to the lack of computing power required to do an accurate 3D modeling in a realistic time-to-solution.

All the applications, introduced before, will easily require the next generation of exascale supercomputing machines that will have unprecedented levels of memory (several petabyte [PB]) and computing power (Exaflop/s). Due to energy consumption constraints, CPU-based exascale machines will aim at reducing their global flop/W ratio by using more and more on-node computing units (many-core processors) with a reduced clock speed and extended data vector register lengths. This will give us access to three levels of parallelism (internode, intranode, and vectorization) that will have to be exploited by our code to perform efficiently on these machines. Unfortunately, most PIC codes are currently not adapted to future exascale computers, which are calling for a significant change of paradigm in the data structures and strategies used to implement and optimize PIC codes.

In this context, the goal of this study is to present all the optimization strategies that were developed in our group in the past 2 years as part of the NERSC Exascale Science Application Program (NESAP) program at NERSC (National Energy Research Scientific Computing Center) to help porting PIC codes to these future exascale architectures (Cori Phase 2).

17.1.2 OUTLINE OF THIS STUDY

In this chapter, we present the high performance library PICSAR ("particle-in-cell application resource"—abbreviated in PXR), which was developed to experiment and test multilayered parallelization strategies that will exploit the three levels of parallelism offered by exascale architectures.

Along this line, this chapter is divided into four sections that are briefly detailed below:

1. In Section 17.1, we shortly review the future evolutions that the high-performance computing (HPC) landscape will undergo in the coming years and its implications on exascale

computing architectures. We then describe how data-structures and parallel implementations used in PIC codes will have to evolve in order to perform efficiently on exascale computers.

2. In Section 17.2, we describe the PIC method, its main hotspot routines and future need for exascale resources in the particular case of the modeling of laser–plasma physics.
3. In Section 17.3, we present all the optimization strategies that were developed as part of the PXR library to fully benefit from the three levels of parallelism that will be offered by exascale architectures.
4. In Section 17.4, we present the benchmarks of the PXR high performance routines on Intel Xeon Phi manycore architectures.

All the results presented in this chapter about PICSAR can be consulted with more details on the official PICSAR website picsar.net.

17.2 CHALLENGES TO PORT PIC CODES ON EXASCALE COMPUTERS

17.2.1 THE HPC LANDSCAPE: PRESENT AND FUTURE EVOLUTIONS

17.2.1.1 The Energy Problem

Achieving exascale computing facilities in the next decade will be a great challenge in terms of energy consumption and will imply hardware and software developments that directly impact our way of implementing PIC codes [1].

Table 17.1 shows the energy required to perform different operations ranging from arithmetic operations (fused multiply add [FMA] or FMADD) to on-die memory/dynamic random access memory (DRAM)/socket/network memory accesses [2]. As 1 pJ/flop/s is equivalent to 1 MW for exascale machines delivering 1 exaflop (10^{18} flops/s), this simple table shows that as we go off the die, the cost of memory accesses and data movement becomes prohibitive and much more important than simple arithmetic operations. In addition to this energy limitation, the draconian reduction in power/flop and per byte will make data movement less reliable and more sensitive to noise, which also push toward an increase in data locality in our applications.

TABLE 17.1
Energy consumption of different operations

Operation	Energy Cost (pJ)	Year
DP FMADD flop	11	2019
Cross-die per word access	24	2019
DP DRAM read to register	4800	2015
DP word transmit to neighbour	7500	2015
DP word transmit across system	9000	2015

Source: K. Bergman et al., In P. Kogge [ed.], *Exascale Computing Study: Technology Challenges in Achieving Exascale Systems*, 2008.

Note: The die hereby refers to the integrated circuit board made of semiconductor materials that usually holds the functional units and fast memories (first levels of cache). This table shows the energy required to achieve different operations on current (Year 2015) and future (Year 2019) computer architectures. DP, double precision; FMADD, fused multiply ADD; DRAM, dynamic random access memory.

17.2.1.2 Impacts at Software and Hardware Levels

At the hardware level, part of this problem of memory locality was progressively adressed in the past few years by limiting costly network communications and grouping more computing ressources that share the same memory ("fat nodes"). There are two paths presently explored by the U.S. Department of Energy (DOE), which is massively investing in both graphics-based (GPU) (Nvidia) and many-core-based (Intel) architectures.

This pushes for a portable way of programming in order to ensure the interoperability of our codes on both types of systems at minor costs. The data structures and strategies developed in this chapter are general and could, in principle, be applied to both GPUs and manycore CPUs providing that a common application program interface (API) between those two architectures is available. With the advent of the target directive in the OpenMP API, which aims at "abstract-ing" the device and makes its use transparent from a user perspective, the code presented here could, in principle, be used with minor revisions on GPUs. However, in the following, we will focus on manycore architectures for which there has been little work to adapt our code and its data structure.

By grouping manycores with a large shared memory, these processors improve the memory local-ity and reduce energy consumption related to data movements. However, grouping more and more of CPU units will imply a reduction of their clock speed. To compensate for the reduction of computing power due to clock speed, future CPUs will, thus, have much wider data registers that can pro-cess or "vectorize" multiple data in a single clock cycle (single instruction multiple data [SIMD]). Along this line, the Intel Xeon Phi manycore processor family was designed with many low-frequency cores sharing an on-chip high-bandwidth memory (HBM or multichannel dynamic random access memory [MCDRAM]). Some of the most powerful existing or upcoming supercomputers are already—or planned to be—equipped with these types of technologies, such as the future Aurora machine at Argonne in the United States [3] or the current Sunway TaihuLight supercomputer in China [4].

At the software level, in order to obtain the best performances on Xeon Phi architectures, algo-rithms will have to be redesigned to achieve good memory locality and efficient vectorization that are required to exploit the full potential of future clusters. Hybrid programming is already seen as the best way to efficiently deal with inner/outer node parallelism and nonuniform memory accesses (NUMA) with the HBM/MCDRAM memory. For efficient vectorization, loops will have to be con-dition free with preferentially FMA operations of the same kind. Vectorization also relies on aligned and contiguous memory accesses: data arrays have to be aligned and accessed contiguously to avoid gather/scatter operations. Cache reuse is also important since multiple load/store from the random access memory (RAM) can inhibit efficient vectorization. The positive outcome is that the result-ing codes will also perform better on current architectures such as the one based on previous Intel processors (Haswell/Ivy Bridge) or Blue Gene architectures.

17.2.2 THE PIC METHOD

PIC methods were first imagined to solve hydrodynamics problems [5,6], until Eulerian hydrocodes matured. They became very popular and are still extensively used in plasma physics since the six-ties, in particular thanks to the pioneering work of J. M. Dawson, O. Buneman, and R. W. Hockney [1,7–9]. The PIC method constitutes a specific way to solve the Vlasov equations in order to sim-ulate kinetic plasma behaviors and interactions with external electromagnetic fields. It is usually opposed to pure Eulerian methods. Its success, still growing, is partly due to its relative efficiency (in opposition with pure Eulerian description) as well as its programming and parallelization simplicity. PIC codes are now used as a simulation tool in a wide range of applications such as astrophysi-cal and space plasma physics, nuclear fusion, electronics, particle acceleration, and high-intensity laser–matter interaction.

17.2.3 Future Needs for Exascale Supercomputers

Multidimensional PIC simulations often require massive parallelism with codes and algorithms that are at the forefront of advanced computing. This increased demand of computational power and the necessity to go to exascale machines are motivated by several factors:

- **The need for large range of time and length scales:** Many simulations are still limited to smaller/simpler cases than in experiments or natural phenomena. For instance, in space plasma physics, the simulation of the Weibel filamentation instability involving the collision of two counter-streaming plasma flows is still a challenge with real ion/electron mass-ratio due to the wide range of space and time scales of the instability.

- **The need for very high resolutions in space and time to accurately describe the physics:** High resolutions are necessary to mitigate numerical artifacts (numerical dispersion, numerical heating, numerical Cerenkov, and instabilities) that could degrade the simulation and lead to unphysical solutions. This is, for instance, the case in high intensity laser–matter interaction for the generation of high-frequency radiations such as harmonics with relativistic plasma mirrors. Here, the interaction of an intense laser with a dense target leads to the creation of a relativistically oscillating plasma mirror at the target surface. Upon reflection of the incident laser on the target, the periodic oscillations of the plasma mirror surface induce a periodic distortion of the reflected field by Doppler effect associated to a high harmonic spectrum in the frequency domain. In order to correctly describe the short harmonic wavelengths with a sufficient number of grid points, therefore a fine discretization is required.

- **The need for 3D simulations:** Many configurations are still limited to two dimension (2D) as a consequence of the two previous points. However, many experimental configurations need 3D effects to be correctly captured. If 2D simulations may give interesting qualitative results, they often lead to incorrect final estimations of the overall energy balance. For instance, in laser–plasma interactions leading to ion acceleration or laser wakefield acceleration of electrons, 2D simulations can help understanding the acceleration mechanisms but typically overestimate the spatial extension and amplitude of the accelerating fields and therefore the final beam energies.

- **The need for additional physical mechanisms:** The physics of the high-intensity laser–matter interactions involves complex intertwined phenomena that sometimes require the use of extraphysical modules. PIC codes are continuously enriched with physical processes, including binary collisions, ionization, radiation emission (Bremsstrahlung, radiation damping, nonlinear inverse Compton scattering), radiative absorption, electron-positron pair creation (Bethe–Heitler, Trident, nonlinear Breit–Wheeler). Many of these multiphysics modules use algorithms from particle physics codes such as Monte-Carlo methods that increase the computational cost of simulations.

17.2.4 The Classical Particle-In-Cell Algorithm

In this section, we present the classical PIC algorithm. We first detail the data structures and parallelization strategies that have been commonly used and highlight their limits on future architectures.

17.2.4.1 The PIC Loop

In a simplified view, the PIC method follows the evolution of a collection of charged macroparticles that evolve self-consistently with their electromagnetic (or electrostatic) fields defined on a grid mesh as schematically described in Figure 17.1. A macroparticle (also referred to as super-particles) represents a group of real particles of the same kind (charge and mass) with the same kinetic properties

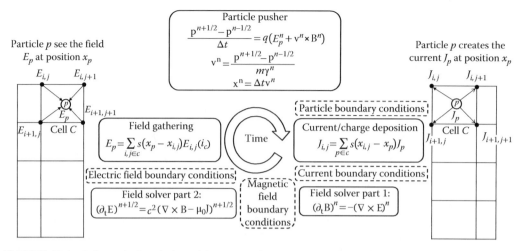

FIGURE 17.1 Schematic description of the core PIC algorithm. Elements around the rotating arrow of the time represent the different steps of the PIC loop. Frames with a plain line correspond to the four computational phases: particle pusher, current/charge deposition, Maxwell solver, and field gathering. Frames with a dashed line correspond to the communication barriers for the parallelized PIC version. The 2D grid on the left illustrates the field gathering process at order 1 (CIC). The field E_p seen by the particle p located in the cell C is determined by interpolation from the four nearest vertices in 2D. The 2D grid on the right illustrates the current deposition process at order 1 (CIC). Here, the current $\mathbf{J}_p = q_p n_p \mathbf{v}_p$ (q_p being the particle charge, n_p the particle weight, and $\mathbf{v}_p$ the particle velocity vector) generated by the particle p located in the cell C is projected on the four nearest vertices.

(speed and propagation direction). With more than 10^6 particles in a hydrogen plasma in a volume of 1 μm^{-3}, and more than 10^{11} in a dense target, following individually all the particles is often not possible and macroparticles are used to allow for a realistic time-to-solution.

The core PIC algorithm involves four steps at each time iteration as shown in Figure 17.1:

1. **Particle pusher:** Advances the velocity and position of macroparticles in time using the relativistic Newton–Lorentz equation of motions.
2. **Current/charge deposition:** Deposits the charge and/or current densities by interpolating particles distributions onto the grid.
3. **Maxwell solver:** Advances Maxwell's equations in time and space (for electromagnetic PIC) or Poisson's equation (for electrostatic PIC) on the grid.
4. **Field gathering:** Interpolates the fields at next time iteration from the grid onto the particles for the next particle push.

Splines of various order [10] can be chosen for interpolation between the macroparticles and the grid during steps 2 and 4 (they are schematically described in Figure 17.1 at order 1 in 2D):

- CIC or order 1 shape factor: interpolation on the four nearest vertices in 2D and the 8 nearest vertices in 3D.
- TSC or order 2 shape factor: interpolation on the nine nearest vertices in 2D and the 27 nearest vertices in 3D.
- QSP or order 3 shape factor: interpolation on the 16 nearest vertices in 2D and the 64 nearest vertices in 3D.

Higher shape factors tend to lead to more numerically stable and accurate simulations, at the expense of higher computational cost. In practice, order 1–3 shape factors are of most common usage.

17.2.4.2 Data Structures Used for Particles and Fields

1. **Particles**. The macroparticles are coded via their properties, such as weights, positions (x, y, z), momenta, energies, and local fields at particle positions.

   On former scalar computer architectures, a common way of storing particle quantities in memory was to use array of structures (AoS), where each structure contained the different properties (positions, momenta, weights, etc.) of one particle contiguously in memory (cf. Fig.17.2a). As the main PIC method (except for the Maxwell solver step) loops over particles, this data structure allows for better memory reuse of particle properties within the different steps of the algorithm (e.g., current deposition, field gathering, and particle pusher) when all steps are performed consecutively for each particle [11].

   However, as we detailed in introduction, upcoming computer architectures will tend to use more and more SIMD vector instructions, for which this data structure significantly hinders vectorization efficiency. Indeed, in order to achieve good vectorization on these machines, data have to be aligned and contiguous in memory to ensure direct single store/load instructions and avoid costly gather/scatter instructions. This has been illustrated on Figure 17.2. In the case of AoS, the properties of successive particles would not be stored contiguously in memory and stored on different cache lines. As most particle routines (particle pusher, current deposition, and field gathering) are vectorized on the particle loop, this will require successive loads/stores from/to memory to fill/empty the data vector register (cf. arrows on Figure 17.2a).

   As we will show later in this study, it is thus much better to use arrays of structure of arrays (SoA)-like data structures for particles where each SoA fits in cache, in order to maximize both cache reuse and vectorization efficiency. Each SoA is called at "tile" and contains arrays of particle quantities (as many arrays as particle quantities, cf. Figure 17.2b). This structure ensures contiguous access of the particle properties in memory and therefore achieves good vectorization on SIMD architectures (direct single store/load to/from memory).

2. **Electromagnetic fields and sources.** Discretized electric E, magnetic B, and current fields J constitute multidimensional grids. The nodes are numerically stored in a multidimensional arrays for each component $(E_x, E_y, E_z, B_x, B_y, B_z, J_x, J_y, J_z)$.

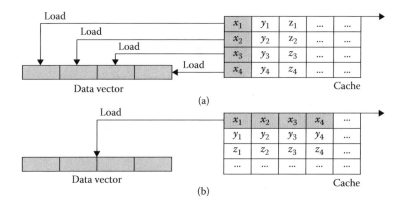

FIGURE 17.2 Data layout of particle arrays in memory. The grid represents cache lines. Quantitites on the same cache line are contiguous in memory. (a) Array of structure (AoS). For each particles p, the particle quantities (x_p, y_p, z_p) are stored contiguously in memory. (b) Structure of arrays (SoA). There is an array per particle quantity. Here, we only illustrated the memory layout for the three position components (x_p, y_p, z_p) of each particle.

17.2.4.3   Distributed Parallelization of the PIC Algorithm

The cartesian domain decomposition is the most popular way to divide the workload between the computing units and has proven to provide good parallel scalings on a large number of cores, and the method is presented in Figure 17.3. The simulation domain is divided into multiple subdomains that are distributed to different message passing interface (MPI) processes. In each subdomain, the field and the current grids have guard regions that extend each subdomain and hold copies from field and currents vertices of adjacent subdomains. These common vertices are located at the boundaries of each subdomain and are necessary for solving the Maxwell equations and depositing/gathering current/fields when particles are located close to the boundaries. Updating the guard cells requires to communicate them from the neighborhood as illustrated in the 2D view of Figure 17.3. For order 2 finite-difference time-domain (FDTD) Maxwell solvers, this method only requires exchange of two guard cells at the margin of each processor subdomain and has demonstrated scaling on up to a million cores.

In 3D, each subdomain has 26 neighbors (8 in 2D). For each subdomain, communications of guard cells can be performed with all neighbors or grouped using the so-called "diagonal technique." The diagonal technique consists in sending guard regions (including corners) in each direction successively, for instance, in the x axis first and then in y and z directions, with MPI synchronizations between each direction. This avoids communications with diagonal neighbors and allows a reduction in the total number of communications from 26 to 6 in 3D (and 8 to 4 in 2D).

The only drawback of the domain decomposition is that it duplicates memory, especially when more guard cells are needed in the case of higher order Maxwell solvers (cf. Section 17.3.3). On future architectures, the available memory per core will tend to decrease and a full MPI implementation of parallel high order Maxwell solvers might run out of memory. In this case, a hybrid MPI/OpenMP parallelization of the PIC algorithm will reveal crucial in order to reduce the total memory footprint of the parallel algorithm as well as reduce the total volume of MPI communications to improve scaling on million cores.

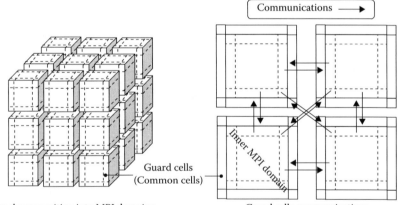

Space decomposition into MPI domains   Guard cells communications

FIGURE 17.3   Schematic of the domain decomposition parallelization method widely used in PIC codes. On the left, a 3D schematic view of the decomposition into subdomains. On the right, the 2D view of the communication pattern. The bold line represents the boundary between the inner subdomain and the guard cell region. The guard cell is an extension of the inner domain that contains data from the neighbors. Parts of the subdomain that are communicated to the neighbor guard cell regions are delimited by the dashed lines. Communications are represented by arrows.

17.2.5 PERFORMANCE ANALYSIS OF THE DIFFERENT PIC STEPS

In the following, we present the PIC loop steps in detail, focusing on their algorithmic characteristics, their advantage, and their issues for computing performances. To illustrate this, we use the PIC codes WARP/PXR.

17.2.5.1 The WARP and PXR PIC Codes

The WARP code is an open-source PIC code dedicated to the modeling of charged particles and electromagnetic fields (fully time dependent or in the static approximation), including the dynamics of charged-particle beams in particle accelerators, and the physics of high-intensity laser–matter interactions [12]. WARP was initially developed at the Lawrence Livermore National Laboratory and is now widely developed and used at the Lawrence Berkeley National Laboratory (LBNL). WARP is written in a combination of (1) Fortran for efficient implementation of computationally intensive tasks, (2) Python for high level specification and control of simulations, and (3) C for interfaces between Fortran and Python.

PICSAR (or PXR) is an open-source full Fortran 90 library designed to provide high-performance PIC subroutines optimized for multi-core and many-integrated core architectures such as Intel Xeon Phi Knights Landing processors (KNL). PXR has been encapsulated in a Python module so that it can be easily interfaced with WARP for production simulations. It can also be run as a stand-alone full Fortran code for benchmarking and performance testing. The domain decomposition parallelization uses MPI for the stencil communications between domains.

As a test case to identify and discuss hotspots, we consider a homogeneous Maxwellian plasma of thermal velocity $0.1c$ where c is the speed of light in vacuum. Using a homogeneous thermalized plasma ensures that the number of particles stays similar and well balanced between processes all along the simulation. The 3D domain is a cube with periodic boundaries in every direction of side length 40 μm and discretized with 100 points in each direction. Each cell contains 40 randomly initialized macroparticles. Computational time ratios will be provided using the new NERSC supercomputer Cori hosted at the LBNL in California. Ranked 5 in the top 500 most powerful supercomputers in November 2016; Cori has two partitions, one equipped with 2004 Intel Haswell computes nodes (two 2.3 GHz 16-core Intel Xeon Processor E5-2698 v3) and the other equipped with 9, 304 Intel Xeon Phi KNL compute nodes (Intel Xeon Phi Processor 7250). The performance results will be given on single Haswell and KNL nodes using the PICSAR Fortran kernel. The code before optimization is used with 32 MPI tasks on a cori Haswell node and 64 on KNL in quadrant or SNC4 flat mode. The full simulation is designed to fit in the MCDRAM memory.

17.2.5.2 Performance Analysis of Particle and Field Routines

The particle pusher—It using the Boris method or the Vay method [13,14], is composed of a single loop over the particles where the momenta are first evolved using the local fields that then enables the computation of the energy, the velocities, and finally the new positions. The loop does not contain memory races (flow dependencies) because particles only use their own properties. Therefore, despite square roots and a few divisions, this step is efficiently vectorized. The algorithm time complexity is in $\mathcal{O}(N_p)$ where N_p is the total number of particles. In the considered test case, the Boris particle pusher represents around 26% of the total simulation time on Haswell node and 3% on KNL node. With the domain decomposition parallelization method, the particles going out of the MPI domain have to be transferred. A communication step, therefore, follows the particle pusher and represents 9.5% of the simulation time on Haswell node and 9% on KNL node.

Below, we consider several Maxwell solvers:

- **Finite difference schemes:** The most popular scheme is the FDTD algorithm [15] that is second-order in space and time and is time-reversible. More recently, nonstandard FDTD have been developed to reduce or suppress the numerical dispersion in some specific propagation direction such as the principal axes of the simulation domain [16–19]. These schemes use extended local stencils when evolving the magnetic field, using an average of the surrounding electric field nodes in the direction orthogonal to the direction of finite-differencing (or vice-versa, i.e., switching "magnetic" and "electric" in the sentence). Here, the quantities are given on staggered (or "Yee") grids, where the electric field components are located between nodes, and the magnetic field components are located in the center of the cell faces. Knowing the current densities at half-integer steps, the electric field components are updated alternately with the magnetic field components at integer and half-integer steps, respectively. These techniques are widely used party because they enable efficient parallelization up to a million cores, as required for the simulation of large-scale problems and because the algorithm is fast and simple. As detailed previously, the distributed parallelization requires the use of guard cells that contain data computed by nearby domains (MPI domains for instance) and therefore a communication step before the magnetic and the electric field updates. These schemes are composed of nested loops over the grid nodes in each direction for each component. Relatively fast with a $\mathcal{O}\left(N_c\right)$ time complexity and $N_c \ll N_p$ where N_c is the number of grid cells, it can nonetheless produce significant unphysical degradation from discretization errors unless using a very high—but costly—resolution. The second-order Yee FDTD solver represents between 1% and 2% of the simulation time including the guard cell communications on both Haswell and KNL. Higher order solvers, by requiring larger stencil order and guard cell regions, do require more computational power.

- **Pseudo-spectral analytical solver:** pseudo-spectral analytical time-domain (PSATD) [20, 21] Maxwell solver, advances analytically Maxwell's equations in the spatio-spectral Fourier and real time domains, assuming that the electromagnetic sources (charge and current) are constant over one time step. Consequently, it is free of numerical dispersion and not subject to a Courant condition. In addition, representing all field values at the nodes of a grid is natural, thus eliminating staggering errors. Nevertheless, despite significant advantages in terms of accuracy, very high-order solvers or pseudo-spectral solvers have not been widely adopted, so far, for large-scale simulations because of their low scalability to a large number of processors, which is due to the requirement of global inter-processor communications in the computation of global Fourier transforms. In Section 17.3.3, a new method has been recently proposed to parallelize the pseudo-spectral solver to an arbitrary number of cores by performing independent local fast Fourier transforms (FFTs) on each subdomain instead of one global FFT on the entire computational domain. This technique, however, implies a small truncation error at the subdomain boundaries that was characterized in a recent study [22].

The current/charge deposition—It is composed of a loop over the particles for which the generated current/charge is computed and projected on the surrounding nodes following a shape factor called S at a given order of interpolation (see Figure 17.1). The Morse–Nielson charge deposition scheme (also referred to as "direct deposition") is described in Listing 17.1. In the code, xp, yp, zp refer to the particle position arrays for each component, xmin, ymin, zmin to the positions of the minimum grid indexes, dxi=1/dx, dyi=1/dy, dzi=1/dz to the inverse of the space steps. The array rho corresponds to the charge grid (ρ). The algorithm complexity is in $\mathcal{O}\left(N_p\right)$ and cannot be vectorized in this form. The main loop over the particles can be divided into two parts: (1) for each particle, the computation of its cell node indices and its weights and (2) the contribution to the vertices of the current/charge arrays. In the second part, memory races occur when two particles are located in the same cells (also see Figure 17.1). Since current/charge arrays also contain guard cells, the current/charge deposition is followed by a communication step.

Listing 17.1   Scalar charge deposition routine for CIC (order 1) particle shape factor.

```fortran
SUBROUTINE simplified_charge_deposition(...)
  ! Declaration and init
  ! Loop on particles
  DO ip =1,np
    ! --- PART 1 -------
    ! Computes current position in grid units
    ixp = (xp(ip)- xmin )* dxi
    iyp = (yp(ip)- ymin )* dyi
    izp = (zp(ip)- zmin )* dzi
    ! Finds node of cell containing particle
    j= floor(ixp)
    k= floor(iyp)
    l= floor(izp)
    ! Computes weigths w1 .. w8
    ........
    ! --- PART 2 -------
    ! Add charge density contributions
    ! To the 8 vertices of current cell
    rho(j,k,l) =rho(j,k,l) + w1
    rho(j+1,k,l) =rho(j+1,k,l) + w2
    rho(j,k+1,l) =rho(j,k+1,l) + w3
    rho(j+1,k+1,l) =rho(j+1,k+1,l) + w4
    rho(j,k,l+1) =rho(j,k,l+1) + w5
    rho(j+1,k,l +1) =rho(j+1,k,l+1) + w6
    rho(j,k+1,l +1) =rho(j,k+1,l+1) + w7
    rho(j+1,k+1,l+1) =rho(j+1,k+1,l+1)+ w8
  ENDDO
END SUBROUTINE simplified_charge_deposition
```

In the tests, both the current and the charge deposition are performed at order 1 (linear interpolation). The current deposition represents 18% of the simulation time on Haswell and 36% on KNL. The charge deposition represents 12% of the simulation time on Haswell and 22% on KNL. Exchange of the guard cells for the current arrays represents a few percent of the simulation time when the number of guard cells is small on Cori both on Haswell and KNL nodes.

The field gathering algorithm—It is described in Listing 17.2. This step is composed of a loop over the particles in which the fields seen by the particles at their positions are determined using the surrounding Maxwell grid nodes with the specified shape factor S. As for the current/charge deposition, the loop can be divided into two parts. The first one is highly similar to the current/charge deposition and consists in determining the indices of the particle on the grids and the weights for the interpolation. The second one is algorithmically the reverse of the current/charge deposition because data is gathered from the grid nodes to the particles. If the inner loop is completely linearized with no if-statements then the entire loop can be vectorized efficiently with minor changes.

Listing 17.2   Electric field gathering routine for CIC (order 1) particle shape factors. Here, the example of the E_x component is considered but the routine is the same for the others and the magnetic field.

```fortran
SUBROUTINE simplified_field_gathering(...)
  ! Declaration and init
  ! Loop on particles
```

```
DO ip =1,np
   ! --- PART 1 -------
   ! Similar as for the charge deposition
   ........
   ! --- PART 2 -------
   ! Determine the field seen by the particle
   ! from the 8 vertices of field cell
   Exp(ip) += w1*Ex(j,k,l)
   Exp(ip) += w2*Ex(j+1,k,l)
   Exp(ip) += w3*Ex(j,k+1,l)
   Exp(ip) += w4*Ex(j+1,k+1,l)
   Exp(ip) += w5*Ex(j,k,l+1)
   Exp(ip) += w6*Ex(j+1,k,l+1)
   Exp(ip) += w7*Ex(j,k+1,l+1)
   Exp(ip) += w8*Ex(j+1,k+1,l+1)
ENDDO
END SUBROUTINE simplified_field_gathering
```

The main slow down for vectorization comes from part 2 in Listing 17.2 that can induce a significant number of **gather instructions**. When the loop on particle is vectorized, field values $Ex(i, j, k)$ (*Ex* variable in Listing 17.2) of successive particles will be loaded in the data vector register. If particles are sorted on a cell basis, vertex indices (j, k, l), etc. of the grid array Ex to load would not change from one particle to another and vectorization will be highly efficient as field values will be loaded only once. If, however, particles are not sorted on a cell basis, vertex indices will be different from one particle to another, therefore potentially involving a significant number of gather operations (consisting of several loads) that could hinder vector performances. The field gathering complexity is in $\mathcal{O}\left(N_p\right)$. It represents 27% of the simulation time on a Haswell and on a KNL node.

17.3 PARALLELISM AND ALGORITHM OPTIMIZATIONS

17.3.1 IMPROVING MEMORY LOCALITY IN PIC CODES USING TILING

Improving the **memory locality** is an essential optimization that can enhance performance on current and future architectures. It consists of manually dividing a large set of data into subsets that are sufficiently small to fit in the different levels of cache when the compiler is not able to do it itself efficiently. Data accesses are rationalized so that blocks are loaded at the most local level, level by level, when used by the computing units. This optimization technique, also known as **cache blocking**, enables a more efficient **cache reuse** when data are used several times in a loop. It helps to suppress **cache misses**, which lead the system to scan the different levels of cache (from the L1 to the double data rate [DRR] and/or the MCDRAM memory) in order to find and load required data. It finally reduces the pressure on the memory bandwidth that slows down memory-bound applications.

The field gathering and the current/charge deposition are composed of a loop over a large group of macroparticles that respectively reads in the vertices of the electric, E, and magnetic, B, field grids and writes in the nearby vertices of the charge, ρ, and current, J, grids. One major bottleneck that arises in these routines and can significantly affect overall performance is cache reuse of the grid quantities.

When particles that are consecutive in memory are localized in the same area in the simulation, blocks of grids loaded in cache for the first particle can be reused for the following ones (see Figure 17.4a). However, the particle distribution often becomes increasingly random as time evolves, leading to numerous cache misses (see Figure 17.4b). Two consecutive particles in memory can then eventually be localized at the two opposite sides of the MPI domain.

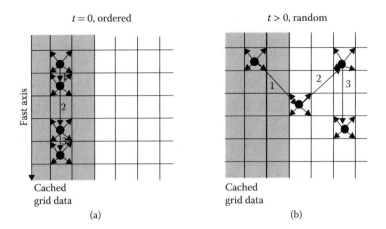

FIGURE 17.4 Importance of cache reuse in deposition routines. Illustration is given in 2D geometry for clarity, with CIC (linear) particle shapes. Panel (a) shows a typical layout at initialization ($t = 0$) where particles are ordered along the "fast" axis of the grid, corresponding to grid cells (gray area) that are contiguous in memory. The loop on particles is illustrated with arrows and index of the loop with numbers 1–3. Using direct deposition, each particle (black point) deposits (arrows) its charge/current to the nearest vertices (4 in 2D and 8 in 3D for CIC particle shapes). Panel (b) illustrates the random case (at $t > 0$) where particles are randomly distributed on the grid. As the algorithm loops over particles, it often requires access to uncached grid data, which then results in a substantial number of cache misses.

In 2D geometry, for a MPI subdomain, grids can usually fit in L2 cache (and usually represents few hundreds of kilobytes [KB]), which is not the case anymore in 3D where an MPI subdomain can represent several MBs for the grids.

To solve this problem and achieve good memory locality, particle arrays for each MPI domain have been divided into smaller subsets called particle tiles. Tiles are SoA (`Type(particle_tile)` in the code) that contain the quantities of the particles localized in the subset space region. MPI domains are therefore represented by an array of tiles, stored in a multidimensional Fortran (of dimension 3 in 3D, 2 in 2D) array `array_of_tiles(:,:,:)` of type `particle_tile` in the code. Using an AoS is more flexible and consumes less memory than using a large multidimensional array with a dimension for the tiles, a dimension for the particles and a last one for their properties (`ppart(1 : nproperties, 1 : nppmax, 1 : ntiles)`). Indeed, when the particle number exceeds the size of the particle property arrays in a tile, only the latter one is resized.

The electromagnetic field and currents arrays used for the Maxwell solvers can be either tiled or not tiled depending on the Maxwell solver used. For low-order Maxwell solver stencils, it has been shown that tiling can strongly benefit stencil computations by increasing memory locality and cache reuse [23]. For pseudo-spectral solvers, it is, however, preferable to keep untiled larger arrays to perform the local FFTs.

In 3D, field and current arrays can fit in L3 cache on Haswell for regular MPI domains on Intel Ivy Bridge or Haswell architectures. Depending on the number of particles per cell, particle arrays can potentially also fit in L3. Field and current blocks local to the tiles are supposed to fit in L2. Another approach has been proposed in Ref. [24] in that the notion of tiling is reduced to a single cell. A more accurate description of the tiling is available in this article [25] and on the PICSAR website: picsar.net/features/tiling-for-better-memory-locality.

17.3.2 Efficient Intranode OpenMP Implementation

OpenMP implementation—Actually, memory locality improvements with tiling also strongly benefits to the efficiency of the OpenMP parallelization of the PIC algorithm. In our implementation, each particle tile is handled by one OpenMP thread running on one core. Choosing a tile size smaller than core caches ensures memory locality of all data handled by each thread. Each thread performs the PIC steps on the particles of the tile it handles. This offers a straightforward parallelization of particle routines. We detail how tiling favors good OpenMP scaling of the PIC loop, as follows:

1. Field gathering/current deposition: each tile deposit currents on a local grid array (grid tile) that have guard cells. In the particular case of current deposition, this avoids memory races in the current deposition step as different tiles hold different cells of the simulation domains. When using pseudo-spectral Maxwell solvers that do not require tiling of the fields and currents, the current deposition is still performed on local grid tiles that will further need to be reduced into the global current arrays before the Maxwell solver step. There are different methods to avoid memory races during the reduction of local arrays with guard cells to the global current array. In PICSAR, we split the initial reduction (single $OMP **DO** LOOP) into three reductions (three $OMP **DO** LOOP) along the three axis (x, y, z) of the simulation with implicit synchronization between each axis. This method ensures that two different threads handling two different tiles would not update current values at the same positions in the global array.

2. Particle sorting: this implementation provides a straightforward parallelization of sorting as different threads handle different cells of the domains.

3. Load balancing: usually, there are many more tiles than OpenMP threads, and intranode load balancing is naturally achieved through the OMP_SCHEDULE clause in OpenMP using the guided or dynamic attribute when the workload is not balanced between the tiles.

Tiling is of particular interest on current multicore and manycore CPU architectures that are usually composed of a high number of cores distributed on several NUMA domains. In PICSAR, the OpenMP parallelization layer benefits from shared memory within NUMA domains and the MPI layer is kept for communications between NUMA domains. This configuration is essential for the scalability on the future machines that will be composed of several millions of lightweight cores. As for MPI subdomains, particles leaving one subdomain have to be transferred into the tiles of the target subdomain. Communications are performed within the shared memory layer. Indeed, leaving macroparticles are directly transferred into the new tile memory that they belong by the tile (OpenMP thread) they are coming from. Considering the case of the homogeneous plasma, the number of MPI subdomains is now reduced to the number of NUMA domains: two on a Cori Haswell node with 16 OpenMP threads per node and four on a KNL node configured in SNC4 with 32 OpenMP threads. The average tile dimension is of $8 \times 8 \times 8$ cells (decomposition into 12 tiles in each direction). With the guard cells, local current and field arrays represent 65 kB and 130 kB approximately. Tiling has led to $\times 1.3$ and $\times 3.6$ speedups on Haswell and KNL node for the current deposition step with order 1 interpolation shape factors. For the field gathering plus the particle pusher, an average of $\times 2.3$ speedups have been obtained on a Haswell and a KNL node. On the whole kernel, tiling has led to a relatively low $\times 1.2$ and $\times 1.1$ speedups on a Cori Haswell node and on a KNL node because of the overhead induced by the particle transfers between tiles. With order 3 interpolation shape factors, the speedup is higher of $\times 1.5$ and $\times 2.8$ on a Cori Haswell node and on a KNL node. It is important to note that the tiling is a cache blocking technique that can also speedup codes without OpenMP parallelization.

Particle exchange—The particle communication routines have been rearranged to avoid **memory races** and to combine the preliminary computations for the shared memory and the MPI transfers. Inside the sender tile, the shared-memory communication process consists of a loop over the macroparticles to determine the ones that will leave the subdomain. A macroparticle that needs to be transferred is placed at the last available index in the property arrays of the receiver tile and suppressed in the sender tile. Note that particle arrays are dynamically resized using an ArrayList-like data structure. When arrays of a tile are full, the size of particle arrays is increased by a factor of 2.

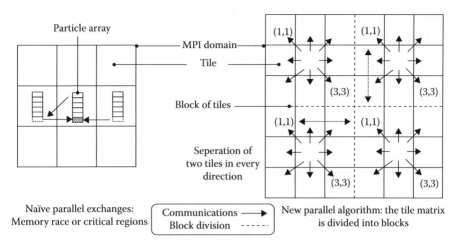

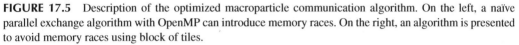

FIGURE 17.5 Description of the optimized macroparticle communication algorithm. On the left, a naïve parallel exchange algorithm with OpenMP can introduce memory races. On the right, an algorithm is presented to avoid memory races using block of tiles.

For X insertions, this allows for a $O(2X)$ insertion time only. Similarly, particle arrays are downsized dynamically when too empty to avoid a too large memory footprint at runtime.

For tiles located at the domain boundary, macroparticles are placed in local buffers (one for each possible direction). Memory races can happen during the shared-memory particle communication when a tile receives two macroparticles at the same time from two different neighbors. With a naïve parallelization, the neighbors can, therefore, potentially write two macroparticles at the same memory location (same index). To solve this problem, the communications are not performed in parallel among the entire tile matrix but using a decomposition into blocks composed of $3 \times 3 \times 3$ tiles in 3D and 3×3 tiles in 2D as described in Figure 17.5. The idea is close to the one described in Ref. [24]. In each block, tiles located at the same local indices (i.e., same coordinates in Figure 17.5) update their particle arrays in parallel in every directions. The consequence of this structure is that there is always a space of two sleeping tiles between working ones in every directions. When a tile transfers some macroparticles to one of its neighbors, the latter will never get macroparticles from another tile at the same time, avoiding memory races. There is no waste of computation units when the number of blocks is still higher than the number of available OpenMP threads. To improve the thread distribution, nested parallel OpenMP regions are used as described in Listing 17.3. The first parallel region applies on the species loop with nthreads_loop1 threads using the NUM_THREADS OpenMP clause. The second region loops over the tiles of similar coordinates inside the blocks with nthreads_loop2 threads. Other solutions are possible to avoid memory races such as having a parallelization on the entire tile matrix and doing communications in each direction step by step (in this case, a synchronization is required after each direction) or using a local buffer per incoming directions (requires more memory).

Listing 17.3 Nested loop algorithm for the macroparticle communications between tiles and MPI domains.

```
SUBROUTINE particle_communications
    ! Number of threads for each parallel region
    nthreads_loop1=MIN(nspecies,nthreads_tot)
```

```
nthreads_loop2=MAX(1_idp,nthreads_tot/nthreads_loop1)

! 1) Shared-memory particle transfers and MPI local buffers

! Loop on the species
!$OMP PARALLEL DO NUM_THREADS(nthreads_loop1)
DO is=1,nspecies

  ! Loop over the tiles in the blocks
  DO ipz=1,3
    DO ipy=1,3
      DO ipx=1,3

        ! Loop over all the tiles of same local indexes
        ! in the blocks
        !$OMP PARALLEL DO NUM_THREADS(nthreads_loop2)
        DO iz=ipz, ntilez,3
          DO iy=ipy, ntiley,3
            DO ix=ipx, ntilex,3

              If the particle goes to another tile
                Paste the properties the destination

              IF the particle goes to another MPI domain
                Paste the properties in local buffers

! 2) Reduction of the local buffers in
! global buffers for the MPI communications
.......
! 3) MPI communication of the buffers
.......
! 4) Ditribution of the particles in the tiles
.......
END SUBROUTINE simplified_charge_deposition
```

The rest of the subroutine is dedicated to the MPI communications. Local buffers are first reduced in parallel into global buffers (one for each direction, 26 in 3D, 8 in 2D). We again use nested parallel region, looping on the particles' species, then on the grid axes, and the communications are then performed. Finally, each tile reads in parallel, the received buffer to collect their macroparticles. As currently written, the particle communication algorithm does not benefit from vectorization because of the large number of if-statements.

17.3.3 HIGHLY SCALABLE PSEUDO-SPECTRAL MAXWELL SOLVERS

17.3.3.1 Limits of Current Low Order FDTD Solver

A large number of 3D plasma-electromagnetic codes including the standard PIC formulation currently uses order $p = 2$ explicit finite differences in both space and time to integrate Maxwell's equations [15]. This means that for calculating the field at a given point of the grid, the field solver uses a stencil of half width $p/2 = 1$, which only requires the fields at nearest grid points. This technique is widely used in part because it achieves efficient parallelization on up to a million cores, as required for the simulation of 3D large-scale problems. Indeed, due to the locality of the Maxwell's solver, the simulation domain is usually divided into independent subdomains with small guard

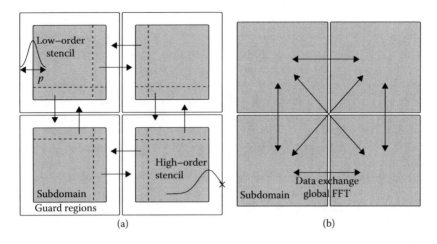

FIGURE 17.6 Parallelization strategies of current Maxwell's solvers. (a) Domain decomposition technique. The sketch represents a 2D layout of cartesian domain decomposition currently used in most low-order FDTD Maxwell's solvers. The whole simulation domain is split into multiple subdomains (gray areas—one per CPU) surrounded by guard regions into which copies of the fields from adjacent subdomains are stored (white regions). Fields are advanced independently using Maxwell's equations on each subdomain. Fields from adjacent subdomains (areas delimited by black dashed lines) are then copied into guard regions to ensure field continuity between subdomains (black arrows). The black curves at subdomain boundaries represent the stencil used to compute the electromagnetic grids. For low-order fields for which order $p/2 < n_{guards}$, the stencil is not truncated at the boundaries. However, for high-order and pseudo-spectral solvers for which $p/2 \gg n_{guards}$, the stencil is truncated at subdomain boundaries. (b) Sketch of the parallelization strategies used in current pseudo-spectral solvers. The domain is also split into multiple subdomains (one per CPU) but the field is not advanced independently on each subdomain. A global FFT is instead performed on the entire computational domain and each subdomain needs to exchange all its data with every other subdomain to perform the global FFT. The cost of this "All to All" operation is computationally much higher than the simple cartesian domain decomposition technique and prevents the scaling of current pseudo-spectral solvers to hundreds of thousands or more cores.

regions of length n_{guards} that are larger than the width of the stencil ($n_{guards} \geqslant p/2$). At each time step, electromagnetic fields are advanced independently on each subdomain and fields at the boundaries (in guard regions) are exchanged between adjacent subdomains (cf. Figure 17.6a).

However, low-order field solvers produce significant unphysical degradation from discretization errors, which reveals to be highly detrimental in the simulation of relativistic laser–plasma interactions. As a consequence, accuracy is practically achievable, but employing them is strongly limited by numerical dispersion and numerical heating or noise, which are particularly critical for laser–plasma accelerator experiments where small unphysical errors can spoil the required high beam quality, or for simulations where accurate description of a large band of frequencies is required (e.g., simulation of high harmonic generation on plasma mirror [26]). In general, the mitigation of all the above-mentioned unphysical effects always requires the use of very high resolution and hence many grid points that considerably increase the required computation time and prevent doing realistic 3D modeling.

17.3.3.2 High Order and Pseudo-Spectral Maxwell's Solvers for Much Higher Accuracy

In contrast, high order field solvers or in the infinite order limit $p \to \infty$ pseudo-spectral solvers offer a number of advantages over standard FDTD schemes in terms of accuracy and stability.

The first advantage is a very high precision. To illustrate this, let's consider a simple one-dimensional (1D) initial value problem $Lu = f(x)$ with L an integral or differential operator, $u(x)$ a function of space x that we want to approximate numerically, and $f(x)$ a known second member.

Furthermore, let's assume that u is defined on a simulation domain of width 1 and discretized by N points (mesh size $\delta x = 1/N$) with spectral solvers, it is possible to approximate the analytical solution u numerically with an error varying as $o((1/N)^N) = o(\delta x^N)$ instead of $o(\delta x^2)$ for order-2 FDTD solvers. The error with spectral solver is, thus, decreasing faster than any finite power of N because the power in the error formula is decreasing too, therefore yielding exponential convergence. When many decimals of accuracy are needed, pseudo-spectral solvers are, thus, the method of choice.

The second advantage of pseudo-spectral solvers is that they drastically reduce memory requirements for a given accuracy. Taking again the simple example described before, a quick calculation shows that we would need $N = 100^{50}$ points with order 2-FDTD solver to reach the same precision as pseudo-sepctral solvers with only $N = 100$ points.

In the particular case of Maxwell's equations, Haber et al. [20] already showed that under certain assumptions, Fourier transforming Maxwell's equation versus space yields an analytical solution for electromagnetic fields in time, called the PSATD solver, which is accurate to machine precision for the electromagnetic modes resolved by the calculation grid. This makes the PSATD solver highly accurate, without any Courant time step limit in vacuum and without any numerical dispersion at any angle/wavelength.

17.3.3.3   Limits to the Scaling of Pseudo-Spectral Solvers to an Arbitrary Large Number of Cores

Nevertheless, despite significant advantages in terms of accuracy and memory-minimization, high order and pseudo-spectral solvers have not been widely adopted so far for large-scale simulations because of their low scalability which is a direct consequence of their spatial nonlocality (large stencils). Indeed, these solvers, most of the time, use distributed FFT-based algorithms that require MPI "all-to-all" global interprocessor communications for matrix transposition (cf. Figure 17.6b) and limits their scaling to a few thousands of cores.

In the MPI version of fastest Fourier transform in the west (FFTW) [27], for instance, the multidimensional grid array to be transformed is first distributed to different processors across its rows (the first dimension) of the data. To perform the FFT of this data, each processor first transforms all the dimensions of the data that are completely local to it (e.g., the rows). Then, the processors have to perform a transpose of the data in order to get the remaining dimension (the columns) local to a processor. This dimension is then Fourier transformed, and the data is (optionally) transposed back to its original order. To perform the matrix transposition (in-place or out-of-place) each processor has to send data to all other processors which is obviously not scalable to a large number of processors.

For many decades, this has prevented the use of pseudo-spectral solvers in very computationally demanding 3D plasma electromagnetic simulations that requires several hundred of thousands of cores. Note that even if the speed of communications would increase in the next decade, it would not be enough to bring realistic time to solution for large-scale problems.

17.3.3.4   A New Method to Scale Pseudo-Spectral Solvers on up to a Million Cores

A new method [21] for solving time-dependent problems proposed to apply the Cartesian domain decomposition technique currently used with low order FDTD solvers to very high order pseudo-spectral solvers. As in the case of low order schemes, the simulation domain would be divided into several subdomains (see Figure 17.6a) and Maxwell's equations solved locally on each subdomain either using local convolution or local FFT's when the order p is very large.

However, there were still two challenges to face before being able to use this technique in production simulation:

1. First, this technique induces small truncation errors that can affect simulation results and that needed to be characterized. The fundamental argument still legitimating this decomposition method is that physical information cannot travel faster than the speed of light.

Choosing large enough guard regions should, therefore, ensure that spurious signal coming from stencil truncations at subdomains boundaries would remain in guard regions and would not enter the simulation domain (see Figure 17.6a) after one time step. Truncations errors were recently extensively characterized and an analytical model [22] was developed, which we can now predict the amount of truncation errors in simulations as a function of the numerical parameters used (solver order, guard cell sizes, and mesh size). In particular, this model shows that one should use very high finite order-p solvers instead of infinite order solvers. Indeed, truncation errors with finite very high order solvers decreases much faster with the number of guard cells while still bringing pseudo-spectral accuracy on almost the whole range of frequency accessible in the simulation. Practically, the truncation error model shows that 10–12 guard cells are sufficient to lower truncation errors below machine precisions with up to order 128 Maxwell solvers, for which pseudo-spectral precision is obtained on almost the whole frequency domain. This model will be especially crucial to determine the number of guard cells to use for a given solver order to minimize truncation errors in production simulations.

2. Then, this technique will need efficient shared-memory implementation of the PIC algorithm. Even though the number of guard cells required ($n_{guards} = 10$) seems low, it can however still affect the scaling of the PIC code to a large number of processors. Therefore, the solution is to enable good shared-memory parallelization of the PIC loop to decrease the total number of MPI-processes (and thus the number of volumes of MPI exchanges) and improve scaling of the pseudo-spectral code to a large number of cores. Shared-memory implementation of particle routines and standard solvers have been addressed in the previous section. For FFTs, we use the FFTW library that provide rather good threaded implementation of FFTs. On Xeon Phi architectures, one might also consider using the Intel Math Kernel Library (MKL) that has been designed to fully exploit hardware features of many-core architectures.

In the recent work, an efficient parallel implementation of pseudo-spectral solvers is now available in WARP/PXR that was recently tested at large scale on the MIRA [28] cluster at Argonne National Lab and benchmarked against standard solvers for relativistic laser–plasma mirror interactions. Scaling tests on MIRA will be presented in the Section 17.4.4 and show efficient scaling of pseudo-spectral solvers on up to 800,000 cores. Benchmarks against standard solvers show significant improvements in terms of time-to-solution and memory footprint required for a given accuracy [29].

17.3.4 AN OPTIMIZED SIMD CHARGE AND CURRENT DEPOSITION ALGORITM

An optimized SIMD algorithm has been developed in PICSAR to target large vector register processors such AVX512 on Intel KNL [25]. Several portable deposition algorithms were developed and successfully implemented on past generations of vector machines (e.g., CRAY, NEC) [30–35]. However, these algorithms do not give good performance on current SIMD architectures that have new constraints in terms of memory alignment and data layout in memory. Most of the vector deposition routines proposed in contemporary PIC codes use compiler-based directives or even C++ Intel intrinsics in the particular case of the Intel compiler, to increase vectorization efficiency [36]. However, these solutions are not portable and require code rewriting for a new architecture. In the following, the new algorithm is presented for the charge deposition at order 1 (CIC) and is shown in Listing 17.4. Note that vectorization methods presented for charge deposition can easily be transposed to current deposition. This algorithm is based on former scheme of Schwarzmeier and Hewit developed for Cray vector machines (see [32]) that demonstrated to give poor performances on Haswell Xeon processors (as demonstrated in Ref. [25]).

Listing 17.4   Optimized SIMD charge deposition algorithm for CIC shape factor.

```fortran
SUBROUTINE optimized_charge_deposition

 ! Declaration and init
 .....
 nnx = ngridx ; nnxy = nnx* ngridy

 moff = (0, 1, nnx, nnx+1, nnxy, nnxy+1, nnxy+nnx, nnxy+nnx+1)
 mx =(1, 0, 1, 0, 1, 0, 1, 0)
 my =(1, 1, 0, 0, 1, 1, 0, 0)
 mz =(1, 1, 1, 1, 0, 0, 0, 0)
 sgn =(-1, 1, 1, -1, 1, -1, -1, 1)


 ! ___ PART 1 _____
 ! Loop over chunks of particles
 DO ip =1,np , LVEC

   ! 1.1) SIMD loop on particles: computes cell index of particle
   ! and their weight on vertices
   ! $OMP SIMD
   DO n=1, MIN(LVEC, np-ip+1)
     nn=ip+n-1

     ! Calculation relative to particle n
     ! Computes current position in grid units
     x = (xp(nn)-xmin)*dxi
     y = (yp(nn)-ymin)*dyi
     z = (zp(nn)-zmin)*dzi

     ! Finds cell containing particles for current positions
     j=floor(x) ; k=floor(y) ; l=floor(z)
     ICELL(n)=1+j+nxguard+(k+nyguard+1)*(nx+2*nxguard) &
     +(l+nzguard+1)*(ny+2*nyguard)

     ! Computes distance between particle and node
     ! for current positions
     sx(n) = x-j ; sy(n) = y-k ; sz(n) = z-l

     ! Computes particles weights
     wq(n)=q*w(nn)*invvol


   ! 1.2) Charge deposition on vertices
   DO n=1, MIN(LVEC ,np-ip+1)
     ic=ICELL(n)

     ! SIMD loop on vertices: Add charge density
     ! contributions to vertices of the current cell
     ! $OMP SIMD
     DO nv=1, 8 !!! - VECTOR
       ww =(- mx(nv)+ sx(n))*( - my(nv)+ sy(n))* &
       (-mz(nv)+ sz(n))* wq(n)* sgn(nv)
       rhocells(nv, ic) = rhocells(nv, ic)+ww
```

```
!   ___  PART 2 _____
! Reduction of rhocells in rho
DO iz=1, ncz
   DO iy=1, ncy
      ! $OMP SIMD
      DO ix =1, ncx !! VECTOR ( take ncx multiple of vector length )
         ic=ix +(iy -1)* ncx +(iz -1)* ncxy
         igrid =ic +(iy -1)* ngx +(iz -1)* ngxy
         rho(orig + igrid + moff (1)) += rhocells (1, ic)
         rho(orig + igrid + moff (2)) += rhocells (2, ic)
         rho(orig + igrid + moff (3)) += rhocells (3, ic)
         rho(orig + igrid + moff (4)) += rhocells (4, ic)
         rho(orig + igrid + moff (5)) += rhocells (5, ic)
         rho(orig + igrid + moff (6)) += rhocells (6, ic)
         rho(orig + igrid + moff (7)) += rhocells (7, ic)
         rho(orig + igrid + moff (8)) += rhocells (8, ic)

END SUBROUTINE optimized_charge_deposition
```

In this new algorithm, a new data structure for the charge, named rhocells, is introduced (for the current: Jxcells, Jycells, Jzcells) that solves the vectorization issue of part 2 in Listing 17.1, and this new data structure is described in Figure 17.7. For each cell icell = (j, k, l) in the main charge array rho, rhocells stores the eight nearest vertices (j, k, l), (j+1, k, l), (j, k+1, l), (j+1, k+1, l), (j, k, l+1), (j+1, k, l+1), (j, k+1, l+1), and (j+1, k+1, l+1) contiguously for CIC (order 1) shape factor. The array rhocells is, thus, of size (8, ncells) with ncells that the total number of cells. This array has to remain sufficiently small to be contained in L2 in addition to rho allowing for an efficient cache reuse. For a tile of $11 \times 11 \times 11$ cells ($8 \times 8 \times 8$ cells with three guard cells in each direction for order 1 shape factor), rhocells has a size of approximately 80 KB. However, the equivalent of rhocells for the current deposition has a size of 250 kB that is close to the L2 cache size on Haswell (256 kB) and half the L2 cache on KNL for a single core (512 kB).

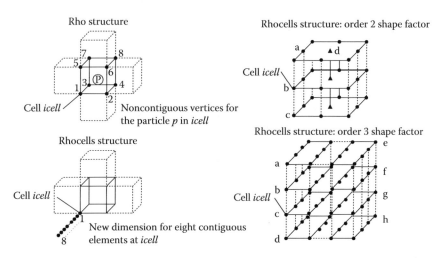

FIGURE 17.7 SIMD charge/current deposition algorithm description. On the left, the rhocells data structure is described and compared to rho for CIC (order 1) particle shape factor. A particle deposits its charge to the eight nearest vertices (black points). For each cell, icell and rhocells stores the eight nearest vertices contiguously. On the right, the rhocells data structure is given for higher shape factor orders.

The particles are divided into chunks of size LVEC where LVEC should be a multiple of the vector register length. The first loop of the code is over the chunks, and the first inner loop (1.1 in Listing 17.4) over the macroparticles of chunk it is vectorized using an OMP SIMD directive. Inside this loop, the cell index of the particles in the charge array (j, l, k and icell) and their weights wq are computed. The second inner loop (1.2 in Listing 17.4) is the charge deposition on rhocells. During this phase, the vectorization is not applied on the particles but on the vertices for each particle. The element rhocells(1, icell) is 64 bytes memory aligned for a given cell icell and the elements rhocells(1:8, icell) entirely fit in one cache line allowing for efficient vector **load/stores**.

After the deposition is done for all particles, rhocells is reduced to the main charge array rho (part 2 in Listing 17.4). This step can be vectorized but does not lead to optimal performances due to the noncontiguous access in rho that induces **gather–scatter** instructions. However, this process is executed N_c times (again, N_c, or ncells, being the total number of cells) against N_p times for the direct or Schwarzmeier and Hewit algorithm (N_p being the number of macroparticles). In many simulations, the number of particles is much higher than the number of cells, $N_p \gg N_c$. Moreover, the reduction is performed just once after all species have contributed to the rhocells data structure.

For higher shape factors that use more than eight vertices, the algorithm is adapted. For TSC (order 2) particle shape factor, the particle deposit their charge to the 27 neighboring vertices as described in Figure 17.7. Increasing the size of rhocells to store 27 vertices would also increase the memory constraint on the L2 leading to inefficient cache reuse. Less memory is left for the macroparticles and the main charge array rho. Instead, the same size is kept for the data structure but the vertices are grouped in a different way. The new structure for rhocells(1:8, 1:ncells) groups 8 points in (y, z) plane for each cell icell (bold line planes in Figure 17.7). Each particle first adds its charge contribution to 24 points in the three planes at icell+1 (a), icell (b), and icell-1 (c). The 3 left points in the central regions (triangle markers) are updated separately. For QSP (order 3) particle shape factor, rhocells is newly adapted and particles deposit their charge to the 64 neighboring vertices in the eight different (y, z) planes as shown in Figure 17.7 (see [25] or the PICSAR website picsar.net/features/optimized-morse-nielson-deposition/ for more details).

Let us first look at the classical charge deposition. If we compare the original WARP subroutine (general order subroutine) implemented in PICSAR and the most optimized subroutine, a speedup of approximatly 3 is obtained on Haswell for an order 1 and order 3 interpolation shape factors. On KNL, the similar comparison shows that we have a speedup of 7 at orders 1 and 6 at order 3. Original WARP deposition subroutines contain branches inside the particle loop that depends on the interpolation order. If we now compare the scalar routines (routines at specific orders) with the optimized subroutines, we have a speedup of 1.7 at orders 1 and 1.4 at order 3 on Haswell. The speedup difference is even higher on KNL, and we obtain around 1.8 at orders 1 and 3. This demonstrates that simple changes like removing branching in intensive loops can provide a significant speedup before any more technical optimization, especially on KNL.

The current deposition is more complex in 3D because of the three components and induces more pressure on the registers. For the latter one, we only compare the order-specific scalar routines to the optimized ones. On Haswell, a speedup of 2 is obtained at orders 1 and 1.3 at order 3. On KNL, we have 2 at orders 1 and 1.7 at order 3.

17.3.5 FIELD GATHERING OPTIMIZATION

Optimization and vectorization of the field gathering are not complicated and are based on simple actions that should be considered at first on many codes:

Remove if-statement from intensive loops: To condense codes, having if-statements inside main loops is common to not have to rewrite the loop entirely for the different branches. Nonetheless, when the loop is executed a large number of iterations, as for the field gathering, the cost of several if-statements become nonnegligible. It is therefore better to externalize the branches when possible by creating several versions of the function than using multicase functions. For the field gathering (also valid for the current deposition), it consists on having a function for each shape factor order.

Linearize loops with constant or known indexes: To enable vectorization of the main loop, small inner loops have to be linearized as done in Listing 17.2 for the reduction in E_x. Temporary variables have to be declared private: `OMP SIMP private(variables...)` to allow correct vectorization.

Computation factorization and FMA: After linearization, redundant calculations have been suppressed by factorizing. Calculation order has been rearrranged to let appear almost only FMA.

Avoid mixing data types: For vectorization again, using same data types prevents conversions that have a nonnegligible cost. Using the smallest needed data type (32-bit integer or float instead of double precision) is also advised to fill vector registers with the maximum number of elements.

This is particularly crucial on MIC architecture that has a reduced speed clock that can only be compensated by vectorization. After being linearized, the field gathering loop can now be vectorized using an OpenMP directive `OMP SIMD`.

Performing the field gathering and the particle pusher in a single loop for each tile increases the flop/bytes ratio and enhances the cache reuse of the macroparticle properties that can not be contained entirely in cache. For cache blocking, the main particle loop is also divided into two loops: a loop over chunks of particles of size `LVEC` and an inner vectorized loop over the particles of the chunk.

All included, these changes provide a speedup of 5.8 on a Haswell node and almost 11 on a KNL node at order 1. At order 3, it gives a speedup of 1.8 on Haswell and again 11 on KNL. More details about the implementation and the speedups of the different optimizations can be obtained at picsar.net/features/optimization-of-the-field-gathering/.

17.3.6 PARTICLE SORTING

As explained in Section 17.2.5, a performance issue for the field gathering, the current and the charge deposition can arise from the fact that two contiguous particles in memory can be spatially far from each other. Therefore, the deposition and the gathering processes use vertices from distant parts of the grids preventing an efficient cache reuse and allowing for potential cache misses if the grids are not in cache. In order to pack the macroparticles according to their location and therefore improve cache reuse for vectorization, a cell sorting algorithm has been implemented as done in Refs. [37,38]. The sorting is performed in the tiles and the granularity (supercells or subcells) can be controlled. When the tiles fit entirely in cache, the sorting provides a relatively low speedup on Haswell node and on KNL for CIC shape factor. The speedup increases at higher orders since more field components are used. With big tiles or without tiling, cache misses increase and can be compensated by the sorting. In the latter case, sorting the macroparticles in supercells is a kind of tiling. The sorting is also limited by the staggered grids: if a sorting is done for one of the grid configurations, it will not match perfectly to the others because of the shifts. The bin sorting algorithm hardly vectorizes due to possible memory races during the sorting process. The bin sorting complexity is in $\mathcal{O}(N_p + N_c)$. This is lower than strict sorting algorithms which usually have in the best case complexity in $\mathcal{O}\left(N_p \log\left(N_p\right)\right)$. Nonetheless, the sorting routine for a given group of particles is still as costly as the nonoptimized charge deposition or field gathering routines. Particles have then to be sorted every given iteration periods so that the gain in computational time exceeds the cost of the sorting process. The sorting period depends on how fast the particles get mixed. It consequently depends on the particle distribution and the local plasma temperature. The sorting efficiency also depends on the number of particles per cells: if too small, the gain will be less interesting with long vector registers.

For the considered test case using the tiling, the most optimized routines and order 1 shape factor, there is no speedup on a Haswell node and on KNL one for the field gathering and the charge deposition whereas the current deposition, obtains a speedup around 1.4 for both architectures. With order 3 shape factor, the speedup on a Haswell node is of 1 (no gain) for the field gathering, 1.8 for the current deposition, and 1.1 for the charge deposition. On KNL, it is about 1.1 for the field gathering, 1.7 for the current deposition, and 1.4 for the charge deposition. These tests have been performed with the tiling. Speedup from sorting is larger without it. In this case, the sorting plays the role of the tiling by grouping particle in memories according to their location.

17.4   ANALYSIS OF THE GLOBAL PERFORMANCE AND BENCHMARKS ON DIFFERENT ARCHITECTURES

17.4.1   SINGLE NODE PERFORMANCES

The overall application gain in performance using all described optimizations (tiling, vectorization, sorting) in this chapter and some additional ones on the PICSAR website is now presented on a single Haswell node and on a single Intel Xeon Phi KNL of the supercomputer Cori. The same simulation case introduced in Section 17.2.5 is used in the following performance studies. The simulations are run the time of 50 iterations approximatly corresponding to $t = 5T_p$ where T_p is the plasma period for a hydrogen plasma density of $n = 10^{25}$ m^{-3}. Simulation times per iteration per particle and per processing unit computed for CIC (order 1) shape factors and QSP (order 3) shape factor are both presented in Figure 17.8. In each figure, both nonoptimized (bars on the left) and optimized code (bars on the right) times are given. The nonoptimized code does not exactly correspond to the original state of PICSAR extracted from WARP since it consists of the scalar routines without internal if-statements.

The first goal of PICSAR is to provide high-peformance subroutine for Intel MIC architectures. The stand-alone version of PICSAR with our test case achieves a speedup of 3.7 on a single KNL

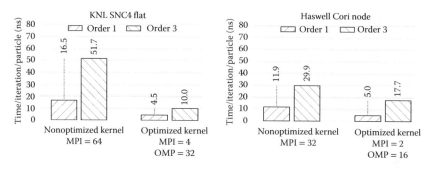

FIGURE 17.8   Times per iteration per particle per computing unit using the nonoptimized and the fully optimized version of PICSAR on a single KNL node and a Haswell node of the supercomputer Cori at NERSC. The nonoptimized code version uses only MPI. We have 64 ranks on KNL and 32 on a Cori Haswell node. With the optimized code, we use both MPI and OpenMP. The considered KNL configuration is SNC4 flat (MCDRAM and DDR are divided into four domains referred to as NUMA domains). We have 1 rank per NUMA domain in this case and 32 OpenMP threads per MPI rank. On Haswell node, we use 1 MPI rank per Haswell processor and 16 openMP thread per MPI.

processor at order 1 interpolation with all optmizations. With an interpolation order of 3, an even higher speedup around 5 have been obtained.

These optimizations that have been implemented for Intel MIC architectures also contribute to speedup the code on previous architectures such as Intel Haswell processors as shown in the Haswell part of Figure 17.8. A speedup of 2.4 is obtained for an shape factor order of 1 and 1.7 at order 3.

We now compare Haswell and KNL results. Without optimization, Figure 17.8 shows that Haswell was much faster than KNL running PICSAR stand-alone. This is the consequence of three main factors:

1. **Presence of the L3 cache on Haswell processors:** The original code is memory bound with a low flop/byte ratio that is compensated by the L3 cache.
2. **MCDRAM shared memory:** Without OpenMP, the large high-bandwidth shared memory is not necessarily fully exploited.
3. **No vectorization:** With a clock speed twice lower and a vector length twice longer than on Haswell processors, vectorization is crucial to get the same performances and the original code is not vectorized on KNL processors.

Using optimized subroutines, KNL becomes faster than Haswell. With a shape fatcor of order 1, the difference is small. Having the same performance is already interesting because it means that the same computation can be achived with less energy. For higher shape factor orders (using the classical current deposition), PICSAR benefits from MIC architectures. At order 3, we have a speedup of almost 2 between KNL and Haswell. Additional studies on a single Cori KNL node can be found on the PICSAR website: http://picsar.net/performance/performance-cori/.

Performance and vectorization efficiency can be further investigated using vendor tools such as Cray tools for Cray systems, Intel tools (Intel Vtune, Advisor, and Inspector), and Allinea tools. In the next section, analysis performed with Intel Advisor and Allinea Map are presented.

17.4.2 PERFORMANCE PROFILING TOOLS

Intel advisor allows to estimate vectorization gains, to analyze vectorization issues such as dependencies and gather/scatter instructions and to find solutions for an efficient vectorization [39]. Estimated efficiency from Intel Advisor are shown in Table 17.2. Simulations were performed on a single core on one node.

Efficiencies do not exactly correspond to speedups that we observe with the nonoptimized routines but they give an estimation of how good is the vectorization. More work will be necessary for the field gathering loop, which concentrates many operations with little data reuse in a single loop (**register spilling**), especially with shape factors of order 3. This applies equally to current deposition loop with shape factors of order 3. Splitting the loop into smaller ones will be explored. The vectorization of the Maxwell solver is also ineffective, due to the stride data access pattern to the field grid points (stencil problem).

Allinea MAP is a profiler for parallel and multithreaded codes [40]. This profiler is used with the original and the optimized version of the code on a Haswell node of Cori (2 MPI tasks and 16 OpenMP threads per task). Part of the profiling is shown in Figure 17.9 for the first five iterations. For both runs, the application activity, the CPU floating-point activity (percentage of time spent executing both scalar and vector floating-point instructions), memory access activity (percentage of time spent memory-related instructions), and the vector floating-point activity averaged over all the cores are given. In the original version of the code, only the particle pusher is automatically vectorized by the compiler. Vectorized instructions therefore account for 34% of the total floating-point intructions. Application activity shows that MPI communications represent a nonnegligible fraction of the iteration time. With the optimized version of the code, vector instructions represent almost 100% of the total floating-point instructions. Time spent in the communication has also been reduced thanks

TABLE 17.2

Vectorization Efficiency Estimated by Intel Advisor for Simulation Analysis on a KNL Core for Order 1 and Order 3 Shape Factors

Subroutine/Loop	Efficiency Order 1 (%)	Efficiency Order 3 (%)
Field gathering	55	15
Current deposition: index computation and weight computation for order 3	59	38
Current deposition: weight computation for order 1 and `jcells` deposition only for order 3	90	100
Current deposition: reduction	10	10
Particle pusher: electric acceleration	100	100
Particle pusher: magnetic rotation	83	83
Particle pusher: position update	100	100
Maxwell solver	18	18

Note: Only the main vectorized loop of the code are presented. Vectorization efficiency is the division of the vectorization gain by the vector length for the longer considered variable type. Vectorization gain is a compiler cost model-based estimate.

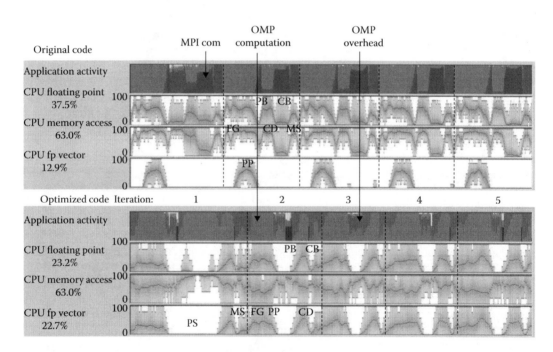

FIGURE 17.9 Allinea MAP analysis of the original and the optimized version of the code showing the application activity, the CPU floating-point activity (percentage of time spent executing both scalar and vector floating-point instructions), memory access activity (percentage of time spent in memory-related instructions) and the vector floating-point activity averaged over all the cores during the first five iterations. Dashed lines demarcate the different iterations. The rest of the percentage not shown here accounts for the CPU integer instruction and branches. The dark gray line is an average of the activity among all processes. The upper and lower light gray bars represent the maximum and minimum thread activity. PP, particle pusher; FG, field gathering; MS, Maxwell solver; CD, current deposition; CB, current boundary conditions; PB, particle boundary conditions; PS, particle sorting.

to MPI communication optimizations. However, application activity shows that OpenMP overheads appeared. In the first iteration, the long interval without vector floating-point instruction corresponds to the particle sorting done at the first iteration (and then every 20 iterations) during the simulation. According to Allinea Map profiling, memory access activity represents 63% of the simulation time in both simulation cases. However, memory accesses are better distributed and more constant accross iteration time with the optimized code. Remaining percentage (not shown in Figure 17.9) is the integer instructions and branches.

Finally, optimization efforts have paid off with a simulation time on KNL similar to Haswell (single node) for order 1 shape factor and even lower by 40% for order 3 shape factor. In the following, the roofline performance model is used to understand the positive effects of these optimizations in a different way.

17.4.3 Application of the Roofline Performance Model

The roofline performance model provides a visual analysis of the computational constraining resources of every systems from single-core to manycores architectures [41,42]. It consists of a 2D graph with information on floating-point performance, operational intensity, and memory performance. It provides a synthesized understanding of the efficiency of a kernel use of system resources, how far the kernel is from reaching the machine peak performance, and what the limitations and tradeoffs are. This model can be applied to the RAM traffic, the cache traffic, and since recently to the HBM that equips MIC architectures. In a roofline graph, the ordinate represents the floating operations per time units (in Gflop/s), whereas the abscissa represents the arithmetic intensity (AI), which is the ratio of total floating-point operations to total data movement (bytes). Roofline graphs are composed of two parts:

1. The first roofline is a rising slope with a steepness that is given by the system peak memory bandwidth, which can be determined from the system itself or the vendor specifications.
2. The second roofline is flat and is given by the peak floating-point performance when the optimum system peak memory bandwidth is reached.

The AI and the Gflop/s can be theoretically determined for simple kernels. Else, the AI (using the total flops and the number of memory access) and the performance (total flops and simulation time) are determined at runtime. If a data point is close to the peak memory bandwidth, the code is mainly memory bound whereas if it is close to the flat roofline and it is considered to be compute bound. Intel Vtune Amplifier is used to calculate the number of accessed bytes from the MCDRAM, and Intel Software Development Emulator (SDE, [43]) for measuring the number of flops [44].

The roofline performance model is applied to the entire kernel and AI points are plotted for different optimization levels in Figure 17.10 for a Haswell node of Cori and Figure 17.11 for KNL. In the figures, only the first roofline is present because of the zoom performed on the points. The optimization levels that are considered are the same as in Section 17.4.1. The triangle marker corresponds to the simulation prior any optimization: without tiling, without vectorized subroutines, and no sorting (level 1). However, in contrast to the previous section, the general-order subroutines are used. On Haswell and KNL, the marker is under the peak stream bandwidth roofline, meaning that the code is essentially memory bound. AI is higher on Haswell than on KNL, with respective values of 0.57 and 0.13 flop/byte and average performances of 17 and 5.6 Gflops/s. The second level of optimization represented by the circle marker corresponds to tiling. By increasing memory management and cache reuse, the number of flops is almost unchanged but the number of RAM accesses is significantly reduced. As a consequence, the ratio flop/byte is increased leading to a translation to the right from the triangle and the circle marker. AI is of 1.1 on Haswell and 0.58 flop/byte of on KNL for respective performances of 32 and 20 Gflops/s.

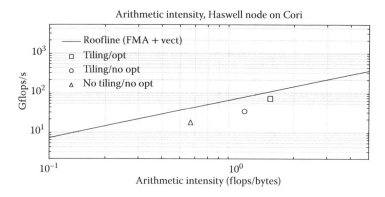

FIGURE 17.10   Roofline performance graph on a Haswell node of Cori. The figure shows arithmetic intensities (flop/byte) and performances (Gflop/s) for three simulations with different levels of optimizations. In the legend, "tiling" means that tiling is activated, "opt" for optimized means that the most optimized subroutines are used. The first simulation (triangle marker) is close to the original situation without tiling (32 MPI tasks/No OpenMP) and with the scalar subroutines. The second simulation (circle marker) uses tiling (2 MPI tasks/16 OpenMP threads per MPI task) and the scalar subroutines. The third simulation (2 MPI tasks/16 OpenMP threads per MPI task) uses the most optimized features: tiling and the vectorized subroutines.

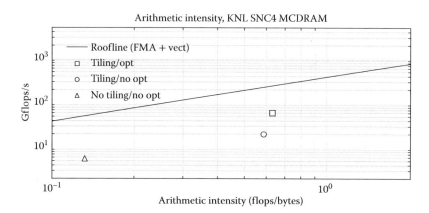

FIGURE 17.11   Roofline performance graph on KNL in flat SNC4 mode. The figure shows arithmetic intensities (flop/byte) and performances (Gflop/s) for three simulations with different levels of optimizations. In the legend, "tiling" means that tiling is activated, "opt" for optimized means that the most optimized subroutines are used. The first simulation (triangle marker) is close to the original situation without tiling (32 MPI tasks/No OpenMP) and with the scalar subroutines. The second simulation (circle marker) uses tiling (2 MPI tasks/16 OpenMP threads per MPI task) and the scalar subroutines. The third simulation (2 MPI tasks/16 OpenMP threads per MPI task) uses the most optimized features: tiling and the vectorized subroutines.

The square marker represents the last level of optimization with vectorized subroutines. Vectorization increases the performance in term of Gflops/s. The flop/byte ratio is also increased and thanks to the memory optimizations made inside the subroutines. AI is finally of 1.5 on Haswell and 0.63 flop/byte of on KNL for respective performances of 68 and 60 Gflops/s. Square markers are still localized under the peak memory bandwidth line meaning that the code is still globally memory bound.

A detailed description of the roofline performance model and its application to PICSAR is available on our website: https://picsar.net/performance/roofline-model-application/.

17.4.4 PERFORMANCES ON LARGE-SCALE SUPERCOMPUTERS

As part of the Director's Discretionary Allocation obtained on 2 June 2016, a series of scaling tests have been performed on the MIRA supercomputer in Argonne. The MIRA machine is a 10-petaflops IBM Blue Gene/Q system [28] composed of 49152 nodes of 16 1600 MHz PowerPC A2 cores interconnected with a five-dimensional (5D) Torus network.

The case of a homogeneous thermalized plasma of thermal velocity $0.1c$ is considered again, with $1024 \times 1024 \times 3072$ grid cells domain and 20 macroparticles per cells. These dimensions allow to divide grid cells equally between MPI domains for all the tests presented in Figure 17.12.

At first, a strong scaling test is performed with MPI only (1 OpenMP thread is used per MPI thread) with a number of cores ranging from 20,000 to 800,000. The FDTD Maxwell solver is used with a stencil of order 2 with two guard cells per MPI domain and the PSATD Maxwell solver is used with a pseudo-stencil of order 128 with 12 guard cells per MPI domain. The results are presented in Figure 17.13. The low-order FDTD Maxwell solver (circle markers) scales very well to the full machine with an efficiency of 98% on approximately 800,000 cores. The high-order FDTD Maxwell solver (square markers) also exhibits a very good scaling with an efficiency of 83% on half of the MIRA machine. The reduction of the efficiency with the high-order FDTD Maxwell solver compared to the low-order FDTD solver is mainly due to the larger number of guard cells that need to be exchanged.

Strong OpenMP scaling tests for the particle and field routines have then been performed with a fixed number of 49 152 MPI tasks. It shows that a speedup of 11 can be achieved with 16 OpenMP threads per MPI process.

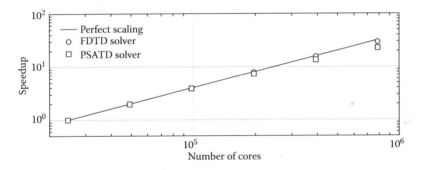

FIGURE 17.12 MPI strong scaling tests of PICSAR on MIRA. The circle markers correspond to scaling data with order 2 FDTD Maxwell solver and two guard cells. The square markers correspond to the PSATD spectral solver at order 128 with 12 guard cells, as used in plasma harmonic simulations.

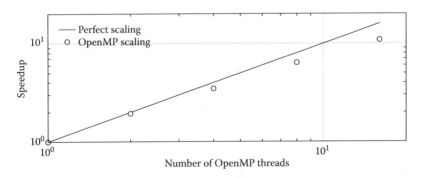

FIGURE 17.13 OpenMP scaling of particle and field routines on MIRA (fixed number of MPI processes). The plain line corresponds to ideal speedup and markers to the routines speedup.

17.5   CONCLUSION AND PROPSECTS

Exascale machines are needed to simulate large-scale and accurate 3D physical problems with PIC codes. Exascale supercomputers may rely on emergent many-integrated core technologies like Intel Xeon Phi KNL processors. To take advantage of these new architectures, existing codes may need to be partially or entirely rewritten to benefit from the manycores, from the large shared HBM and from vectorization. Codes using the PIC method are examples of codes that may perform more poorly without adaptations on MIC than on previous architectures. This has been illustrated using the open-source Fortran library PICSAR, dedicated to ultimately provide a kernel of high-performance sub-routines for PIC elementary operations. The main PICSAR optimizations for MIC architectures that have been presented in this chapter are

- Tiling: cache blocking optimization for L2 cache
- OpenMP: inner node parallelization
- MPI and tile communications
- Vectorization of the current deposition
- Vectorization of the field gathering
- Particle sorting

The efficiency and scalings of input/output (IO) have not been discussed in this chapter. Yet, in large-scale 3D simulation, IO data volume workflow can represent several TB if all particles are dumped at regularly close iteration intervals, and the optimization of IO subroutines can be very challenging, especially in a massively parallel environment. Advantage of emerging burst buffer technologies will be the subject of a forthcoming study for PICSAR, for standard IO, as well as in-situ or in-transit data analysis and visualization. Future studies may also involve the optimization of additional physical algorithms to the classical PIC loop kernel presented here.

ACKNOWLEDGMENTS

We thank Rebecca Hartman Baker, Tuomas Koskela, Andrey Ovsyannikov, Tareq Malas, Brian Friesen, Doug Jacobsen, Alice Koniges, Thorsten Kurth, Dong Li, Nigel Tan, and Helen He for interesting and fruitful discussions. We thank Zakhar Matveev and Anthony Nguyen from Intel for their support in using Intel tools (SDE, Advisor) and their advice in optimization. We acknowledge Doug Doerfler for useful support on AI calculations. This work was supported by the European Commission through the Marie Skłodowska-Curie actions (Marie Curie IOF fellowship PICSSAR grant number 624543) and U.S. Department of Energy (US-DOE) High Energy Physics program CAMPA and SciDAC program ComPASS, supported by the Office of Science of the under Contract No. DE-AC02-05CH11231. This material is based upon work supported by the Advanced Scientific Computing Research Program in the US-DOE, Office of Science, under Award Number DE-AC02-05CH11231. This research used resources of the National Energy Research Scientific Computing Center, a DOE, Office of Science User Facility supported by the Office of Science of the US-DOE under Contract No. DE-AC02-05CH11231.

GLOSSARY

Arithmetic intensity: The number of floating operations performed per loaded bytes.

Cache blocking: Cache blocking is an optimization technique that consists on decomposing a memory pattern into smaller structures adequatly chosen to fit in some specific levels of cache and then to work on these subsets in cache. Cache blocking enables to achieve an efficient cache reuse.

Cache hit: Cache hit at a given cache level means that data are found at this level.

Cache miss: Cache miss at a given cache means that data are not located at this level.

Cache reuse: Cache reuse happens when data are loaded at a given level of cache and reused several times for multiple computations without being released. Cache reuse means that less data are moved for a set of instructions leading to better performances.

Floating operation performance: Number of floating operations performed per time unit.

Gather: A gather consists on several loads to form a contiguous vector when a vector operation is performed on noncontiguous data in memory.

Load instruction: Data transfer from the L1 cache level to the vector register.

Memory bandwidth: Amount of data transfered per time unit from one memory to another or from one device to another.

Memory latency: Latency refers to the time delay between a command to tranfer some data and the execution.

Memory locality: Memory locality refers to where and how the data are accessible by the processing unit when the latter one needs to access them. Good memory locality means that data should be at the closer cache level to the processing unit when required.

Register spilling: Moving a variable from a register to main memory. A spilled variable must be loaded in and out of main memory for every read/write operation, resulting in poorer performance. (Intel definition, https://software.intel.com/en-us/articles/intel-advisor-vectorization-advisor-glossary.)

Scatter: A scatter consists on several store to update noncontiguous data after a vector operation has been performed.

Store instruction: Data transfer from the vector register to the cache.

Vectorization: An operation is said to be vectorized when the same instruction is applied simultaneously on multiple data. This can be opposed to scalar/sequential operations in which each data is treated one by one.

REFERENCES

1. C. K. Birdsall and A. B. Langdon. *Plasma Physics via Computer Simulation. Series in Plasma Physics.* New York, NY: Taylor & Francis, 2005.
2. K. Bergman, S. Borkar, D. Campbell, W. Carlson, W. Dally, M. Denneau, P. Franzon, et al. *Exascale Computing Study: Technology Challenges in Achieving Exascale Systems.* P. Kogge, editor and study lead, Arlington County, VA: DARPA, 2008.
3. Aurora Super-computer. http://aurora.alcf.anl.gov/. Accessed: 2016-07-15.
4. Sunway TaihuLight Super-computer. https://www.top500.org/system/178764/.
5. F. H. Harlow and M. W. Evans. A machine calculation method for hydrodynamics problems, Los Alamos Scientific Laboratory report LAMS-1956, 1955.
6. A. Amsden. The particle-in-cell method for the calculation of the dynamics of compressible Fluids, Los Alamos Scientific Laboratory report LA 3466, 1969.
7. R. W. Hockney and J. W. Eastwood. *Computer Simulation using Particles.* London, UK: CRC Press, 1988.
8. O. Buneman. Computer space plasma physics. In *Simulation Techniques and Software*, H. Matsumoto and Y. Omura editors, page 67, Tokyo, Japan: Terra Scientific Publishing Company, 1993.
9. J. M. Dawson. Particle simulation of plasmas. *Reviews of Modern Physics*, 55:403–447, April 1983.
10. H. Abe, N. Sakairi, R. Itatani, and H. Okuda. High-order spline interpolations in the particle simulation. *Journal of Computational Physics*, 63(2):247–267, 1986.
11. K. J. Bowers, B. J. Albright, L. Yin, B. Bergen, and T. J. T. Kwan. Ultrahigh performance three-dimensional electromagnetic relativistic kinetic plasma simulation. *Physics of Plasmas*, 15(5):055703, 2008.

12. Warp pic code. http://warp.lbl.gov/. Accessed: 2016-07-15.

13. J. P. Boris. *The Acceleration Calculation from a Scalar Potential*. Princeton, NJ: Plasma Physics Laboratory, March 1970.

14. J. -L. Vay. Simulation of beams or plasmas crossing at relativistic velocity). *Physics of Plasmas* (1994–present), 15(5):056701, 2008.

15. K. S Yee. Numerical solution of initial boundary value problems involving Maxwells equations in isotropic media. *IEEE Transactions on Antennas and Propagation*, 14(3):302–307, 1966.

16. J. B. Cole. A high-accuracy realization of the Yee algorithm using non-standard finite differences. *IEEE Transactions on Microwave Theory and Techniques*, 45(6):991–996, June 1997.

17. M. Karkkainen, E. Gjonaj, T. Lau, T. Weiland. Low-dispersion wake field calculation tools. In *Proceeding of the International Computational Accelerator Physics Conference*, Chamonix, France, 2006.

18. B. M. Cowan, D. L. Bruhwiler, J. R. Cary, E. C. Michel, and C. G. R. Geddes. Generalized algorithm for control of numerical dispersion in explicit time-domain electromagnetic simulations. *Physical Review Special Topics-Accelerators and Beams*, 16(4):041303, 2013.

19. R. Lehe, A. Lifschitz, C. Thaury, V. Malka, and X. Davoine. Numerical growth of emittance in simulations of laser-wakefield acceleration. *Physical Review Accelerators and Beam*, 16:021301, February 2013.

20. I. Haber, R. Lee, H. H. Klein, and J. P. Boris. Advances in electromagnetic simulation techniques. In *Proceedings of the 6th Conference on Numerical Simulation of Plasmas*, pp. 46–48, Berkeley, CA, 1973.

21. J. -L. Vay, I. Haber, and B. B. Godfrey. A domain de-composition method for pseudo-spectral electromagnetic simulations of plasmas. *Journal of Computational Physics*, 243:260–268, June 2013.

22. H. Vincenti and J. -L. Vay. Detailed analysis of the effects of stencil spatial variations with arbitrary high-order finite-difference Maxwell solver. *Computer Physics Communications*, 200:147–167, March 2016.

23. M. McCool, A. D. Robison, and J. Reinders. *Structured Parallel Programming: Patterns for Efficient Computation*. Burlington, MA: Morgan Kaufmann, 2012.

24. I. A. Surmin, S. I. Bastrakov, E. S. Efimenko, A. A. Gonoskov, A. V. Korzhimanov, and I. B. Meyerov. Particle-in-cell laser-plasma simulation on xeon phi coprocessors. *Computer Physics Communications*, 202:204–210, 2016.

25. H. Vincenti, R. Lehe, R. Sasanka, and J. -L. Vay. An efficient and portable SIMD algorithm for charge/current deposition in particle-in-cell codes. *Computer Physics Communications*, 210: 145–154, 2016.

26. H. Vincenti, S. Monchoce, S. Kahaly, G. Bonnaud, P. H. Martin, and F. Quere. Optical properties of relativistic plasma mirrors. *Nature Communications*, 5, 3403 EP 2014.

27. M. Frigo and S. G. Johnson. The fastest Fourier transform in the west. Technical Report MIT-LCS-TR-728, Massachusetts Institute of Technology, September 1997.

28. Mira Team. Mira super-computer. https://www.alcf.anl.gov/mira. Accessed: 2016-07-22.

29. G. Blaclard, H. Vincenti, R. Lehe, and J. L. Vay. An efficient and portable SIMD algorithm for charge/current deposition in particle-in-cell codes. arXiv preprint, arXiv:1601.02056, 2016.

30. A. Nishiguchi, S. Orii, and T. Yabe. Vector calculation of particle code. *Journal of Computational Physics*, 61(3):519–522, 1985.

31. E. J. Horowitz. Vectorizing the interpolation routines of particle-in-cell codes. *Journal of Computational Physics*, 68(1):56–65, 1987.

32. J. L. Schwarzmeier and T. G. Hewitt. In *Proceedings of the 12th Conference on Numerical Simulation of Plasmas*, San Francisco, CA, 1987.

33. A. Heron and J. C. Adam. Particle code optimization on vector computers. *Journal of Computational Physics*, 85(2):284–301, 1989.

34. G. Paruolo. A vector-efficient and memory-saving interpolation algorithm for pic codes on a Cray X-MP. *Journal of Computational Physics*, 89(2):462–482, 1990.

35. J. U. Brackbill, J. W. Eastwood, D. V. Anderson, and D. E. Shumaker. Particle simulation methods hybrid ordered particle simulation (hops) code for plasma modelling on vector-serial, vector-parallel, and massively parallel computers. *Computer Physics Communications*, 87(1):16–34, 1995.

36. R. A. Fonseca, J. Vieira, F. Fiuza, A. Davidson, F. S. Tsung, W. B. Mori, and L. O. Silva. Exploiting multi-scale parallelism for large scale numerical modelling of laser wakefield accelerators. *Plasma Physics and Controlled Fusion*, 55(12):124011, December 2013.

37. V. K. Decyk and T. V. Singh. Particle-in-cell algorithms for emerging computer architectures. *Computer Physics Communications*, 185(3):708–719, 2014.

38. H. Nakashima. Manycore challenge in particle-in-cell simulation: How to exploit 1–TFlops peak performance for simulation codes with irregular computation. *Computers and Electrical Engineering*, 46:81–94, 2015.

39. Intel Advisor. https://software.intel.com/en-us/intel-advisor-xe. Accessed: 2016-07-22.

40. Allinea Map. http://www.allinea.com/products/map. Accessed: 2016-07-22.

41. S. Williams, A. Waterman, and D. Patterson. Roofline: An insightful visual performance model for multicore architectures. *Communications of the ACM*, 52(4):65–76, 2009.

42. Roofline Performance Model. http://crd.lbl.gov/departments/computer-science/PAR/research/roofline/. Accessed: 2016-07-22.

43. Intel Software Development Emulator. https://software.intel.com/en-us/articles/intel-software-development-emulator. Accessed: 2016-07-22.

44. NERSC Team. Arithmetic intensity. http://www.nersc.gov/users/application-performance/measuring-arithmetic-intensity/. Accessed: 2016-07-22.

18 Extreme-Scale *De Novo* Genome Assembly

Evangelos Georganas, Steven Hofmeyr, Leonid Oliker,
Rob Egan, Daniel Rokhsar, Aydin Buluc,
and Katherine Yelick

CONTENTS

18.1 Overview of *De Novo* Genome Assembly...409
18.2 The Meraculous Assembly Pipeline 410
18.3 The Partitioned Global Address Space Model in UPC..413
18.4 Distributed Hash Tables in a PGAS Model 414
    18.4.1 Basic Implementation of a Distributed Hash Table 414
    18.4.2 Use Cases of Distributed Hash Tables in the HipMer Pipeline 415
18.5 Parallel Algorithms in HipMer 417
    18.5.1 Parallel *k*-mer analysis..417
    18.5.2 Parallel Contig Generation 418
    18.5.3 Parallel Read-to-Contig Alignment 420
    18.5.4 Parallel Scaffolding and Gap Closing..421
    18.5.5 Summary of Communication Patterns and Costs 422
18.6 Performance Results 422
    18.6.1 Strong Scaling Experiments 423
    18.6.2 I/O Caching..424
    18.6.3 Performance Comparison with Other Assemblers 425
18.7 Challenges for Future Architectures..425
18.8 Related Work 427
18.9 Conclusions...428
References 428

18.1 OVERVIEW OF *DE NOVO* GENOME ASSEMBLY

Genomes are the fundamental biochemical elements underlying inheritance, represented by chemical sequences of the four DNA "letters" A, C, G, and T. Genomes encode the basic software of an organism, defining the proteins that each cell can make, and the regulatory information that determines the conditions under which each protein is produced, allowing different organs and tissues to establish their distinct identities and maintain the stable existence of multicellular organisms like us. Sequences that differ by as little as one letter can cause the expression of proteins that are defective or are inappropriately expressed at the wrong time or place. These differences underlie many inherited diseases and disease susceptibility.

Each organism's genome is a specific sequence, ranging in length from a few million letters for typical bacterium to 3.2 billion letters for a human chromosome to over 20 billion letters for some plant genomes, including conifers and bread wheat. Genomes differ between species, and even between individuals within species; for example, two healthy human genomes typically differ at more than 3 million positions, and each can contain over 10 million letters that are absent in the other.

Genomes mutate between every generation and even within individuals as they grow, and some of those mutations can drive cells to proliferate and migrate inappropriately, leading to diseases such as cancer. We do not yet know which sequence differences are important but hope to learn these rules by sequencing millions of healthy and sick people, and comparing their genomes.

Determining a genome sequence "*de novo*" (i.e., without reference to a previously determined sequence for a species) is a challenging computational problem. Modern sequencing instruments can cost-effectively produce only short sequence fragments of 100–250 letters, read at random from a genome (so-called "shotgun" sequencing). A billion such short reads can be produced for around $1000, enough to redundantly sample the human genome 30 times in overlapping short fragments. The computational challenge of "genome assembly" is then to reconstruct chromosome sequences from billions of overlapping short sequence fragments, bridging a six order of magnitude gap between the length of the individual raw sequence reads and a complete chromosome.

Reconstructing a long sequence from short substrings is in general a nondeterministic polynomial time (NP) hard problem and must rely on heuristics and/or take advantage of specific features of genome sequences [1,2]. Current genome assembly algorithms typically rely on single node, large-memory (e.g., 1 TB) architectures and can take a week to assemble a single human genome, or even several months for larger genomes like loblolly pine [3]. These approaches clearly do not scale to the assembly of millions of human genomes. While some distributed-memory parallel algorithms have been developed, they do not scale to massive concurrencies as they exhibit algorithmic bottlenecks and the irregular access patterns that are inherent to these algorithms amplify the distributed-memory parallelization overheads.

The work presented in this chapter addresses the aforementioned challenges by developing parallel algorithms for *de novo* assembly with the ambition to scale to massive concurrencies. The result of this work is HipMer [4], an end-to-end high-performance *de novo* assembler designed to scale to massive concurrencies. HipMer uses (1) high-performance computing (HPC) clusters or supercomputers for both memory size and speed, (2) a global address space programming model via Unified Parallel C (UPC) [5] to permit random accesses across the aggregate machine memory, and (3) parallel graph algorithms and hash tables optimized for the statistical characteristics of the assembly process to reduce communication costs and increase parallelism. Our work is based on the Meraculous [6,7] assembler, a state-of-the-art *de novo* assembler for short reads developed at the Joint Genome Institute. Meraculous is a hybrid assembler that combines aspects of de Bruijn-graph-based assembly with overlap-layout-consensus approaches and is ranked at or near the top in most metrics of the Assemblathon II competition [8]. The original Meraculous used a combination of serial, shared-memory parallel, and distributed-memory parallel code. The size and complexity of genomes that could be assembled with Meraculous were limited by both speed and memory size. Our goal was a fast, scalable parallel implementation that could use the combined memory of a large-scale parallel machine and our work [4,9–11] has covered all aspects of the single genome assembly pipeline.

The rest of this chapter is organized as follows. Sections 18.2 through 18.4 describe the fundamental concepts that we build upon in this chapter and provide the necessary background. Section 18.5 details the parallelization in the HipMer algorithms. Section 18.6 presents performance results of HipMer on large scale. Section 18.7 highlights the main challenges for porting HipMer to manycore architectures. Section 18.8 briefly overviews related works and finally Section 18.9 concludes this chapter.

18.2 THE MERACULOUS ASSEMBLY PIPELINE

In this section, we review the Meraculous [6,7] single genome assembly pipeline and its main algorithmic components. This pipeline constitutes the basis for our parallel algorithms. Even though the description of the pipeline is specific to Meraculous, the high-level algorithmic techniques are relevant to any *de novo* genome assembler, which is based on de Bruijn graphs.

The input to the genome assembler is a set of short, erroneous sequence fragments of 100–250 letters, read at random from a genome (see Figure 18.1). Note that the genome is redundantly sampled at a depth of coverage *d*. Typically, these reads fragments come in pairs and this information will be further exploited in the pipeline. Paired reads are also characterized by the *insert size*, the distance between the two distant ends of the reads. Thus, given the read lengths and the corresponding insert size, we have an estimate for the gap between the paired reads. Typically, the reads are grouped into libraries and each library is characterized by a nominal insert size and its standard deviation. Libraries with different insert sizes play a significant role in the assembly process, as will be explained later in this section. The Meraculous pipeline consists of four major stages (see Figure 18.2a).

1. *k*-mer analysis: The input reads are processed to exclude errors. First, the reads are chopped into *k*-mers, which are overlapping sequences of length *k*. Figure 18.2b shows the *k*-mers (with *k* = 3) that are extracted from a read. Then, the *k*-mers extracted from all the reads are counted and those that appear fewer times than a threshold are treated as erroneous and discarded. Additionally, for each *k*-mer we keep track of the two neighboring bases in the original read it was extracted from (henceforth, we call these bases *extensions*). The result of *k*-mer analysis is a set of *k*-mers and their corresponding extensions that with high probability include no errors.

The redundancy *d* in the read dataset is crucial in the process of excluding the errors implicitly. More specifically, an error at a specific read location yields up to *k* erroneous *k*-mers. However, there

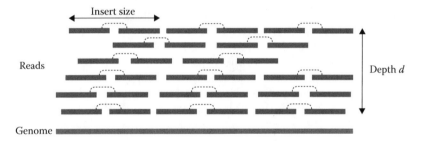

FIGURE 18.1   Reads extracted from a genome with a depth of coverage *d*.

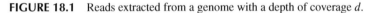

FIGURE 18.2   (a) The Meraculous assembly pipeline. (b) Extracting *k*-mers (*k* = 3) from the read GATCTGAACCG.

are more reads covering the same genome location due to the redundancy d. More precisely, given the read length L we expect to find a true, error-free k-mer on average $f = d \cdot (1 - (k-1)/L)$ times in the read dataset where f is the mean of the Poisson distribution of key frequencies [12] and most of these k-mer occurrences will be error-free. Therefore, if we find a particular k-mer just one or two times in our read dataset, then we consider that to be erroneous. On the other hand, k-mers that appear a number of times proportional to d are likely error-free.

2. Contig generation: The resulting k-mers from the previous step are stored in a de Bruijn graph. This is a special type of graph that represents overlaps in sequences. In this context, k-mers are the vertices in the graph, and two k-mers that overlap by $k-1$ consecutive bases are connected with an undirected edge in the graph (see Figure 18.3 for a de Bruijn graph example with $k = 3$).

Due to the nature of DNA, the de Bruijn graph is extremely sparse. For example, the human genome's adjacency matrix that represents the de Bruijn graph is a $3 \cdot 10^9 \times 3 \cdot 10^9$ matrix with between two and eight nonzeros per row for each of the possible extensions. In Meraculous, only k-mers which have unique extensions in both directions are considered, thus each row has exactly two nonzeros.

Using a direct index for the k-mers is not practical for realistic values of k, since there are 4^k different k-mers. A compact representation can be leveraged via a hash table: A vertex (k-mer) is a key in the hash table and the incident vertices are stored implicitly as a two-letter code [ACGT][ACGT] that indicates the unique bases that immediately precede and follow the k-mer in the read dataset. By combining the key and the two-letter code, the neighboring vertices in the graph can be identified.

In Figure 18.3, all k-mers (vertices) have unique extensions (neighbors) except from the vertex GAA that has two "forward neighbors," vertices AAC and AAT. From the previous k-mer analysis results, we can identify the vertices that do not have unique neighbors. In the contig generation step, we exclude from the graph all the vertices with nonunique neighbors. We define *contigs* as the connected components in the de Bruijn graph. Via construction and traversal of the underlying de Bruijn graph of k-mers, the connected components in the graph are identified. The connected components have linear structure since we exclude from the graph all the "fork" nodes or equivalently the k-mers with nonunique neighbors. The contigs are (with high probability) error-free sequences that are typically longer than the original reads. In Figure 18.3 by excluding the vertex GAA that doesn't have a unique neighbor in the "forward" direction, we find three linear connected components that correspond to three contigs.

3. Aligning reads onto contigs: In this step, we map the original reads onto the generated contigs. This mapping provides information about the relative ordering and orientation of the contigs and will be used in the final step of the assembly pipeline.

The Meraculous pipeline adopts a seed-and-extend algorithm in order to map the reads onto the contigs. First, the contig sequences are indexed by constructing a seed index, where the seeds are all substrings of length k that are extracted from the contigs. This seed index is then used to locate candidate read-to-contig alignments. Given a read, we extract seeds of length k, look them up in the seed index and as a result we get candidate contigs that are aligned with the read because they share

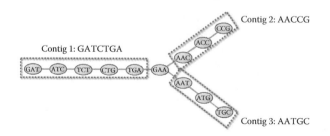

FIGURE 18.3 A de Bruijn graph of k-mers with $k = 3$.

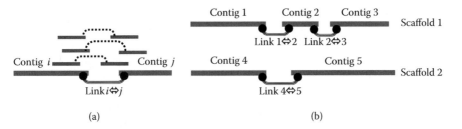

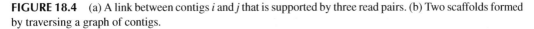

FIGURE 18.4   (a) A link between contigs *i* and *j* that is supported by three read pairs. (b) Two scaffolds formed by traversing a graph of contigs.

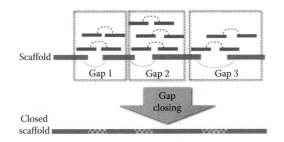

FIGURE 18.5   The gap closing procedure.

common seeds. Finally, an extension algorithm (e.g., Smith–Waterman [13]) is applied to extend each found seed and local alignments are returned as the final result.

4. Scaffolding and gap closing: The scaffolding step aims to "stitch" together contigs and form sequences of contigs called *scaffolds* by assessing the paired-end information from the reads and the reads-to-contigs alignments. Figure 18.4a shows three pairs of reads that map onto the same pair of contigs *i* and *j*. Hence, we can generate a link that connects contigs *i* and *j*. By generating links for all the contigs that are supported by pairs of reads we create a graph of contigs (see Figure 18.4b). By traversing this graph of contigs, we can form chains of contigs which constitute the scaffolds. Note that libraries with large insert sizes can be used to generate long-range links among contigs. Additionally, scaffolding can be performed in an iterative way by using links generated from different libraries at each iteration.

After the scaffold generation step, it is possible that there are gaps between pairs of contigs. We then further assess the reads-to-contigs mappings and locate the reads that are placed into these gaps (see Figure 18.5). Ultimately, we leverage this information and close the contig gaps by performing a mini-assembly algorithm involving only the localized reads for each gap. The outcome of this step constitutes the result of the Meraculous assembly pipeline.

In Sections 18.3 through 18.5, we examine the programming model, the main distributed data structure, and the parallel algorithms that are employed in the HipMer pipeline.

18.3   THE PARTITIONED GLOBAL ADDRESS SPACE MODEL IN UPC

The Partitioned Global Address Space (PGAS) programming model is employed in parallel programming languages. In this model, any thread is allowed to directly access memory on other threads. In the PGAS model, two threads may share the same physical address space or they may own distinct physical address spaces. In the former case, remote-thread accesses can be done directly using load and store instructions, while in the latter case, a remote access must be translated into a communication event, typically using a communication library such as GASNet [14] or hardware specific layers such as Cray's DMAPP [15] or IBM's PAMI [16].

An alternative communication mechanism typically employed in parallel programming languages is message passing, where the communication is done by exchanging messages between threads (e.g., see the Message Passing Interface [MPI] [17]). In such a communication model, both the sender and the receiver should explicitly participate in the communication event and therefore requires coordinating communication peers to avoid deadlocks. The programmer's burden in such a two-sided communication model can be further exaggerated in situations where the communication patterns are highly irregular as in distributed hash table construction. On the other hand, the PGAS model requires the explicit participation only of the peer that initiates the communication and as a result parallel programs with irregular accesses are easier to implement. Such a communication mechanism is typically referred to as *one-sided communication*. In addition to PGAS languages like UPC [5], there are programming libraries such as SHMEM [18] and MPI 3.0 [19] with one-sided communication features.

UPC is an extension of the C programming language designed for HPC on large-scale parallel machines by leveraging a PGAS communication model. UPC utilizes a single program multiple data (SPMD) model of computation in which the amount of parallelism is fixed at program startup time. On top of its one-sided communication capabilities, UPC provides global atomics, locks, and collectives that facilitate the implementation of synchronization protocols and common communication patterns. In short, UPC combines the programmability advantages of the shared-memory programming paradigm and the control over data layout and performance of the message passing programming paradigm. According to the memory model of UPC, each thread has a portion of shared and private address space. Variables that reside in the shared space can be directly accessed by any other thread and typically the program should employ synchronization mechanisms in order to avoid race conditions. On the other hand, variables that live in the private space can be read and written only by the thread owning that particular private address space.

Overall, the global address space model and the one-sided communication capabilities of UPC simplify the implementation of distributed data structures and highly irregular communication patterns. Such communication patterns are ubiquitous in our parallel algorithms described in Section 18.5.

18.4   DISTRIBUTED HASH TABLES IN A PGAS MODEL

A common data structure utilized in subsequent parallel algorithms is the distributed hash table. There is a wide body of work on concurrent hash tables [20–26] that focuses on shared-memory architectures. There is also a lot of work on distributed hash tables (see [27,28] and survey of Zhang et al. [29]) specially designed for large-scale distributed environments that support primitive `put` and `get` operations. Such implementations do not target dedicated HPC environments and therefore have to deal with faults, malicious participants, and system instabilities. Such distributed hash tables are optimized for execution on data centers rather than HPC systems with low-latency and high-throughput interconnects. There are some simple distributed-memory implementations of hash tables in MPI [30] and UPC [31], but they are used mainly for benchmarking purposes of the underlying runtime and do not optimize the various operations depending on the use case of the hash table. In this section, we describe the basic implementation of a distributed hash table using a PGAS abstraction. We also identify a handful of use cases for distributed hash tables that enable numerous optimizations for HPC environments.

18.4.1   Basic Implementation of a Distributed Hash Table

We present the vanilla implementation of a distributed hash table by following an example of a distributed de Bruijn graph. Figure 18.6a shows a de Bruijn graph of k-mers with $k = 3$ and Figure 18.6b

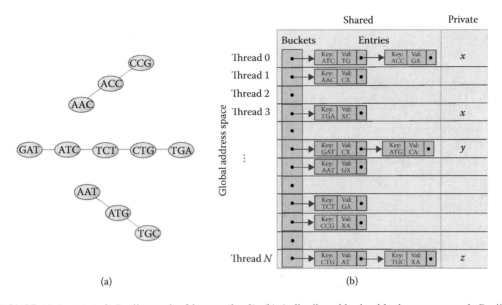

FIGURE 18.6 (a) A de Bruijn graph of k-mers ($k = 3$). (b) A distributed hash table that represents de Bruijn graph at left.

illustrates its representation in a distributed hash table. A vertex (k-mer) in the graph is a key in the hash table and the incident vertices are stored implicitly as a two-letter code [ACGTX][ACGTX] that indicates the unique bases that follow and precede that k-mer. This two-letter code is the value member in a hash table entry. Note that the character X indicates that there is no neighboring vertex in that direction. By combining the key and the two-letter code, the neighboring vertices in the graph can be identified. More specifically, by concatenating the last $k - 1$ letters of a key and the first letter of the value, we get the "forward" neighboring vertex. Similarly, by concatenating the second letter of the value and the first $k - 1$ letters of that key, we get the "preceding" neighboring vertex.

In our example, all the hash table entries are stored in the shared address space and thus they can be accessed by any thread. The buckets are distributed to the available threads in a cyclic fashion to achieve load balance. Our hash table implementation utilizes a chaining rule to resolve collisions in the buckets (entries with the same hash value). We emphasize here that the hash tables involved in our algorithms can be gigantic (hundreds of gigabytes up to tens of terabytes) and cannot fit in a typical shared-memory node. Therefore, it is crucial to distribute the hash table buckets over multiple nodes and in this quest the global address space of UPC is convenient.

In Section 18.4.2, we list different use-case scenarios of distributed hash tables. These use-case scenarios are encountered in our parallel algorithms described in Section 18.5 and are presented up front in order to highlight the optimization opportunities.

18.4.2   Use Cases of Distributed Hash Tables in the HipMer Pipeline

Here we identify a handful of use cases for the distributed hash tables that allow specific optimizations in their implementation. These use cases will be used as points of reference in the section that details our parallel algorithms.

Use Case 1—Global Update-Only (GUO) phase: The operations performed in the distributed hash table are only global updates with commutative properties (e.g., inserts only). The global hash table will have the same state regardless of insert order, although it might possibly have different underlying representation due to chaining. The GUO phase can be optimized by dynamically aggregating fine-grained updates (e.g., inserts) into batch updates. In this way, we can reduce the number of messages and synchronization events. We can also overlap computation/communication or pipeline communication events to further hide the communication overhead.

A typical example of such a use case is a producer/consumer setting where the producers operate in a distinct phase from consumers, for example, all consumers insert items in a hash table before anything is consumed/read.

Use Case 2—Global Reads and Writes (GRW) phase: The operations performed during this phase are global reads and writes over the *already inserted entries*. Typically, we can't batch reads and/or writes since there might be race conditions that affect the control flow of the governing parallel algorithm. However, we can use global atomics (e.g., compare-and-swap) instead of fine-grained locking in order to ensure atomicity. The global atomics might employ hardware support depending on the platform and the corresponding runtime implementation. We can also build synchronization protocols at a higher level that do not involve the hash table directly but instead are triggered by the results of the atomic operations. Finally, we can implement the delete operation of entries with atomics and avoid locking schemes.

For example, consider the consumers in a producer/consumer scenario that compete for the entries of the hash table. The entries may have utilization signatures (i.e., "used" binary flags) that can be accessed via global atomics and indicate whether the corresponding entries have been consumed or not. An orthogonal optimization for this use-case scenario is to adopt locality sensitive hashing schemes to increase locality and decrease communication volume/latency overhead of global atomics.

Use Case 3—Global Read-Only (GRO) phase: In such a use case, the entries of the distributed hash table are read-only and a degree of data reuse is expected. The optimization that can be readily employed is to design software caching schemes to take advantage of data reuse and minimize communication. These caching frameworks can be viewed as "on demand" copying of remote parts of the hash table. Note that the read-only phase guarantees that we do not need to provision for consistency across the software caches. Such caching optimizations can be used in conjunction with locality-aware partitioning to increase effectiveness of the expected data reuse. Initially even if the data is remote, it is likely to be reused later locally.

A typical example of this use case is a lookup-only hash table that implements a database/index. This is a special case of the consumer side in a producer/consumer setting where the entries can be consumed an infinite number of times.

Use Case 4—Local Reads and Writes (LRW) phase: In this use case, the entries in the hash table will be further read/written only by the processor owning them. The optimization strategy we employ in such a setting is to use a deterministic hashing from the sender side and local hash tables on the receiver side. The local hash tables ensure that we avoid runtime overheads. Additionally, high-performance, serial hash table implementations can be seamlessly incorporated into parallel algorithms.

For example, consider items that are initially scattered throughout the processors and we want to send occurrences of the same item to a particular processor for further processing (e.g., consider a "word-count" type of task). Each processor can insert the received items into a local hash table and further read/write the local entries from there.

We emphasize that this is not an exhaustive list of use cases for distributed hash tables. Nevertheless, it captures the majority of the computational patterns we identified in our parallel algorithms that are detailed in Section 18.5. Table 18.1 summarizes the optimizations we can employ for the various use cases of the distributed hash tables. Multiple of the aforementioned use cases can be encountered during the lifetime of a distributed hash table; in most of the cases, the optimizations can be easily composed (e.g., by having semantic barriers to signal the temporal boundaries of the phases). For example, the GUO phase can be followed by a GRO phase in a scenario where a database is first built via insertion of the corresponding items into a hash table and later the distributed data structure is reused as a global lookup table.

TABLE 18.1
Distributed Hash Table Optimizations for Various Use-Case Scenarios

Use Case	Dynamic Message Aggregation	Remote Global Atomics	Caching of Remote Entries	Locality Sensitive Hashing	Serial Hash Table Library
GUO	✓	✓		✓	
GRW		✓		✓	
GRO			✓	✓	
LRW	✓			✓	✓

18.5 PARALLEL ALGORITHMS IN HipMer

In this section, we detail the parallelization of the Meraculous pipeline presented in Section 18.2. In our description, we refer to ideas from Sections 18.3 and 18.4 regarding the PGAS programming model and distributed hash tables.

18.5.1 PARALLEL *k*-MER ANALYSIS

Counting the frequencies of each distinct *k*-mer involves reading the input DNA short reads, parsing the reads into *k*-mers, and keeping a count of each distinct *k*-mer that occurs more than ϵ times ($\epsilon \approx 1, 2$). The reason for such a cutoff is to eliminate sequencing errors. *k*-mer analysis additionally requires keeping track of all possible extensions of the *k*-mer from either side. This is performed by keeping two short integer arrays of length four per *k*-mer, where each entry in the array keeps track of the number of occurrences of each nucleotide [ACGT] on either end. If a nucleotide on an end appears more times than a threshold t_{hq}, it is characterized as high-quality extension. One of the difficulties with performing *k*-mer analysis in distributed memory is that the size of the intermediate data (the set of *k*-mers) is significantly larger than both the input and the output, since each read is subsequenced with overlaps of $k - 1$ base pairs.

As each processor reads a portion of the reads and extracts the corresponding *k*-mers, a deterministic map function maps each *k*-mer to a processor id. Once the *k*-mers are generated, we perform an irregular all-to-all exchange step in order to communicate the *k*-mers among the processors based on the calculated processor ids. This deterministic mapping assigns all the occurrences of a particular sequence to the same processor, thus eliminating the need for a global hash table; instead, each processor maintains a local hash table to count the occurrences of the received *k*-mers. We refer to this model of computation as "owner-computes." Note that this computational pattern fits the Use Case 4 (LRW) of the distributed hash tables. Given the genome size G, the coverage d, and the read length L, the total number of *k*-mers that have to be communicated is $\Theta(\frac{Gd}{L}(L - k + 1))$.

In this parallel algorithm, memory consumption quickly becomes a problem due to errors because a single-nucleotide error creates up to k erroneous *k*-mers. It is not uncommon to have over 80% of all distinct *k*-mers erroneous, depending on the read length and the value of *k*. We ameliorate this problem using Bloom filters, which were previously used in serial *k*-mer counters [32]. A Bloom filter [33] is a space-efficient probabilistic data structure used for membership queries. It might have false positives, but no false negatives. If a *k*-mer was not seen before, the filter can accidentally report it as "seen." However, if a *k*-mer was previously inserted, the Bloom filter will certainly report it as "seen." This is suitable for *k*-mer counting as no real *k*-mers will be missed. If the Bloom filter reports that a *k*-mer was seen before, then the corresponding processor inserts that *k*-mer to the actual local hash table in order to perform the counting. Our novelty is the discovery that localization of *k*-mers via the deterministic *k*-mer to processor id mapping is *necessary and sufficient* to extend the benefits of Bloom filters to distributed memory.

The false positive rate of a Bloom filter is $\Pr(e) = (1 - e^{-hn/m})^h$ for m being the number of distinct elements in the dataset, n the size of the Bloom filter, and h the number of hash functions used. There is an optimal number of hash functions given n and m, which is $h = \ln 2 \cdot (m/n)$. In practice, we achieve approximately 5% false positive rate using only 1%–2% of the memory that would be needed to store the data directly in a hash table (without the Bloom filter). Hence, in a typical dataset where 80% of all k-mers are errors, we are able to filter out 76% of all the k-mers using almost no additional memory. Hence, we can effectively run a given problem size on a quarter of the nodes that would otherwise be required.

We have so far ignored that Bloom filters need to know the number of distinct elements expected to perform optimally. While dynamically resizing a Bloom filter is possible, it is expensive to do so. We therefore use a cardinality estimation algorithm to approximate the number of distinct k-mers. Specifically, we use the Hyperloglog algorithm [34], which achieves less than $1.04/\sqrt{m}$ error for a dataset of m distinct elements. Hyperloglog requires a only several kilobytes of memory to count trillions of items. The basic idea behind cardinality estimators is hashing each item uniformly and maintaining the minimum hash value. Hyperloglog maintains multiple buckets for increased accuracy and uses the number of trailing zeros in the bitwise representation of each bucket as the estimator.

The observation that leads to minimal communication parallelization of Hyperloglog is as follows. Merging Hyperloglog counts for multiple datasets can be done by keeping the minimum of their final buckets by a parallel reduction. Consequently, the communication volume for this first cardinality estimation pass is *independent of the size of the sequence data* and is only a function of the Hyperloglog data structure size. In practice, we implement a modified version of the algorithm that uses 64-bit hash values as the original 32-bit hash described in the original study [34] is not able to process our massive datasets.

One downside of this parallel counting approach is that highly complex plant genomes, such as wheat, are extremely repetitive and it is not uncommon to see k-mers that occur millions of times. Such high-frequency k-mers create a significant load imbalance problem, as the processors assigned to these high-frequency k-mers require significantly more memory and processing times. We improve our approach by first identifying frequent k-mers (i.e., "heavy hitters" in database literature) and treating them specially [4]. In particular, the "owner-computes" method is still used for low-to-medium-frequency k-mers but the high-frequency k-mers are accumulated locally on each processor, followed by a final global reduction. Since an initial pass over the data is already performed to estimate the cardinality (the number of distinct k-mers) and efficiently initialize our Bloom filters, running a streaming algorithm for identifying frequent k-mers during the same pass is essentially free.

18.5.2 PARALLEL CONTIG GENERATION

Once we have performed the k-mer analysis step, it is necessary to store the resulting k-mers in a distributed hash table that represent the de Bruijn graph in a compact way. A vertex (k-mer) is a key in the hash table and the incident vertices are stored implicitly as a two-letter code [ACGT][ACGT] that indicates the unique bases that immediately precede and follow the k-mer in the read dataset. These graphs typically are huge and require hundreds of gigabytes or even terabytes for large genomes in order to be stored in memory. Therefore, we employ the global address space of UPC in order to transparently store the distributed hash table in distributed memory and overcome the limitations of requiring specialized, large shared-memory machines.

During the parallel hash table construction, we consider only the k-mers that have unique high-quality extensions in *both directions*. These k-mers are hashed and sent to the proper (potentially remote) bucket of the hash table by leveraging the one-sided communication capabilities of UPC. We recognize this computational pattern as the Use Case 1 (GUO) of the distributed hash tables, therefore we can mitigate the communication and synchronization overheads by leveraging dynamic message aggregation. In particular, we designed a dynamic aggregation algorithm [9] where the k-mers are aggregated in batches before being sent to the appropriate processors. The pattern deployed here is also an irregular all-to-all communication. However, unlike k-mer analysis, the total number of k-mers that have to be communicated is $\Theta(G)$, since multiple occurrences of k-mers have been collapsed during the k-mer analysis stage and this condensed k-mer set should have size proportional to the genome size G.

Once the distributed de Bruijn graph (hash table) has been constructed, we traverse it in parallel and identify the connected components that represent the *contig* sequences. Typically such de Bruijn graphs have extremely high-diameter (the connected components in theory can have size up to the length of chromosomes) and therefore traditional parallelization strategies of the graph traversal would not scale to extreme concurrencies.

In order to form a contig, a processor p_i chooses a random k-mer from its own part of the distributed hash table as seed and creates a new *subcontig* (incomplete contig) data structure which is represented as a string and the initial content of the string is the seed k-mer. Processor p_i then attempts to extend the subcontig toward both of its endpoints using the high-quality extensions stored as values in the distributed hash table. To extend a subcontig from its right endpoint, processor p_i uses the $k-1$ last bases and the right high-quality extension R from the right-most k-mer in the subcontig. It therefore concatenates the last $k-1$ bases and the extension R to form the next k-mer to be searched in the hash table. Processor p_i performs a lookup for the newly formed k-mer and if it is found successfully, the subcontig is extended to the right by the base R. The same process can be repeated until the lookup in the hash table fails, meaning that there are no more UU k-mers that could extend this subcontig in the right direction. A subcontig can be extended to its left endpoint using an analogous procedure. If processor p_i cannot add more bases to either endpoint of the subcontig, then a contig has been formed (or equivalently a connected component in the de Bruijn graph has been explored) and is stored accordingly.

Figure 18.7 illustrates how the parallel algorithm works with three processors. Processor 0 picks a random traversal seed (vertex CTG) and initializes a subcontig with content CTG. Then, by looking

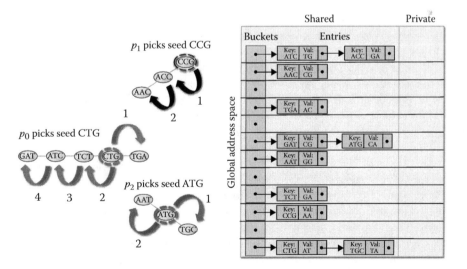

FIGURE 18.7   Parallel de Bruijn graph traversal. Processor 0 picks a k-mer called "traversal seed" (vertex CTG) and with four lookups in the distributed hash tables it explores the four remaining vertices of that connected component. The numbered arrows indicate the order in which processor 0 looks up the corresponding vertices in the distributed hash table. In an analogous way, processors 1 and 2 pick seeds CCG and ATG, respectively, and explore in parallel with processor 0 different connected components of the underlying de Bruijn graph.

in the distributed hash table the entry CTG it gets back the value AT, meaning that the right extension is A and the left extension is T. After that, processor 0 forms the next k-mer to be looked up (TGA) by concatenating the last two bases of CTG and the right extension A—this lookup corresponds to the arrow 1 of processor 0. By following the analogous procedure and three more lookups in the distributed hash table, processor 0 explores all the vertices of that connected component that corresponds to the contig GATCTGA. The numbered arrows indicate the order in which processor 0 looks up the corresponding vertices in the distributed hash table. In an analogous way, processors 1 and 2 pick seeds CCG and ATG, respectively, and explore in parallel with processor 0 different connected components of the underlying de Bruijn graph.

All processors independently start building subcontigs and no synchronization is required unless two processors pick initial k-mer seeds that eventually belong in the same contig. In this case, the processors have to collaborate and resolve this conflict in order to avoid redundant work and race conditions. The high-level idea of the synchronization protocol for conflict resolution is that one of the involved processors backs off, and the other processor takes over the computed "subcontig" from the processor that backed off. We designed a lightweight synchronization scheme [9] based on remote atomics and in our previous work we proved (under some modeling assumptions) that our synchronization algorithm is scalable to massive concurrencies. Finally, the parallel traversal is terminated when all the connected components in the de Bruijn graph are explored.

The access pattern in the distributed hash table consists of highly irregular, fine-grained lookup operations. The size of the de Bruijn graph is proportional to the genome size, thus the traversal involves visiting $\Theta(G)$ vertices via atomics and irregular lookup operations. The computational task of the graph traversal is to visit all the already inserted k-mers in the distributed hash table. During this parallel procedure, we cannot batch reads and/or writes since there might be race conditions that affect the control flow of the synchronization algorithm. However, we use global atomics instead of fine-grained locking and we build synchronization protocols at a higher level that do not involve the distributed hash table directly but instead are triggered by the results of the atomic operations on the objects stored inside the hash table. We recognize this computational pattern as the Use Case 2 (GRW) of the distributed hash tables.

18.5.3 PARALLEL READ-TO-CONTIG ALIGNMENT

Here we describe the parallel algorithm that maps the original reads onto the contig sequences. First, each processor reads a distinct portion of the contig sequences and stores them in global address space such that any other processor can access them. Every contig sequence of length C contains $C - k + 1$ distinct seeds (substrings) of length k. We extract in parallel seeds from the contigs and associate with every seed the contig from which it was extracted. Since the contigs constitute a fragmented version of the genome, in total we extract $\Theta(G)$ seeds.

Once the seeds are extracted from the contigs, they are stored in a global hash table, referred to as the seed index where the key is a seed and the value is a pointer to the contig from which this seed has been extracted. The seed index is distributed and stored in global shared memory such that any processor can access and lookup any seed. Essentially, the seed index data structure provides a mapping from seeds to contigs. The seeds are stored in the global seed index via an irregular all-to-all communication step similar to the hash table construction in the contig generation phase. Again, we recognize this computational pattern as the Use Case 1 (GUO) of the distributed hash tables, therefore we can mitigate the communication and synchronization overheads by leveraging dynamic message aggregation. Figure 18.8 illustrates how two contigs are indexed by using seeds with length $k = 3$.

After the seed index construction, we proceed with the aligning phase where every read is mapped onto contigs. Initially, each processor is assigned an equal number of reads. For each read of length L, a processor extracts all $L - k + 1$ seeds of length k contained in it. Given a seed s from a read, the processor performs a fine-grained lookup in the global seed index and locates the candidate

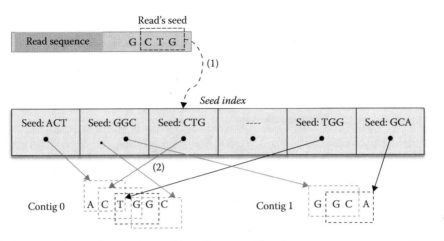

FIGURE 18.8   Locating read-to-contig candidate alignments. First, the processor extracts a seed from the read sequence (CTG seed). Next, the processor looks up the distributed seed index (arrow 1) and finds that a candidate contig sequence is Contig 0 (arrow 2). Finally, the Smith–Waterman algorithm is executed using as inputs the read and the Contig 0 sequences.

contigs that have in common the seed s with that read. Thus, each one of the read-to-contig candidate alignments can be located in $\Theta(1)$ time. Figure 18.8 exhibits an example of how we can locate a read-to-contig candidate alignment by leveraging the seed index. We emphasize that in the alignment phase, all processors operate concurrently on distinct reads.

If we naively execute an exhaustive lookup of all possible seeds, in total we have to perform $O(\frac{Gd}{L}(L-k+1))$ lookups. Our optimized parallel algorithm though [10] identifies properties in the contigs during the hash index construction that reduce significantly the number of lookups.

We also made the observation that our parallel alignment phase makes no writes/updates in the distributed seed index or the distributed data structure that stores the contig sequences after their construction phase; it just uses them for lookups/reads. We recognize this computational pattern as the Use Case 3 (GRO) of the distributed hash tables and our parallel algorithm [10] exploits software caches to maximize data reuse and avoids off-node lookups.

Finally, after locating a candidate contig that has a matching seed with the read we are processing, the Smith–Waterman algorithm is executed with input the read and contig sequences in order to perform local sequence alignment. The output of this stage is a set of reads-to-contig alignments.

18.5.4   PARALLEL SCAFFOLDING AND GAP CLOSING

The first part of scaffolding involves processing of the reads-to-contig alignments (Figure 18.4a) and generating links among contigs. In order to parallelize this operation, we index only the relevant alignments (those that indicate that two contigs should be connected) via a distributed hash table. This distributed hash table construction employs an irregular all-to-all communication pattern similar to the contig generation stage (Use Case 1 of distributed hash tables). We emphasize that the graph of contigs (and consequently the number of links among them) is orders of magnitude smaller than the k-mer de Bruijn graph because the connected components in the k-mer graph are now contracted to single vertices in the contig-graph. According to the Lander–Waterman statistics [35], the expected number of contigs is $\Theta(dG/L \cdot e^{-d})$.

Then, we process the contigs to identify properties (e.g., average k-mer depth, termination states) that will help us further simplify the contig-graph. This step necessitates looking up k-mer info in a global hash table of k-mers with $\Theta(G)$ size. Afterward we introspect the contig-graph to identify bubble structures via a parallel traversal which requires irregular lookups in the distributed contig-graph representation and global atomics (Use Case 2 of the distributed hash tables). After

TABLE 18.2
Major Communication Operations in the HipMer Pipeline

Stage	Communication Pattern	Volume of Data
k-mer analysis	All-to-all exchange	$\Theta(Gd \cdot (L - k + 1)/L)$
Contig	All-to-all exchange	$\Theta(G)$
generation	Irregular lookups	$\Theta(G)$
	Global atomics	$\Theta(G)$
Sequence	All-to-all exchange	$\Theta(G)$
alignment	Irregular lookups	$O(Gd \cdot (L - k + 1)/L)$
	All-to-all exchange	$\Theta(G)$
Scaffolding	Irregular lookups	$\Theta(G)$
	Global atomics	$\Theta(dG/L \cdot e^{-d})$
Gap closing	All-to-all exchange	$\Theta(\gamma Gd/L)$

the bubble removal step, we traverse the simplified graph and generate scaffolds (Figure 18.4b). The last traversal is done by selecting starting vertices in order of decreasing length (this heuristic tries to stitch together first "long" contigs) and therefore it is inherently serial. We have optimized this component and found that its execution time is insignificant compared to the previous pipeline operations—this behavior is expected as the input contig-graph is relatively small as explained earlier.

The gap closing stage uses the read-to-contig alignments, the scaffolds, and the contigs to attempt to assemble reads across gaps between the contigs of scaffolds (see Figure 18.4b). To determine which reads map to which gaps, the alignments are processed in parallel and projected into the gaps. We utilize a distributed hash table to localize the unassembled reads onto the appropriate gaps via irregular all-to-all communication. Assuming that a fraction γ of the genome is not assembled into contigs, then this communication step involves $\Theta(\gamma Gd/L)$ reads. Finally, the gaps are divided into subsets and each set is processed by a separate processor, in an embarrassingly parallel phase.

18.5.5 SUMMARY OF COMMUNICATION PATTERNS AND COSTS

Given the genome size G, the read length L, the coverage d, and the fraction γ of the reads that are not assembled into contigs, Table 18.2 summarizes for each stage the main communication patterns along with the corresponding volume of communication. These communications patterns govern the efficiency of the parallel pipeline at large scale, where most of the stages are communication-bound. The different communication patterns have, however, vastly different overheads. For example, the all-to-all communication exchange is typically bounded by the bisection bandwidth of the system assuming that the partial messages are large enough and there is enough concurrency to saturate the available bandwidth. On the other hand, fine-grained, irregular lookups, and global atomics are typically latency-bound and their efficiency relies upon the ability of the interconnect to serve those fine-grained, irregular request efficiently at high concurrencies.

18.6 PERFORMANCE RESULTS

Parallel performance experiments are conducted on Edison, the Cray XC30 located at NERSC. Edison has a peak performance of 2.57 Pflops/sec, with 5576 compute nodes, each equipped with 64 GB RAM and two 12-core 2.4 GHz Intel Ivy Bridge processors for a total of 133,824 compute cores, and interconnected with the Cray Aries network using a Dragonfly topology. For our experiments, we use Edison's parallel Lustre /scratch3 file system, which has 144 Object Storage

Servers providing 144-way concurrent access to the I/O system with an aggregate peak performance of 72 GB/sec.

To analyze HipMer performance behavior, we examine a human genome for a member of the CEU HapMap population (identifier NA12878) sequenced by the Broad Institute. The genome contains 3.2 Gbp (billion base pair) assembled from 2.9 billion reads (290 Gbp of sequence), which are 101 bp in length, from a paired-end insert library with mean insert size 395 bp. Additionally, we examine the grand-challenge hexaploid wheat genome (*Triticum aestivum* L.) containing 17 Gbp from 2.3 billion reads (477 Gbp of sequence) for the homozygous bread wheat line "Synthetic W7984." Wheat reads are 150–250 bp in length from five paired-end libraries with insert sizes 240–740 bp. Also, for the scaffolding phase, we leveraged (in addition to the previous libraries) two long-insert paired-end DNA libraries with insert sizes 1 and 4.2 kbp. This important genome was only recently sequenced for the first time [36] and requires high-performance analysis due to its size and complexity.

18.6.1 STRONG SCALING EXPERIMENTS

Figures 18.9 and 18.10 show the end-to-end strong scaling performance of HipMer (including I/O) with the human and the wheat datasets, respectively, on the Edison supercomputer. For the human dataset at 15,360 cores, we achieve a speedup of 11.9 times over our baseline execution (480 cores). At this extreme scale, the human genome can be assembled from raw reads in just ≈8.4 minutes. On the complex wheat dataset, we achieve a speedup up to 5.9 times over the baseline of 960 core execution, allowing us to perform the end-to-end assembly in 39 minutes when using 15,360 cores. In the end-to-end experiments, a significant fraction of the execution time is spent in parallel sequence alignment, scaffolding, and gap closing (e.g., 68% for human at 960 cores); *k*-mer analysis requires less runtime (28% at 960 cores) and contig generation is the least expensive computational component (4% at 960 cores).

The *k*-mer analysis and the contig generation steps scale efficiently for both datasets up to 15,360 cores, while the *combined* step of alignment, scaffolding, and gap closing exhibits better scaling on the human dataset. Even though the alignment and gap closing modules for the wheat dataset exhibit similar scaling to the human test case, the scaffolding step consumes a significantly higher fraction of the overall runtime. There are two main reasons for this behavior. First, the highly repetitive nature

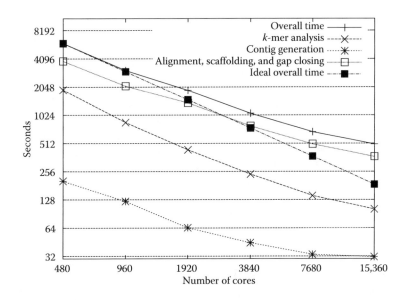

FIGURE 18.9 End-to-end strong scaling for the human genome. Both axes are in log scale.

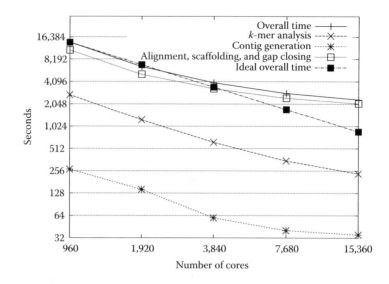

FIGURE 18.10   End-to-end strong scaling for the wheat genome. Both axes are in log scale.

of the wheat genome leads to increased fragmentation of the contig generation compared with the human DNA, resulting in contig graphs that are contracted by a smaller fraction. Hence, the serial component of the scaffolding module requires a relatively higher overhead compared with the human dataset. Second, the execution of the wheat pipeline as performed in our previous work [37] requires four rounds of scaffolding with libraries of different insert sizes, resulting in even more overhead within the serial module.

The strong scaling results presented here contradict the conventional wisdom that algorithms with highly irregular accesses (like *de novo* genome assembly) are prohibitive for distributed-memory systems. We showed that as long as the parallel algorithms are highly scalable and do not exhibit algorithmic/serialization bottlenecks, they perform fewer irregular operations on the critical path as the concurrency increases, therefore decreasing eventually the overall execution time.

18.6.2   I/O CACHING

Our modular design of the pipeline enables flexible configurations that can be adapted appropriately to meet the requirements of each assembly. For instance, one might want to perform multiple rounds of scaffolding to facilitate the assembly of highly repetitive regions or to iterate over the k-mer analysis step and contig generation multiple times (with varying k and other parameters) in order to extract information that is latent within a single iteration. These configurations imply that the input read datasets should be loaded multiple times. Even in a typical, single pass execution of the pipeline, the reads constitute the input to multiple stages, namely k-mer analysis, alignment, and gap closing. Reloading the reads multiple times from the parallel file system, imposes a potential I/O bottleneck for the pipeline. However, at scale we have the unique opportunity to cache the input reads and all the intermediate results onto the aggregate main memory, thus avoiding the excessive I/O and concurrent file system accesses. In order to achieve the I/O caching in a transparent way, we leverage the POSIX shared-memory infrastructure and thus all the subsequent input loads are streamed through the main memory.

Figure 18.11 shows the end-to-end strong scaling performance of HipMer on the human dataset up to 23,040 Edison cores. We present this experiment in order to highlight the importance of the I/O caching. Note that the baseline concurrency is 1920 cores; we need at least 80 Edison nodes, each with 24 cores, to fit all the data structures *and* cache the input datasets in memory ($\approx$5 TB). The line

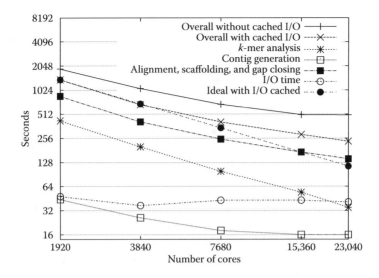

FIGURE 18.11 Strong scaling of the human dataset with I/O caching. Both axes are in log scale.

with × ticks shows the end-to-end execution time *including* the I/O, which is cached in main memory once the input reads are loaded. The ideal strong scaling is illustrated by the line with solid circles. At the concurrency of the 23,040 cores, we completely assemble the human genome in 3.91 minutes and obtain a strong scaling efficiency of 48.7% relative to the baseline of 1,920 cores. In order to illustrate the effectiveness of the I/O caching, we performed the same end-to-end experiments where the input reads are loaded from the Lustre file system five times (solid line). This experiment does not exhibit any scaling from 15,360 to 23,040 cores due to the I/O overhead, thus demonstrating that I/O caching is crucial for scaling to massive concurrencies. At the scale of 23,040 cores, the version with I/O caching is almost two times faster than the version without this optimization.

The efficiency of the I/O (reading the input reads once) is illustrated by the line with empty circles. We observe that the I/O is almost a flat line across the concurrencies and yields a read bandwidth of ≈16 GB/sec (the theoretical peak of our Lustre file system is 48 GB/sec). With 80 Edison nodes, we are able to saturate the available parallelism in the Lustre file system and further increasing the concurrency does not help to improve the I/O performance. The lines with ∗, □, ■ ticks show the partial execution time for (1) the *k*-mer analysis, (2) contig generation, and (3) sequence alignment, scaffolding, and gap closing, respectively. We conclude that all the components scale up to 23,040 cores and do not impose any scalability impediments.

18.6.3 PERFORMANCE COMPARISON WITH OTHER ASSEMBLERS

To compare the performance of HipMer relative to existing parallel *de novo* end-to-end genome assemblers, we evaluated Ray [38,39] (version 2.3.0) and ABySS [40] (version 1.3.6) on Edison using 960 cores. Ray required 10 hours and 46 minutes for an end-to-end run on the human dataset. ABySS, on the other hand, took 13 hours and 26 minutes solely to get to the end of contig generation. The subsequent scaffolding steps are not distributed-memory parallel. At this concurrency on Edison, HipMer is approximately 13 times faster than Ray and *at least* 16 times faster than ABySS.

18.7 CHALLENGES FOR FUTURE ARCHITECTURES

With the advent of exascale computing architectures expected within the next few years, many challenges arise into porting efficiently the HipMer *de novo* assembly pipeline to larger and more complex systems. In this section, we briefly describe these challenges and their implications for highly irregular algorithms, like our *de novo* assembly pipeline, and the underlying runtime support.

The architectural trends dictate that the degree of parallelism within the system's node will be increased considerably compared to contemporary supercomputing systems. For instance, the Edison supercomputer (used for our experimental evaluation) is equipped with 24-core nodes, while NERSC's newest supercomputer, named Cori, features Intel Xeon Phi "Knight's Landing" nodes, each having 68 cores. We expect this trend to hold on the way to exascale. Also, the number of nodes on exascale systems is expected to rise significantly. The combination of the increased number of cores per node and the large number of nodes will yield an unprecedented level of parallelism that should be exploited by the algorithms. In such a massively parallel environment, it is crucial to adopt asynchronous algorithmic approaches that do not suffer from load imbalance and system performance fluctuations. The parallel hash table construction and the parallel de Bruijn graph traversal algorithms described in Section 18.5.2 are examples of such asynchronous algorithms that do not exhibit synchronization bottlenecks on the critical path of execution. On the other hand, parallel algorithms which rely on bulk synchronous communication will most likely be inadequate for applications with highly irregular accesses.

In Section 18.5, we highlighted the different communication patterns that are stressed throughout the HipMer pipeline, namely all-to-all exchanges, irregular lookups, and global atomics. Accommodating these communication operations efficiently as the system size increases is critical into porting HipMer to exascale architectures. More specifically, the all-to-all exchange primitives should be mapped efficiently on the underlying network topologies in order to maximize the attainable bandwidth, and ideally should avoid excessive synchronization. Additionally, the operations which are latency-bound like the irregular lookups and the global atomics should exploit efficient protocols and routing algorithms that avoid hot spots on a large-scale system. Furthermore, taking advantage of network capabilities like remote direct memory access (RDMA) and hardware atomics will play a tremendous role in obtaining low-latency and low-overhead communication primitives. The aforementioned communication optimizations would be preferably applied at the runtime level and therefore could be seamlessly employed at the HipMer application level.

All the parallel algorithms in Section 18.5 are detailed in the context of a flat SPMD execution model, where each UPC thread is mapped onto a compute core of the system. However, the way these UPC threads are instantiated during execution time has implications for the overall performance and the memory footprint of the runtime. For instance, one could use one process per UPC thread; alternatively, one could opt for hierarchical approaches where multiple UPC threads are mapped onto a single process. Both approaches have advantages and disadvantages, but with the arrival of exascale it is imperative to take into account the scale of the systems and reevaluate the design space. Designs where the runtime's data structures scale in size and complexity quadratically with the number of nodes and/or cores per node are prohibitive. With this in mind, it is more likely that highly optimized hierarchical designs would be suitable for runtimes that target exascale systems. One could additionally adopt analogous hierarchical strategies at the HipMer's application level. However, dealing with this issue upfront at the runtime/communication library level would provide a more robust ecosystem and make the HipMer codebase more portable.

Even though the performance of the HipMer pipeline is mostly dominated by communication and subsequently by the way the communication is orchestrated within the parallel algorithms, it is crucial to optimize the core computations for the underlying architectures. Such computations include mostly string operations (e.g., k-mer extraction, reverse complementation of sequences, local alignment of sequences, string comparisons) and calculations of hash values. These computations can take advantage of vectorization and hence it is important to leverage vectorized implementations of these core computations throughout the pipeline. This necessity is even more emphasized within the context of the current architectural trends, where the single-thread performance heavily depends on efficient utilization of the vector units.

The process of porting our assembly pipeline to exascale can be tackled on multiple fronts. In the previous paragraphs, we explained how some of the key performance factors lie within the UPC runtime level. From this point of view, effective portability of the pipeline is translated into efficient UPC runtime implementation for exascale systems. An additional opportunity to facilitate the porting to exascale systems emerges within the context of the distributed data structures described in Section 18.4. We could capitalize on the level of abstraction offered by our distributed data structures and their Use Cases (see Table 18.1) and optimize the core operations of the pipeline at the library level of the data structures. The benefit of such an approach is that the distributed data structure library utilized in HipMer could be specialized for each target system (e.g., with appropriate communication optimizations, topology, and hierarchical considerations) while the core codebase will remain unmodified. Finally, we highlight that the parallel algorithms in the HipMer pipeline are designed to scale to massive concurrencies and do not exhibit fundamental impediments in porting them to exascale systems.

18.8 RELATED WORK

As there are many *de novo* genome assemblers and assessment of the quality of these is well beyond the scope of this chapter, we refer the reader to the work of the Assemblathons I [41] and II [8] as examples of why Meraculous [6] was chosen to be scaled, optimized, and reimplemented as HipMer. For performance comparisons, we primarily refer to parallel assemblers with the potential for strong scaling on large genomes (such as plant, mammalian, and metagenomes) using distributed computing or clusters.

Ray [38,39] is an end-to-end parallel *de novo* genome assembler that utilizes MPI and exhibits strong scaling. It can produce scaffolds directly from raw sequencing reads and produces timing logs for every stage. One drawback of Ray is the lack of parallel I/O support for reading and writing files. As shown in Section 18.6, Ray is approximately 13 times slower than HipMer for the human dataset on 960 cores.

ABySS [40] was the first *de novo* assembler written in MPI that also exhibits strong scaling. Unfortunately only the first assembly step of contig generation is fully parallelized with MPI and the subsequent scaffolding steps must be performed on a single shared-memory node. As shown in Section 18.6, ABySS' contig generation phase is approximately 16 times slower than HipMer's entire end-to-end solution for the human dataset on 960 cores.

PASHA [42] is another partly MPI-based de Bruijn graph assembler, though not all steps are fully parallelized as its algorithm, like ABySS, requires a large memory single node for the last scaffolding stages. The PASHA authors do claim over two times speedup over ABySS on the same hardware.

YAGA [43] is a parallel distributed memory that is shown to be scalable except for its I/O, but the authors could not obtain a copy of this software to evaluate. HipMer employs efficient, parallel I/O so is expected to achieve end-to-end performance scalability. Also, the YAGA assembler was designed in an era when the short reads were extremely short and therefore its runtime will be much slower for current high-throughout sequencing systems.

SWAP [44] is a relatively new parallelized MPI-based de Bruijn assembler that has been shown to assemble contigs for the human genome and performs strong scaling up to about one thousand cores. However, SWAP does not perform any of the scaffolding steps, and is therefore not an end-to-end *de novo* solution. Additionally, the peak memory usage of SWAP is much higher than HipMer, as it does not leverage Bloom filters.

There are several other shared-memory assemblers that produce high-quality assemblies, including ALLPATHS-LG [45] (pthreads/OpenMP parallel depending on the stage), SOAPdenovo [46] (pthreads), DiscovarDenovo [47] (pthreads), and SPADES [48] (pthreads), but unfortunately each of these requires a large memory node and we were unsuccessful at running these experiments using our datasets on a system containing 512 GB of RAM due to lack of memory. This shows the importance of strong scaling distributed-memory solutions when assembling large genomes.

18.9 CONCLUSIONS

In this chapter, we presented HipMer, the first end-to-end highly scalable, high-quality *de novo* genome assembler demonstrated to scale efficiently on tens of thousands of cores. HipMer is two orders of magnitude faster than the original Meraculous code and at least an order of magnitude faster than other assemblers, including those with incomplete pipelines and lower quality. Parts of the HipMer pipeline were used in the first whole-genome assembly of the grand-challenge wheat genome [37]. HipMer is so fast, that by using just 17% of Edison's cores, we could assemble 90 Tbases/day, or all of the 5400 Tbases in the Sequence Read Archive [49] in just 2 months. Also, the HipMer technology makes it possible to improve assembly quality by running tuning parameter sweeps that were previously prohibitive in terms of computation.

Obtaining this scalable pipeline required several new parallel algorithms and distributed data structures which take advantage of a global address space model of computation on distributed-memory hardware, remote atomic memory operations, and novel synchronization protocols. Additionally, we developed runtime support to reduce communication cost through dynamic message aggregation and statistical algorithms that reduced communication through locality aware hashing schemes. We showed that high-performance distributed hash tables with various optimizations constitute a powerful abstraction for this type of irregular data analysis problems.

We believe our results will be important both in the application of assembly to health and environmental applications and in providing a conceptual framework for scalable genome analysis algorithms beyond those presented here. The code for HipMer is open source and can be downloaded at: `https://sourceforge.net/projects/hipmer/`.

REFERENCES

1. JT Simpson and M Pop. The theory and practice of genome sequence assembly. *Annual Review of Genomics and Human Genetics,* 16:153–72, 2015.
2. JD Kececioglu and EW Myers. Combinatorial algorithms for DNA sequence assembly. *Algorithmica,* 13:751, 1995.
3. A Zimin, KA Stevens, et al. Sequencing and assembly of the 22-gb loblolly pine genome. *Genetics,* 196(3):875–90, March 2014.
4. E Georganas, A Buluc, et al. HipMer: An extreme-scale de novo genome assembler. In *Proceedings of the International Conference for High Performance Computing, Networking, Storage and Analysis*, page 14. New York, NY: ACM, 2015.
5. T El-Ghazawi and L Smith. UPC: Unified parallel C. In *Proceedings of the 2006 ACM/IEEE Conference on Supercomputing*, page 27. Tampa, FL: ACM, 2006.
6. JA Chapman, I Ho, et al. Meraculous: De novo genome assembly with short paired-end reads. *PLOS ONE,* 6(8):e23501, August 2011.
7. JA Chapman, I Ho, et al. Meraculous2: Fast accurate short-read assembly of large polymorphic genomes. *PLOS ONE,* arXiv:1608.01031 [cs.DS] 2016.
8. K Bradnam, J Fass, et al. Assemblathon 2: Evaluating de novo methods of genome assembly in three vertebrate species. *Gigascience,* 2(1):10, 2013.
9. E Georganas, A Buluc, et al. Parallel de Bruijn graph construction and traversal for de novo genome assembly. In *Proceedings of the International Conference for High Performance Computing, Networking, Storage and Analysis (SC'14)*, New Orleans, LA: IEEE, 2014.
10. E Georganas, A Buluc, et al. merAligner: A fully parallel sequence aligner. In *Proceedings of the International Parallel and Distributed Processing Symposium*. Hyderabad, India: IEEE, 2015.

11. E Georganas. Scalable Parallel Algorithms for Genome Analysis. PhD thesis, University of California, Berkeley, CA, 2016.

12. B Liu, Y Shi, et al. Estimation of genomic characteristics by analyzing *k*-mer frequency in de novo genome projects. arXiv preprint arXiv:1308.2012, 2013.

13. TF Smith and MS Waterman. Identification of common molecular subsequences. *Journal of Molecular Biology*, 147(1):195–197, 1981.

14. D Bonachea. Gasnet specification, v1.1. http://gasnet.lbl.gov/CSD-02-1207.pdf, 2002. Accessed April 10, 2017.

15. M ten Bruggencate and D Roweth. DMAPP-an API for one-sided program models on baker systems. In *Proceedings of the 2010 Cray User Group Conference*, 2010.

16. S Kumar, AR Mamidala, et al. PAMI: A parallel active message interface for the blue gene/q supercomputer. In *IEEE 26th International on Parallel and Distributed Processing Symposium (IPDPS)*, pages 763–73. Shanghai, China: IEEE, 2012.

17. W Gropp, E Lusk, et al. A high-performance, portable implementation of the MPI message passing interface standard. *Parallel Computing*, 22(6):789–828, 1996.

18. B Chapman, T Curtis, et al. Introducing OpenSHMEM: SHMEM for the PGAS community. In *Proceedings of the Fourth Conference on Partitioned Global Address Space Programming Model*, New York, NY page 2. ACM, 2010.

19. J Dinan, P Balaji, et al. An implementation and evaluation of the MPI 3.0 one-sided communication interface. *Concurrency and Computation: Practice and Experience*, 28(17):4385–4404, 2016.

20. J Shun and GE Blelloch. Phase-concurrent hash tables for determinism. In *Proceedings of the 26th ACM Symposium on Parallelism in Algorithms and Architectures*, pages 96–107. Prague, Czech Republic: ACM, 2014.

21. M Herlihy, N Shavit, and M Tzafrir. Hopscotch hashing. In *International Symposium on Distributed Computing*, Arcachon, France pages 350–64. Berlin, Heidelberg: Springer, 2008.

22. M Hsu and W-P Yang. Concurrent operations in extendible hashing. In *Proceedings of the 12th International Conference on Very Large Database*, Kyoto, Japan Vol. 86, pages 25–8, 1986.

23. CS Ellis. Concurrency in linear hashing. *ACM Transactions on Database Systems (TODS)*, 12(2):195–217, 1987.

24. V Kumar. Concurrent operations on extendible hashing and its performance. *Communications of the ACM*, 33(6):681–94, 1990.

25. MM Michael. High performance dynamic lock-free hash tables and list-based sets. In *Proceedings of the 14th Annual ACM Symposium on Parallel Algorithms and Architectures*, pages 73–82. Winnipeg, Manitoba: ACM, 2002.

26. O Shalev and N Shavit. Split-ordered lists: Lock-free extensible hash tables. *Journal of the ACM (JACM)*, 53(3):379–405, 2006.

27. H Balakrishnan, MF Kaashoek, et al. Looking up data in P2P systems. *Communications of the ACM*, 46(2):43–8, 2003.

28. S Ratnasamy, P Francis, et al. A scalable content-addressable network. In *Proceedings of the 2001 Conference on Applications, Technologies, Architectures, and Protocols for Computer Communications*, Vol. 31. San Diego, CA: ACM, 2001.

29. H Zhang, Y Wen, et al. A Survey on Distributed Hash Table (DHT): Theory, Platforms, and Applications, http://www3.ntu.edu.sg/home/ygwen/Paper/ZWYX-13.pdf. 2013. Accessed: April 10, 2017.

30. R Gerstenberger, M Besta, and T Hoeer. Enabling highly-scalable remote memory access programming with MPI-3 one sided. In *2013 SC-International Conference for High Performance Computing, Networking, Storage and Analysis (SC)*, pages 1–12. Denver, CO: IEEE, 2013.

31. C Maynard. Comparing one-sided communication with MPI, UPC and SHMEM. In *Proceedings of the Cray User Group (CUG)*, Stuttgart, Germany, Cray User Group, Incorporated, 2012.

32. P Melsted and JK Pritchard. Efficient counting of *k*-mers in DNA sequences using a Bloom filter. *BMC Bioinformatics*, 12(1):333, 2011.

33. BH Bloom. Space/time trade-offs in hash coding with allowable errors. *Communications of the ACM*, 13(7):422–6, 1970.

34. P Flajolet, E Fusy, et al. Hyperloglog: The analysis of a near-optimal cardinality estimation algorithm. In *DMTCS Proceedings*, AH:137–156, 2008.

35. RC Deonier, S Tavare, and M Waterman. *Computational Genome Analysis: An Introduction*. New York, NY: Springer Science and Business Media, 2005.

36. KFX Mayer, J Rogers, et al. A chromosome-based draft sequence of the hexaploid bread wheat (*Triticum aestivum*) genome. *Science*, 345(6194):1251788, 2014.

37. J Chapman, M Mascher, et al. A whole-genome shotgun approach for assembling and anchoring the hexaploid bread wheat genome. *Genome Biology*, 16:26, 2015.

38. S Boisvert, F Laviolette, and J Corbeil. Ray: Simultaneous assembly of reads from a mix of high-throughput sequencing technologies. *Journal of Computational Biology*, 17(11):1519–33, 2010.

39. S Boisvert, F Raymond, et al. Ray meta: Scalable de novo metagenome assembly and profiling. *Genome Biology*, 13(R122), 2012.

40. JT Simpson, K Wong, et al. Abyss: A parallel assembler for short read sequence data. *Genome Research*, 19(6):1117–23, 2009.

41. D Earl, K Bradnam, et al. Assemblathon 1: A competitive assessment of de novo short read assembly methods. *Genome Research*, 21(12):2224–41, December 2011.

42. Y Liu, B Schmidt, and DL Maskell. Parallelized short read assembly of large genomes using de Bruijn graphs. *BMC Bioinformatics*, 12(1):354, 2011.

43. BG Jackson, M Regennitter, et al. Parallel de novo assembly of large genomes from high-throughput short reads. In *IEEE International Symposium on Parallel and Distributed Processing (IPDPS' 10)*. Atlanta, GA: IEEE, 2010.

44. J Meng, B Wang, et al. SWAP-assembler: Scalable and efficient genome assembly towards thousands of cores. *BMC Bioinformatics*, 15(Suppl 9):S2, 2014.

45. S Gnerre, D MacCallum, et al. High-quality draft assemblies of mammalian genomes from massively parallel sequence data. In *Proceedings of the National Academy of Sciences USA*, 108(4):1513–18, 2010.

46. R Li, H Zhu, et al. De novo assembly of human genomes with massively parallel short read sequencing. *Genome Research*, 20(2):265–72, 2010.

47. D Jaffe. Discovar: Assemble genomes and find variants. http://www.broadinstitute.org/software/discovar/blog/, 2014. Accessed April 10, 2017.

48. A Bankevich, S Nurk, et al. SPAdes: A new genome assembly algorithm and its applications to single-cell sequencing. *Journal of Computational Biology*, 19(5):455–77, May 2012.

49. Sequence Read Archive (SRA). Database Growth. http://www.ncbi.nlm.nih.gov/sra/docs/sragrowth/. Accessed on July 18, 2016.

19 Exascale Scientific Applications

Programming Approaches for Scalability, Performance, and Portability: KKRnano

*Paul F. Baumeister, Marcel Bornemann, Dirk Pleiter,
and Rudolf Zeller*

CONTENTS

19.1 Introduction..431
19.2 Scientific Methodology...432
    19.2.1 Sparse Dyson Equation...434
    19.2.2 Linear-Scaling $\mathcal{O}(N)$ Mode...434
19.3 Algorithmic Details and Performance Characteristics 435
19.4 Programming Approach...437
    19.4.1 Scalability 439
    19.4.2 I/O 440
    19.4.3 GPU Acceleration 441
    19.4.4 Portability 441
    19.4.5 External Libraries 441
19.5 Software Practices 442
    19.5.1 Conditional Compilation and Metaprogramming............................442
    19.5.2 Automatic Documentation...442
    19.5.3 Verification and Validation 442
    19.5.4 Distribution and Licensing 443
    19.5.5 Build System...443
19.6 Performance Results 443
    19.6.1 Scaling on Blue Gene/Q...443
    19.6.2 Performance on GPU-Accelerated POWER8 Nodes 443
    19.6.3 I/O Performance using SIONlib 444
19.7 Energy Considerations..445
19.8 Summary and Outlook..446
Acknowledgments 447
References 447

19.1 INTRODUCTION

Materials development for advanced twenty-first century applications is supported more and more by understanding of the atoms at the nanoscale origin of their mesoscopic and macroscopic properties.

For this goal, beyond empirical concepts and experimental data, quantum mechanical calculations within density functional theory (DFT) have proved to be an increasingly powerful tool. DFT [1,2] treats the many-electron system by single-particle equations using the electron density instead of the many-electron wavefunction as the basic quantity. Although this represents an obvious simplification, calculations for systems with many atoms still remain a serious computational challenge because the computational effort in conventional density functional calculations increases with the third power of the number of atoms in the system. Systems with a few hundred atoms can be treated routinely today, but larger systems require enormous computer resources. Therefore, considerable effort has been spent in recent years to reduce the effort by novel techniques, which exploit the so-called near-sightedness of electronic matter [3] by utilizing the exponential decay of the density matrix. These techniques are continuously improved and with the advent of the next generation of supercomputers it will become possible to simulate the quantum effects in entire semiconductor devices. These devices are currently scaled down below the 10 nm range and thus contain about 10^5 atoms.

One of the new computer codes, which avoids the unfavorable cubic scaling of the computational effort with system size, is KKRnano [4], which was designed from the outset to run efficiently on supercomputers with the IBM Blue Gene/P and Blue Gene/Q architectures. KKRnano, an all-electron code, which makes no use of pseudopotentials, is based on the full-potential Korringa–Kohn–Rostoker (KKR) multiple-scattering, Green-function method as described in Ref. [5]. It avoids the cubic increase of computing time by adopting the techniques explained in Ref. [6] and has been used so far for systems containing as many as 10^5 atoms. Such calculations could be processed with up to 1.8 million parallel tasks on the 5.9-Petaflop installation of the Blue Gene/Q architecture at the Jülich Supercomputing Centre [7].

In the following section, we provide background information on the methodology of KKRnano and explain differences compared with other approaches in the field. This is followed by a description of the algorithmic details of KKRnano and an overview on performance characteristics in Section 19.3. We continue with explaining our implementation (Section 19.4) and software practices (Section 19.5). Additionally, we present in Section 19.6 various performance results with particular focus on a scalability analysis on Blue Gene/Q and on graphics processing unit (GPU) accelerated POWER 8 nodes. Finally in Section 19.7 we report an analysis of energy efficiency, before we provide a summary and outlook toward exascale in Section 19.8.

19.2 SCIENTIFIC METHODOLOGY

Contrary to most other DFT codes, which determine the Kohn–Sham wavefunctions by solving the differential Kohn–Sham wave equation, in KKRnano the Green function $G(\mathbf{r}, \mathbf{r}'; \epsilon)$ for the Kohn–Sham equation is obtained by solving the integral equation

$$G(\mathbf{r}, \mathbf{r}'; \epsilon) = G^{\text{ref}}(\mathbf{r}, \mathbf{r}'; \epsilon) + \int d\mathbf{r}'' G^{\text{ref}}(\mathbf{r}, \mathbf{r}''; \epsilon) \left[v_{\text{eff}}(\mathbf{r}'') - v^{\text{ref}}(\mathbf{r}'') \right] G(\mathbf{r}'', \mathbf{r}'; \epsilon) \quad (19.1)$$

where the integral extends over the space covered by the system. Here, $v_{\text{eff}}(\mathbf{r})$ is the Kohn–Sham effective potential and $G^{\text{ref}}(\mathbf{r}, \mathbf{r}'; \epsilon)$ is the known Green function of a suitable reference system with potential $v^{\text{ref}}(\mathbf{r})$. In KKRnano, the reference system consists of constant repulsive potentials within nonoverlapping spheres around the atoms. With this choice, the reference Green function has two important properties: it decays exponentially with the distance $\mathbf{r} - \mathbf{r}'$ for ϵ values relevant in DFT calculations and its angular-momentum representation around the atoms is easily obtained from the analytically known Green function for the system with zero potential.

By dividing space into nonoverlapping cells around the atoms, the angular-momentum representation of the Green functions is used to separate the solution of Equation 19.1 into intracell and intercell parts. The intracell part requires the solution of single-site integral equations where the integral extends only over the volume of a single cell. Because these equations are independent of each other, the computational effort for this part scales linearly with system size and can be parallelized in an efficient way. The intercell part requires the solving of a large linear matrix equation

$$G_{LL'}^{nn'}(\epsilon_i) = G_{LL'}^{\text{ref},nn'}(\epsilon_i) + \sum_{n''} \sum_{L''} G_{LL''}^{\text{ref},nn''}(\epsilon_i) \sum_{L'''} \Delta t_{L''L'''}^{n''}(\epsilon_i) G_{L'''L'}^{n''n'}(\epsilon_i) \qquad (19.2)$$

for the Green function matrix elements $G_{LL'}^{nn'}(\epsilon_i)$. Here, $\Delta t_{L''L'''}^{n}$ is the difference between single-cell scattering matrices in the system and the reference system. $G_{LL'}^{\text{ref},nn'}$ is the known matrix element of the reference Green function. The indices n and $L = (\ell, m)$ label cells and angular-momentum components, whereas ϵ_i denotes integration mesh points used in the complex ϵ plane to obtain the fundamental quantity in DFT, the electronic density $n(\mathbf{r})$. In KKRnano, $n(\mathbf{r})$ can be calculated very precisely, if the effective potential around each atom is understood as a nonlocal, angular-projection potential [8] because for such potentials the dependence of $n(\mathbf{r})$ on the angular variables $\hat{\mathbf{r}} = \mathbf{r}/|\mathbf{r}|$ can be represented exactly [9] so that only the dependence on the radial variable $|\mathbf{r}|$ is calculated numerically.

Massively parallel calculations with KKRnano have been used so far to investigate a variety of large disordered multicomponent alloys. Among others, these are dilute magnetic semiconductors, phase change materials, and high-entropy alloys.

Dilute magnetic semiconductors are semiconductors doped with magnetic impurities. These systems are technologically interesting because they can be used for novel electronic devices, which utilize not only the electron charge but also the electron spin. For such so-called spintronic devices, it is important that the materials are magnetic with Curie temperatures T_C as high as room temperature or above. Gadolinium-doped gallium-nitride is a system that has attracted significant attention because magnetic moments of 4000 μ_B per Gd atom and ferromagnetism above room temperature were claimed in experiments (μ_B is the Bohr magneton). KKRnano was used to study realistic models of Gd-doped GaN with codoping by nitrogen or oxygen interstitials or Ga vacancies. It was found that only Ga vacancies provide a robust path to magnetic percolated clusters, which can explain the ferromagnetism with rather large moments.

Phase change materials, in particular alloys of germanium, antimony, and tellurium, are basic materials for DVD and BluRay technology because laser heating can be used to switch their structure between crystalline and amorphous phases, which have considerably different optical properties. KKRnano was used to investigate how vacancies and vacancy clusters can explain the experimentally observed metal–insulator transition in $GeSb_2Te_4$. It was found that the presence of vacancy clusters, which appear in the insulating phase and are dissolved in the metallic phase, likely plays an important role in the electronic resistance because it can act as a strong scattering center. Other interesting systems are phase change materials doped by magnetic impurities because not only the electronic resistivity but also the magnetic state can be changed by switching between the two phases, which could be exploited for new multivalued memory devices. In calculations with KKRnano, it was found that doping of $Ge_2Sb_2Te_5$, particularly with chromium impurities, leads to a strong tendency for ferromagnetism with T_C values close to room temperature for large impurity concentrations.

High-entropy alloys consisting of four or more metallic elements, which crystallize in simple face-centered cubic (FCC) lattices, constitute a relatively new class of materials with favorable properties like high hardness, wear resistance, and corrosion resistance. With KKRnano, energetics and magnetism in CrFeCoNi alloys were investigated. From the calculated local energies and their static fluctuations, it can be concluded that completely random solutions are not stable. Instead, an L_{12} structure is energetically preferred where the Cr atoms, which show the largest environmental effects with magnetic moments varying between -1.7 μ_B to $+0.8$ μ_B, are at the cube corners and Fe, Co, and Ni atoms are randomly distributed at the other sites.

19.2.1 Sparse Dyson Equation

Formally, without indices, the Dyson Equation 19.1 is equivalent to

$$X = \Delta t + \Delta t G^{\text{ref}} X \tag{19.3}$$

where the matrix $X = \Delta t + \Delta t G \Delta t$ is known as the scattering path operator in the KKR method. Contrary to conventional KKR programs, in KKRnano this matrix equation is not solved by Gaussian elimination, but iteratively as

$$X^{(i+1)} = \Delta t + \Delta t G^{\text{ref}} X^{(i)} \tag{19.4}$$

using the transpose-free quasi-minimal residual (TFQMR) method of Freund and Nachtigal [10]. The use of an iterative solution method is associated with several advantages. The first advantage is that the matrix elements $X_{LL'}^{nn'}$ can be determined separately for each atom n' and for each angular-momentum component L', which makes straightforward parallelization possible. The second advantage is that the exponential decay of the reference Green function $G^{\text{ref}}(\mathbf{r}, \mathbf{r}', \epsilon)$ can be utilized easily. By neglecting exponentially small matrix elements of G^{ref} in (19.4) this matrix is turned into a sparse matrix where the number of nonzero matrix elements is proportional to $N_{\text{cl}} N_{\text{at}}$ instead of proportional to N_{at}^2 as for a dense matrix. Here, N_{cl} is the number of atoms in a cluster in the vicinity of each atom. Usually, clusters consisting of 20–50 atoms are sufficient [11]. The sparse matrix multiplication in (19.4) requires order N_{at}^2 operations, which leads to a computational effort that increases only quadratically with the number of atoms and not cubically as in conventional DFT codes. The third advantage is that one can terminate the iterations, if the desired precision of the results is achieved. For instance, in DFT-based, total-energy calculations, the energy error is a few millielectron volts per atom, if the bound for the quasi-minimal residual (QMR) residual norm is set to $||r|| = 10^{-3}$, and a few tenths of millielectron volts for $||r|| = 10^{-4}$ as shown in Ref. [6] for periodic model systems consisting of supercells with 16,384 copper or palladium atoms.

19.2.2 Linear-Scaling $\mathcal{O}(N)$ Mode

The computational effort in KKRnano, which scales quadratically with system size, if no compromise on precision is made, can be reduced to a linear scaling by exploiting that the Green function $G(\mathbf{r}, \mathbf{r}', \epsilon)$ decays with the distance $|\mathbf{r} - \mathbf{r}'|$. This decay is exponential for the complex values of ϵ used in KKRnano, but considerably slower than the decay of the reference Green function $G^{\text{ref}}(\mathbf{r}, \mathbf{r}', \epsilon)$, so that matrix elements $G_{LL'}^{nn'}$ in (19.2) or $X_{LL'}^{nn'}$ in (19.4) are more important for larger distances between atoms n and n' than matrix elements $G_{LL'}^{\text{ref},nn'}$. Nevertheless, by trading some precision for computational speed, the matrix elements $X_{LL'}^{nn'}$ can be neglected beyond truncation regions of about $N_{\text{tr}} > 1,000$ atoms. This reduces the overall computational complexity in KKRnano to $N_{\text{it}} N_{\text{cl}} N_{\text{tr}} N_{\text{at}}$, where N_{it} is the number of TFQMR iterations and makes calculations possible for large systems involving up to 100,000 atoms. In order to assess which values of N_{tr} are reasonable, the dependence of the ground-state total energy on the number N_{tr} of atoms in the truncation region around each atom was studied with KKRnano. Because highly ordered pristine metallic systems are more demanding than insulators or disordered systems, the model systems used were constructed by a 32-times-repetition of simple cubic unit cells with four copper or palladium atoms in all three space directions. Using N_{tr} values between 55 and 34,251 for the constructed periodic supercells with 131,072 atoms, it was found that for small truncation regions with 55 atoms the error in the total energy can be as large as 0.1 eV per atom. Although this accuracy can be acceptable for certain applications, the usual goal in DFT calculations is to determine the total energy with an accuracy of a few millielectron volts per atom. In KKRnano, this can be achieved even for difficult ordered systems with truncation regions containing a few thousand atoms [6,12].

19.3 ALGORITHMIC DETAILS AND PERFORMANCE CHARACTERISTICS

The following listing gives an overview on the sequence of tasks, which is executed during one run of KKRnano:

```
.1 read input.
.1 read potentials.
.2    compute general atom quantities.
.3       compute E-dependent atom quantities.
.3       setup reference system.
.4          setup scattering path operator.
.4          invert operator using TFQMR.
.5             invoke sparse operator times block vector.
.4          use diagonal elements of inverse.
.3       .
.2    compute density.
.2    solve Poisson equation for new potential.
.1 store potential.
```

In the Kohn–Sham DFT scheme, the effective potential and the density must be determined self-consistently because they depend on each other in a nonlinear manner. Usually, they are calculated alternately in about 50–200 self-consistency steps. A typical run of KKRnano begins with reading some control parameters, the coordinates of the cell centers, the radial integration mesh in each cell and an initial guess for the effective potential. Before the self-consistency steps are started, it is necessary to calculate the so-called shape functions, which describe the geometric shape of the cells in an angular-momentum representation, and the Madelung coefficients, which are used to determine the electrostatic part of the effective potential from the charge multipole moments in each cell. These parts of the calculation require no communication and are easily parallelized over the cells.

During the self-consistency steps, the intracell part requires to solve systems of $(\ell_{max} + 1)^2$ coupled radial integral equations, as given in Ref. [8], in order to obtain $\Delta t^n_{LL'}(\epsilon_i)$ and to solve systems of linear equations of dimension $N_{cl}(\ell_{max} + 1)^2$ to obtain $G^{r,nn'}_{LL'}(\epsilon_i)$. Here, ℓ_{max} is a cutoff parameter, which essentially determines the angular resolution of the projection potential in each cell by restricting the projection to a subspace of spherical harmonics with $\ell \leq \ell_{max}$. The standard choice in KKRnano is $\ell_{max} = 3$ leading to $(\ell_{max} + 1)^2 = 16$ because common experience with the KKR method has shown that this is enough for most purposes. The intracell part requires no communication and is easily parallelized over the cells.

For the subsequent intercell part, the matrices $\Delta t^n_{LL'}(\epsilon_i)$ and $G^{ref,nn'}_{LL'}(\epsilon_i)$ with N_{cl} values of n' must be communicated to those other processors where they are needed to solve the Dyson Equation 19.4 by the TFQMR method. The matrices X, Δt, and G^{ref} in (19.4) are sparse matrices with X containing $N_{at}N_{tr}(\ell_{max} + 1)^4$ nonzero elements and G^{ref} containing $N_{at}N_{cl}(\ell_{max} + 1)^4$ nonzero elements, whereas Δt contains nonzero blocks of size $(\ell_{max} + 1)^4$ only on the diagonal. During the TFQMR iterations, which require no communication and are easily parallelized over the cells, the sparsity structure of the matrices is exploited so that multiplication with zeros is avoided (see Section 19.4). After the TFQMR iterations, the density and the charge multipole moments in each cell are calculated by simply summing up the Green-function contributions at the mesh points ϵ_i with appropriate integration weights.

For the calculation of the effective potential, the charge multipole moments of each cell, a data set of $(2\ell_{max} + 1)^2 = 49$ numbers, must be communicated to all other cells. This small amount of data together with the $256(N_{cl} + 1)$ numbers, which must be transferred before the TFQMR iterations but only to the participating cells, indicates that the principle communication time in KKRnano is, in many situations, of not much importance.

In order to enable the use of more processors than atoms at the expense of additional communication, KKRnano optionally employs two other levels of parallelization besides the natural parallelization over the cells. One is over the two magnetic spin directions in ferromagnetic systems and the other is over the energy points ϵ_i. Because for physical reasons quite different numbers of TFQMR iterations are needed for different values of ϵ_i, a simple distribution to parallel tasks is inefficient. Instead, the ϵ_i is pooled in two or three groups where each group is treated by one thread. Load balancing is achieved by dynamically updating this grouping during the self-consistency steps.

Besides the coarse-grained parallelism described earlier, KKRnano can exploit additional parallelization levels, particularly for the most time-consuming part, the TFQMR iterations. For the Blue Gene/P JUGENE, which was installed at the Jülich Supercomputing Centre from 2008 to 2012, an additional parallelization over the 16 L components in the Dyson equation was implemented, which allowed to utilize all 294,912 processors available on this machine for a ferromagnetic nickel-palladium system containing 3,072 atoms. For the Blue Gene/Q JUQUEEN, which has been operational at the Jülich Supercomputing Centre since 2012, the parallelization strategy for the iterative calculation of the matrix elements $X_{LL'}^{nn'}$ by using (19.4) was changed. Instead of calculating the columns labeled by L' by using several message passing interface (MPI) processes, the block rows labeled by n are calculated by several OpenMP threads. The advantages of the OpenMP implementation are simple use of the hybrid programming model of the Blue Gene/Q, increased flexibility, because the number of block rows, which equals the number of atoms, is much larger than the number of columns per atom, reduced memory requirements, because the shared memory on the nodes is exploited, and increased flop rate, because matrix–vector operations are replaced by matrix–matrix operations. This enabled efficient use of all 1,835,008 parallel threads available on Blue Gene/Q for a phosphorus-silicon system with 57,344 atoms and 57,344 empty cells.

In typical runs, most of the time is spent in the TFQMR algorithm. Each iteration involves two matrix multiplications. The amount of floating point operations and the amount of data, which needs to be loaded and stored, are denoted by I_{fp}, I_{ld}, and I_{st}, respectively:

$$I_{\mathrm{fp}} = 2N_{\mathrm{it}} \cdot \frac{N_{\mathrm{at}}}{N_{\mathrm{MPI}}} \cdot N_{\mathrm{tr}} N_{\mathrm{cl}} \cdot b^3 \cdot 8\,\mathrm{f\,lop} \tag{19.5}$$

$$I_{\mathrm{ld}} = 4N_{\mathrm{it}} \cdot \frac{N_{\mathrm{at}}}{N_{\mathrm{MPI}}} \cdot N_{\mathrm{tr}} N_{\mathrm{cl}} \cdot b^2 \cdot 16\,\mathrm{byte} \tag{19.6}$$

$$I_{\mathrm{st}} = 2N_{\mathrm{it}} \cdot \frac{N_{\mathrm{at}}}{N_{\mathrm{MPI}}} \cdot N_{\mathrm{cl}} \cdot b^2 \cdot 16\,\mathrm{byte} \tag{19.7}$$

where $b = (\ell_{\max} + 1)^2$. For more details see [13]. From these quantities, the arithmetic intensity (AI) can easily be determined as $\mathrm{AI} = I_{\mathrm{fp}}/(I_{\mathrm{ld}} + I_{\mathrm{st}})$.

The original implementation supporting the $\mathcal{O}(N)$-mode by truncation was restricted to exactly one source atom per MPI process. In terms of programming, this involved much less bookkeeping than the current version that supports $n_{\mathrm{at}} = N_{\mathrm{at}}/N_{\mathrm{MPI}}$ source atoms per MPI process. One MPI process per atom also infers a large memory overhead related to global arrays, associated with the MPI library requirements and, in particular, the scattering path operator. However, a lower total memory consumption can be achieved with the flexibility of several source atoms per MPI process because elements of the scattering path operator can be shared to a large extent. This also yields an improved performance of the sparse matrix multiplication as the AI of multiplying two square complex matrices of dimension 16 is 4 flop/byte. The multiplication of a square matrix and a rectangular matrix with the long dimension $16\,n_{\mathrm{at}}$ leads to an $\mathrm{AI}(n_{\mathrm{at}})$ that converges toward 5.32 flop/byte and is as high as 5.0 for five source atoms, c.f. Figure 19.1.

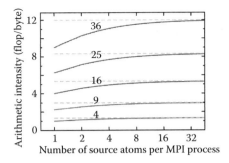

FIGURE 19.1   The nominal arithmetic intensity (AI) for a matrix multiplication $A \times X$ increases for larger matrix dimensions. Here, $A \in \mathbb{C}^{b \times b}$ and $X \in \mathbb{C}^{b \times nb}$. The solid lines show the AI as a function of n for $b \in \{4, 9, 16, 25, 36\}$.

19.4   PROGRAMMING APPROACH

The main programming language for KKRnano is Fortran and, here, the usage of Fortran 90 or Fortran 95 is encouraged over Fortran 77, in particular when adding new functionalities or restructuring old parts of the code. The character of Fortran as a domain-specific language is particularly useful to avoid many explicit loops (Fortran 90 (:)-syntax), natural handling of matrix data layout and matrix operations and intrinsic support of complex numbers. Furthermore, Fortran implies reasonable assumptions beneficial for compiler optimization, as assuming no pointer-aliasing between arguments. In KKRnano, derived data types assist to encapsulate data that belong to one module. This leads to relatively short argument lists and, hence, readable code. Especially for the less performance critical parts, this infers a structure that is easy to maintain. Each derived data type is defined in a Fortran module that exposes a creation and a destruction routine. By keeping memory management statements at well-defined locations, it becomes easier to maintain the code and adapt data layout to new architectures.

The following code block gives an example of a simple derived data type:

```
module AtomicCoreData_mod
  private
  public :: AtomicCoreData, create, destroy

  type AtomicCoreData
    integer                      :: ellcore
    double precision             :: Ecore
    double precision             :: corecharge
    integer                      :: irmd
    double precision, allocatable :: rhocat(:,:)
  endtype

  interface create
    module procedure createAtomicCoreData
  endinterface

  interface destroy
    module procedure destroyAtomicCoreData
  endinterface

  contains
```

```fortran
subroutine createAtomicCoreData(self, irmd)
  type(AtomicCoreData), intent(inout) :: self
  integer,              intent(in)    :: irmd

  self%irmd = irmd

  self%ellcore = -1
  self%Ecore = 1.d9

  allocate(self%rhocat(irmd,2)) ! both spin directions
  self%rhocat = 0.d0
endsubroutine

elemental subroutine destroyAtomicCoreData(self)
  type(AtomicCoreData), intent(inout) :: self

  integer :: ist
  deallocate(self%RHOCAT, stat=ist)
endsubroutine

endmodule
```

KKRnano exploits the Fortran 90 feature of overloading names of routines (generic names) using the `interface` statement. With that, any derived data type in KKRnano can, for example, always be destructed by

```fortran
call destroy(core_state)
```

which causes all dynamically sized member fields to be deallocated. The `elemental` keyword even adds more comfort because that allows a single call onto an array of core state data items:

```fortran
call destroy(core_states(:))
```

In general, name overloading, encapsulation and the Fortran-module syntax for namespacing with the `only`-syntax allow for a well-structured approach toward large software packets written in Fortran 90 that can be maintained with relatively low efforts and facilitate porting to new architectures.

The truncation approach which is made in KKRnano asks for a matrix storage format that reflects the resulting block sparsity pattern. To this aim, we employ the variable block row (VBR) format where only the elements in nonzero blocks alongside four pointer arrays for structure description are stored [14]. The pointer array KVST contains the global row index and simultaneously the global column index of each block row because in KKRnano only square blocks occur. The block-wise column index of each block is stored in JA. KA indexes into the beginning of each block in the array A where all elements from nonzero blocks are saved. Elements in IA point to the beginning of each block row in both JA and KA. KVST, IA, and KA are supplemented by one concluding element to indicate the end of the matrix structure. In the example in Figure 19.2, KVST has three elements plus the concluding element. The first three elements point to the beginning of each block row, whereas the fourth element points behind the last block row to indicate where the matrix structure ends. As the example matrix is composed of five nonzero blocks, JA contains five entries and each entry gives the block column index of the corresponding block in row-major order, for example, first block can be found in first block column, second block in third block column, third block in second block column, and so on. KA also holds as many entries as nonzero blocks plus one concluding entry similar to that in KVST. Because in this example, all blocks have equal size (2×2), there is a constant increment of 4 for the entries in KA. Finally, IA accommodates four pointer entries (three block rows plus concluding element) and each, apart from the last one, points to the start of a new block row in JA and KA.

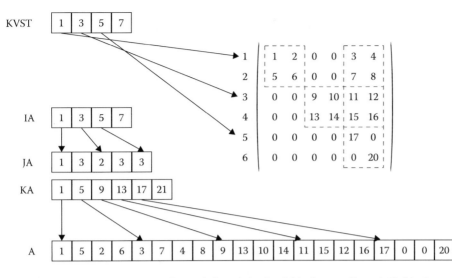

FIGURE 19.2 A 6×6 sparse matrix consisting of nine 2×2-blocks stored in variable block row (VBR) format. The five nonzero blocks are highlighted by dashed lines.

19.4.1 SCALABILITY

KKRnano makes use of the MPI for exploiting the distributed memory parallelism. After each operator inversion, information related to atoms, which belong to the same truncation cluster but are processed by different MPI ranks, needs to be exchanged. Therefore, point-to-point communication operations dominate. For typical work loads, the time spent in the inverter is significantly larger than the time required for communication.

For the implementation of more than one source atom per MPI process ($n_{at} > 1$), the bookkeeping during communication becomes difficult, in particular using the truncation mode. Therefore, the communication of the ingredients required to construct the scattering path operator makes use of one-sided MPI communication routines. A call to `MPI_Win_create` allows the other MPI processes to access the locally stored quantities.

For testing the scalability of the $\mathcal{O}(N)$-mode, we selected an input deck with 2197 atoms. It features a supercell of 13^3 unit cells of pristine FCC crystal. This is large enough to perform truncation with a truncation cluster size N_{tr} of 1289, that is, there are six shells of target atoms around each source atom (Figure 19.3).

The test indicated that about 32 % of the runtime for an energy point was spent in communication for the reference Green function. Here, 2197 MPI processes exchanged $13 \cdot 16 \cdot 16$ `double complex` with 1288 other processes, that is, 280.66 GiByte per energy point were transferred over the network. The high percentage means that other ways of communication should be explored.

An experiment with direct point-to-point, nonblocking communication showed that the communication time can be reduced significantly. Whereas the memory-saving, one-sided communication routine took 2.2 seconds, direct MPI messages were faster by 5×, so that only about 8.4 % of the iteration time was spent for communication.

In order to analyze both runs, an instrumentation with ScoreP [15,16] generates profile summaries (or traces) that can be analyzed using Scalasca [17–19], see Figure 19.4. The number of visits to each function, the time spent in it, and, very important for the analysis of the communication pattern, the number of bytes sent and received. However, the latter numbers are not given in the case of one-sided communication. Scalasca was used to extract the time spent in the communication functions. As each process delivers a timing result for each function call that has not been filtered out, the display of the distribution of time values is particularly useful, see the rightmost panel in Figure 19.4.

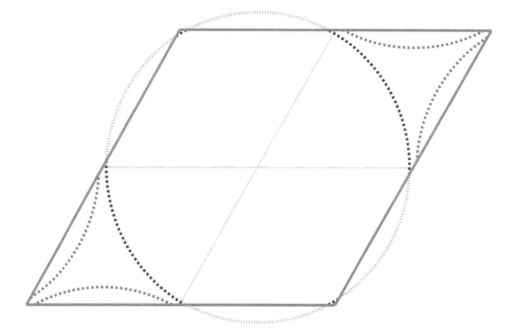

FIGURE 19.3 The truncation sphere with a radius of six lattice constants contains 1289 target atoms in three-dimensions using a supercell of 13^3 face centered cubic unit cells, that is, 2197 source atoms. This input deck is designed so that truncation spheres do not overlap with their periodic images and is used for profiling the application in $\mathcal{O}(N)$-mode.

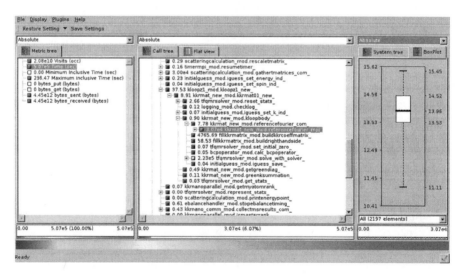

FIGURE 19.4 Cube viewer for profiles analyzed by the Scalasca tool. Metrics are selected in the left panel, the investigated function is shown embedded into its call tree in the center panel and statistics about the time per call or other quantities are shown on the right.

19.4.2 I/O

It is often underestimated how I/O can limit scalability of an application. KKRnano performs tasking-local POSIX I/O operations, which on currently available parallel file systems do not scale. To mitigate this problem we use SIONlib [20], which is a library that combines accesses to one file per MPI task into accesses to shared files. It requires only minimal changes to the application as it allows to continue using the most popular functions for reading and writing.

19.4.3   GPU ACCELERATION

One of the most promising approaches for realizing exascale computers is the use of compute devices, which are operated at relatively low clock frequency but are capable of performing an extremely larger number of floating-point operations each clock cycle. One increasingly popular example of such devices are GPU. Due to the low clock frequency they feature a throughput of floating-point operation versus power consumption ratio that is much higher compared with standard CPUs. Efficient exploitation of such devices requires applications with a high level of parallelism and large AI. These are exactly the features of the TFQMR solver in KKRnano. For this reason, this solver has been ported to GPUs [13] in order to be able to exploit the performance of GPU-accelerated high performance computing (HPC) systems.

This custom-tailored solver is written in C++ employing the CUDA toolkit provided by NVIDIA. The basic structure of the existent TFQMR can be adopted but data transfer from host to device and vice versa has to be taken care of and matrix operations must be implemented in a CUDA-conform manner. The latter can be easily achieved by using cuBLAS routines as they are included in the CUDA toolkit. However, tests showed that in our area of application, where sparse matrices are essential, self-written algorithms can outperform those routines. Therefore, KKRnano does not rely on library function calls but is equipped with its own custom-built linear algebra routines that handle matrices stored in a block sparse format.

19.4.4   PORTABILITY

As discussed earlier, a typical work load based on KKRnano spends most of its time in the TFQMR solver. The default implementation of this solver comprises 400 lines-of-code. For this reason, we can apply the following portability strategy:

- The bulk of the code is kept Fortran 95 compliant.
- Parallelization of the application is compliant to MPI version 3 and OpenMP version 3.
- Optionally, KKRnano can be linked to specialized versions of the TFQMR solver.

The application can thus easily be ported to any platform supporting Fortran 95, MPI, and OpenMP. These requirements are sufficiently easy to fulfill such that KKRnano runs both, on almost any HPC system and Linux laptop computers. In fact, it has been ported to a variety of HPC systems, including IBM Blue Gene, different clusters based on Intel Xeon processors or IBM POWER processors (with or without GPUs). As this does not guarantee the required high level of performance for HPC systems, we do foresee specialized versions of the solver, which might be implemented in a nonportable way. One example for this approach is our GPU-accelerated version of this solver, which is implemented in CUDA (see section 19.4). The efforts for creating such specialized code versions is typically moderate and thus a good balance between additional efforts and performance gains can be obtained.

19.4.5   EXTERNAL LIBRARIES

Wherever possible, standard linear algebra is performed in BLAS [21] and LAPACK [22] routines. Here, vendor-tuned libraries, where available (Intel Math Kernel Library [MKL] or IBM Engineering and Scientific Subroutine Library [ESSL]), deliver good performance. In particular, matrix–matrix multiplications for complex matrices (zgemm) are used intensively for small and mid-sized matrices. Another essential ingredient are inversion routines like zgetrf, zgetrs, or zgetri, which make use of efficient vendor-provided implementations on matrices of dimension $b \cdot N_{cl}$.

19.5 SOFTWARE PRACTICES

The development of KKRnano is facilitated by a software repository running the version control system `git`. The corresponding `git`-server [23] allows distributed development and a simple management of source code versions and documentation. The software project management system `TRAC` [24] is used as a web front end for meeting minutes and ticketing besides uncomplicated access to single files from different uploaded branches.

19.5.1 CONDITIONAL COMPILATION AND METAPROGRAMMING

KKRnano makes use of the preprocessor. This means that most source files carry the suffix `.F90` rather than `.f90`. This invokes the Fortran-internal preprocessor. It is used here to deactivate code that was included for testing purposes with `-D NDEBUG` and activates machine-specific solutions in selected routines.

As mentioned earlier, Fortran 90 supports generic functions that allow for well-readable, high-level routines. Nevertheless, C++-like template programming is not supported. Some KKRnano code structure, however, would benefit from such language capabilities. The workaround for this missing feature of the programming language is code generation with small Python scripts or text replacement with the Linux command line tool `sed`. The latter is used for communication routines. Besides all the advantages to the ease of programming, the drawback of an explicit routine interface that is being checked during compilation is that interfaces for many data types need to be generated. Due to the absence of template programming, KKRnano generates communication routines for `double precision`, `double complex`, `integer` using text replacement with `sed`. Although solutions using the preprocessor are viable, `sed` has the advantage that the other preprocessor directives survive the manipulation.

Furthermore, a Python script is used to generate the routine that reads the input file. Because various data types and combinations (with or without default value) are required for the user control of KKRnano, the input file reader is re-generated by the Python code when an input file keyword is added or changed or default values are adjusted.

19.5.2 AUTOMATIC DOCUMENTATION

The KKRnano documentation is twofold: Most features are described for practical reasons in the source code following the Doxygen convention [25]. However, some issues that should be addressed in more detail as well as small tutorials are available in a "DokuWiki" [26]. The importance of a good documentation cannot be overestimated because in KKRnano a lot of parameters have to be introduced whose meaning and significance in the bigger context are not necessarily clear at first sight. Therefore, parameters that are in KKRnano commonly defined within F90 types are extensively described and the purposes of subroutines are explained if instructive.

19.5.3 VERIFICATION AND VALIDATION

The correctness of the code is verified using a set of regression tests collected in `tests.py`. Here, the test cases inherit from `unittest` and compare the total energy results of serial runs on small input decks (up to 16 source atoms) with precalculated values. Furthermore, verification of the MPI parallelization can be switched on checking if the results stay unchanged when source atoms are distributed to up to a single source atom per process.

The script `tests.py` also provides validation: The convergence with respect to ℓ_{max} is checked for a single atom of copper in FCC symmetry. At the same time, this test is used to reveal implementation errors when compiled with the `checkbounds` option.

The full set of regression tests runs about 20 minutes on a workstation. Single subtests can be as fast as 2 seconds.

19.5.4 DISTRIBUTION AND LICENSING

Access to the KKRnano-repository is currently available in the context of scientific collaboration with the Institute for Advanced Simulation because the code has not yet been released under a public license.

19.5.5 BUILD SYSTEM

KKRnano uses traditional makefiles. A single `Makefile` can take various configuration options to specify the flavor of executable one wants to build. The options control the machine type to compile for the compilers used and the compiler flags. In particular, the preprocessor macros are controlled here, whereas some of them will be set depending on the machine type. The dependency table of the source files is explicitly expressed in the makefile. This is necessary to ensure that Fortran modules are already refreshed and present in the module directory before compiling a source file that includes them via `use`-statement. The dependency table can be constructed using a tool invoking `grep` on all source files. A prerequisite here is that developers have to stick to the rule that only one module is defined inside one source file and the name of the module matches the name of file (except for the `.F90` suffix and case sensitivity). Parallel compilation (`make -j`) allows a full recompilation plus linking in about 15 seconds (`-O3`) on a workstation with four cores.

19.6 PERFORMANCE RESULTS

19.6.1 SCALING ON BLUE GENE/Q

A weak scaling analysis for KKRnano has been conducted on the Blue Gene/Q installation JUQUEEN at Jülich Supercomputing Centre. Its peak performance is 5.9 Petaflop/sec. Here, the code has been applied to a host crystal of silicon with a single impurity atom of phosphorous that adds a shallow donor state into the bandgap of the semiconductor. Meaningful input decks can easily be generated for any size of the host material. The lattice structure of Si is the diamond structure, which exhibits relatively large void spaces in between the atoms. In order to describe it properly, a body-centered cubic structure is set up so that there are two Voronoi cells per atom, one containing an atomic core and one for the void. The results of the measurements with four OpenMP threads per MPI process is shown in Figure 19.5. As the characteristic scaling behavior of the $\mathcal{O}(N)$ mode starts when the system is larger than the truncation cluster, the measurements begin at 2000 cells. Despite some deviations from the linear behavior that might stem from suboptimal job partitions on the machine, the solver time per cell is around 645 seconds for $1k$, $8k$, and $65k$ atoms.

19.6.2 PERFORMANCE ON GPU-ACCELERATED POWER8 NODES

An in-depth performance analysis of the GPU-accelerated TFMQR solver on server based on POWER8 processors and NVIDIA Tesla K40m GPUs has been published in [13]. To improve utilization of the GPU it is required to use the multiprocess service (MPS) feature to run multiple solvers concurrently. Benchmark data show that at least four GPU tasks are required to maximize performance on a K40m GPU for a typical problem size (see Figure 19.6). It can also be observed that performance is best when the number of atoms in the truncation cluster N_{tr} is 1000 or more. This

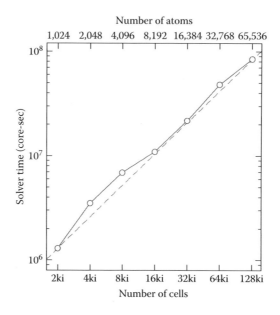

FIGURE 19.5   Blue Gene/Q weak scaling of a the multiple scattering solver for a single atom of phosphorous in a host crystal of silicon. Two cells per atom have been used to describe the diamond structure with body-centered cubic Voronoi cells, one message passing interface (MPI) process per cell and four OpenMP threads per MPI process. Around 2000 cells were inside the truncation cluster, 1.7 % for the largest calculation.

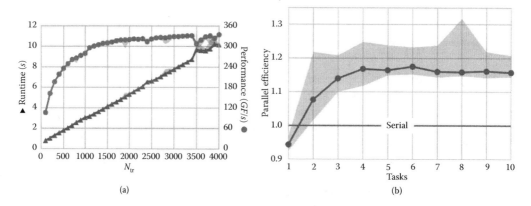

FIGURE 19.6   (a) Performance of graphics processing units-transpose-free quasi-minimal residual solver using 1000 iterations with varying truncation cluster size N_{tr}. (b) Efficiency of the multiprocess service on an NVIDIA K40m.

can be ascribed to the GPU architecture, which favors large matrices as they appear when working with a big truncation cluster.

19.6.3   I/O PERFORMANCE USING SIONlib

The standard output of KKRnano consists of the effective potential, the charges (in angular-momentum decomposition), the total-energy contributions, and the forces on the atoms. In the original version of KKRnano, the output to the file system was implemented using task-local, unformatted Fortran writes to nonshared files. The individual record for each cell was written by the MPI process responsible for that cell. Although charges, total-energies, and forces are needed only after the last

TABLE 19.1

Times t_{POSIX} and $t_{SIONlib}$ (in seconds) on the Blue Gene/Q JUQUEEN Used to Write the Effective Potential Task-Locally to Nonshared Files and to a Shared SIONlib File for a System with 114,688 Cells Using 114,688 Message Passing Interface (MPI) Tasks and a Varying Number of OpenMP Threads per MPI Task

N_{OpenMP}	N_{tasks}	t_{direct}	$t_{sionlib}$
2	229,376	101	10.1
4	458,752	166	11.1
8	917,504	434	7.3
16	1,835,008	831	8.1

self-consistency step, the situation is different for the effective potential. Usually, for monitoring and improving convergence of the self-consistency steps about 15 steps are done in one production job, which requires that the potential is written to the file system at least once per job. It is, however, desirable to write the potential after every self-consistency step to obtain checkpoint data. They can be used to restart the calculation, for instance, after an unexpected crash of a production job or if the self-consistency process was manually terminated because its behavior changed from convergent to divergent.

Although the potential comprises only a few hundred kBytes per cell, for large systems the use of separate files for each MPI rank severely limits the scalability. This is obvious in Table 19.1 where the third column shows the time, which was used to write about 10 GByte for a system with 114,688 cells using between 2 and 16 OpenMP threads per MPI process. The time for output approximately doubles if the number of processing units is doubled unless SIONlib is used.

To alleviate this problem we use SIONlib [12], which is a library that prevents file-system overhead arising from hundreds of thousands of task-local files by using a small number of shared files. SIONlib employs a communication layer to aggregate metadata among the tasks and exploits the I/O infrastructure and the file-system properties. The use of SIONlib in KKRnano, which required only minimal changes to the code by replacing the standard functions for reading and writing by the appropriate SIONlib functions, gave impressive improvements for the output time as shown in the fourth column of Table 19.1. The time for output remains almost constant at a level far below the total time used in a self-consistency step.

The current version of KKRnano is expected to scale to a significantly larger number of tasks without being limited by I/O performance. There is more room for improvements by better tuning of the SIONlib parameters. No fine-tuning was attempted for Table 19.1 where only a single, shared SIONlib file was used. Tests using eight SIONlib files gave reduced times of 3.6 and 2.7 seconds for 229,376 and 458,752 tasks.

19.7 ENERGY CONSIDERATIONS

The combination of a throughput-optimized compute accelerator with a general-purpose CPU makes the choice of clock speeds of both chips an important task. As shown by Hater et al. [27] on a heterogeneous compute node with two POWER8 CPUs and two K40m GPUs, dynamic voltage and frequency scaling (DVFS) can be tuned to minimize the energy-to-solution. Here, the POWER8 processors are featuring a rich cache hierarchy and also offer a substantial floating point performance, about 20 % that of the GPUs. The energy-to-solution becomes a function of the CPU frequency.

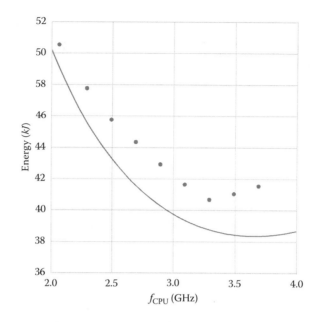

FIGURE 19.7   Energy to solution for the quasi-minimal residual (QMR) solver running on POWER8 processors as a function of the CPU frequency. Dots show the results of power measurements, whereas the solid line represents the energy model (With kind permission from **Springer Science+Business Media**: Hater, T. et al., *Exploring Energy Efficiency for GPU-Accelerated POWER Servers*, 2016.)

Despite the high base power consumed by the main memory, the best choice of CPU frequency is not the highest possible. Measurements of the total power consumption of the compute node (dots in Figure 19.7) infer that the POWER CPUs should solve the iterative inversion problem at 90 % of the max. CPU frequency to achieve the highest energy efficiency. The findings can be explained by an energy model (solid line in Figure 19.7) that is derived from a performance model extended by prefactors for the power consumption, which are extracted from device-resolved power measurements. For the GPU solver, the optimal frequency is the maximum of 875 MHz and the energy-to-solution for the input parameters used in the investigation is lower by about 40 % compared with the CPU solver.

19.8   SUMMARY AND OUTLOOK

In this chapter, we provided an introduction into the KKR method and described its features, which allows for addressing challenging problems in materials science. One important feature is the extreme scalability that can be achieved on massively parallel computers, which was demonstrated with one specific implementation of this method, namely KKRnano. We consider it as a showcase that a systematic approach to increase the parallelism of an application pays off.

For KKRnano we presented details of the implementation and, in particular, aspects on how it is parallelized. Furthermore, we discuss our programming approach aiming for both, maintainability and portability of the code. This allows us to run the code on desktop systems for code development and on HPC systems of the highest performance class when using KKRnano for research. On the latter systems, a very high level of scalability could be demonstrated, for example, on an IBM Blue Gene/Q system at Jülich Supercomputer Center with 458,752 cores.

Based on an analysis of the properties of the key algorithms and their implementation in KKRnano, we expect that the level of parallelism can be further significantly increased. This will be crucial to exploit future architectures, for example, architectures including GPUs as compute accelerators. These devices feature an extremely large throughput of floating-point operations per clock cycle. As it can be expected that this hardware level parallelism will further increase, KKRnano is in a good position to exploit these future compute technologies efficiently. This could already be demonstrated using node architectures, which are similar to those of the Summit supercomputer, which will be installed at Oak Ridge National Lab. The simple properties of the most performance critical kernel, namely the operator solver based on the TFMQR algorithm, makes this application also suitable for unconventional architectures. An interesting example is processing-in-memory architectures, where data transport is avoided by integrating compute capabilities into the memory, which can help to significantly reduce energy-to-solution [28]. As power consumption is expected to become the major limiting factor for further increase of HPC system performance, it will become more and more important to explore such unconventional architectures toward exascale computing.

ACKNOWLEDGMENTS

We would like to thank all contributors to the application, in particular Alexander Thiess and Elias Rabel, who proved the linear scaling and brought the code into shape. For support on the Blue Gene systems, especially focussed to I/O, we thank Wolfgang Frings and Kay Thust from the Jülich Supercomputer Center. We would also like to acknowledge the efforts toward the GPU implementation by Thorsten Hater in the context of the Exascale Innovation Center and the POWER Acceleration and Design Center. Finally, we thank Jiri Kraus (NVIDIA) for consultancy.

REFERENCES

1. Hohenberg, P., Kohn, W. Inhomogeneous electron gas. *Phys. Rev.* 136, B864–B871 (November 1964), http://link.aps.org/doi/10.1103/PhysRev.136.B864.
2. Kohn, W., Sham, L.J. Self-consistent equations including exchange and correlation effects. *Phys. Rev.* 140, A1133–A1138 (November 1965), http://link.aps.org/doi/10.1103/PhysRev.140.A1133.
3. Prodan, E., Kohn, W. Nearsightedness of electronic matter. *Proc. Natl. Acad. Sci. USA* 102, 11635–11638 (2005).
4. Thiess, A., Zeller, R., Bolten, M., Dederichs, P.H., Blügel, S. Massively parallel density functional calculations for thousands of atoms: KKRnano. *Phys. Rev. B* 85, 235103 (June 2012).
5. Papanikolaou, N., Zeller, R., Dederichs, P.H. Conceptual improvements of the KKR method. *J. Phys. Condens. Matter* 14(11), 2799–2824 (2002).
6. Zeller, R. Towards a linear-scaling algorithm for electronic structure calculations with the tight-binding Korringa-Kohn-Rostoker Green function method. *J. Phys. Condens. Matter* 20(29), 294215 (2008).
7. Forschungszentrum Jülich. High-Q Club. http://www.fz-juelich.de/ias/jsc/EN/Expertise/High-Q-Club_node.html/, accessed: 2016-08-21.
8. Zeller, R. Projection potentials and angular momentum convergence of total energies in the full-potential Korringa-Kohn-Rostoker method. *J. Phys. Condens. Matter* 25(10), 105505 (2013).
9. Zeller, R. The Korringa-Kohn-Rostoker method with projection potentials: Exact result for the density. *J. Phys. Condens. Matter* 27(30), 306301 (2015).
10. Freund, R.W., Nachtigal, N. QMR: A quasi-minimal residual method for non-Hermitian linear systems. *Numer. Math.* 60(1), 315 (1991).
11. Zeller, R. Evaluation of the screened Korringa-Kohn-Rostoker method for accurate and large-scale electronic-structure calculations. *Phys. Rev. B* 55, 9400–9408 (April 1997).
12. Zeller, R. Linear scaling for metallic systems by the Korringa-Kohn-Rostoker multiple-scattering method. In: Papadopoulos, M.G., Zalesny, R., Mezey, P.G., Leszczynski, J. (eds.), *Linear-Scaling Techniques in Computational Chemistry and Physics: Methods and Applications. Challenges and Advances in Computational Chemistry and Physics*, pp. 475–505. Dordrecht: Springer Netherlands (2011).

13. Baumeister, P.F., Bornemann, M., Bühler, M., Hater, T., Krill, B., Pleiter, D., Zeller, R. *Addressing Materials Science Challenges Using GPU-accelerated POWER8 Nodes*, In: Dutot, P.-F., Trystram, D. (eds.), Euro-Par 2016: Parallel Processing, Lecture Notes in Computer Science, Vol. 9833, pp. 77–89. Cham: Springer International Publishing (2016).

14. Saad, Y. Sparskit: A basic tool kit for sparse matrix computations - version 2 http://www-users.cs.umn.edu/~saad/software/SPARSKIT/ (1994).

15. RWTH Aachen: Score-P. http://www.vi-hps.org/projects/score-p/, accessed: 2016-07-21.

16. Knüpfer, A., Rössel, C., Mey, D.a., Biersdorff, S., Diethelm, K., Eschweiler, D., Geimer, M., Gerndt, M., Lorenz, D., Malony, A., Nagel, W.E., Oleynik, Y., Philippen, P., Saviankou, P., Schmidl, D., Shende, S., Tschüter, R., Wagner, M., Wesarg, B., Wolf, F. Score-P: A joint performance measurement runtime infrastructure for Periscope, Scalasca, TAU, and Vampir. In: Brunst, H., Müller, M.S., Nagel, W.E., Resch, M.M., (eds.), *Tools for High Performance Computing 2011* pp. 79–91. Berlin, Heidelberg: Springer Verlag (2012).

17. German Research School of Simulation Sciences. Scalable performance measurement infrastructure for parallel codes. http://www.scalasca.org/, accessed: 2016-07-21.

18. Geimer, M., Hermanns, M.A., Siebert, C., Wolf, F., Wylie, B.J.N. *Scaling Performance Tool MPI Communicator Management*, In: Cotronis, Y., Danalis, A., Nikolopoulos, D.S, Dongarra, J. (eds.), Recent Advances in the Message Passing Interface, Lecture Notes in Computer Science, Vol. 6960, pp. 178–187. Berlin: Springer Verlag (2011).

19. Geimer, M., Wolf, F., Wylie, B.J.N., Ábrahám, E., Becker, D., Mohr, B. The scalasca performance toolset architecture. *Concurr. Comput.* 22(6), 702–719 (2010), http://dx.doi.org/10.1002/cpe.1556.

20. Frings, W., Wolf, F., Petkov, V. Scalable massively parallel I/O to task-local files. In: *Proceedings of the Conference on High Performance Computing Networking, Storage and Analysis (SC '09).* pp. 17: 1–17:11. New York, NY: ACM (2009), http://doi.acm.org/10.114/1654059.1654077.

21. University of Tennessee and Oak Rigde National Laboratory. BLAS Basic Linear Algebra Subprograms. http://www.netlib.org/blas/, accessed: 2016-07-20.

22. University of Tennessee and Oak Rigde National Laboratory. LAPACK–Linear Algebra PACKAGE. http://www.netlib.org/lapack/, accessed: 2016-07-20.

23. Hamano, J. Git. https://git-scm.com/, accessed: 2016-07-20.

24. Edgewall Software: The Trac project. https://trac.edgewall.org/, accessed: 2016-07-20.

25. van Heesch, D. Doxygen. http://www.doxygen.org/, accessed: 2016-07-20.

26. Gohr, A. DokuWiki. https://www.dokuwiki.org/.

27. Hater, T., Anlauf, B., Baumeister, P., Bühler, M., Kraus, J., Pleiter, D. Exploring energy efficiency for GPU-accelerated POWER servers. In: Taufer, M. (ed.), *High Performance Computing, Lecture Notes in Computer Science*, Vol. 9945, pp. 207–227. Cham: Springer International Publishing (2016).

28. Baumeister, P.F., Hater, T., Pleiter, D., Boettiger, H., Maurer, T., Brunheroto, J.R. *Exploiting In-Memory Processing Capabilities for Density Functional Theory Applications*. In: Desprez, F., Dutot, P.-F., Kaklamanis, C., Marchal, L., Molitorisz, K., Ricci, L., Scarano, V., Vega-Rodriguez, M.A., Varbanescu, A.L., Hunold, S., Scott, S.L., Lankes, S., Weidendorfer, J. (eds.), *Euro-Par 2016: Parallel Processing Workshops, Lecture Notes in Computer Science*, Vol. 10104, pp. 750–762. Cham: Springer International Publishing (2017).

20 Real-Space Multiple-Scattering Theory and Its Applications at Exascale

Markus Eisenbach and Yang Wang

CONTENTS

20.1 Introduction..449
20.2 MST and Linear Scaling...450
20.3 Performance of Original LSMS...453
20.4 Code Refactoring for Multithreaded Architectures 455
    20.4.1 Communication...455
    20.4.2 GPU Acceleration...456
20.5 Performance and Scaling of LSMS 457
20.6 Future Needs and Challenges for Exascale 459
Acknowledgments 459
References 460

20.1 INTRODUCTION

In recent decades, the *ab initio* methods based on density functional theory (DFT) (Hohenberg and Kohn 1964, Kohn and Sham 1965) have become a widely used tool in computational materials science, which allows theoretical prediction of physical properties of materials from the first principles and theoretical interpretation of new physical phenomena found in experiments. In the framework of DFT, the original problem that requires solving a quantum mechanical equation for a many-electron system is reduced to a one-electron problem that involves an electron moving in an effective field, while the effective field potential is made up of an electrostatic potential, also known as Hartree potential, arising from the electronic and ion charge distribution in space and an exchange–correlation potential, which is a function of the electron density and encapsulates the exchange and correlation effects of the many-electron system. Even though the exact functional form of the exchange-correlation potential is formally unknown, a local density approximation (LDA) or a generalized gradient approximation (GGA) is usually applied so that the calculation of the exchange–correlation potential, as well as the exchange–correlation energy, becomes tractable while a required accuracy is retained. Based on DFT, *ab initio* electronic structure calculations for a material generally involve a self-consistent process that iterates between two computational tasks: (1) solving an one-electron Schrödinger equation, also known as Kohn–Sham equation, to obtain the electron density and, if needed, the magnetic moment density, and (2) solving the Poisson equation to obtain the electrostatic potential corresponding to the electron density and constructing the effective potential by adding the exchange–correlation potential to the electrostatic potential. This self-consistent process proceeds until a convergence criteria is reached.

At the atomistic length scale, modern *ab initio* electronic structure calculation techniques are capable of determining fundamental electronic and magnetic properties, such as electrical conductivity, magnetic moments, exchange interactions, and magneto crystalline anisotropy of bulk materials or atom clusters. There exist a number of DFT based *ab initio* methods. They are usually classified according to the technique used to solve the one-electron Schrödinger equation or according to the representation used for the electron density, potential, and wave function solutions (the Kohn–Sham orbitals). The representation refers to the kind of the basis functions used to the Kohn–Sham orbitals. These usually fall into two broad classes: local atomic-like orbitals (e.g., Gaussian-type functions, linear muffin-tin orbitals) and plane-waves. Of course, it is possible to avoid using a basis function representation altogether by numerically solving the one-electron Schrödinger equation on grids or finite elements. The choice of the technique and the basis functions is made to minimize the computational costs and ease the programming efforts, while maintaining a sufficient numerical accuracy. Currently, the most popular *ab initio* method for the electronic structure calculation is probably the plane-wave based pseudopotential method.

A theoretical approach to materials simulation usually starts with constructing a unit cell that repeats itself along x-, y-, and z-directions to fill the entire space. The unit cell consists of the constituent atoms in a predetermined proportion and in a real-space distribution to mimic the atomic composition and spatial arrangement in the actual material. A fundamental problem arises, however, with conventional *ab initio* methods when applied to unit cells containing a large number of atoms (~ 100) that the amount of computational work, or more precisely, the number of floating-point operations, increases as $O(N_a^3)$, the third power of the number of atoms (N_a) in the unit cell. In plane-wave based *ab initio* methods, for example, the Kohn–Sham orbital wave function orthogonalization step scales as $O(N^2 N_p)$, where N is the number of Kohn–Sham orbitals, N_p is the number of plane waves, and both N and N_p are proportional to N_a, and the electron density calculation step using FFT scales as $O(N \cdot N_p \log N_p)$. For a moderate system size ($N_a \sim 100$ or less), the prefactor for the $O(N \cdot N_p \log N_p)$ scaling processes dominates so that the electronic structure calculation scales approximately as $O(N_a^2 \log N_a)$. For larger system sizes, however, the computing time spent on the orthogonalization step, which essentially scales as $O(N_a^3)$, will become dominant. Another problem with conventional *ab initio* methods is the lack of efficient schemes for parallel implementation, mainly due to the fact that the dominating computational tasks are global in nature. Especially, when the number of atoms is large, very few **k**-points are actually needed for the Brillouin zone integration and the parallelization over the **k**-points for the band structure calculation and the Brillouin zone integration no longer shows any advantage.

Because of the problems mentioned above, applying conventional *ab initio* methods to the electronic structure calculation for materials with complex structures (e.g., nanostructures, interfaces, defects, etc.) that require large unit cell sizes is obviously prohibitive. To overcome this computational bottleneck, many efforts have been made since early 1990s to develop approximate methods to solve the electronic structure for systems involving large unit cell with an acceptable computational cost. The result is the so-called order-N methods, for which the computational effort of the methods scale linearly, that is, $O(N_a)$, with respect to the number of atoms in the unit cell, rather than cubically like the conventional *ab initio* methods. For the rest of this chapter, we present a linear scaling *ab initio* method based on multiple scattering theory (MST) that shows clear advantage over other *ab initio* methods, we demonstrate its petascale computing capability in the *ab initio* calculation of magnetic and crystal structure phase transition temperatures, and finally we discuss its potential applications at exascale.

20.2 MST AND LINEAR SCALING

As described in the Introduction section, a major task in a DFT based *ab initio* electronic structure calculation is to solve the one-electron Schrödinger equation, in atomic units ($\hbar = 1$ and $m_e = \frac{1}{2}$), as follows,

$$\left[-\nabla^2 + \underline{V}_{\text{eff}}(\mathbf{r}) \right] \underline{\Psi}_\alpha(\mathbf{r}) = \epsilon_\alpha \underline{\Psi}_\alpha(\mathbf{r}) \tag{20.1}$$

where the effective potential $\underline{V}_{eff}(\mathbf{r})$ is a spin-polarized LDA or GGA function of electron density, and it is in general, a 2×2 matrix in spin space. Subscript α is a label identifying the quantum state of the Kohn–Sham orbital wave function $\underline{\Psi}_\alpha(\mathbf{r})$, associated with energy eigenvalue ϵ_α. The valence-electron density is a summation of the probability density for both spin up ($\sigma = 1$) and spin down ($\sigma = 2$) states of the valence band:

$$\rho(\mathbf{r}) = \sum_{\substack{\alpha \\ \epsilon_\alpha \leq \epsilon_F}} \sum_{\sigma=1}^{2} |\Psi_{\sigma,\alpha}(\mathbf{r})|^2 \tag{20.2}$$

provided that the Kohn–Sham orbital wave functions are orthonormal and the Fermi energy ϵ_F is known. Developed by Korringa (1947) and independently by Kohn and Rostoker (1954), MST offers a powerful tool for solving Equations 20.1 and 20.2. In the MST approach, also known as the Korringa–Kohn–Rostoker (KKR) method, a crystal is considered as made up of nonoverlapping and space filling atomic cells (or Voronoi polyhedra), $\{\Omega_n, n = 1, 2, \ldots\}$, each of which is centered at an atomic site described by a position vector $\mathbf{R}_n$ in the crystal and each atomic cell is considered as an electron scattering center with scattering potential $v^n(\mathbf{r}_n)$ identical to the effective potential inside the cell and zero outside, that is,

$$\underline{v}^n(\mathbf{r}_n) = \begin{cases} \underline{V}_{eff}(\mathbf{r}), & \text{if } \mathbf{r} \in \Omega_n \\ 0, & \text{otherwise} \end{cases} \tag{20.3}$$

where $\mathbf{r}_n = \mathbf{r} - \mathbf{R}_n$ is a coordinates vector centered at $\mathbf{R}_n$. Solving the one-electron Schrödinger (20.1) can thus be cast into a multiple-scattering problem, in which the Bloch wave is essentially a standing wave solution of the multiple-scattering processes.

A multiple-scattering process in a crystal can be described in terms of so-called scattering path matrix $\underline{\tau}^{nm}(\epsilon)$, which depicts all possible scattering processes of an electron traveling from site n to site m. In real-space formalism, the scattering path matrix is given by

$$\begin{aligned} \underline{\tau}^{nm}(\epsilon) &= \underline{t}^n(\epsilon)\delta_{nm} + \underline{t}^n(\epsilon) \sum_{k \neq n} \underline{g}^{nk}(\epsilon)\underline{t}^k(\epsilon) + \underline{t}^n(\epsilon) \sum_{k \neq n} \underline{g}^{nk}(\epsilon)\underline{t}^k(\epsilon) \sum_{j \neq k} \underline{g}^{kj}(\epsilon)\underline{t}^j(\epsilon) \\ &\quad + \underline{t}^n(\epsilon) \sum_{k \neq n} \underline{g}^{nk}(\epsilon)\underline{t}^k(\epsilon) \sum_{j \neq k} \underline{g}^{kj}(\epsilon)\underline{t}^j(\epsilon) \sum_{i \neq j} \underline{g}^{ji}(\epsilon)\underline{t}^i(\epsilon) + \ldots \\ &= \underline{t}^n(\epsilon)\delta_{nm} + \underline{t}^n(\epsilon) \sum_{k \neq n} \underline{g}^{nk}(\epsilon)\underline{\tau}^{km}(\epsilon) \end{aligned} \tag{20.4}$$

where, $\underline{g}^{nk}(\epsilon)$ is free-electron propagator matrix, also known as real-space structure constant matrix, that describes the propagation of a free-electron, with kinetic energy ϵ, from site n to site k, $\underline{t}^n(\epsilon)$ is the single-site scattering t-matrix associated with site n due to potential $\underline{v}^n(\mathbf{r}_n)$. Obviously, the second equation for $\underline{\tau}^{nm}(\epsilon)$ can be cast into a matrix inverse problem as follows,

$$\underline{\tau}^{nm}(\epsilon) = \begin{pmatrix} [\underline{t}^1(\epsilon)]^{-1} & \underline{g}^{12}(\epsilon) & \underline{g}^{13}(\epsilon) & \cdots \\ \underline{g}^{21}(\epsilon) & [\underline{t}^2(\epsilon)]^{-1} & \underline{g}^{23}(\epsilon) & \cdots \\ \underline{g}^{31}(\epsilon) & \underline{g}^{32}(\epsilon) & [\underline{t}^3(\epsilon)]^{-1} & \cdots \\ \vdots & \vdots & \vdots & \ddots \end{pmatrix}^{-1}_{nm} \tag{20.5}$$

where the subscript nm on the right hand side implies the subblock at the nth row and mth column of the big matrix after the inverse is taken. For a crystal with infinite number of atoms, the number subblocks of the big matrix is of course infinite. However, if the crystal has a periodic structure, the matrix inverse in (20.5) can be reduced to the inverse of a matrix with $N_a \times N_a$ subblocks.

One important advantage of the MST method is its convenient access to the Green function for one-electron Schrödinger (20.1). In the vicinity of atomic site n, the Green function can be expressed in terms of the single-site regular solutions $Z_L^n(\mathbf{r}_n; \epsilon)$ and irregular solutions $J_L^n(\mathbf{r}_n; \epsilon)$ due to single-site potential $v^n(\mathbf{r}_n)$ as follows,

$$G_{\sigma\sigma'}(\mathbf{r}_n, \mathbf{r}_n'; \epsilon) = \sum_{LL'} Z_{L\sigma}^n(\mathbf{r}_n; \epsilon) \tau_{L\sigma L'\sigma'}^{nn}(\epsilon) Z_{L'\sigma'}^{n*}(\mathbf{r}_n; \epsilon) - \sum_L Z_{L\sigma}^n(\mathbf{r}_n; \epsilon) J_{L\sigma}^{n*}(\mathbf{r}_n; \epsilon) \delta_{\sigma\sigma'} \tag{20.6}$$

where matrix index L simply represents a combination of angular momentum quantum number l and magnetic quantum number m. This allows us to calculate the electron density and magnetic moment density associated with the valence states in atomic cell Ω_n by taking the imaginary part of Green function trace integrated in valence energy band,

$$\rho^n(\mathbf{r}_n) = -\frac{1}{\pi} \mathrm{ImTr} \int_{\epsilon_B}^{\epsilon_F} \underline{G}(\mathbf{r}_n, \mathbf{r}_n; \epsilon) d\epsilon$$

$$\mathbf{m}^n(\mathbf{r}_n) = -\frac{1}{\pi} \mathrm{ImTr} \int_{\epsilon_B}^{\epsilon_F} \underline{\sigma} \cdot \underline{G}(\mathbf{r}_n, \mathbf{r}_n; \epsilon) d\epsilon \tag{20.7}$$

In Equation 20.7, $\underline{\sigma} = (\underline{\sigma}_x, \underline{\sigma}_y, \underline{\sigma}_z)$ is the Pauli matrix vector, ϵ_B is the bottom energy of the valence band, and the energy integration can be conveniently carried along a contour in the complex energy plane so that only few tens of complex energy points are necessary. Obviously, this Green function approach makes the calculation of the band structures and Kohn–Sham orbital wave functions unnecessary and consequently, the time-consuming procedure for orthogonalizing and normalizing the wave functions can be entirely avoided. It is necessary to note that the MST method is an all-electron approach to the *ab initio* electronic structure calculation. All-electron means that electronic states for both valence and core electrons are treated on equal footing—in contrast to pseudopotential methods where core electrons are not explicitly treated. Another important feature of the MST method is that the only global operation required for obtaining the Green function is the calculation of a multiple-scattering matrix for each atom. It is this step that accounts for the major portion of the floating-point operations of the entire electronic structure calculation. Specifically, like other conventional *ab initio* methods, the MST method still suffers cubic scaling limitation since the calculation of multiple-scattering path matrices in (20.5) requires inverse of a matrix whose size is proportional to the number of atoms in the unit cell, and therefore its application is essentially restricted to problems whose size is within less than a thousand atoms.

Moreover, the MST method becomes linear scaling if one approximates the calculation of $\tau_{L\sigma L'\sigma'}^{nn}(\epsilon)$, $n = 1, 2, \ldots, N_a$, by neglecting the multiple-scattering processes that involve atoms at a distance greater than a cut-off radius R_{LIZ} from atom n. This is the essence of the locally self-consistent multiple-scattering (LSMS) method (Wang et al. 1995). The idea behind this approximation is based on the observation that the scattering processes involving far away atoms influence the local electronic states less and less as the distance from the scatter under study is increased, an example of nearsightedness proposed by W. Kohn. In the LSMS method, the space within R_{LIZ} centered at an atom is called local interaction zone (LIZ) of the atom. If there are M_a atoms in the LIZ at atom n, the time cost for calculating $\tau_{L\sigma L'\sigma'}^{nn}(\epsilon)$, and thus the Green function, for atom n does not depend on N_a, rather it depends on M_a. Since we only have to repeat the Green function calculation for each atom, the total time cost for the entire electronic structure calculation will scale linearly with respect to N_a, the number of atoms in the unit cell. In addition, parallelism is intrinsic to the method since the Green function calculation for each atom and each energy point along the complex contour is essentially independent. Consequently, there are no global operations involved in the process of calculating the Green function other than A few trivial global sum operations, for

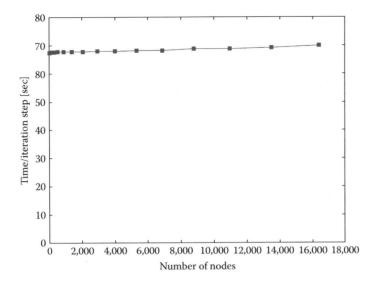

FIGURE 20.1 Weak scaling of LSMS. This figure demonstrates the weak scaling of LSMS in the number of atoms for the GPU-accelerated version on Titan. Thus, 16 atoms on 4 nodes require 67 seconds per iteration step and 65,536 atoms on 16,384 nodes require 70 seconds, resulting in a parallel scaling efficiency of 96% across Titan.

example, the summation of the net charge in each atomic cell for the determination of the electron chemical potential, and therefore the code is highly parallel, as demonstrated in Figure 20.1. The LSMS method has proved to be a very useful tool for performing the *ab initio* calculation for complex structures involving tens of thousands of atoms. It was the first scientific application to pass the teraflop computing speed barrier while investigating the magnetic properties of a non collinear magnetic structure of 1458 iron atoms and was one of the very few scientific applications that demonstrated early petascale computing capability.

20.3 PERFORMANCE OF ORIGINAL LSMS

The original design of LSMS (Wang et al. 1995) targeted parallel distributed memory architectures that became the standard for high-performance computing architectures during the 1990s and early 2000s. The success of the LSMS approach for scaling of first principles calculations is demonstrated by its success in winning the Gordon-Bell prize in 1998 (Ujfalussy et al. 1998) and it being one of the first codes to surpass 1 TFlop/s sustained performance. The performance potential of the LSMS method is largely based on the observation that for typical calculations, 95% of floating-point operations are spent in dense linear algebra for double precision complex numbers. These floating-point operations are dominated by the matrix inversion to calculate the scattering path matrix τ as shown in Equation 20.5.

To calculate the energy of the system and the charge and magnetization densities, the LSMS algorithm only requires the diagonal block of the τ-matrix corresponding to the central site of the LIZ. This observation allows us to effectively reduce the number of operations required to calculate the matrix inverse by considering subblocks of the matrices and employing the Schur complement (Schur 1905)

$$\begin{pmatrix} A & B \\ C & D \end{pmatrix}^{-1} = \begin{pmatrix} U & V \\ W & Y \end{pmatrix} \tag{20.8}$$

where A, D, U, and Y are square matrices and

$$U = (A - BD^{-1}C)^{-1} \tag{20.9}$$
$$V = -(A - BD^{-1}C)^{-1}BD^{-1} \tag{20.10}$$
$$W = -D^{-1}C(A - BD^{-1}C)^{-1} \tag{20.11}$$
$$Y = D^{-1} + D^{-1}C(A - BD^{-1}C)BD^{-1} \tag{20.12}$$

In particular, we utilize the expression for the upper diagonal block U and apply it multiple times to formulate an iterative procedure to obtain the site diagonal block of the scattering path matrix τ. The diagonal blocks A and U have size $k \times k$ and the blocks D and Y have size $k' \times k'$. Our algorithm successively reduces the size of the matrix to be inverted from $k + k'$ to k and thus reduces the sizes of the matrices to be multiplied and inverted. While this algorithm retains the computational complexity of the underlying matrix inversion and multiplication algorithms of $O(N^3)$, the total number of operations, and thus the prefactor of the computational cost is reduced with respect to a full inversion of the multiple-scattering matrix. The algorithm that results from this is detailed in Algorithm 20.1. Here, the size N of the matrix is divided into a sequence os subblocks of size k_i. While the top right subblock of size k_1 has to contain the desired subblock of the inverse matrix, the size of all the other subblocks k_2 to k_n is arbitrary and their number and size provides opportunities for tuning and adaptation to different architectures.

Algorithm 20.1   Block inversion algorithm to calculate the top diagonal $k_1 \times k_1$ block of an $N \times N$ matrix A

1:   set block sizes $k_{1...n}$ such that k_1 is the size of the desired sub block of the inverse of matrix A which has size $N = \sum_i k_i$.

2:   $K \leftarrow N$ initialize the start of the subblock to be inverted

3:   **for** $i = n$ **down to** 2 **do**

4:     $K \leftarrow K - k_i$

5:     $A_{K...K+k_i,K...K+k_i} \leftarrow A_{K...K+k_i,K...K+k_i}^{-1}$

6:     $A_{1...K-1,1...K-1} \leftarrow$
     $A_{1...K-1,1...K-1} - A_{1...K-1,K...K+k_i}A_{K...K+k_i,K...K+k_i}A_{K...K+k_i,1...K-1}$

7:   **end for**

8:   $A_{1...k_1,1...k_1} \leftarrow A_{1...k_1,1...k_1}^{-1}$

This algorithm can be readily implemented using standard dense linear algebra libraries such as Basic Linear Algebra Subprograms (BLAS) and Linear Algebra Package (LAPACK). As our matrices are general double precision complex matrices, the matrix multiplications in step six use the BLAS routine zgemm. Multiple options exist for the matrix inversion of the lower block in step five and for the final inversion in step eight. Since the τ^{-1} matrix is generally a non-Hermitian complex matrix, many efficient iterative solver algorithms that rely on Hermitian matrices are not available. Thus, we use either a quasi-minimal residual (QMR) iterative algorithm (Freund and Nachtigal 1991) or a direct LU solver, which we utilize in the implementation of the matrix block inversion for the Graphics processing unit (GPU) accelerators due to its deterministic behavior. The resulting Fortran routine that implements this algorithm in the original version of LSMS is shown in listing 20.1. For typical calculations with a LIZ with ≈ 100 atoms and an angular momentum cutoff $l_{max} = 3$, this single subroutine accounts for $\approx 90\%$ of the runtime of the CPU version of LSMS.

Listing 20.1 CPU version of the matrix block inversion function `zblock_lu`

```
      subroutine zblock_lu(a,lda,blk_sz,nblk,ipvt,mp,idcol,k)
 ...
c Do block LU
      n=blk_sz(nblk)
      joff=na-n
      do iblk=nblk,2,-1
      m=n
      ioff=joff
      n=blk_sz(iblk-1)
      joff=joff-n
c invert the diagonal blk_sz(iblk) x blk_sz(iblk) block
      call zgetrf(m,m,a(ioff+1,ioff+1),lda,ipvt,info)
c calculate the inverse of above multiplying the row block
c blk_sz(iblk) x ioff
      call zgetrs('n',m,ioff,a(ioff+1,ioff+1),lda,ipvt,
     &    a(ioff+1,1),lda,info)
      if(iblk.gt.2) then
      call zgemm('n','n',n,ioff-k+1,na-ioff,cmone,a(joff+1,ioff+1),
         lda,
     &    a(ioff+1,k),lda,cone,a(joff+1,k),lda)
      call zgemm('n','n',joff,n,na-ioff,cmone,a(1,ioff+1),lda,
     &    a(ioff+1,joff+1),lda,cone,a(1,joff+1),lda)
      endif
      enddo
      call zgemm('n','n',blk_sz(1),blk_sz(1)-k+1,na-blk_sz(1),cmone,
     &    a(1,blk_sz(1)+1),lda,a(blk_sz(1)+1,k),lda,cone,a,lda)
      end
```

20.4 CODE REFACTORING FOR MULTITHREADED ARCHITECTURES

The original implementation of LSMS (Wang et al. 1995) only anticipated distributed memory parallelism and a message passing paradigm. Hence, in preparing the code to efficiently utilize multicore architectures and to make effective use of the accelerators provided on Oak Ridge National Laboratory's (ORNL) Titan and similar and future machines, major restructuring of the data structures in our LSMS code was required. Additionally, a complete rewrite of the original LSMS code proved to be too time intensive and therefore impractical, so we had to find a compromise balance between retaining original Fortran 77 subroutines and newly written functions. This endeavor was facilitated by the structure of the Fortran code that avoided the use of global variables, common blocks, and saved values in subroutines, while passing all input parameters and results explicitly through subroutine arguments, thus minimizing side effects throughout the code.

20.4.1 COMMUNICATION

The distribution of atoms among the compute nodes is determined at the start of the calculation and remains fixed throughout a run. Since the computational effort per atom in a regular lattice structure is nearly the same for all atoms in the system, a distribution that places the same number of atoms per compute node is sufficient and does not cause any load balancing issues. In noncrystalline systems, where the size of the LIZ changes significantly between sites, this is not necessarily the case and further efforts to achieve satisfactory load balancing will be needed.

Once the atoms are distributed between the compute nodes, the communication pattern can be determined. Algorithm 20.2 constructs these communication lists based on the atoms contained in the LIZ of each atomic site. As the data required to construct the communication pairs is available locally, this does not require any communication by itself. The algorithm constructs two lists for each node: a list `tmatFrom` of remote sites that will send data to the local node and a list of the remote nodes `tmatTo` that will expect to receive data for sites from the local node. Note that Algorithm 20.2 scales quadratic with the number of atom sites in the system, as it has to loop over all atoms in the system, not only the local atoms and build their LIZ. The construction of the LIZ itself requires a loop over all atom sites in our naive implementation. Since the construction of the communication lists is performed only once at startup of the code, this does not represent a serious performance concern.

Algorithm 20.2   The construction of the LIZ communication lists

---

    for all atoms i in the crystal **do**
        build the local interaction zone $LIZ_i = \{j | dist(\mathbf{x}_i, \mathbf{x}_j) < r_{LIZ}\}$ of atom i
        for all atoms j in LIZ_i **do**
            add atom j to the list R_i of data to receive for atom i (`tmatFrom`)
            add atom i to the list S_j of data to send from atom j (`tmatTo`)
        end for
    end for
    remove duplicate entries from S_j and R_i

Each compute node i hosts a fixed number of atoms N_{atom}^i. The single-site t-matrices that are needed in a MPI rank i to construct the τ-matrices are stored in a linear array `tmatStore`, such that `tmatStore[j]` contains the t-matrices for the atoms local to this node if $j < N_{atom}^i$ and remote scatterers otherwise. The local data is filled at the beginning of each self-consistency step by the single-site scattering solver and the single-scatterer t-matrices are exchanged according to the communication lists built at start-up using nonblocking MPI-point to point communication. Even in the GPU-accelerated version of LSMS, this communication is infrequent enough (from once every few second to about once a minute), that communication overheads do not impact the scalability of the code.

20.4.2   GPU Acceleration

As significant amount of computational effort of LSMS is spent in inverting the multiple-scattering matrix, a non-Hermitian complex matrix, this formed the main focus for porting to GPUs and the experience is guiding the portability to future platforms. The algorithm described above for obtaining the top diagonal block of the τ-matrix enabled us to perform the matrix inversion completely within the GPU memory using multiple kernels without the need for data transfer between the CPU host and the GPU side of the node. After the inverse τ-matrix is constructed in the device memory, the code follows the same outline as the CPU algorithm with minor modification. The calls to the BLAS routine (ZGEMM) for matrix multiplication are readily replaced by their cuBLAS equivalents (`cublasZgemm`). The matrix inversion of the diagonal blocks that arise in Algorithm 20.1 has been implemented in a custom kernel that was optimized for the Kepler architecture used on Titan. This inversion kernel performs an LU factorization and back-substitution inside a single kernel that does not require any data transfer between the GPU and CPU. Due to the restrictions of the GPU architecture, especially the available register memory, the inversion kernel is restricted to a maximum double complex matrix size of ≈ 130, thus determining the block size for the block inversion algorithm.

After calculating the matrix inverse to obtain the τ-matrix, the second most time consuming operation in the nonaccelerated version of the code was the construction of the matrix to be inverted. This construction of the KKR matrix accounts for approximately 40% of the remaining time if the matrix is constructed on the CPU side of the Titan compute nodes and transferred to the GPU memory for inversion. The amount of data transfer can be significantly reduced by constructing the KKR matrix on the GPU. The matrix to be inverted on the right hand side of Equation 20.5 for the scattering path matrix τ depends only on the relation of the positions of the atoms that are represented by the components of the free space Green's function $g_{nm}(\epsilon)$ and the single-site scattering matrices t that are calculated by solving independent initial value problems for each atomic site. As the atom positions can be transferred to the accelerator memory during setup, the positions usually do not change during the calculation. The single-site scattering t-matrices for spherical scattering potentials are calculated on the CPU for the local atoms and the t matrices for remote atoms that contribute to the LIZ are received from the remote nodes that own these atoms. The individual blocks in the inverse τ matrix on the right-hand side of Equation 20.5 can then be independently calculated on the accelerator inside a single kernel that threads over the subblocks that are associated with atom pairs in the LIZ, thus providing substantial opportunity for parallelism, resulting in M^2 independent blocks in τ^{-1} for a LIZ with M atoms. Constructing τ^{-1} on the accelerator significantly reduces the amount of data that needs to be transfered from the CPU to the GPU to be linear in the number of sites in the LIZ from being quadratic if τ^{-1} was constructed on the CPU.

20.5   PERFORMANCE AND SCALING OF LSMS

The LSMS code has been known in the past for its performance and scalability on HPC architectures from the TFlop/s range (Ujfalussy et al. 1998) into the PFlop/s range, such as the previous Jaguar system at Oak Ridge (Eisenbach et al. 2009). By implementing the main computational routines for the construction and inversion of the multiple-scattering matrices as GPU kernels and using the updated communication scheme outlined above, we are able to extend the scalability of LSMS to current accelerated systems such as Titan while utilizing the performance capabilities of these systems (Eisenbach et al. 2017). In addition to the work on the performance and scalability of the multiple-scattering method itself to deal with large systems and to reduce the amount of approximations needed, we are also including the capability to sample over multiple configurations of the system. This has allowed us to implement Wang–Landau Monte Carlo (Wang and Landau 2001) calculations using first principles calculations to obtain finite temperature behavior of magnets (Eisenbach et al. 2011) and alloy phase transitions (Khan and Eisenbach 2016) without the need for approximate models. This additional layer of parallelism that is currently implemented using a master–slave approach (Eisenbach et al. 2009), provides further scalability for these combined first principles and classical statistical physics Wang–Landau Locally Self-consistent Multiple Scattering (WL-LSMS) calculations.

We evaluate the performance and scalability of both the LSMS and the WL-LSMS versions of the code on Titan and compare the GPU-accelerated version with the performance of the CPU only version. First, we demonstrate that the utilization of GPU accelerators in the computationally intensive parts of the code for calculating the multiple-scattering matrix together with a significant restructuring of the high-level architecture of the code and of the communication pattern enabled us to maintain the excellent scalability of LSMS. In Figure 20.1 we show the near-perfect weak scaling of the LSMS code in the number of atom, while maintaining the number of atom per compute node over five orders of magnitude from 16 iron atoms to 65,536 atoms. With four iron atoms per node on the Titan system at Oak Ridge we find that for a LIZ of 113 atoms and angular momentum cutoff of $l_{max} = 3$, the weak scaling parallel efficiency is 96%. This performance puts calculations of million atom size systems within reach for the next generation of supercomputers such as the planned Summit system at the Oak Ridge Leadership Computing Facility.

To measure the performance improvement of the GPU-accelerated version of LSMS, we performed identical WL-LSMS calculation for 1024 iron atoms with 290 walkers on 18,561 nodes on Titan for 20 Monte Carlo steps per walker. We record the runtime of both versions and we count the number of floating-point operations for the CPU-only version using Papi counters. The CPU version executes at a rate of 1.86 PFlop/s and the GPU-accelerated WL-LSMS shows a speedup of the total code of 8.6x resulting in a performance of 14.5 PFlop/s. Additionally, we measured the instantaneous power consumption of Titan for this full machine run of WL-LSMS. The measurements were performed both for the CPU-only calculation and the GPU-accelerated version of the code. The results are shown in Figure 20.2. The difference in power consumption between the compute intensive LSMS calculations, that take most of the time and the Monte Carlo steps that are marked by a significant drop in the power consumption is obvious and this allows a clear comparison of the two runs.

The scaling of the WL-LSMS code is shown in Figure 20.3. The system simulated is the same 1024-iron atom system that was used for the performance and power consumption measurements using 128 compute nodes per Monte Carlo walker. Note that WL-LSMS requires an additional compute node for the master process that controls the Wang–Landau calculation. The code shows good scalability in the number of walkers reaching nearly one Monte Carlo sample per second on 12,289 Titan nodes.

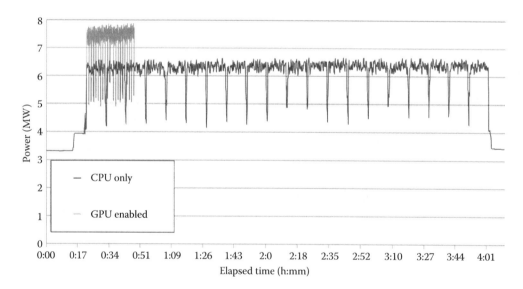

FIGURE 20.2 Comparison of power consumption of LSMS on Titan. This figure shows the power trace for identical runs of WL-LSMS on Titan using only the CPUs (black) and utilizing the GPUs for the scattering matrix calculations described in the text (gray). The calculation performed 20 Wang–Landau steps per walker for 1024 Fe atoms on 18,561 Titan nodes. The GPU code yields 14.5 PF-sustained performance versus 1.86 PF for the CPU-only version. Thus the accelerated code shows a 8.6x speedup and a 7.3x reduction in power consumption. (3,500 kWh vs. 25,700 kWh). (From Eisenbach, M. et al., *Comput. Phys. Commun.*, 211, 2–7, 2017.)

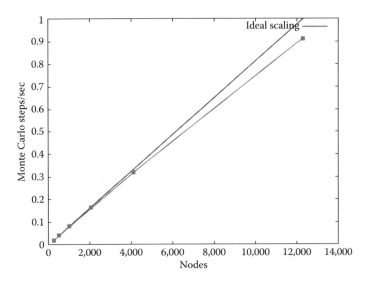

FIGURE 20.3   Scaling of WL-LSMS performance on Titan. We show the scaling of the Monte Carlo Wang–Landau sampling for a 1024-Fe-atom simulation cell using 128 compute nodes per Monte Carlo walker. The code shows good scaling in the number of states sampled. With two walkers on 257 nodes WL-LSMS generates 0.0208 samples/s and with 96 walkers on 12,289 nodes 0.9128 samples/s resulting in a parallel efficiency of 92%.

20.6   FUTURE NEEDS AND CHALLENGES FOR EXASCALE

Our current effort lies in expanding the capabilities of the LSMS code to enable the energy and force calculations for atomic displacements, thus allowing us to perform relaxation and first principles molecular dynamics simulations at length scales that approach what is currently possible with classical force field methods. We expect that our experience with the dense linear algebra part of the code will continue to translate to future architectures and that the use of libraries in the spirit of BLAS and LaPACK will continue to provide the bulk of the performance in the multiple-scattering part of the LSMS code together with few and optimized kernels. Thus, the main new challenges we will face are related to the solution of the single-scatterer problem that has not been an issue for the performance of the LSMS code to date.

In the currently implemented multiple-scattering formalism in LSMS, we assume spherical atomic scattering potentials and a constant interstitial potential between the atomic sites, which is computationally cheap to treat and does not create any performance concerns. To be able to relax lattice structures, perform first-principles molecular dynamics or include lattice vibration within statistical physics calculations, it is necessary to go beyond this approximation and consider space filling nonspherical scattering potentials. The differential equations for these potentials (Liu et al. 2016) are significantly more expensive to solve, approaching the cost of the matrix inversion for typical LIZ sizes. Consequently, it will be important to port these full-potential scattering solvers to GPUs in future versions of the code.

Additionally, large-scale calculations for nonspherical potentials will require a scalable approach to solving the Poisson equation for space-filling charge densities on the grids used by the multiple-scattering method.

ACKNOWLEDGMENTS

This work has been sponsored by the U.S. Department of Energy, Office of Science, Office of Advanced Scientific Computing Research (software optimization and performance measurements) and Basic Energy Sciences, and Materials Sciences and Engineering Division (basic theory). The work presented here used resources of the Oak Ridge Leadership Computing Facility, which is supported by the Office of Science of the U.S. Department of Energy under contract no. DE-AC05-00OR22725.

REFERENCES

Eisenbach, M., Larkin, J., Lutjens, J., Rennich, S., and Rogers, J. H. (2017). GPU acceleration of the locally selfconsistent multiple scattering code for first principles calculation of the ground state and statistical physics of materials. *Comput. Phys. Commun.*, 211:2–7.

Eisenbach, M., Nicholson, D. M., Rusanu, A., and Brown, G. (2011). First principles calculation of finite temperature magnetism in Fe and Fe3C. *J. Appl. Phys.,* 109(7):07E138.

Eisenbach, M., Zhou, C.-G., Nicholson, D. M., Brown, G., Larkin, J., and Schulthess, T. C. (2009). A scalable method for ab initio computation of free energies in nanoscale systems. In *Proceedings of the Conference on High Performance Computing Networking, Storage and Analysis, SC '09*, pages 64:1–64:8, New York, NY: ACM.

Freund, R. and Nachtigal, N. (1991). QMR: A quasi-minimal residual method for non-Hermitian linear systems. *Numer. Math.*, 60:315–339.

Hohenberg, P. and Kohn, W. (1964). Inhomogeneous electron gas. *Phys. Rev.*, 136:B864–B871.

Khan, S. N. and Eisenbach, M. (2016). Density-functional Monte-Carlo simulation of CuZn order-disorder transition. *Phys. Rev. B*, 93(2):024203.

Kohn, W. and Rostoker, N. (1954). Solution of the Schrödinger equation in periodic lattices with an application to metallic Lithium. *Phys. Rev.*, 94:1111–1120.

Kohn, W. and Sham, L. J. (1965). Self-consistent equations including exchange and correlation effects. *Phys. Rev.*, 140:A1133–A1138.

Korringa, J. (1947). On the calculation of the energy of a Bloch wave in a metal. *Physica*, 13:392–400.

Liu, X., Wang, Y., Eisenbach, M., and Stocks, G. M. (2016). A fullpotential approach to the relativistic single-site Green's function. *J. Phys: Condens. Matter*, 28(35):355501.

Schur, I. (1905). Neue Begründung der Theorie der Gruppencharaktere. *Sitzungsberichte der Königlich Preussischen Akademie der Wissenschaften*, pages 406–432, Berlin: Verlag der königlichen Akademie der Wissenschaften.

Ujfalussy, B., Wang, X., Zhang, X., Nicholson, D., Shelton, W., Stocks, G., Canning, A., Wang, Y., and Gyorffy, B. (7–13 November 1998). High performance first principles method for complex magnetic properties. In *Proceedings of the ACM/IEEE Supercomputing 98 Conference*, Orlando, FL, IEEE Computer Society, Los Alamitos, CA, 90720.

Wang, F. and Landau, D. P. (2001). Efficient, multiple-range random walk algorithm to calculate the density of states. *Phys. Rev. Lett.*, 86(10):2050–2053.

Wang, Y., Stocks, G. M., Shelton, W. A., Nicholson, D. M. C., Temmerman, W. M., and Szotek, Z. (1995). Order-n multiple scattering approach to electronic structure calculations. *Phys. Rev. Lett.*, 75:2867.

21 Development of QMCPACK for Exascale Scientific Computing

Anouar Benali, David M. Ceperley, Eduardo D'Azevedo,
Mark Dewing, Paul R. C. Kent, Jeongnim Kim,
Jaron T. Krogel, Ying Wai Li, Ye Luo,
Tyler McDaniel, Miguel A. Morales, Amrita Mathuriya,
Luke Shulenburger, and Norm M. Tubman

CONTENTS

21.1 Introduction and Scientific Methodology 461
21.2 Algorithms 463
21.3 Data Layout Optimizations for SMP Nodes 465
    21.3.1 AoS-to-SoA Transformation of Outputs...467
    21.3.2 AoSoA Transformation (Tiling) 468
    21.3.3 Exposing Greater Parallelism 470
21.4 High-Performance Determinant Update Algorithms..471
    21.4.1 Delayed Rank-k Update...471
    21.4.2 Performance Results 472
21.5 Adoption of Workflow Tools 473
21.6 Development Practices and Strategies 476
    21.6.1 Unit Testing 476
21.7 Summary...477
Acknowledgments 478
References 478

21.1 INTRODUCTION AND SCIENTIFIC METHODOLOGY

QMCPACK is an open source code for performing electronic structure calculations using quantum Monte Carlo (QMC) techniques [1–3]. Why are electronic structure computations important? The electronic structure describes the behavior of the electrons in a material, and consequently, underlies many scientific disciplines such as condensed matter physics, materials science, biophysics, chemistry, and astrophysics. In electronic structure calculations, one solves the equations of quantum mechanics to determine the electronic structure for a set of atoms. In principle, all the properties of the system, whether gas, solid, or molecular, can be determined from this solution. The underlying physics and chemistry can therefore be elucidated and routes to improving desired properties be identified.

QMC methods provide the most accurate general computational approach to obtain the energy of electronic systems. Although these methods require significantly more computational effort than other popular methods such as density functional theory, in general the results have smaller and potentially quantifiable errors. The ability to quantify errors is a key distinction from other methods.

In addition, the methods can more readily take advantage of highly parallel architectures than most other electronic structure methods as we describe below. For these reasons, QMC methods are expected to be included among the major applications on exascale resources.

The scientific possibilities for QMC with exascale computing will lead to many new applications. With petascale supercomputing resources, we have already seen advances in the simulation of molecular systems with multideterminant wavefunctions [4], simultaneous optimization of joint electronic and ionic degrees of freedom in wavefunctions [5] and simulations of strongly correlated solids such as transition metal oxides and cuprate systems [6,7]. In many cases, these simulations show close agreement with experiment, and yet many questions about the electronic structure of molecular systems and strongly correlated solids remain unraveled. With the next generation of computing resources, we expect to be able to treat larger, more complex systems at a higher level of accuracy than is possible today.

What distinguishes QMC methods? Stated simply, QMC refers to a family of methods that solve the many-body Schrödinger equation using statistical sampling algorithms. In the next section and the rest of this chapter, we describe two of the most commonly used methods implemented in QMC-PACK, namely variational Monte Carlo (VMC) and diffusion Monte Carlo (DMC). These algorithms construct random walks of electrons treated as particles, that is, they produce a series of samples, or snapshots, of electron positions in the material or molecule. Taken together, these snapshots represent the solution to Schrödinger's equation–the wavefunction–from which physical properties can be derived. To achieve high accuracy, tabulated trial wavefunctions on 3D grids are used both for importance sampling and also to enforce the fermionic nature of the electronic wavefunction. QMC methods therefore have computational requirements that are different from purely particle based methods such as classical molecular dynamics. For a text book description of these methods see [8]. Many methodological details and example applications to molecules and solids are also discussed in review articles [9,10].

The VMC and DMC methods are closely related to each other: VMC is typically a preliminary step to a DMC calculation and shares most of the code with common procedures and data. We can characterize the size of a computation by the number of electrons (N), though in reality many more variables enter into consideration for the resources, a calculation will require (e.g., type of atoms, desired accuracy and properties, etc.). Typical calculations today use N between one hundred and one thousand electrons and for most purposes, this can be considered the effective system size. This number is limited by available memory, overall computational cost, and floating-point requirements. Availability of exascale hardware will naturally alleviate the floating-point limitation on system size, N, but the memory limitations are expected to become more complex. Although increase in total memory storage is expected, this will be accompanied by more complex and deeper memory architectures. These developments are forcing reevaluation of the established algorithms used in QMC that we outline in Section 21.2, and we therefore focus on optimization targeted at improving on-node efficiency in this article.

In Section 21.3 we present data layout optimizations that are specifically targeted for improving memory and vectorization efficiency on multicore nodes, such as Intel® Xeon Phi™ processors. These optimizations also facilitate increased on-node parallelism. In Section 21.4 we present an improved algorithm for the determinant updates that are the leading computational cost in QMC. Current science investigations are also motivating the adoption of workflow systems to treat the complexity, ease reproducibility and data sharing. In Section 21.5 we describe the Nexus workflow system that we have developed.

The likely greater heterogeneity of supercomputer architectures and supercomputer systems is creating a pressure for increased code complexity. This requires the adoption of modern software engineering approaches which has not always been the case in this scientific domain. In Section 21.6 we review these approaches in the context of QMCPACK and expect that they will be useful when planning the evolution of other science codes toward the exascale.

21.2 ALGORITHMS

In this section, we outline what drives the choice of algorithms used in QMC and QMCPACK. Internally, QMC codes are similar to particle-based classical molecular dynamics codes with the addition of a limited amount of simple dense linear algebra.

Given a Hamiltonian that describes electrons in a material,

$$H = \sum_{i=1}^{N} [-\frac{1}{2}\nabla_i^2 + v_{\text{ext}}(r_i)] + \sum_{i<j}^{N} \frac{1}{r_{ij}} \tag{21.1}$$

our goal is to find the lowest eigenvalue in the 3N-dimensional space of electron coordinates consistent with particle symmetry: $H\Psi(R) = E\Psi(R)$ where $\Psi(r_1, ..., r_N)$ is the wavefunction and E the energy. The first two terms in the Hamiltonian are the kinetic energy and the single particle external potential, respectively. The last term in the Hamiltonian is the electron–electron interaction. Without this term, the Schrödinger equation would reduce to solving a 3D partial differential equation; with this term, the problem is in 3N dimensions. The wavefunction must be antisymmetric (change sign) under exchange of electron coordinates. This requirement makes the problem particularly difficult; in the context of QMC algorithms, it is known as the fermion sign problem.

One of the simplest QMC methods is VMC, which is a direct application of the variational method. If the trial wavefunction $\Psi_T(R)$ satisfies the requisite symmetries and boundary conditions, then the ratio $E_V = \int dR\Psi_T^*(R)H\Psi_T(R)/\int |\Psi_T(R)|^2$ is an upper bound to the exact ground state energy. Instead of performing the integrals by explicit quadrature, one performs a random walk in the 3N-dimensional coordinate space of the electrons using Monte Carlo sampling. By avoiding the need to use forms that facilitate numerical quadrature, use of Monte Carlo sampling allows the use of much more complex trial wavefunctions, and hopefully, ones that encapsulate electronic correlation and essential physics succinctly. To illustrate VMC within this article, we consider only the commonly-used generalized Slater–Jastrow function:

$$\Psi_T(R) = \exp(-U(R)) \sum_k c_k \det_k(\phi_i(r_j)) \tag{21.2}$$

Here $U(R)$ is a real function (Jastrow function) of all the electron coordinates, which is symmetric under electron exchange and serves to correlate the electrons (keep them away from each other). On the other hand, the determinants make the wavefunction antisymmetric. The 3D functions ϕ_i in the determinants are the orbitals, and the determinants differ by which orbitals are included. The Jastrow function $U(R)$ and the orbitals are parameterized and their values optimized to minimize E_V. Most commonly, the Jastrow function is a sum of short-ranged functions of combinations of electron–electron and electron–atom distances.

The overwhelming preponderance of time in QMC calculations is spent in evaluating the trial function during the random walk. During the random walk, electrons are moved one at a time and the new trial move is accepted or rejected based on the changes in the trial wavefunction. This means that the new orbitals in the determinants are evaluated and the new value of the determinant computed. The change in the Jastrow function part of the trial wavefunction is also computed by evaluating the terms in the Jastrow factor that depend on the moved electron.

In Figure 21.1, we show execution profiles obtained for a recent research calculation [11]. Three quarters of the runtime is spent evaluating data needed to update the trial wavefunction. This fraction is expected to increase with larger problem sizes (N).

The largest cost in computing the trial wavefunction is the evaluation of the determinants. In order to accelerate the evaluation, the matrix inverses are kept and updated as the electronic positions evolve. Hence, each electron step requires $\mathcal{O}(N^2)$ computation, while a full recomputation of the

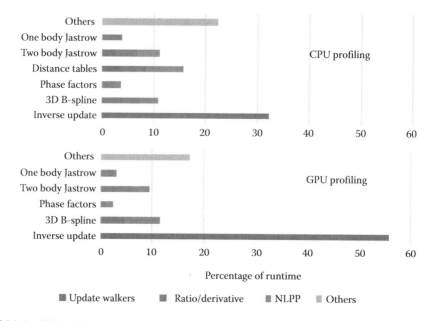

FIGURE 21.1 CPU and GPU profiles for a 108-atom $(TiO_2)_{36}$ supercell with 864 electrons. The CPU profile was obtained on IBM BlueGene/Q platform and the GPU profile was obtained using NVIDIA K20X GPUs.

determinant would require $\mathcal{O}(N^3)$ computation. Details of these updates are discussed in more detail below. The memory requirements are dominated by the storage of the orbitals. In a material with N electrons, values of $N/2$ orbitals are kept on a 3D grid with a resolution of approximately 0.2 nm. Given typical system sizes (N) and grid sizes, the total memory requirements are therefore measured in GBytes. To evaluate the orbitals quickly these must be kept local to where the computation is done, that is, on each compute node. Optimizing memory layouts for speed of evaluation is critical. We will discuss approaches to this in Section 21.3.

The VMC algorithm described above gives the lowest upper bound consistent with the assumed trial function. DMC takes it a step further by finding the lowest energy consistent with the nodes (or phase if it is complex) of the assumed trial function. That is, let us define the many-body phase of the trial function by $\Theta(R) = Im[\ln(\Psi_T(R))]$. Then in the class of functions $\exp[-U(R) + i\Theta(R)]$, we minimize the ground-state energy with respect to $U(R)$. This can be mapped to a random walk problem where instead of working in a space of a single system, it is formulated in an ensemble $\{R_i\}$ of P walkers. Each walker executes a VMC random walk for a certain number of steps, then based on the error of the trial wavefunction, branches by either dying or making one or more copies of itself. DMC has been found to lower the VMC error by roughly an order of magnitude. Although a more complex algorithm, the computational operations are very similar to VMC which enables considerable code reuse.

One method to run DMC in parallel is obvious: each processor holds a few walkers, between a few and a hundred. During a simulation cycle, a fixed number of walkers are advanced one *pass* (moves of each electron are attempted once) and the branching factor of each new configuration is evaluated. To maintain an overall normalization of the population, we make use of the average energy of all the walkers, which implies a synchronization between processors. Optimal ways to perform this and the related load-balancing steps have been explored [12]. In addition, global averages of physical properties need to be performed and outputted.

There are many additional opportunities for parallelism: to reduce effects of the small number of electrons in the simulation cell, one often averages over boundary conditions [13]; each boundary condition is essentially an independent calculation with results averaged together in a postprocessing step. Although improvements in parallelization will likely be needed in the future, efficiencies of over 95% are already obtainable on 1 million Blue Gene cores. This is possible due to the relatively large computational cost of evaluating the trial wavefunctions, often leading to periods of seconds between communication. Speeding up the trial wavefunction evaluation to reduce time to solution is, therefore, the primary challenge for exascale computation, and the focus of this chapter.

21.3   DATA LAYOUT OPTIMIZATIONS FOR SMP NODES

One of the main trends of HPC systems is the increasing parallelism available in a single multicore shared-memory processor (SMP) node. The average number of nodes of big clusters has not increased in recent years but the compute capacity of a node has been increasing through using more cores, multiple hardware threads per core, and wider SIMD units. The manycore architecture expands the parallelism on a node dramatically. For example, the second generation Intel® Xeon Phi™ processors, formerly codenamed Knights Landing (KNL) [14], have up to 72 cores, four threads per core, and a double-precision SIMD width of eight. Its theoretical peak is more than 10 times of that of an IBM Blue Gene/Q (BG/Q) node of 16 cores, four threads per core and a double-precision SIMD width of four [15].

In QMCPACK, most of the computational time is spent on evaluations of ratios, derivatives for quantum forces, and local energies. The choice of the single particle orbital (SPO) representation (the ϕ_i functions in Equation 21.2) is crucial for QMC efficiency and scalability with problem size. For a proposed electron move, the value, gradient, and laplacian of N SPOs are evaluated at a random particle position and $N_{ion}N$ values at random positions are evaluated to compute the pseudopotential contribution with N_{ion} ions (v_{ext} in Equation 21.1). The cost to compute the value of ϕ scales linearly with the number of basis function evaluations which tends to grow with the system size. This amounts to $\mathcal{O}(N^2)$ cost for each particle move and in total SPO evaluations scale as $\mathcal{O}(N^3)$ per MC step.

For this reason, it is efficient to use a localized basis with compact support. In particular, 3D tricubic B-splines provide a basis in which only 64 elements are nonzero at any given point in space [16,17]. This allows the rapid evaluation of each orbital in constant time. Furthermore, this basis is systematically improvable with a single parameter so that accuracy can be easily checked. Extensive analysis of QMCPACK simulations reveals that SIMD efficiency is low except for B-spline evaluations in intrinsics and BLAS/LAPACK routines that are available in optimized libraries, such as Intel MKL [18]. The abstractions for 3D physics, encapsulated in generic `ParticleAttrib<DT>` for N particles, are responsible for the low SIMD efficiency. Here, `DT` denotes elemental datatype per particle, such as a position (x,y,z). AoS representations for physical entities are logical and advantageous in expressing physical concepts at high levels and have been widely adopted in HPC applications. However, the computations using AoS datatypes are difficult for compilers to optimize. Optimizing *hotspots* using architecture-specific intrinsics is possible but it is not scalable or portable. Therefore, portable optimization of the code, in preparation of the exascale era followed a step-by-step process on a single SMP node : (1) data layout transformations from AoS to structure-of-arrays (SoA) of particle abstractions in 3D for SIMD efficiency, (2) grouping B-spline orbitals by using array of SoA (AoSoA) technique for memory/cache utilization, and (3) using threads over the orbital groups to reduce the time required to generate each new Monte Carlo sample.

These optimizations were supported and realized through an Intel Parallel Computing Center at Argonne National Laboratory and Sandia National Laboratories and a full article describing the methods can be found in [19].

In the following, we describe briefly these three steps, as implemented in QMCPACK and exposed through the B-spline evaluation routine. QMCPACK implements three routines for B-spline evaluations namely, V, VGL, and VGH. The pseudocode in Figure 21.2 shows the kernel of VGH, which computes N values, gradients (real 3D vectors), and Hessians (symmetric 3D tensors). A total of

```
class Bspline {
  T b[Nx][Ny][Nz][N];
  // other data
  void VGH(T x, T y, T z, T* v, T* g, T* h) {
    //compute the lower-bound index x0,y0,z0
    //compute prefactors using (x-x0,y-y0,z-z0)
    for(int i=0; i<4; ++i)
      for(int j=0; j<4; ++j)
        for(int k=0; k<4; ++k) {
        T* C=b[i+x0][j+y0][k+z0];
        for(int n=0; n<N; ++n) {
         val[n]    += F(C[n]);
         g[3*n+0]+= Gx(C[n]);    g[3*n+1]+= Gy(C[n]);
         h[9*n+0]+= Hxx(C[n]);   h[9*n+1]+= Hxy(C[n]);
         ...
      }}}
};
```

FIGURE 21.2 A pseudo `Bspline` class and `VGH` function.

$13N$ output components are evaluated. V and VGL differ from VGH by how many components are computed, while having the same computational and data access patterns. Depending on the simulation cell type, either VGH or VGL is used with the corresponding adopter class to apply phase factors for computations of the first and second derivatives of Ψ_T. V is used with pseudopotentials for the local energy computation. For solids, VGH is used during the drift-diffusion phase. Both V and VGH routines, each roughly contribute 50%. The evaluations of V are correlated and this property can be exploited by QMC drivers to improve cache utilization as shown in [20].

The outputs v, g, and h are the starting addresses of the particle attributes, V[N] (values) and AoS types of G[N][3] (gradients) and H[N][3][3] (Hessians). They are encapsulated in the `ParticleAttrib<DT>` class. Gradients (Hessians) are the first (second) derivatives with respect to the particle position and are used to compute quantum forces and Laplacians. The allocation of the b coefficient array uses an aligned allocator, `posix_memalign` or `_mm_malloc`, and includes padding to ensure the alignment of b[i][j][k] to a 256/512-bit cache-line boundary. In 4D memory space, accesses to b start at a random input grid point b[x0][y0][z0] that satisfies $x_{x0} \le x < x_{x0} + \Delta_x$ for Δ_x the grid spacing in x-direction. The same conditions apply to y and z. Thus, 64 input streams are issued to access N coefficient values. In total, $64N$ stride-one reads and $13N$ mixed-strided writes are executed for each random input. The arithmetic intensity is low at roughly one fused multiply-add (FMA) for each component and the bandwidth plays a critical role in deciding the throughput of the B-spline routines, especially for v. The cost of prefactor computations at (x, y, z) is amortized for N, which had big impacts on the performance on scalar processors.

Our experiments*[†] are carried out on four different shared memory multi/manycore processors, 18-core single-socket Intel® Xeon® processor E5-2697v4 (Broadwell, BDW), the first

---

* Software and workloads used in performance tests may have been optimized for performance only on Intel microprocessors. Performance tests, such as SYSmark and MobileMark, are measured using specific computer systems, components, software, operations and functions. Any change to any of those factors may cause the results to vary. You should consult other information and performance tests to assist you in fully evaluating your contemplated purchases, including the performance of that product when combined with other products. For more complete information visit http://www.intel.com/performance. Intel, Xeon, and Intel Xeon Phi are trademarks of Intel Corporation in the U.S. and/or other countries.

† Optimization Notice: Intel's compilers may or may not optimize to the same degree for non-Intel microprocessors for optimizations that are not unique to Intel microprocessors. These optimizations include SSE2, SSE3, and SSSE3 instruction sets and other optimizations. Intel does not guarantee the availability, functionality, or effectiveness of any optimization on microprocessors not manufactured by Intel.

generation Intel® Xeon Phi™ coprocessor 7120P (KNC), the second generation Intel® Xeon Phi™ processor 7250P (KNL), and IBM Blue Gene/Q BG/Q.

21.3.1   AoS-to-SoA Transformation of Outputs

The data layout of output arrays follows an AoS pattern. For example, g assumes the internal ordering of the gradients in [xyz|xyz|..|xyz] sequence. A better way to store this data is in a SoA format and use three separate streams for each component. It lets us align the individual output streams for efficient loading and storing, removes strided accesses, and in turn eliminates the need to use gather/scatter instructions.

Figure 21.3 shows a pseudocode with an AoS-to-SoA data layout transformation of the gradient and Hessian arrays. BsplineSoA::VGH uses three separate streams for the gradient vector and nine streams for the Hessian tensor. Only a few components are shown for brevity. A minor change is made to exploit the symmetric nature of the Hessian, which results in a total of 10 (1+3+6) output streams instead of the 13 of the baseline. To help the compiler autovectorize and ignore any assumed dependencies, we place #pragma omp simd (OpenMP 4.0) on top of the inner most loop. These changes lead to efficient vectorization over N and ensure stride-one accesses to both read-only b and output arrays. For the VGL function, in addition to the AoS-to-SoA transformation, we unroll the loop over the z-dimension and move the allocation of temporary arrays of the baseline out of the kernel.

Data-layout transformations from AoS to SoA are commonly employed by HPC applications on Intel® Xeon Phi™ [21,22] and have been shown to be essential on the processors with SIMD units. The same transformation boosts the performance of other critical computational steps involving distance tables, and Jastrow factors and are keys to modernize QMCPACK on processors with wide SIMD units. Such transformations, however, require fundamental changes in the core physics abstractions and affect high-level QMC implementations. Care must be taken to minimize their impacts on theoretical development and users.

Figure 21.4 presents the performance before and after the AoS-to-SoA transformation of output arrays, showing 2–4x speedups for small- to medium-sized problem on the Intel CPUs. We can estimate the vector efficiency of a routine by comparing Ps on KNL. For a small problem such as 256, the vector efficiency is low at 1.2x with the AoS datatypes. In contrast, this efficiency with SoA objects is greater than four. The read bandwidth utilization increases to sustained 238 GB/s from 60–98 GB/s. KNC gets the biggest boost with the AoS-to-SoA transformation being an in-order processor with high memory bandwidth. This SoA technique is currently implemented in the

```
class BsplineSoA {
  void VGH(...) {
    //compute starting addresses for all the components
    for (i,j k) {
      T* C=b[i+x0][j+y0][k+z0];
      #pragma omp simd
      for(int n=0; n<N; ++n) {
        v[n]   += F(C[n]);
        gx[n]  += Gx(C[n]);   gy[n]+= Gy(C[n]);
        hxx[n]+= Hxx(C[n]); hxy[n]+= Hxy(C[n]);
        . . .
  }}}
};
```

FIGURE 21.3   A pseudo *BsplineSoA* class and *VGH* function.

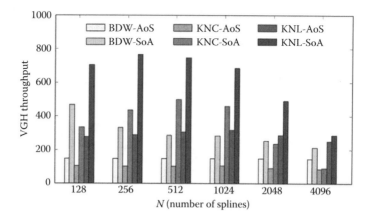

FIGURE 21.4 VGH Performance with AoS-to-SoA data layout transformation.

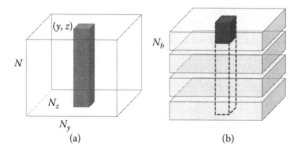

FIGURE 21.5 Data access pattern of read-only B-spline coefficients b at a random position (y, z): (a) previous einspline library and (b) AoSoA implementation. The outermost x-dimension is not shown.

QPX version for BG/Q, leading to 2.2x speedup. However, this solution is not portable and, more importantly, not needed with the transformation.

21.3.2 AoSoA Transformation (Tiling)

Efficient cache utilization is critical for obtaining high performance on large cache-based architectures. As described earlier, accesses to b are random, limiting cache reuse. Typically, the size of b is too big to fit in to even last-level caches. Also, the output arrays grow with N and they move out of the caches for large problems. To overcome these challenges, we use array of BsplineSoA objects, effectively splitting the coefficients along N in M groups, as shown in Figure 21.5b. We use $N_b = N/M$ the tile size. Figure 21.6 is a pseudo user code using M BsplineSoA and WalkerSoA objects of N_b size. We use the restructured data layout of Section 21.3.1 without any modification.

The Array-of-SoA data layout transformation is a cache-blocking optimization. We break up the problem into M *active* working sets to make them fit in to caches. We divide the working set across the spline dimension N, which is common to both the inputs and outputs. Output arrays get split and fit in to caches for efficient reduction operations. Input four dimensional (4D) coefficient array b which is shared among all the threads, also gets split across its innermost dimension. This allows it to fit in the big last-level caches. This AoSoA data layout transformation helps reduce the size of the working set for each evaluation and exposes parallelism that will be utilized to shorten computational time in the strong-scaling sense. Tiling of a multidimensional grid achieve similar effects in grid-based applications [21]. We use the AoSoA data-layout transformation and tiling interchangeably.

```
void Compute( const int N, const int M) {
  int Nb = N/M;
  class WalkerSoA {T v[Nb], g[3*Nb], h[6*Nb];};

  BsplineSoA soa[M](Nb);
  WalkerSoA w[M](Nb);

  for(int t=0; t<M; ++t)
    soa[t].VGH(x, y, z, w[t].v, w[t].g, w[t].h);
}
```

FIGURE 21.6 Pseudocode using *BsplineSoA* and *WalkerSoA* objects.

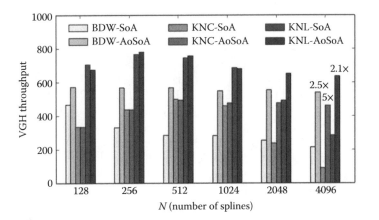

FIGURE 21.7 VGH Performance with SoA to AoSoA transformation (tiling).

Starting at $N_b = 16$, we explore tile sizes in the multiple of two till $N_b = N$ and call N_b with the highest throughput as the optimal tile size. Figure 21.7 presents VGH performance before and after the AoSoA transformation, showing significant improvement for $N = 2048$ and 4096. We use the obtained optimal N_b on each platform: $N_b = 64$ on BDW and 512 on Knights Corner (KNC) and Knights Landing (KNL). With tiling, we obtain sustained throughput across the problem sizes on all the cache-based architectures we have tested.

On BDW and BG/Q with the shared last level cache (LLC), $N_b = 64$ gives the speedups of 1.2–2.5 (BDW) and 1.2–1.5 (BG/Q) across all the problem sizes. At $N_b = 64$, the entire working set, including the input and output arrays fit in their LLC and hence, gives the best speedup. The speedups on KNC and KNL are obtained for $N > 512$. With the optimal $N_b = 512$ on KNC and KNL, output arrays stay in L1/L2 caches and the write-bandwidth decreases from 177 GB/s to 43 GB/s even for $N = 4096$. The vector efficiency also increases from 2.5x to 4.2x.

In contrast to BDW and BG/Q with the shared LLC, KNC, and KNL do not show peak performance at small N_b values. Although L2 caches are kept fully coherent among cores, they are private to cores on KNC and to tiles with two cores on KNL. Due to random accesses of the large b array, private L2 caches may contain duplicate copies of elements at any point of time. The duplication reduces the effective size for L2 cache. Even for $N_b = 16$ tile size, the input working set size is 7 MB, which may not allow it to fit into their L2. For KNC and KNL, a performance peak is obtained at $N_b = 512$. The improvement comes from fitting output arrays in cache, allowing efficient reduction operations over 64 grid points. In QMCPACK, each thread owns a `ParticleSet` object and should keep it in cache during the entire run. This AoSoA approach will facilitate the same cache-blocking efficiency in the full application for any problem size.

The optimal tile size N_b is independent of problem size N for a particular architecture and can be estimated based on L1 or LLC cache size, if present. For production runs, N_b can be tuned for each architecture, making the algorithm cache-oblivious for any N and N_b. The information of the cache subsystems narrows the range of optimal N_b.

21.3.3 EXPOSING GREATER PARALLELISM

Exposing parallelism within a walker is crucial to extend scaling further and to reduce the memory footprint. The AoSoA data layout transformation described in the previous subsection naturally exposes parallelism of B-spline operations. Each `BsplineSoA` object shown in Figure 21.6 can be managed independently, without synchronizing with the other objects. Figure 21.8 shows a method for this nested threading built upon the SoA objects. Alternative approaches include placing `#pragma omp parallel for` or the equivalent within the single object to compute N values. We found the parallelization over M objects is better than over N within an object. It minimizes OpenMP overhead and reaps the benefit of smaller working set sizes of both inputs and outputs.

We implement this step with an explicit data partitioning scheme by assigning nth threads for each walker and distributing M objects among nth threads. The OpenMP thread ID and tile IDs are used to distribute the work and allocations, initializations and computations are done in an OpenMP `parallel` region, instead of using `#pragma omp parallel for` as shown in Figure 21.8. This removes any overhead from OpenMP run time for the nested parallelism and sets the ideal performance that can be obtained by cache and threading optimizations.

Enabling nested parallelism in QMCPACK will follow our approaches to optimize B-spline evaluations by introducing AoSoA transformations in the core physical abstractions, such as `ParticleSet`. The users can choose nth to maximize their resources and science productivity as shown in Figure 21.9, which summarizes the performance improvement on all the platforms.

```
#pragma omp parallel for
for(int t=0; t<M; ++t)
    soa[t].VGH(x, y, z, w[t].v, w[t].g, w[t].h);
```

FIGURE 21.8 Equivalent B-spline kernel using nested OpenMP.

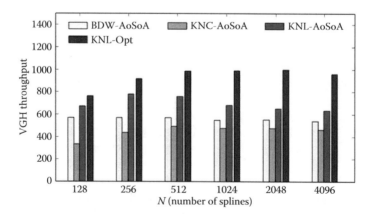

FIGURE 21.9 Highest measured performance for BDW, KNC, and KNL, including nested threading performance on KNL.

21.4 HIGH-PERFORMANCE DETERMINANT UPDATE ALGORITHMS

As described in Section 21.2 the essential numerical and computational steps in QMC are all associated with the evaluation of the trial wavefunction. For large enough system sizes or sufficiently complicated trial wavefunctions, the wavefunction evaluation step dominates the execution profile. Identifying efficient but accurate wavefunction forms, and algorithms to evaluate them that are well mapped onto the execution hardware, are the key to obtaining high performance. As the hardware changes, it is important to reexamine established algorithms and determine if new algorithms can be adopted that will run more efficiently.

To evaluate the transition probability in the Monte Carlo, we require the ratio of the old and new wavefunction values. For single-electron moves, this requires computing the ratio of determinants with a single row or column changed. Computing and updating the value of the determinants is formally an $\mathcal{O}(N^3)$ scaling operation and is the leading cost in large QMC calculations.

Within the last seven years, significant progress has been made improving the efficiency of evaluating the determinantal parts of wavefunctions used in high-accuracy molecular calculations. For molecules, the electron count is often small, under 100 electrons, but the determinants in the configuration interaction-type wavefunctions may number in the millions. The new algorithms exploit the similarities between these determinants [4,23] and require only a single determinant to be fully updated while computing changes from a single reference determinant. By only storing and updating a single determinant, the memory and speed improvements are substantial.

For solid-state calculations of materials, the largest calculations today may have thousands of electrons, but use only a single determinant. Electron counts measured in the thousands are expected at the exascale. For these runs, the $\mathcal{O}(N^3)$ matrix update cost is much more severe than for the molecular case. Unfortunately, the standard update algorithm has relatively low compute intensity and is memory bandwidth limited. As we explain below, these can be updated for significantly higher performance, particularly on graphic processing units (GPU) systems. By deferring matrix updates for several Monte Carlo steps, we are able to introduce a blocked matrix update algorithm that uses higher level BLAS functions and runs significantly faster overall—the algorithm becomes less memory-bandwidth limited and more compute limited. The underlying Monte Carlo moves are unmodified.

21.4.1 DELAYED RANK-K UPDATE

QMC codes avoid full recalculation of the determinant values by exploiting the matrix determinant lemma, as described in [24]. For a Slater matrix $\mathbf{A}$, and its updated form $\mathbf{A}_{\mathbf{new}}$ following a proposed electron move, the ratio $\det(\mathbf{A}_{\mathbf{new}})/\det(\mathbf{A})$ can be found via the matrix determinant lemma:

$$\det(\mathbf{A} + \mathbf{ue}'_k) = (1 + \mathbf{e}'_k \mathbf{A}^{-1}\mathbf{u})\det(\mathbf{A}) \qquad (21.3)$$

The change in $\mathbf{A}$ due to the movement of the electron coordinate r_k is given by $\mathbf{A} + \mathbf{ue}'_k$, where $\mathbf{ue}'_k$ is the outer product of the electron change vector $\mathbf{u}$ and the kth basis vector. Dividing this equation on both sides by $\det(\mathbf{A})$, the existing determinant, yields the desired ratio on the right-hand side. Hence, the ratio can be combined with an initially computed determinant value to obtain the value of the new determinant. The computational challenge is to efficiently compute the ratio using this equation and to efficiently apply the accepted changes to the Slater matrix. Conventionally, these updates utilize the Sherman–Morrison scheme which can be evaluated at $\mathcal{O}(N^2)$ cost.

The main problem with the conventional Sherman–Morrison scheme is that it utilizes only level 1 and 2 BLAS, which do not run optimally even on current CPU architectures. Our *delayed update* scheme is designed to evaluate multiple accepted Monte Carlo moves before any updates are made to the Slater matrix. This strategy requires that after an electron move is proposed and accepted, the move's application to the Slater matrix is delayed, and any needed ratios are obtained using the matrix determinant lemma. We term this *determinant lookahead*. A group of accepted changes are queued until a preset limit on unapplied moves p is reached or a move is denied. The accepted moves are applied to the Slater matrix together as a single rank-k update. Because this involves multiple column changes, higher-level BLAS than the conventional algorithms can be used, and because the column changes are adjacent, memory accesses are optimized. For multiple move probability estimation, we use a generalization of the matrix determinant lemma:

$$\det(\mathbf{A} + \mathbf{U}\mathbf{V}') = \det(\mathbf{I_m} + \mathbf{V}'\mathbf{A}^{-1}\mathbf{U})\det(\mathbf{A}) \tag{21.4}$$

where $m \leq p$ is the number of contiguous changes to consider, $\mathbf{I_m}$ is the identity matrix of size m, and $\mathbf{U}$ is an $N \times m$ matrix containing a proposed electron change vector in each column. Note $\mathbf{A}^{-1}$ is $N \times N$ and $\mathbf{V}$ is formed by appending m standard basis vectors columnwise. The matrix $\mathbf{I_m} + \mathbf{V}'\mathbf{A}^{-1}\mathbf{U}$ is, at most, $p \times p$, and the calculation of its determinant is much less burdensome than recomputing $\det(\mathbf{A})$ because $p^3 \ll N^3$.

Up to p of these smaller determinants are computed prior to each update to $\mathbf{A}$. Because the $\mathbf{I_m} + \mathbf{V}'\mathbf{A}^{-1}\mathbf{U}$ matrices are relatively small (for low values of p), this work is performed on CPU, even for our GPU-accelerated implementation.

We stress that our delayed update approach is entirely equivalent to multiple contiguous electron-by-electron moves and not a single multiple-electron move. The handling of rejected moves is a matter of acceptance ratios and details of the underlying hardware: for high enough acceptance ratios, it can be fastest to treat any rejections as a simulated acceptance, substituting an unchanged column.

21.4.2 Performance Results

We have obtained performance results from a standalone test application that assumes 100% acceptance probability. The performance is therefore representative of what would be obtained in a DMC calculation with a typical acceptance rate of over 99%. This QMC algorithm utilizes most of our computer time. Other QMC algorithms with lower acceptance probabilities are expected to obtain lower performance depending on the acceptance ratio and the how rejected moves are treated. Numerical stability, not shown here, is similar to the conventional algorithm.

Figure 21.10 shows representative results for updating 16 size 2048×2048 matrices on both conventional CPU and hybrid CPU+GPU systems. For the hybrid GPU+CPU calculation, only the full matrix updates are performed on the GPU, using separate CUDA streams and CUBLAS. The determinant lookahead is performed on the CPU, although for large enough k, this work could be scheduled on the GPU. The full determinants are stored on the GPU to minimize transfers, consistent with the production QMCPACK code.

Recognizing that $k = 1$ corresponds to the conventional algorithm, we find that the new delayed update algorithm performs well, with greater than an order of magnitude speedup obtained by $k = 16$. The performance continues to improve with greater k, but begins to taper off due to the increasing cost of the determinant lookahead operations. Based on the current profiles of the CPU and GPU codes (Figure 21.1), a substantial speedup is expected when the algorithm is implemented in production.

The new update algorithm addresses the demands for greater compute intensities and more efficient memory hierarchy utilization of current and future architectures by enabling use of matrix–matrix BLAS operations. This also facilitates use of nested parallelism and threaded BLAS, compatible with the proposal in Section 21.3. Based on the observed performance, we expect that the delayed update algorithm will become standard in large QMC calculations with single determinants. We are currently exploring other variants of the algorithm that might have further reduced cost as well as more refined implementations.

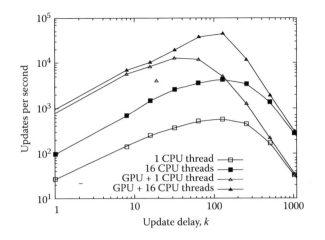

FIGURE 21.10   Performance of matrix update algorithm for 16 size 2048×2048 matrices as a function of delay k. The conventional algorithm is equivalent to $k = 1$. The delayed update algorithm performance peaks near $k = 100$, but turns over due to the increased CPU work in the determinant look ahead. Timings were obtained on OLCF's Titan with up to 16 OpenMP threads and a single K20X GPU.

21.5   ADOPTION OF WORKFLOW TOOLS

Exascale machines will make an unprecedented level of computational resources available. As these machines become available, science applications of QMC will expand in size as not only larger problems, previously infeasible, come within reach, but also the breadth of questions that can be explored simultaneously. From a task management point of view, moving to the exascale will present challenges of handling scientific workflows in a scalable, efficient, and robust manner.

Individual QMC calculations are not monolithic, they have substructure including the generation of single particle orbitals with DFT, orbital file format conversion, optimization of Jastrow correlation factors, and a variety QMC total energy calculations such as variational, diffusion, and reptation Monte Carlo. The natural progression within a research project involves many layers of calculations ranging from exploratory to production—possibly spanning multiple machine environments and scales—and even these project phases are connected through data dependencies.

Currently, this complex of management tasks falls primarily to the individual computational scientist, which at the exascale results in both increased human effort and increased opportunity for the injection of human error into the calculation process. Increased production of knowledge in physics, chemistry, and materials science via QMC hinges on the ability to reduce error propagation and accelerate the totality of the calculation. In anticipation of these challenges, we have developed a workflow management system [25], called Nexus, which has been tailored to the needs of QMC in the context of large-scale scientific workflows.

Specifically, the qualities and capabilities we require of such a system are

1. Manage an arbitrary number of workflows, each represented as a directed acyclic graph (DAG).
2. Be simple enough for new researchers to use effectively.
3. Be robust to interruption, operating only on uncompleted tasks upon restart.
4. Operate with all codes in the dependency chain: Quantum Espresso, GAMESS, QMC-PACK, and conversion tools.
5. Manage tasks in a variety of computing environments, from personal workstations to managed high-performance computing resources including those at NERSC, Oak Ridge Leadership Computing Facility (OLCF) and Argonne Leadership Computing Facility (ALCF)

6. Be sufficiently flexible to support new codes and machine environments with minimal development effort.
7. Generate and write arbitrary input files for supported codes, guarding against faulty input to the extent possible.
8. Parse and organize output data, including performing statistical analysis on QMC data.
9. Submit simulation jobs and assess them for successful completion, including the quality of expected output data.
10. Be capable of collecting multiple jobs at the same level in the workflow hierarchy for batched submission.
11. Process dependencies between simulation codes such as orbital and Jastrow information.

Nexus fulfills these needs by partitioning the required capabilities between high-level Python classes, as shown in Figure 21.11 and providing a minimal scripting interface to users.

The `Simulation` class is ultimately responsible for handling all the demands of a single application instance prior to, during, and following its execution. These demands include writing input files, copying in associated input data (such as pseudopotential files), requesting job submission, monitoring output for completion and correctness, copying lightweight data files for storage, preprocessing output data for later analysis, and providing information (such as an optimized structure or Jastrow factor) to dependent simulations to incorporate as input. The majority of these actions are handled directly by the `Simulation` class. It operates as a heavy base class for more specialized simulation classes (such as `Qmcpack`) that handle the production and incorporation of specific data dependencies. The lightweight `SimulationInput` and `SimulationAnalyzer` (and their heavier derived classes) support the Simulation class in the specific actions of input file reading/generation/writing and output data processing, respectively.

The `PhysicalSystem` class represents the physical system under study (e.g., crystalline diamond) including coarse electronic information and complete details of the atomic structure (via the `Structure` class). Separating physical system information into its own class is useful because this information is generally shared in common among different simulations in a QMC workflow (e.g., DFT orbital generation, Jastrow optimization, and production DMC). The `Job` class contains

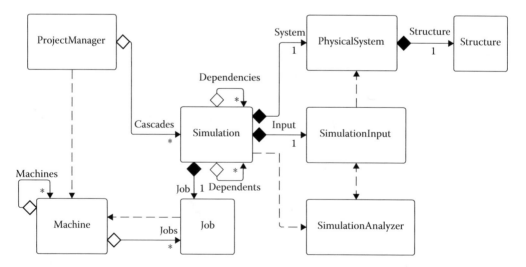

FIGURE 21.11 UML diagram of central Nexus classes (Reproduced from J. T. Krogel, *Computer Physics Communications*, 198, 154, 2016.).

information regarding job submission for the local machine, such as the target queue, the number of nodes required, and the wall-clock time of the job.

Similar to `Simulation`, the `Machine` class is a heavy base class that handles actual job submission and queue monitoring. Lightweight derived classes allow for local variations in job submission and monitoring needs, for example, working on machines at OLCF (`Titan`) and ALCF (`Mira`) differ somewhat, while sharing most details.

The `ProjectManager` class oversees the execution of simulation workflows by driving the `Simulation` and `Machine` classes at regular update intervals, in between which the Nexus host process sleeps. At each update interval, the status of the job queue is checked and the simulation workflows (DAG's) are traversed with each active simulation progressing, as appropriate, through the various stages/actions as described for the `Simulation` class. If the host process is interrupted and resumed, the distributed workflow state is loaded from the disk ensuring that only remaining management tasks are performed, that is, completed jobs are not resubmitted.

The user interface for Nexus is made available within the Python scripting environment. A basic Nexus script resembles input files familiar to practitioners. More complex workflows can be generated either by copying simulation elements in the input script, or through the use of standard programming constructs such as `for` loops and **if** statements. An example of a single QMC workflow that Nexus might execute is shown in Figure 21.12. In this case, a single DFT calculation is performed with Quantum Espresso and its outputted orbitals are converted into the ESHDF file format used by QMCPACK. Two- and three-body Jastrow factors are optimized in sequence with QMCPACK. Finally, diffusion Monte Carlo calculations using either the locality approximation or T-moves are performed with QMCPACK for each Jastrow factor. For greater detail about Nexus' implementation and user interface, consult [25].

Use of Nexus offers key advantages to QMC practitioners that we anticipate will increase in importance at the exascale. A primary benefit is increased time efficiency for the computational scientist, offsetting the demands of inevitably heavier workflow tasks. The workload of performing all steps manually is traded for the much lighter task of workflow specification, which is further facilitated by programmability of the workflows. An additional advantage is that commonplace errors made while composing input or handling data dependencies can be minimized as the collected understanding of experienced practitioners are continually encoded into safeguarding mechanisms. Preprocessed and internally stored numerical simulation output allows more efficient data analysis, with the potential to perform the desired analysis tasks within the workflow script itself. Finally, the intrinsic

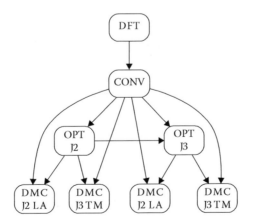

FIGURE 21.12 Example QMC workflow diagram. DFT orbital generation is followed by orbital conversion, optimization of two and three body Jastrows, and DMC performed with the locality approximation or T-moves.

transferability of Nexus workflow scripts enables the provenance, sharing, and reproducibility of otherwise transient research processes.

21.6 DEVELOPMENT PRACTICES AND STRATEGIES

Although the fundamental QMC algorithms implemented in QMCPACK are relatively simple, the source code is comparatively large: approximately 400K source lines of C++ in 1800 files. Nearly 100K of this total results from the different wavefunction converters and input/output functionality required to interface with other electronic structure packages, both wavefunctions and pseudopotentials. Furthermore, as a mature QMC package, QMCPACK implements a large variety of wavefunctions and QMC methods. Refactoring and updating the source for new architectures, for example, as described in Section 21.3, while maintaining developer productivity is therefore an increasing challenge.

CMake [26] is used as the configuration and build system. The source code is currently managed in Subversion (SVN) using Assembla, a commercial project hosting service. It provides source code management, issue tracking, a project wiki, and more.

For much of its lifetime, QMCPACK, in common with many scientific domain codes, was developed using only informal testing. A suite of informal reference or *gold* results was used to catch errors. This method does not scale as developers are added, because it requires that the new developers have the domain knowledge to interpret the results, and it also does not provide the fine-grained level of testing needed to refactor the code.

The challenge of scaling the development effort has led to three important goals: significantly increasing the level of testing, reducing the number of source lines where possible, and establishing standards for contributed code and adopting code reviews. Although these goals are worthwhile for any software project, the emerging increased diversity in compilers, libraries, and architectures provides additional motivation to adopt them.

First, we have significantly increased the level of testing, adding integration, and also unit tests. The different types of testing are managed through CTest, which is a testing tool distributed with CMake. There is a nightly build and test cycle that is monitored with CDash [27], a software testing server. Several HPC machines are tested on a regular basis, including a Cray XC30 (Eos) and a Cray XK7 (Titan) at Oak Ridge and a BG/Q (Cetus) at Argonne. The types of automated testing include system level tests, converter tests, and unit tests. System level tests run the code and compare the results against known values. There are short (few minute) and long (half an hour) versions of the system tests. Because of the basic nature of the Monte Carlo algorithms, the results have statistical errors. The converter tests test the conversion of wavefunctions from Quantum Espresso and GAMESS. Correct conversion at the program interfaces is subtle and the converters are often a source of problems.

Second, we would like to reduce the number of source lines. For example, the wavefunction converters could be written in a higher level language than C++, which would add flexibility and ease of further development. Before doing so, we require an extensive test suite.

Third, in order to improve maintainability of contributed code, we are instituting minimum code standards that encompass format, Doxygen comments, testing, and documentation. To facilitate this, QMCPACK will move to a git-based version control system and adopt the *pull request* model. This workflow has the benefit of preserving discussion around individual requests that are currently not captured in SVN commit messages.

21.6.1 UNIT TESTING

Unit testing enables projects to scale to a larger number of developers, and provides assurances to developers that changes are not breaking the code. Unit tests also help design by driving refactoring, and can lead to a less tightly-coupled architecture [28].

Two unit test frameworks were evaluated for possible use in the project: Google Test [29] and Catch [30]. These have similar capabilities, but the decision was made to use Catch, partly because it has a more natural syntax for assertions. Catch is also smaller and easier to integrate into the project—it is distributed in a single include file about 400KB in size. Google Test takes about 3.7 MB consisting of 220 files in 25 directories.

See Listing 21.1 for a sample test using Catch. For different types, there is only a single assertion clause (REQUIRE) and it uses normal comparison expression syntax. Compare with the Google Test version in Listing 21.2. There are different assertion macros for different types (EXPECT_EQ, EXPECT_DOUBLE_EQ), with each macro taking two arguments.

Listing 21.1   Unit test example using Catch

```
TEST_CASE("vector size one", "[vector]")
{
    std::vector<double> v;
    v.push_back(1.0);

    REQUIRE(v.size() == 1);
    REQUIRE(v[0] == Approx(1.0));
}
```

Listing 21.2   Unit test example using Google Test

```
TEST(Numerics, vector_size_one)
{
    std::vector<double> v;
    v.push_back(1.0);

    EXPECT_EQ(v.size(), 1);
    EXPECT_DOUBLE_EQ(v[0], 1.0);
}
```

21.7   SUMMARY

There are many outstanding scientific challenges that QMC is poised to address on the next generation of supercomputers—direct simulation of complex superconducting materials or exotic quantum phases on a fully *ab initio* basis, creating a predictive tools for catalysis on surfaces or providing a fundamental reference database for high-accuracy materials properties for materials innovation.

To address these challenges on the next generation of supercomputers requires more than simple weak scaling of the application. We have presented data layout optimizations and improved algorithms that yield significant performance gains on the latest architectures. Our main focus has been improving on-node performance by improving utilization of the memory hierarchy and enabling more compute bound operations. Although these developments were targeted at the next generation, by virtue of their effectiveness today, we may consider that the next generation of computer architectures has already arrived.

ACKNOWLEDGMENTS

The data layout optimizations for Intel® Xeon Phi™ were realized through an Intel Parallel Computing Center. Over the last decade, support for QMCPACK has come from many grants, in particular from NSF DMR-0325939 and DOE OCI-0904572. Currently, the code is supported as part of the Computational Materials Sciences Program funded by the US Department of Energy, Office of Science, Basic Energy Sciences, Materials Sciences and Engineering Division. An award of computer time was provided by the INCITE program. This research used resources of both the Argonne and Oak Ridge Leadership Computing Facilities, which are DOE Office of Science User Facilities supported under contracts DE-AC02-06CH11357 and DE-AC05-00OR22725.

REFERENCES

1. QMCPACK. http://www.qmcpack.org, 2016.
2. K. Esler, J. Kim, D. Ceperley, and L. Shulenburger. Accelerating quantum Monte Carlo simulations of real materials on GPU clusters, *Computing in Science Engineering*, vol. 14, no. 1, p. 40, 2012.
3. J. Kim, K. P. Esler, J. McMinis, M. A. Morales, B. K. Clark, L. Shulenburger, and D. M. Ceperley. Hybrid algorithms in quantum Monte Carlo, *Journal of Physics: Conference Series*, vol. 402, no. 1, p. 012008, 2012.
4. B. K. Clark, M. A. Morales, J. McMinis, J. Kim, and G. E. Scuseria. Computing the energy of a water molecule using multideterminants: A simple, efficient algorithm, *Journal of Chemical Physics*, vol. 135, no. 24, p. 244105, 2012.
5. N. M. Tubman, I. Kylänpää, S. Hammes-Schiffer, and D. M. Ceperley. Beyond the Born-Oppenheimer approximation with quantum Monte Carlo methods, *Physical Review A*, vol. 90, p. 042507, 2014.
6. K. Foyevtsova, J. T. Krogel, J. Kim, P. R. C. Kent, E. Dagotto, and F. A. Reboredo. Ab *initio* quantum Monte Carlo calculations of spin super exchange in cuprates: The benchmarking case of Ca_2CuO_3, *Physical Review X*, vol. 4, p. 031003, 2014.
7. A. Benali, L. Shulenburger, J. T. Krogel, X. Zhong, P. R. C. Kent, and O. Heinonen. Quantum Monte Carlo analysis of a charge ordered insulating antiferromagnet: The Ti4O7 Magneli phase, *Physical Chemistry Chemical Physics*, vol. 18, p. 18323, 2016.
8. R. M. Martin, L. Reining, and D. M. Ceperley. *Interacting Electrons*. Cambridge: Cambridge University Press, 2016.
9. W. M. C. Foulkes, L. Mitas, R. J. Needs, and G. Rajagopal. Quantum Monte Carlo simulations of solids, *Reviews of Modern Physics*, vol. 73, p. 33, 2001.
10. M. Dubecky, L. Mitas, and P. Jurecka. Noncovalent interactions by quantum Monte Carlo, *Chemical Reviews*, vol. 116, pp. 5188–5215, 2016.
11. Y. Luo, A. Benali, L. Shulenburger, J. T. Krogel, O. Heinonen, and P. R. C. Kent. Phase stability of TiO_2 polymorphs from diffusion quantum Monte Carlo, *arXiv:1607.07361*, 2016.
12. J. T. Krogel and D. M. Ceperley. Population control bias with applications to parallel diffusion Monte Carlo, in *Advances in Quantum Monte Carlo. ACS Symposium Series* (S. Tanaka, S. Rothstein, and W. A. Lester, eds.), vol. 1094, p. 13. Washington, DC: American Chemical Society, 2012.
13. C. Lin, F. H. Zong, and D. M. Ceperley. Twist-averaged boundary conditions in continuum quantum Monte Carlo algorithms, *Physical Review E*, vol. 64, p. 016702, 2001.
14. Knights Landing (KNL). 2nd Generation Intel® Xeon Phi™ Processor. http://www.hotchips.org/, 2016.
15. Top 500 list, June 2013. http://www.top500.org/lists/2013/06/.
16. D. Alfè and M. J. Gillan. An efficient localized basis set for quantum Monte Carlo calculations on condensed matter, *Physical Review B*, vol. 70, no. 16, p. 161101, 2004.
17. K. P. Esler. Einspline B-spline Library. http://einspline.sf.net.
18. Intel® Math Kernel Library (Intel® MKL). https://software.intel.com/en-us/intel-mkl, 2016.
19. A. Mathuriya, Y. Luo, A. Benali, L. Shulenburger, and J. Kim. Optimizations of Bspline-based orbital evaluations in quantum Monte Carlo on multi/many-core shared memory processors, *Submitted to IPDPS 2017, arXiv:1611.02665*, 2016.

20. Q. Niu, J. Dinan, S. Tirukkovalur, A. Benali, J. Kim, L. Mitas, L. Wagner, and P. Sadayappan. Global-view coefficients: A data management solution for parallel quantum Monte Carlo applications, *Concurrency and Computation: Practice and Experience*, vol. 28, p. 3655, 2016.
21. J. Jeffers and J. Reinders, eds. *High Performance Parallelism Pearls: Multicore and Many-core Programming Approaches*, vol. 1. Boston, MA: Morgan Kaufmann Publishers, 2015.
22. J. Reinders and J. Jeffers, eds. *High Performance Parallelism Pearls: Multicore and Many-core Programming Approaches*, vol. 2. Boston, MA: Morgan Kaufmann Publishers, 2015.
23. P. K. V. V. Nukala and P. R. C. Kent. A fast and efficient algorithm for Slater determinant updates in quantum Monte Carlo simulations, *Journal of Chemical Physics*, vol. 130, no. 20, p. 204105, 2009.
24. S. Fahy, X. W. Wang, and S. G. Louie. Variational quantum Monte Carlo nonlocal pseudopotential approach to solids: Formulation and application to diamond, graphite, and silicon, *Physical Review B*, vol. 42, no. 6, p. 3503, 1990.
25. J. T. Krogel. Nexus: A modular workflow management system for quantum simulation codes, *Computer Physics Communications*, vol. 198, p. 154, 2016.
26. CMake. https://cmake.org, 2016.
27. CDash. http://www.cdash.org/, 2016.
28. A. Tarlinder. *Developer Testing: Building Quality into Software*. Boston, MA: Addison-Wesley Professional, 2016.
29. Google Test. https://github.com/google/googletest, 2016.
30. Catch (C++ Automated Test Cases in Headers). https://github.com/philsquared/Catch, 2016.

22 Preparing an Excited-State Materials Application for Exascale

Jack Deslippe, Felipe H. da Jornada, Derek Vigil-Fowler, Taylor Barnes, Thorsten Kurth, and Steven G. Louie

CONTENTS

22.1 Introduction to the *ab initio* GW Approach and BerkeleyGW 481
22.2 Target System and Optimization Strategy 484
    22.2.1 Cori at NERSC 484
    22.2.2 Optimization Approach 486
22.3 MPI+OpenMP Scaling on MPP Systems 487
22.4 Computational Bottlenecks in GW 488
    22.4.1 Bottleneck A: Computing Orbital Transition Probabilities 489
    22.4.2 Bottleneck B: Construction of the Electronic Polarizability 489
    22.4.3 Bottleneck C: Computation of Electron Self-Energy 490
    22.4.4 Bottleneck D: Computation of Electron Self-Energy with Full Energy
        Dependence 491
22.5 GW Optimization Process for Cori 491
    22.5.1 Bottleneck A: Optimization Strategy 491
        22.5.1.1 Arithmetic Intensity and Absolute Performance of FFTs on KNL 493
    22.5.2 Bottleneck B: (Polarizability): Optimization Strategy 493
    22.5.3 Bottleneck C: (GPP Self-Energy): Optimization Strategy 497
    22.5.4 Bottleneck D: Optimization Strategy 501
22.6 Summary 503
Acknowledgments 504
References 504

22.1 INTRODUCTION TO THE *AB INITIO* GW APPROACH AND BERKELEYGW

Many important properties of materials can be ascertained through a description of their *electronic-structure*—a description of the quantum orbitals available to electrons in the system (in analogy with the 1S, 2S, 2P, etc. orbitals in atoms) and transition probabilities among the orbitals in the presence of an external perturbation (e.g., shining light on the material). Density functional theory (DFT) within the Kohn–Sham approach [1] provides an accurate formalism for computing the ground state properties of materials (those properties of a material where the electrons are occupying the lowest energy orbitals) through the solution of the coupled self-consistent Kohn–Sham equations:

$$\left\{ \frac{p^2}{2m} + V(\mathbf{r}) + V_{\mathrm{H}[\rho]}(\mathbf{r}) + V_{\mathrm{xc}}[\rho](\mathbf{r}) \right\} \varphi_i(\mathbf{r}) = \varepsilon_i \, \varphi_i(\mathbf{r}) \tag{22.1}$$

with

$$\rho(\mathbf{r}) = \sum_{i}^{occ} |\varphi_i(\mathbf{r})|^2 \qquad (22.2)$$

Here, $V(\mathbf{r})$ is the potential from the ions (nuclei plus core electrons), V_H is the Hartree potential (the electrostatic interaction between an electron and the average charge density of all the valence electrons in the system), V_{xc} is the exchange-correlation potential (which represents the many-body electron interactions beyond the Hartree term, including forces associated with the Pauli exclusion principle, and must be approximated in practical calculations), and ε_i is the eigenvalue of the Kohn–Sham equations. Both V_H and V_{xc} are functions of the electron density ρ. These equations have the form of a coupled eigenvalue problem, where ρ is computed self-consistently.

The solution to Equation 22.1 is known to predict accurate material structures, total energies, chemical reaction pathways, and vibrational properties of a crystal without relying on additional constants or parameters determined from experiment—hence DFT is known as an *ab initio* approach. However, calculations beyond Kohn–Sham DFT are required for the computation of excited-state material properties such as bandgaps (i.e., the energy of the system with one extra electron, minus the energy with one electron removed), band-alignments at interfaces between materials, conductivity, and optical absorption spectra. These properties are of central importance in the efficiency and ultimate performance characteristics underpinning optoelectronic, energy conversion, energy storage, and other functional materials of great interest for designing next generation energy materials. The *ab initio* GW plus Bethe–Salpeter equation (GW-BSE) approach is regarded as the gold standard for the determination of these properties, and, as we demonstrate in this chapter, is particularly well-suited for exascale high performance computing (HPC) systems—more so, for example, than Kohn–Sham DFT.

One of the main limitations of DFT within the Kohn–Sham formalism with typical approximations for V_{xc}, such as the local density approximation (LDA) or general gradient approximation (GGA), is that ε is rigorously only a Lagrange multiplier and should not be interpreted formally as electron excitation energies. Indeed, the interpretation of these values as single electron excitation energies leads to miss-prediction of important material properties (like the bandgap of semiconductors) that are both quantitatively and qualitatively wrong—in some cases, predicting materials to be metallic instead of semiconducting (see Figure 22.1).

The *ab initio* GW method [2,3] and its implementation within the BerkeleyGW software package [4,5] captures the excited-state properties of materials in a quantitatively accurate way, which fundamentally cannot be accomplished with standard Kohn–Sham DFT methods. The GW method is named after the approximation to the many-body electron–electron interacting equation of motion in materials, where the electron self-energy (the change in the electron's energy due to the interaction with other electrons in the system) is written as $\Sigma = iGW$, where G stands for the electron Green's function, and W for the Coulomb interaction between two charges in the presence of electronic screening. Screening is an effect wherein charges in the material rearrange themselves in the presence of an external field, which can originate from an external electromagnetic perturbation such as light, or due to the addition/removal of an electron to/from the system. GW calculations, based on the framework of many-body theory, contain an accurate description of this screening phenomena, which is missing in DFT calculations (even those employing more advanced hybrid functional approximations to V_{xc}). In neutral two-particle excitations, an electron is promoted to a conduction orbital leaving a hole behind in the valence orbital (see Figure 22.2), and in this case the two particles (i.e., the electron and the hole) also interact with each other via a screened Coulomb interaction.

In BerkeleyGW and other common GW applications, the excitation energies of the system are computed perturbatively starting from the Kohn–Sham eigenvalues and eigenvectors. Even though, as mentioned above, the Kohn–Sham eigenvalues shouldn't formally be interpreted as excitation energies they are typically good starting points for more accurate theories, such as the GW approximation. For this reason, BerkeleyGW is typically used in conjunction with DFT codes such as

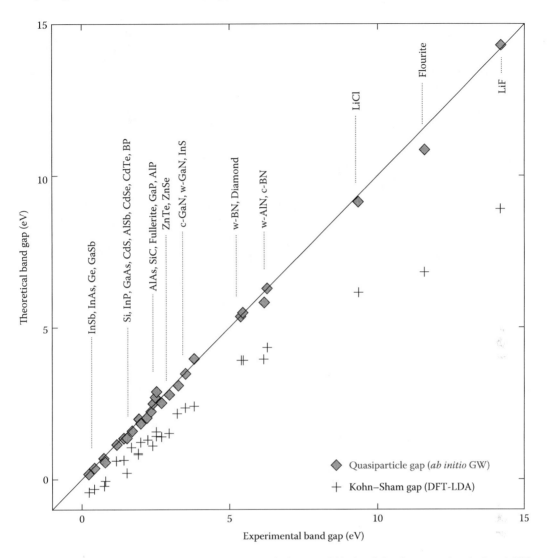

FIGURE 22.1   Comparison of the Density Functional Theory (within local density approximation) and GW calculations with the experimentally observed bandgap for a number of materials. The line $y = x$ corresponds to perfect agreement. (Adapted from Cohen, M. L. and Louie, S. G., *Fundamentals of Condensed Matter Physics.* Cambridge, UK: Cambridge University Press, 2016 [24].)

PARSEC [6], PARATEC [7], Quantum ESPRESSO [8], SIESTA [9], ABINIT, and OCTOPUS. Other common DFT codes like VASP incorporate their own GW methods.

The fundamental equations of motion in many-body perturbation theory can be cast into a Kohn–Sham-like form [2,3]:

$$\left[-\frac{1}{2}\nabla^2 + V_{\text{loc}} + \Sigma(E_n) \right] \phi_n = E_n \phi_n \tag{22.3}$$

where V_{loc} is the *local* ionic and Hartree potentials, Σ is the electronic self-energy in the GW approximation, $\Sigma = iGW$, which depends on the electronic orbitals and energies E_n. The orbitals are either expressed in real space, as $\phi_n(\mathbf{r})$, or, as in BerkeleyGW and other *plane-wave* codes, in Fourier space as $\phi_n(\mathbf{G})$, stored as complex double-precision arrays.

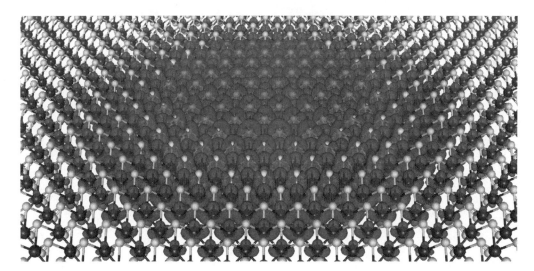

FIGURE 22.2   Schematic of the interacting electron-hole state in monolayer MoS_2.

GW applications, for example, BerkeleyGW [5], Yambo [10], WEST [11], are becoming increasingly used at large-scale high performance computing (HPC) facilities around the world, such as the National Energy Research Supercomputing Center (NERSC) [12]—the DOE Office of Science production HPC center at the Lawrence Berkeley National Laboratory in Berkeley, California. NERSC recently deployed a new HPC system, named Cori (named after the American scientist Gerty Cori), powered by more than 9600 nodes with Intel's Xeon-Phi Processors (code named Knights Landing or KNL for short), which are based on Intel's recent Many Integrated Core (MIC) architecture. The GW method (with many layers of parallelism to exploit) is much more suited for this type of system than common Kohn–Sham DFT approaches, which tend to parallelize over only the plane-wave basis set and one band index. In a typical GW calculation, we can parallelize over k,q (kpoints and kpoint differences), band pairs, plane-wave pairs, and energy values. See the discussion in Section 22.4.2, for example.

BerkeleyGW [4,5] is a massively parallel software package for GW and related calculations of materials ranging from bulk insulators, metals, and semiconductors to nanostructured systems and molecules. The package takes advantage of all the levels of parallelization available mentioned earlier. The code is written primarily for the FORTRAN 2003 standard. Like many applications at NERSC, it was originally optimized for HPC like those at NERSC via message passing interface (MPI). In the remainder of this chapter, we discuss the process for porting and optimizing the BerkeleyGW application for state-of-the-art energy-efficient HPC systems, namely the Cori system at NERSC with an eye toward exascale. We discuss our optimization strategy and how each novel feature in the Xeon-Phi architecture is exploited and ultimately affects BerkeleyGW performance.

22.2   TARGET SYSTEM AND OPTIMIZATION STRATEGY

22.2.1   CORI AT NERSC

As our target pre-exascale system for GW optimizations, we discuss here the recently installed Cori Cray XC40 system at NERSC [13]—shown in Figure 22.3. The Cori system has two partitions using the same Cray Aries interconnect. The largest of which (the partition we target for optimization) has

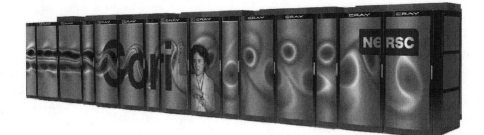

FIGURE 22.3 The Cori system at NERSC. (Courtesy of NERSC.)

over 9600 compute nodes, each powered by Intel's KNL Xeon-Phi processor. The KNL Xeon-Phi processor is commonly referred to as a *manycore* processor and is on Intel's exascale roadmap. Each compute node on the Cori KNL partition has one KNL processor with 68 compute cores—each of which supports up to four hardware threads.

The Phase-1 Cori system consists of over 1500 compute nodes powered by two Intel Xeon (Haswell generation) processors, each with 16 cores (supporting two hardware threads). The Haswell processors are a recent generation of Intel's traditional data-center processor. Despite having 16 cores per part, they are not typically referred to as *manycore*.

The differences between a compute node in the Haswell and KNL partition are summarized in Table 22.1.

NERSC users have traditionally been able to compile their existing codes on the latest HPC system at the center and achieve higher performance on the new system than its predecessor. When migrating between the Cori Phase 1 system (the Cori Haswell partition) and the *Phase 2* KNL partition, however, it is often the case that work is required to match and exceed the performance of the earlier partition. As you can see in the table, the frequency of the cores on the KNL is about half that of the Xeons. In order to make up for this, a code needs to effectively use one or more of the features where KNL outstrips Haswell by (1) scaling well to 68 cores per node via either MPI or a shared memory programming model; (2) accelerating memory bandwidth sensitive code with effective use of the

TABLE 22.1

Node-Level Difference between the Cori Haswell and KNL Partitions

Category	Haswell	KNL
Cores	16	68
Hardware threads	64	272
AVX frequency	2.3 GHz	1.2 GHz
FLOPs per cycle	16	32
Vector Width	256 Bit	512 Bit
VPUs	2	2
Fused multiply-add support	Yes	Yes
Off-package memory	128 GB	96 GB
Off-package memory bandwidth	120 GB/sec	102 GB/sec
On-package memory		16 GB
On-package memory bandwidth		460 GB/sec
Total L3 cache	80 MB	
Total L2 cache	8 MB	34.8 MB
L2 cache per core	256 KB	512 KB

16 GB of on-package MCDRAM (Multi-Channel DRAM) with a bandwidth 4× that of the DRAM on Haswell; and (3) taking advantage of the eight double-precision wide vector units on the KNL processors (2× the width of Haswell). Typically speaking, it is necessary for a code to satisfy case (1) and at least one of case (2) and case (3) in order to achieve better performance on the KNL partition. Note, the potential 32 FLOPs per cycle is the result of each KNL sporting two vector processing units (VPU) with eight wide vectors that support fused multiply-add (FMA) (two FLOP) instructions.

As shown in the table, the KNL processor lacks an L3 cache. However, the MCDRAM on the chip may be optionally configured as a direct mapped cache. We discuss some of the impact of the lack of the L3 cache in the results below. The on-chip mesh (the 2D layout of the 68 cores arranged as 34 tiles) may be configured in multiple Nonuniform Memory Architecture (NUMA) configurations, where the tag-directories are localized to half or quarters of the mesh and the hemispheres and quadrants are optionally exposed to the operating system (OS) as NUMA domains. In the results in the following sections, we typically find only small performance differences between these modes (less than 10%). However, we note that when performing multinode runs, having multiple MPI tasks per node is required to saturate the bandwidth available to the node—so running with multiple MPI ranks per node in a hybrid application (regardless of NUMA configuration) is typically optimal.

22.2.2 Optimization Approach

Our approach to optimizing BerkeleyGW for KNL and the Cori system has been to profile the code to identify four key bottlenecks and to generate stand-alone mini-apps or kernels that represent the single node-level work within these bottlenecks performed during a calculation at scale [14]. In Section 22.4, we discuss these four code bottlenecks and expand on the discussion in Ref. [14].

As discussed in the previous section, obtaining high performance on KNL typically requires both the ability to scale to 68 cores on each node and either effective use of the MCDRAM or the wide VPUs or both. Knowing which of these (or both) to target in a given kernel or code bottleneck is essential for defining an optimization path. Toward this end, we utilize heavily the roofline performance model [15] developed at the Berkeley lab.

Figure 22.4 shows the performance bounds for a 64 core (7210) KNL node. The *x*-axis is the arithmetic-intensity (FLOP/bytes-from-memory) for a given algorithm that is typically measured empirically (with the vtune [16] (bytes) and Intel Software Development Emulator [17] [FLOPs]

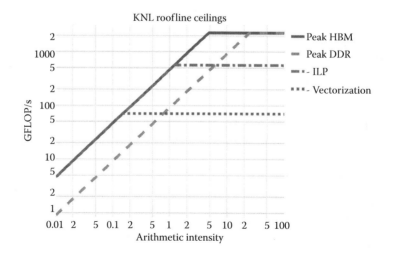

FIGURE 22.4 Roofline performance ceilings for a 7210 Knights Landing node used for node-level optimizations targeting Cori. See text for details.

tools) in a given code or kernel. For a given arithmetic intensity, the value of the piece-wise linear curves represent the performance ceiling. The two diagonal lines correspond to calculations performed by running out of MCDRAM versus running out DDR (Double Data Rate Memory). The position of each diagonal line corresponds to the memory bandwidth, the line to left with a higher ceiling corresponding to running out of MCDRAM.

For codes with a low arithmetic intensity (poor data reuse), the performance ceiling is low—the product of the arithmetic intensity x available memory bandwidth. For codes with large arithmetic intensity (high data reuse in the registers and or L1 and L2 cache levels), the ceiling eventually flattens out—meaning the code is no longer limited in practice or in theory by the available memory bandwidth. We show three ceilings for such codes, corresponding to different scenarios. The first line, labeled Peak (HBM), and which flattens at just beyond 2 Teraflops is the ceiling for a code that is implemented optimally. DGEMM (Matrix-Matrix multiply routines in is the BLAS library) could be expected to perform near this ceiling. The second line, labeled "-ILP" (Instruction Level Parallelism), corresponds to a code that vectorizes well but does not utilize FMA instructions and does not have enough work or instruction-level parallelism to keep both VPUs busy—together these account for a factor of 4 in performance. The third line corresponds to an additional lack of vectorization (a factor of 8 below line 2). Thus, if your code, despite having a high arithmetic intensity, does not vectorize or take advantage of FMA instructions and both VPUs, your expected performance is well under 100 GFLOP/sec.

We use the roofline model to visualize the absolute performance of our code bottlenecks. This is illustrated in detail in the BerkeleyGW bottleneck C example below.

22.3   MPI+OPENMP SCALING ON MPP SYSTEMS

Before discussing the optimization of the GW bottlenecks on the KNL architecture, let us have a brief aside to discuss the strong scaling of the application on a traditional HPC system—in our case, the Edison system at NERSC. For this purpose, we consider a system containing over 1000 atoms in the simulation box, shown in Figure 22.5, and which represents a silicon crystal with a vacancy defect.

As shown in Figure 22.6, the performance of the code at large scales is significantly better with fewer MPI ranks and more OpenMP threads per rank. From the timings, we see that this appears to be related largely to the communication step. The total amount of data transferred during communication is independent of the ratio of threads to MPI tasks. However, the number of messages goes up, and the size of each message goes down as we increase the number of MPI tasks (by reducing the number of threads). The difference that we see between the performance with 12 threads versus 3 threads can be largely attributed to MPI latency. This demonstrates one of the major benefits of moving from a pure MPI model to a MPI+OpenMP model at large scales. Collective communication time (particular when the number of collectives scales linearly with the number of ranks) can be reduced. For example, when running on 40,000 cores in pure MPI mode, the average message size in our example would be 435 kB. With 12 threads, the message size increases to 5 MB and the number of messages decreases by a factor of 12.

BerkeleyGW performs IO on a few large datastructures, namely the polarizability and dielectric matrices described below. For the above system, these files are 17 GB per frequency, where on order 100 frequencies are typically used, but IO is done per frequency. We use collective HDF5 writes for these files, which are distributed among the MPI ranks in a block cyclic pattern, and see performance between 1 and 5 GB/sec for these writes on the Cori lustre parallel filesystem at the node counts discussed. The performance is nearly independent of the stripe size and count. For the benchmark we presented earlier, one can see this is not a significant fraction of the walltime. However, it is interesting to note that performance of writes in this block cyclic distributed data structure are far from ideal—the peak performance of the filesystem being orders of magnitude higher and, when writing noncollectively, one file-per-process (FPP), we achieve approximately 60 GB/sec bandwidth (an order of magnitude better performance).

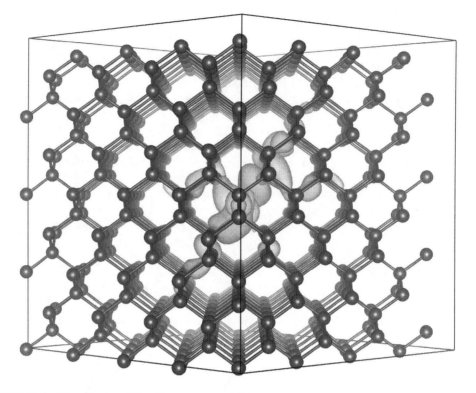

FIGURE 22.5   Visualization of the silicon vacancy defect system, containing more than 1000 atoms, used for the strong scaling study.

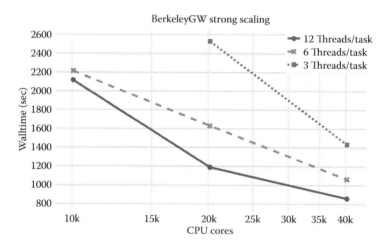

FIGURE 22.6   Strong scaling of silicon defect calculations for a single k-point and frequency (additional trivial near linear scaling over k-points and frequencies is possible) on the NERSC Edison system. The calculation includes a complete run of the `epsilon.x` executable including both I/O and communication.

22.4   COMPUTATIONAL BOTTLENECKS IN GW

Simulations of materials with BerkeleyGW contain several well-separated steps that (depending on the problem size) require similar large amounts of walltime and CPU time. When targeting optimizations for the Cori Phase 2 system, we created representative stand-alone mini computational kernels or benchmarks to run on one node but represent the node level work for a problem at scale (1000's of nodes) for each step or bottleneck. These steps and their derived kernels are described in this section. In the following section, we discuss the performance and optimization process on KNL for each of these steps.

22.4.1 BOTTLENECK A: COMPUTING ORBITAL TRANSITION PROBABILITIES

Materials respond to perturbations (e.g. exposure to light) by transitioning its electrons from low- to high-energy orbitals. The associated transition probabilities are a main ingredient in a GW calculation. The computation of these transition probabilities (often referred to as *matrix elements*) between electron orbitals is our first computational bottleneck in a typical GW calculation. Since BerkeleyGW operates in a plane-wave basis, we construct a general perturbations as a superposition of plane-wave perturbations, and compute transition probabilities as

$$M_{nn'}(\mathbf{G}) = \int \phi_n(\mathbf{r}) * \phi_{n'}(\mathbf{r}) e^{i\mathbf{Gr}} d\mathbf{r} \tag{22.4}$$

where $\mathbf{G}$ is a wave-vector and $\phi_n(\mathbf{r})$ is an electronic orbital (complex double precision arrays) generated from a DFT application like Quantum ESPRESSO. In BerkeleyGW, we compute these matrix elements by noting that the integral represents a Fourier transform of the produce of states:

$$M_{nn'}(\{\mathbf{G}\}) = \mathrm{FFT}^{-1}\left(\phi_n(\mathbf{r})\phi_{n'}^*(\mathbf{r})\right) \tag{22.5}$$

Therefore, we must compute a three-dimensional (complex double precision) FFT over the simulation cell of the material for each pair of electronic states (orbitals). The number of electron orbitals that is occupied or unoccupied within a given energy window scales linearly with the number of atoms N in the simulation cell. Therefore, the computational complexity of this calculation is $O(N^3 \log N)$.

In the BerkeleyGW package, this computation is parallelized via MPI over the band pairs and via OpenMP over the plane-waves/real-space-grid in the FFTs themselves. The FFTs are treated locally rather than the vast distributed FFTs in plane-wave DFT applications, which leads to limitations on strong scaling in plane-wave DFT apps. The difference is that DFT apps either cannot or do not express enough parallelism over bands to saturate the MPI parallelism available on a large scale HPC system. By parallelizing over not just bands but band-pairs (order 1 million for a system with a few hundred atoms), BerkeleyGW does not need to utilize MPI to parallelize the FFTs themselves—leaving this work to be performed by OpenMP on the node and avoiding costly all-to-all communication. The limitation of this approach is that the memory for an entire 3D box must fit within a node, however even for system with many 10s of thousands of atoms (well beyond the largest target systems) this is not an issue on the Cori system. We discuss this more in the next section.

22.4.2 BOTTLENECK B: CONSTRUCTION OF THE ELECTRONIC POLARIZABILITY

The electronic polarizability, one of the main ingredients in the self-energy, is a matrix that combines the various orbital transition probabilities computed in Bottleneck A to describe how the electron charge density as a whole changes in response to a perturbation. This is the important quantity that describes how electrons reorient to *screen* electric fields in the material. In particular, the previously computed *matrix elements* from Bottleneck A are combined to compute the response of the system via

$$\chi_{\mathbf{GG'}}^0(E) = \sum_n^{occ} \sum_{n'}^{emp} M_{nn'}(\mathbf{G}) M_{nn'}(\mathbf{G'}) \frac{1}{E_n - E_{n'} - E}, \tag{22.6}$$

where E_n are the input orbital energies and E is the perturbation energy (or frequency) typically computed on a grid of around 50–100 values. The computation is implemented within the BerkeleyGW package as a large parallel (complex double precision) matrix–matrix multiply (ZGEMM in BLAS notation):

$$\chi^0_{\mathbf{GG'}}(E) = \mathbf{M}(\mathbf{G}, (n, n'), E) \cdot \mathbf{M}^{\mathrm{T}}(\mathbf{G'}, (n, n'), E) \tag{22.7}$$

for each of about 50–100 E values. Here (n, n') can be thought of as a combined index summed over during the matrix-multiply. The matrices $\mathbf{M}$ are constructred from the matrix elements M:

$$\mathbf{M}(\mathbf{G}, (n, n'), E) = M_{nn'}(\mathbf{G}) \cdot \frac{1}{\sqrt{E_n - E_{n'} - E}} \tag{22.8}$$

This expression contains an explicit transpose of data over the MPI ranks. As mentioned earlier, the $M_{nn'}$ are distributed to the MPI tasks via the band pairs nn'. The matrix-multiply itself scales as $O(N^4)$, with the number of atoms N - and generally should dominate the walltime. After the matrix-multiply, we distribute out the $\chi^0_{\mathbf{GG'}}$ matrix in block cyclic manner for the purposes of performing an inversion with ScaLAPACK [18]. The inversion scales as $O(N^3)$ and is typically not a significant computational bottleneck. Writing the matrix to disk from the distributed block cyclic layout was discussed in the previous section.

22.4.3 BOTTLENECK C: COMPUTATION OF ELECTRON SELF-ENERGY

The next two bottlenecks represent the solution to Equation 22.3 itself, which can be done in a variety of approximations. The self-energy operator, Σ, in Equation 22.3 is often approximated as diagonal in the DFT orbital basis—requiring only an update of the electron energies (from the input Kohn–Sham eigenvalues) to be computed within first-order perturbation theory, rather than a rediagonalization of the Hamiltonian. We also often employ the so-called Hybertsen–Louie generalized plasmon-pole (GPP) approximation [2,3], which requires us to compute Equation 22.6 at only a single value of E and extrapolates the energy dependence of χ using a model. This simplifies the energy integrals needed to compute Σ into the following form:

$$\Sigma_n = \sum_{n'} \sum_{\mathbf{GG'}} M^*_{n'n}(-\mathbf{G}) M_{n'n}(-\mathbf{G'}) \frac{\Omega^2_{\mathbf{GG'}}}{\tilde{\omega}_{\mathbf{GG'}} \left(E - E_{n'} - \tilde{\omega}_{\mathbf{GG'}} \right)} v(\mathbf{G'}) \tag{22.9}$$

where Ω and $\tilde{\omega}$ represent complex double precision arrays that can be quickly computed from $\chi^0_{\mathbf{GG'}}(0)$, $v = 1/\mathbf{G}^2$ is the Coulomb potential in Fourier space and E is the energy at which we evaluate the material response—typically at or near the energy of the orbital ϕ_n. The operation is a matrix reduction (often called tensor contraction in the chemistry community), which is an often memory-bandwidth bound application, but the arithmetic intensity improves when we consider multiple E values surrounding the exact ϕ_n orbital energy (the default is 3, but many applications require more). To compute a band structure, a user typically compute Σ_n for all n in a given energy region. In this scenario, the computation cost scales with the number of atoms, N, in the system as $O(N^4)$. If one only wants to calculate a gap requiring a fixed two energy states, n, then the scaling is only $O(N^3)$. The MPI parallelization here is performed over n, n', and G—the latter index also being used for OpenMP. We discuss vectorization in the next section.

22.4.4 Bottleneck D: Computation of Electron Self-Energy with Full Energy Dependence

One may relax the GPP approximation described earlier and instead obtain Σ as an integral over all frequencies for which we computed χ^0:

$$\Sigma_n = \frac{i}{2\pi} \sum_{n'} \sum_{GG'} M^*_{n'n}(-\mathbf{G}) M_{n'n}(-\mathbf{G'}) \int_0^\infty dE' \frac{\left[\epsilon_{GG'}\right]^{-1}(E')}{E - E_{n'} - E'} v(\mathbf{G'}) \tag{22.10}$$

where $\epsilon_{GG'}$ is a complex double-precision array derived from $\chi^0_{GG'}(E')$. This has formally the same complexity as the GPP sums earlier (and it should be noted that the integral on the right is independent of n). The bottleneck is typically the evaluation of an intermediate array:

$$A_{nn'}(E') = \sum_{GG'} M^*_{n'n}(-\mathbf{G}) M_{n'n}(-\mathbf{G'}) \left[\epsilon_{GG'}\right]^{-1}(E') v(\mathbf{G'}) \tag{22.11}$$

for all n'. The MPI parallelization here is the same as in bottleneck C above.

22.5 GW OPTIMIZATION PROCESS FOR CORI

In this section, we will discuss the optimization process for the BerkeleyGW bottlenecks as described above on the Cori manycore HPC system. The bottlenecks naturally stress different aspects of the system and require different optimization strategies that in many ways represent a diverse workload.

In order to test single-node performance on KNL, we utilized the *Carl* test system at NERSC sporting Xeon-Phi 7210 (64-core) processors. The frequency of these parts is 1.1 GHz—which is slightly lower than the 68-core Xeon-Phi processors in the cori system. The 64 cores are laid out on a 2D mesh connected to four channels of MCDRAM and 6 channels of DDR. Each node has 16GB of on chip MCDRAM and 96GB of off chip DDR memory. The MCDRAM can be configured either as a allocatable memory (with the option of splitting the MCDRAM and the distributed tag directory into 1, 2 or 4 NUMA domains) or as transparent direct mapped cache. In our studies below, we use the *quadrant* NUMA mode (which localized the distributed tag-directory to cores in quadrants of the 2D mesh but exposes each memory system as only a single NUMA domain).

The Intel Haswell performance numbers are based on calculations run on a single node of the Cori-Phase 1 system with two 16-core Xeon E5-2698 processors per node and two NUMA domains. OpenMP scaling numbers are typically obtained with two MPI ranks per node unless otherwise stated. Ivy Bridge numbers are computed on NERSC's Edison system with two 12-core E5-2695 processors across two NUMA domains per node. Sandy Bridge numbers are computed on NERSC's babbage system with two 8-core E5-2670 processors per node again spread across two memory domains. We compare the most optimized build on each architecture using the 2016 update 2 suite of the Intel compiler and libraries. When projecting to Cori, we assume the ability to strong scale to a sufficient number of nodes to run entirely out of MCDRAM as discussed in the MPI scalability section.

We compile to the highest ISA (instruction set) on each architecture using AVX instructions on Sandybridge and Ivy-Bridge, AVX2 on Haswell and AVX-512 on KNL.

22.5.1 Bottleneck A: Optimization Strategy

As mentioned earlier, the $O(N^2)$ three-dimensional FFTs (one for each pair or orbitals) is distributed via band pairs over MPI ranks, leaving each individual FFT (typically many FFTs) to be performed on node. This approach is well suited for a manycore system like Cori for a number of reasons:

(1) We minimize internode communication, allowing near perfect strong scaling and (2) we leave a significant amount of work for each node to exploit with OpenMP and vector parallelism. This is superior to typical DFT applications that parallelize each individual FFT among the nodes.

The threaded three-dimensional FFTs are performed by simple calls to the threaded FFT implementations within the MKL library, which optimize the thread and vector level performance on each architecture. Therefore, for this particular computational bottleneck, significant hand tuning was not required. Nevertheless, we developed a runnable kernel to compare performance on the KNL architecture against Knight's Corner (KNC) and various Xeon architectures. The kernel typically makes a single threaded `fftw_many` [19] call on complex double-precision arrays. We use a single call to mimic the lack of data-reuse in the real application as we cycle through band pairs though we initialize MKL ahead of time.

In Figure 22.7, we show the performance of FFT library calls in two limits: the limit of a single very large ($960 \times 960 \times 480$—6 GB) FFT box, and the limit of many (400) small FFTs of dimension 135^3. For the single FFT, we estimate throughput on the dual socket Xeon systems by using half the socket and multiplying the performance by 2—a somewhat unfair comparison (in favor of the Xeons) but consistent with our real application use case and generally more fair than comparing Xeon performance on a single three-dimensional FFT spread across two NUMA domains—except when running at scale, where MPI latency considerations (as shown earlier) may drive us toward one MPI rank per node. In the case of the FFT many runs, we choose an optimal number of MPI ranks versus OpenMP threads in each case—two in the case of Xeon and four in the case of Xeon-Phi. In both cases, we run entirely out of MCDRAM on the KNL node.

In the large limit, we can see that the kernel is memory bandwidth bound running out of DDR on the KNL. Running out of MCDRAM (utilizing the `numactl` Linux command) provides a 2× performance advantage. This speedup does not appear in the many-small FFT limit, where data for each individual FFT have better reuse out of the lower levels of cache. More details on the thread scaling can be found in Ref. [14].

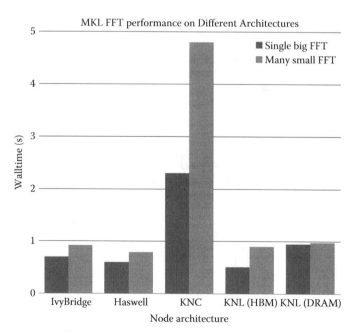

FIGURE 22.7 Walltime comparison (lower is better) for the single large and many small FFTs limits on multiple different Intel processors.

Runs of the large FFT limit on the KNL outperform runs on Haswell by about 20%, whereas the many smaller FFTs perform at near the same performance on Haswell and our 64 core KNL part. The KNL drastically outperforms the previous generation Xeon-Phi, KNC by a factor of 4. It is interesting to note that if MPI rank reduction is advantageous when running at scale (as was shown above), running with a single rank across both Xeon sockets drops the Xeon performance by nearly a factor of 2, meaning NUMA affects signficantly hamper threaded FFT performance.

In analogy to the BerkeleyGW FFT bottleneck (bottleneck A), exact-exchange calculations within a DFT application such as Quantum ESPRESSO [8] requires the computation of FFTs for each band pair. We additionally applied our optimizations to Quantum ESPRESSO the details of which can be found in refs. [14,20].

22.5.1.1   Arithmetic Intensity and Absolute Performance of FFTs on KNL

It is interesting to consider the absolute performance of FFTs on the KNL. We performed a series tests varying both the number of FFTs within a three-dimensional `fft_many` call and the size of the box. Figure 22.8 shows the measured arithmetic intensity for these calls in both the fftw [19] and Intel FFT interface. The arithmetic intensity saturates at around 2–3 (with peaks and valies for special dimensions).

Figure 22.9 shows the absolute performance of the FFTs on KNL. One can see that the performance peaks around 150 GFLOP/sec with nearly a factor of 2 speed up at the peaks corresponding to FFTs of special dimensions typically products of powers of 2, 3, and 5.

22.5.2   Bottleneck B (Polarizability): Optimization Strategy

As described in the earlier section, the computation of the polarizability is dominated by a large parallel complex matrix–matrix multiply, ZGEMM. However, another computationally intensive step exists in the kernel—the evaluation of Equation 22.8. In this operation, we stream through the M matrices and multiply each element by a precomputed energy denominator. In our benchmark, we turn the combined code into a stand-alone runnable microkernel representing the node level work on a much bigger problem with 4000 planewave basis elements, 2000 orbitals, and 20 energy values for the polarizability. The matrix size per node in our mini-app runs is (565×4800) typical of a medium size problem on an HPC cluster like Cori.

Looking at thread scaling beyond a handful of threads—in the case of this example, 64 threads and beyond—one quickly identifies serial bottlenecks in the application. One of the common discoveries in our FORTRAN applications is that seemingly harmless array initialization statements like the following:

```
example_array = 0D0
```

which actually represent loops, are not threaded by the compiler, and become a serial bottleneck. The above being a short notation for

```
do i = 1, n
  do j = 1, m
    example_array(j,i) = 0D0
  enddo
enddo
```

Because a single thread can only use a small fraction of the available MCDRAM bandwidth, these statements really hinder performance. Mitigation for this is to explicitly write the loops for such statements and associated OMP directives or, in recent compilers versions, to use the OMP WORKSHARE construct.

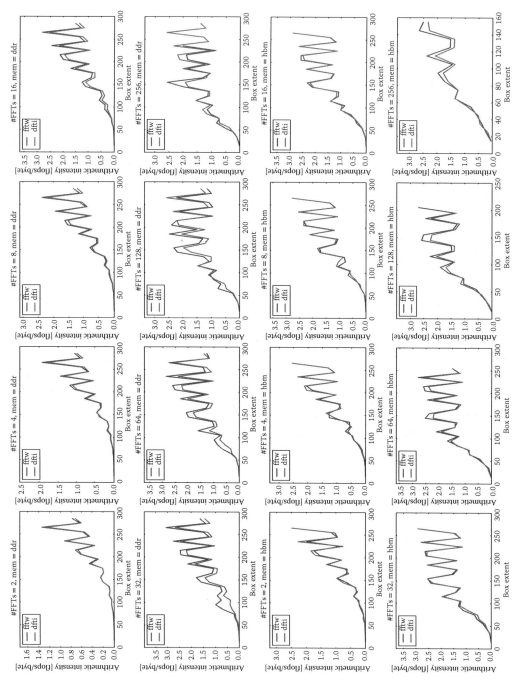

FIGURE 22.8 Measured arithmetic intensity (AI) for fft_many in the FFTW and intel FFT interfaces to MKL.

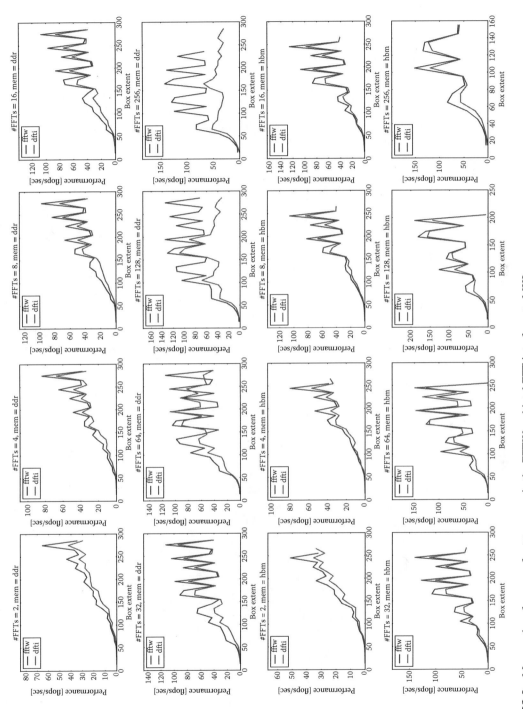

FIGURE 22.9 Measured performance for `fft_many` in the FFTW and intel FFT interfaces to MKL.

Another issue with FORTRAN array statements is that they are considered by the Intel compiler as the innermost loops in a code. Consider the following code:

```
mylength=2
allocate(example_array(mylength))
do i = 1, 100000
  example_array(:) = 0
  ... lots of work
enddo
```

Chances are that the programmer would have hoped that the explicit loop would be vectorized by the compiler; instead, the compiler will actually consider the array assignment statement as the innermost loop, will vectorize that code, and yield poor performance because of the low trip count.

We additionally found a surprising amount of time being spent in the *stream* like step described earlier (multiplying array *M* by a complex denominator). This was essentially due to the fact that we precomputed **M** for every *E* ahead of the first ZGEMM. We replace this with a computation **M** on the fly for each *E* before the ZGEMM for that energy. The same data structure (or data container) can then be reused in L2 cache across the KNL tiles for this operation.

Figure 22.10 shows the performance of the polarizability kernel across our test Xeon and Xeon-Phi architectures. The optimal performance on the KNL node is about 33% faster than the best Haswell node result with 80% of the runtime coming from the ZGEMM operations. On the Xeon nodes, we run with two MPI ranks (one per socket) compared with a single MPI rank on KNL (where the problem fits entirely within the MCDRAM). More details on the thread scaling can be found in ref [14].

On KNL, the above choice of a pure OpenMP run mode represents the optimal mix of MPI tasks versus OpenMP threads. Using a mix with two MPI tasks results in performance that is roughly 30%

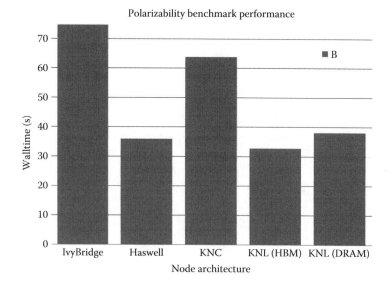

FIGURE 22.10   Performance of the polarizability benchmark on multiple dual-socket Xeon and Xeon-Phi architectures (single node comparisons).

slower and more MPI tasks continue to hurt performance. The slowdown is predominantly from the ZGEMM stage—where performing a single large ZGEMM on node outperforms multiple concurrent ZGEMM across multiple MPI ranks.

It is interesting to note that the trend does not change when booting the KNL into the SNC4 NUMA configuration—where the data for each MPI task can be allocated explicitly into the nearest quadrant of MCDRAM. MKL still performs best with a single large ZGEMM per node.

The optimized performance for the entire kernel on KNL and Haswell is 385 and 350 GFLOP/sec, respectively.

22.5.3   BOTTLENECK C (GPP SELF-ENERGY): OPTIMIZATION STRATEGY

Of the various computational bottlenecks in the BerkeleyGW application, the computation of the self-energy within the GPP approximation provides the richest optimization space. The code is comprised of hand-tuned array reduction-like operations with an arithmetic intensity (AI) between 1 and 5 as we discuss below. Because its AI is near the vertex of the roofline curve on KNL, maximizing performance requires using all the KNL hardware features appropriately.

This bottleneck represents a new bottleneck in the BerkeleyGW application based on more intense use-cases (for computing complete electronic-bandstructures on system with many atoms) that increase the computational complexity to $O(N^4)$. In recent years, this step has become a large fraction of a typical GW run. However, only recently, through the NESAP program, has it received optimization attention. We once again create a single-node benchmark for benchmarking that represents the node level work of a problem running at scale.

In order to discuss the impact of various code changes, we break our optimizations into a series of 5 steps (6 code versions including the original) and analyze the performance at each step. We describe the 6 code versions below:

Version 1: We start with an MPI-only code where little attention had been given to performance.

Version 2: We restructure the code targeting MPI, OpenMP, and Vector parallelism. We explicitly leave loops with large trip counts (see code snippets under Version 4) targeting OpenMP with thread counts on the order of 100 or higher, over the G' loop and vectorization, with trip counts in the thousands, over the ig loop.

Version 3: Like many applications running on KNL for the first time, we find that the loops we were targeting for vectorization were not being autovectorized by the compiler. In order to allow the compiler to auto-vectorize, we made a number of code changes in the following simplified code block

```
!$OMP DO reduction(+:achtemp)
do igp = 1, ngpown
   ...

      do iw=1,3 ! Original Inner Loop Bad for Vectorization

       ...

         do ig = 1, ncouls

            !Below loop prevents compiler from auto-vectorizing
```

```
!if (abs(wtilde_array(ig,my_igp) * eps(ig,my_igp))
    .lt. TOL) cycle

delw = array2(ig,igp) / (wxt(iw) - array1(ig,igp))
...
scht = scht + ... * array2(ig) * delw * array3(ig,
    my_igp)

  enddo ! loop over g

  achtemp(iw) = achtemp(iw) + 0.5D0*scht*vcoul(igp)

enddo

enddo
```

We move the innermost loop with a trip count of 3 out to ensure the *ig* loop is considered by the compiler for vectorization. We additionally remove cycle statements added to the code to save work on 10%–30% of the iterations but prevents the 8x speedup possible from vectorization. This is an example of how branching within loops, intended to save work when executing sequentially needs to be removed in order to generate efficient vector code. The new large inner loop, with trip counts of over 1000, allows us to ignore vectorization overhead due to alignment as well as peel and remainder loops, which account for trip-counts not being an exact multiple of the vector width.

Version 4: Despite improving the performance of the code (Figure 22.12) by a large factor, a number of mysteries remained. First, the performance on KNL was not outperforming Haswell by a large margin, and second, the performance was significantly worse running out of DDR versus MCDRAM—which was not expected for a FLOP heavy kernel.

Using the VTune [16] performance analysis tool, we were able to discover that the culprit was a high number of L2 misses on both Xeon and Xeon-Phi. This is because accessed rows of arrays 1–3 in the above code block require just over 1.5 MB of data per thread. This is 3× larger than the available L2 per core on KNL and about 6× larger than the L2 available on Haswell. Therefore, these rows must be streamed into the registers past L2 for each iw iteration. However, the required data sizes do fit into the two 40 MB L3 caches on the dual socket Xeon node, where we run 16 threads per socket. However, the lack of the L3 on Xeon-Phi leads us to add a level of cache blocking with blocks targeting the L2 cache on both the Haswell and KNL architectures. We make the following code modification:

```
do igp = 1, ngpown ! OpenMP
  do iw = 1 , 3
    do ig = 1, igmax
      load array1(ig,igp) !512KB per row
      load array2(ig,n1) !512KB per row
      load array3(ig,igp) !512KB per row
```

to

```
do igp = 1, ngpown ! OpenMP
  do igbeg = 1, igmax, igblk
```

```
do iw = 1 , 3
    do ig = igbeg, min(igbeg + igblk,igmax)
        load array1(ig,igp) !512KB per row
        load array2(ig,n1) !512KB per row
        load array3(ig,igp) !512KB per row
```

This improves our data reuse out of L2 by a factor of 3, which is significant in a situation where the arithmetic intensity is around 1—see Figure 22.11. The effect of this change is nearly negligible on Haswell because of the presence of an L3 cache.

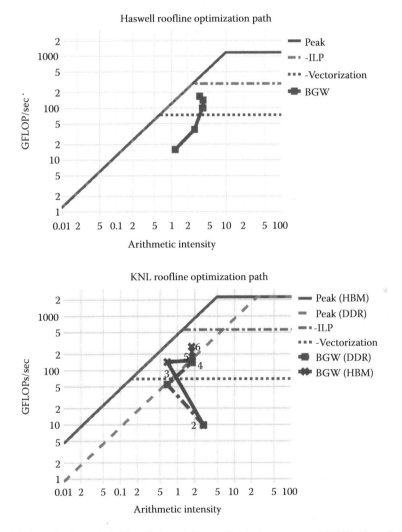

FIGURE 22.11 (See color insert.)   Visualization of the optimization process of GW bottleneck C within the roofline [15,22,23] model on Haswell and Knights Landing (KNL). The two separate diagonal lines on the KNL roofline represent the different ceilings when running out of DDR and MCDRAM with the KNL booted into flat memory mode. The three ceilings correspond to peak performance, peak minus the use of fused multiply-add instructions and instruction saturation on dual VPUs, and the additional loss of vector parallelism (a factor of 4 for Haswell and a factor of 8 for KNL). See text for more details.

However, the change makes a noticeable difference on KNL, particularly when running out of DDR. The cache-blocking speedup when running out of MCDRAM is a nonnegligible (5%) but significantly smaller than the speedup when running out of DDR.

Version 5: We remove some unnecessary if statements in the *ig* vector loop to improve vector performance.

Version 6: We utilize hyperthreads (256 total threads on KNL and 64 on Haswell). The speedup from using multiple threads per core suggests we are achieving some level of latency hiding within the application. See the discussion later.

The roofline model [21,22] provides a useful way to visualize this optimization process on both Haswell and KNL. Figure 22.11 shows the self-energy kernel for each of the six optimization steps described earlier.

On Haswell, the process shows a consistent march toward higher arithmetic intensity and higher performance (GFLOP/sec). The optimized (step 6) performance is at 170 GFLOP/sec with an arithmetic intensity of around 3.

It is natural to ask why the performance does not hit the predicted ceiling in the roofline model. In general, the main utility of the roofline model is its ability to force programmers ask and answer this question, which nearly always leads to important discoveries about the apps performance. In this case, there are a combination of factors: there is a divide in the final loop that has a multicycle latency, there is a lack of balance between multiplies and adds and there is branching (if statements) still remaining in the code. The inner, vector loop, has the form:

```
do ig = igbeg, min(igend,igmax)
  wdiff = wxt - array1(ig,igp)
  delw = array1(ig,igp) / wdiff
  ...
  scha(ig) = array2(ig,n1) * delw * array3(ig,igp)
  if (wdiff.gt.limittwo .and. delw.lt.limitone) then
    scht = scht + scha(ig)
  ...
```

Therefore on Haswell, we find ourselves at just below 50% of the ceiling for our given arithmetic intensity. Note that the hyperthreading slightly decreases the arithmetic intensity, likely due to sharing of the caches between hyperthreads.

The optimization path is a bit more circuitous on KNL. Step 3, which greatly improved the overall performance (GFLOP/sec), has the unintended side-effect of also reducing the arithmetic intensity. When running out of DDR, the performance at step 3 is pinned to the roofline ceiling, meaning the kernel is memory bandwidth bound at this point. The reduction in AI is due to the fact that we lost a factor of 3 in data reuse by moving the iw loop outward. This is later resolved by the cache blocking in step 4, which gives us the best of both worlds: high data reuse and an inner loop with a high trip count for vectorization.

Because compilers began supporting the OpenMP 4.x standard, however, a simpler solution would be to place an OMP SIMD directive on the original outer *G* loop to force vectorization even though the inner iw loop exists. It is important to note though that the lack of an L3 cache on the KNL can definitely be felt by applications.

Although the optimization steps improve the performance on both Haswell and KNL, we note that the performance delta is higher on KNL than Haswell for all steps. Even after the cache blocking optimization, the performance on KNL when running out of DDR remains pinned to the roofline ceiling. Therefore, no further optimizations (including hyperthreading) improve the performance. However, we have additionally tried explicitly modifying the source code with FASTMEM directives and find we can obtain a slight speedup compared to running the entire problem out of the MCDRAM by explicitly placing the three arrays (`array1`, `array2`, and `array3`) in fast memory. The small boost appears to be coming from the ability to draw the combined bandwidth of both MCDRAM and DDR.

Figure 22.12 shows the overall speedup of the full BerkeleyGW application as a result of these improvements. The improvements to Sigma.x demonstrate the bottleneck C improvements while the improvements to Epsilon.x general come from bottlenecks A and B.

We additionally note in Figure 22.11 that step 6, simply using more threads than the number of physical cores on the nodes, makes a significant difference in performance—nearly 30% speedup on both KNL (MCDRAM) and Haswell. Hyperthreads can typically help an application when they hide latency in instructions generated by a single thread or drive higher bandwidth. In our case, the divides are a potential source of latency and general lack of balanced instructions may allow multiple threads to better keep the various ports busy.

Details on the thread scaling of this benchmark can be found in Ref. [14]. There is, however, a slight speedup to running 4 MPI tasks (in either SNC4 mode or quadrant mode) of 10% over a pure threaded run. At high concurrencies, we see that the scaling for runs on KNL out DDR saturate due to memory bandwidth limitations as shown in the roofline model.

This section has shown that a very rich optimization space exists for applications with arithmetic intensities between 1 and 10, it is essential to effectively use all the hardware features the KNL has to offer in order to achieve maximum performance.

22.5.4   BOTTLENECK D: OPTIMIZATION STRATEGY

Our example run for the time sensitive equation in Bottleneck D

$$A_{nn'}(E') = \sum_{\mathbf{GG'}} M_{n'n}^*(-\mathbf{G})M_{n'n}(-\mathbf{G'})\left[\epsilon_{\mathbf{GG'}}\right]^{-1}(E')v(\mathbf{G'}) \tag{22.12}$$

has the following parameters: $Nn' = 1$, $Nn = 128$, $NG = 10{,}000$, $NG' = 200$, $NE = 200$ per node. Therefore, ϵ has size 6.1 GB, M^* has size 20 MB, M has size 400 kB, and v has size 3 kB.

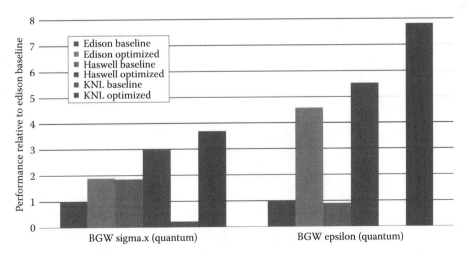

FIGURE 22.12   Comparison of original and optimized walltimes for BerkeleyGW on Edison, Cori Haswell and Cori KNL when running at scale (42 and 420 nodes for sigma.x and epsilon.x). We compare the same number of nodes on all three systems.

The original code had the form:

```
do inp = 1, Nnp
  do iE = 1, NE
    do iGp = 1, NGp
      do iG = 1, NG
        temp = temp + M(iGp,inp)*Ms(iG,inp)*eps(iG,iGP,iE)
```

One can clearly see that we stream in eps, the variable for ϵ, Nnp times in this original construction. In our optimized code, we strive to take advantage of the fact that $M \ll M^* \ll \epsilon$ and restructure the loops as

```
do iE = 1, NE
  do iGt = 1, NG, nblock
    do iGp = 1, NGp
      do inp = 1, Nnp
        do iG = igt, min(igt+nblock,NG)
          temp = temp + M(iGp,inp)*Ms(iG,inp)*eps(iG,iGP,iE)
```

In this new construction, we reuse nblock elements of eps out of cache Nnp times. We reuse nblock ∗ Nnp elements of Ms out of cache NGp times. And, each element of M is reused nblock times.

It should be pointed out that the loop reordering in the production code is not nearly as simple as the above code snippets make it seem. The inp loop in the actual code was actually several hundred lines above the iE loop. Thus changing the loop order actually involved significant code changes.

We auto-tune the selection of nblock and find a value of 50 to be optimal on KNL. On Haswell, the performance is fairly insensitive to the blocking due to the fact that Ms fits in the L3 cache on the chip. This is yet another example, where more care to reuse data in L2 cache is required on KNL than Haswell.

Figure 22.13 shows the walltime before and after optimization on Haswell and KNL. We notice something unusual, the speedup from optimization is significantly greater on Haswell than it is on KNL. This can be explained by the fact that the original code was essentially doing data streaming

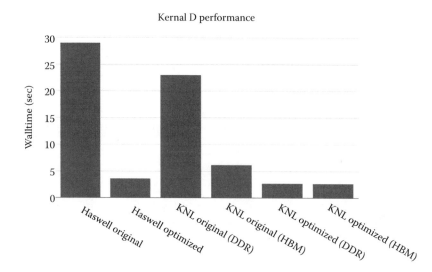

FIGURE 22.13 Walltime for BerkeleyGW bottleneck D on dual-socket Haswell and Knights Landing before and after optimization.

from DRAM. Because the problem fits in MCDRAM on KNL, the performance of the original code on KNL is actually greater than on Haswell. One might have been tempted to declare success at this point, but by reordering the loops and adding a level of cache-blocking, we not only improve the performance dramatically on Xeon (by close to a factor of 10) but also improve the performance by over a factor of 2 on KNL. The final code runs about 35% faster on KNL than Haswell. It is also interesting to note that after the optimizations, the performance on KNL out of MCDRAM (HBM) and DDR is nearly identical, showing that we are effectively using the L2 caches.

22.6   SUMMARY

In this chapter, we have described the optimization strategy and optimization process for the BerkeleyGW software package toward performant execution on exascale systems. Toward this end, we used the NERSC Cori system as our pre-exascale target system. The Cori system represents many of technology trends on the way to exascale—manycores (68 per chip) and hardware threads (272 per chip), vector units that support many FLOPs per cycle (32), and a deeper memory hierarchy (16 GB of MCDRAM and 96 GB of DDR per node).

We demonstrate that codes (unoptimized Sigma FF benchmark code) with low arithmetic intensity (and high bandwidth requirements) that fit in the fastest memory tier will perform relatively better on the manycore KNL architecture than the traditional Xeon (Haswell) architecture with little effort. However, in the case of the Bottleneck D code, further optimizations improving data-reuse improve the performance on both architectures.

In general, though, out-performing Haswell on KNL requires at least two of the following qualities: efficient thread scaling, efficient vectorization and effective use of the high-bandwidth memory, which two depends on the arithmetic intensity of the application. For codes with arithmetic intensities between 1 and 10 (e.g., our GPP sigma benchmark) have an extremely rich optimization space. For such codes, optimal performance ultimately requires all of the following: effective thread/task scaling, effective use of vector-units and effective use of MCDRAM.

Ultimately, what is the payoff of all these optimization efforts? Figure 22.12 shows the overall application speedups and Figure 22.14 shows the ultimate reward, a 30% reduction in energy to

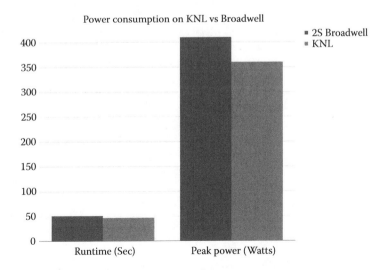

Power consumption on KNL vs Broadwell

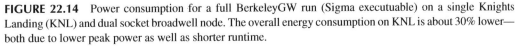

FIGURE 22.14   Power consumption for a full BerkeleyGW run (Sigma executuable) on a single Knights Landing (KNL) and dual socket broadwell node. The overall energy consumption on KNL is about 30% lower—both due to lower peak power as well as shorter runtime.

solution for a complete BerkeleyGW run on KNL versus Broadwell (the Haswell Xeon successor). As we move toward exascale, both in terms of architecture and in terms of application optimization, we expect this gap will widen.

ACKNOWLEDGMENTS

This work was supported by the Center for Computational Study of Excited-State Phenomena in Energy Materials at the Lawrence Berkeley National Laboratory, which is funded by the U.S. Department of Energy, Office of Science, Basic Energy Sciences, Materials Sciences and Engineering Division under Contract No. DE-AC02-05CH11231, as part of the Computational Materials Sciences Program. Previous work supported by the SciDAC Program on Excited-State Phenomena in Energy Materials funded by the US Department of Energy (DOE), Office of Basic Energy Sciences and of Advanced Scientific Computing Research, under Contract No. DE-AC02-05CH11231 at Lawrence Berkeley National Laboratory. Derek Vigil-Fowler is support by NREL's LDRD Director's Postdoctoral Fellowship. This research used resources of the National Energy Research Scientific Computing Center, a DOE Office of Science User Facility supported by the Office of Science of the DOE under Contract No. DE-AC02-05CH11231.

We acknowledge helpful conversations with Mike Greenfield, Paul Kent, David Prendergast, and Pierre Carrier.

REFERENCES

1. Kohn, W. and L. J. Sham. Self-consistent equations including exchange and correlation effects. *Physical Review* 140, no. 4A (1965): A1133.
2. Hybertsen, M. S. and S. G. Louie. Electron correlation in semiconductors and insulators: Band gaps and quasiparticle energies. *Physical Review B* 34, no. 8 (1986): 5390.
3. Hybertsen, M. S. and S. G. Louie. First-principles theory of quasiparticles: Calculation of band gaps in semiconductors and insulators. *Physical Review Letters* 55, no. 13 (1985): 1418.
4. BerkeleyGW: http://www.berkeleygw.org.
5. Deslippe, J., G. Samsonidze, D. A. Strubbe, M. Jain, M. L. Cohen, and S. G. Louie. BerkeleyGW: A massively parallel computer package for the calculation of the quasiparticle and optical properties of materials and nanostructures. *Computer Physics Communications* 183, no. 6 (2012): 1269–1289.
6. Kronik, L., A. Makmal, M. L. Tiago, M. M. G. Alemany, M. Jain, X. Huang, Y. Saad, and J. R. Chelikowsky. PARSEC the pseudopotential algorithm for real-space electronic structure calculations: Recent advances and novel applications to nanostructures. *Physica Status Solidi (B)* 243, no. 5 (2006): 1063–1079.
7. Pfrommer, B., D. Raczkowski, A. Canning, and S. G. Louie. PARATEC (PARAllel Total Energy Code), Lawrence Berkeley National Laboratory (with contributions from Mauri F, Cote M, Yoon Y, Pickard C, Heynes P). For more information, see www.nersc.gov/projects/paratec.
8. Giannozzi, P., S. Baroni, N. Bonini, M. Calandra, R. Car, C. Cavazzoni, D. Ceresoli, G. L. Chiarotti, M. Cococcioni, I. Dabo, A. Dal Corso, S. Fabris, G. Fratesi, S. de Gironcoli, R. Gebauer, U. Gerstmann, C. Gougoussis, A. Kokalj, M. Lazzeri, L. Martin-Samos, N. Marzari, F. Mauri, R. Mazzarello, S. Paolini, A. Pasquarello, L. Paulatto, C. Sbraccia, S. Scandolo, G. Sclauzero, A. P. Seitsonen, A. Smogunov, P. Umari, and R. M. Wentzcovitch. QUANTUM ESPRESSO: A modular and open-source software project for quantum simulations of materials. *Journal of Physics: Condensed Matter* 21, no. 39 (2009): 5502 doi:10.1088/0953-8984/21/39/395502.
9. Soler, J. M., E. Artacho, J. D. Gale, A. Garca, J. Junquera, P. Ordejn, and D. Snchez-Portal. The SIESTA method for ab initio order-N materials simulation. *Journal of Physics: Condensed Matter* 14, no. 11 (2002): 2745.

10. Marini, A., C. Hogan, M. Grning, and D. Varsano. Yambo: An ab initio tool for excited state calculations. *Computer Physics Communications* 180, no. 8 (2009): 1392–1403.

11. Govoni, M. and G. Galli. Large scale GW calculations. *Journal of Chemical Theory and Computation* 11, no. 6 (2015): 2680–2696.

12. NERSC: http://www.nersc.gov.

13. NERSC: Cori, http://www.nersc.gov/systems/cori/.

14. Deslippe, J., F. H. da Jornada, D. Vigil-Fowler, T. Barnes, N. Wichmann, K. Raman, R. Sasanka, and S. G. Louie. Optimizing excited-state electronic-structure codes for Intel Knights Landing: A case study on the BerkeleyGW software. In *International Conference on High Performance Computing*, pp. 402–414. Cham: Springer International Publishing, 2016.

15. Williams, S., Auto-tuning performance on multicore computers. PhD thesis, EECS Department, University of California, Berkeley, CA (December 2008).

16. Intel Vtune: https://software.intel.com/en-us/intel-vtune-amplifier-xe.

17. Tal, A., Intel software development emulator, https://software.intel.com/en-us/articles/intel-software-development-emulator.

18. Blackford, L. S., J. Choi, A. Cleary, E. D'Azevedo, J. Demmel, I. Dhillon, J. Dongarra, S. Hammarling, G. Henry, A. Petitet, K. Stanley, D. Walker, and R. C. Whaley. *ScaLAPACK Users' Guide*, Philadelphia, PA: Society for Industrial and Applied Mathematics, 1997.

19. Frigo, M. and S. G. Johnson. FFTW: An adaptive software architecture for the FFT. In *Proceedings of the 1998 IEEE International Conference on Acoustics, Speech and Signal Processing*, vol. 3, pp. 1381–1384. IEEE, 1998.

20. Barnes, T. A., T. Kurth, P. Carrier, N. Wichmann, D. Prendergast, P. R. C. Kent, J. Deslippe et al. Improved treatment of exact exchange in Quantum ESPRESSO. Computer Physics Communications, May 31, 2017.

21. Williams, S., Roofline performance model, http://crd.lbl.gov/departments/computer-science/PAR/research/rooine/.

22. Williams, S., A. Watterman, and D. Patterson. Rooine: An insightful visual performance model for floating-point programs and multicore architectures. *Communications of the ACM* (April 2009).

23. CS Roofline Toolkit: https://bitbucket.org/berkeleylab/cs-rooine-toolkit.

24. Cohen, M. L. and Louie, S. G. *Fundamentals of Condensed Matter Physics*. Cambridge, UK: Cambridge University Press, 2016.

23 Global Gyrokinetic Particle-in-Cell Simulation

William Tang and Zhihong Lin

CONTENTS

23.1 Introduction...507
23.2 Scientific Methodology...508
    23.2.1 Kinetic Electron Models...510
    23.2.2 Resistive Tearing Mode Model..510
    23.2.3 Collisionless Tearing Mode Model...510
    23.2.4 Global PIC Geometric Models 511
    23.2.5 Global PIC Grid Considerations 513
23.3 Algorithmic Details 513
23.4 Programming Approach...515
    23.4.1 Scalability 515
    23.4.2 Performance...516
    23.4.3 Portability 517
    23.4.4 External Libraries 519
23.5 Software Practices 519
23.6 Benchmarking Results...520
23.7 Time-to-Solution and Energy-to-Solution Comparative Studies....................................525
23.8 Concluding Comments 526
References 527
Research Team...528

23.1 INTRODUCTION

As the global energy economy makes the transition from fossil fuels toward cleaner alternatives, fusion becomes an attractive potential solution for satisfying the growing needs. Fusion energy, which is the power source for the sun, can be generated on earth, for example, in magnetically confined laboratory plasma experiments (called *tokamaks*) when the isotopes of hydrogen (e.g., deuterium and tritium) combine to produce an energetic helium *alpha* particle and a fast neutron—with an overall energy multiplication factor of 450:1. Building the scientific foundations needed to develop fusion power demands high-physics-fidelity predictive simulation capability for magnetically confined fusion energy (MFE) plasmas. To do so in a timely way requires utilizing the power of modern supercomputers to simulate the complex dynamics governing MFE systems—including International Thermonuclear Experimental Reactor (ITER), a multibillion dollar international burning plasma experiment supported by seven governments representing over half of the world's population. Currently, under construction in France, ITER will be the world's largest tokamak system, a device that uses strong magnetic fields to contain the burning plasma in a doughnut-shaped vacuum vessel. In tokamaks, unavoidable variations in the plasma's ion temperature profile drive microturbulence—fluctuating electromagnetic fields, which can grow to levels that can significantly increase the transport rate of heat, particles, and momentum across the confining magnetic field. Because the balance between these energy losses and the self-heating rates of the actual fusion reactions will ultimately determine the size and cost of an actual fusion reactor, understanding and possibly controlling the underlying physical processes is key to achieving the efficiency needed to help ensure the practicality of future fusion reactors. The associated motivation drives the pursuit of sufficiently realistic calculations of turbulent transport that can only be achieved through advanced simulations. The present paper on advanced particle-in-cell (PIC) global simulations of plasma microturbulence at the extreme scale is accordingly associated with this fusion energy science (FES) grand challenge [1,2].

The research and development (R&D) described in this document targets new physics insights on MFE confinement scaling by making effective use of powerful world class supercomputing systems. Specifically, the long-time behavior of turbulent transport in ITER scale plasmas is studied using simulations with unprecedented phase-space resolution to address the reliability/realism of the well-established picture of the influence of increasing plasma size on confinement in tokamaks and the associated physics question of how/if the aforementioned turbulent transport changes with the size of laboratory plasmas up to the ITER scale.

Associated knowledge gained addresses the key question of how turbulent transport and associated confinement characteristics scale from present generation devices to the much larger ITER plasmas. This involves the development of modern software capable of using leadership class supercomputers to carry out reliable first-principles-based simulations of multiscale tokamak plasmas.

Particle dynamics can in general be described either by a five-dimensional (5D) gyrokinetic equation (for low-frequency turbulence) or six-dimensional (6D) fully kinetic equation (for high-frequency waves). The flagship gyrokinetic toroidal code (GTC) and its *codesign* partner GTC-P are massively parallel PIC codes designed to carry out first principles, integrated simulations of thermonuclear plasmas, including the future burning plasma ITER. These codes solve the 5D gyrokinetic equation in full, global toroidal geometry to address kinetic turbulence issues in magnetically confined fusion experimental facilities. GTC is the key production code for the fusion SciDAC project Gyrokinetic Simulation of Energetic Particle Turbulence and Transport (GSEP) Center and the Accelerated Application Readiness (CAAR) program at the Oak Ridge Leadership Computing Facility (OLCF). It is the only PIC code in the world fusion program capable of multiscale simulations of a variety of important physics processes in fusion-grade plasmas including microturbulence, energetic particle dynamics, collisional (neoclassical) transport, kinetic magnetohydrodynamic (MHD) modes, and nonlinear radio-frequency (RF) waves. GTC interfaces with MHD equilibrium solvers (e.g., in codes such as equilibrium and reconstruction fitting code (EFIT), VMEC, and M3D-C1) for addressing realistic toroidal geometry features that include both axisymmetric tokamaks and nonaxisymmetric stellarators. A recent upgrade enables this code to carry out global PIC simulations covering both the tokamak core and scrape-off layer (SOL) regions. It should also be noted that the current flagship version of GTC is capable of both perturbative (δf) and nonperturbative (full-f) simulations with capability of dealing with kinetic electrons, electromagnetic fluctuations, multiple ion species, collisional (neoclassical) effects using Fokker–Planck collision operators, equilibrium current and radial electric field, plasma rotation, sources/sinks and external antennae for auxiliary wave heating.

Beyond the conventional application domain of gyrokinetic simulation for microturbulence, the GTC code has a long history in pioneering the development and application of gyrokinetic simulations of mesoscale electromagnetic Alfven eigenmodes excited by energetic particles (EP) in toroidal geometry. This is one of the most important scientific challenges that must be addressed in future burning plasma experiments such as ITER. Accordingly, the GTC work-scope has recently been extended to include simulation of macroscopic kinetic MHD modes driven by equilibrium currents. The associated importance is that such efforts could ultimately lead to key knowledge needed to systematic analyze and possibly help avoid or mitigate highly dangerous reactor relevant thermonuclear disruptions.

23.2   SCIENTIFIC METHODOLOGY

The GTC and GTC-P codes include all of the important physics and geometric features captured in numerous global PIC simulation studies of plasma size scaling over the years—extending from the seminal work in the *Phys. Rev. Letter* (PRL) by Lin et al. [3] up to the more recent PRL paper by McMillan et al. on "system size effects on gyrokinetic turbulence" [4]. The current generally supported picture is that size-scaling follows an evolution from a *Bohm-like* trend where the confinement degrades with increasing system size to a *Gyro-Bohm-like* trend where the confinement for Joint European Torus (JET)-sized plasmas begins to *plateau* and then exhibits no further confinement degradation as the system size further increases toward ITER-sized plasmas. A number of physics papers over the past decade have proposed theories—such as turbulence spreading—to account for this transition to gyro-Bohm scaling with plasma size for large systems. From a physics perspective, the main point in this paper is that this key decade-long fusion physics picture of the transition or *rollover* trend associated with toroidal ion temperature gradient microinstabilities that are highly prevalent in tokamak systems—should be re-examined by modern supercomputing-enabled simulation studies, which are now capable of being carried out with much higher phase-space resolution and duration. With a focused approach based on performance optimization of key functions within PIC codes in general, GTC-P, the *codesign* focus, has demonstrated the effective usage of the full power of current leadership class computational platforms worldwide at the petascale and beyond to produce efficient nonlinear PIC simulations that have advanced progress in understanding the complex nature of plasma turbulence and confinement in fusion systems for the largest problem sizes. Unlike fluid-like computational fluid dynamics (CFD) codes, GTC-P has concentrated on the fact that PIC codes are characterized by having less than 10 key operations, which can then be an especially tractable target for advanced computer science performance optimization methods. As illustrated in Ref. [5], these efforts have resulted in accelerated progress in a discovery-science-capable global PIC code that models complex physical systems with unprecedented resolution and produces valuable new insights into reduction in *time-to-solution* as well as *energy-to-solution* on a large variety of leading supercomputing systems.

The *flagship* GTC code is—as noted earlier—especially comprehensive with respect to the complex physics included. Its productivity over the years is well illustrated in Table 23.1.

TABLE 23.1
Demonstration of GTC Productivity and Impact → Delivery of Scientific Advances with Use of Increasingly Powerful Supercomputing Systems

GTC Simulation	Computer Name	PE # Used	Speed (TF)	# Particles Used	Time Steps	Physics Discovery (Publication)
1988	Cray T3E NERSC	10^2	10^{-1}	10^8	10^4	Ion turbulence zonal flow (*Science*, 1998)
2002	IBM SP NERSC	10^3	10^0	10^9	10^4	Ion transport size scaling (*PRL*, 2002)
2007	Cray XT3/4 ORNL	10^4	10^2	10^{10}	10^5	Electron turbulence (*PRL*, 2007); EP transport (*PRL*, 2008)
2009	Jaguar/Cray XT5 ORNL	10^5	10^3	10^{10}	10^5	Electron transport scaling (*PRL*, 2009); EP-driven MHD modes
2012 to present	Cray XT5 Titan ORNL Tianhe-1A (China)	10^5	10^4	10^{11}	10^5	Kinetic-MHD (*PRL*, 2012); Turbulence + EP + MHD TAE Modes (*PRL*, 2013)
2018 (future)	Path to Exascale HPC Resources	TBD	10^6	10^{12}	10^6	Turbulence + EP + MHD + RF

*** GTC is first FES code to deliver production run simulations @ TF in 2002 and PF in 2009.

Several key associated computational methodologies will be elaborated upon as follows.

23.2.1 Kinetic Electron Models

The small electron mass presents a numerical difficulty for simultaneously treating the dynamics of ions and electrons in long time simulations. A fluid-kinetic hybrid electron model [6] currently implemented in GTC overcomes this difficulty by expanding the electron drift kinetic equation using the electron–ion mass ratio as a small parameter. The model accurately recovers low-frequency plasma dielectric responses and faithfully preserves linear and nonlinear wave-particle resonances. Maximum numerical efficiency is achieved by overcoming the electron Courant condition and suppressing tearing modes and high-frequency modes thus effectively suppressing electron noise. The fluid-kinetic hybrid electron model avoids the well-known *cancellation problem* in some gyrokinetic particle and continuum codes [7]. The *cancellation problem* arises when solving a particular form of the Ampere's law, where two large terms are artificially added to the original Ampere's law. These two terms are needed because canonical momentum is used as an independent variable to overcome a numerical difficulty of calculating the inductive electric field by an explicit time derivative. Analytically, these two terms should cancel with each other exactly. However, a small error in numerically evaluating these two large terms can give rise to a residue, which leads to a large error in solving the Ampere's law.

Moving into the near future, the very high-resolution capability in advanced PIC codes (such as GTC-P) hold strong promise of being further improved on the 200-petaflop near future leadership-class systems such as Summit and Aurora. Accordingly, progress toward delivering the skin-depth grid resolution capability [8] to comprehensively avoid the aforementioned *cancellation problem* in fully kinetic electromagnetic codes will likely be realistically achievable with access to the 200 PF class of supercomputers.

23.2.2 Resistive Tearing Mode Model

The fluid-kinetic hybrid electron model incorporating equilibrium current enables global gyrokinetic PIC simulations of both pressure-gradient-driven and current-driven instabilities—as well as their nonlinear interactions in multiscale simulations [9]. However, the fluid-kinetic hybrid electron model removes the tearing modes in order to improve numerical properties for facilitating production runs. In order to simulate classical tearing modes, the current GTC capabilities have recently been extended for this model with the inclusion of a resistivity term (due to friction between electrons and ions) in the electron momentum equation

$$\frac{\partial \delta A_{\parallel}}{\partial t} = -c\mathbf{b}_0 \cdot \nabla \delta \phi + \frac{cT_e}{e}\mathbf{b}_0 \cdot \nabla \, \partial n_e + \eta \frac{c^2}{4\pi}\nabla_{\perp}^2 \delta A_{\parallel}$$

This nonideal term introduces the resistive tearing mode physics into the GTC formulation [10].

23.2.3 Collisionless Tearing Mode Model

Collisionless tearing mode (e.g., microtearing mode), which is not considered important for conventional tokamak plasmas such as ITER, has been observed in some high-spherical tokamaks such as National Spherical Torus Experiment (NSTX), Mega Amp Spherical Tokamak (MAST). To simulate the collisionless tearing mode, we have implemented in GTC two new kinetic electron models: the split-weight scheme that treats the full physics of the electron drift kinetic equation including the collisionless tearing mode, and the finitemass electron fluid model. In the finite-mass electron fluid model, the electron momentum equation in the original fluid-kinetic hybrid electron model is upgraded to include the electron inertia term using the following electron parallel momentum equation:

$$\frac{\partial \delta \xi}{\partial t} = \frac{\omega_{pe}^2}{c} \left(\frac{T_e}{n_0 e} \mathbf{b}_0 \cdot \nabla \delta n_e - \mathbf{b}_0 \cdot \nabla \delta \phi \right) + \frac{4\pi e}{m_e c} \frac{\delta B_\perp}{B} \cdot \nabla n_0 T_e$$

Here, $\delta \xi = - \left(\nabla_\perp^2 = -\frac{1}{D_e^2} \right) \delta A_{||}$. The rest of the set of equations of the original fluid-kinetic hybrid electron model remain unchanged. Electron kinetic effects appear in higher order equations by using the nonadiabatic part of the drift-kinetic equation. This finite-mass kinetic-fluid hybrid electron model has been verified for simulating the theoretically predicted collisionless tearing mode instability in the slab geometry [11] and has been recently implemented and verified in GTC [12].

The formulation of the resistive and collisionless tearing has been combined to simulate the transition from collisionless to resistive tearing mode when the resistivity is increased, and recover the correct scaling of the collisionless tearing mode growth rate on the skin depth when the skin depth is much shorter than the macroscopic length. This complete formulation will be used for self-consistent simulation of the interactions between microturbulence, tearing mode, and neoclassical transport proposed in this project. This fluid electron models for collisionless tearing mode has recently been extended and verified to fully incorporate electron kinetic effects using the fluid-kinetic electron models, which is being applied to study the microtearing mode in high beta plasmas.

23.2.4 GLOBAL PIC GEOMETRIC MODELS

In plasma turbulence studies, the standard approach is to divide the physical quantities into an equilibrium part and a fluctuating part. The GTC code uses two set of meshes—one for the specification of the equilibrium and the other to represent fluctuating turbulent fields. In particular, the turbulence mesh is an unstructured field-aligned mesh for finite difference or finite element in three-dimensional (3D) space.

The equilibrium quantities are governed by the Grad–Shafranov equation for toroidal geometry, while the fluctuating part is driven by various instabilities that lead to turbulent transport. Equilibrium magnetic configurations typically used in gyrokinetic simulations come from (1) analytic models such as the simple circular cross section or the Miller equilibrium and (2) numerical equilibrium codes such as EFIT or variational moments equilibrium code (VMEC). For the rapidly evolving optimization studies that deliver very high-resolution results from investigations of plasmas with increasing problem size on the most powerful supercomputing systems, the practical choice—as exemplified by the *codesign* GTC-P code—is the category (1) analytically based equilibria. On the other hand, comprehensive production runs carried out by the *flagship* GTC code demand interfacing with the numerical equilibria of category, (2) that properly represent the actual experimental conditions.

The most accurate representation of the equilibrium in tokamaks is by using magnetic flux coordinates rather than Cartesian coordinates. This is because that most important equilibrium quantities, such as plasma temperature and density, can be shown to depend on the magnetic flux only. The flagship GTC code employs magnetic flux coordinates (ψ, θ, ζ) to represent the electromagnetic fields and plasma profiles, where ψ is the poloidal magnetic flux, θ is the poloidal angle, and ζ is the toroidal angle. Specifically, the inputs come from the numerical magnetic equilibrium and plasma profiles are obtained from EFIT/VMEC by transforming the equilibrium quantities defined in the cylindrical coordinates (R, ϕ, Z) to those defined in the magnetic coordinates (ψ, θ, ζ). The equilibrium data are provided by MHD equilibrium codes for the magnetic field strength B, and cylindrical coordinates (R, Φ, Z) of points forming magnetic flux surfaces. Additionally, the flux functions representing poloidal $g(\psi)$ and toroidal $I(\psi)$ currents, magnetic safety factor $q(\psi)$, and minor radius $r(\psi)$—defined as a distance from the magnetic axis along the outer mid-plane—are provided. First-order continuous B-splines are implemented for the one-dimensional (1D), two-dimensional (2D), and 3D functions to interpolate the complicated magnetic geometry and plasma profiles, which provide a good compromise between high numerical confidence and reasonable computational efficiency.

The GTC capability to carry out simulations of problems with general toroidal geometry has recently been extended to also include nonaxisymmetric configurations. For nonaxisymmetric devices, the equilibrium data are presented on the uniform (ψ, θ) grid for all $n = (1, 2, \dots N)$ toroidal harmonics. To reduce the computational load and memory usage, the transformation of nonaxisymmetric variables into spline functions of ζ is chosen for implementation in GTC, with spline coefficients associated with a particular grid point ζ_i being stored by processors with corresponding toroidal rank using message passing interface (MPI) parallelization. An example of GTC results for a stellarator plasma is shown in Figure 23.1 [13].

The GTC-P code deploys the so-called *large aspect ratio equilibrium*, which is an analytical model describing a simplified toroidal magnetic field with a circular cross-section. The associated model takes into account the key geometric and physics properties needed to carry out a meaningful study of the influence of increasing plasma size on magnetically confined fusion plasmas. Such an approach enables working with a sufficiently straightforward but nevertheless discovery-science-capable physics [3,4] code that makes more tractable the formidable task of developing the algorithmic advances needed to take advantage of the rapidly evolving modern platforms featuring, for example, both homogenous and hybrid architectures. The associated physics approach is to deploy GTC-P plasma size-scaling studies because it is a fast streamlined modern code with the capability to efficiently carry out computations at extreme scales with unprecedented resolution and speed on present-day multipetaflop computers [5]. The corresponding scientific goal is to accelerate progress toward capturing new physics insights into the key question of how turbulent transport and associated confinement characteristics scale from present generation laboratory plasmas to the much larger ITER-scale burning plasmas. This includes a systematic characterization of the spectral properties of the turbulent plasma as the confinement scaling evolves from a *Bohm-like* trend where the confinement degrades with increasing system size to a *Gyro-Bohm-like* trend where the confinement basically *plateaus* exhibiting no further confinement degradation as the system size further increases. *Lessons learned* achieved in a timely way from this codesign effort can be expected to expedite associated advances in the flagship GTC code in particular as well as to generally providing valuable information on PIC performance modeling advances to ongoing and future efforts in improving PIC code deployment on multipetaflop supercomputers on the path to exascale and beyond.

FIGURE 23.1 Illustration of GTC 3D mode structure for the $n = 1$ global Alfvén eigenmode in the large helical device (LHD) stellarator. (From D.A. Spong, *Phys. Plasmas* 22, 055602, 2015.)

23.2.5 GLOBAL PIC GRID CONSIDERATIONS

In accurately tracking the key physics in magnetically confined toroidal plasmas, the GTC and GTC-P codes utilize a highly specialized grid that follows the magnetic field lines as they twist around the torus (see Figure 23.1). This allows the code to retain the same accuracy while using fewer toroidal planes than a regular, non–field-aligned grid. From relevant physics considerations, because short wavelength waves parallel to the magnetic field are suppressed by Landau damping, increasing the grid resolution in the toroidal dimension will leave the results essentially unchanged. Consequently, a typical production simulation run usually consists of a constant number of poloidal planes (e.g., 32 or 64) wrapped around the torus. Each poloidal plane is represented by an unstructured grid, where the grid sizes in the radial and poloidal dimensions correspond approximately to the size of the gyro-radius of the particles. As we consider larger plasma sizes (e.g., 2× in major and minor radius), the number of grid points in each 2D plane increases 4×. The number of grid points for a 3D grid increases 4× as well because the number of planes in the toroidal dimension remains the same for all problem sizes. For a modest-sized fusion device (e.g., the DIII-D tokamak at General Atomics in San Diego, CA), the associated plasma simulation typically uses ~128,000 grid points in a 2D plane. As we move to the larger JET device and then eventually to the ITER size plasmas, the number of grid points increases 4× and 16×, respectively. Using a fixed number of 64 toroidal planes, the total number of grid points for an ITER-sized plasma will be ~131 million. With 100 particles per cell resolution, an ITER-sized simulation will accordingly involve ~13 billion particles. Tracking the dynamics of this large number of particles would of course be an extremely daunting task without access to leadership-class supercomputers.

23.3 ALGORITHMIC DETAILS

The basic PIC method is of course a well-established computational approach that simulates the behavior of charged particles interacting with each other through pair-wise electromagnetic forces. At each time step, the particle properties are updated according to these calculated forces. For applications on powerful modern supercomputers with deep cache hierarchy, a pure particle method is very efficient with respect to locality and arithmetic intensity (compute bound). Unfortunately, the $O(N^2)$ complexity makes a particle method impractical for plasma simulations using millions of particles per process.

Rather than calculating $O(N^2)$ forces, the PIC method, which was introduced by J. Dawson and N. Birdsall in 1968, employs a grid as the media to calculate the long range electromagnetic forces. This reduces the complexity from $O(N^2)$ to $O(N+M\log M)$, where M is the number of grid points and is usually much smaller than N. Specifically, the PIC simulations are being carried out using *macro* particles ($\sim 10^3$ times the radius of a real charged ion particle) with characteristic properties, including position, velocity, and weight. However, achieving high parallel and architectural efficiency is very challenging for a PIC method due to potential fine-grained data hazards, irregular data access, and low arithmetic intensity. The issue gets more severe as the high performance computing (HPC) community moves into the future to address even more radical changes in computer architectures as the multicore and manycore revolution progresses.

In this chapter, the computational approach involves the advanced development of a comprehensive *ab initio* PIC global (3D) code GTC and its codesign partner GTC-P, which cover equations underlying gyrokinetic theory. As highly scalable PIC codes used for studying microturbulent transport in tokamaks, GTC and GTC-P solve the gyro-phase-averaged Vlasov–Poisson set of equations (*gyrokinetic equations*) using discrete, charged particles. Particles interact with each other through a self-consistent field evaluated on a grid that covers the whole simulation domain. The charge of each particle is deposited on the grid by interpolating to its nearest grid points, resulting in a charge density that is then used in the evaluation of the field by solving the Poisson field equations. At the position of each particle, the field is then evaluated, again by interpolation, and used in the equations of motion to advance the particles.

The parallel algorithms in GTC-P are implemented with MPI and OpenMP. The original implementation included a 1D domain decomposition and a particle decomposition. This design had shown nearly perfect scaling with the number of particles. However, as the grid size is increased when simulating large fusion devices, the 1D domain decomposition produces a significant memory footprint. To address this issue, an extra dimension domain decomposition feature was added to GTC-P [15,24]. As a result, the particles are now fully distributed across all processes while the grid is split with the implementation of an appropriate 2D domain decomposition scheme—thereby greatly reducing the memory footprint and improving cache reuse. This algorithm was developed specifically for the Blue Gene systems in order to handle the limited amount of memory per node. It is a capability introduced in GTC-P to simulate very large fusion devices on BG/P with unprecedented efficiency—a key feature that was exercised in current studies to greatly facilitate examining the key question of how plasma microturbulence properties might be affected as the plasma size increases from that of existing experiments to the future very large plasmas characteristic of ITER. Finally, the GTC-P code has data parallelism at the loop level through the use of OpenMP directives. All the loops over the particles and grid points are fairly large and contain data parallelism. This method has been used very successfully on the multicore processors and has contributed to the excellent scaling of the GTC-P code on large-scale homogenous architecture supercomputers including the BG/Q *Mira* and *Sequoia* systems in the United States and on the Fujitsu K Computer in Japan (Figures 23.2 and 23.3).

Using resources from INCITE, previous early Science Projects (ESP) at the Argonne Leadership Computing Facility (ALCF), and Director's Discretionary allocations from both the ALCF and OLCF in the past few years, GTC-P has demonstrated excellent scalability to more than 100,000 cores on leadership computing facilities at Argonne National Laboratory (ANL) and Oak Ridge National Laboratory (ORNL). It has been successfully deployed for major scientific production runs on the IBM BG/Q/*Mira*—where the excellent weak scaling performance was carried over to much larger scale on LLNL's more powerful Sequoia system. These results are illustrated in Figure 23.2. In addition, it is relevant to note that the GTC-P code was the featured U.S. code in the G8 international exascale project in nuclear fusion energy (NuFuSE) (http://www.nu-fuse.com/) that was supported in the United States by the National Science Foundation (NSF) [14]. The G8 program helped provide

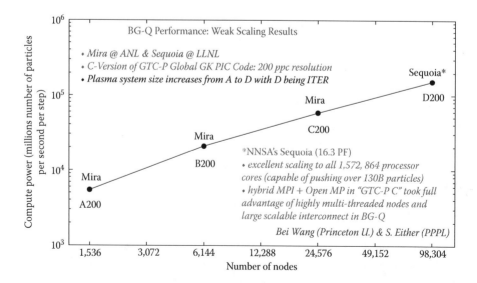

FIGURE 23.2 GTC-P code performance on world-class IBM BG/Q systems.

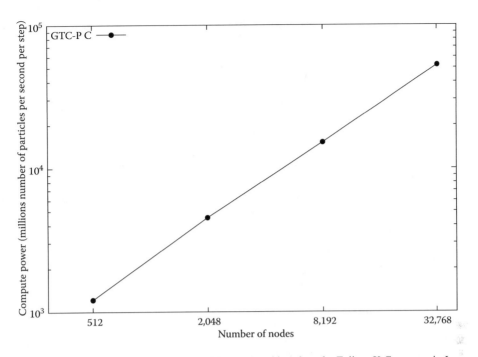

FIGURE 23.3 Excellent weak scaling of the GTC-P code achieved on the Fujitsu-K Computer in Japan.

unique access to a variety of international leadership class computational facilities such as the Fujitsu K Computer in Japan. As seen from subsequent stimulating new results [5], substantive impact can be expected to help stimulate progress in preparing for actual research engagement on ITER. In order to do so in a timely way, it is critically important that new software for extreme concurrency systems that demand increasing data locality be developed to help accelerate progress toward the ultimate goal of computational fusion research—a predictive simulation capability that is properly validated against experiments in regimes relevant for practical fusion energy production.

23.4 PROGRAMMING APPROACH

The basic parallel programming approach for global PIC codes such as GTC and GTC-P include (1) explicit message passing using MPI; (2) quasishared memory models such as Global Arrays for internode communication; (3) architecture-specific models such as CUDA for computing on graphic processing units (GPUs) and (4) directive-based compiler options such as OpenMP and OpenACC with possible promise of being more cross-machine portable between architectures.

In the course of describing global PIC code characteristics/considerations with respect to scalability, performance, portability, modern computational platforms, and external libraries, associated discussions will touch on the rationale for the chosen programming approach, and the associated balance between performance and portability. In particular, attention will be focused on specific challenges for global PIC applications in achieving efficiency on exascale architectures.

23.4.1 SCALABILITY

This section focuses on a description of efforts to improve the scalability of the global PIC codes—well represented by GTC and GTC-P. In particular, key topics highlighted include (1) on-node thread scaling and (2) between node scaling.

The GTC/GTC-P codes have been designed with four levels of parallelism: (1) an internode distributed memory domain decomposition via MPI; (2) an internode distributed memory particle decomposition via MPI, (3) an intranode shared memory work partition implemented with OpenMP; and (4) a single instruction, multiple data (SIMD) vectorization within each core. This approach was shown to lead to nearly perfect scaling with respect to the number of particles [18].

In order to efficiently address large grid sizes and the associated significant memory increase, the domain decomposition in GTC-P is further extended in the radial dimension (beyond the toroidal dimension). This leads to a 2D domain decomposition and enables carrying out true weak scaling studies, where both particle and grid work are appropriately scaled. The multilevel particle and domain decompositions provide significant flexibility in distributed-memory task creation and layout. Although the ranks in the toroidal dimension are usually fixed as 32 or 64 due to Landau damping physics, there is freedom to choose any combination of process partitioning along the radial and particle dimensions. For scaling with a fixed problem size, the procedure involves first partitioning along the radial direction and then switching to particle decomposition for additional scalability. The decompositions were implemented with three individual communicators in MPI (toroidal, radial, and particle communicator), and further tuning is made available via options to change the order of MPI rank placement.

A gyrokinetic PIC simulation typically has highly anisotropic behavior, with the velocity parallel to the magnetic field being an order of magnitude larger than that in the perpendicular direction. Consequently, the message sizes in the toroidal dimension can be 10 times larger than those in the radial dimension at each time step. On Blue Gene systems with explicit process mapping, it is convenient and effective to group processes to favor the MPI communicator in the toroidal dimension [15]. For other systems, assigning consecutive ranks for processes within each toroidal communicator generally leads to improved performance.

To maximize on-node performance and efficiency, modern processor architectures have evolved with more cores and wider vector units in a single node. In order to fully exploit the emerging architectures on the path to exascale, it is important that application scientists design their software such that the algorithms and the implementations map well on the hardware for maximum scalability. In GTC/GTC-P, multicore parallelism is further exploited using shared-memory multithreading, which provides an additional multiplicative source of speed up. In an earlier version of GTC-P, *holes* were used to represent nonphysical *invalid* particles, that is, in a distributed environment, at every time step, the particles that are being moved to other processes are marked as *holes* and considered to be *invalid* in the local particle array. These invalid particles are then removed from the array periodically to empty memory space for new incoming particles. In this type of implementation, two particles in consecutive memory locations may have different operations in *charge* and *push* depending on if they belong to the same type of particles (valid or invalid) or not. This accordingly introduces difficulty for automatic vectorization. To maximize the usage of vector units, the latest version of GTC-P and GTC removes the holes completely for *charge* and *push* by filling the holes at the end of *shift* and using the new incoming particles sent from neighboring processors at every time step. If a process has sent more particles than received, then the remaining holes are filled with the last particles in the array. A similar strategy has been applied for the GPU implementation to remove the branch statement caused by the *holes*.

23.4.2 PERFORMANCE

Global PIC code performance considerations involve a proper description of the GTC/GTC-P approach to achieving high performance on advanced architectures with focus on (1) improving data locality and vectorization; (2) improving thread scalability; and (3) making appropriate algorithmic changes.

Particle in cell algorithms is challenging to optimize on modern computer architectures due to issues such as data conflict and data locality. In GTC and GTC-P, parallel binning algorithms have been developed to improve data locality for *charge* and *push*. More specifically, several choices are provided to bin the particles, that is, along the radial dimension and along the poloidal dimension. The best binning strategy will be used for production runs by first running a few benchmarks. In GTC-P, the additional use of intrinsics has helped improve the vectorization of the binning implementation. On GPUs, the CUDA version of the binning algorithm was implemented using the Thrust Library.

To address the data conflict issue in *charge*, optimization strategies have been explored via static replication of grid segments that are coupled with synchronization via atomics, where the size of the replica may be traded for increased performance [16]. The best performance is often obtained by employing the full poloidal grid for each OpenMP thread. In GTC with only toroidal domain decomposition, the full poloidal grid replication dramatically increases the temporary grid-related storage for large size grid on manycore architectures such as the Intel Xeon Phi systems. As such, static replication of grid segments that are coupled with synchronization via atomics will likely be the best strategy. In GTC-P, the radial domain decomposition solves locality and memory pressure without resorting to costly atomics. In essence, because only a small segment of the full poloidal grid is required for a hazard-free charge deposition, the private grid replication strategy can be readily employed on a per thread basis for the best performance.

23.4.3 PORTABILITY

Global PIC codes such as GTC and GTC-P have demonstrated increasing capability for portability over the past few years across different architectures. In this section, the associated techniques applied for doing so are discussed along with examples of success achieved. In general, a high priority is being placed on portability in HPC because of the significant differences between quite different main-line approaches receiving heavy emphasis and by government investments—a prominent example being the major architectural differences between the upcoming 200 PF systems: the SUMMIT system at the OLCF and the AURORA system at the ALCF. Because both approaches have significant exciting potential for enabling accelerated performance at scale, most advanced applications—including prominent global PIC codes such as GTC/GTC-P—will continue to focus attention on achieving BOTH performance enhancement and portability. For example, performance portability of these advanced codes helps ensure—in a risk mitigation sense—the capability to perform very well on whichever platform proves to provide the greater eventual computing at extreme scale advantage.

GTC-P has been particularly successful in porting modern optimized versions across a wide range of multi petaflop platforms at full or near-to-full capability. Benefit is associated in part from the fact that GTC-P is not critically dependent on any third-party libraries. For example, this effort was initiated with the implementation a highly optimized Poisson solver with multithreading capability. Additional performance enhancement for both GTC and GTC-P has been obtained by utilizing a specialized damped Jacobi iterative solver [17]. In this iterative solver, the damping parameter was carefully chosen to favor the desired range of wavelengths for the fastest growing modes in plasma turbulence simulations. As a result, a small and fixed iteration count is sufficient to achieve the desired accuracy.

Although achieving the best performance on each explored architecture requires platform-specific optimization strategies, a *pluggable* software component approach in architecting the GTC/GTC-P application codes has proven to be a quite successful approach. Specifically, the interface is preserved across all implementations targeting CPU-based codes as well as GPU (or Xeon Phi) hybrid implementations. Components are chosen based on the target platform during the application build process. This enables having a unified code base with the best-possible performance, without sacrificing portability. Behind the unified interface, platform-specific optimization strategies are systematically investigated.

Some optimizations, such as sorting particles and vectorization, are common to all platforms, but implementation details differ. Other optimizations, such as handling non-uniform memory access (NUMA) issues and load imbalance, are specific to certain platforms. Designing routine interfaces is of crucial importance to allow portability without compromising performance-tuning opportunities.

GTC-P uses the MPI-3 standard for distributed-memory communication, including the exploration of explore one-sided communication. The motivation here is again to provide better portability for diverse architectures and programming models.

Significant advances in GTC-P on manycore processors with respect to portability and scalability have been recently achieved by porting the code to GPU systems with OpenACC 2.0 as a viable option instead of CUDA. This has led to the very recent success in porting and optimizing an OpenACC 2.0 version of GTC-P on the Sunway TaihuLight Supercomputer—the new No. 1 system on the international Top500 as of June 2016. The only approach for achieving good performance on TaihuLight requires software compatibility with their SWACC compiler—a customized OpenACC 2.0 syntax supported software.

In more generally considering the question of portability onto a broad variety of modern computational platforms, experiences with GTC-P indicate that machines such as the IBM BG/Q Mira demand at least 49,152-way MPI parallelism and up to 3 million-way thread-level parallelism in order to fully utilize the system. Although distributing particles to at least 49,152 processes is straightforward, the distribution of a 3D torus-shape grid among those processes is certainly a nontrivial task. For example, first considering the 3D torus as being decomposed into subdomains of uniform volume, the subdomains close to the edge of the simplest circular geometry system will contain more grid points than the core. This leads to potential load imbalance issues for the associated grid-based work.

Through a close collaboration with the Future Technologies Group at the Lawrence Berkeley National Laboratory, a new version of GTC-P has been developed and optimized to address the challenges in the PIC method for leadership-class systems in the multicore/manycore regime [16,22, 23]. As noted earlier in this document, GTC-P includes multiple levels of parallelism, a 2D domain decomposition, a particle decomposition, and a loop level parallelism implemented with OpenMP— all of which help enable this modern global PIC code to efficiently scale to the full capability of the largest extreme scale homogeneous HPC systems currently available [5]. Special attention has been paid to the load imbalance issue associated with domain decomposition [15]. To improve single node performance, a *structure-of-arrays* (SOA) data layout has been chosen for particle data. This is accompanied by aligning memory allocation to facilitate SIMD intrinsic binning of particles to improve locality and the use of *loop fusion* to improve arithmetic intensity. Irregular nested loops have been manually flattened to expose more parallelization to OpenMP threads. GTC-P features a 2D topology for point-to-point communication. On the IBM BG/Q system with 5D torus network, communications have been optimized with customized process mapping. Data parallelism has also been continuously exploited through SIMD intrinsics (e.g., QPX intrinsics on IBM BG/Q) and by improving data movement through software prefetching.

Overall, GTC-P has incorporated four levels of parallelism including (1) an internode distributed memory 2D domain decomposition via MPI, (2) an internode distributed memory particle decomposition via MPI, (3) an intranode shared memory work partition implemented with OpenMP, and (4) an SIMD vectorization within each core. In common with the large majority of codes in the fusion energy science/plasma physics application domain, GTC-P was originally written in Fortran language. However, to better facilitate interdisciplinary collaborations with computer science and applied math colleagues, modern versions of this code have been developed in C language as well as a CUDA implementation for dealing with GPUs. As just noted, this capability has recently been further advanced with the development and implementation of an OpenACC 2.0 version of GTC-P. Although the original Fortran version of this code is still used for verification purposes in cross-checking and benchmarking results, the primary utilization has involved the C and CUDA versions for performance studies and physics production runs on supercomputing systems such as the ALCF's *Mira* and the OLCF's *Titan*.

In dealing with heterogenous supercomputing platforms such as *Titan*, the approach followed in the deployment of global PIC codes involves off-loading the computationally intensive and highly scalable subroutines to GPUs, whereas the communication-dominant subroutines remain on CPUs. Performance, however, is known to be impeded due to the synchronization of atomic operations and the unavoidable memory transpose associated with the structure-of-array to array-of-structure data layout. To address this issue, the time-consuming global memory atomic operations have been replaced with local shared memory atomic operation. This R&D activity falls generally in the category of advances and challenges involving heterogeneous architectures.

23.4.4 EXTERNAL LIBRARIES

A multilevel parallelization using MPI/OpenMP has been designed in GTC to scale up to millions of cores and to take advantage of the memory hierarchy of current generation parallel supercomputers. GTC is the first fusion code to reach the tera-scale in 2001 on the Seaborg computer at the National Energy Research Scientific Computing Center (NERSC) [3] and the petascale in 2008 on the Jaguar computer at ORNL in production simulations [19]. Through collaborations with computer scientists from hardware vendors, GTC was the first large-scale fusion code to fully utilize the heterogeneous architectures using GPU accelerators on the Tianhe-1A [20] and Titan, and using Intel Many Integrated Core (MIC) accelerators on Tianhe-2 [21]. As the codesign partner to the electromagnetic GTC code, the electrostatic GTC-P code has demonstrated its high-resolution portability capabilities on the top seven supercomputers worldwide [15]. These advances were enabled in significant measure by a well-established collaboration between Princeton University's Institute for Computational Science and Engineering (PICSciE) members Bei Wang and W. Tang and SciDAC SUPER Institute members L. Oliker and S. Williams and their colleagues in LBNL's Future Technology Group.

The flagship GTC code was originally written in Fortran 90. The CPU version has been parallelized using an MPI/OpenMP hybrid programming model with GPU and Intel Xeon Phi acceleration. Fortran 90 modules are used in this code to manage global data—with every class of global data (e.g., field data, particle data) having its own module. GTC has previously relied upon the Department of Energy (DOE)-funded PETSc toolkit at ANL to implement the electromagnetic parallel solvers. Advanced third-party packages, such as LLNL's HYPRE multigrid solver, have more recently been implemented in GTC as part of the current CAAR GTC Project at the OLCF. HYPRE can be used via the PETSc framework with a simple parameter change in a runtime configuration file.

23.5 SOFTWARE PRACTICES

The GTC code, which was originally written in Fortran 90, has a CPU version, which has been parallelized using an MPI/OpenMP hybrid programming approach with GPU and Intel Xeon Phi acceleration. This code uses Fortran 90 modules to manage global data—with every class of global data (e.g., field data, particle data) having its own module. GTC currently uses the DOE-funded PETSc toolkit to implement and utilize the electromagnetic parallel solvers. Advanced third-party packages, such as LLNL's HYPRE multigrid solver have recently been implemented in GTC and can be used via the PETSc framework with a simple parameter change in a runtime configuration file. The GTC features multiple levels of parallelism:

1. First, a 1D domain decomposition is implemented in the symmetric toroidal direction using MPI. The particles are divided between MPI processes in the domain decomposition

wherein each process owns a fraction of the total particles in that domain as well as a private copy of the local toroidal grid.

2. In order to further increase MPI parallelism, a second level of decomposition, a particle decomposition, is introduced. Particles are divided, but fields are shared between MPI processes in the particle decomposition. Field solvers are parallelized using all MPI processes in the particle decomposition.

3. The third level of parallelism is an intranode shared memory partitioning (via OpenMP) of both particle and grid-related computation. This results in a near-perfect scaling with the number of particles.

Moving forward, advanced radial domain decomposition methodology developed and successfully deployed in the codesign GTC-P code to efficiently reduce the memory footprint in larger problem size challenges is introduced into the flagship GTC code.

23.6 BENCHMARKING RESULTS

In this section, a description is provided of performance and portability of GTC and GTC-P, using the porting approach performed on these codes using the described programming approach. The discussion includes the following: (1) What has worked well and what has not; (2) the level of effort that was needed to do the refactoring and porting—with associated description of the benchmarking and profiling results; (3) illustration of parallel scaling behavior, displayed on a logarithmic scale such that linear or ideal scaling is shown as a straight line for either weak or strong scaling; and (4) a description of *time to solution*, which is a metric more meaningful than parallel efficiency, as the latter is often based on an arbitrary data point. Of course, *time to solution* is the key metric that counts the most for computational domain scientists. Finally, while exascale resources are, obviously, not as yet available, the topic of extrapolation to exascale resources, based on expected architectural roadmaps, is be discussed.

Two sets of weak scaling studies for the comprehensive GTC code have been carried out on Titan up to nearly the full system 16,384 nodes. However, at the time of this study, a significant number of the Titan nodes were unavailable, making it impossible to run on all 18,688 nodes. The first test set is called *particle weak scaling study*, where the grid size is held fixed, but the total number of particles is scaled up. The second set of test is called *hybrid weak scaling study*, where both the grid size and the total number of particles are scaled. The first study holds the number of particles per MPI rank and the number of grid cells per MPI rank nearly constant, thus representing a conventional weak scaling study. However, the second study is a more realistic performance scaling study based on a typical production run of the code: grid size is proportional to the square root of number of nodes. For both sets of weak scaling study, the number of particles per processor is fixed at 3.2 million. Compared with CPU (16 cores AMD 6274), GPU (NVIDIA K20x) has boosted the overall performance by $1.6 - 3.0\times$. The decrease of the performance speedup in large processor counts is mainly due to the increased portion of non-GPU accelerated subroutines as well as MPI time (Figures 23.4 and 23.5).

The GTC Poisson solver currently runs on the CPU. Though it is presently not the most time-consuming part of GTC simulations, the solver time requirements have become more significant since other parts of the code have been accelerated using GPUs. The standard PETSc solver has accordingly been replaced with a HYPRE multigrid solver as part of the CAAR GTC Project. This solver has the clear advantage of being threaded to effectively use the CPUs while also being scalable to many compute nodes. Figures 23.6 and 23.7 show comparative timings of the PETSc solver and the HYPRE multigrid solver for a representative set of GTC test cases. The HYPRE solver for these cases is $\sim4\times$ faster than the standard PETSc solver and has better scaling properties.

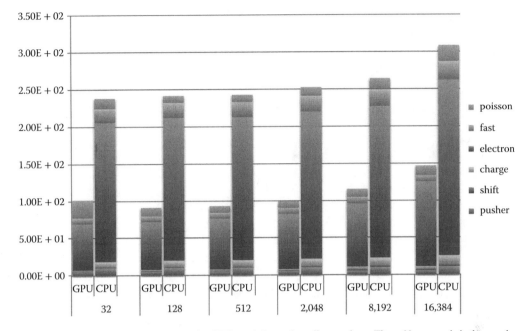

FIGURE 23.4 The timing breakdown for GTC particle weak scaling study on Titan. *Note*: *x*-axis is the number of nodes and *y*-axis the total wall-clock time. GPUs are shown to deliver up to 3× speedup compared with CPUs.

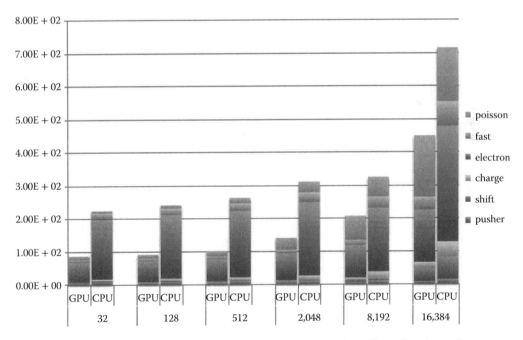

FIGURE 23.5 The timing breakdown for GTC hybrid weak scaling study on Titan. Here the work per processor is increased as node count is increased. *Note*: *x*-axis is the number of nodes and *y*-axis the total wall-clock time. GPU delivers up to 3.0× speedup compared with CPU.

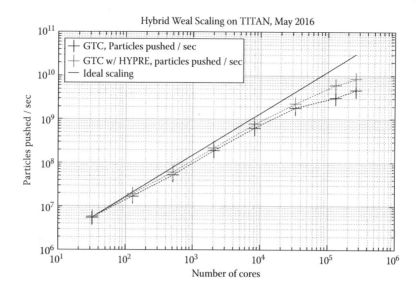

FIGURE 23.6 Comparative performance of the PETSc versus HYPRE solvers for representative GTC Code cases on Titan (May, 2016).

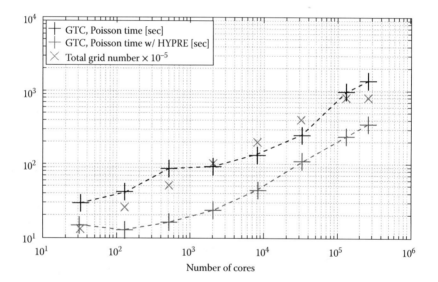

FIGURE 23.7 Weak scaling of GTC on Titan (top), with the number of nodes ranging from 32 to 16,384 (88% of the whole machine). Both grid number and total particle number are increased, but the number of particles per core remains constant. The Poisson time (bottom) shows the improved performance due to the new deployment of the HYPRE multigrid solver. Total grid number for this scaling study is also shown.

The benchmarking results associated with GTC-P—a highly scalable code developed to efficiently utilize modern parallel computer architectures at the petascale and beyond—features simulations of confinement physics for large-scale MFE plasmas that have been carried out for the first time with very high phase-space resolution and long temporal duration to deliver important new scientific results that are cross-benchmarked versus results achieved on various top supercomputing systems. This was enabled by deployment of this code on world-class *homogeneous* architecture systems such as the IBM BG/Q *Mira* at the ALCF and also *Sequoia* at LLNL and the Fujitsu K Computer at RIKEN, Kobe, Japan. Some specific accomplishments are summarized in Figures 23.2 and 23.3.

The success of these efforts was greatly facilitated by the fact that true interdisciplinary collaborative effort with Computer Science and Applied Math scientists has accelerated completion of the modern C as well as GPU-compatible versions of the GTC-P code. The demonstrated capability to run at scale on the largest open-science BG/Q system (*Mira* at the ALCF) opened the door to obtain access to the National Nuclear Security Administration's (NNSA) *Sequoia* system at LLNL, which then produced the outstanding weak scaling performance results shown on Figure 23.2.

With regard to a more global perspective, the G8-sponsored international program on computing at extreme scale [14], has enabled this project to gain collaborative access to a number of the top international supercomputing facilities—including the Fujitsu K Computer, previously Japan's top-ranked system worldwide. Results from weak-scaling studies carried out on the K-computer are illustrated in Figure 23.3.

Having demonstrated the ability to effectively utilize the most powerful *homogeneous* supercomputing platforms worldwide, GTC-P R&D efforts have also examined performance characteristic on *heterogeneous* architectures. More generally, these studies represent productive investigations of extreme scale science across advanced scientific computing basic research programs with fusion energy science as an illustrative application domain. Here, the focus was on developing new algorithms for advanced *heterogeneous* supercomputing systems such as the GPU/CPU *Titan* at DOE's OLCF and the Intel Xeon Phi/Intel Xeon *Stampede* at the National Science Foundation's (NSF), Texas Advanced Computing Center (TACC). In doing so, a new version of GTC-P code was developed that features algorithms, which include new heterogeneous capabilities for deployment on hybrid GPU (NVIDIA K20)/CPU as well as the Intel Xeon Phi/Intel Xeon systems such as Stampede and also TH-2 in China. From a verification perspective, this research effort includes systematic comparison of new results against the successful work described in earlier in studies that featured high-resolution, long temporal scale simulation results obtained on world-class *homogeneous* systems such as the IBM BG/Q *Mira* at the ALCF, *Sequoia* at LLNL, and the K-Computer in Kobe, Japan.

The development of a GPU version of GTC-P suitable for accelerator-based architectures powered by GPUs started with focus on off-loading the particle-based phases—charge deposition, particle push, and particle shift—to the GPUs, while keeping the grid-based phases of the code on the CPUs. Consequently, the GPU implementation includes three programming models: CUDA and OpenMP within a node and MPI between nodes. The particle-based phases are good candidates for GPU implementation because they are especially computation intensive, taking at least 80% of the total computational time of the code [15].

Exploiting massive fine-grained parallelism on GPUs is nevertheless a nontrivial challenge for PIC codes, which feature fine-grained data hazards, irregular data access, and low computational intensity. For example, memory locality that typically *improves* the performance of most routines actually *degrades* the performance for atomics because of access conflicts. These conflicting requirements for locality and conflict avoidance have made previous [16]—as well as continuing—efforts to optimize the performance of modern codes on GPU systems both interesting and challenging.

The particle shift phase of the code consists of four steps: (1) finding the particles on the GPUs that need to be sent to other MPI processes, (2) copying these particles from GPUs to CPUs through a PCI bus, (3) sending/receiving particles with MPI_Sendrecv on CPU, and (4) copying the received particles from CPUs to GPUs. Performance is found to be impeded in finding and buffering those particles on GPUs and in explicit memory transfers between multiple memory spaces within a compute node.

Several optimization strategies to boost the code performance on different GPU architectures have been developed. On the older Fermi chip, the performance degraded mainly due to slow global atomic operation. We have accordingly replaced the associated global memory operation with a shared memory atomic operation—but with more memory usage. On the newer generation Kepler chip, where the performance of global atomic operations is dramatically improved, we have kept the global memory access with less memory usage. In addition, we have optimized the data layout such that the access of the global memory is achieved in a more coalesced way. In addition, a binning-based shift algorithm is being developed to improve the performance in finding and buffering shifted particles on GPUs with massive parallelism. This is the first step of the particle shift phase of the particle-based operations.

More recent developments of advanced algorithms addressing the programming challenges on hybrid GPU/CPU systems have produced some quite encouraging results (enabled by OLCF Director's Discretionary time allocation) on Titan—a Cray XK7 system with 299,008 CPU cores and 18,688 K20 NVIDIA GPUs. Quite favorable particle-scaling performance results have been obtained for fixed plasma size with the starting point being the porting of a significantly improved GPU version of GTC-P to Titan. For example, excellent performance of the GPU-version of GTC-P has recently been demonstrated on the *Titan* system at the OLCF. The results of associated very favorable weak-scaling studies are illustrated in Figure 23.8.

In ongoing investigations of the comparative capabilities of powerful *heterogeneous* system versus those of the petascale *homogeneous* systems, it is of course necessary to move beyond particle scaling studies to carry out true weak-scaling studies, where the plasma size increases.

In the GPU implementation in the GTC-P code, the GPU and CPU operations were not overlapped. For example, when the GPU is busy with charge deposition, the CPU remains idle. To utilize the full node power, two different methods are now being developed in this code. One approach is to distribute the particles among CPUs and GPUs such that they work concurrently. The ratio of the number of particles on each device (CPU or GPU chip) is carefully selected such that the work on CPUs and GPUs are well balanced. This is also the strategy we are using on heterogeneous systems with MIC coprocessors (which will be described in more detail below). Another approach is to utilize the idle processors for in situ data analysis [25]. In view of the fact that for applications with great code complexity where the accelerator/coprocessor provides only a moderate speedup compared

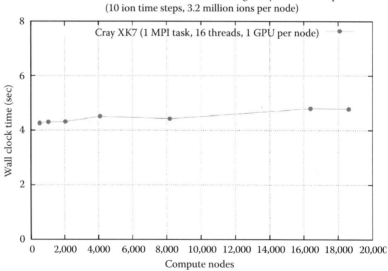

FIGURE 23.8 Particle weak scaling results obtained from the graphic processing unit (GPU)-version of the GTC-P code as carried out for a fixed D3D-size tokamak plasma on the Titan LCF at ORNL.

with commodity CPUs, both of the promising alternative approaches just noted here can better utilize CPUs together with the accelerator/coprocessor and thus significantly improve the performance in terms of *time to solution*.

In order to fully utilize the parallelism in the coprocessor/accelerator, it is important that algorithms be designed such that the code can exploit the SIMD vectorization. On GPUs, this translates to coalescent memory access and we accordingly structured the data array layout to facilitate SIMD vectorization. These developments can be expected to have a strong impact in dealing with the multithreading challenges for efficient deployment of a large number of processors on modern GPU/CPU heterogeneous systems. Continuing engagement with key personnel at NVIDIA is a significant *codesign* asset in these R&D efforts.

With regard to the deployment experience on NSF's *Stampede* system with Intel Xeon Phi (*MIC*) coprocessors at the TACC, each coprocessor is considered at present as a separate node, and the code has been run in symmetric mode. In dealing with 2D domain decomposition, we distribute each subdomain along with its associated particles to a single *Stampede* server node. We further distribute the particles in one subdomain among 2 Intel Xeon E5 and Intel Xeon Phi coprocessor by turning on the particle decomposition. This means that each Intel Xeon E5 and Intel Xeon Phi in a node carries a fraction of the total number of particles in that subdomain. Having the host and the coprocessor in a node share the same subdomain (for particle decomposition only) can avoid running the grid-based subroutines on coprocessors redundantly.

Currently, the particles are divided evenly among the host and the coprocessor. Later, we will distribute the particles such that the work on the host and the coprocessor is well balanced. The chip level parallelism is exploited with OpenMP and SIMD vectorization. For example, on a node with 2 Intel Xeon E5, we use OpenMP with 16 threads. Similarly, on a coprocessor with the Intel Xeon Phi, we use OpenMP with 240 threads.

The data conflict issue in the charge deposition phase is addressed by providing each thread a private copy of the local grid following by a reduction operation to merge all copies together. The summation order is carefully arranged such that no costly synchronization is required. It should be noted here that the private copy strategy is applicable when the code has operational 2D domain decomposition needed to efficiently deal with large-size plasmas.

From the perspective of *codesign R&D*, it is appropriate to note here that ongoing discussions with lead designers at Intel can be expected to provide access to more advanced Intel MIC systems. In addition, we will have access to and be able to test our newly developed hybrid algorithms on the powerful heterogeneous TH-2 Intel MIC hybrid supercomputer in Guangzhou, China—currently the #2-rated system worldwide.

23.7 TIME-TO-SOLUTION AND ENERGY-TO-SOLUTION COMPARATIVE STUDIES

Energy is becoming an increasingly large impediment to advances in supercomputing. The net energy efficiency of large scientific simulations can be particularly nonintuitive as one moves from one processor or network architecture to the next as the interplay between performance and power is highly dependent on algorithm and architecture. Using power measured under actual load via system instrumentation, Table 23.2 shows the energy per time step on 4K nodes of Mira, Titan, and Piz Daint when using 80M grid points, 8B ions, and 8B electrons. We observe that although Mira required the most wall clock time per time step, it also required the least power per node. When combined, this ensured Mira required the least energy per time step of all platforms considered. Conversely, using the host-only configurations on Titan and Piz Daint required between 2× and 4× the energy with the difference largely attributable to the lack of scalability on Titan. Interestingly, although accelerating the code on

TABLE 23.2

ENERGY-TO-SOLUTION **ESTIMATES (for Mira, Titan, and Piz Daint)**

	CPU-Only			CPU+GPU	
	Mira	**Titan**	**Piz Daint**	**Titan**	**Piz Daint**
Nodes	4,096	4,096	4,096	4,096	4,096
Power/node (W)	69.7	254.1	204.9	269.4	246.5
Time/step (s)	13.77	15.46	10.00	10.11	6.56
Energy (kWh)	1.09	4.47	2.33	3.10	1.84

- Energy per ion time step (kWh) by each system/platform for the weak-scaling, kinetic electron studies using 4 K nodes.
  (Watts/node) * (#nodes) * (seconds per step) * (1 kW/1,000 W) * (1 h/3,600 sec)
- Power/Energy estimates obtained from system instrumentation including compute nodes, network, blades, AC to DC conversion, etc.

these platforms significantly reduces wall clock time per time step, it only slightly increased power. As such the energy required for the GPU-accelerated systems was reduced nearly proportionally with run time. Optimizing for energy-to-solution could be at odds with time-to-solution when we use technologies such as dynamic voltage and frequency scaling (DVFS). When internode communication dominates the execution time by scaling down the CPU frequency we observed up to 25% reduction in energy consumption (for the C problem size) for an increase in the execution time by 28%. We conducted such an experiment on an Intel Haswell-based Cray XC40 system. Although supported by much hardware, DVFS control is not enabled on most of the systems studied. As such, we could not explore such optimization on all systems. Another impediment to adopting such technology is the policy adopted for resource allocation by HPC compute facilities, which is based on CPU-hours rather than energy consumption. This policy makes the most performant solution the best from the user perspective.

With regard to energy-efficient scientific computing, instrumenting scientific applications to measure energy when running on large supercomputing installations today can be cumbersome and obtrusive—requiring significant interaction with experts at each center. As such, most applications have little or no information on energy-to-solution across the architecture design space spectrum. In order to affect energy-efficient codesign of supercomputers, energy measurement must be always-on by default with, at a minimum, total energy and average power reported to the user at the end of an application. By reporting energy by component (memory, processor, network, storage, etc), scientists and vendors could codesign their applications and systems to avoid energy hotspots and produce extremely energy-efficient computing systems.

23.8   CONCLUDING COMMENTS

As a final comment, it is appropriate to note that a broader impact of the work presented in this chapter is the delivery of benefits to PIC codes in general because the associated codes share a common algorithmic foundation. For example, the continuing developments targeted in the current project can be expected to have a strong impact in dealing with the multithreading challenges for efficient deployment of a large number of processors on modern heterogeneous systems with coprocessors (GPU or MIC)—advances that should prove beneficial to any particle-mesh algorithm for such systems.

REFERENCES

1. W. Tang et al. *Scientific Grand Challenges: Fusion Energy Sciences and the Role of Computing at the Extreme Scale*, PNNL-19404, 212 pp. (March, 2009). http://www.er.doe.gov/ascr/ProgramDocuments/Docs/FusionReport.pdf.

2. R. Rosner et al. Opportunities & challenges of exascale computing—DoE Advanced Scientific Computing Advisory Committee Report, Office of Science, U.S. Department of Energy Fall (November, 2010).

3. Z. Lin et al. Size scaling of turbulent transport in magnetically confined plasmas, *Phys. Rev. Lett.* 88, 195004 (2002).

4. B. F. McMillan, et al. System size effects on gyrokinetic turbulence, *Phys. Rev. Lett.* 105, 155001 (2010).

5. W. Tang et al. Extreme scale plasma turbulence simulations on top supercomputers worldwide, *SC16 International Conference on High Performance Computing, Networking, Storage and Analysis*, Salt Lake City, UT; oral presentation, paper 399, 14–18 November, 2016.

6. Z. Lin et al. A fluid-kinetic hybrid electron model for electromagnetic simulations, *Phys. Plasmas* 8, 1447–1450 (2001).

7. Y. Chen and Parker, S. E. A δf particle method for gyrokinetic simulations with kinetic electrons and electromagnetic perturbations, *J. Comput. Phys.*, 189, 463–475 (2003).

8. E. Startsev and W. W. Lee. Finite beta simulations of microinstabilities, *Phys. Plasmas*, 21, 022505 (2014).

9. W.J. Deng et al. Linear properties of reversed shear Alfvén eigenmodes in the DIII-D tokamak, *Nucl. Fusion*, 52, 043006 (2012).

10. D. Liu et al. Verification of gyrokinetic particle simulation of current-driven instability in fusion plasmas. II. Resistive tearing mode, *Phys. Plasmas*, 21, 122520 (2014).

11. D. Liu et al. A finite-mass fluid electron simulation model for low-frequency electromagnetic waves in magnetized plasmas, *Plasma Phys. Control. Fusion*, 53, 062002 (2011).

12. D. Liu et al. Verification of gyrokinetic particle simulation of current-driven instability in fusion plasmas. III. Collisionless tearing mode, *Phys. Plasmas*, 23, 022502 (2016).

13. D. A. Spong. 3D toroidal physics: Testing the boundaries of symmetry breaking, *Phys. Plasmas*, 22, 055602 (2015).

14. G8 Research Council's "Exascale Computing for Global Scale Issues" Project in Fusion Energy "NuFuSE" [http://www.nu-fuse.com/]—An international HPC collaborative research project involving the U.S., France, Germany, Japan, and Russia (2011–2014). It is supported in the United States by the National Science Foundation (NSF)—with W. Tang as the US PI.

15. B. Wang. Kinetic turbulence simulations at extreme scale on leadership-class systems, oral presentation at SC13, published in SC13, *Proceedings of 2013 International Conference on High Performance Computing, Networking, Storage and Analysis,* Denver, CO. http://www.hpcwire.com/2013/11/16/sc13-research-highlight-extreme-scale-plasma-turbulence-simulation/ (November, 2013).

16. K. Madduri, et al. Memory-efficient optimization of gyrokinetic particle-to-grid interpolation for multi-core processors, *Supercomputing*, (SC), 2009.

17. Z. Lin et al. Method for solving the gyrokinetic Poisson equation in general geometry, *Phys. Rev. E*, 52, 5646 (1995).

18. S. Ethier et. al. Gyrokinetic particle-in-cell simulations of plasma microturbulence on advanced computing platforms. *Journal of Physics: Conference Series*, 16, 1–15 (2005).

19. Y. Xiao and Lin, Z. Turbulent transport of trapped electron modes in collisionless plasmas, *Phys. Rev. Lett.*, 103, 085004 (2009).

20. X. Meng et al. Heterogeneous programming and optimization of gyrokinetic toroidal code and large-scale performance test on th-1a, *International Supercomputing Conference*, Leipzig, Germany, pp. 81–96. Springer (2013).

21. J. Dongarra. Visit to the National University for Defense Technology Changsha, China (2013). http://www.netlib.org/utk/people/JackDongarra/PAPERS/tianhe-2-dongarra-report.pdf.

22. K. Madduri et al. Gyrokinetic toroidal simulations on leading multi- and many-core HPC systems, *The International Conference for High Performance Computing, Networking, Storage, and Analysis (SC'11)*, Seattle, WA (2011).

23. K. Madduri et al. Gyrokinetic particle-in-cell optimization on emerging multi- and many-core platforms, *Parallel Comput.* 37(9), 501–520 (2011).

24. S. Ethier et al. Petascale parallelization of the gyrokinetic toroidal code. *Proceedings of 2010 High Performance Computing for Computational Science (VECPAR'10)*, Berkeley, CA, 2010.
25. F. Zheng et al. Gold Rush: Resource efficient *In Situ* scientific data analytics using fine-grained interference-aware execution. *Proceedings of 2013 International Conference on High Performance Computing, Networking, Storage and Analysis*, Denver, CO, 2013.

RESEARCH TEAM

- Princeton University: *Prof. William Tang, Dr. Bei Wang, Princeton Institute of Computational Science & Engineering (PICSciE), Dr. Stephane Ethier, Princeton Plasma Physics Laboratory (PPPL)*
- University of California, Irvine: *Prof. Zhihong Lin), Dr. Wenlu Zhang, Dr. Jian Bao, Dr. Animesh Kuley, Dept. of Physics & Astronomy*
- ETH Zurich: *Prof. Torsten Hoefler, Grzegorz Kwasniewski, Intel Parallel Computing Center*
- Lawrence Berkeley National Laboratory (LBNL): *Dr. Leonid Oliker, Dr. Samuel Williams, Dr. Khaled Ibrahim, Computer Science Department*
- Penn State University: *Prof. Kamesh Madduri, Department of Computer Science & Engineering*
- Argonne Leadership Computing Facility (ALCF): *Dr. Hal Finkel, & Dr. Venkat Vishwanath*
- Texas Advanced Computing Center (TACC): *Dr. Carlos Rosales-Fernandez*
- Oak Ridge National Laboratory: *Dr. Scott Klasky, Dr. Ed D'Azevedo, Dr. Wayne Joubert*
- NVIDIA, Inc.: *Dr. Peng Wang, Dr. Tom Gibbs*

24 The Fusion Code XGC

Enabling Kinetic Study of Multiscale Edge Turbulent Transport in ITER

*Eduardo D'Azevedo, Stephen Abbott, Tuomas Koskela,
Patrick Worley, Seung-Hoe Ku, Stephane Ethier,
Eisung Yoon, Mark S. Shephard, Robert Hager,
Jianying Lang, Jong Choi, Norbert Podhorszki,
Scott Klasky, Manish Parashar, and Choong-Seock Chang*

CONTENTS

24.1 Introduction..529
24.2 Scientific Methodology...530
24.3 Algorithmic Details and Programming Approach......................................532
    24.3.1 Hybrid Lagrangian δf Scheme 534
    24.3.2 Electron Push Kernel 534
        24.3.2.1 Triangle Searching..535
        24.3.2.2 Particle Sorting..536
        24.3.2.3 GPU Specific Optimizations..537
        24.3.2.4 Portability 537
    24.3.3 Multispecies Collision Kernel 538
    24.3.4 Dynamic Load Balancing 540
    24.3.5 Parallel and Staged I/O 542
24.4 Performance Methodology 543
24.5 Performance Results...543
    24.5.1 Weak and Strong Scalability of Electron Push Kernel 543
    24.5.2 GPU Optimization of the Electron Push Kernel.............................544
    24.5.3 Multispecies Collision Kernel 546
    24.5.4 Load Balancing..547
    24.5.5 Parallel I/O..548
    24.5.6 ITER Simulation 548
24.6 Scientific Implications 549
Acknowledgments 551
References 551

24.1 INTRODUCTION

Magnetic fusion experiments are essential for next-generation burning plasma experiments such as the International Thermonuclear Experimental Reactor (ITER).* The success of ITER is critically

---

* www.iter.org

dependent on sustained high-confinement (H-mode) operation, which requires an edge pedestal of sufficient height for good core plasma confinement without producing deleterious large-scale, edge-localized instabilities. The plasma edge presents a set of multiphysics, multiscale problems involving a separatrix and complex three-dimensional (3-D) magnetic geometry. Perhaps the greatest computational challenge is the lack of scale separation; for example, temporal scales for drift waves, Alfvén waves, and edge localized mode (ELM) instability dynamics have a strong overlap. Similar overlap occurs in the spatial scales for the ion poloidal gyro-radius, drift wave, and plasma pedestal width. Microturbulence and large-scale neoclassical dynamics self-organize together nonlinearly. The traditional approach of separating fusion problems into weakly interacting spatial or temporal domains clearly breaks down in the edge. A full kinetic model (total-f nonperturbative model) must be applied to understand and predict the edge physics, including nonequilibrium thermodynamic issues arising from the magnetic topology (e.g., the open field lines producing a spatially sensitive velocity hole), plasma wall interactions, neutral and atomic physics [1,2].

The goal of this effort is to develop the XGC advanced simulation software for utilizing extreme parallelism on leadership computing resources. XGC is based upon a first-principles kinetic particle-in-cell (PIC) approach to address the challenges associated with understanding the edge region of magnetically confined plasmas. XGC obtained large allocations of leadership computing resources in the U.S. Department of Energy (DOE) Innovative and Novel Computational Impact on Theory and Experiment (INCITE) program.* XGC has benefited from performance engineering from the Institute for Sustained Performance, Energy, and Resilience (SUPER†) in the Scientific Discovery through Advanced Computing (SciDAC) Program in DOE.‡ XGC has also been selected for the Exascale Science Applications Program (NESAP) at the National Energy Research Scientific Computing Center (NERSC) and the Center for Accelerated Application Readiness (CAAR) program at the Oak Ridge Leadership Computing Facility (OLCF) to be optimized for the next generation of supercomputers.

Section 24.2 provides an outline of the type of scientific challenges and current state-of-the-art in gyrokinetic modeling. The many algorithmic enhancements and innovations for achieving high performance on leadership computing hardware are described in Section 24.3. Section 24.3.1 describes a hybrid Lagrangian δf scheme that reduces the number of particles by a factor of 2. The electron particle push and multispecies collision kernels account for over 80% of the computational time. Thus, these two kernels have been the focus of performance optimization activities and are described in Section 24.3.2 and Section 24.3.3. Dynamic load balancing for redistributing the computational load in the collision kernel (proportional to number of grid cells) and the push kernel (related to number of particles) is described in Section 24.3.4. XGC achieves high performance in parallel I/O via the Adaptable IO System (ADIOS) library. ADIOS uses distributed memory of dedicated nodes via the DataSpace library as a large data cache for improving performance of parallel I/O and in performing in situ data analysis and reduction. This key innovative technology is described in Section 24.3.5. The details on the numerical experiments and how performance was measured are described in Section 24.4, and the results are listed in Section 24.5. Finally, the scientific implications and conclusions are described in Section 24.6.

24.2 SCIENTIFIC METHODOLOGY

Few energy sources promise as much as plasma fusion, the reaction that powers the stars. Fusion is essentially inexhaustible, uses only abundant low atomic number elements, does not produce long-lived nuclear waste as in fission reactors, is not subject to meltdown, does not produce greenhouse

---

* http://www.doeleadershipcomputing.org/incite-program/
† www.super-scidac.org
‡ www.scidac.gov

gases, is not tied to a particular geographical location, and can deliver more energy per gram of fuel than any other sources. However, designing a reliable device that can sustain continuous and economic fusion reaction conditions, and produce electricity more efficiently than current low-carbon methods, is still an ongoing and very active research goal. The most promising method for generating fusion energy is that of magnetic confinement, where high-temperature plasma is confined in a torus-shaped device using strong magnetic fields to control the trajectories of the charged ions and electrons forming the plasma. The most successful design so far is that of the *tokamak*, an axis-symmetric magnetic toroidal device that has been the focus of fusion research for several decades and has nearly achieved break-even between fusion energy produced and input energy to sustain the plasma. This feat was accomplished in two of the largest tokamaks ever built, the Tokamak Fusion Test Reactor (TFTR) in the United States[*] and the Joint European Torus (JET) in the United Kingdom.[†]

Empirical models based on decades of experiments in tokamaks of various sizes indicate that a larger tokamak would certainly break that barrier and produce an amount of fusion energy far greater than the energy needed to sustain the plasma. Based on this promise, seven countries are currently collaborating to build the world's largest tokamak, ITER, in the South of France.[‡] This regime of very high-yield fusion reactions at large physical scale has, however, never been achieved in any of the previous tokamaks, so many physics questions still remain. One general question is "will extrapolation through data regression of phenomena found in today's tokamaks still hold true in ITER?"

One of the most critical questions needing to be answered is the one concerning the level of energy flux impinging on the so-called *divertor plate* at the bottom of the device. This is the exhaust-energy collection method in current tokamaks where the magnetic field is shaped in such a way that the ions and electrons escaping the hot core of the plasma are directed toward actively cooled, localized divertor plates where heat may be dissipated using various methods. For ITER, this exhaust power on the divertor will be extremely high, about 100 MW. A simple data extrapolation, assuming that the same physics, governs the edge plasma between the present tokamaks and ITER, predicts that the 100 MW exhaust power will be mostly deposited on a ~1 cm width toroidal band on the divertor plates. If this prediction is correct, the power deposition density will exceed the limit 10 MW/m^2 imposed by known materials on earth and damage the divertor. Will the prediction be correct in ITER? Will the ITER divertor be vaporized prematurely unless some highly expensive and technically difficult measure is taken? Or, will it survive? The answer depends on many complex and interrelated nonlinear plasma physics mechanisms that cannot be answered unless an extreme scale computing tool can be used.

This is why fusion scientists must rely on high-fidelity, extreme-scale simulations for predictive understanding of what will happen to the divertor heat-load footprint during the operation of ITER. The work presented in this chapter aims at addressing this question with XGC, the most comprehensive kinetic plasma code ever developed for studying the multiscale transport physics in tokamaks, both in contact with the wall and in diverted magnetic field geometry. Before the innovative improvements presented in this chapter, an ITER simulation was not possible even on the largest U.S. open-science Cray XK7 Titan supercomputer. Several state-of-the-art innovative improvements in the high-performance computational and algorithmic technologies now allow the simulation of the ITER device using realistic shape of the vacuum vessel walls and divertor using over 300 billion total marker particles.

Among the plasma equations that can be solved on today's computers for strongly magnetized plasmas in a toroidal confinement device, the gyrokinetic equation can describe the most fundamental physics. The six-dimensional (6-D) Vlasov equation for the particle distribution function $f(\mathbf{x}, \mathbf{v}, t)$

---

[*] See https://en.wikipedia.org/wiki/Tokamak_Fusion_Test_Reactor

[†] See https://www.euro-fusion.org/jet/

[‡] See http://www.iter.org

can be expressed as

$$\frac{\partial f}{\partial t} + \frac{d\mathbf{x}}{dt} \cdot \frac{\partial f}{\partial \mathbf{x}} + \frac{d\mathbf{v}}{dt} \cdot \frac{\partial f}{\partial \mathbf{v}} = C \tag{24.1}$$

The five-dimensional (5-D; 3-D in configuration space and two-dimensional (2-D) in velocity space) gyrokinetic equation is derived from the 6-D Vlasov Equation 24.1 by limiting the physics of interest to time scales much slower than the ion cyclotron frequencies (slower than 0.1 μs), but still keeping all the fundamental physics such as the wave-particle Landau resonance, nonlinear dynamics, and the particle trapping without an empirical model. Solving the original 6-D Vlasov equation requires at least exascale computers for a tokamak size simulation of ions and electrons.

Even with the gyrokinetic reduction of the Vlasov equation to five-dimensions, further simplifications have been necessary to fit both the kinetic ions and electrons into today's leadership class computers. The simplifications have been in either using the perturbative scheme, called the δf scheme that is based on the perturbative assumption that the background plasma is fixed at Maxwellian distribution function (hence giving up the possible multiscale interaction of turbulence with the background plasma and neutral atomic physics), or using only the kinetic ions assuming a Boltzmann type adiabatic electron fluid response to the electrostatic potential dynamics. Another further simplification that has been adopted is a simplified coordinate system called *flux coordinate system* that does not allow the simulation domain to approach the magnetic separatrix surface in the edge and the magnetic axis in the central core. All the existing U.S. gyrokinetic codes (e.g., GTC [3]) use these assumptions in order to enable the turbulence simulations on today's leadership class computers. For these reasons, even though a fundamental understanding of the edge plasma physics has been recognized for over 30 years as one of the most critical topics in the magnetic fusion problem and for ITER, a GTC could not attack this problem. In the edge plasma, the background plasma dynamics, the turbulence dynamics, and the atomic physics interactions with neutral particles all are important and exist with nonlinear multiscale interactions among them. Moreover, the edge plasmas are in nonthermal states and thus have a non-Maxwellian particle distribution function that requires a fully nonlinear Fokker–Planck collision operation satisfying the conservation properties.

24.3   ALGORITHMIC DETAILS AND PROGRAMMING APPROACH

In order to overcome these critical deficiencies in the current state-of-the-art codes and to address the complicated edge specific problems, a new modern GTC XGC has been developed over two funding periods in the SciDAC program. XGC is a whole-volume, full-function 5-D modeling code without the perturbation assumption, with a unique capability in modeling the edge plasma in realistic diverted magnetic geometry (see Figure 24.1), together with neutral particle recycling at the material wall surface and their Monte-Carlo transport using atomic charge-exchange and ionization cross-sections. In order to handle the complicated edge geometry, XGC uses an unstructured triangular mesh. XGC is based upon the conventional PIC ordinary differential equations (ODE) solver method. Unlike the conventional PIC codes developed in the United States, (1) the cylindrical coordinate system is used in XGC to include the magnetic separatrix surface and the magnetic axis in the simulation domain, and (2) all the non-Vlasov, dissipative physical processes are solved on 5-D configuration-velocity space grid using finite element, finite difference schemes, which include the nonlinear Fokker–Planck collisions, charge exchange, ionization, and radiation. Due to the enormity of the particle number required for full-function multiscale gyrokinetic simulation of tokamak plasma, the majority of the computing time is spent on the ODE particle push.

The original XGC before the innovative improvements presented in this work was not capable of simulating the ITER edge plasma, which has been the main goal of the XGC project. We would have occupied the entire Titan for almost 2 months to finish one physics case even if the hardware fail-

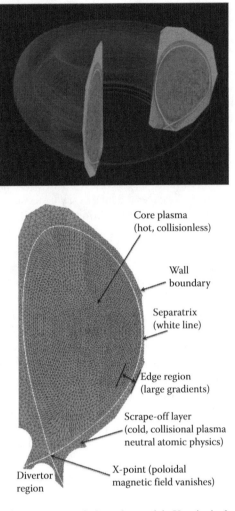

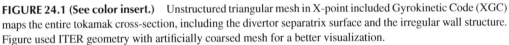

Core plasma
(hot, collisionless)

Wall
boundary

Separatrix
(white line)

Edge region
(large gradients)

Scrape-off layer
(cold, collisional plasma
neutral atomic physics)

Divertor
region

X-point (poloidal
magnetic field vanishes)

FIGURE 24.1 (See color insert.) Unstructured triangular mesh in X-point included Gyrokinetic Code (XGC) maps the entire tokamak cross-section, including the divertor separatrix surface and the irregular wall structure. Figure used ITER geometry with artificially coarsed mesh for a better visualization.

ure did not occur. The combined techniques presented here shortened the time-to-solution by well over six times even without the dramatic improvement in the I/O speed. If we also consider that, using adaptive checkpointing with ADIOS, the I/O time became negligible compared to the more than an hour-long checkpoint writing using the Message Passing Interface Input/Output (MPI-IO) library, the time-to-solution is effectively improved by well over a factor of 10. The simulation wall clock time is shortened to about 5 days. As a result, XGC is making scientific discoveries on ITER which have not been possible before. XGC achieves near perfect scalability on leadership computing platforms by integrating several state-of-the-art techniques for taking advantage of advanced architecture such as effectively exploiting graphics processing unit (GPU), when available, for the most intensive particle push operation, performing dynamic balancing of computational load, enhanced algorithm for multispecies collision, and advanced techniques to perform parallel I/O using ADIOS. ADIOS [4] has made a major improvement in the XGC project and is widely used in other application codes. The partial differential equation (PDE) solver parts of the code, which encapsulate the long-range data dependencies in the PIC method, are more demanding in extreme scale computing.

One advantage of the full-function method used in the improved XGC is that it accounts for these global data dependencies in the PDE solver with only a small fraction of the total work.

Following independent particle trajectories in the PIC method is ideally suited for acceleration on GPUs. GPU acceleration has been used in modeling beam dynamics in accelerator modeling [5], kinetic simulation of space plasma [6], and the GTC code for modeling of plasma in core region of tokamak [3]. However, a successful implementation of particle pushing on GPUs for XGC must overcome two unique challenges. Among various fusion GTCs, only XGC uses an unstructured triangular mesh to extend beyond the last closed flux surface and include the X-point to model the realistic complicated geometry of the tokamak fusion device. Thus, charge deposition and interpolation of field quantities require a more sophisticated search algorithm. Another challenge in XGC is due to the higher charge to mass ratio of electrons compared with ions so that electrons have speeds about 60 times faster than ions. A simple approach of domain decomposition and passing particles to neighbor processors once a particle leaves a domain may be reasonable for pushing ions but will incur a very high communication cost in pushing electrons, especially when particle data are resident on GPU device memory. The techniques taken to mitigate these challenges are described in the next section.

24.3.1   Hybrid Lagrangian δf Scheme

In the edge region of a tokamak, the distribution function of edge plasma deviates significantly from the Maxwellian distribution. However, a significant part of the distribution function represents the slow-varying background plasma compared with the turbulence dynamics. To handle the global-type non-Maxwellian distribution function more efficiently, XGC uses a hybrid-Lagrangian δf scheme, which is mathematically equivalent to the full-f scheme and which utilizes the combination of PIC and continuum grid method [7]. In this scheme, the Lagrangian particle push is used for the fast-varying turbulent Vlasov dynamics and the continuum grid is used for the slow-varying irreversible physics of Boltzmann equation, such as Coulomb collision and neutral atomic physics. Also, the scheme slowly converts particle weights into the coarse-grained phase space grid and reduces the growth of the particle weights. One advantage of this new innovation, not used in any of the GTCs, is the reduction in the particle number by at least factor of 2, thus, leading to the reduction of computing time by a similar factor.

24.3.2   Electron Push Kernel

The algorithm for pushing electrons has a deeply nested call graph and needs access to derived type data structures in modules. To avoid frequent interruptions to GPU (and CPU) for MPI communication, the data for the background electric field of the entire simulation volume are replicated and stored on the GPU. This allows the electron push kernel with multiple subcycling time steps, usually 60, to proceed without any MPI communication. MPI communication is used only at the end of the subcycling to scatter particles to their final destination.

Within each subcycling step, each electron is advanced in time using a fourth order Runge–Kutta (RK4) integrator. The basic operations needed for each RK4 substep are shown in Algorithm 24.1. Equilibrium magnetic field data are read from an input file on a regularly spaced Cartesian grid spanning the poloidal cross-section and interpolated to the particle positions using a bicubic spline. In order to interpolate the electric field (which lives on the discrete poloidal meshes) to the true particle position, the field line must be traced to find its intersection with the nearest triangular mesh. Finding the correct triangular element from the intersection coordinates is challenging and will be discussed in Section 24.3.2.1. The electric field and potential can then be interpolated from the mesh to the particle position and the electron advanced. Performance across all architectures can be improved by sorting the electrons at some period of subcycles, as will be discussed in Section 24.3.2.2.

Algorithm 24.1   The Electron Push Algorithm

---

for each subcycle **do**
   for each particle **do**
      for each 4th order Runge–Kutta substep **do**
         1) calculate magnetic field at particle position by evaluating a bicubic spline interpolant
         2) follow the field line to the footpoint on a nearby poloidal plane
         3) locate the correct triangle for this footpoint
         4) interpolate the electric potential ϕ and field $\mathbf{E}$ from the triangular mesh to the particle position
         5) advance the particle position and momentum based on the fields
   end for
   end for
end for

PGI CUDA Fortran is used to port this important kernel onto GPUs. The initial GPU implementation greatly expanded XGCs capabilities, but in Section 24.3.2.3, we will discuss specific optimizations that produced a significant improvement in performance. A user-determined fraction of particles is assigned to the GPU (about 70% in the original port, and 85% for the optimized code), the remaining are processed concurrently by the multiple CPU cores using OpenMP. This arrangement makes full concurrent use of both the GPU and multiple CPU cores on Titan. As more computing power shifts toward the GPUs on next generation platforms, the particle push is likely to shift entirely to the accelerator.

At present, we maintain separate versions of the electron push routine for GPUs, conventional CPUs, and many-threaded vector architectures like the Intel Xeon Phi. In Section 24.3.2.4 we discuss our (still ongoing) efforts to use directive-based languages such as OpenACC and OpenMP to create a performance portable electron push routine.

24.3.2.1   Triangle Searching

Because XGC uses a carefully generated unstructured triangular mesh that is aligned to magnetic field lines and has mesh vertices located on magnetic flux surfaces, geometry searching presents a significant challenge. The problem can be described as follows: given the 2-D coordinates of a particle on a poloidal plane, find the enclosing triangle and the barycentric coordinates.* In XGC, this search process is implemented using a lookup table method. A fine background rectangular grid is overlayed on the unstructured triangular mesh. At initialization, a list is generated for each rectangular cell that contains the overlapping triangles. We order this list by triangle area, as a proxy for the probability that a particle in the cell will be in a given triangle. On average, only four triangles overlap with each grid cell. Identifying the resident triangle for a given particle then involves hashing into the rectangular grid and then iterating through the list of triangles until the correct domain is found.

Although significantly faster than other methods (such as a directed walk across the poloidal plane), this search suffers from several weaknesses. First, the performance of the lookup table is extremely sensitive to the size of the overlayed rectangular mesh and the correct dimensions may need to be empirically determined. Second, to enable fast lookup, the cells of the rectangular grid must be evenly spaced, which means its size cannot be tuned evenly across a mesh with varying element sizes (see, e.g., Figure 24.1). Finally, the full process of accessing both the list of triangles

---

* See `http://mathworld.wolfram.com/BarycentricCoordinates.html` for definition of barycentric coordinates of a triangle.

and the triangular mesh coordinates has proved to be punishing on hardware caches, particularly on GPUs but even on larger and deeper memory hierarchies like the Intel Xeon Phi Knights Landing (KNL) processor.

These drawbacks could be addressed by more sophisticated algorithms but analysis of the particle motions reveals a more effective solution. Numerical accuracy constrains how far particles can travel in a given timestep, meaning their motion across the mesh per RK4 step per cycle is small compared to the cell size. Furthermore, particle motion is predominantly parallel to the magnetic field. The sophisticated XGC1 field following method ensures that an electron moving along its field line will stay tied to the footpoint triangle on a given poloidal plane until it travels far enough in the toroidal (ϕ) direction to cross to a different mesh (and therefore triangle). Taken together, these two constraints mean the probability is very high that for any given instance where the search routine is run, the triangle found will be the same as the last known resident triangle. For test runs of the full electron push routine (including the full subcycling process), over 99% of triangle searches *do* reside in the last known triangle. By storing the last known triangle for each particle and checking it before running the search, we have reduced the cost of the triangle search to a negligible time (<5% of the kernel time), with the triangle check being the majority of that time.

24.3.2.2 Particle Sorting

Frequent sorting of particle data is widely acknowledged as an effective means to decrease scattered memory access and increase cache usage in PIC codes [6]. In XGC, we use a simple counting sort algorithm [8], which has linear complexity $O(n + k)$ where n is the number of elements (particles) and k is the number of keys (the choice of keys will be discussed shortly). The basic counting sort algorithm is as follows: (1) calculate (nonunique) keys for each particle; (2) count the number of occurrences for each key; (3) prefix-sum the list of counts to find the starting position for each key; (4) reorder the particles into their correct memory locations. To provide an implementation that is agnostic to Struct-of-Arrays (SoA) or Array-of-Structs (AoS) particle layouts, we use a list of calculated keys to sort an index array and generate a permutation vector. The permutation vector *iperm* is a one-dimensional index array of length given by the number of particles, where element $iperm[j] = i$ gives the source index i in the original particle array that must be moved to location j in the sorted array. In other words, *iperm* stores the mappings for a gather operation, which we then apply to the arrays storing particle components (in the SoA storage type) or entire particles (in the AoS storage type).

On GPUs, we have explored using the `thrust` library [9] to accelerate the sorting process. In principle, only step 1 is unique to a given code; steps 2–4 could make use of library calls. Direct sorting of particle data is not, however, feasible due to our requirement that several arrays be sorted using the same keys. This multiarray reordering can be accomplished using `thrust` in two ways: an iterator can be used to encapsulate multiple arrays so that they can be sorted by one call to the library or multiple calls to a stable sorting method can be used to sort each array using the same key sequence. The former uses significant memory, as it requires (at the minimum) a full duplicate of each array to coexist on the device. The latter is limited by the stable sort algorithms implemented in `thrust`, which have significantly worse complexity than counting or radix sorts. In practice, we use `thrust` only for the prefix-sum calculation (step 3). The other steps are simple enough that they can easily be implemented and hand tuned.

The correct choice of keys is still an area of active research. Initially, we sorted by cell of the rectangular lookup table (see Section 24.3.2.1). The NESAP and CAAR teams discovered that sorting by the last known resident triangle resulted in much better caching and better consistency across vector lanes or block threads. This choice results in a slight increase in the number of keys (k above) but this resulted in marginal increase in the sorting time. Although the domain covered by each key is smaller in this method (triangle vs. lookup table cell), the timescale for the sort to degrade is not much shorter. This is likely due to the field following method of XGC, where particles remain anchored to the same triangle until they move far enough in the ϕ direction, at which point their footpoint will move several triangles away. This is not likely to be true in other PIC models.

The ideal sorting period must be determined empirically on a given architecture based on the cost of the sorting routine and the gains in electron push kernel performance. On NVIDIA K20X GPU, we find the optimal period to be ~five cycles. For Intel KNL, in contrast, it is only advantageous to sort at the beginning of the electron push routine. The optimal sorting period for NVIDIA Pascal P100 GPU is still being evaluated.

24.3.2.3 GPU Specific Optimizations

Minimally, pushing each particle through each Runge–Kutta substep of each cycle involves the steps shown in Algorithm 24.1. Only the final step (5) involves writing data to device memory; steps 1, 3, and 4 involve read-only access; step 2 is pure calculation using the Runge–Kutta integrator. Full optimization of the XGC kernel was accomplished by minimizing traffic to GPU global memory. Profiling results revealed that the overwhelming majority of memory traffic came from the read-only accesses. In principle, if the PGI compiler can determine that a given variable is read-only for the duration of the kernel execution (via the use of `intent(in)` attributes), it will generate `__lgd` CUDA intrinsics to access these data through the read-only texture cache. Because the read-only cache is not kept coherent with global memory, this path can only be used if the compiler can determine that it is safe to do so. We have found that these are not generated for the XGC electron push kernel, likely because of the complicated call tree. Instead, we explicitly use the texture capabilities in CUDA Fortran and C. The electric potential and field information in (4), as well as data structures related to the field following and search in (2) and (3), are all exclusively accessed through CUDA Fortran `texture` pointers. The bicubic spline interpolation in step (1) is inherently a 2-D process as the interpolant representing the equilibrium magnetic flux function is constructed from regularly spaced inputs on r, z space. This structure is ideal for binding to a 2-D layered texture reference; a capability that can only be accessed in CUDA C/C++. We have found this 2-D CUDA C implementation of the spline interpolation to result in a 5% decrease in the execution time of the total kernel compared with a CUDA Fortran implementation bound to linear memory.

The use of texture pointers produced a substantial decrease in the run-time of the electron push kernel. Additional improvement was achieved defining module scalars and short arrays with the CUDA Fortran `constant` attribute. At the compile time, PGI provides several useful flags that provided minor gains to performance; specifically, the little known `madconst` flag places Fortran module array descriptors into constant memory. With this final set of improvements, we were able to eliminate all spurious loads from global memory from the XGC electron push kernel.

24.3.2.4 Portability

Large XGC science runs require the full resources of the OLCF Titan supercomputer and as a consequence significant resources have been dedicated to the above performance optimizations on NVIDIA GPUs, both before and as part of the CAAR program. The electron push routine has also been tuned for the Intel Xeon Phi architecture by the XGC NESAP team at NERSC. At present, the GPU and vectorized versions of the routine require separate code, with a third version for conventional CPUs. With leadership computing supporting both the vector and GPU tracks for the foreseeable future, we have spent significant effort attempting to unify the various electron push kernels using directive-based models. Our approach has largely focused on porting both the conventional CPU and vectorized versions of the code to GPUs using OpenACC directives, using the highly tuned CUDA code as the performance target. Both approaches have encountered significant challenges.

The greatest limitation in porting the conventional CPU is the deep call tree. As discussed earlier, a key component of PUSHE performance on the GPU is the use of read-only memory pathways in *leaf* routines deep within the call tree. The same compiler limitations discussed in Section 24.3.2.3 also impact the OpenACC kernel. The directive-based model does not, however, permit the architecture specific memory mappings that we use to bypass these issues in the CUDA code. As a consequence, the OpenACC implementation is significantly slower. We are able to alleviate the problem somewhat by replacing the leaf routines, such as the bicubic spline interpolation, with optimized CUDA versions on the device using the `bind` clause on the OpenACC `routine` directive. Even exploiting this capability, the OpenACC code is still only half as fast as the optimized CUDA Fortran kernel. As compilers mature, this performance gap is expected to close but at present the difference is too substantial for the OpenACC version of the conventional CPU kernel to be considered viable.

We have had even less success porting the vectorized kernel developed for Intel Xeon Phi by the NESAP team, where the deep call tree has also presented problems. For automatic vectorization by the Intel compiler, the vector parallelism must be visible as close to the instruction level as possible, even using OpenMP4 `simd` directives. We block the top-level particle loop into chunks of ~64 particles and repack the particles into SoA tiles. These tiles or *Particle vectors* are then passed into modified *vector* forms of subroutines, which loop over the particles in a tile. These tiles must be passed all the way through to the low-level *leaf* routines and iterated over there, otherwise the compiler cannot detect sufficient parallelism to generate vector instructions. Although this vectorization challenge has the same root cause as the read-only memory issue on the GPU (i.e., the deep call tree), the solution is at odds with optimized GPU code. Efficient execution on the GPU relies on convincing the compiler, as much as possible, that computations for the particles are completely independent and that the read-only data pipelines are safe to use. Vectorization, however, requires ensuring the compiler that groups of particles are sufficiently tightly bound to allow processing *in lockstep* using single instruction multiple data (SIMD) vector instructions. Currently, compilers targeting GPUs view this binding as a decrease in parallelism from thread level to block level independence, resulting in an overuse of shared memory and numerous block private memory allocations that destroy performance. Overcoming these challenges and creating a unified performance portable kernel on both GPUs and Intel Xeon Phi will require either significant advances in compiler technology or a redesign of the electron push algorithm.

24.3.3 MULTISPECIES COLLISION KERNEL

The Fokker–Planck–Landau (FPL) collision operator is an advection-diffusion type integro-differential equation describing time evolution of a probability distribution function through collisions in velocity space. The effects of collisions, viz. advection and diffusion, on species s against species s' are introduced through the drag and diffusion coefficients $E_{ss'}$ and $D_{ss'}$ given by moments of the distribution function $f_{s'}$ for species s'. The system of equations for all species governing collisions has intrinsically nonlinear nature in that colliding and target particles are not manifestly separable because its not passive (one-way) interaction to one species from the other species but mutual interactions among multiple species.

Although most other kinetic codes in magnetic confinement fusion have adopted linearized collision operators, XGC employs a fully nonlinear, Eulerian, multispecies FPL operator [10,11] for an accurate but more compute-intensive description of nonequilibrium plasma in the tokamak edge. The main reason for the popular use of linearized collision operator is due to its analytic and numerical tractability. The reduced linearized operator from the nonlinear one is based on the assumption that distribution function $f_{s/s'}$ does not deviate significantly from the assumed equilibrium distribution function, (shifted-) Maxwellian $f_{M,s'}$, as $\delta f_{s/s'}/f_{M,s/s'} \ll 1$ where $f_{s/s'} = f_{M,s/s'} + \delta f_{s/s'}$. This allows one to calculate $E_{ss'}$ and $D_{ss'}$ analytically using a Maxwellian $f_{M,s'}$, instead of numerically using the actual $f_{s'}$ with the same density, temperature, and mean flow. However, the plasma edge and scrape-off layer, where XGC mainly focuses, involve ion orbit effects and strong sources and sinks, which strongly distorts the distribution function. Therefore, the assumption made for the linearization is usually invalid in the targeting simulation region. Also, collision processes among multiples species

---

Algorithm 24.2   The Nonlinear Collision Algorithm

> **for each** interaction pair of species (s, s') **do**
>> **for each** cell of $f_s(I, J)$ **do**
>>> **for each** cell of $f_{s'}(i', j')$ **do**
>>>> Compute $\langle\langle M_{s/m}\rangle\rangle(I, J, i', j')$
>>> **end for**
>> **end for**
> **end for**
> **while** error of conserving physical quantities $>$ tolerance **do**
>> **for each** species s **do**
>>> **for each** cell of $f_s(I, J)$ **do**
>>>> Compute $E(I, J)$ and $D(I, J)$ from reduction of $\langle\langle M_{s/m}\rangle\rangle(I, J, i', j')$ over (i', j')
>>> **end for**
>> **end for**
>> Construct system $Ax = b$; solve for x {using LAPACK band solver, SuperLU, or PETSc solvers}
>> For each species s: update f_s
> **end while**

---

are dominant in the edge region. This is why XGC adopts the costly multispecies nonlinear collision operator.

Numerically, the collision operator is solved on an Eulerian velocity grid. In order to use this method in conjunction with XGC, a Lagrangian PIC code, the distribution functions on the velocity grid are reconstructed by marker particles sampled from each Voronoi cell around a mesh vertex of configuration space. Given the distribution functions, an implicit time advance using an iterative Picard method is performed to get changes of the distribution function through collisions as shown in Algorithm 24.2. Each nonlinear iteration requires direct solution of a sparse linear system that has a sparsity pattern associated with a regular stencil on a rectangular grid. If a $N_u \times N_v$ velocity grid is used, then each Voronoi cell needs to solve a $(N_u N_v)$ by $(N_u N_v)$ sparse linear system. Subsequently, marker particles' weights are adjusted according to the changed distribution. Note that in this algorithm, collision operations in different collision cells are independent with each other. In other words, these computations do not require coordination or communication at all and are all local to each MPI process.

Originally, the calculation of the drag and diffusion coefficients of the nonlinear collision operation was highly compute intensive and more expensive than the electron push kernel, taking over 70% of the total compute time. Accordingly, we undertook an intensive optimization effort (see Table 24.3). Physics-based optimization of this calculation by using symmetries of the FPL operator offers a major performance improvement and better cache performance. Thorough analysis of memory access patterns with the symmetries motivated a reordering of the dimensions of a handful of large arrays resulting in enhanced data reuse and reduced memory consumption. Additional performance improvement is achieved by modifying the algorithms for the calculation of the drag and diffusion coefficients such that they can be vectorized efficiently by the compiler.

The original implementation of the FPL operator followed conventional parallelization strategies in that each MPI process solved the Fokker–Planck equation sequentially on the vertices of a patch of the configuration space mesh, whereas long loops were parallelized with OpenMP. The leadership computing architectures, however, offer a higher degree of parallelism per compute node than one can use effectively following this simple strategy and using more MPI tasks per compute node instead may be prohibitive due to memory constraints and communication overhead. The alternative is to use nested parallelism as allowed by the OpenMP standard. Instead of applying nested OpenMP parallelism to nested loops, we use the fact that the evaluation of the collision operator on every vertex of the configuration space mesh is independent and parallelize the whole collision operator subroutine with all its internal subroutine calls in addition to the OpenMP parallelized loops inside the collision operator. Thus, each MPI process can solve the Fokker–Planck equation on several mesh vertices in parallel. The outer level of our nested OpenMP implementation proved to be more efficient than the inner level (16.6% on Titan, 44.8% on Edison; see Section 24.5 and Figure 24.5), but more threads in the outer level also require more memory. Because the number of threads assigned to each of those two levels of parallelization is flexible, the available OpenMP threads can be arranged to optimize the balance between time-to-solution and memory consumption on different high performance computing (HPC) architectures.

Performance portability on the collision operator has proven to be far more achievable than for the electron push kernel. The regular structure and high degree of parallelism in the method map well to both vector architectures and GPUs. With the help of NVIDIA representatives at a GPU Hackathon, OpenACC directives were added to the existing OpenMP and vector optimized code. The most significant bottleneck involved copies of the multispecies collision matrix M to the device and copies of the drag and diffusion coefficients $E_{ss'}$ and $D_{ss'}$ back to the host. Matrix M can, however, be constructed purely on the device, allowing the use of an `!$acc data create` directive to avoid data transfer. Thus, only the final coefficients need to be copied back.

The high degree of parallelism also allows the use of asynchronous data transfers and kernel launches with the OpenACC `async()` clause but encounters the same memory constraints as the top-level OpenMP threading. This issue is particularly prominent on K20X, where the device memory is limited to 6 GB. On Titan, we are forced to limit the collision routine to the PGI OpenACC maximum of eight streams. The correct number of streams will need to be determined empirically on newer architectures.

Furthermore, the collision routine may clash with the electron push routine for device memory. As a performance feature, the PGI OpenACC runtime implements a memory management back-end that does not release device memory once allocated. As such, in large runs on Titan the OpenACC collision kernel would compete with the CUDA Fortran electron push kernel for GPU memory, resulting in a crash. Our current (suboptimal) solution is to force the OpenACC runtime to release memory with the PGI `acc_clear_freelists()` routine, although this may be replaced by OpenACCv2.5 `finalize` directives and calls as they are implemented in compilers. A better long-term solution is to manage all device memory through the same runtime, either OpenACC or CUDA. This code-wide memory management structure will, however, require significant development work.

The collision and particle push routines also compete for computational resources. Due to the fact that the mesh-based collision operator favors even distribution of configuration space vertices among processes, whereas the particle push favors an even distribution of particles, the dynamic load balancing algorithm described in Section 24.3.4 is designed to find a compromise between balancing mesh and particle-related work.

24.3.4   Dynamic Load Balancing

The XGC computational mesh for a toroidal domain consists of uniformly spaced poloidal planes (toroidal cross-sections) and an unstructured triangular tessellation within each poloidal plane (e.g., Figure 24.1), where each poloidal plane tessellation is identical. Note that each processor has a copy of this 2-D mesh but the load balancing scheme allocates only a subdomain for managing marker-particles. As described in [12], mesh vertices in a poloidal plane are ordered first by *flux surface*, starting from the inner surface. Within each flux surface vertices are ordered by poloidal angle. This results in an approximate *spiral* space-filling curve ordering that preserves locality of the associated triangles and is roughly monotonic in the radial direction (see Figure 24.2).

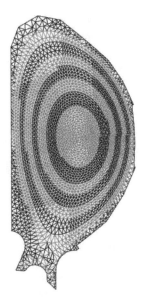

FIGURE 24.2   The spiral decomposition on 2-D grid for partitioning work to processors.

The full computational mesh is partitioned over a virtual 2-D processor array, with a *column* subset of processes sharing responsibility for a poloidal plane, and a *row* subset of processes sharing responsibility for one element of a 1-D partition of the mesh vertex index space. Load imbalances are controlled by adjusting the 1-D partition of the vertex indices, resulting in particle and vertex-related data being moved between processes in the same *column* for each adjustment. Although some load imbalances occur in the toroidal direction (between poloidal planes), these develop and dissipate very quickly and are difficult to address effectively. Fortunately, addressing load imbalances in the radial and poloidal directions has been sufficient to enable good performance scalability.

In XGC simulations, there are two significant sources of load imbalance currently: (1) the distribution of particles, which primarily shows up in load imbalance when calculating the electron *push* that determines where electrons move during a timestep, and (2) the temperature profile across the mesh cells, which determines the computational cost per mesh cell of the iterative algorithm that calculates the effect of collisions between particles (from the particle distribution functions in each cell). Simple 1-D optimization algorithms to adjust the partition of the mesh vertex index space are very effective for equidistributing particle count or empirically measured per cell collision cost across the processes. However, optimal partitions for the particle count are typically not efficient for load balancing the collision cost, and vice versa.

Particle-related computation cost is not a simple function of the particle distribution, and attempts to define a weighted objective function that captures both particle and collision cost have not been successful heretofore. The current solution is to first solve the collision cost optimization problem with a constraint on the maximum particle count load imbalance. An outer Golden Section Search algorithm is then used to find the constraint that minimizes wallclock time. This wallclock time is determined empirically, so the Golden Section Search algorithm is implemented as part of the timestepping algorithm, with a new constraint evaluated at some user specified frequency of timesteps. As both the particle distribution and temperature profile evolve with the simulation, this Golden Section Search has been modified to allow it to adapt to such changes and to be resilient to performance perturbations.

The relative costs of the collision operator and the electron push, and thus the importance of this hybrid load balancing scheme, are dependent on the details of the experiment. Such details include, for example, total particle count, mesh size, and how frequently the collision operator is calculated. It also depends on the target platform, for example, whether electron push is accelerated on a GPU or not. In one production experiment, this hybrid load balancing scheme doubled the performance as compared to just load balancing the particle count per process, as described in Section 24.5.4.

24.3.5 PARALLEL AND STAGED I/O

The I/O overhead of XGC comes from several output data sets with various I/O conditions. The largest overhead comes from checkpointing, about 20 TB in this ITER simulation case. We switched to ADIOS for I/O many years ago when the existing solution took well over 1 hour to write the checkpoint data for every hour of computation, practically doubling the time-to-solution even if no hardware fault occurs. If hardware failure occurs once in 2 hours, the time-to-solution can become infinity. Now it takes about 4 minutes to write this amount. The minimal physical analytics data is much smaller (500 GB), but it is written more frequently. A surprisingly large overhead for this data comes from the metadata collection in ADIOS. Finally, one of the small diagnostic data sets takes disproportionally long time to write because every process writes a small amount of data to a single file, which does not scale well. In comparison, another diagnostic file using one process per simulation plane, that is, 32 or 64 processes, causes a negligible overhead.

We have addressed these different overheads using three optimizations designed for the ITER simulation. First, ADIOS scales well in terms of storage targets when writing large amount of data from many processors. Atlas, the Lustre-based parallel file system at the OLCF, is partitioned into two subfile systems to balance the user load in general. However, our project was allowed to write to both subfile systems and the scalability of ADIOS enabled us to use them efficiently. We were able to double the I/O bandwidth this way, writing from 16,384 Titan nodes to 2,000 object storage targets (OSTs) and to decrease the time to write 20 TB data to about 4 minutes.

Second, the ADIOS metadata are duplicated when using I/O methods that output to multiple files, for example, aggregation. Each aggregator writes data into a separate file and it also writes the corresponding local metadata into that file. The large overhead comes from gathering all metadata to a single process, which merges and writes it into a final single file. This global metadata are used when a data set is opened for reading, and it is, therefore, essential to have it. Because it can be created from the information stored in the subfiles, we created a mechanism in ADIOS to generate this global metadata post-mortem from the subfiles. The creation process is slower this way because all subfiles have to be opened and the local metadata have to be read, processed, and merged to create the global metadata file. However, in the improved technique this is now a single-node postprocessing action and thus it does not impact the simulation runtime.

Third, a small diagnostics data set takes too much time to write. The access from all processes to the file system's metadata server, the creation of an output file, and the write access from all processes to that file, represent most of the I/O time. The idea of buffering such small data quickly and writing them to permanent storage asynchronously has been around for a long time and the next-generation supercomputers will be equipped with burst buffers to accommodate such scenarios. At this time, we rely on the staging capability of ADIOS provided by DataSpaces [13]. ADIOS applications can switch to the DataSpaces method instead of a file-based method to stage the output instead of writing to disk. There is no need to change the application source to do this [14]. DataSpaces ran as a data server on a separate set of nodes. Although one extra node would be sufficient for storing the 2 GB diagnostics data set, in the presented runs, we needed many staging nodes to handle the large number of connections from all processes because the Gemini network interface has strong limits in handling a large number of connections and memory registrations for remote access. We used 128 nodes as servers to handle 16,834 clients sending the data. Another 64 node application was running along with the simulation and the server, pulling the data set from the server and writing to the file system.

24.4 PERFORMANCE METHODOLOGY

XGC was instrumented manually using timer start/stop routines from a slightly modified version of the GPTL library [15] to capture wallclock time per process (and per thread in threaded regions) for the full code, for the main timestep loop, and for important phases, including initialization, electron push, collision operator, and I/O. All performance data presented here used MPI_WTIME as the underlying timer.

Note that the XGC code is structured as (1) initialization, (2) timestep loop, (3) finalization, including any end of job I/O. So the timestep loop represents essentially all wallclock time in production runs.

Scaling studies, both weak and strong, were for relatively short simulations (e.g., 10 ion timesteps), and did not include initialization or significant I/O because these would not have been weighted appropriately for a production run. Results are presented in terms of maximum wall time over processes, either directly or as a normalized inverse time rate. Note that there are numerous synchronization points within the timestepping algorithm and there is little per process variability in the main timestep loop timer.

Weak scaling studies are weak in particle count only and strong in mesh size. Only two (both production) meshes are used in the results presented here, one for the DIII-D device and one for the ITER device.

I/O overhead was studied with the ITER production run with relatively short simulations but with all I/O turned on. This was weak scaling in both particle count and mesh size. The overhead of writing each data set was calculated from the profile data of the I/O routines and the main timestep loop of XGC.

Production run performance is presented for the main timestepping loop and for the important computational kernels, all measured using MPI_WTIME via the GPTL library.

Most of the performance data were collected on the Cray XK7 Titan system at the OLCF at the ORNL. Titan consists of 18,688 compute nodes. Each compute node has 32 GBytes of memory, one 16-core AMD Opteron 6200 Interlagos processor with eight floating point units and a NVIDIA K20X Kepler GPU with 6 GBytes of device memory. Each Interlagos processor has eight 256-bit floating point compute units shared by 16 integer cores. Two compute nodes are connected to a Cray Gemini network device (NIC) that has over 160 GBytes/s of routing capacity. The global network is arranged as a 3-D torus. The Portland Group pgf90 CUDA Fortran compiler version 15.7 was used with the *-fast* optimization option.

Other performance data presented here were collected on (1) the IBM BG/Q Mira system at the ALCF at the ANL; (2) the Cray XC30 Edison system at the National Energy Research Scientific Computing Center (NERSC) at the Lawrence Berkeley National Laboratory (LBNL); and (3) the Cray XC30 Cori (phase 1) system in NERSC at LBNL.

24.5 PERFORMANCE RESULTS

24.5.1 WEAK AND STRONG SCALABILITY OF ELECTRON PUSH KERNEL

XGC achieves near perfect weak scaling on most leadership computing systems and Figure 24.3 describes performance improvement in the main XGC time stepping loop on the Cray XK7 Titan system after the initial CUDA Fortran port of the electron push kernel. The performance data are for a weak scaling in particle count per compute node but a fixed size mesh. This run did not include the new multispecies collision kernel.

Optimization consisted of both optimizations to the associated MPI communication algorithms and porting the kernel to the GPU. The GPU port included the option to compute some particles on the

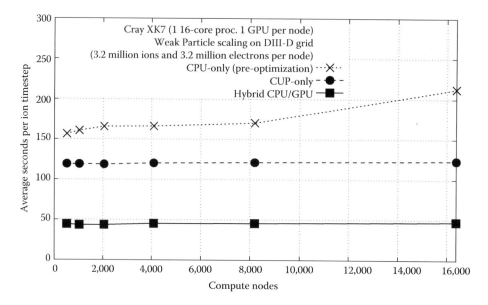

FIGURE 24.3 Weak scaling plot of XGC on DIII-D tokamak grid on Cray XK7 Titan immediately after introduction of drift kinetic electron capability, after optimizations to MPI communication algorithms, and after porting of electron push kernel to graphics processing unit (GPU).

CPU as well and the percentage of particles assigned to the CPU instead of the GPU is a performance tuning option. Note that this does not capture the performance improvement from waiting to shift particles (between processes) until after all electron push subcycles were complete. Note also that the original implementation of the CUDA Fortran kernel was used for these scaling results, rather than the highly optimized version described in Section 24.3. Both the optimized push and the multispecies collision operator are currently used in production runs but the pace of development has been rapid enough that they are newer than any large scaling tests.

Figure 24.4 shows good performance of XGC for (realistic) strong scaling experiments on a variety of leadership computing architectures, including Cray XK7 Titan, IBM BlueGene/Q Mira at ALCF, and Cray XC30 Edison at NERSC. Note that each data point on these curves represents potentially different performance tunings, for example, number of MPI processes per node, number of OpenMP threads per process, whether to and how much hyperthreading to use (XC30 and BG/Q), mapping of virtual processor grid onto physical topology.

The degradation in the strong scaling on Cray XK7 Titan (Figure 24.4) is a combination of communication overhead becoming more evident as computation cost decreases (due to strong scaling) and a decrease in the effectiveness of the GPU acceleration as the workload per node decreases. Further communication optimization targeting full large system runs and further CPU/GPU load balancing could mitigate some of this. Note that BG/Q results do not show this because it is compute bound out to the largest node count. The Cray XC30 Edison system used was not large enough to exhibit a communication overhead effect for this example (as well as due to having a higher performance interconnect).

24.5.2 GPU OPTIMIZATION OF THE ELECTRON PUSH KERNEL

The impact of the algorithmic and GPU-specific optimizations discussed in Section 24.3 is shown in Table 24.1. Results are given for a test case of 5×10^6 electrons pushed through a full ion step with 40 sub-cycles. To better illustrate the impact of the changes, we compare two different resolution

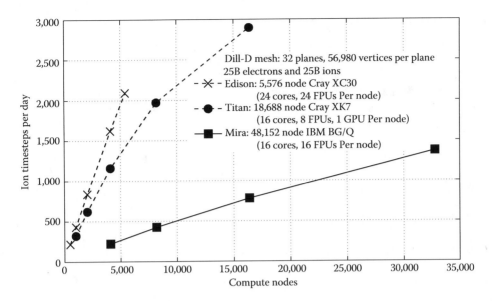

FIGURE 24.4   Strong scaling plot of XGC on DIII-D tokamak grid on Cray XK7 Titan, IBM BG/Q Mira, and Cray XC30 Edison.

TABLE 24.1

Improvements Achieved by Optimization of XGC's Electron Push Routine, Evaluated Using 5×10^6 Particles on Two Different Meshes

Opt.	3 mm ITER	12 mm ITER
Initial	38 sec	23 sec
Improved searching	15 sec	12 sec
Improved sorting	13 sec	11 sec
Read-only mapping	7.8 sec	7.2 sec
Compiler options	7.7 sec	7.0 sec
Current OpenACC	14 sec	12 sec

Note: Each line includes all prior optimizations (as described in Section 24.3), with the exception of the OpenACC result, which is a separate code.

---

meshes for the same ITER relevant geometry; a 3-mm mesh with $\sim 10^6$ triangles and a 12-mm mesh with $\sim 10^4$ triangles. In the initial code, it is significantly faster to push the electrons on the lower resolution mesh. Each successive optimization brings the execution time close together; the better exploitation of GPU hardware allows true particle parallelism to determine execution, rather than the bottlenecks involved in read-only memory access.

We also include, for reference, the current best performance of the OpenACC kernel adapted from the conventional CPU code. This version includes both the algorithmic search and sort improvements, and bound versions of leaf routines written in CUDA C, allowing for some exploitation of the GPU read-only hardware. Note that performance of this version is significantly better than the initial CUDA Fortran implementation but far short of the fully optimized kernel.

24.5.3 MULTISPECIES COLLISION KERNEL

The performance characteristics of XGC's nonlinear FPL operator have been evaluated in smaller scale XGC simulations (to check efficiency of nested OpenMP parallelism) and in a stand-alone OpenMP/OpenACC-only version of the collision operator (other optimizations). Because the collision operation is completely local, a full-scale test is not needed. We tested different combinations of threads in the outer and inner OpenMP level ranging from all threads assigned to the outer level to all threads assigned to the inner level. The results are shown in Figure 24.5. Shifting threads from the inner OpenMP level to the outer level results in a performance improvement of 16.6% on Titan and 44.8% on Edison (Figure 24.5). Table 24.2 shows the runtime for the collision kernel using flat and cache mode on KNL.

The improvements achieved by exploiting symmetries, optimizing memory access, facilitating automatic vectorization by the compiler, and porting to GPUs with OpenACC are summarized in Table 24.3. We tested the initial implementation (*Initial*), a version after exploiting symmetries and optimizing memory access (*V1*), and a version with optimized vectorization (*V2*) in addition to the optimizations in *V1*. We chose the same thread configuration of 16 outer and 1 inner OpenMP thread in all three versions. The velocity space mesh used for this test has 64×32 vertices and the collision operator was evaluated for five collision cells per OpenMP thread and for five timesteps. The largest speedup (3.3) is observed on Titan, whereas the improvement on Mira (2.1) and Cori Phase I (2.9) is more modest. We also include the results of the OpenACC GPU implementation on Titan using

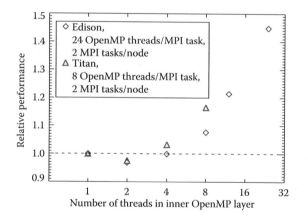

FIGURE 24.5   Relative performance of XGC's collision operator on Edison (Cray XC30, NERSC), and Titan (Cray XK7, OLCF) with different partitions of OpenMP threads between the outer and inner layer of our nested OpenMP parallelization (see also Ref. [10]). The horizontal axis indicates the number of OpenMP threads assigned to the inner OpenMP layer. The case in which all threads are assigned to the inner OpenMP layer is equivalent to conventional OpenMP strategies that parallelize only long loops. In contrast, our nested OpenMP approach parallelizes calls to complex subroutines, which themselves contain OpenMP parallelized loops.

TABLE 24.2
Performance of Original, Cache Optimized, and Vectorized Collision Kernels for Flat and Cache Mode in KNL Using 64 Outer Threads

	Original	Cache Optimized	Vectorized
Flat	114.5 sec	61.8 sec	29.8 sec
Cache	108.2 sec	45.0 sec	17.5 sec

TABLE 24.3

Performance Improvements Achieved by Optimization of XGC's Collision Operator

	Initial	V1	V2	ACC
Titan	178.3 sec	72.3 sec	54.8 sec	10.5 sec
Mira	214.2 sec	121.2 sec	103.2 sec	N/A
Cori Phase I	41.0 sec	17.7 sec	14.4 sec	N/A

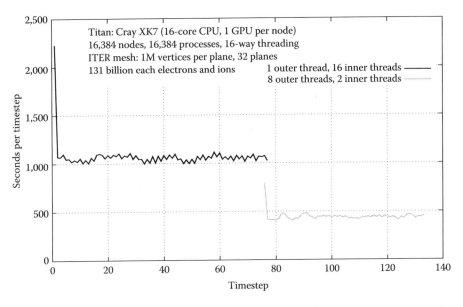

FIGURE 24.6 Wall-clock time per timestep for two checkpoint/restarts for a production run on the Cray XK7 Titan. The number of inner and outer OpenMP threads used in the collision operator were changed between the two restarts. The first timestep in each used particle-only load balancing. Subsequent steps used the hybrid scheme, also taking into account load imbalance in the collision calculation.

eight GPU streams to process the same test case. The OpenACC kernel shows an impressive 17× speedup compared with the original code on Titan.

24.5.4 Load Balancing

Figure 24.6 demonstrates the impact of both the hybrid load balancing scheme and the nested OpenMP threading in the multispecies collision operator. During a science run on the Cray XK7 Titan, it was observed that the collision performance was poor because of a poor choice in the number of inner and outer OpenMP threads. The job was killed and restarted after a checkpoint. The first timestep after a restart uses particle-only load balancing as there is no collision cost history preserved across the restart (in the current implementation). Once collision cost data become available, the load balancing scheme can then address both sources of load imbalance. This resulted in a doubling performance in this case. Not shown here, but after adjusting the threading and after load balancing, the maximum collision cost per process was approximately 40% of the cost of the main timestepping loop. Note that MPI_WTIME was used directly to collect these data, not the GPTL (General Purpose Timing Library).

24.5.5 Parallel I/O

Figure 24.7 shows the improvements on file-based I/O overhead by using the latest ADIOS software. We ran the ITER production case with weak-scaling both in the number of particles and the grid granularity. Using both subfile systems with 2,000 OSTs and turning off of global metadata collection more than doubled the performance of writing 20 TB restart data. Skipping global metadata collection also improved the overhead for the more frequent minimal physics data (of 500 GB) by a factor of 2. Figure 24.8 shows that staging of the 3-D diagnostic data set (of 2 GB) using DataSpaces eliminated the associated high overhead to the total runtime of the application.

24.5.6 ITER Simulation

In the production run of an ITER simulation, 131 billion ions and 131 billion electrons are pushed utilizing 16,384 nodes of the Cray XK7 Titan system. In addition, there is a smaller number of Monte

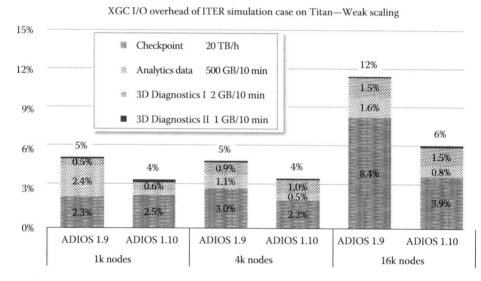

FIGURE 24.7 Input/Output (I/O) overhead of the production ITER simulation with XGC. ADIOS 1.10 used two file systems to improve the speed of writing checkpoint data and skipped metadata writing of analysis data. One diagnostic was negligible in all cases but the other diagnostics data still represented a considerable amount of overhead.

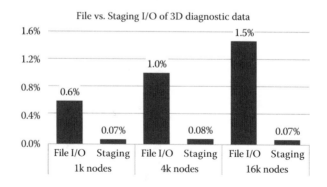

FIGURE 24.8 File versus DataSpaces-based staging I/O overhead of the diagnostic data in the production ITER simulation with XGC. The file-based information is the same as the diagnostic overhead in the ADIOS 1.10 runs in Figure 24.7. The staging information is from identical simulation runs except for using DataSpaces for staging the diagnostic data.

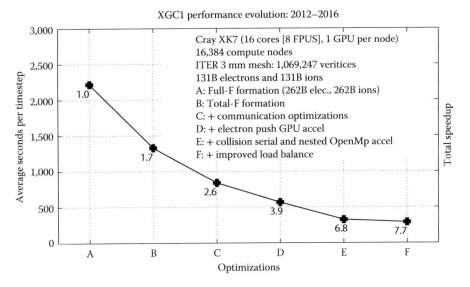

FIGURE 24.9 Cumulative performance improvement of ITER simulation with the enhancements described in this work. Note that point D represents the initial Compute Unified Device Architecture (CUDA) Fortran port of the electron push routine and point E represents the optimized Central Processing Unit (CPU)-only version of the collision routine (*V2* in Table 24.3). The optimized push and OpenACC collision kernel are not represented here.

Carlo neutral particles. The ITER mesh used in the simulation has 32 million vertices in configuration space (1 million vertices per plane × 32 poloidal planes) and each configuration vertex has 31×30 velocity vertices. Hence, the number of particle per velocity vertex is 4.4 for each species, which is a reasonable number to resolve the non-Maxwellian distribution function of edge plasmas.

Figure 24.9 shows the performance improvement of an ITER simulation with the enhancements described in Section 24.3. The stages of optimization correspond to the following improvements: A—the original XGC simulation code; B—the Hybrid Lagrangian δf scheme (Section 24.3.1); C—communication optimizations around the subcycling scheme (Section 24.3.2); D—initial (but not optimized) CUDA Fortran port of the push routine (Section 24.3.2); E—improved (but not GPU enabled) collision routine (Section 24.3.3); F—load balancing between particles and collisions (Section 24.3.4). A total factor of 7.7 performance improvement is obtained in the main timestep loop.

I/O improvement (Section 24.3.5) was not included in this result because the data are collected with very short simulations. Even though very short, tests of this scale are costly and we have not generated a matching point for the improved GPU electron push and OpenACC collision routines because they are still being optimized. Considering the fact that the MPI-only checkpointing would take well over 1 hour for every hour of simulation, and that the adaptive I/O reduces the checkpointing time to 4 minutes, the effective reduction in time-to-solution is much greater than 10.

24.6 SCIENTIFIC IMPLICATIONS

The scientific implication of the "greater than 10 times acceleration in the time-to-solution," achieved here from the combined state-of-the-art computational and algorithmic technology, is priceless. It enabled the ITER edge plasma simulation on Titan in about five wallclock days. Experimentalists maps the divertor heat-load width to the outboard midplane (giving a quantity λ_q) along the magnetic field lines, which usually results in a 10:1 compression in the width compared with that measured at the divertor plate. Using the experimental data from the present tokamak devices and assuming that the edge physics in ITER is the same, a simple regression analysis predicted $\lambda_q \simeq 1$ mm, or the heat-load width on the divertor target is ~1 cm. If correct, this would result in a heat flux density that is a few times higher than the strongest known material can withstand, and an expensive and difficult restructuring and redesign of the divertor structure is needed without any assurance. The XGC simulation enabled by the present endeavor predicts that the heat-load width is more than five times wider: $\lambda_q > 5$ mm, as shown in Figure 24.10, and the edge physics in ITER is different from the present tokamaks. $\lambda_q > 5$ mm could allow ITER to move ahead with the present divertor design without further delay. The physics details of this study had been chosen as an invited talk at the 26th IAEA Fusion Energy Conference, October 2016.

In preparation for the large ITER simulation, we validated our approach by carrying out several numerical experiments of two existing tokamaks: the conventional DIII-D device and the small aspect ratio National Spherical Torus Experiment (NSTX). Good agreement with the XGC1 simulation results is obtained for both cases, giving us confidence that XGC contains the right physics to tackle ITER. As an example, Figure 24.11 shows the comparison of XGC simulated divertor heat-flux widths λ_q with published experimental data for NSTX plasmas.

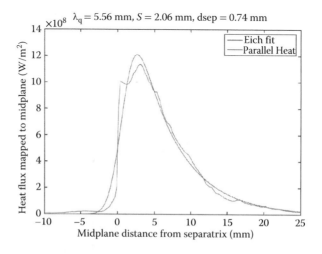

FIGURE 24.10   Heat flux footprint predicted from XGC on the outer divertor plate in a model ITER plasma edge at 15MA, mapped to the outboard midplane. λ_q is ~ 5.6 mm, which is more than five times greater than prediction $\lambda_q \simeq 1$ mm based on the data from present experiment. The input model plasma profile is obtained in collaboration with the ITER headquarter team.

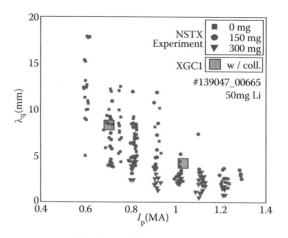

FIGURE 24.11   Comparison of the XGC simulated λ_q (large square) against National Spherical Torus Experiment experimental data (small dots).

ACKNOWLEDGMENTS

Support provided through the SciDAC program funded by U.S. DOE Office of Advanced Scientific Computing Research and Office of Fusion Energy Sciences. The work was performed at Princeton Plasma Physics Laboratory under Contract No. DE-AC02-09CH11466, at Oak Ridge National Laboratory under Contract No. DE-AC05-00OR22725, at Rensselaer Polytechnic Institute under Grant DE-SC0008449, and at Rutgers University under Grants DE-SC0007455 and DE-FG02-06ER5485. Awards of computer time were provided by the Innovative and Novel Computational Impact on Theory and Experiment (INCITE) program. This research used resources of OLCF, ALCF, and NERSC, which are U.S. DOE Office of Science User Facilities supported under contracts DE-AC05-00OR22725, DE-AC02-06CH11357, and DE-AC02-05CH11231, respectively.

REFERENCES

1. D. P. Stotler, C. S. Chang, S. H. Ku et al. Energy conservation tests of a coupled kinetic plasma-kinetic neutral transport code, *Computational Science and Discovery*, 6(1), 1–17, 2013.

2. D. P. Stotler, C. S. Chang, S. H. Ku et al. Pedestal fueling simulations with a coupled kinetic plasma-kinetic neutral transport code, *Journal of Nuclear Materials*, 438, S1275, 2013.

3. B. Wang, S. Ethier, W. Tang et al. Modern gyrokinetic particle-in-cell simulation of fusion plasma on top supercomputers, *Arxiv preprint*, 2015, arXiv:1510.05546v1 [cs.DC] [Online]. Available: http://arxiv.org/abs/1510.05546.

4. Q. Liu, J. Logan, Y. Tian et al. Hello ADIOS: The challenges and lessons of developing leadership class I/O frameworks, *Concurrency and Computation: Practice and Experience*, 26(7), 1453–1473, 2014.

5. Q. Lu and J. Amundson. Synergia CUDA: GPU-accelerated accelerator modeling package, in 20th International Conference on Computing in High Energy and Nuclear Physics (CHEP2013), *Journal of Physics: Conference Series*, 513 (052021), 2014. IOP Publishing.

6. K. Germaschewski, W. Fox, N. Ahmadi et al. The plasma simulation code: A modern particle-in-cell code with load-balancing and GPU support, *Arxiv preprint*, 2015, arXiv:1310.7866v2 [Online]. Available: http://arxiv.org/abs/1310.7866.

7. S. Ku, R. Hager, C. Chang et al. A new hybrid-Lagrangian numerical scheme for gyrokinetic simulation of tokamak edge plasma, *Journal of Computational Physics*, 315, 467–475, 2016.

8. W. Sun and Z. Ma. Count sort for GPU computing, in *Proceedings of the 2009 15th International Conference on Parallel and Distributed Systems, ser. ICPADS '09*. Washington, DC: IEEE Computer Society, 2009, pp. 919–924. [Online]. Available: http://dx.doi.org/10.1109/ICPADS.2009.

9. N. Bell and J. Hoberock. Thrust: A productivity-oriented library for CUDA, in *GPU Computing Gems Jade Edition*, 1st ed., Chap. 26, W.-M. W. Hwu, Ed., San Francisco, CA: Morgan Kaufmann Publishers, 2011, pp. 359–372.

10. R. Hager, E. Yoon, S. Ku et al. A fully non-linear multi-species Fokker–Planck–Landau collision operator for simulation of fusion plasma, *Journal of Computational Physics*, 315(15), 644–660, 2016.

11. E. S. Yoon and C. S. Chang. A Fokker–Planck–Landau collision equation solver on two-dimensional velocity grid and its application to particle-in-cell simulation, *Physics of Plasmas (1994–present)*, 21(3), 032503, 2014 [Online]. Available: http://scitation.aip.org/content/aip/journal/pop/21/3/10.1063/1.4867359.

12. M. Adams, S. Ku, E. D'Azevedo et al. Scaling to 150k cores: Recent algorithm and performance engineering developments enabling XGC1 to run at scale, *Journal of Physics: Conference Series*, 180(1), 012036, July 2009.

13. C. Docan, M. Parashar, and S. Klasky. DataSpaces: An interaction and coordination framework for coupled simulation workflows, *Cluster Computing*, 15(2), 163–181, 2012 [Online]. Available: http://dx.doi.org/10.1007/s10586-011-0162-y.

14. N. Podhorszki, J. Logan, H. Abbasi et al. Flexible I/O programming API for big data analytics, in *Proceedings of the Fifth International Workshop on Big Data Analytics: Challenges, and Opportunities (BDAC-14)*, New Orleans, LA, November 2014.

15. J. Rosinski. General purpose timing library (GPTL): A tool for characterizing performance of parallel and serial applications, *Cray User Group*, 2009, also Available: http://jmrosinski.github.io/GPTL.

Index

A

Ab initio electronic structure, 449–450
Ab initio GW method, 482
Ab initio molecular dynamics (AIMD)
    overall timings for, 172–174
    simulation, 166–167
Abstract-device interface for parallel I/O (ADIO), 291
ABySS assembler, 427
Accelerated Climate Modeling for Energy (ACME), 188
    GPU refactoring, *see* GPU refactoring, of ACME
        atmosphere
    nested OpenMP, *see* OpenMP
    ongoing codebase changes and future directions, 206
    portability considerations, 203–206
Accelerator memory, 57
"acc loop gang" directives, 204, 205
"acc loop vector" directives, 204, 205
"acc parallel loop" construct, 204, 205
Aces4
    coupled-cluster singles and doubles, 162
    domain scientists and, 155
    fault tolerance, 162–163
    performance, 159–160
    portability, 160–161
    QC methods implemented in, 152
    runtime system, 156, 161
    system development, 155
        barriers, 158–159
        graphical processing unit, 159
        load balancing, 158
        servers, 157–158
        Super Instruction Architecture, 155–157
        workers, 157
ACES III software project, 81, 151, 156
ACME, *see* Accelerated Climate Modeling for Energy
Adaptable IO System (ADIOS), 530, 533, 542, 548
Adaptable Seismic Data Format (ASDF), 282, 287, 294–296
Adaptive mesh refinement (AMR), 8, 99, 329–330
    block structured, 307
    Cartesian structured, 307–308
    embedded boundary, *see* Embedded boundary-adaptive
        mesh refinement
    partitioning and, 332–333
AddressSanitizer, 41, 42
ADIO, *see* Abstract-Device interface for parallel I/O
ADIOS, *see* Adaptable IO System
Adjoint methods, 280
Adjoint tomography, 280
Aerodynamic flow control, 320
    algorithmic details, 330, 334
        adaptive mesh control and partitioning, 332–333
        flow solver, 330–332
    computational approach, 328
        adaptive mesh control, 329–330
        parallel flow solver, 328–329
    discretization choices, 321

    programming approach, 334–335
        external libraries, 336–337
        performance, 335
        portability, 335–336
    in situ visualization and computational steering,
        337–340
    software practices, 337
    synthetic jet actuators, 321–322
    turbulence modeling, 320–321
    vertical tail/rudder assembly, *see* Vertical tail/rudder
        assembly
Aggregated stencils, 310–311
Agile development, 301
AI, *see* Arithmetic intensity
AIMD, *see* Ab initio molecular dynamics
Airline manufacturers, 323
Alanine dipeptide, 123
ALCF, *see* Argonne Leadership Computing Facility
Allinea DDT, 37–38
Allinea MAP, 399, 400
ALLPATHS-LG assembler, 427
AMR, *see* Adaptive mesh refinement
Amsterdam density functional (ADF) package, 148
AoS, *see* Array of structures
AoSoA, *see* Array of structure of arrays
Application programming interface (API), 127, 291, 355,
    378
Application-Specific Energy Template (ASET)
    approach, 4–5
    dynamic power steering, 7–8
    with global information, 2–3
    for load-imbalanced workloads, 10–11
    with local information, 2
    per-core energy, 5–6
    power-saving mechanisms, 3–4
    principles, 6
    vs. dynamic load balancing, 11–15
    wave-front algorithms, 8–10
Application state, 4
Argonne Aurora, 137
Argonne Leadership Computing Facility (ALCF), 337, 339,
    361, 514
Arithmetic intensity (AI), 401, 402
Array of structure of arrays (AoSoA), 361, 465
    transformation, 468–470
Array of structures (AoS), 381
    datatypes, 465
    to SoA transformation, 467–468
ASDF, *see* Adaptable Seismic Data Format
ASET, *see* Application-Specific Energy Template
Asynchronous message-driven execution, 126
Asynchronous sampling, 20
Asynchronous thread control, 39
Atomic event, 22–23
Atomic velocities, 121
Aurora Early Science Program, 327–328
Autobuild tests, 134
Automatic documentation, KKRnano, 442

B

Bandwidth, in MPI communication, 192–193
Barcelona Supercomputing Center (BSC), 211, 212
Basic Linear Algebra Subprograms (BLAS), 454
BCs, *see* Boundary conditions
Benedict–Webb–Rubin (BWR) equation, 235
BerkeleyGW, 482, 484
    FFT bottleneck, 491–493
    GPP self-energy, 497–501
    polarizability, 493–497
    walltime computation, 501–503
BFB, *see* Bit-for-bit
BFD, *see* Binary File Descriptor
Biased calculations, 123
BiCGSTAB, 61, 351
Big data, 68, 280, 293, 303
Binary File Descriptor (BFD), 23
Binary instrumentation, 23
Bin sorting algorithm, 397
Biogeochemistry, 188
Biomolecular systems, simulations of, 121
Bit-for-bit (BFB), 202
BLAS, *see* Basic Linear Algebra Subprograms
BLAS-1/-2/-3, 53–55, 60, 61
Blocked distribution
    pattern, 11
    of tensors, 77
Block inversion algorithm, 454
Block-sparse tensor, 78
Block structured adaptive mesh refinement, 307
Block-threaded GPU version, 113
Bloom filter, 417, 418
Blue Gene/P JUGENE, 436
Blue Gene/Q JUQUEEN, 436
    scaling on, 443, 444
BOAST-generated kernels, 283
Body integrals, 189
Bonded terms, molecular mechanics, 121
Born–Oppenheimer approximation, 152
Boundary conditions (BCs), 259
Boundary exchange, for bandwidth/latency, 197
Breaking up element loops, 194, 204
BSC, *see* Barcelona Supercomputing Center
Bspline class, 466
BsplineSoA object, 469, 470
Build system, KKRnano, 443
Bulk injection velocity, 243
Burn unit, 100
Burst buffer, in-transit processing mechanisms, 315–317

C

CAAR program, *see* Center for Accelerated Application
    Readiness program
C++ abstractions, 310
Cache blocking, 386, 388, 397
Cache misses, 386, 397
Cache reuse, 378, 386
Caches, 56–57, 59
Call stack unwinding, 21–22
CAM, *see* Community Atmosphere Model
CAM-SE, *see* Community Atmosphere Model-Spectral
    Element

CANDECOMP/PARAFAC, 75
Canonically conjugate momenta, 350
Canonical polyadic (CP) decomposition, 75
Cartesian structured adaptive mesh refinement, 307–308
CASK framework, *see* Cray Adaptive Sparse Kernel
    framework
Catch, 477
CC, *see* Coupled cluster
CCSD, *see* Coupled-cluster singles and doubles
CCSD(T) approach, 177, 178
CCSNe, *see* Core-collapse supernovae
CCT, *see* Characteristic computational time
CCV, *see* Classical covisualization
CDC, *see* Clock delta compression
Center for Accelerated Application Readiness (CAAR)
    program, 53, 99
CESM, *see* Community Earth System Model
CESM-Ensemble Consistency Test (CESM-ECT), 212–213
CESM Large Ensemble (CESM-LE) Project, 209
CEU HapMap population, 423
CFD, *see* Computational fluid dynamics
CFG, *see* Control flow graph
Characteristic computational time (CCT), 237
Chare arrays, 126
Chares, 126
Charge/current deposition, 380, 384
Charm++, 120–121, 124, 134–135
Chombo-Crunch, 308, 313–316
Chroma code, 346
    generic code through base class defaults, 358–359
    levels, 355–356
    performance libraries, 360–361
        QPhiX Library, 361–362
        QUDA library, 363–364
    polymorphism, 359–360
    QDP++, 356–358
    scripting via XML and Command Pattern, 360
CIG, *see* Computational Infrastructure for Geodynamics
CkIO, 130–131
Classical *a posteriori* visualization workflow, 337, 338
Classical covisualization (CCV), 338
Classical particle-in-cell (PIC) algorithm, 376
    data structures, 381
    distributed parallelization for, 382
    PIC loop, 379–380
CLI, *see* Command-line interface
Climate simulation, 191
Clock delta compression (CDC), 43
Cluster operator, 69
CMake-based system, 162, 476
CMIPs, *see* Coupled Model Intercomparison Projects
CMOS, *see* Complementary metal–oxide–semiconductor
Code generators, 365
Code maintainers, 301
Code refactoring, for multithreaded architectures, 455
    communication, 455–456
    graphics processing unit acceleration, 456–457
Code structure, 191
    bandwidth and latency, 192–193
    data and loops, 191–192
    OpenMP, 192
Collapsing/pushing if-statements, 204–205
Collective variables, 136–137

Collisionless tearing mode model, 510–511
Color, QDP++, 356
Columbia Physics System (CPS) code, 349
Colvars, *see* Collective variables
Command-line interface (CLI), 39
Command pattern, 360
Communication-avoidant iterative methods, 64
Community Atmosphere Model (CAM), 209, 211, 213
Community Atmosphere Model-spectral Element
    (CAM-SE), 188–192, 195, 197, 198
Community Earth System Model (CESM)
    computational platforms, 213, 214
    data analytics, 225–227
    Ensemble Consistency Test, 212–213
    exascale computations, issues with, 210
    folding analysis, 212
    high-resolution simulations, 208–210
    HOMME dynamical core, *see* High-Order Method
        Modeling Environment
    kernel extraction, 211–212
    performance analysis, 211
Compilation tests, 302
Compiler-based instrumentation, 23
Complementary metal–oxide–semiconductor (CMOS),
    3–4
Complete neglect of differential overlap (CNDO) method,
    147
Computational chemistry, 145
    computers and uses in, 145–147
    quantum chemistry programs, 147–149
Computational data, I/O optimization
    in adjoint-based seismic workflows, 292
    overview of, 289–290
    parallel file formats and I/O libraries, 290–291
Computational domain, 242–243
Computational fluid dynamics (CFD), 258
    codes, 509
    turbulence modeling and, 320–321
Computational Infrastructure for Geodynamics (CIG), 301
Computational steering, *in situ* visualization and, 337–340
Computation factorization, FMA and, 397
Compute flow, of NAMD, 125
Conditional compilation, KKRnano, 442
Configuration interaction (CI) method, 69, 146
Connection Machine, 147
Context event, 23
Contigs
    aligning reads onto, 412–413
    generation, 412
        parallel, 418–420
Continuum extrapolation, 354
Contraction/correlation function construction, 354–355
Contravariant indices, 68
Control Data Corporation (CDC), 146
Control flow graph (CFG), 30
CORAL systems, 63
Core-collapse supernovae (CCSNe), 95, 98–99, 103
Cori Phase 1 system, 313, 491
Cori system, 426, 484–486
Correlation function, 353
Counters, hardware, 19–20
Coupled cluster (CC), 165, 174
    tensors, 176
    theory, 69

Coupled-cluster singles and doubles (CCSD), 69, 70,
    162, 163
    formulation, 175, 176
Coupled Model Intercomparison Projects (CMIPs), 208
Courant–Friedrichs–Lewy (CFL) condition, 189
Covariant indices, 68
Covered box pruning, 312, 313
C/O white dwarfs (C/O WDs), 96, 97
C preprocessor (CPP), 202
CPS code, *see* Columbia Physics System code
CPU, power draw analysis, 12, 13
Cray Adaptive Sparse Kernel (CASK) framework, 66
Cray DataWarp burst buffer, 316
Cray LibSci, 53
Cray Scientific and Math Libraries (CSML), 66
Cray XK7 Titan system, 68, 160, 246, 476, 531, 543–545,
    547, 548
Cryosphere system, 188
CSML, *see* Cray Scientific and Math Libraries
CTest, 476
Cubic equations, 235
CUDA Debugger API, 18, 25
CUDA FORTRAN, 192, 193
CUDA-MEMCHECK, 26
CUDA MPS server, 192, 197
Current deposition, 388, 393–396
cuTT library, 79
CyberShake developers, 298
Cyclic distribution of tensors, 77

D

DAG, *see* Directed acyclic graph
Darshan, 35–36
Data-centric sampling (DCS), 21
Data, code structure, 191–192
Data layout flexibility, 365
DataSpaces, ADIOS applications, 542
Data structures, for particles and fields, 381
DCS, *see* Data-centric sampling
De Bruijn graphs, 410, 412, 415, 419–420
Debugging tools, 18, 36–37
    Allinea DDT, 37–38
    message-passing interface, 43–44
    Stack Trace Analysis Tool, 42–43
    TotalView, 39–41
    Valgrind, 41–42
DEC, *see* Digital Equipment Corporation
Decoupling, 299
Degenerate electrons, 95
Delayed rank-k update, 471–472
δf scheme, 532, *see also* Hybrid-Lagrangian δf scheme
Dennard scaling, 210
De novo genome assembly, 409–410, 427
Dense linear algebra operations, 53–54
Density functional theory (DFT), 146, 432, 449
    equations, 167
    Kohn–Sham formalism, 481–482
    vs. ab initio GW, 483
Density matrix renormalization group (DMRG) procedure,
    70
deriv, 195–196
Detached eddy simulation (DES), 321, 326
devel branch, 301

`developer` repositories, 240–241
Device-to-host data transfers (DtoH), 115
DFT, *see* Density functional theory
Diagonal technique, 382
Diffusion Monte Carlo (DMC) methods, 462, 464
Digital Equipment Corporation (DEC), 146
DIRAC, 80, 149
Direct approach, tensor contraction, 78
Directed acyclic graph (DAG), 226, 299
Direct numerical simulation (DNS), 232, 258–259, 320
DiscovarDenovo assembler, 427
Discretization
    choices, 321
    errors, 350
Distributed hash table, in PGAS model
    in HipMer pipeline, 415–417
    implementation of, 414–415
Distributed-memory message-passing, 236
Distributed memory model, 236
Distributed parallelization, of PIC algorithm, 382
Distributed parallel linear algebra software for multicore
        architectures (DPLASMA), 65
DMC methods, *see* Diffusion Monte Carlo methods
DMRG procedure, *see* Density matrix renormalization
        group procedure
DNS, *see* Direct numerical simulation
Domain decomposition, in molecular dynamics
        module, 180
Domain-specfic virtual processor (DSVP), 82–86
Domain-specific instruction set (DSIS), 82
Domain-specific language (DSL), 86, 259, 264, 283
Domain specific library (DSL), 166
    force and energy evaluation, 181
Domain-specific virtual I/O service (DSVIOS), 89
Domainsplit, in EBIndexSpace generation, 308–309
Double-degenerate scenario, 97
Double-detonation scenario, 97
DPLASMA, *see* Distributed parallel linear algebra
        software for multicore architectures
DPS, *see* Dynamic power steering
DRAM, *see* Dynamic random-access memory
DSL, *see* Domain-specific language; Domain specific
        library
DSL SIAL, 155
DSOs, *see* Dynamic shared objects
DSVIOS, *see* Domain-specific virtual I/O service
DSVP, *see* Domain-specfic virtual processor
DtoH transfers, *see* Device-to-host data transfers
DVFS, *see* Dynamic voltage and frequency scaling
Dynamical core, 198
Dynamic load balancing, 7
    Charm++, 127
    migration of task, 11
    per-core execution time for, 12, 14
    XGC, 540–542, 547
Dynamic power steering (DPS), 3
    ASET-enabled, 12–15
    concept of, 7–8
Dynamic random-access memory (DRAM), 189, 190
Dynamic shared objects (DSOs), 23, 24
Dynamics simulations, 120
Dynamic voltage and frequency scaling (DVFS), 3–4,
        12, 526
Dyson equation, 434

E

Early Science Program (ESP), 322, 327–328, 514
Earth-system models (ESMs), 207–208, *see also*
        Community Earth System Model
EB-AMR, *see* Embedded boundary-adaptive mesh
        refinement
EBIndexSpace generation, 308–309
`EBISLayout` operation, 309
EBS, *see* Event-based sampling
EBStencil object, 311
ECT, *see* Ensemble Consistency Test
Edison system, 313–315, 422, 426
Electromagnetic fields, 381, 387
Electron momentum equation, 510
Electron push kernel
    algorithm, 534, 535
    GPU optimization of, 537, 544–545
    particle sorting, 536–537
    portability, 537–538
    triangle searching, 535–536
    weak/strong scaling, 543–545
Electron self-energy, computation of, 490
    with full energy dependence, 491
Electronic polarizability, 489–490
Electronic structure theory, 73
Electrostatic interactions, 12
Element-based partitioning, 331
Element boundary averaging, 190
Element loops, breaking up, 194, 204
`element_t`, 192
Embedded boundary-adaptive mesh refinement (EB-AMR),
        307–308
    aggregated stencils, 310–311
    burst buffer and in-transit workflow, 315–317
    Chombo-Crunch, 308, 313–316
    covered box pruning, 312
    EBIndexSpace generation, 308–309
    Edison and Cori Phase 1 system, 313
    feedback-based load-balancing scheme, 310
    hybrid MPI+OpenMP model, 312–313
    local geometry caching effects, 309
    sparse file format, 311–312
EMSL, *see* Environmental Molecular Sciences
        Laboratory
Engineering and Scientific Subroutine Library (ESSL), 66
Ensemble Consistency Test (ECT), 212–213
Entry methods, 126
Environmental Molecular Sciences Laboratory
        (EMSL), 166
Equation of state (EoS), 98, 108–109, 114–115
ESMs, *see* Earth-system models
ESP, *see* Early Science Program
ESSL, *see* Engineering and Scientific Subroutine Library
Euler equations, 101
Event-based sampling (EBS), 20, 21
Event-based tracing, 24, 25
Exascale programming models, 259
Exascale supercomputers, future needs for, 379
ExaTENSOR project, 71, 78, 80, 83
Exchange-correlation potential, 482
Exchange, data, 192
Exotic particles, 346
Extensive knowledge, 241

External libraries, 441
    Fastest Fourier Transform in the West, 131–132
    Tcl interpreter, 132

F

Factorization operation, 54
Fastest Fourier Transform in the West (FFTW), 131–132,
    392–393
Fast Fourier transform (FFT), 122, 384, 392–393
    algorithm, 168–170
    bottleneck, 491–493
    3D, 173, 174
Fault tolerance, 90, 139–140
FDTD algorithm, *see* Finite-difference time-domain
    algorithm
Feedback-based load-balancing scheme, 310
Fermion doubling problem, 348
Fermion sign problem, 463
FFM matrix product, 170, 171
FFT, *see* Fast Fourier transform
FFTPACK library, 170
FFTW, *see* Fastest Fourier Transform in the West
Field gathering, 380, 385, 388, 396–397
FIFO, *see* First-in-first-out
Fifth-Generation computing systems, 124
Fine-grain computations, 138
Fine-grained parallelism, 129
Finite difference schemes, 383–384
Finite-difference time-domain (FDTD) algorithm, 382
First-in-first-out (FIFO), 129
FLASH algorithm
    equation of state, 108–109
    multiphysics implementation, 102–103
    nuclear burning module, 104–108
    OpenMP threading, 100, 111–112
    partial differential equation solvers, 99
    physics modules, 100–102
Flattening arrays, for reusable subroutines, 194–195
FLEXWIN program, 296
Flight Reynolds number flow control
    Aurora ESP, 327–328
    estimating resources for, 326
    five-fold rise in, 326–327
Floating point operations per second (FLOPS), 272
Fluid-kinetic hybrid electronmodel, 510
Flux coordinate system, 532
FMA, *see* Fused multiplyadd
FMF matrix product, 170, 171
Fokker–Planck–Landau (FPL) collision operator, 538–541,
    546, 547
Folding analysis, 212
Force-gradient integrators, 352
Fortran, 146
    CUDA FORTRAN, 192, 193
    FORTRAN II, 147
    high performance, 355
    for KKRnano, 437
    and MPI implementation, 261
Fortran 90, 437, 438, 442
Fortran 95, 437
FPL collision operator, *see* Fokker–Planck–Landau collision
    operator
Fuel jet, bulk injection velocity of, 243

Functional integrals, 347
Functional tests, 303
Function filtering, 22
Fused multiplyadd (FMA), 56
Fusion energy, 507

G

Galerkin's method, 331
GA/MPI programming model, 166
`gang`, 196
GAs, *see* Global arrays
Gas parallel toolkit, 181–183
Gather instructions, 386
Gather–scatter instructions, 396
Gauge configuration, 348
Gauge generation, 349–352, 354
Gaussian-type orbitals (GTOs), 146
Gauss–Legendre–Lobatto (GLL) points, 189
GEMM, *see* General matrix-matrix multiplication
Generalized gradient approximation (GGA), 449, 482
Generalized minimal residual (GMRES) method, 55, 331
General matrix-matrix multiplication (GEMM), 54, 58, 59
General-purpose graphical processing units (GPGPUs),
    25–26
General Purpose Timing Library (GPTL), 201, 211
General tensor–tensor contraction (GETT) kernel, 79
Generic code, through base class, 358–359
Genomes, 409–410
Geodesic integration methods, 122
GGA, *see* Generalized gradient approximation
Git repositories, 302
`glob_1`, 195–196
Global adjoint tomography
    Adaptable Seismic Data Format, 294–296
    preprocessing stage, 282
    scientific methodology of, 280
    Seismic Analysis Code, 293
    SPECFEM3D_GLOBE, *see* SPECFEM3D_GLOBE
    workflow, 281
    workflow management strategies, 297–300
Global arrays (GAs), 166
    operations, 182
    toolkit, 153, 181
Gluons, 345
GMRES method, *see* Generalized minimal residual method
Godunov method, 101
Golden Section Search algorithm, 541
Google Test, 477
GPGPUs, *see* General-purpose graphical processing units
GPP self-energy, 497–501
GPTL, *see* General Purpose Timing Library
GPU, *see* Graphics processing unit
GPU Manager, 129–130
GPU refactoring, of ACME atmosphere, 188–189
    code structure
        bandwidth and latency in MPI communication,
        192–193
        data and loops, 191–192
        OpenMP, 192
        CUDA FORTRAN, 193
    mathematical considerations and computational impacts
        element boundary averaging, 190
        mathematical formulation, 189

OPENACC refactoring
  boundary exchange for bandwidth, 197
  breaking up element loops, 194
  flattening arrays for reusable subroutines, 194–195
  loop collapsing and reducing repeated array accesses, 195
  thread master regions, 193–194
  using shared memory and local memory, 195–196
optimizing for pack, exchange, and unpack, 197–198
runtime characterization, throughput and scaling, 191
testing for correctness, 198–199
Grad–Shafranov equation, 511
Graphical user interface (GUI), 39, 40
Graphics processing unit (GPU), 18, 104, 159, 189, 261
  acceleration, 447, 454, 456–457, 534
  kernels, roofline analysis of, 273
  KKRnano, 441
  optimization
    electron push kernel, 537, 544–545
    on Titan, 105–108, 113–114
  performance tools and interfaces, 25–26
  per node, 262
  refactoring, *see* GPU refactoring, of ACME atmosphere
  simulations, 283–284
  streaming multiprocessor in, 21
  usage by Titan, 272
  in XGC project, 533
Green function, 432, 433, 452
Grid, 189
GTC, *see* Gyrokinetic toroidal code
GTC-P code, 508, 511, 512, 517–518, 523, 524
GTOs, *see* Gaussian-type orbitals
GUI, *see* Graphical user interface
GW, *see also* BerkeleyGW
  ab initio method, 482
  computational bottlenecks in, 488–489
    electronic polarizability, 489–490
    electron self-energy, 490–491
    orbital transition probabilities, 489
  Cori system, 484–486
  roofline model, 486–487
  silicon vacancy defect system, 487, 488
GW-Bethe–Salpeter-Equation (GW-BSE) approach, 482
Gyrokinetic toroidal code (GTC), 508
  algorithmic details, 513–515
  benchmarking results, 520–525
  collisionless tearing mode, 510–511
  fluid-kinetic hybrid electron model, 510
  global PIC grid considerations, 513
  PIC geometric models, 511–512
  Poisson solver, 520
  programming approach, 515
    external libraries, 519
    performance, 516–517
    portability, 517–519
    scalability, 515–516
  software practices, 519–520
  time-to-solution and energy-to-solution, 525–526

H

Hadoop ecosystem, 298
Hardware performance counters, 19–20
Hardware reconfiguration, 141
Hartree–Fock equations, 146, 147
Hartree–Fock–Slater (HFS), 146
Hartree potential, 449, 482
Haswell, KNL partition and, 485
HBM, *see* High bandwidth memory
HDF5, 31, 100, 240, 291, 294, 311, 487
Hellman–Feynman theorem, 152
Helmholtz equation of state, 101
Hessenberg matrix, 60
Heterogeneous compute node architecture, 247
Heterogeneous debugging, 39–40
Heterogeneous load balancing, 129–130
Heterogeneous system architecture (HSA) project, 80
Heterogenous environment, 298
HFS, *see* Hartree–Fock–Slater
Hierarchical domain-specfic virtual processor (H-DSVP), 85
Hierarchical virtualization, 82
High bandwidth memory (HBM), 138, 378
High-energy physics, 346
High-level quantum chemistry methods
  benchmarks, 178–179
  Intel Xeon Phi KNC coprocessor, 177–178
  tensor contraction engine, 174–176
High-Order Method Modeling Environment (HOMME), 199–201, 213
  algorithm details, 214–216
  CAM-like configuration, 218–220, 223–224
  computational phases, 215
  KNL-based system, 224–225
  parallelization improvements, 216–218
  single-core optimization, 218
  WACCM-like configuration, 219–224
High performance computing (HPC), 1, 52, 258
  dense linear algebra operations, 53–54
  energy problem, 377
  hierarchical virtualization of, 82
  platforms, 335
  processors of, 56
  programming for, 259
  scale abstraction, 84–85
  software and hardware levels impacts, 378
  sparse linear algebra operations, 55
  system decomposition tree, 85–88
High Performance Fortran (HPF), 355
High-resolution simulations, CESM, 208–210
HipMer pipeline, 410
  challenges for future architectures, 425–427
  communication operations in, 422
  distributed hash tables in, 415
    global update-only phase, 415–416
    local reads and writes phase, 416
    for various use-case scenarios, 417
  parallel algorithms in, 417
    communication patterns and costs, 422
    contig generation, 418–420
    parallel *K*-mer analysis, 417–418
    parallel read-to-contig alignment, 420–421
    scaffolding and gap closing, 421–422
  performance results, 422–423
    comparison with other assemblers, 425
    I/O caching, 424–425
    strong scaling experiments, 423–424
HOMME, *see* High-Order Method Modeling Environment

Host-to-device data transfers (HtoD), 115
HPC, *see* High performance computing
HPCToolkit, 30–31
HSA project, *see* Heterogeneous system architecture
      project
HtoD, *see* Host-to-device data transfers
Human genome, end-to-end strong scaling for, 423
Hybrid decomposition of patches, 125
Hybrid-Lagrangian δf scheme, 534
Hybrid Monte Carlo (HMC), 350
Hybrid MPI+OpenMP model, 312–313
Hybrid parallelism, 239
    shared memory, 246–251
Hybrid weak scaling study, 520, 521
Hydrodynamics solver, 104, 116
Hyperloglog algorithm, 418
Hyperthreading, 56
Hyperviscosity operator, 190
HYPRE software, 66
HYPRE solver, 520, 522

I

IBS, *see* Instruction-based sampling
icell, 395–396
If-statements, 396–397
    collapsing and pushing, 204–205
ILP, *see* Instruction level parallelism
ILU, *see* Incomplete LU factorization
Improved intra/internode communication, 128
INCITE, *see* Innovative and Novel Computational Impact
      on Theory and Experiment
Incomplete LU factorization (ILU), 55, 62–63
Index variance, 68
Indirect approach, tensor contraction, 78
Innovative and Novel Computational Impact on Theory and
      Experiment (INCITE), 322, 326–327
In situ data extracts (ISDE), 338–339
Instruction-based sampling (IBS), 20–21
Instruction level parallelism (ILP), 487
Instrumentation, 22–23
Intel Math Kernel Library, 66
Intel MKLlibrary, 170
Intel® Language Extensions for Offloading, 177
Intel Parallel Computing Center, 465
Intel Phi processors, 60, 484, 485
Intel Xeon Phi hardware, 166
Intel Xeon Phi KNC coprocessor, implementation for,
      177–178
Intel Xeons microprocessor, 284
Interconnect hardware, 57–58
International Business Machines (IBM), 146
    Power microprocessor, 21, 284
International Thermonuclear Experimental Reactor (ITER),
      507, 508, 529, 531, 548–549
Interoperable Technologies for Advanced Petascale
      Simulations (ITAPS), 328
Interval event, 23
In-transit processing mechanisms, 315–317
I/O system
    caching, 424–425
    KKRnano, 440–441
    performance using SIONlib, 444–445
    systems, 251–252

Isoefficiency analysis, 123–124
Isometric compression, 76
Isosurface of instantaneous value of Q, 325
ITAPS, *see* Interoperable Technologies for Advanced
      Petascale Simulations
ITER, *see* International Thermonuclear Experimental
      Reactor
Ivy-Bridge, 491

J

Jacobi preconditioning, 61
Jenkins, 134
JET, *see* Joint European Torus
Jet-in-cross-flow (JICF), 241–242
JIT, *see* Just-in-time
Joint European Torus (JET), 509, 531
Just-in-time (JIT), 365

K

Kernel extraction, 211–212
Kernel Generator (KGEN), 211–212
kernels, 195
Kernels *vs.* parallel loop, 205–206
KGEN, *see* Kernel Generator
Kinetic electron models, 510
Kinetic energy, 95
Kitware Inc., 337, 338
KKR method, *see* Korringa–Kohn–Rostoker method
KKRnano, 432, 447
    algorithmic details, 435–437
    energy considerations, 445–446
    performance results, 443–445
    programming language for, 437–439
        I/O, 440–441
        portability, 441
        scalability, 439–440
    scientific methodology, 432–434
        linear-scaling (N)mode, 434
        sparse Dyson equation, 434
    software practices, 442–443
k-mer analysis, 411–412
    parallel, 417–418
KNC Intel Xeon Phi implementation, 178–179
Knights Landing (KNL), 166, 465
    FFTs, absolute performance of, 493
    and Haswell partitions, 485
    node, 383, 386, 388, 397
    overall timings for, 172–174
    systems, 135–136
Kohn–Sham method
    DFT scheme, 166, 435, 481–482
    equation, 449
    orbitals, 450, 451
    wavefunctions, 432
Kokkos package, 65
Kokkos programming model, 249, 250
Korringa–Kohn–Rostoker (KKR) method, 432, 451
Kronecker delta, 76
Krylov solver methods, 55, 61, 66
KVST, 438
K20x GPU (Kepler) *vs.* 16 CPU-cores (AMD Interlagos),
      284

L

Lagrange multiplier algorithms, nonlocal pseudopotential and, 170–172
Lanczos method, 55
LAPACK, *see* Linear algebra package
Large aspect ratio equilibrium, 512
Large eddy simulation (LES), 258, 320–321, 326
    approach, 232
    external libraries, 240
    performance, 239–240
    RAPTOR, *see* RAPTOR
    scalability, 236–239
    software practices, 240–241
Large-scale supercomputers, performances on, 403
Last level cache (LLC), 141
Latency
    boundary exchange for, 197
    in MPI communication, 192–193
Lattice gauge fields, 347
Lattice quantum chromodynamics (LQCD), 346, 347–349
    continuum extrapolation, 354
    gauge generation, 349–352
    methodology and computational challenges, 354–355
    observable calculation, 352–353
    Schur even–odd preconditioning, 353–354
L-BFGS, *see* Limited-memory Broyden-Fletcher-Goldfarb-Shanno
LDA, *see* Local density approximation
LeanMD, 139
Leap-frog integrator, 122
Legion, 263
    programming model, 262
LEM, *see* Linear eddy model
Lennard–Jones parameters, 122
LES, *see* Large eddy simulation
Library interposition, 23–24
LibSci library, 66
Limited-memory Broyden-Fletcher-Goldfarb-Shanno (L-BFGS), 282
Limiting oscillations, 190
Linear algebra operations
    applications, 52–53
    changing memory hierarchies, 64
    dense, 53–54, 58–60
    growing resilience concerns, 64–65
    higher thread count, 63–64
    sparse, *see* Sparse linear algebra operations
    third-party libraries, 65–66
    vendor libraries, 66
Linear algebra package (LAPACK), 454
    routine, 441
    standards, 54, 65
Linear combination of atomic orbitals-molecular orbital (LCAO-MO) method, 146, 147
Linear eddy model (LEM), 239
Linearize loops, with constant indexes, 397
Linear scaling
    multiple scattering theory and, 450–453
    N mode, 434
Link fields, 347
Link variables, 347
Liquid fuel jet (n-decane), 242, 243
LIZ, *see* Local interaction zone

LLC, *see* Last level cache
Load balancing, 388, *see also* Dynamic load balancing
    feedback-based, 310
    heterogeneous, 129–130
    XGC, 540–542, 547
Load-imbalanced workloads, 10–11
Load instruction, 381
Load/stores, 396
Local density approximation (LDA), 449, 482
Local geometry caching, 309
Local interaction zone (LIZ), 452, 456
Locally selfconsistent multiple-scattering (LSMS) method, 452
    communication, 455–456
    CPU version of, 454–455
    graphics processing unit acceleration, 456–457
    multiple-scattering formalism in, 459
    performance of, 453–455
    scaling of, 457–459
    weak scaling of, 453
Local memory, 195–196
Logical node, 128
Long recurrence methods, 61
Loop collapsing repeated array accesses, 195
Looping, pushing down callstack, 205
Loop reducing repeated array accesses, 195
Lowering/raising tensor indices, 67, 68
Lower/upper triangular matrix factorization (LU), 60
LQCD, *see* Lattice quantum chromodynamics
LS-DALTON, 70
LSMS method, *see* Locally selfconsistent multiple-scattering method
L3 cache, on Haswell processors, 399

M

Machine class, 475
Mach number, 320
Macroparticles, 381, 388–389
Magnetically confined fusion energy (MFE), 507
Magnetic confinement, 531
Makefile, 443
Manual loop fissioning the callstack, 205
Many-body perturbation theory, 483
Mapping algorithms, 64
Markov state models, 123
master branch, 301
master repositories, 240–241
Mathematical formulation, 189
Matrix algebra for GPU and multicore architectures (MAGMA), 65
Matrix elements, 489–490
Matrix–matrix multiplication, 54, 78
Matrix product state (MPS), 70
Maximum stable CFL (MSCFL), 190
Maxwell solver, 380
MCDRAM, 486, 487, 491, 501, 503
    shared memory, 399
MCPT, *see* Molecular-cluster perturbation theory
MD, *see* Molecular dynamics
Memcheck, in Valgrind, 41–42
Memory bandwidth, 386, 401, 402
Memory hierarchy, 56

Memory locality, 386–387
Memory races, 388, 389
Meraculous assembler, 410
Meraculous assembly pipeline, 410
    aligning reads onto contigs, 412–413
    contig generation, 412
    k-mer analysis, 411–412
    scaffolding and gap closing, 413
Message passing interface (MPI), 6, 26–27, 100,
        111–112, 125
    applications, 259
    communication, 191, 355, 382
    GA/MPI model, 166
    implementation
        Fortran and, 261
        hybrid OpenACC, 261–262
    KKRnano, 439
    MPI+OpenMP model, 487–488
    MPI+X model, 38, 44, 238
    SIA in C++, 155
    and thread debugging, 43–44
    vs. OpenMP
        benchmarking, 199, 201, 203
        efficiency, 203
Metaprogramming, KKRnano, 442
Methods and techniques in computational chemistry
        (METECC), 148
Metric tensor, 67, 68
MFE, see Magnetically confined fusion energy
MG2, see Morrison Gettelman Microphysics version 2
Migratability, 126–127
Milestoning, 123
Minimal coupling prescription, 347
Mixing data types, 397
MM, see Molecular mechanics
MMM matrix product, 170, 171–172
Model development, 231
Model for Prediction Across Scales-Ocean (MPAS-O),
        188
Model validation, 231
Modern Techniques in Computational Chemistry
        (MOTECC), 148
Molecular-cluster perturbation theory (MCPT), 152, 162,
        163
Molecular dynamics (MD), 120, 179, 350
    hierarchical ensemble methods, 181
    synchronization and global reductions, 180
Molecular mechanics (MM), 120
Molecular simulations, 12
Monotone limiter, 190
Monte Carlo integral, 351
Monte Carlo methods, 349
Monte Carlo sampling, 463
Moore's law era, 52
Morrison Gettelman Microphysics version 2 (MG2), 212
Morton number, 111
MPAS-Ocean model, 206
MPI, see Message passing interface
MPP system, MPI+OPENMP scaling on, 487–488
MRNet, see Multicast/Reduction Network
MST, see Multiple scattering theory
Multiblock domain decomposition, 244
Multicast/Reduction Network (MRNet), 40

Multichannel out-of-order instruction queue (MC-OOO-IQ),
        83
Multicopy algorithms, 123
Multicore compute nodes, 128
Multidimensional models, 102
Multigrid methods, 55, 63
Multiple independent simulations, 123
Multiple instruction, multiple data (MIMD) machine, 147
Multiple scattering theory (MST), 450
    linear scaling and, 450–453
Multipole/multigrid algorithm, 102
Multispecies collision kernel, 538–540

N

NAMD, see Nanoscale Molecular Dynamics
NamedObject, 360
Nanoscale Molecular Dynamics (NAMD), 120, 132–133
    algorithmic details, 123–126
    benchmarking results, 135–139
    energy, power, and variation, 140–141
    fault tolerance, 139–140
    programming approach, see Programming approach,
        NAMD
    scientific methodology, 120–123
    software practices, 132–135
National Center for Atmospheric Research (NCAR), 208
    Yellowstone system, 209, 213, 219, 222
National Energy Research Scientific Computing Center
        (NERSC), 519
    Carl test system, 491
    Cori system, 484–486
    Edison and Cori Phase 1 system, 313
Native mode, 177
Navier–Stokes equations, 233, 320, 330, 331
NCAR, see National Center for Atmospheric Research
nelemd, 191
NERSC, see National Energy Research Scientific
        Computing Center
Nested loops, 191–192
NetCDF, 291
Network interface cards (NICs), 19
Network switches, 19
Nexus, 473–475
NICs, see Network interface cards
NINJA tool, 44
nlev, 191
Nonbonded computes, 125
Nonbonded electrostatic interactions, 122
Nonbonded terms, molecular mechanics, 121–122
Noncommercial quantum chemistry, 149
Nonlinear collision algorithm, 539
Nonlinear matrix eigenvalue equations, 146
Nonlocal pseudopotential, and Lagrange multiplier
        algorithms, 170–172
Nonuniform memory access (NUMA), 57, 138, 378, 388
Nonuniform memory architecture (NUMA), 486
Nonvolatile memory (NVM)-based storage devices, 226
Nonvolatile RAM (NVRAM), 57, 64, 89, 316, 317
NorthWest chemistry (NWChem) modeling software, 165,
        183
    parallel infrastructure of, 166
No-slip boundary conditions, 243

np×np, 191
ntimelevels, 191
Nuclear burning module, 104–108
  benchmarking, 109–111
Nucleosynthesis, postprocessing analysis of, 102
NUMA, *see* Nonuniform memory access; Nonuniform
              memory architecture
NVIDIA
  GPU-accelerated libraries, 66
  proprietary tools, 25
NVLINK hardware, 57, 59
NVRAM, *see* Nonvolatile RAM
NWChem AIMD program, 170

O

Oak Ridge Leadership Computing Facility (OLCF), 237,
              280
  Jaguar system, 52, 53
  Spider file system, 292
Oak Ridge National Laboratory (ORNL), 68
Oak Ridge Summit, 136, 137
Object Server Target (OST), 292
ODE, *see* Ordinary differential equation
Offload model, 177
OFI, *see* Open Fabrics Interfaces
OLCF, *see* Oak Ridge Leadership Computing Facility
OMPCOLLAPSE parameter, 177
OMPT, 27
One-dimensional (1-D) sweeps, 189
One-sided communication, 414
OpenACC, 81, 537–538
  directives, 239, 248
  implementation, 115
    of S3D, 262
  refactoring, 193
    breaking up element loops, 194
    flattening arrays for reusable subroutines,
              194–195
    loop collapsing and reducing repeated array
              accesses, 195
    boundary exchange for bandwidth, 197
    shared and local memory, 195–196
    thread master regions, 193–194
  vector loop, 205
OpenBLAS, 65
Open Fabrics Interfaces (OFI), 131
OpenMP, 81, 177, 192, 199, 470
  algorithmic structure, 199–201
  benchmarking results, 203
  code structure, 192
  constructs, 166
  directives, 239
  implementation, 387–390
  OpenMP4.5, 115
  performance tool interface for, 27
  programming approach, 201–202
  software practices, 202
  threading, 100, 105–108, 111–112
Optimal isometric tensor, 76
Optimization technique, 58
  communication, 129
  pack, exchange, and unpack, 197–198

per-core energy, 5–6
  related to wide nodes, 138–139
Optimized SIMD charge deposition algorithm, 393–396
Orbital transition probabilities, 489
Ordinary differential equation (ODE), 190
  solvers, 99, 104
ORNL, *see* Oak Ridge National Laboratory
Out-of-core algorithms, 57, 88
Overdecomposition, 126

P

Packing and unpacking (PUP) framework, 126
Packing, data, 192
PAPI, *see* Performance Application Programming Interface
Parallel BLAS (PBLAS), 65
Parallel debuggers, 37
Parallel efficiency, 124
Parallel Engineering and Scientific Subroutine Library
              (PESSL), 66
Parallel flow solver, 328–329
Parallel hierarchic adaptive stabilized transient analysis
              (PHASTA), 324, 326, 328–329, 331
  forms of, 332
  scaling of, 333
Parallel I/O, 130–131
  libraries, 291
Parallelism, algorithm optimizations and
  efficient intranode OpenMP implementation, 387–390
  field gathering optimization, 396–397
  highly scalable pseudo-spectral Maxwell solvers,
              390–393
  improving memory locality, 386–387
  particle sorting, 397–398
  SIMD charge and current deposition algoritm, 398–399
Parallel linear algebra software for multicore architectures
              (PLASMA), 65
parallel loop, 195
Parallel optimization, HOMME dynamical core, 216–218
Parallel read-to-contig alignment, 420–421
Parallel scaffolding, and gap closing, 421–422
Parallel Thread eXecution (PTX) code, 365
Parallel unstructured mesh infrastructure (PUMI), 330
Parallel Virtual Machine (PVM), 125
Parallel workloads, power allocation scheme, 6–8
Parameterized models, 97
ParaProf tool, 31, 32
ParaView Coprocessing Library, 338
pardo statement, 158
ParMA partitioning tool, 330
Partial differential equation (PDE), 189
  solvers, 99
Particle exchange, 388
Particle-in-cell (PIC) code, 7–8, 376, 378
  classical PIC algorithm, 379
    data structures, 381
    distributed parallelization for, 382
    PIC loop, 379–380
  different steps
    particle and field routines performance analysis,
              383–386
    WARP and PXR, 383
  future needs for exascale supercomputers, 379

HPC landscape
  energy problem, 377
  impacts at software and hardware levels, 378
  large-scale supercomputers, 403
  limits of current low order FDTD solver, 390–391
  parallelism and algorithm optimizations
    efficient intranode OpenMP implementation,
      387–390
    field gathering optimization, 396–397
    improving memory locality, 386–387
    particle sorting, 397–398
    pseudo-spectral Maxwell solvers, 390–393
    SIMD charge and current deposition algoritm,
      393–396
  performance profiling tools, 399–401
  PIC methods, 378
  roofline performance model, 401–402
  single node performances, 398–399
  XGC, 530, 532
Particle-in-cell (PIC) loop, 379–380
Particle-mesh Ewald (PME) method, 122
Particle pusher, 380
Particle sorting, 388, 397–398
  electron push kernel, 536–537
Particle weak scaling study, 520, 521
Partitioned global address space (PGAS), 131
  distributed hash tables in, 414–417
  software, 166
  in Unified Parallel C, 413–414
Partition function, 347
Partition graph, 332
PASHA assembler, 427
Path integrals, 347
PBLAS, *see* Parallel BLAS
PCI, *see* Peripheral component interconnect
PDE, *see* Partial differential equation
PE, *see* Processing element
PEBS, *see* Precise event-based sampling
Peng–Robinson (PR) equations, 235
Per-core energy, optimization of, 5–6
Performance Application Programming Interface (PAPI), 29
Performance-portable implementation, 80
Performance tools, 18, 28
  Darshan, 35–36
  HPCToolkit, 30–31
  Performance Application Programming Interface, 29
  Score-P, 32–33
  TAU, 31–32
  Vampir, 33–35
Peripheral component interconnect (PCI), 191
Persistent communication, 129
PESSL, *see* Parallel Engineering and Scientific Subroutine
  Library
PETE, *see* Portable Expression Template Engine
PETSc, *see* Portable, extensible toolkit for scientific
  computation
PGAS, *see* Partitioned global address space
PGI CUDA Fortran, 535
PHASTA, *see* Parallel hierarchic adaptive stabilized
  transient analysis
`PhysicalSystem` class, 474
Physics, 189
Phys. Rev. Letter (PRL), 509

PIC code, *see* Particle-in-cell code
PICSciE, *see* Princeton University's Institute for
  Computational Science and Engineering
Pilot approach, 298
Planetary dynamics, 120
Plane-wave DFT methods, 166–168
Plaquette variable, 348
PME method, *see* Particle-mesh Ewald method
PMI, *see* Process management interface
PMI-Exascale (PMIx), 28
PMIx, *see* PMI-Exascale
Poisson equation, 102, 449
Polarizability, computation of, 489–490, 493–497
Polymorphism, 359–360
Portability considerations
  in ACME, 203–204
    breaking up element loops, 204
    collapsing and pushing if-statements, 204–205
    kernels *vs.* parallel loop, 205–206
    manual loop fissioning and pushing looping down,
      205
  in NAMD, 131
Portable Expression Template Engine (PETE), 357
Portable, extensible toolkit for scientific computation
  (PETSc), 66, 331
Portable programming model, 52
POSIX I/O interface, 290
Power-aware job scheduling, 140
Power gating, 3, 4
POWER8 nodes, 445
  graphics processing unit-accelerated, 443–444
  quasi-minimal residual solver running on, 446
Power-saving mechanisms, 3–4
Power scaling, 3–4
Power shifting, 3, 4
Precise event-based sampling (PEBS), 20
PREM, on Titan, 285–286, 288
Pressure–volume–temperature (PVT), 235
PRF, *see* Primary reference fuel
Primary reference fuel (PRF), 258, 259, 262
Princeton University's Institute for Computational Science
  and Engineering (PICSciE), 519
Principle of persistence, 127
Prioritized messages, 129
PRL, *see* Phys. Rev. Letter
Processing element (PE), 127
Process management interface (PMI), 28
Program Database Toolkit (PDT), 24
Program instrumentation, 22–23
Programming approach, NAMD
  external libraries
    fastest Fourier transform in the west, 131–132
    Python interpreter, 132
  migratability, 126–127
  overdecomposition, 126
  performance and scalability, 127
    GPU manager and heterogeneous load balancing,
      129–130
    optimizing communication, 129
    parallel I/O, 130–131
    SMP optimizations, 128–129
    topology aware mapping, 127–128
  portability, 131

Programming model, 57, 248, 249
`ProjectManager` class, 475
Propagator computation, 354
Property evaluation scheme, 235
PRs, *see* Pull requests
Pruning effect, covered box, 312, 313
PSATD, *see* Pseudo-spectral analytical time-domain
Pseudofermion fields, 351
Pseudo-spectral analytical time-domain (PSATD), 384
Pseudo-spectral Maxwell solvers
    current low order FDTD solver limits, 390–391
    high order FDTD solver limits, 391–392
    limits to scaling, 392
    methods, 392–393
p-state, 3, 5–6
PTX code, *see* Parallel Thread eXecution code
Pull requests (PRs), 202
PUMI, *see* Parallel unstructured mesh infrastructure
PUP framework, *see* Packing and unpacking framework
PVM, *see* Parallel Virtual Machine
PXR code, 383
PYAVERAGER, 225–226
PYCONFORM, 226
PYRESHAPER, 225–226
Python, 211, 225
    interpreter, 132
    script, 442

Q

QCD, *see* Quantum chromodynamics
QCD Linear Algebra (QLA), 355
QCD message passing (QMP), 355
QDP++, 355–358
QDP-JIT/LLVM, 366–369
QDP-JIT/PTX, 364–366
QLA, *see* QCD Linear Algebra
QM calculations, *see* Quantum mechanical calculations
QMCPACK, 461
    algorithms, 463–465
    AoSoA transformation, 468–470
    AoS-to-SoA transformation of outputs, 467–468
    B-spline evaluations, 465–466
    computational time, 465
    development practice and strategies, 476
        unit testing, 476–477
    exposing greater parallelism, 470
    high-performance determinant update algorithms, 471
        delayed rank-k update, 471–472
        performance results, 472–473
    workflow tools, adoption of, 473–476
QMC techniques, *see* Quantum Monte Carlo techniques
QM–MD calculations, 121
QMP, *see* QCD message passing
QPhiX Library, 361–362
Quadratic isoefficiency, 124
Quantum chemical software, 146
Quantum Chemistry Program Exchange (QCPE) system, 148
Quantum chromodynamics (QCD), 345
Quantum mechanical (QM) calculations, 121
Quantum Monte Carlo (QMC) techniques, 461, 462
    algorithms, 463, 472, 476

calculation, 473
    scientific possibilities for, 462
    workflow diagram, 475
Quark–gluon interaction matrix, 352
Quark line diagrams, 352
Quark propagator, 352
Quarks, 345
Quasi-minimal residual iterative algorithm, 454
QUDA library, for GPUs, 363–364

R

Radiation hydrodynamics (RHD) simulations, 99
RANS, *see* Reynolds-averaged Navier–Stokes
Rapid Radiation Transport Model (RRTMG), 212
RAPTOR, 233
    with graphics processing units, 238–239, 251
    hybrid shared memory parallelism in, 246–251
    spatial scheme in, 236
    third-party libraries, 240
Rational Hybrid Monte Carlo (RHMC), 350
Ray assembler, 427
RCCI, *see* Reactivity controlled compression ignition
RDMA, *see* Remote direct memory access
Reactivity controlled compression ignition (RCCI)
    combustion results, 266–270
    simulation setup, 264–265
Reality, QDP++, 356
Recursive bisection, in EBIndexSpace generation, 309
Reduced memory footprint, 128
Reduction in launch time, 128–129
Registers, hardware considerations, 56–57, 59
Register spilling, 399
Remote direct memory access (RDMA), 129, 426
ReMPI, 43–44
Rensselaer Polytechnic Institute, 322
Research and development (R&D), 508
Resistive tearing mode model, 510
Resolution rank, 75
Restrictive power budgets, 7
Reverse debugging, 39
Reynolds-averaged Navier–Stokes (RANS), 258
    approximation, 232
    models, 320, 325, 326
Reynolds number, 320, 326–327
Reynolds stress tensor, 243
RHD simulations, *see* Radiation hydrodynamics simulations
rho, 395–396
rhocells, 395–396
RK integrator, *see* Runge–Kutta integrator
Roofline model, 401–402, 486–487
Routine-level instrumentation, 23
RRTMG, *see* Rapid Radiation Transport Model
RTS, *see* Runtime system
Runge–Kutta (RK) integrator, 190, 534
Runtime system (RTS), 124, 140–141
    characterization, 191
    enhancements, 138

S

SAC, *see* Seismic Analysis Code
Sample-based tracing, 24, 25

Sampling, 20–21
Scaffolding, gap closing and, 413
ScaLAPACK, 65, 66
Scalasca tool, 439, 440
Scale-adaptive tensor algebra, 72
Scaling, 191
Scattering path matrix, 451
Schrödinger equation, 152, 153, 159, 449, 450, 463
Schur even–odd preconditioning, 353–354
SciDAC, *see* Scientific Discovery through Advanced
        Computing
Science goals, 137
Scientific Discovery through Advanced Computing
        (SciDAC), 328, 346, 508
Score-P, 32–33
SDAG, *see* Structured Dagger
Seaborg computer, 519
Seismic Analysis Code (SAC), 282
        data format, 293
Seismic tomography, 279, 280
Seismograms, 303
Seismology, 293
Seismometers, 293
Selective power boosting, 7
SE method, *see* Spectral element method
Servers, 157–158
Shadow Hamiltonian methods, 351
Shared-memory model, 195–196, 247
Shared-memory processor (SMP) nodes, data layout
        optimizations for, 465–467
Sherman–Morrison scheme, 471
Shifted conjugate gradients, 352
Short-range repulsive interactions, 122
Short recurrence methods, 61
SIA, *see* Super Instruction Architecture
SIA/Aces4 programming model, 153, 154
SIAL programs, 154–156, 158, 161, 162
Silicon vacancy defect system, 487, 488
SIMD, *see* Single instruction multiple data
Simulated years per day (SYPD), 191, 199, 203
`SimulationAnalyzer`, 474
`Simulation` class, 474
`SimulationInput`, 474
SimVascular, 329
Singe, 263–264
Single-core performance optimization, 218
Single-degenerate scenario, 97
Single instruction multiple data (SIMD), 131, 264
        vectorization pattern, 177, 178
Single node performances, 398–399
Single particle orbital (SPO) representation, 465
Single program multiple data (SPMD) model, 308, 313, 414
Singular value decompositions (SVDs), 54
SIONlib, I/O performance using, 444–445
`sip_barrier`, 157, 158
Six-dimensional (6-D) Vlasov equation, 531–532
Slack time, 2, 5
Slater–Jastrow function, 463
Slater-type orbitals (STOs), 146
SM, *see* Streaming multiprocessor
Smagorinsky eddy viscosity model, 235
SMP, *see* Symmetric multi-processing
SOAPdenovo assembler, 427

Soave–Redlich–Kwong (SRK) equations, 235
Software practices
        Charm++, 134–135
        Nanoscale Molecular Dynamics, 132–133
Sorting efficiency, 398
Source-code instrumentation, 23
Space–time subfilter variance, 232
SPADES assembler, 427
Sparse linear algebra operations, 55
        incomplete factorization, 62–63
        iterative linear solvers and eigensolvers, 61
        multigrid methods, 63
        preconditioners, 62
        sparse matrix–vector products, 61–62
SparseMaps, 78
Sparse matrix–vector products (SpMVs), 61–62
Sparse plot file, 311–312
Spatial decomposition, 124
SPECFEM3D_GLOBE
        bandwidth for, 292
        computational data, *see* Computational data, I/O
                optimization
        overview and programming approach, 282–283
        in postprocessing stage, 289
        scalability and benchmarking, 282–289
Spectral element (SE) method, 188–190, 195
Speed-aware load balancing, 140
Spherically symmetric models, 98
Spin, QDP++, 356
Spintronic devices, 433
SPMD model, *see* Single program multiple data model
SpMVs, *see* Sparse matrix–vector products
Spontaneous ignition front, 267
SQLite database, 299, 300
SRK equations, *see* Soave–Redlich–Kwong equations
SSP method, *see* Strong stability preserving method
Stack Trace Analysis Tool (STAT), 42–43
Staggered fermion, 348
Standard Template Library (STL), 39
STAT, *see* Stack Trace Analysis Tool
Stencils, in EB-AMR approach, 310–311
S3D-Legion code, 259–261, 275–276
        chemical mechanisms, 261
        combustion results, 266
                reactivity controlled compression ignition
                        simulation, 266–270
                temporal jet simulation results, 270–271
        contributions, 259–260
        direct numerical simulation, 258–259
        high-performance computing programming, 259
        hybrid OpenACC, 261–262
        legion, 262, 263
        performance
                performance bottlenecks, 272–274
                strong scaling, 274–275
                weak scaling, 274
        simulation setup, 264
                reactivity controlled compression ignition
                        simulation, 264–265
                temporal jet simulation, 265–266
        singe, 263–264
        weak scaling, 274
Store instruction, 381

STOs, *see* Slater-type orbitals
Streaming multiprocessor (SM), 21
Streamline upwind/Petrov–Galerkin (SUPG) stabilization
        method, 330
Strong scaling test, 403
Strong stability preserving (SSP) method, 190
Structured Dagger (SDAG), 134
Subfillter contributions approach, 233
Sunway TaihuLight, 80
Supercomputer centers, 146
Super Instruction Architecture (SIA), 151–152
        distributed arrays, 153
        programming model, 154–155
        structure of, 155–157
Supernovae, 95
Super-parametrization, 206
SVDs, *see* Singular value decompositions
SWAP assembler, 427
Symmetric multi-processing (SMP), 127
        optimization, 128–129
Synchronous sampling, 20
Synthetic jet
        actuators, 321–322
        periodic flow, 324
SYPD, *see* Simulated years per day

T

TACC, *see* Texas Advanced Computing Center
Task-based programming model, 58, 81
TAU Performance System®, 31–32
TCE, *see* Tensor contraction engine
Tcl interpreter, 132
Temporal filtering techniques, 232
Temporal integration scheme, 235
Temporal jet simulation, 265–266
        combustion results, 270–271
        simulation setup, 265–266
Tensor algebra
        decompositions and higher level operations, 74–76
        in different scientific disciplines, 67–71
        hardware abstraction scheme and virtual processing,
                80–84
        operations, 71–73
        parallel algorithms for, 77–79
        projected exascale computing hardware roadmap, 79–80
Tensor blocks, 72
Tensor calculus, 68
Tensor contraction engine (TCE), 174–176
Tensor contraction operation, 71–72, 78
Tensor network (TN) theory, 70
Tensor order, 67
Tensor rank, 67
Tensor signature, 89
Tensor train (TT), 70
Tensor transpose compiler (TTC), 79
Tensor valence, 67
Testing for correctness, 198–199
tests.py, 442
Texas Advanced Computing Center (TACC), 213
TFQMR method, *see* Transpose-free quasi-minimal residual
        method
TFTR, *see* Tokamak Fusion Test Reactor

Theoretical–numerical framework, 242
Thermal-aware load balancing, 140
Thermal bomb models, 98
Thermodynamic properties kernel, 250
Thermonuclear supernovae, 96, 97
Thermophysical properties kernel, 249
Thinking Machines, 147
Third-party tools, 18
ThreadBlockList_GPU loop structure, 106
Thread control, in TotalView, 39
Thread debugging, message passing interface and, 43–44
Threading, hardware considerations, 56, 58
Thread master regions, 193
ThreadWithinBlock_GPU version, 107–109, 111
3D FFT, 168–170, 173, 174
3D numerical solvers, 279
3D seismic wave equation solver, 280
3D simulations, 379
3D tricubic B-splines, 465
Throttling, 22
Throughput, 191
tilesize, 176
Time discretization, 190
timelevels, 191
Time to solution, 520
Titan system
        heterogeneous compute node architecture, 247
        JAGUAR and, 237
        PREM on, 285–286, 288
        RAPTOR on, 238
Tokamak, 507, 531
Tokamak Fusion Test Reactor (TFTR), 531
Topology aware mapping, 127–128
TotalView debugger
        architecture, 40
        requirements, 40–41
        thread-control features of, 39
Tpetra subpackage, 65
Tracers, 189
Tracing approach, 24–25
Transition probabilities, in GW calculation, 489
Transport methods, 291
Transport properties kernel, 250
Transpose-free quasi-minimal residual (TFQMR) method,
        434, 435–436, 441, 447
Triangle searching, electron push kernel, 535–536
Trilinos, 65
Truncations errors, 393
Tucker kernel, 76
Tucker tensor decomposition, 75
Turbulent combustion, 258
2D distribution pattern, 11
2D simulations, 379
Type Ia supernovae (SNe Ia), 96–98

U

Unified Parallel C (UPC), 410, 414, 426
        Partitioned Global Address Space model in, 413–414
Unit testing, 302, 476–477
Unix profiling tools, 20
Unpack, data, 192
UPC, *see* Unified Parallel C

V

Valgrind, 41–42
Validation, KKRnano, 442–443
Vampir, 33–35
Van der Waals attractive interactions, 122
Variational moments equilibrium code (VMEC), 511
Variational Monte Carlo (VMC)
    algorithm, 464
    methods, 462, 463
Vector hardware, 56, 58
Vectorization, 378, 381, 386, 393, 397, 399
Verification, KKRnano, 442–443
Vertical tail/rudder assembly
    flow control on, 322–323
    simulations, 324
        Aurora Early Science Program, 327–328
        flight reynolds number flow control, 326–327
        past simulations at Re = 0.35 million, 324–325
VGH function, 466, 467
    with AoS-to-SoA data layout transformation, 468
    with SoA to AoSoA transformation, 469
Viscous aerodynamic flows, 320
Viscous stress tensor, 234
Vlasov equation, 378, 531–532
VMC, *see* Variational Monte Carlo
VMEC, *see* Variational moments equilibrium code

W

WalkerSoA object, 468
Wang–Landau Locally Self-consistent Multiple Scattering
        (WL-LSMS) calculations, 457–459
WARP code, 383
Water cycle, 188
Wave-front algorithms, 6, 8–10
Wave propagation theory, 279

Wheat genome, end-to-end strong scaling for, 424
Wilson fermions, 348
WL-LSMS calculations, *see* Wang–Landau Locally
        Self-consistent Multiple Scattering calculations

X

Xeon Phi architecture, QPhiX Library for, 361–362
Xeon Phi processor/coprocessor, 65, 484, 485
Xeon, QPhiX Library for, 361–362
XGC, 530
    Adaptable IO System in, 533, 542
    electron push kernel, *see* Electron push kernel
    Fokker–Planck–Landau collision operator, 538–541,
        546, 547
    graphics processing unit for, 533–534
    hybrid-Lagrangian δf scheme, 534
    ITER simulation, 531, 548–549
    load balancing, 540–542, 547
    multispecies collision kernel, 538–540, 546–547
    parallel and staged I/O, 542, 548
    particle in-cell approach, 530, 532
    performance data, collection of, 543
    scientific implications, 549–550
    unstructured triangular mesh in, 532–533
XML, 360

Y

YAGA assembler, 427
Yellowstone system, 209, 213, 219, 222

Z

ZMAT file, 155
Zone-threaded GPU version, 113